PRINCIPLES OF CHEMISTRY

A MODELS APPROACH

PRINCIPLES OF CHEMISTRY

A MODELS APPROACH

Terrence J. Swift
Case Western Reserve University

D. C. HEATH AND COMPANY
Lexington, Massachusetts / Toronto / London

Published simultaneously in Canada.

Printed in the United States of America.

International Standard Book Number: 0-669-93021-0

Library of Congress Catalog Card Number: 74-11822

Preface

Instructors in college introductory chemistry courses are faced with an extremely formidable task. Their students have a wide variety of interests, ranging from those of the prospective health scientist to those of the future chemical researcher. Because the career interests of most of these students require them to take chemistry courses beyond the introductory level, basic chemistry courses must present the fundamental facts and concepts in such a way that students not only learn them but can use and retain them. When one considers the wealth of facts and concepts most chemists consider basic, it is easy to conclude that the instructor's task is almost impossible.

Added to the problem is the great diversity of preparation in chemistry with which students enter college. Some enter with excellent chemistry backgrounds and are fully prepared for a sophisticated treatment of the subject at the college level. Unfortunately, the number is usually relatively small. In fact, a larger number of students have had almost no preparation at all in chemistry and in instances where there has been preparation retention by the student is extremely poor. Thus, the diversity of background and the lack of retention on the part of students in the introductory chemistry course demand that the instructor begins at a very elementary level and proceeds in a very systematic fashion. The material must be presented in such a way that students—including the well-prepared—are interested and challenged.

How does one construct a chemistry course which is both highly structured for the poorly-prepared student and highly flexible for the well-prepared student? The author of *Principles of Chemistry: A Models Approach* believes that a solution lies in centering the course around the systematic development of the physical intuition of students. It is generally conceded that there are at least three factors involved in mastering a science. These are acquiring a knowledge of the basic facts of the science, acquisition of the quantitative techniques and developments associated with the concepts of the science, and understanding those concepts in physical terms. The latter factor must be considered the most important of the three, because it ties together the other two and serves as a guide for further pursuits in the science. Furthermore, without a good physical understanding, students soon forget the facts and concepts.

How can physical intuition be systematically developed? One way is by concentrating on the construction and application of models. Models in chemistry are mental pictures of matter which chemists formulate in order to understand the properties of matter. They are useful approximations of material objects that can be developed systematically from very crude approximations to extremely sophisticated ones.

The process of model building is fundamental to the thinking of most chemists and is without doubt the best way yet devised to tie together chemical facts and concepts. Furthermore, model building can be done in such a way as to involve the student in every step of the process. It is an approach that all students will find interesting and challenging, including the well prepared ones.

Instructors have had considerable experience in constructing and using models of matter; students have not. As a consequence, there are two forewords to this text, one addressed to the instructor and the other to the student. However, before the forewords a brief discussion of the process of model building and some general comments on the philosophy of the models approach as reflected in the text are in order.

Model construction begins with experimental observations. The model is thus a collection of proposed features of the matter under investigation which are consistent with the experimental observations. One of the most useful characteristics of models is their predictive value. Once constructed, the model leads to predictions of additional facts. The appropriate experiments are then devised and performed and the predictions checked to provide corroboration for the model. If the experimental results are in disagreement with the predictions, the model is either discarded and a new one constructed or the original model is refined. Thus, the process of model building involves a never-ending interplay between experiment and theory.

Generally, college students in introductory chemistry courses should not have to accept a concept on faith. Concepts must make sense to them in physical terms or they will neither be able to use the concepts nor retain them.

Because model building is a highly physical process (as contrasted with mathematical procedures) it is very desirable to minimize the mathematical level and the number of mathematical operations used in the process. Consequently, the calculus is not used in this text, and even the amount of algebra is kept to a minimum. Important equations are developed, but in general they follow from models rather than lead to them.

I wish to acknowledge my debt of gratitude to two former teachers who have served as great examples to me of what a scholar and teacher should be. They are Professor George Schulte of Loras College and Professor R. E. Connick of the University of California, Berkeley. I wish also to acknowledge with gratitude the assistance of a number of reviewers, whose contributions to the development of the manuscript were far above the ordinary. They are Professor Robert Brasted, University of Minnesota; Professor George Fleck, Smith College; Professor Michael Henchman, Brandeis University; and Professor Frank Rioux, St. John's University, Minnesota. I would be remiss if I did not express my appreciation to Professor G. R. McMillan of Case Western Reserve University for his many helpful discussions and suggestions, and also to Mr. T. S. Viswanathan for his meticulous efforts in correcting proofs.

It has been a great pleasure working with Mr. Jeff Holtmeier, Dr. Paul Bryant, and others of the staff at D. C. Heath and Company. They all have been completely dedicated to making this the best text possible. A textbook which has the development of visual images as its theme must contain outstanding artwork. Mr. Leonard Preston, the artist, is responsible for several key developments in the artwork. My thanks also go to Mrs. Eileen Green and Mrs. Frances Novak, who typed the manuscript on some schedules that were often nearly impossible. Finally, I would like to acknowledge the patience and understanding of my wife, Therese, and my daughters, Mary Margaret and Elizabeth, for tolerating a preoccupied husband and father.

TERRENCE J. SWIFT

Contents

Foreword to the Instructor

This text has been successfully used in manuscript form by me in a large introductory chemistry course at Case Western Reserve University in the 1973–74 academic year, the manuscript having been organized from my lectures in the course in previous years. Each chapter as it appears in the book was put in final form only after it had been classroom tested. In this foreword I hope to explain the rationale for the approaches taken in key chapters and to indicate areas where I believe special emphasis should be placed. Both explanations are presented in more detail in an instructor's manual that is available from the publishers.

As a general comment I would point out that there is no such thing as a correct model. Students should be made aware that it is pointless for them to expect to be told the "truth," as far as explanation of chemical phenomena are concerned. The reason, of course, being that there is no such thing as absolute truth in any explanation. One of the virtues of the process of model building is that it shows the limitations of explanations. You can aid immensely in making this point by using alternative models to the ones developed in the text. Students thus can see that there is more than one valid way to view the same phenomenon.

The first six chapters are almost exclusively concerned with model building. I have put special emphasis on the process in Chapter 2. The gaseous state, of itself, does not merit the lengthy treatment given it. However, Chapter 2 is the chapter in which the discussions of model building begin in earnest and is the first place where the student is given examples of models and shown how they relate to actual observations.

I believe that to make sense out of chemistry in physical terms it is necessary to have an adequate grasp of such elementary physical quantities as momentum and kinetic energy. Hence, basic concepts in mechanics are developed briefly in Chapter 2, with other elementary physical concepts being presented as they are needed in succeeding chapters. In these developments no previous knowledge of physics on the part of the student is assumed.

The principal model developed in the first half of the text is the wave model of the atom (Chapter 4). The concepts of energy levels, quantum numbers, and orbitals seem like magic to students unless these concepts are developed. It is not necessary to go into Schrödinger wave mechanics for this development. Waves are easy to picture and energy levels, quantum numbers, and atomic orbitals can be derived in a nonmathematical fashion which is relatively easy to visualize.

Chapter 5 is concerned with chemical bonding. Students need to have, early in the course, a very clear picture of why any given chemical bond exists. In addition they need to have a model with which to predict the numbers, kinds, and something of the strengths of bonds between most atoms. Such a model is that of molecular orbitals formed, in Chapter 5, by the pictorial overlap of atomic orbitals. Molecular orbitals are an anathema to many instructors and consequently are often de-emphasized in their teaching. However, I believe these instructors are overlooking an extremely useful model, which can be (but admittedly often is not) constructed in a way which is easy to visualize.

Chapter 6 contains a variation on the valence-shell, electron-pair repulsion method for predicting molecular geometry. The variation has proven to be very effective in improving students' predictive ability and it is used often in succeeding chapters.

A major part of the descriptive chemistry in the text appears in Chapters 7 through 11. This is rather unusual, in that it precedes the discussions of thermodynamics and equilibrium. My experience has been that descriptive chemistry is best approached on two different levels. The first is a qualitative level, based on the concepts of chemical bonding, and the second is a quantitative level based on thermodynamics. Most students experience great difficulties with chemical equilibria unless they are familiar with the molecular structures and reactions involved.

By placing the major part of descriptive chemistry in the text immediately following the development of atomic and molecular models we can demonstrate the predictive and correlative power of models. Large amounts of descriptive material are difficult to present successfully, because it is easy to lose the interest of the student confronted with a mass of detail. As a consequence, descriptive chemistry (particularly non-metallic inorganic chemistry) is often neglected, a situation approved of by some chemical educators but deplored by others. I believe that the models approach provides some solution to this problem. By pointing out (Chapters 7 through 11) just how successful they are in predicting and explaining molecular structures and reactions, models provide a theme for relating otherwise apparently unrelated facts.

While the predictive approach to descriptive chemistry has its advantages in maintaining student interest, it also has one potential drawback that must be recognized. Since the models precede the facts, students can easily conclude that the facts are subservient to the models, and that something is true because we predicted the something to be true. They, therefore, must be reminded in this game we are playing that a proper perspective must be kept. Models come and go, but facts remain facts.

In recent years, many high school chemistry courses have become very concept centered, at the expense of the consideration of chemical compounds and chemical reactions. Consequently, most of the compounds and chemical reactions discussed in Chapters 7 through 11 of this text may be unfamiliar to many of your students. I recommend emphasis on classroom demonstrations, particularly in connection with Chapters 8 and 9. These two chapters have been written so that the students can master them through their own reading. The challenge is to bring them to life, and this can be done through classroom demonstrations.

Chapter 12 is, for an introductory text, a rather thorough treatment of spectroscopy. We, as instructors, confidently state all sorts of detailed information about molecular structures, and students certainly wonder about the source of our information. With the judicious use of a few simple models spectroscopic methods can be introduced which will allow the students themselves to deduce molecular structures.

The approach to thermodynamics in Chapter 15 is probably the most important feature of this text. It has been my experience that of all the areas of chemistry thermodynamics is the most difficult for students to grasp, use, and retain. This is particularly true of anything having to do with the second law. I believe that the problem exists because the student is not given any easily visualizable model which relates to the second law. To some extent this problem exists because, in a strict sense, thermodynamics is model independent. One can use the laws of thermodynamics without relating them to atoms and molecules in any way. As a chemist I have always found this way of using the laws self-defeating, and pedagogically the model-independent approach has been a failure in every instance that I have ever seen it used. As with other physical entities, entropy and Gibbs free energy must be made visualizable or students will have little feeling for them. The models of matter spreadedness and energy spreadedness developed in Chapter 15 permit the visualization of entropy and Gibbs free energy. The use of these two models has permitted my students to grasp, use, and retain the concepts of thermodynamics. In fact, it was the success in this area of the use of visualizable models that led to the development of the model approach in this text.

A mathematical analysis of a simple process involving matter spreading is presented in the text. The detailed analysis of energy spreading is more complicated and is not presented. Norman Craig has given such an analysis (N.C. Craig, J. Chem. Educ., **47**, 342 (1970)) in a particularly lucid and concise treatment.

The models of matter spreadedness and energy spreadedness adequately introduce the concept of equilibrium, and the general concept of equilibrium is consequently developed in Chapter 16. Chapters 17, 18, 19, and 20 are a return to descriptive chemistry. The treatment in these chapters is quantitative in terms of specific equilibria. The student learns the standard procedures for handling equilibria quantitatively. However, the models of matter and energy spreadedness are woven throughout these four chapters, so that the student can visualize why specific equilibria are as they are.

Chapter 21 is devoted to chemical kinetics. The focus of this chapter is on the question of why some reactions are rapid and others are slow. In order to answer that question specific reaction mechanisms are proposed. There probably are no more debatable models in chemistry than specific reaction mechanisms. Except for a few cases, space limitations in the text prevent proposals of alternate mechanisms for reactions and discussing the pros and cons of each. It would help the student to think of reaction mechanisms in terms of the visualizable encounters of atoms and molecules, if you supply alternative mechanisms.

Chapter 22 is a lengthy treatment of biochemistry, a topic that my students have usually voted the most interesting in the course. The chapter has far more detail than the student is expected to master. However, it contains several very important concepts, which should be presented in a framework of real structures and reactions, even though students will not learn and retain these structures and reactions. The chapter is unique among the others in that it contains sections that are identified as optional material. These sections include detailed analyses of some important life processes, the degree of detail exceeding that normally presented in an introductory text. It is my intention that students use these sections as they see fit to help them understand the processes discussed. As a minimum they will see the types of molecular structures and chemical reactions involved. As a maximum they can verify, detail for detail, important statements that are made about life processes.

The length of the text has been kept to one which can be completely covered in two semesters or three quarters. This is particularly important in a text that is as highly structured as this one. If students are to accept nothing on faith, chapters must build on each other and the last chapters are a culmination of the whole development.

The traditional drawback of the thematic approach, exemplified by this text, is the inflexibility of textual presentation. I have tried to minimize this problem by organizing the chapters in an order which I believe is most common in introductory courses.

The problems at the ends of chapters emphasize explanations and predictions rather than substitutions into equations, although some of the latter are included to acquaint students with units and their uses. Particularly challenging problems are indicated by asterisks. The problems are rarely repetitive and their numbering follows the order of topic development in each chapter. Numerical answers and other answers which are not lengthy are included after the text's appendices. Detailed solutions to all problems are included in the instructor's manual. It is my hope that those problems which are particularly illustrative of important concepts be discussed in class.

Foreword to the Student

The first question to be addressed in any course is, "What can you, the student, expect to obtain from this course?" The answer in the case of a course in which this text is used is that you may expect to be able to view the properties of matter in essentially the same way as the practicing chemist does. This, of course, raises a second question. "What is unique about the way in which the practicing chemist views the properties of matter?" When observing a steel bar, for example, a chemist might note that it is a solid at room temperature; it is shiny, hard, and a good conductor of heat and electricity; it undergoes chemical changes (e.g., it rusts when exposed to air and water). The practicing chemist can account for these properties, although it is by no means obvious to the nonchemist what it is about the steel that gives it such properties and makes it different from wood.

The ability of chemists to account for the properties of matter lies in their ability to construct **models** of any given substance. These models are mental images of the tiny units of which all matter is composed. Mentally, chemists visualize atoms and molecules, their structures and their motions, and picture these structures and motions giving rise to measurable properties.

I like to view models of matter as abstract paintings. Any given substance is incredibly complex in its structure, and if we are to understand even one facet of the substance's behavior we must focus on that one facet, almost to the exclusion of all others. This is what the abstract artist does, and it is also what the scientist does in constructing models. The result in both cases is a distorted view of the object as a whole but, hopefully, one that provides greater insight into the particular facet under study. Thus, in the case of the steel bar we would employ a different mental picture in considering rusting than we would in considering hardness. However, the two pictures would not be contradictory, any more than two artists' conceptions of the same object would be contradictory.

Because models are actually inexact views of matter, two different scientists will often employ different models in viewing the same phenomenon. There is no such thing as a correct model. There are only models with varying degrees of usefulness. In this text I have tried to introduce those models most commonly employed by chemists. Your instructor will undoubtedly wish to introduce his or her own models —views of certain objects and processes. It is most desirable that you also be an active participant in this process of model building. The process is one of the most stimulating activities of a scientist, and you can participate in it even as your scientific background is only beginning to develop. The problems at the ends of chapters place a strong emphasis on model development, as well as on the explanation of phenomena in terms of those models.

The first six chapters of the text are devoted to building up pictures of atoms and molecules, their structures, and something of their motions. Chapters 7, 8, 9, 10, and 11 are largely concerned with the application of the models previously developed to the chemical behavior of matter in its myriad of forms. Our models allow us both to understand the behavior of matter and to predict that behavior in many cases. Predictive ability is one of the most desirable skills you can acquire.

Chapters 12, 13, and 14 are concerned with physical properties. These chapters contain a mixture of model development and application. Chapters 15 and 16 are devoted to the development of a general model for spontaneous change. In Chapters 17, 18, 19, and 20 this model is applied to a wide variety of physical and chemical changes. Chapter 21 deals with the rate at which chemical changes occur. We shall want to be able to see precisely why a given change occurs rapidly or slowly and to be able to devise ways to speed it up or slow it down. Chapter 22 is a lengthy application of many of our models to the chemistry of life processes.

If I may be permitted to rephrase the answer to the question which opened this foreword, the aim in this text is to have the subject of chemistry make sense to you in physical terms which you can visualize.

1/ Atoms and Molecules

Chemistry is sometimes loosely defined as the study of matter. However, almost all other branches of knowledge and inquiry fit into that sweeping definition. The biologist and psychologist treat the structure and behavior of living matter; the engineer deals with the properties of matter that can be put to useful purposes; and the humanist is concerned exclusively with that most important collection of matter, man. What is it about chemistry that distinguishes its approach to studying matter? If we were to eavesdrop on any two chemists discussing their subject, the answer would be readily apparent, for the words *atom* and *molecule* would repeatedly enter the conversation, much more so than in conversations between humanists, engineers, or scientists in other branches of knowledge.

The chemist carries with him a more or less well-developed picture (depending on the chemist) of tiny fundamental units of which all matter is composed, whose structures and motions determine the multitude of different properties we associate with different types of matter. This picture of atoms and molecules began to be developed in an effective fashion about 1800. In the years since 1800, it has been developed into an extremely strong basis for our understanding of matter. In this chapter, we shall consider the following questions.

1. How do we know there are such things as atoms and molecules?
2. How do we know the composition of any given molecule?
3. How much do atoms and molecules weigh?
4. What insights do the answers to the first three questions give us about the properties of matter?

1–1. Measurements, Foundations of Model Construction

There is perhaps nothing more puzzling to the layman about chemistry than the chemists' constant referral to atoms. How do we know there are atoms? Have you ever seen one? Seeing is, of course, only one piece of evidence for the existence of things (and not a completely reliable one at that, as witness the mirage in the desert). In the case of the mirage in the desert, hearing and smelling would be much more reliable than sight because sight is more easily tricked under such circumstances.

The most valuable observations to the scientist who investigates matter are quantitative measurements. One of the most important distinguishing features of the scientist's approach to the properties of matter is his reliance on precise quantitative measurements. We shall see shortly which type of measurement provided the first convincing evidence for the existence of atoms. However, before we can do that it will be necessary to know some fundamental properties that are measurable and how the measurements are made and reported.

Units

The two most commonly measured properties of any object are its **mass** and its **volume.** The volume of an object is defined as the space the object occupies. The mass is a measure of resistance of an object to being moved (this resistance is termed **inertia**). The mass of an object is normally measured from the force exerted on the object by the earth's gravity since this force has been shown to be directly proportional to the mass of an object.

The gravitational force on an object is defined as the **weight** of the object, and the weight is given by the statement

$$\text{Weight} = mg \tag{1-1}$$

where m is the mass of the object and g is termed the **earth's gravitational force constant.** The constant g has the same value for all objects on earth, so that when two objects have the same mass they also have the same weight. This allows us to measure masses in a comparative fashion. First, some object is chosen to be the standard mass with which all other masses will be compared. The mass of this standard is defined to be one unit of mass. The unit is given a name and in the **metric system** of units, the system that scientists use, the unit of mass is the **gram.**

Once we have a standard mass, the mass of any object can be measured by subjecting the object and a number of standard masses (or known fractions or multiples of standard masses) to the earth's gravity and adjusting the number of standard masses until the two weights are the same. This is the principle of the balance. Figure 1–1 shows the essential features of a balance. A beam connected to two suspended pans is balanced on a sharp-edged support so that the balance indicator (shown in color) is exactly vertical. The object whose mass is to be determined is placed in the left pan and standard masses (gram masses made of brass, or fractions or multiples of those masses) are added to the right until the indicator is vertical again. At this point the mass of the object is equal to the sum of the standard masses. Determining the mass of an object by using a balance is commonly called **weighing** the object. Thus, *weight* in common usage means *mass*

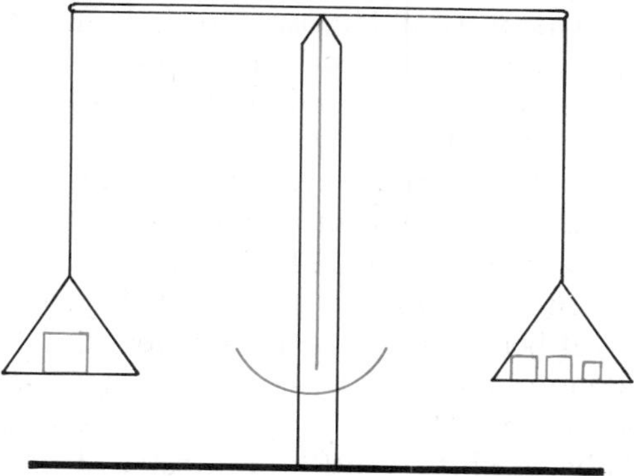

Figure 1–1
Schematic determination of the mass of an object (*left*) by balancing its weight with that of standard masses (*right*).

determined by weighing. We shall use this definition of weight throughout the text, and the terms *mass* and *weight* interchangeably.

The preceding discussion of mass and weight has led to the information that mass is measured by comparison with an arbitrarily defined unit mass. Such arbitrary standards or units are defined for all other measurements also. In addition to mass, the most common measurements used in science are volume and length. In the metric system, the units of volume and length are the **liter** and the **meter,** respectively.

Measurements reported in grams, liters, and meters have little meaning for us unless we have some feeling for the sizes of these three units. Every American has a feeling for the units of the English system of measurements. If told the weight of something in pounds, an American can form a good mental image of the amount of matter represented by that measurement. The same can be said of a distance reported in feet or a volume in gallons. To convert this feeling for the English units into a feeling for the metric units, we require **conversion factors,** giving the English equivalent of each metric unit. These conversion factors are given in the column at the right in Table 1–1.

Measurement	Metric Units	Conversion Factor
Mass	gram (g) 1 kg = 10^3 g 1 mg = 10^{-3} g	1 lb = 453.6 g
Length	meter (m) 1 km = 10^3 m 1 cm = 10^{-2} m 1 mm = 10^{-3} m	1 m = 3.28 ft
Volume	liter (l) 1 l = 10^3 cm^3 1 ml = 10^{-3} l	1 l = 1.057 qt

Table 1–1
Metric Units and Conversion Factors to English Units

In any system of measurement, the basic unit is related to several smaller and larger units, and there are numerical factors that relate these units to the basic unit. For example, an English weight unit is the ounce and there are 16 ounces in a pound.

The great advantage to the metric system is that, unlike the English system, units that measure the same quantity are related by powers of ten. Thus, a given bar of copper weighs 332.6 grams (g) or 0.3326 kilogram (kg). The same conversion in the English system might be from pounds to ounces, and it is more difficult because we must remember that there are 16 ounces in a pound. Conversions are much easier to remember for metric units, since we only have to remember a few prefixes. The prefix **kilo** (k) always indicates 1000 (10^3) basic units, **centi** (c) indicates $\frac{1}{100}$ (10^{-2}) of a basic unit, and **milli** (m) indicates $\frac{1}{1000}$ (10^{-3}) of a basic unit. These conversions within the metric system are also shown in Table 1–1. Since volume equals length cubed, the unit of volume (the liter) is related to the unit of length through 1 liter $= 10^3$ cm$^3 = 10^{-3}$ m^3.

Figure 1–2 is an illustration of one of the conversion factors in Table 1–1, namely that between the English and metric units of length. The length is measured at the top of the figure with a foot ruler, which is subdivided into

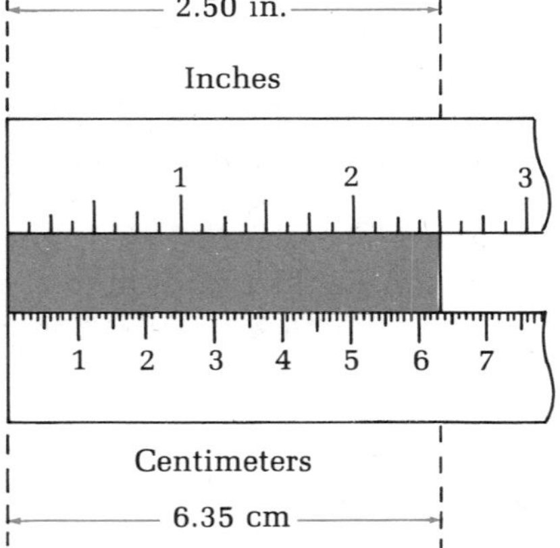

Figure 1–2
Measurement of a length in inches (*top*) and centimeters (*bottom*).

inches and fractions of inches. At the bottom is a meter stick, which is subdivided into centimeters and millimeters. The length of the object is both 2.50 in. and 6.35 cm. Thus, the conversion factor between centimeters and inches is

$$1 \text{ inch} = \frac{6.35}{2.50} \text{ cm} = 2.54 \text{ cm}$$

The International Bureau of Weights and Measures established, in 1960, a new system of units based on the metric system. This International System (SI) of Units has only slowly and partially been adopted by the chemical community. As a consequence, we shall not observe strict use of the new units in this text. Generally, we shall use the units chemists commonly use. They are the metric units of Table 1–1 and others to be introduced subsequently in this text.

Significant Figures

One of the most obvious things associated with any measurement is its inherent uncertainty. For example, one way to measure the mass of this text is to lift it to experience the force exerted on your arm. Since you have lifted known masses before, you can use that experience to estimate the mass of this text. You might estimate that the text weighs 5 pounds, but you are aware that it could weigh 4 pounds or 6 pounds. On the other hand, you are quite confident that it doesn't weigh less than 2 pounds or more than 8 pounds. Thus, you would report the mass as 5 plus or minus 3 (5 ± 3) pounds. All measurements, even those made with very expensive instruments, involve uncertainty, and the scientist must always be aware of the approximate limits of that uncertainty, just as in the hefting of the text.

Let us now consider a measurement, namely the length of the object in Figure 1–2. How do we incorporate uncertainty into our reporting of the length? We could report the length as 6 cm, but this would be foolish because we can read the meter stick more precisely than to the nearest cm. In fact, we are certain that the length is between 6.3 and 6.4 cm. We can even estimate the length better than that since it appears to be approximately halfway between 6.3 and 6.4. Thus, it is reasonable to report the length as 6.35 with the implication that there is some degree of uncertainty in the last digit, 5.

It would be foolish to report the length as 6.350, for example. This number implies that the digit 5 is certain and the uncertainty lies in the zero. The reporting of a number conveys definite information about the uncertainty in the measurement of that number. The digits reported are termed **significant figures,** meaning numbers that either are certain or are good estimates. If the mass of this text is reported as simply 5 lb, some correct conclusions are inescapable. The uncertainty is not as large as 10 lb, or we should not have reported a number less than 10 lb. However, the uncertainty is at least one pound or we should have estimated and reported another significant figure, for example 5.3 lb. Thus, the last significant figure reported establishes an approximate uncertainty.

Suppose that we were asked to estimate the diameter of the earth. A reasonable guess is eight thousand miles. How should we report that number in digital form to indicate that the uncertainty of the estimate is in thousands of miles? Surely we should not report it as 8000 miles, since this implies four significant figures. All we are using the zeros for in 8000 is to indicate the power of ten in the number. We could just as well report the number as 8×10^3, a notation of the form that scientists use to denote separately the significant figures and the power of ten. There is no mistaking the number of significant figures in 8×10^3 because only one figure is given.

The purpose of measurements is to allow us to draw meaningful conclusions about what is measured. Often calculations are performed between the measurements and the conclusion. The validity of the conclusion always depends on the uncertainties of the final numbers used to draw that conclusion. If the uncertainties are too large, no valid conclusions can be drawn. Thus, an important consideration in every calculation is that the uncertainty in the measurements be carried through the calculation properly, so that we do not fool ourselves into drawing unwarranted conclusions.

As an example, let us consider a company that proposes to sell a 12-ounce soft drink, which they plan to market at a price that should yield a 5 percent profit if it sells as well as expected. How precisely should the soft drink be measured so that the company does not cheat itself out of its profit? Suppose that the measurement is only to the nearest ounce so that every bottle could contain as much as 13 oz or as little as 11 oz. This is a possible error of

$$\left(\tfrac{1}{12}\right)(100) = 8.5 \text{ percent}$$

and this could more than offset the proposed profit. A proper control might be to the nearest tenth of an ounce.

Suppose that instead of doing our calculation with the uncertainty in one bottle, we worked with the total volume to be produced in one year. Let this be 1,000,000 bottles. If we multiply 12 by 1,000,000, we obtain 12,000,000 oz as the volume to be produced. One ounce appears to be insignificant compared with this large number, and we might be tempted to say that the measurement of soft drink into each bottle need not be done carefully. This is an incorrect conclusion, and it arose because we didn't carry through the uncertainty in 12 in its multiplication by 1,000,000. Since each bottle can contain 13 ounces, the maximum possible output per year is 13,000,000 and the possible error is

$$\frac{1{,}000{,}000}{12{,}000{,}000}(100) = 8.5 \text{ percent}$$

Uncertainty is neither created nor destroyed in a calculation. We did not gain any greater degree of certainty by multiplying the uncertain number 12 by the highly certain number 1,000,000. The result of the multiplication should have been reported as 12×10^6 oz per year. **In any multiplication or division, the percentage of uncertainty of the answer is equal to the percentage of uncertainty of the more uncertain of the two numbers that go into the calculation.**

Another common error that sometimes creeps into calculations occurs in addition and subtraction. Suppose that we are asked to add the two numbers 263 and 3.2×10^{-2}. The temptation is to do the operation simply because we are told to and to report the answer as 263.032. To perform the addition, we have implicitly converted 263 to 263.000, adding three more significant figures that are not given. **Only significant figures may be added to or subtracted from other significant figures.** The correctly reported sum of 263 and 3.2×10^{-2} is 263.

Conversion of Units

We shall have many occasions within this text for using conversion factors between units such as those given in the second and third columns of Table 1–1. A recurrent problem for many students is whether to multiply or divide by the conversion factor in a given unit conversion. This problem does not arise if careful attention is paid to the units involved. For example, suppose that we wish to express 2.64 pounds in grams. The conversion factor that relates these units is 1 lb = 453.6 g; the pound is 453.6 times as large as the gram. If both sides of the conversion factor equation are divided by 1 lb, we obtain

$$1 = \frac{453.6\ \text{g}}{1\ \text{lb}}$$

We can now multiply 2.64 lb by (453.6 g/1 lb) without changing its value, because this is the same as multiplication by unity. The result is

$$\begin{aligned} 2.64\ \text{lb} &= 2.64\ \cancel{\text{lb}} \left(\frac{453.6\ \text{g}}{1\ \cancel{\text{lb}}}\right) \\ &= (2.64)(453.6\ \text{g}) \\ &= 1.20 \times 10^{3}\ \text{g} \end{aligned}$$

If we multiply as shown, the units of pounds cancel and the desired unit grams is obtained. If we had chosen to multiply instead by (1 lb/463.6 g), the resulting units would have been lb^2/g, and we should have known that the conversion factor was used improperly.

1–2. The Existence of Atoms

The question why matter takes different forms and displays different properties has interested thinking people for centuries. However, it was only in the latter part of the eighteenth century that the pertinent measurements were made that allowed the first firm conclusions that matter is composed of atoms. There is a large and interesting history of the thinking prior to 1800 on the composition of matter.*

*The interested reader is directed in particular to the introductory text by Dickerson, Gray, and Haight, cited at the end of this chapter, for an excellent, concise discussion of that history.

Law of Definite Proportions

One of the ways in which a wide variety of properties among the different substances in nature can arise is the combining of some substances to form other substances, or mixtures. It was only natural that scientists should try to separate substances that they encountered into simpler substances in the hope of arriving at a set of elemental substances that form the basis for all other substances. A common way to accomplish separation, or decomposition, is by heating. For example, the English scientist Joseph Priestley found that a certain red solid decomposed when heated, yielding a silvery liquid and a gas. However, neither the silvery liquid nor the gas could be decomposed to yield simpler substances.

Substances that could not be decomposed into simpler substances were termed **elements,** and each element was given a name and a symbol. Priestley's red solid yields the silvery liquid element **mercury**, symbolized by Hg, and the gas **oxygen,** symbolized by O. By 1800, many such elements had been identified. They included most of the common metals we know today, together with such nonmetallic species as hydrogen, oxygen, nitrogen, and sulfur. The complete set of elements now known is listed on the inside front cover of this text, along with the symbols for each.

Certain elements combine with one another to produce substances that have different properties from either of the elements from which they were formed. For example, liquid mercury and gaseous oxygen combine to form a red solid. The new substance formed by combining elements is termed a **compound.** A key discovery was made in the late eighteenth century concerning such chemical transformations and this is that **the combined weights of the product substances are equal to the combined weights of the reactant substances.** For example, if 10.000 g of the red solid derived from mercury and oxygen is heated, the products are 9.261 g of mercury and 0.739 g of oxygen.

Scientists in the late eighteenth century recognized the importance of this discovery largely through the efforts of the French chemist Antoine Lavoisier. Lavoisier's work showed how important weight measurements were in chemical transformations, and chemists then began to focus primarily on the ratios of weights with which elements or compounds combined with each other to form new substances. They discovered two kinds of combinations. Sugar and water are an example of the first kind, in which the combination can occur over a wide range of weight ratios. The combination may be written in the following manner:

$$\text{Water} + \text{sugar} \rightarrow \text{syrup}$$

$$x \text{ grams} + y \text{ grams} \rightarrow (x + y) \text{ grams.}$$

The ratio of y to x can be anything from nearly zero up to that for thick syrup. Such combinations that occur over a continuous range of compositions are called **mixtures.**

The second type of combination often occurs only in one weight ratio. For example, lime and the gaseous product of beer making (now known as carbon dioxide) react to produce the principal constituent of limestone in only a single weight ratio:

$$\text{Carbon dioxide} + \text{lime} \rightarrow \text{limestone}$$
$$x \text{ grams} + (1.274x) \text{ grams} \rightarrow (2.274x) \text{ grams}$$

The carbon dioxide reacts with an amount of lime precisely 1.274 times its own weight. Any smaller amount of lime will leave some unreacted gas, and any more will result in unreacted lime. Many combinations—elements and elements, elements and compounds, and compounds and compounds—have been found to combine in only a single ratio by weight. Some of these weight ratios are given in Table 1–2.

Substance A	Substance B	Product	Weight Ratio (B/A)
Sodium	Chlorine	Table salt	1.542
Hydrogen	Chlorine	Hydrogen chloride	35.172
Calcium	Oxygen	Lime	0.3990
Calcium	Sulfur	Calcium sulfide	0.8000
Hydrogen	Sulfur	Hydrogen sulfide	15.902
Potassium	Bromine	Potassium bromide	2.043

Table 1–2
Weight Ratios for the Combination of Pairs of Substances A and B

When many experiments on many different substances yield results that display the same theme, those results form the basis of an **experimental law.** The law that the definite weight ratios in Table 1–2 show is the **law of definite proportions.** This law represents the first substantial evidence for the idea that elements and compounds are composed of fundamental units. If each element or compound consists of tiny units, each with a distinctive weight, there will undoubtedly be transformations in which some number of units of one substance combine with a fixed number from another substance. This picture must yield a combination of the two substances in a fixed proportion by weight. Figure 1–3 represents this combination for lime and carbon dioxide: the extremely small, fundamental unit of each substance is represented by a box, and the size of the box is proportional to the weight of the unit. If we take a 1.000-g sample of carbon dioxide, it will contain a fixed number of fundamental units of carbon dioxide. If the lime unit weighs 1.274 times as much as the carbon dioxide unit, 1.274 grams of lime will contain the same number of units as 1.000 gram of carbon dioxide. Such a situation would result in an example of the law of definite proportions.

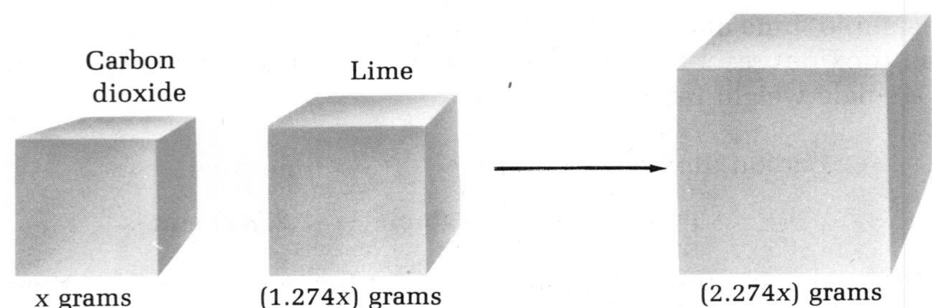

Figure 1–3 Graphic representation of the reaction between lime and carbon dioxide, in terms of the combination of very small units of given weights.

Law of Multiple Proportions

There are alternative explanations to the law of definite proportions that do not require the existence of fundamental units. Perhaps the lime contains holes and only a fixed amount of gas can fill in the holes. Such a hole theory could explain the law of definite proportions in all cases.

Let us design an experiment that can allow us to decide between the hole hypothesis and the fundamental-unit hypothesis. If fundamental units exist, they may not have to combine in one fixed ratio. Perhaps under one set of conditions one fundamental unit of a substance A combines with one unit of a substance B to form a new substance, which we could symbolize as AB. It is not inconceivable that a second set of conditions can be found in which one unit of A combines with two units of B to produce a new substance, symbolized as AB_2. The two combinations are depicted in Figure 1–4 in the box representation analogous to Figure 1–3. In this case, there are two definite proportions for the combination of A and B, but the two proportions are related to each other in a simple fashion. For a given weight of A, the weight of B required to produce AB_2 is exactly twice that to produce AB. Thus, the ratio of the two definite proportions is two. Many cases of such multiple combinations are known, and the small number of definite weight ratios found are indeed related to each other in a simple way. The simple relation between the definite weight ratios is the **law of multiple proportions.** An example of this law occurs with the combination of carbon and oxygen. Carbon burns in a limited supply of oxygen to form carbon monoxide (CO) and the weight ratio of carbon to oxygen in CO is 0.750. When oxygen is in excess, carbon and oxygen form carbon dioxide (CO_2) and the weight ratio (C/O) is 0.375.

The hole theory is hard pressed to explain multiple proportions. Any fractional number of holes can be filled. Why should there be a simple relation between the numbers of holes filled under two different sets of circumstances? From the law of definite proportions, the English scientist John Dalton, early in the nineteenth century, formulated the atomic hypothesis of matter: **Each element contains a single type of fundamental unit,**

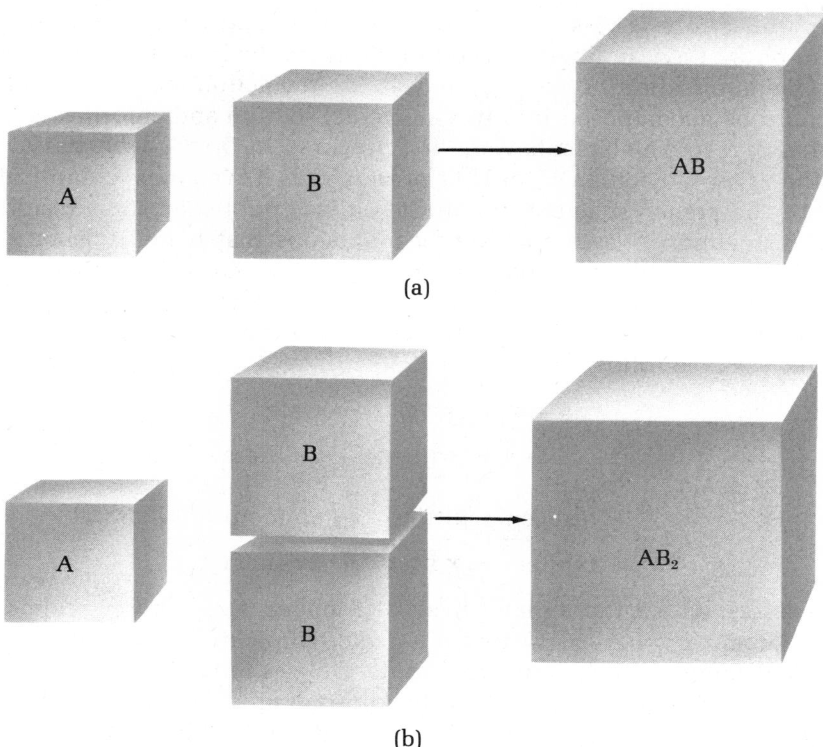

Figure 1–4
Two ways for fundamental units of A and B to combine: (a) as AB, (b) as AB_2.

the atom, and the atoms of different elements possess different weights. Different kinds of atoms can combine to form molecules, which are the fundamental units of compounds.

1–3. Molecular Formulas and Molecular Weights

The laws of combining weights are compelling evidence for the existence of atoms and molecules. However, we do not know from those laws the weights of individual atoms and molecules; nor do we know the numbers and kinds of atoms in given molecules (**molecular formulas**). Determination of molecular and atomic weights and molecular formulas was the central problem in chemistry in the first half of the nineteenth century.

The Avogadro Hypothesis

We can illustrate this central problem by a concrete example. Water can be converted to the elements hydrogen (H) and oxygen (O) by passing an electrical current through it. The chemical conversion that passing an electrical current through a substance brings about is termed **electrolysis.** The technique of electrolysis was introduced and widely applied in the laboratory of Humphrey Davy in England about the time of the Dalton atomic hypothesis. One application of electrolysis was the revelation that common table salt could be decomposed into two elements, sodium and chlorine. A number of very reactive elements were first prepared in Davy's laboratory.

Is the water molecule OH, or H_2O, or HO_2, or is it even more complicated? In 1808, the French scientist Joseph Gay-Lussac published some results on the combination of gases and some conclusions that were to become important in solving the formula of water and molecular formulas in general. If the pressures and temperatures of different gases are made the same, the relation between the volumes of the reacting gases is a simple one. Some examples are the following:

$$\begin{array}{rcl}\text{Hydrogen} + \text{oxygen} & \rightarrow & \text{water} \\ 2x \text{ liters} + x \text{ liters} & \rightarrow & 2x \text{ liters}\end{array}$$

$$\begin{array}{rcl}\text{Hydrogen} + \text{nitrogen} & \rightarrow & \text{ammonia} \\ 3x \text{ liters} + x \text{ liters} & \rightarrow & 2x \text{ liters}\end{array}$$

These results led the Italian physicist Amadeo Avogadro to propose, in 1811, that equal volumes of gases at a given temperature and pressure contain equal numbers of fundamental units. He was forced to make a further hypothesis, however, to account for the volume proportions in the reaction of hydrogen and oxygen to produce water and the reaction of hydrogen and nitrogen to produce ammonia. Dalton had proposed that all gaseous elements consisted of separate atoms. If this were true, and if equal gas volumes contained equal numbers of atoms in the case of elements and molecules in the case of compounds, the two reactions would be formulated from their combining volumes as

$$\begin{array}{rcl}\text{Hydrogen} + \text{oxygen} & \rightarrow & \text{water} \\ 2 \text{ volumes} + 1 \text{ volume} & \rightarrow & 2 \text{ volumes} \\ 2 \text{ units} + 1 \text{ unit} & \rightarrow & 2 \text{ units} \\ 2\,\text{H} + \text{O} & \rightarrow & 2 \text{ units}\end{array}$$

$$\begin{array}{rcl}\text{Hydrogen} + \text{nitrogen} & \rightarrow & \text{ammonia} \\ 3 \text{ volumes} + 1 \text{ volume} & \rightarrow & 2 \text{ volumes} \\ 3 \text{ units} + 1 \text{ unit} & \rightarrow & 2 \text{ units} \\ 3\,\text{H} + \text{N} & \rightarrow & 2 \text{ units}\end{array}$$

If two identical units are to be formed from the combinations H_2O and NH_3, it would be necessary to split the oxygen and nitrogen atoms. Avogadro's additional hypothesis was that hydrogen, oxygen, and nitrogen consist of the diatomic molecules, H_2, O_2, and N_2. The volume analysis then yields

Hydrogen +	oxygen	→ water
2 volumes +	1 volume	→ 2 volumes
2 units +	1 unit	→ 2 units
$2\,H_2$ +	O_2	→ $2\,H_2O$

Hydrogen +	nitrogen	→ ammonia
3 volumes +	1 volume	→ 2 volumes
3 units +	1 unit	→ 2 units
$3\,H_2$ +	N_2	→ $2\,NH_3$

The Cannizzaro Procedure

Unfortunately the Avogadro hypothesis that hydrogen, oxygen, and nitrogen are diatomic had no independent support, and the whole problem of molecular formulas remained controversial until the middle of the nineteenth century, when the Italian chemist Stanislao Cannizzaro published the definitive answer.

The steps in Cannizzaro's procedure are as follows:

1. At a predetermined temperature and pressure, which are made the same for all gases, measure the densities of a large number of gases that contain hydrogen. The **density** of any substance is defined as the weight of any given amount of that substance divided by the volume occupied by that amount. If the Avogadro hypothesis of "equal volumes–equal numbers" of fundamental units is correct, the ratio of any two of these densities is the ratio of the **molecular weights** of the two substances.
2. Analyze each hydrogen-containing gas and report the number of grams of hydrogen in one liter of the gas at the temperature and pressure of Step 1.
3. In a large number of gases, some gases will consist of molecules containing one hydrogen atom; others will have two, three, and so forth. Take the weight of hydrogen in one liter of each gas and divide it by the weight of hydrogen in one liter of pure hydrogen gas. Results of steps 1, 2, and 3 are reported in Table 1–3 for a number of hydrogen-containing gases. The table lists the ratio of the weight of hydrogen in one liter of each gas to the weight of hydrogen in one liter of pure hydrogen.

No matter how many substances are listed in this table no ratio in the last column is ever found below 0.5. Surely, Cannizzaro concluded, this

Table 1–3 Ratios of the Densities of Hydrogen-Containing Gases to the Density of Hydrogen Gas. The Ratios of the Weights of Hydrogen in One Liter of Hydrogen-Containing Gases to the Weight of Hydrogen in One Liter of Pure Hydrogen

Compound	Density Ratio (Compound/Hydrogen)	Hydrogen Weight Ratio (Compound/Hydrogen)
Water vapor	9.0	1.0
Natural gas (methane)	8.0	2.0
Ammonia	8.5	1.5
Hydrogen chloride	18.2	0.5
Hydrogen bromide	40.5	0.5
Hydrogen sulfide	17.0	1.0

means that Avogadro was correct and gaseous hydrogen is H_2. Furthermore, each molecule of water contains two hydrogen atoms, natural gas four, ammonia three, hydrogen chloride one, and so forth.

He could then define a system of molecular weights based on the arbitrary definition that the H_2 molecule weighs two **atomic mass units** (amu). This makes the H atom the standard atom in this system and the H atom has a mass of one amu.

With the molecular formula and weight of H_2 established, Cannizzaro could then determine other molecular formulas and atomic and molecular weights in a manner that we now consider. Table 1–4 lists the molecular weights of oxygen-containing substances. In each case the molecular weight in amu was calculated as twice the ratio of the density of the gas to the density of hydrogen gas. The weight percentage of each element in each compound was then determined, and the table reports the weight of oxygen in amu in one molecule of each of the compounds. For example, the molecular weight of water is listed as 18.0 because the density ratio, water/hydrogen, in Table 1–3 is 9.0. Water contains 88.9 percent oxygen by weight, so the weight of oxygen in each water molecule is $(18.0)(0.889) = 16.0$ amu. This number appears in the last column of Table 1–4. This comparison of hydrogen and water is also depicted in Figure 1–5. No matter how many oxygen-containing compounds are listed in this table, the lowest number found in

Table 1–4 Weights of Oxygen in Various Molecules

Molecule	Molecular Weight (amu)	Weight of Oxygen in One Molecule (amu)
Water	18.0	16.0
Fermentation gas	44.0	32.0
Ozone	48.0	48.0
Oxygen	32.0	32.0
Burned sulfur	64.0	32.0
Laughing gas	44.0	16.0

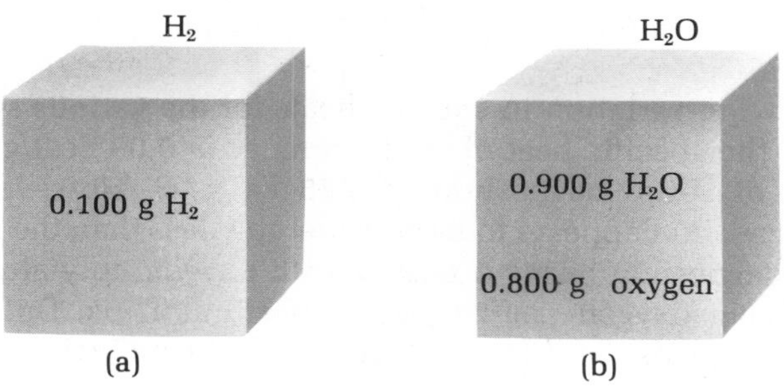

Figure 1–5
Data useful in the determination of the atomic weight of oxygen and the molecular formula of water. Given are the weights of (a) hydrogen and (b) water and oxygen in one-liter containers of (a) gaseous hydrogen and (b) gaseous water at the same temperature and pressure.

the right-hand column is 16.0, and surely 16.0 amu is the atomic weight of oxygen.

This procedure may then be applied to nitrogen-containing compounds, or to any other element found in a sizable number of gaseous species. In this manner, fermentation gas is found to be CO_2 and the atomic weight of carbon is 12.0. The acrid-smelling gas produced by the burning of sulfur is SO_2 and the atomic weight of sulfur is 32.0; and laughing gas is N_2O with the atomic weight of nitrogen equal to 14.0. From Table 1–4, we note that elemental gaseous oxygen may exist either in the normal form, O_2, or in the more unusual form O_3, ozone.

Rule of Dulong and Petit

The Cannizzaro method was invaluable because it firmly established atomic weights for many elements and formulas for many molecules. Unfortunately, it is limited to those elements from which gaseous molecules can be formed under normal conditions, and this limitation excludes most of the heavier elements, whose compounds are typically solids except at elevated temperatures. Thus, for example, 16 g oxygen combines with 200 g liquid mercury to yield the red solid oxide of mercury. If this solid has the formula HgO, the atomic weight of mercury is 200 (we take the atomic weight of oxygen as 16.0); but perhaps it is HgO_2, in which case the atomic weight is 400, or Hg_2O with an atomic weight of 100.

In 1819, two French scientists, Pierre Dulong and Alexis Petit, discovered a property of solids that resolves the ambiguity in the atomic weight of mercury and other heavy elements. They measured the **specific heats** of many solids. The specific heat involves two physical measurements that we have not yet talked about. The first is temperature (see Section 2–1). You are already well acquainted with the measurement of temperature in **Fahrenheit** degrees (°F). Scientists use the **centigrade** temperature scale and temperatures are reported in **Celsius** degrees (°C). We are all aware that the addition of heat to a substance normally increases the temperature of the substance. The specific heat of a substance is defined as the amount of heat required to raise the temperature of one gram of the substance by one degree C. The unit of heat is the **calorie,** which is defined as the amount of heat required to raise the temperature of one gram of liquid water by one degree Celsius.

There is a wide variation in specific heats for the various solid elements. For example, the specific heat of solid mercury is 0.033 cal/g-°C, while the specific heat of solid magnesium is 0.25 cal/g-°C. However, the atomic weight of magnesium appears to be considerably less than the atomic weight of mercury. Magnesium burns if heated with oxygen, to yield a solid oxide containing 16.0 g oxygen per 24.3 g magnesium. Thus, Dulong and Petit reasoned that perhaps the difference in specific heats is the consequence of the different atomic weights of mercury and magnesium. From the combining weights of these substances with a given weight of oxygen, we may conclude that the mercury atom is considerably heavier than the magnesium atom. Thus, one gram of mercury contains many fewer atoms than one gram of magnesium. If a heat input increases the motion of atoms (a correct assumption, as we shall see in Chapter 2), perhaps equal numbers of atoms in a solid absorb equal amounts of heat in a given temperature rise, regardless of the nature of those atoms. Thus, we wish to compare the heats required to raise equal numbers of atoms by one degree C. We can do this by multiplying the specific heat by the atomic weight, and this arithmetic product should be the same for all solid elements. In the cases of mercury and magnesium, we do not know the atomic weights from the combining weight data given in this section, but there are only a limited number of possibilities. By examining the specific heats of many solid elements and the possible atomic weights, Dulong and Petit became convinced that there is such a constant product and it is approximately 6 cal-amu/g-°C. This value would indicate that the approximate atomic weights of Hg and Mg are

$$\text{Hg: } \frac{6}{0.033} = 180$$

$$\text{Mg: } \frac{6}{0.25} = 24$$

and that the atomic weights indicated by the combination of these elements with oxygen are 200 and 24.3 for Hg and Mg, respectively.

Table 1–5 shows that the rule of Dulong and Petit is a remarkably good one, applicable to a wide range of elements. It does fail with the very lightest elements, however. We can explain this failure through a model to be developed in Chapter 4, and the failure itself is discussed in terms of that model in Chapter 12.

The table shows that the rule also applies to solid compounds that do not contain light elements such as hydrogen or oxygen. Thus for HgI_2 the average atomic weight is

$$\frac{200.59 + 2(126.90)}{3} = 151.46 \text{ amu}$$

When this number is multiplied by the specific heat of HgI_2, the result is

	Specific Heat	Atomic Weight	Product
Element			
Boron	0.31	10.8	3.3
Carbon (graphite)	0.17	12.0	2.0
Magnesium	0.25	24.3	6.1
Aluminum	0.21	27.0	5.7
Sulfur	0.18	32.0	5.8
Iron	0.11	55.8	6.1
Copper	0.093	63.5	5.9
Arsenic	0.080	74.9	6.0
Silver	0.056	107.9	6.0
Iodine	0.050	126.9	6.3
Platinum	0.032	195.1	6.2
Mercury	0.033	200.6	6.6
Lead	0.031	207.2	6.4
Compound			
H_2O (ice)	0.50	6.0	3.0
MgO	0.22	20.1	4.4
$MgCl_2$	0.19	31.7	6.0
FeS	0.14	43.9	6.1
CuI	0.066	95.2	6.3
AgCl	0.088	71.7	6.3
HgI_2	0.041	151.5	6.2
PbS	0.050	119.6	6.0

Table 1–5 Products of Atomic Weight and Specific Heat for Some Solid Elements and Compounds

6.2 cal-amu/g-°C, and the agreement with the Dulong-Petit rule is excellent. Surely the rule, in addition to being very useful in obtaining atomic weights, indicates a great similarity in the motions of atoms in a wide variety of solids.

A list of the known elements with their symbols and atomic weights is printed inside the front cover of this text. The reader may notice that hydrogen is listed as 1.008, instead of exactly 1, as previously indicated in the discussion of the Cannizzaro method. This small change arises from a change in the arbitrarily chosen standard. (We shall treat this question in more detail in Chapter 3 when we begin to treat the structures of atoms themselves.)

1–4. Reaction Equations and the Mole Concept

Reaction Equations

Atomic and molecular weights and also molecular formulas were originally obtained largely from data on the definite combining weights of elements and compounds in reactions. The molecular formulas allow us to formulate any chemical reaction in terms of a **reaction equation,** which gives the numbers and kinds of reactant molecules that yield fixed numbers and kinds of product molecules. Examples of reaction equations are the following:

$$H_2(g) + Cl_2(g) \rightarrow 2\,HCl(g) \qquad \textbf{1–2}$$

$$H_2O(l) + CO_2(g) \rightarrow H_2CO_3(aq) \qquad \textbf{1–3}$$

$$2\,Na(s) + Cl_2(g) \rightarrow 2\,NaCl(s) \qquad \textbf{1–4}$$

Reaction 1–2 says that one molecule of gaseous hydrogen (gaseous species are indicated by a lowercase g in parentheses) reacts with one molecule of chlorine gas to produce two molecules of hydrogen chloride. In reaction 1–3, one molecule of liquid water reacts with one molecule of gaseous carbon dioxide to produce a molecule of carbonic acid dissolved in water (the symbol *aq*, for "aqueous," indicates a solution in water). In reaction 1–4, two atoms of solid sodium react with a chlorine molecule to produce solid sodium chloride.

There are two requirements for writing a proper equation for a reaction. **First,** it must properly represent the reactant and product species. Thus, it would be improper to write reaction 1–2 as

$$H(g) + Cl(g) \rightarrow HCl(g) \qquad \textbf{1–5}$$

since gaseous hydrogen and chlorine consist almost exclusively of diatomic molecules. The designations *g*, *l*, *s*, and *aq* are often placed after each species in parentheses to indicate its physical state. **Second,** the reaction equation must be balanced: the same numbers and kinds of atoms must appear on both sides of the equation.

The chemist thinks about a given chemical reaction in terms of individual molecules and the reaction equation. It is an experimental fact that gaseous hydrogen and oxygen in the presence of a spark combine explosively to yield water and a large quantity of heat. The chemist wants to know what it is about the structure of H_2O that leads to such a large energy release when it is formed. Why should hydrogen be H_2; why not H or H_3 or something else? Clearly the O_2 molecule must be ruptured during production of H_2O, because H_2O contains only one oxygen atom. How can the fission of a molecule take place so rapidly as it obviously must in an explosion? Once the existence and formulas of molecules had been demonstrated in all matter, chemists set about the task that continues to this day—explaining all properties of matter in terms of the structures and motions of the constituent molecules. We want to find out, in a logical fashion, how those structures and motions were discovered and how they relate to specific properties of bulk substances.

The Mole

In focusing on molecules, we shall refer repeatedly to reaction equations and think of them as applying to individual molecules. However, an equation like reaction 1–2 tells us not only how molecules combine, but also how large collections of molecules in bulk matter combine. Equation 1–2 still holds when multiplied through by any constant.

It would be very convenient in calculating the weight of hydrogen that reacts with some weight of chlorine if we had a unit that corresponds to the number of molecules contained in the quantity of matter encountered in typical laboratory experiments. Equation 1–2 tells us that a unit of H_2 combines with another unit Cl_2, to yield two units of HCl. An appropriate unit is the weight in grams, equal to the sum of the atomic weights (in atomic mass units), of elements in the formula of each substance in the reaction equation. For H_2 this is 2.016 g, for Cl_2 it is 70.90 g, and for HCl it is 36.46 g. Each of these quantities contains the same number of molecules. This is the **formula weight** of the element or compound.

The unit we have defined is the **mole.** Our definition makes the mole a given number of molecules, a number contained in the quantities of matter normally encountered in the chemical laboratory. This number of molecules in a mole is termed **Avogadro's number** (symbolized by N) in honor of Amadeo Avogadro, and it has been found to be 6.023×10^{23} molecules per mole. There are many independent ways to measure this number. It is also called the Loschmidt number after the Austrian school teacher who first calculated it in 1865.

Let us do some elementary conversions involving the mole and Avogadro's number. How many moles does 10.62 g of solid silver chloride, AgCl, represent? How many atoms does this sample of AgCl contain? In the first problem, we need the conversion factor from grams of AgCl to moles of AgCl, and this is the molecular weight.

$$\begin{aligned} 1 \text{ mole of AgCl} &= 1 \text{ mole of Ag} + 1 \text{ mole of Cl} \\ &= 107.87 \text{ g of Ag/mole of Ag} + 35.45 \text{ g of Cl/mole of Cl} \\ &= 143.32 \text{ g of AgCl.} \end{aligned}$$

Thus, the conversion is

$$10.62 \text{ g of AgCl} \left(\frac{1 \text{ mole of AgCl}}{143.32 \text{ g of AgCl}}\right) = 7.41 \times 10^{-2} \text{ mole of AgCl}$$

The number of atoms in this sample is twice the number of AgCl units, or

$$\begin{aligned} &(7.41 \times 10^{-2} \text{ mole}) \left(\frac{2 \text{ atoms}}{1 \text{ molecule}}\right) \left(\frac{6.023 \times 10^{23} \text{ molecules}}{1 \text{ mole}}\right) \\ &= 8.9 \times 10^{22} \text{ atoms} \end{aligned}$$

Empirical Formulas from Chemical Analyses

One of the most immediately useful applications of the mole concept is in obtaining proportions of atoms in molecules from weight proportions of the constituent elements of any compound. The measurement of the weight percentages of the elements in substances is termed **stoichiometry,** and it forms a sizable segment of the subject **analytical chemistry.** Techniques exist for identification and quantitative measurement of all of the elements, either alone or in compounds, and the bases of many of the qualitative (identification) and quantitative analyses will subsequently be demonstrated.

With these procedures, it is not difficult to determine the elemental composition by weight of any given compound. Suppose that a liquid compound were analyzed and found to contain 75.9 percent carbon, 6.3 percent hydrogen, and 17.8 percent nitrogen by weight. A 100.0-gram sample contains 75.9 grams = 75.9/12.0 = 6.32 moles of carbon atoms. The sample also contains 6.3/1.008 = 6.3 moles of hydrogen atoms and 17.8/14.0 = 1.27 moles of nitrogen atoms. The atom ratio in each molecule of this compound is (6.32/1.27) : (6.3/1.27) : (1.27/1.27), or 5.0 : 5.0 : 1.0. The molecule must have the formula C_5H_5N or an integral multiple of it: $C_{10}H_{10}N_2$, $C_{15}H_{15}N_3$, or something even more complex. The formula C_5H_5N, which is the simplest of all the possibilities, is known as the **empirical formula** of the compound. Such an empirical formula can always be obtained by analysis of the compound.

Determining the molecular formula necessitates a molecular weight measurement for the compound in addition to elemental analysis. Suppose that our liquid can be vaporized at some convenient temperature and pressure and that the vapor is 39.5 times as dense as gaseous hydrogen under the same conditions. This shows the molecular weight to be 79.0. Therefore the molecular formula is C_5H_5N. It is a relatively easy task to obtain the molecular formula of any pure compound that is easily vaporized.

The mole concept can also be used to relate weights of species to each other through molecular formulas and reaction equations. For example, we might ask what weight of $Cl_2(g)$ is required to convert 0.562 g of $H_2(g)$ to HCl(g) according to equation 1–2.

$$H_2(g) + Cl_2(g) \rightarrow 2\ HCl(g)$$

This weight of H_2 represents 0.562/2.016 = 0.279 mole of H_2. The reaction equation reads that 0.279 mole of H_2 reacts with 0.279 mole of Cl_2 to produce (2)(0.279) = 0.558 mole of HCl. Thus, (0.279)(70.90) = 19.8 g of Cl_2 are required for complete reaction of the H_2.

Mole Concept in Quantitative Analysis

The mole concept is particularly useful in the calculations that are normally performed in the quantitative analysis of substances. A common quantitative analysis for fluorine takes advantage of the very low solubility of solid CaF_2, calcium fluoride, in water. Suppose we take a 1.000-g sample of a compound known to contain fluorine and heat it in a sealed container with Na_2O_2, sodium peroxide. Such treatment is known to convert the fluorine to NaF, sodium fluoride, which is highly soluble in water. The fluorine is then quantitatively removed from solution as CaF_2 by adding the highly soluble $Ca(NO_3)_2$, calcium nitrate. The solid CaF_2 is filtered off and dried. Finally the CaF_2 is treated with sulfuric acid and thereby converted quantitatively to $CaSO_4$, calcium sulfate, which is then thoroughly dried by heating and weighed. From the weight of the $CaSO_4$, the weight percentage of fluorine in the original compound can be directly calculated through the mole concept.

The changes involved in the sequential steps for converting solid NaF to $CaSO_4$ are

$$\mathrm{NaF(s) \rightarrow NaF(aq)} \qquad \textbf{1–6}$$

$$\mathrm{2\,NaF(aq) + Ca(NO_3)_2(aq) \rightarrow CaF_2(s) + 2\,NaNO_3(aq)} \qquad \textbf{1–7}$$

$$\mathrm{CaF_2(s) + H_2SO_4(l) \rightarrow CaSO_4(s) + 2\,HF(g)} \qquad \textbf{1–8}$$

Let the number of moles of fluorine atoms in the 1.000-g sample of compound be x. The first step of the procedure will then yield x moles of NaF, because the NaF formula unit contains one fluorine atom. We can use equations 1–6, 1–7, and 1–8 to obtain the number of moles of each of the substances involved in the analysis.

$$\mathrm{NaF(s) \rightarrow NaF(aq)}$$

$$\mathrm{x\ moles \rightarrow x\ moles}$$

$$\mathrm{2\,NaF(aq) + Ca(NO_3)_2(aq) \rightarrow CaF_2(s) + 2\,NaNO_3(aq)}$$

$$\mathrm{x\ moles} + \frac{\mathrm{x}}{2}\,\mathrm{moles} \rightarrow \frac{\mathrm{x}}{2}\,\mathrm{moles} + \mathrm{x\ moles}$$

$$\mathrm{CaF_2(s) + H_2SO_4(l) \rightarrow CaSO_4(s) + 2\,HF(g)}$$

$$\frac{\mathrm{x}}{2}\,\mathrm{moles} + \frac{\mathrm{x}}{2}\,\mathrm{moles} \rightarrow \frac{\mathrm{x}}{2}\,\mathrm{moles} + \mathrm{x\ moles}$$

Suppose that the final weight of $CaSO_4$ is 0.781 g, so that $x/2$ is $0.781/136.1 = 5.74 \times 10^{-3}$ mole, where 136.1 is the molecular weight of $CaSO_4$. Thus, the number of moles of fluorine in the original sample was

$$\mathrm{x} = (2)(5.74 \times 10^{-3}) = 1.148 \times 10^{-2}\ \mathrm{mole}$$

The atomic weight of fluorine is 19.0, and consequently the weight of fluorine in the original sample was $19.0\mathrm{x} = 0.218$ g. The compound in question contains 21.8 percent fluorine by weight, and we have one of the first numbers required in determining its molecular formula.

Let us consider a second example. A mixture of aluminum metal and copper metal is easily analyzed for aluminum because aluminum reacts with an aqueous solution of HCl and copper does not. The reaction is the following:

$$Al(s) + 3\,HCl(aq) \rightarrow AlCl_3(aq) + \tfrac{3}{2}\,H_2(g) \qquad \textbf{1–9}$$

A 0.3196-g sample of the Al-Cu mixture yielded 0.0225 g of hydrogen gas. What was the percentage by weight of Al in the mixture? From equation 1–9, we see that x moles of Al yield (3x/2) moles of H_2. The number of moles of H_2 in 0.0225 g is given by

$$\text{Moles of } H_2 = \frac{(0.0225\text{ g})}{(2.016\text{ g/mole})} = 0.0112 \text{ mole}$$

The number of moles of Al in the original sample was $\frac{2}{3}$ of 0.0112 mole = 7.47×10^{-3} mole. The weight of Al in the original sample was

$$\text{Grams of Al} = (7.47 \times 10^{-3}\text{ mole})(26.98\text{ g/mole})$$

$$= 0.201 \text{ g}$$

The weight percentage of Al in the original sample was

$$\text{Wt. \%} = \frac{(0.201)}{(0.3196)}(100) = 63.0\%$$

The Limiting Reagent

Chemical reactions are seldom carried out in the exact molecular proportions given in the reaction equation. At least one of the reacting species is ordinarily present in excess. Suppose that we were to mix 5.16 g of $H_2(g)$ and 10.53 g of $Cl_2(g)$ and allow them to react. How many grams of HCl are produced? The numbers of moles of H_2 and Cl_2 involved are 5.16/2.016 = 2.56 and 10.53/70.90 = 0.1485, respectively. The reaction equation is the following:

$$H_2(g) + Cl_2(g) \rightarrow 2\,HCl(g)$$

According to this reaction equation, the number of moles of HCl produced is twice the number of moles of H_2 reacting, which in turn equals the number of moles of Cl_2 reacting. In our case, the numbers of moles of the reactants are unequal, and 0.1485 mole represents the maximum amount of H_2 that can react. This limitation arises because each H_2 requires a Cl_2 for reaction and there is only 0.1485 mole of Cl_2 available. Thus, Cl_2 is termed the **limiting reagent** since it limits the amount of product formed. In this case, the number of grams of HCl formed is (0.1485)(2)(36.46) = 10.83 g.

Mole Concept Applied to Substances in Solution

Reactions are often carried out in which the reacting species are dissolved in an appropriate solvent. Water is the most common solvent. In such reactions, adding the reacting species by measured solution volumes rather than weight proves convenient. For this reason, the amount of a given substance in a solution is commonly reported in terms of the **molarity** of the solution. The molarity of a substance in solution is defined as the number of moles of that substance in one liter of solution.

Let us consider some common sample calculations associated with the concept of molarity. In the first calculation, we wish to prepare 250 ml of a 1.25-molar (M) solution of NaCl. How many grams of NaCl are required? One liter of a 1.25M solution would require 1.25 moles of NaCl. Since only 0.250 l is being prepared, we require $(1.25)(0.250) = 0.313$ mole of NaCl $= (0.313)(58.44) = 18.29$ g of NaCl.

A common solution reaction is one in which aqueous solutions of NaCl and silver nitrate ($AgNO_3$) are mixed. The reaction is

$$NaCl(aq) + AgNO_3(aq) \rightarrow AgCl(s) + NaNO_3(aq) \qquad \textbf{1-10}$$

How many ml of a 0.1262M solution of $AgNO_3$ are required to react completely with 43.73 ml of a 0.0894M solution of NaCl? How many grams of AgCl are produced in the complete reaction? Since reacting species are always simply related by numbers of moles through the reaction equation, we must first convert the ml of NaCl solution to moles of NaCl. The result is

$$\text{Moles of NaCl} = \left(\frac{0.0894 \text{ mole}}{\text{liter}}\right)(0.04373 \text{ liter})$$

$$= 3.909 \times 10^{-3} \text{ mole NaCl}$$

The reaction equation shows that one mole of NaCl reacts with one mole of $AgNO_3$, so that 3.909×10^{-3} mole of $AgNO_3$ is required in the present case. The number of liters of $AgNO_3$ solution required is shown in the equation

$$\text{Liters of solution} = \left(\frac{\text{liter}}{0.1262 \text{ mole}}\right)(3.909 \times 10^{-3} \text{ mole})$$

$$= 0.03097 \text{ liter}$$

$$= 30.97 \text{ ml}$$

The number of moles of silver chloride (AgCl) produced is 3.909×10^{-3}, which converted equals $(3.909 \times 10^{-3})(143.32) = 0.560$ g.

SUGGESTIONS FOR ADDITIONAL READING

Kieffer, W. F., *The Mole Concept in Chemistry*, Reinhold, New York, 1962.

Nash, L. K., *Stoichiometry*, Addison-Wesley, Reading, Mass., 1969.

Benfey, O. T., *Classics in the Theory of Chemical Combination*, Dover, New York, 1963.

Dickerson, R. E., Gray, H. B., and Haight, G. P., *Chemical Principles*, Benjamin, Menlo Park, Calif., 1974.

PROBLEMS

(Answers to selected end-of-chapter problems are given at the back of the text, following the Appendices.)

1. The force of gravity on the moon is one-sixth the force of gravity on the earth. The mass of an object is determined to be 5.00 grams by weighing it on a balance on earth. The balance, object, and standard masses are transported to the moon, and the object is reweighed. What mass will be measured for the object in this new determination?
2. Five people are told to measure the length of a room, and they report the following values, all in feet: 10.53, 10.41, 10.58, 10.62, and 10.40. If you were to report the average of these five measurements, how many significant figures would you give in the number you report?
3. Give the results of each of the following operations. Give only the significant figures in each case.
   a. (2.968)(1.92)/36.8932.
   b. 1.241 + 26.34 + 0.3784.
   c. 9.34 − 5.6.
4. Make each of the following unit conversions.
   a. 5.69 gallons to liters.
   b. 132 grams to pounds.
   c. 5280 ft (1 mile) to km.
   d. 100 meters to yards.
   e. 25 miles per hour to meters per second.
5. How many liters are in a cubic foot? A liter of liquid water weighs one kilogram. What is the weight in pounds of a cubic foot of water?
6. The molecular formula of table salt is NaCl. Use the atomic weights listed on the inside front cover of this text to calculate the percentage by weight of sodium in table salt.
7. Two compounds are known that contain only hydrogen and oxygen. The first, water, contains 11.19 percent hydrogen by weight. The second, hydrogen peroxide, contains 5.93 percent hydrogen by weight. Show that these two compounds represent an example of the law of multiple proportions. The molecular formula of water is H_2O. The hydrogen peroxide molecule also contains two hydrogen atoms. What is the molecular formula of hydrogen peroxide?
8. Table 1–6 contains density and composition information about four

nitrogen-containing gases, identified only by the code letters A, B, C, and D. The first column of numbers in the table lists the ratio of the gas density of the compound to the density of oxygen gas of known molecular weight equal to 32.0. The second column lists the percentage by weight of nitrogen in each of the four compounds. Given the fact that the molecules of at least one of the gases contain only one nitrogen atom, find the number of N atoms in molecules of each of the four compounds. What is the atomic weight of nitrogen?

Compound	Density Ratio (Compound/O_2)	Weight Percent N
A	1.375	63.6
B	0.875	100.0
C	0.9375	46.7
D	0.844	51.9

Table 1-6

9. Solid iron reacts vigorously with gaseous chlorine to produce a solid iron chloride that contains 44.06 percent iron. If the atomic weight of Cl is known to be 35.45, and the atomic weight of Fe is supposedly unknown at the time of this measurement, what do we calculate for the atomic weight of Fe if we assume the compound to be FeCl? $FeCl_2$? $FeCl_3$?

The specific heat of solid iron is 0.11 cal/g-°C. Use the rule of Dulong and Petit to decide which of the three formulas for iron chloride is correct.

10. A solid oxide of manganese is found to contain 63.19 percent manganese by weight. The specific heat of solid manganese is 0.12 cal/g. What is the most precise value of the atomic weight of Mn obtainable from these data?

11. Which of the following solid species should you expect to obey the Dulong-Petit rule: aluminum chloride, $AlCl_3$; aluminum sulfate, $Al_2(SO_4)_3$; ammonium iodide, NH_4I; potassium chloroplatinate, K_2PtCl_6? The specific heats of the four compounds are 0.188, 0.180, 0.115, and 0.112 cal/g respectively. Are your predictions borne out by calculation?

12. Balance each of the following reaction equations:

a. $H_2O_2(l) \rightarrow O_2(g) + H_2O(l)$.
b. $NH_4NO_3(s) \rightarrow H_2O(l) + N_2O(g)$.
c. $C_6H_6(l) + O_2(g) \rightarrow CO_2(g) + H_2O(g)$.

13. How many water molecules are there in a 10-ounce glass of liquid water? The density of liquid water is 1.000 g/ml.

14. The solid element calcium burns in oxygen to form a white solid oxide. It is found that 0.532 g of Ca yields 0.744 g of oxide. Use the known atomic weights of Ca and O to obtain the atomic ratio of Ca to O in the solid oxide.

15. A solid compound contains the following composition by weight: 35.6 percent potassium, 17.0 percent iron, 21.9 percent carbon, and 25.5 percent nitrogen. What is the empirical formula of the compound?

16. Protein contains a sizable percentage of nitrogen. The common method for determining the approximate protein content of a produce such as cattle feed is the Kjeldahl method for the quantitative determination of nitrogen. In this method all of the nitrogen is converted to ammonia, NH_3, whose quantity is easily measured. Suppose that a 0.1531-g sample of feed yields 0.0262 g of ammonia. What is the weight percentage of nitrogen in the cattle feed?

17. A mixture of NaCl and KCl is found to contain 53.6 percent Cl by weight. What is the percentage by weight of NaCl in the mixture?

18. Analyses for carbon and hydrogen are particularly easy to perform because these species are easily converted to CO_2 and H_2O by combustion with oxygen. A given compound was found to contain only oxygen, hydrogen, and carbon. A 0.2863-g sample of it was combusted with oxygen to yield 0.4199 g of CO_2 and 0.1718 g of H_2O. What is the empirical formula of the compound?

19. One method for cleaning up spilled liquid mercury is to spread powdered sulfur on it. The reaction that takes place is

$$Hg(l) + S(s) \rightarrow HgS(s)$$

and the solid HgS, mercuric sulfide, is easily swept up. Suppose equal weights of mercury and sulfur are allowed to react. What is the limiting reagent under these circumstances? What percentage by weight of the other substance remains, not reacting?

20. A compound possesses the molecular formula $C_3H_6O_2$. You wish to perform an analysis for carbon and hydrogen by combusting a 0.2969-g sample of the compound with O_2. The products are CO_2 and H_2O. What is the minimum weight of O_2 required? Actually you would use a large excess of O_2. Why?

21. A certain concentrated aqueous solution of sulfuric acid (H_2SO_4) contains 90 percent H_2SO_4 by weight. What weight of this solution would you use to prepare 5.00 liters of a 0.100-molar solution of H_2SO_4?

22. Ammonia (NH_3) dissolves in aqueous solutions of hydrochloric acid (HCl) according to the following reaction:

$$NH_3(g) + HCl(aq) \rightarrow NH_4Cl(aq)$$

What volume of 0.1016*M* HCl solution is required to react with the NH_3 produced in problem 16?

In practice an excess of HCl solution is employed to ensure that all of the NH_3 reacts. The excess HCl is then allowed to react with a solution of NaOH according to the following reaction:

$$HCl(aq) + NaOH(aq) \rightarrow NaCl(aq) + H_2O(l)$$

In a given analysis, 23.48 ml of 0.08640*M* NaOH was required for complete reaction with the excess HCl in an HCl solution containing NH_4Cl. Before addition of the ammonia, the HCl solution was 50.00 ml in volume and it was 0.1153*M* in HCl. How much ammonia was added to this solution?

23. A 0.2386-g sample of a mixture of NaCl and KCl is dissolved in water. A 0.1294*M* solution of $AgNO_3$ is added slowly until all of the dissolved chloride has been converted to solid silver chloride (AgCl). It was found that 30.55 ml of the $AgNO_3$ solution was required for complete reaction. What was the weight percentage of NaCl in the original sample?

2/ Models of Gaseous Substances

The concepts of atoms and molecules have now been introduced, but as yet you probably cannot form any picture in your mind of atoms and molecules from the data and conclusions so far presented. You don't know about their sizes, shapes, and motions. As we form models of substances in the gaseous state, we should be able to use them to explain common phenomena associated with gases. For example, you are probably aware that someone speaking in a room in which most of the air is replaced with helium speaks in a high-pitched voice. Why does this occur? Hydrogen is a much better conductor of heat than air. Why is this so? Why does the moon have no atmosphere? What is the principal factor controlling the speed of sound in air? These are questions that we should be able to answer by using models of gases.

2–1. The Gas Laws

The three states in which we commonly find matter are the solid, liquid, and gaseous states. The gaseous state differs markedly from the other two in that gases are normally much less dense than either solids or liquids. This low relative density is manifested in the large volume a gas occupies compared with the volume for the same number of molecules in the liquid or solid state at the same temperature and pressure. The volumes of substances in any of the three states can be conveniently compared, when there are equal numbers of molecules per volume, by reporting the **molar volume,** that is, the volume occupied by one mole of the substance. For example, the molar volume of water at 100 °C is 0.0187 liter/mole, while the molar volume of the steam produced by boiling this water at 100 °C is 30.6 liters/mole.

A second marked difference between gases and substances in the other two states is in their compressibilities. If liquid water is contained in a cylinder, as shown in Figure 2–1, and a sizable pressure applied to a piston, the volume of liquid water will, in general, decrease only slightly. The same will be true for any other liquid and for any solid. Gases, however, are

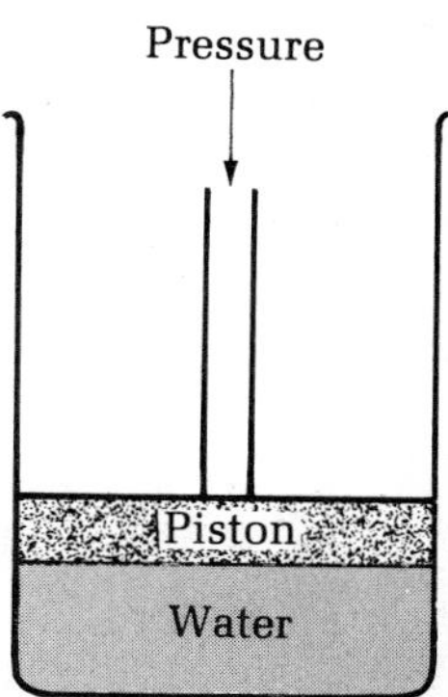

Figure 2–1
Apparatus for demonstrating the low compressibility of water.

"springy," and even a moderate pressure applied to a piston when a gas is enclosed in a cylinder produces a large fractional volume change.

A popular idea at one time to explain the low density and springiness of gases was that molecules became large and feathery in the boiling process. Thus, one model of the gaseous state is of large, fluffy molecules all in contact with one another, like popcorn in a pan. Liquids and solids are the unpopped kernels and boiling is the popping process. It is difficult to reconcile this model with the quantitative molar volume data, however. Different solids and liquids possess different molar volumes, but all gases possess the same molar volume according to the Avogadro hypothesis. Surely if the kernels are of different sizes, the popcorn ought to be of different sizes too. Thus, we must abandon a model of fluffy molecules all in contact with each other in gases.

The Mercury Barometer

Molar volumes of gases depend on both pressure and temperature. A useful model of the gaseous state must account quantitatively for changes in molar volume with pressure and temperature changes. First we must learn what the quantitative relation among gas volume, pressure, and temperature is. A mercury barometer can be used for measuring the pressure of air in the atmosphere. The Italian scientist Evangelista Torricelli first demonstrated this simple device in the middle of the seventeenth century; it is depicted in Figure 2–2. A glass tube is filled with mercury and inverted in a beaker of mercury in such a way that no air enters the tube. Air pressure causes the mercury to be held up in the column at height h, which varies typically between 70 and 80 cm depending on the day the measurement is made and the altitude at which it is made. The height of the mercury is independent of the cross-sectional area of the tube. Mercury is a convenient barometer liquid because of its high density of 13.6 g/ml. Liquid water, with a density of 1.0 g/ml, would be held up at least 70(13.6)/(1.0) = 952 cm = 31.2 ft.

The distance h = 76 cm = 760 mm of Hg has been chosen as the **standard atmosphere** of pressure, and most pressures we shall be reporting will be in atm (meaning standard atmospheres). Small pressures are commonly reported in torr, meaning mm of Hg, a unit named in honor of Torricelli. Thus, 760 torr = 1 atm.

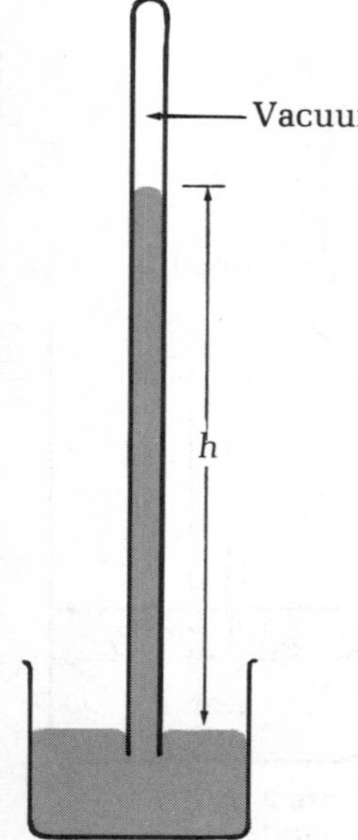

Figure 2–2
A Torricelli barometer.

Boyle's Law

In 1662, Robert Boyle published the results of his quantitative study of the compressibility of air. He used a U-shaped glass tube, as shown in Figure 2–3. He poured mercury into the open end, trapping air in the closed end, as shown. The pressure of the enclosed gas in atm is

$$P = \frac{h + h'}{760} \qquad \textbf{2–1}$$

where h is the distance in mm shown in Figure 2–3 and h' is the atmospheric pressure in torr at the time of the measurement of h. The tube used by Boyle (Figure 2–3) was graduated so that the volume of enclosed air could be easily determined as the pressure varied.

The pressure of the enclosed air was easily increased by adding additional mercury through the open end of the tube. Boyle discovered that if the pressure of enclosed air is doubled, the air volume halves. In general, the pressure P and volume V are related through

$$PV = \text{constant} \tag{2-2}$$

Boyle also discerned that the constant in equation 2–2 depends on the mass of air trapped in the tube and the temperature. At a given temperature, the constant in the equation increases with the quantity of trapped air, and with a given quantity of air the constant increases with the temperature.

When individual gases such as nitrogen and oxygen were discovered in air and other pure gaseous compounds were discovered, their compressibilities were also measured by the Boyle method, and equation 2–2 was found to hold for all these gases. The equation was then elevated to the status of **Boyle's law.**

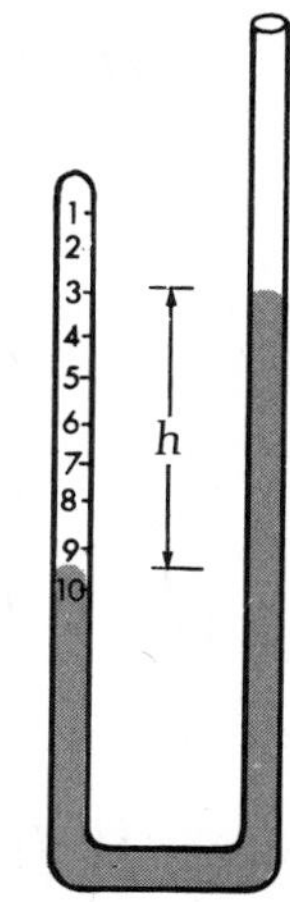

Figure 2–3
U-shaped tube for measuring the compressibility of gases and gas mixtures.

Intensive and Extensive Properties

Boyle did not pursue the quantitative dependence of the constant in equation 2–2 on the quantity of air or its temperature. Before we pursue these dependences, we must consider the nature and measurement of temperature. In the language of the scientist, temperature is an **intensive property** of matter, as contrasted with mass and volume which are **extensive properties.** To understand this distinction, consider Figure 2–4. In part (a) of this figure are shown two beakers of water that are in the same room at the same temperature and pressure. Each beaker contains 100 ml of water weighing 100 g. In part (b), the contents of one beaker have been emptied into the other. In combining the two water samples, some measurable properties of the water in beaker 2 have doubled and some have remained the same. The combined water has twice the mass and twice the volume of the sample originally in beaker 2. Any property that changes in going from the condition of (a) to (b) is, by definition, an extensive property. Thus, mass and volume are extensive properties. The pressure and temperature did not change between (a) and (b). Temperature and pressure are therefore intensive properties.

Measurement of Temperature

Temperature is an elusive property of matter. One of our goals is to relate temperature to our fundamental model of atoms and molecules. Everyone has a sense of the hotness and coldness of matter, but it is difficult to define just exactly what hotness and coldness are. Long before the molecular

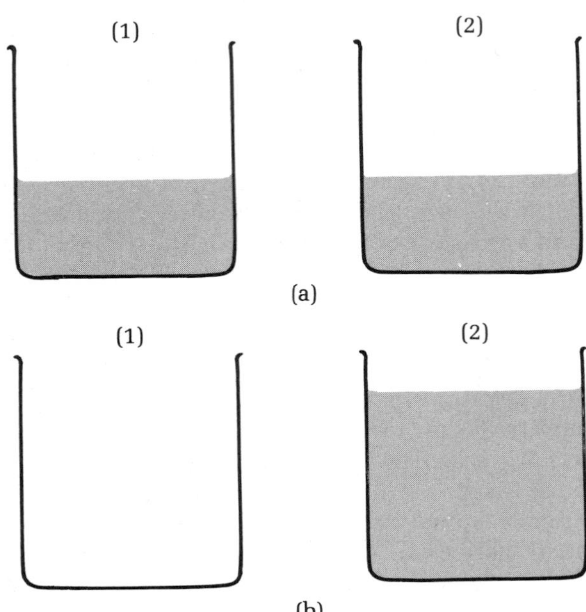

Figure 2–4
An illustration of extensive and intensive properties. (a) Identical beakers containing equal volumes of water; (b) the contents of one beaker emptied into the other.

model was developed, people found it desirable to report the hotness or coldness of the air on any given day in a way that could be generally understood. This is the sense in which temperature is used by most people even today. One requires no knowledge of the molecular basis of temperature to know how the air outside will feel, having heard the temperature given on a televised weather report.

To report reliable temperatures, we need a **reference temperature.** We need a system with a reproducible temperature that can serve as a standard for reporting all others. In turn, we must have a way of verifying the reproducibility of this standard and of reporting differences between any other temperature and the standard. A convenient standard is the temperature of ice and liquid water in contact with each other (in an insulated container so that the temperature is uniform throughout the mixture). The reproducibility of this temperature is verified by measuring the volume of any substance immersed in this water-ice mixture, since the volume of any sample of matter is found to vary with its hotness and coldness. Thus, a mercury thermometer is a volume-measuring device, involving changes in the volumes of mercury and glass. The fact that the mercury level of a given thermometer reaches the same height in all water-ice mixtures shows that the freezing point of water is indeed a reproducible temperature and suitable as a standard.

When the mercury thermometer is placed in air on a hot summer day, the mercury level rises, indicating that mercury, like most liquids and solids, expands as its temperature increases. We could measure this temperature by designating a standard thermometer (one based on the freezing point of pure water) and reporting the temperature on the hot summer day as 10.3 cm, meaning that the mercury height is 10.3 cm above the height for the water-ice mixture. This would not be satisfactory, since the standard thermometer can only be in one place at a time and many thermometers are required for the many measurements that must be made on any given day. This problem is surmounted by locating a second reproducible reference temperature, namely the boiling point of water at a pressure of one standard atmosphere. Suppose the mercury level on our hot summer day is 35.0 percent of the way from the first standard level (water-ice) to the second (boiling water). It will be at this relative level on all mercury-in-glass thermometers no matter how they are constructed, and we have a measure of temperature that everyone can use and understand.

The 35.0 percent in the preceding discussion is converted to degrees by defining a temperature scale using our two standards. Actually, such a scale is used, and it is called the Celsius scale after Anders Celsius, its originator. The water-ice temperature is defined as 0 °C and the boiling water temperature as 100 °C, so the temperature on the hot summer day is 35 °C. The Celsius scale is used in science in preference to the more awkward Fahrenheit scale, which is based on the definitions of the water freezing point as 32 °F and the water boiling point as 212 °F. The converting equation between Celsius (°C) and Fahrenheit (°F) temperature is

$$°F = \tfrac{9}{5}\,°C + 32\,° \qquad \textbf{2–3}$$

so that the 35 °C day is also a 95 °F day. This conversion between °C and °F is depicted in Figure 2–5.

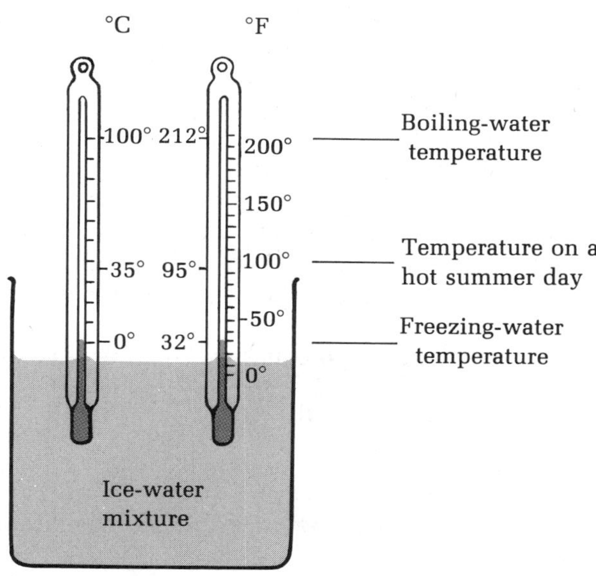

Figure 2–5 Comparative temperatures on Celsius and Fahrenheit thermometers.

Law of Charles and Gay-Lussac

It is relatively easy to modify the **U** tube of Figure 2–3 to determine the temperature dependence of gas volume at a given pressure. At a given measured temperature, such as room temperature, the gas volume is noted along with the mercury height h. The tube is then placed in a hot water bath of measured temperature and the gas is observed to expand, increasing h. Mercury is then removed from the tube until h, and thus the gas pressure, returns to its original value, and the gas volume at the new temperature recorded.

The French scientists Charles and Gay-Lussac determined that all the gases they studied, within the uncertainties of their measurements, showed the same temperature dependence of volume, namely that given in equation 2–4:

$$\frac{V}{V_0} = 1 + \alpha t \qquad \textbf{2–4}$$

In this equation, V_0 is the gas volume at $t = 0\,°C$ and it depends on the quantity of gas and the pressure. The constant α (Greek lowercase *alpha*) is the fractional increase in gas volume resulting from a temperature rise of 1 °C. It is termed the **coefficient of thermal expansion.** Charles and Gay-Lussac found, experimentally, that α has the same value for all gases irrespective of their identities, masses, and pressures. Actually, α shows a small dependence on pressure at pressures from zero to 1 atm. However, these changes were beyond the range of precision of the measurements of Charles and Gay-Lussac. As the gas pressure approaches zero, the value of α approaches $3.6609 \times 10^{-3}\,°C^{-1}$. Note that this strange unit of reciprocal temperature is required to make αt unitless as 1 and V/V_0 are in equation 2–4.

The existence of this reciprocal temperature that applies to all gases suggests that α can be converted to a new standard temperature of much greater fundamental significance than the freezing point of water. Suppose that we define a new temperature scale by

$$T = t + \frac{1}{\alpha} = t + 273.15, \qquad \textbf{2–5}$$

where T is termed the **absolute temperature** and is reported in degrees Kelvin (°K), a unit named in honor of the English scientist, Lord Kelvin. The word *absolute* indicates that the standard of this new scale has fundamental significance for all matter.

If equation 2-5 is substituted into equation 2–4, the result is

$$\begin{aligned} V &= V_0(1 + \alpha t) = V_0\alpha\left(t + \frac{1}{\alpha}\right) \qquad \textbf{2–6} \\ &= V_0\alpha T. \end{aligned}$$

The volume of every gas at a very low pressure can be extrapolated to zero at a common temperature of $T = 0\ ^\circ K = -273.15\ ^\circ C$. (We use the word *extrapolated* because this statement can only be a projection for the substance as a gas—all gases will liquefy if the temperature is lowered sufficiently.) Figure 2–6 shows a plot of the volume of one mole of gaseous nitrogen at a very low constant pressure as a function of temperature in °C. The solid straight line terminates before V reaches zero due to nitrogen liquefaction. The dashed line is the extrapolation to $0\ ^\circ K = -273.15\ ^\circ C$.

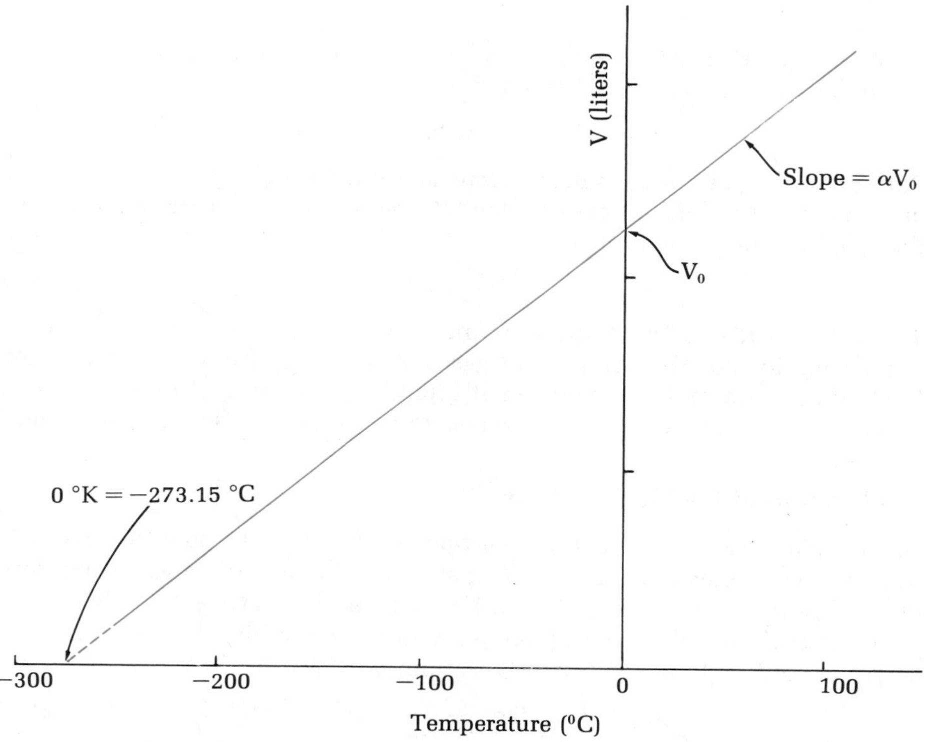

Figure 2–6
The volume of one mole of gaseous N_2, at a constant low pressure, as a function of temperature.

Further evidence for the fundamental significance of 0 °K comes from the fact that as any substance is cooled to near 0 °K, it becomes increasingly difficult to cool it further. Practically, 0 °K appears to represent some lower limit of obtainable temperatures that cannot be reached but only approached. The reason why nature possesses such a limit is considered in Chapter 15.

Ideal Gas Law

The constant V_0 of equation 2–6 is related to the gas pressure by Boyle's law—

$$V_0 = \frac{(\text{constant}_0)}{P} \qquad \textbf{2–7}$$

where the subscript zero indicates the value of the constant of equation 2–2 at 0 °C. This constant_0 still depends on the quantity of gas measured. The Avogadro hypothesis demands that at a given temperature and pressure the volumes of gases be directly proportional to the number of molecules they contain (and thus to the number of moles n of gas present). Thus equation 2–7 may be rewritten as

$$V_0 = \frac{n(\text{constant}'_0)}{P}, \qquad \textbf{2–8}$$

where $\text{constant}'_0$ has the same value for all gases. Equation 2–8 may now be combined with equation 2–6 to yield

$$PV = n(\text{constant}'_0)\alpha T. \qquad \textbf{2–9}$$

Both $\text{constant}'_0$ and α are independent of the nature of the gas, so they may be combined to yield a new constant R, **the ideal gas constant,** as well as the **ideal gas law,** equation 2–10:

$$PV = nRT \qquad \textbf{2–10}$$

The value of R is 0.08206 liter-atm/mole-°K. Equation 2–10 is an excellent approximation for the behavior of gases at pressures from zero to approximately one atm and we shall use it often. The meaning of the word *ideal* will become much clearer when we examine models of the gaseous state.

Applications of the Ideal Gas Law

Chemical analyses are sometimes carried out in which the quantity of a solid in a mixture is measured by having that solid react to yield a gas. A mixture of solid copper and solid zinc can be analyzed by adding the solids to an aqueous solution of HCl. Copper does not react with HCl, but zinc dissolves according to the following reaction:

$$Zn(s) + 2\,HCl(aq) \rightarrow ZnCl_2(aq) + H_2(g). \qquad \textbf{2–11}$$

A 0.3624-g sample of the solid mixture was observed to yield 271.2 ml of gaseous H_2 at 25.3 °C and 255 torr. What is the percentage by weight of zinc in the original sample? From equation 2–10, we can convert these data to the number of moles of H_2 produced.

$$\begin{aligned} n_{H_2} &= \frac{PV}{RT} \\ &= \frac{(255\text{ torr})(1\text{ atm}/760\text{ torr})(0.2712\text{ l})}{(0.08206\text{ l-atm/mole-°K})(298.5\text{ °K})} \\ &= 3.71 \times 10^{-3}\text{ mole of } H_2 \end{aligned}$$

According to equation 2–11, this must be produced by 3.71×10^{-3} mole of

Zn, or $(3.71 \times 10^{-3})(65.37) = 0.242$ g of Zn. Thus, the weight percentage of Zn in the original sample was $(0.242/0.3624)(100) = 66.8$ percent.

A common application of equation 2–10 is in the determination of molecular weights. We have learned that molecular weights of gaseous species can be determined by comparing the gas density with the density of hydrogen at the same temperature and pressure. With equation 2–10, the density of the gas alone suffices as long as the pressure and temperature are known. We shall illustrate this calculation by showing a method for the molecular weight determination of low-boiling liquids. As an example we pick carbon tetrachloride (CCl_4), with a boiling point of 76.8 °C. A small amount of liquid CCl_4 is introduced into a weighed glass bulb of known volume through a narrow opening in the tip. This bulb is then immersed in boiling water, as shown in Figure 2–7. The liquid CCl_4 will quickly evaporate, leaving the bulb filled with vapor at atmospheric pressure. The narrow glass tip is then

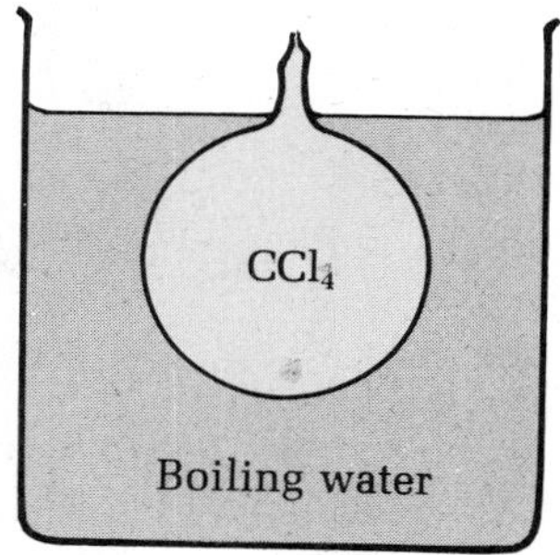

Figure 2–7
Apparatus for determining the molecular weight of a liquid having a low boiling point.

sealed with a torch and the bulb is cooled and weighed. The gain in weight over the empty bulb is the weight of CCl_4 vapor present before the cooling. Suppose that the volume of the bulb is 243 ml, the atmospheric pressure at the time of the measurement was 710 torr, and consequently the boiling point of water was 98.5 °C = 371.7 °K. Under these conditions, the bulb must contain the following number of moles of CCl_4:

$$n_{CCl_4} = \frac{PV}{RT} = \frac{(710/760)(0.243)}{(0.0821)(371.5)} = 7.44 \times 10^{-3} \text{ mole of } CCl_4$$

Suppose that the weight of vapor was found to be 1.14 g. The molecular weight is

$$MW_{CCl_4} = \text{grams/mole} = 1.14 \text{ g}/7.44 \times 10^{-3} \text{ mole} = 154 \text{ g/mole}$$

Dalton's Law of Partial Pressures

From equation 2–10 we can deduce another result, which was discovered experimentally by John Dalton. Equation 2–10 shows that the pressure exerted by any gas in a given volume at a given temperature is independent of the nature of the gas and depends only on the number of moles of gas present. In a gas mixture, such as air, the total pressure must be the sum of partial pressures that each of the gases present exerts independently. The partial pressure of a gas A in a gas mixture is the pressure that would be exerted on the walls of the container if all molecules except those of A were removed. Each partial pressure is given by equation 2–10, where n is the number of moles of the species whose partial pressure is being calculated.

Let us illustrate partial pressures with an example depicted in Figure 2–8. Suppose that we have 10.0 g of O_2 and 15.0 g of N_2 in a 10.0-liter container at 25 °C = 298 °K. What is the total pressure and what fractions of it do O_2 and N_2 exert individually? The numbers of moles of the components are 10.0/32.0 = 0.312 for n_{O_2} and 15.0/28.0 = 0.536 for n_{N_2}. The total pressure is given by

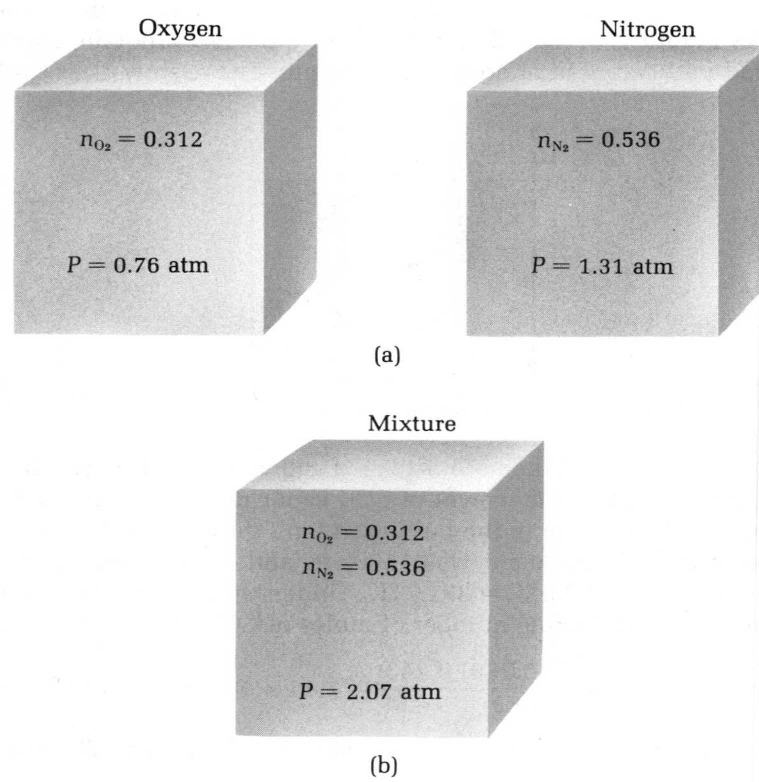

Figure 2–8 An illustration of Dalton's law of partial pressures. (a) Separate 10.0-liter containers of O_2 and N_2 at 25 °C, (b) the gases combined in the same 10.0-liter container.

$$P = \frac{(n_{N_2} + n_{O_2})}{V} RT = \frac{(0.848)(0.0821)(298)}{(10.0)} = 2.07 \text{ atm}$$

The partial pressure of oxygen is given by

$$P_{O_2} = \frac{n_{O_2}RT}{V} = \frac{(0.312)(0.0821)(298)}{(10.0)} = 0.76 \text{ atm}$$

and similarly $P_{N_2} = 1.31$ atm. The partial pressure P_{O_2} is related to the total pressure as

$$\frac{P_{O_2}}{P} = \frac{n_{O_2}}{n_{O_2} + n_{N_2}} = x_{O_2}$$

where x_{O_2} is the standard symbol for the mole fraction. This last equation is the mathematical statement of Dalton's law, that is, **the fractional contribution a given gaseous species makes to the total pressure of a gas mixture equals the mole fraction of that species in the gas.**

2–2. A Kinetic Model of Gases

What is the source of the pressure an enclosed gas exerts? The pressure that anything exerts is defined as the force it effects on a given area divided by that area. A good illustration of the distinction between force and pressure is provided by the Torricelli barometer in Figure 2–2. If the cross-sectional area of the mercury column of height h is A, the mass of the column is $hA(13.6)$, where 13.6 g/ml is the density of mercury. The force that the column exerts on the base A is $(13.6)Ahg$, where g is the earth's gravitational force constant. The equation for pressure exerted on this base area is

$$\text{Pressure} = \frac{\text{force}}{A} = (13.6)hg$$

This pressure equals the atmospheric pressure and it does not depend on the value of A.

Figure 2–9 shows three models of an enclosed gas that accommodate exertion of force by the gas on the enclosing walls. Figure 2–9(a) considers the molecules as compressed elastic spheres. This model we previously discarded. Model (a) would surely yield different molar volumes for different gases at a given temperature and pressure, because of different molecular

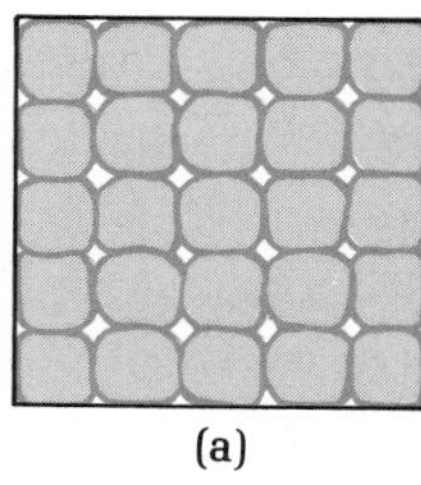

(a)

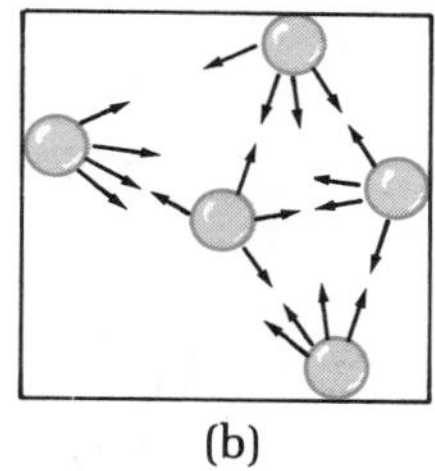

(b)

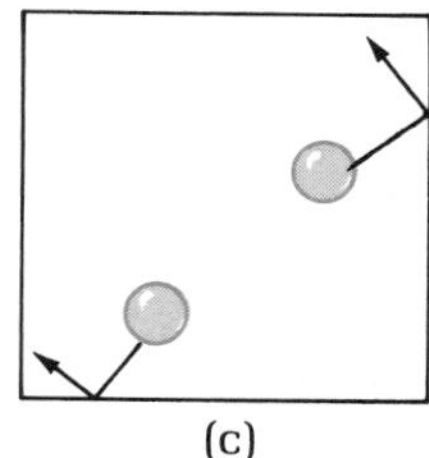

(c)

Figure 2–9
Three models of the gaseous state.

sizes between gases. The Avogadro hypothesis demands that the large volumes occupied by gases consist mainly of empty space between well-separated molecules. Only then can the molar volume be independent of molecular size.

In Figure 2–9(b), the molecules are shown with repulsive forces (shown by arrows) between each other, so that the molecules at the wall are being forced outward. This model yields pressure; but it cannot be reconciled with the existence of solids and liquids, which involve molecules in close proximity and which exist even at quite low pressures. If molecules repel each other strongly, why don't they do so in liquids and solids, in which they are closest together?

Figure 2–9(c) shows well-separated noninteracting molecules exerting forces on the walls through collisions arising from their rapid motion inside the container. The model avoids the pitfalls of (a) and (b) and it accounts for pressure; but does it relate pressure to volume and mole number in the correct way (that is, by equation 2–10), and does it allow us to appreciate the nature of temperature?

To answer these questions, we must focus on the mechanics of collisions at the walls. Molecular motion is one of the key features of matter, calling for an understanding of the physical laws of motion. There are no laws that belong exclusively to either physics or chemistry; understanding chemistry requires a grasp of what are commonly termed the laws of physics. Hence, many so-called physical concepts are pertinent material to chemists, and we shall consider them as the need arises.

Some Elementary Mechanics

If we are to understand the consequences of molecular motion, we must characterize that motion. The most obvious characteristic of the motion of anything is its **velocity.** Velocity is a **vector** property, indicating that it possesses a given direction as well as a magnitude. Arrows symbolize vector quantities; the length of the arrow indicates the magnitude of the quantity and the direction indicates the direction of the vector quantity. Figure 2–10

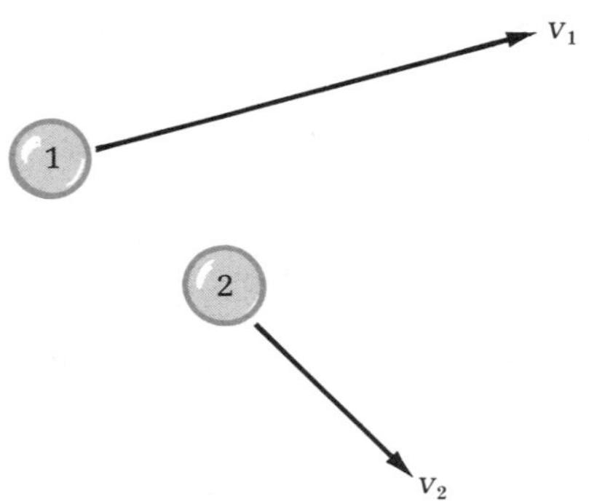

Figure 2–10 Vector representations of the velocities v_1 and v_2 of two gas molecules.

shows two gas molecules and the vector arrows that represent their velocities. The arrow for molecule 2 is exactly half the length of the arrow for molecule 1, indicating that the **speed** of molecule 2 is half the speed of molecule 1. Speed is defined as the magnitude of velocity.

Associated with the velocity is a second vector quantity, the **momentum,** given by the product of the mass and the velocity. The momentum of an object determines the force that the object exerts in a collision. Sir Isaac Newton discovered that the force that an object exerts in a collision equals the rate of change of the object's momentum with time:

$$F = \frac{\Delta(mv)}{\Delta t} = \frac{m\Delta v}{\Delta t} \qquad \textbf{2–12}$$

Greek uppercase *delta* (Δ) is the standard symbol for a change of anything. It is always the value after the change minus the value before the change. In equation 2–12, $\Delta(mv)$ is the change in momentum occurring in a time interval Δt, and Δv is the change in velocity occurring in that same time interval. The mass m was factored out in equation 2–12 because it is a constant. There are two units of force in common use in science. The first is the **dyne,** which is defined as a momentum change of one gram-cm/sec in a time of one second, or

$$1 \text{ dyne} = 1 \text{ g-cm/sec}^2$$

The second is the **newton,** represented as follows:

$$1 \text{ newton} = 1 \text{ kg-m/sec}^2$$

Let us calculate the conversion factor between dynes and newtons to gain additional practice in unit conversions:

$$1 \text{ newton} = \frac{(1 \text{ kg})(10^3 \text{ g/1 kg})(1 \text{ m})(10^2 \text{ cm/1 m})}{\text{sec}^2} = \frac{10^5 \text{ g-cm}}{\text{sec}^2} = 10^5 \text{ dynes}$$

If a force is exerted, **work** can be performed because work W is defined as the force F exerted by an object times the distance d through which the object exerts the force. For example, the lifting of a 100-gram weight a vertical distance of 10.0 cm is the performance of an amount of work given by

$$W = Fd = (mg \text{ dynes})(h \text{ cm})$$
$$= (100)(10.0)g \text{ dyne-cm}$$

An object that can perform work is said to possess **energy,** since energy is defined as the ability to perform work. That energy may be a consequence of the motion of the object. For example, a moving automobile may perform work on a pedestrian if the driver isn't careful enough. The energy of motion is termed **kinetic energy** (T), and the relation is given by

$$T = \frac{mv^2}{2} \qquad \textbf{2–13}$$

Just as there are two common units of force, there are two common units of energy associated with them. The first is the **erg**, which is defined as the work done by a force of one dyne exerted through a distance of one cm. Thus,

$$1 \text{ erg} = 1 \text{ dyne-cm}$$

The second unit is the **joule**, which is defined as the work done by a force of one newton exerted through a distance of one meter. Thus,

$$1 \text{ joule} = 1 \text{ kg-m}$$

(In a problem at the end of the chapter, you are asked to derive the conversion factor 1 joule = 10^7 ergs.)

Suppose that a molecule collides with a wall at the angle θ (Greek lowercase *theta*), as shown in Figure 2–11. The angle is measured between the

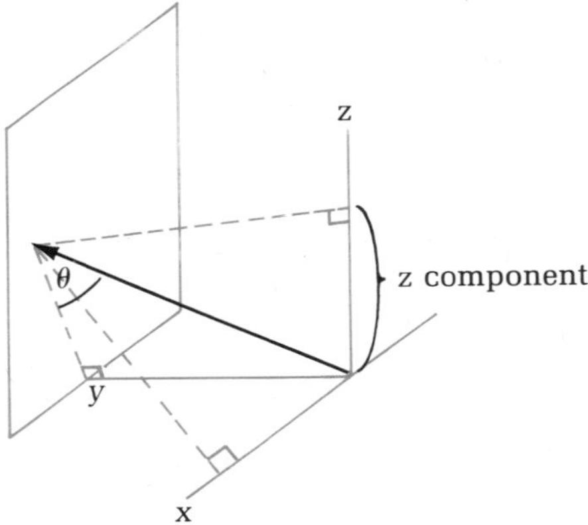

Figure 2–11 Resolution of a momentum vector into three components.

flat wall and the momentum vector of the molecule. A system of mutually perpendicular coordinate axes is also shown in the figure, with its origin at one end on the momentum vector. As shown in the figure, the x, y, and z coordinates of the molecule are all increasing as the molecule moves in the direction of the momentum vector. Another way of saying this is that the momentum has **components** in the x, y, and z directions. These components are obtained by drawing the dashed lines shown from the vector tip perpendicular to each of the three axes. Each component is measured by the distance from the origin of the axial system to the dashed line along the appropriate axis. The z component is indicated in the figure. The force on the wall due to the collision of the molecule in Figure 2–11 depends only on the y component of momentum. The x and z components are parallel to the wall, while y is perpendicular.

Gas Pressure and Collisions at the Walls

If the pressure of a gas is exerted because of molecular collisions at the walls, the pressure must fluctuate with time because there will be short time intervals during which no collisions occur. However, if our model yields a great many collisions in one second, we will find it nearly impossible to detect the fluctuations, and only a time average pressure will be observed. It is this average pressure that we shall attempt to calculate.

Let us take a section, A (shown in Figure 2–12), of the wall enclosing a gas. Since the gas pressure is the force exerted on A divided by A, we are concerned with the calculation of the force on the section due to the collisions of molecules with it. If we pick a time interval Δt, the force is given by $\Delta mv/\Delta t$, where Δmv is the total molecular momentum change at the section A in the interval Δt.

We need to know how many collisions occur at A in Δt and the momentum change associated with each. This, in turn, requires that we know in which directions the molecules move and how fast they move. Since the pressure of an enclosed gas is observed to be the same at all of the walls, the molecules must be moving randomly with equal probabilities for motion in all possible directions. This indicates that if we sum up all of the momentum components for all of the molecules, we shall find that the components are equally divided into six parts, corresponding to motion along $+x$, $-x$, $+y$, $-y$, $+z$, and $-z$ directions. The pluses and minuses are included because there are, for example, as many particles moving from right to left as from left to right. Thus, the distribution of momentum as it concerns direction is exactly the same as though one-sixth of the molecules at any given time were traveling exclusively in each of these six directions (that is, with no intermediate directions allowed). In the calculation of gas pressure, we are perfectly justified in using this six-directional model even though we know it to be an oversimplification. That is the nature of a model.

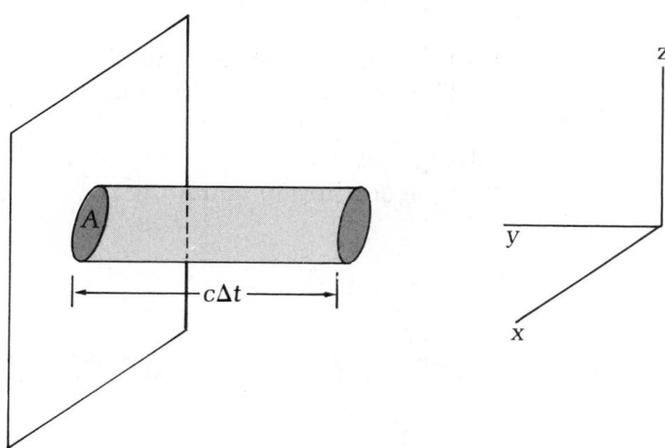

Figure 2–12
Imaginary cylinder containing all the gas molecules that will collide with wall area A in time Δt, in the six-directional model of gases.

In the six-directional model, only molecules moving along $+y$ (one-sixth of the total molecules) can collide with A. The others are moving away from A or parallel to A.

Do all of the molecules move with the same speed? For the moment we shall assume that they do. This cannot be a correct assumption, because if the molecules collide with the walls, some must speed up and some must slow down as a consequence of those collisions. There must be a distribution of speeds. For the sake of clarity, we shall first carry through the pressure derivation with the assumption of a uniform speed c. Then we shall see how this simplistic picture transforms for a realistic speed distribution.

Figure 2–12 shows a cylinder with A as the base and sides parallel to y. The height of the cylinder is shown as $c\Delta t$. This cylinder contains every molecule that will strike A in the interval Δt. Those near the wall will collide well within Δt and those at the distance $c\Delta t$ away will arrive just as the interval ends. The number of collisions at A in Δt is one-sixth the number of molecules inside the cylinder because only these are moving towards A. The number of molecules in the cylinder is represented by

$$\text{Molecules in cylinder} = \left(\frac{\text{molecules}}{\text{volume}}\right)(\text{volume of cylinder})$$

$$= \left(\frac{\text{moles}}{\text{volume}}\right)N(\text{volume of cylinder})$$

$$= \left(\frac{n}{V}\right)N(Ac\Delta t)$$

where N is Avogadro's number.

The expression for the number of collisions at A in the time interval Δt is

$$\text{Collisions} = (\text{molecules in cylinder})/6$$

$$= \frac{nNAc\Delta t}{6V} \qquad \textbf{2–14}$$

Each of these molecules hits A and bounces straight back with the same speed it had before it hit. The momentum goes from mc to $-mc$, since momentum is a vector quantity and the direction of the vector has been reversed. The momentum change per collision is $2mc$, and the total momentum change at A in Δt (which, according to equation 2–12, equals the force on A) is given as follows:

$$F = \frac{nNAmc^2}{3V} \qquad \textbf{2–15}$$

Since pressure equals force per unit area, the pressure relation is shown by

$$P = \frac{nNmc^2}{3V} \qquad \textbf{2–16}$$

Equation 2–16 is the pressure the gas would exert if all of the molecules moved with speed c. The only effect of a speed distribution is to add another source of fluctuation to the pressure. Not only does the pressure fluctuate because there are short periods in which no collisions occur, but successive collisions yield different forces. This means that equation 2–16 must be averaged over the molecular speeds:

$$P = \frac{nNm\langle c^2 \rangle}{3V} \tag{2–17}$$

In equation 2–17 the brackets indicate the average of the square of the molecular speed.

The average square of a set of numbers often causes confusion for students. Let us take the numbers 1, 2, 2, and 3 and calculate the average square of the numbers. It is

$$\langle n^2 \rangle = (1^2 + 2^2 + 2^2 + 3^2)/4 = \tfrac{18}{4}$$

This value of $\langle n^2 \rangle$ is not equal to the square of the average of n, which is

$$\langle n \rangle^2 = \left(\frac{1 + 2 + 2 + 3}{4}\right)^2 = 4$$

The Nature of Temperature

Equation 2–17 relates the pressure to n and V exactly as they are related in equation 2–10. Let us equate PV/n from equations 2–10 and 2–17 to obtain an expression for temperature in terms of molecular properties:

$$\frac{PV}{n} = RT = \frac{Nm\langle c^2 \rangle}{3} = \frac{2N}{3}\left(\frac{m\langle c^2 \rangle}{2}\right) \tag{2–18}$$

The term $m\langle c^2 \rangle/2$ is the average kinetic energy per gas molecule. Therefore, $Nm\langle c^2 \rangle/2$ is the average kinetic energy of a mole of gas molecules. From equation 2–18 this molar kinetic energy is also given by $3\,RT/2$, and the average kinetic energy per gas molecule is shown as follows:

$$\frac{\text{Average kinetic energy}}{\text{Molecule}} = \frac{3RT}{2N} = \frac{3kT}{2} \tag{2–19}$$

In equation 2–19, the constants R and N have been combined to yield a new constant, $k = R/N$, termed **Boltzmann's constant** in honor of Ludwig Boltzmann, who contributed significantly in the late nineteenth century to the kinetic molecular model of the gaseous state. From equation 2–19 we see that the intensive property T is a measure of the average kinetic energy of gas molecules.

Equation 2–19 also gives us insight into what happens when heat is added to a gas. According to the equation, an increase in temperature of a

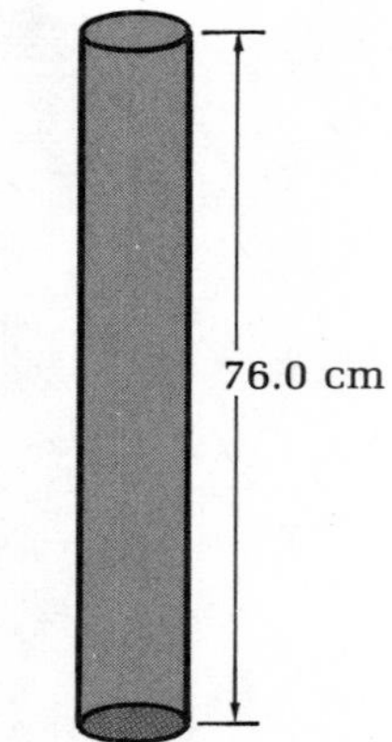

Figure 2–13 An illustration of a pressure of 1 atmosphere being exerted on base area A cm² by a mercury column of height 76.0 cm.

gas is accompanied by an increase in the energy of molecular motion. And, in connection with the rule of Dulong and Petit, an input of a fixed amount of heat also accompanies any particular temperature increase. Thus, heat is converted to the energy of molecular motion, and heat input is the input of energy. There are conversion factors between the unit of heat, the calorie, and the two units of energy already introduced. These conversion factors are

$$1 \text{ calorie} = 4.184 \text{ joules} = 4.184 \times 10^7 \text{ ergs}$$

Units of R and k

According to equations 2–18 and 2–19, the units of R and k are energy/mole-°K and energy/molecule-°K, respectively, where we can use any of several units of energy. Up to this point we have encountered R only in connection with equation 2–10 as $R = 0.08206$ liter-atm/mole-°K. The unit liter-atm does not immediately suggest energy. Let us show that the liter-atm is a unit of energy by converting it to ergs.

Our first task is to convert the standard atmosphere to an equivalent pressure in dynes/cm², that is, force per unit area. Figure 2–13 shows a column of mercury 76.0 cm high with a base area of A cm². The force that the mercury column exerts on the base area is the mass of mercury in the column times g, the earth's gravitational force constant. This constant g is 980.7 dynes/g. The mass of mercury is equal to the volume of the column times the mercury density of 13.59 g/cm³. The complete conversion from 1 atm to equivalent dynes/cm² is written:

$$1 \text{ atm} = \frac{(13.59 \text{ g/cm}^3)(76 \text{ cm})(980.7 \text{ dynes/g})\, A \text{ cm}^2}{A \text{ cm}^2}$$

$$= 1.013 \times 10^6 \text{ dynes/cm}^2$$

Since 1 liter $= 10^3$ ml $= 10^3$ cm³,

$$1 \text{ liter-atm} = (10^3 \text{ cm}^3)\left(\frac{1.013 \times 10^6 \text{ dynes}}{\text{cm}^2}\right)$$

$$= 1.013 \times 10^9 \text{ dyne-cm}$$

$$= 1.013 \times 10^9 \text{ ergs}$$

Thus, an alternative set of units for R can be shown:

$$R = \left(\frac{0.08206 \text{ liter-atm}}{\text{mole-°K}}\right)\left(\frac{1.013 \times 10^9 \text{ ergs}}{\text{liter-atm}}\right) = \frac{8.313 \times 10^7 \text{ ergs}}{\text{mole-°K}}$$

and a value of k is given as

$$k = \frac{8.313 \times 10^7 \text{ ergs/mole-°K}}{6.023 \times 10^{23} \text{ molecules/mole}}$$

$$= \frac{1.380 \times 10^{-16}\ \text{erg}}{\text{molecule-°K}}$$

Because the conversion from ergs to joules is so simple (10^7 ergs = 1 joule), we can easily obtain the further values, $R = 8.313$ joules/mole-°K and $k = 1.380 \times 10^{-23}$ joule/molecule-°K. Finally R and k can be expressed in terms of calories as 1.987 cal/mole-°K and 3.299×10^{-24} cal/molecule-°K, respectively. It is in these last units that we shall most commonly use R. For convenience in subsequent referral, these R and k values are listed in Table 2–1. A convenient conversion factor to remember is

$$1\ \text{liter-atm} = 24.2\ \text{cal}$$

Energy	R (energy/mole-°K)	k (energy/molecule-°K)
Liter-atm	0.08206	1.366×10^{-25}
Ergs	8.313×10^7	1.380×10^{-16}
Joules	8.313	1.380×10^{-23}
Calories	1.987	3.299×10^{-24}

Table 2–1
Values of R and k in Different Energy Units

We can now use the value of k in ergs to obtain a value of $\langle c^2 \rangle$, the average square molecular speed, for a typical gas at 25 °C. From equation 2–19, the average kinetic energy of a gas molecule at 25 °C is 6.17×10^{-14} erg. Let us take N_2 as our typical gas. The mass m of a nitrogen molecule is obtained as follows:

$$m = \left(\frac{28.0\ \text{g/mole}}{6.023 \times 10^{23}\ \text{molecules/mole}} \right) = 4.65 \times 10^{-23}\ \text{g/molecule}$$

Many times we shall require an atomic or molecular mass in grams. This mass is always obtained by dividing the mass in amu by Avogadro's number.

With this value of m, $\langle c^2 \rangle$ is given by

$$\langle c^2 \rangle = \frac{2(6.17 \times 10^{-14})}{(4.65 \times 10^{-23})} = 2.65 \times 10^9\ \text{cm}^2/\text{sec}^2.$$

The square root of this number, $\langle c^2 \rangle^{1/2}$, is a form of average speed termed the **root mean square speed,** and it is symbolized as c_{rms}. In the case of nitrogen, c_{rms} is 5.15×10^4 cm/sec. This speed is relatively large. The world's fastest human can run 100 m in 10 sec, while the average N_2 molecule can travel the same distance in 0.194 sec.

With the value of c_{rms} for N_2, we can check our earlier assumption that wall collisions in a typical gas are so frequent that pressure fluctuations are unobservable. We shall calculate the number of collisions in 1 sec at a wall area A of 1 cm^2 in nitrogen gas at 1 atm pressure and 25 °C. The number of collisions is given by equation 2–14. The number of nitrogen molecules per liter nN/V is given by PN/RT, according to equation 2–10 (N is Avogadro's number). In this case,

$$\frac{\text{molecules}}{\text{liter}} = \frac{(1.000)(6.02 \times 10^{23})}{(0.08206)(298)} = 2.46 \times 10^{22}$$

If A and c in equation 2–14 are expressed in cm^2 and cm, respectively, this number must be converted to 2.46×10^{19} molecules/cm^3.

From equation 2–14, the number of collisions/sec at A is shown by the relation:

$$\frac{\text{Collisions}}{\text{sec}} = \frac{(2.46 \times 10^{19})(1.00)(5.15 \times 10^{4})}{6} = 2.11 \times 10^{23}$$

or the average time between collisions is $1/2.11 \times 10^{23} = 4.74 \times 10^{-24}$ sec. This time is so short that no instrument could come near to detecting it and as far as we can see, the surface A appears to be undergoing continuous collisions.

2–3. Speed Distribution in Gases

The general model of the gaseous state that we have developed is known as the **kinetic (motional) theory of gases.** Its basic assumptions are:

1. Molecular sizes are negligible compared with the average distance between molecules.
2. There are no attractions or repulsions between molecules.
3. There is rapid and chaotic motion of the gas molecules.
4. There are vast numbers of individual, independently moving molecules.

The model properly accounts for gas pressure and reveals the nature of temperature. In addition it yields, in equation 2–19, two predictions, which can be checked experimentally as tests of the model. We consider them successively.

Graham's Law

Consider the representation shown in Figure 2–14. The circles represent nitrogen molecules at 1 atm pressure in a container that has a pinhole at the top so that molecules can slowly escape into the evacuated space surrounding the container. The passage of N_2 molecules through the pinhole is termed **effusion.**

What is the rate at which N_2 molecules effuse? If the pinhole has a cross-sectional area A, the rate of loss is simply the collision rate of N_2 molecules with the area A (although the molecules pass through the hole instead of

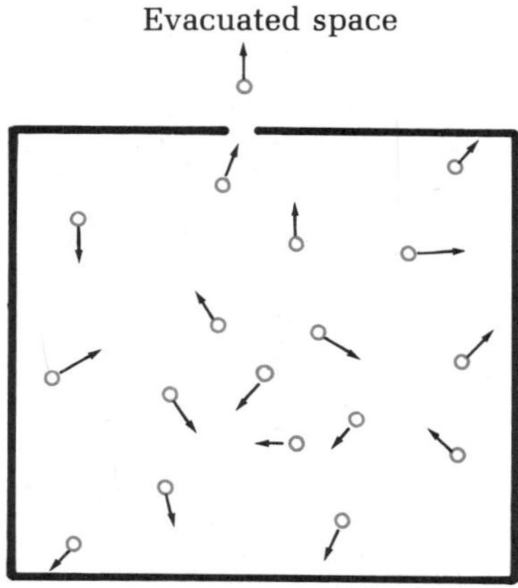

Figure 2–14
The effusion of gaseous molecules from a container.

colliding with it). Thus the effusion rate in molecules/sec is given by the collision rate in equation 2–14.

Now suppose that we repeat the experiment with hydrogen instead of nitrogen, but keep the same temperature and initial pressure. How does H_2 compare with N_2 in effusion rate? According to equation 2–14, the ratio of effusion rates is represented by

$$\frac{\text{Rate}_{H_2}}{\text{Rate}_{N_2}} = \frac{(n/V)_{H_2} c_{rms(H_2)}}{(n/V)_{N_2} c_{rms(N_2)}} \qquad \textbf{2–20}$$

in which we have replaced c in equation 2–14 by c_{rms}. The term n/V is given from the ideal gas law as P/RT. Since P and T are the same for H_2 and N_2, these terms cancel in equation 2–20. By equation 2–19, we have

$$\frac{m_{H_2} c_{rms(H_2)}^2}{2} = \frac{3kT}{2}$$

$$\frac{m_{N_2} c_{rms(N_2)}^2}{2} = \frac{3kT}{2}$$

or

$$\frac{c_{rms(H_2)}}{c_{rms(N_2)}} = \left(\frac{m_{N_2}}{m_{H_2}}\right)^{1/2} = \left(\frac{Nm_{N_2}}{Nm_{H_2}}\right)^{1/2} = \left(\frac{MW_{N_2}}{MW_{H_2}}\right)^{1/2} \qquad \textbf{2–21}$$

where MW denotes molecular weight. Substitution of equation 2–21 into equation 2–20 yields a ratio of effusion rates of

$$\frac{\text{Rate}_{H_2}}{\text{Rate}_{N_2}} = \left(\frac{MW_{N_2}}{MW_{H_2}}\right)^{1/2} \qquad \textbf{2–22}$$

This result that the ratio of effusion rates at constant P and T is given by the inverse ratio of the square roots of the molecular weights was discovered experimentally in the mid–nineteenth century, and it is termed **Graham's**

law. It is a direct consequence of the fact that **the average kinetic energy of gaseous molecules depends only on the temperature of the gas and not on the nature of the gas.** Thus lighter molecules travel faster and escape faster. Graham's law is a substantial verification of the kinetic model of gases.

An Experimental Determination of Speed Distribution

A far stricter test of the kinetic model than Graham's law derives from the calculated values of c_{rms} obtainable from equation 2–19 for any gas of known molecular weight. It is an enormous triumph for the model if it can predict these precise numbers. However, to perform this test we must be able to measure experimentally the distribution of speeds in a gas so that we can obtain c_{rms}. This can be done by modifying the effusion experiment. The molecules pass through the pinhole as a narrow beam with a variety of speeds characterized by the speed distribution. Suppose that the pinhole is in the side of the container so that the beam leaves the container horizontally. As the molecules travel horizontally, they are pulled downward by gravity. Suppose that the beam consists of silver atoms obtained by vaporizing silver in an oven. The beam is then allowed to pass through an otherwise evacuated space to impinge on a cold, flat surface, as shown in Figure 2–15. The downward deflection of any atom increases with the length

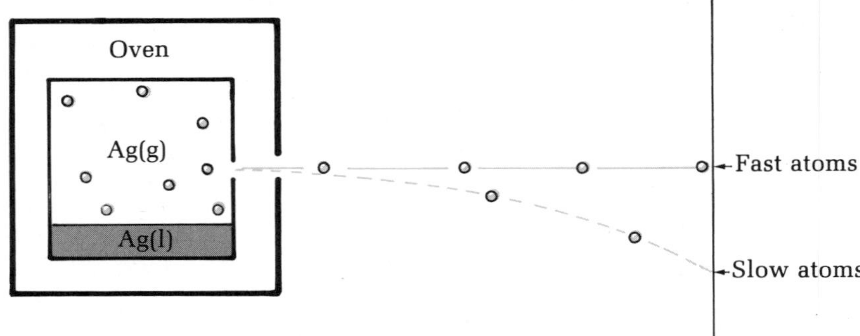

Figure 2–15 Apparatus for determining the distribution of atomic speeds in silver vapor.

of time it takes to pass from the pinhole to the flat surface. Thus, as the figure shows, slow atoms are deflected more than fast ones. The flat surface will be coated with solid silver in a vertical line and the speed distribution can be obtained from the measurable amounts of silver in various distance intervals along the line. Since the gravitational force constant is known as well as the pinhole-surface distance, any given vertical deflection corresponds to a precisely calculable atomic speed, and, therefore, any distance interval along the silvered line corresponds to a precisely calculable range of atomic speeds.

By this and other techniques, the speed distribution in almost any gas can be measured over a wide range of temperatures. Figure 2–16 shows the fraction of nitrogen molecules in speed intervals of width equal to 1.0×10^4 cm/sec in a gas at 25 °C. The fraction is shown as a vertical line drawn at the upper limit of the interval. Thus, for example the fraction of molecules with speeds between zero and 1.0×10^4 cm/sec is approximately 0.02 and is indicated in the figure by the vertical line at a speed of 1.0×10^4 cm/sec. One mole of N_2 gas contains $(0.02)(6.02 \times 10^{23}) = 1 \times 10^{22}$ molecules with speeds between zero and 1.0×10^4 cm/sec.

We used equation 2–19 previously to calculate c_{rms} for N_2 molecules at 25 °C as 5.15×10^4 cm/sec. Figure 2–16 shows that the greatest concentration of speeds is indeed in the region near 5×10^4 cm/sec. A plot such as Figure 2–16 can yield only an approximation to c_{rms}. However, as we make the size of the speed intervals smaller, the approximate value approaches the exact value as a limit, the exact experimental value being 5.15×10^4 cm/sec in agreement with the prediction from the kinetic model.

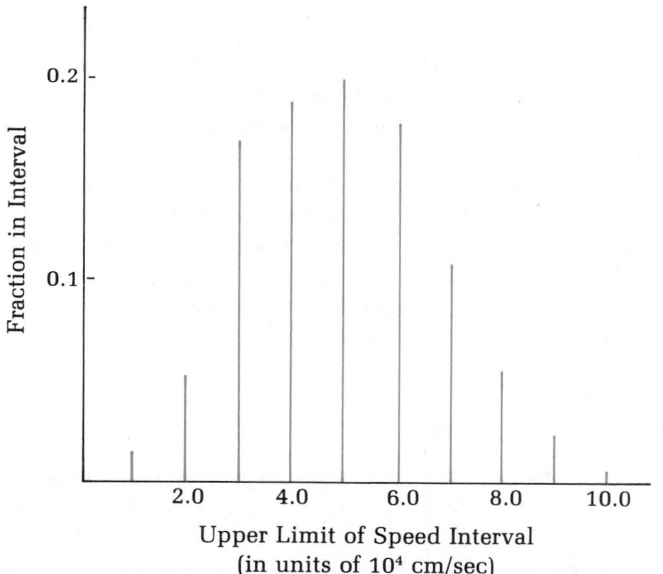

Figure 2–16 Speed distribution for N_2 gas molecules at 25 °C plotted as the fraction of molecules within speed intervals of width equal to 1.0×10^4 cm/sec. The fraction within a given speed interval is plotted as the height of the vertical line at the upper limit of that speed interval.

Equation 2–19 also predicts the temperature dependence of c_{rms}. In all cases, c_{rms} must increase with increasing temperature. In the case of N_2, for example, c_{rms} must be 10.64×10^4 cm/sec at 1000 °C. The experimental speed distribution for N_2 at 1000 °C is shown in Figure 2–17. As predicted, the greatest concentration of speeds is in the region around 10×10^4 cm/sec. Figure 2–17 reveals another aspect of the heating of gases that we shall pursue subsequently. As the gas is heated, the speed distribution broadens;

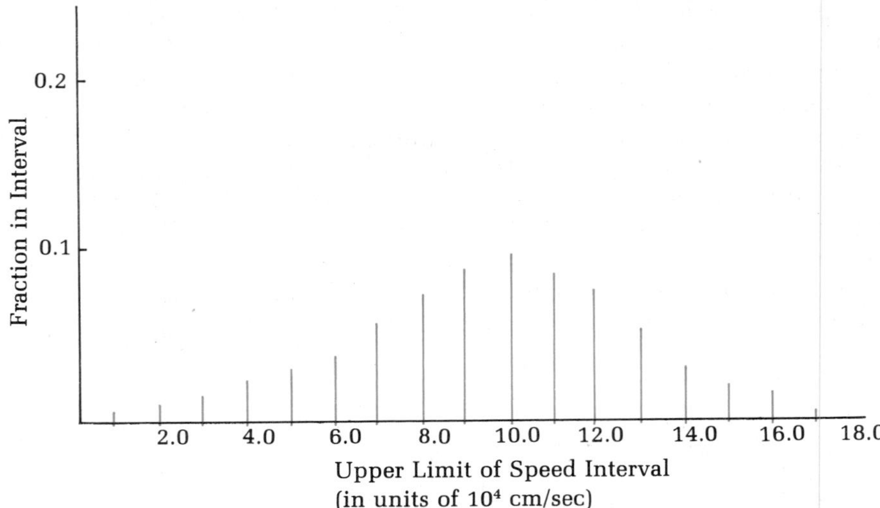

Figure 2–17
Speed distribution for N_2 gas molecules at 1000 °C plotted as the fraction of molecules within speed intervals of width equal to 1×10^4 cm/sec.

the distribution shown in Figure 2–16 is much more sharply peaked than the distribution in Figure 2–17. As more energy is supplied to the gas molecules, higher kinetic energies become practically accessible and the motional possibilities available to the molecules increase. Thus, temperature is a measure not only of the average kinetic energy of gas molecules but also of the extent to which the total kinetic energy of all the molecules is spread among different speed intervals.

Source of Speed Distribution

The kinetic model is very successful at predicting c_{rms} for gas molecules at any temperature, but it also should explain other features suggested by Figures 2–16 and 2–17. For example, why are relatively few molecules found with speeds near zero or with speeds greater than approximately $2c_{rms}$? Suppose that somehow we could obtain a sample of N_2 gas at 25 °C in which all of the molecules moved with a speed of 5×10^4 cm/sec at some given time. This uniformity would soon be destroyed by collisions of molecules with each other and with the walls. In the derivation of equation 2–15, we assumed that any molecule striking a wall at speed c would bounce back with speed c, and this is an incorrect assumption for every collision. It is true only for the collisions that are the average of all collisions. However, as the applicability of c_{rms} shows, averages suffice in that derivation. In the present context, we require a more detailed molecular picture of a collision at a wall.

The wall itself is composed of atoms, each of which vibrates about a fixed site in the wall, and the motion of these atoms must somehow be related to the wall temperature. At the moment a gas molecule hits the wall, there should be equal probabilities that the wall atom is moving away from the gas molecule as in Figure 2–18(a), or toward it, as in Figure 2–18(b). The "rear-end" collision in (a) results in a loss of speed of the gas molecule and a transfer of energy from the gas to the wall. The two parts of Figures 2–18(a) and (b) depict the two situations immediately before (left) and after (right) collision. The arrows are velocity vectors. The "front-end" collision in (b) results in energy transfer from the wall to the gas and a gain in the speed of the gas molecule. If the wall and the gas are at the same temperature, the time average transfer of energy is zero; that is, in one second the energy transferred from the wall by front-end collisions equals the energy transferred to the wall by rear-end collisions. In the collision of any given molecule with the wall under this circumstance of equal temperatures, there are exactly equal probabilities that the molecule will be speeded up or slowed down. Consequently, the average collision involves no speed change. This was our assumption made in deriving equation 2–15.

With this more detailed picture of wall collisions in mind, let us return to the imaginary nitrogen gas with uniform molecular speeds of 5×10^4 cm/sec. At this speed, collisions with a wall at 25 °C will yield very nearly equal probabilities of a speedup or a slowdown, since this speed is very close to the average one for N_2 molecules at 25 °C. Suppose that a given molecule undergoes 100 wall collisions. What is its most probable speed after these collisions, and what are the probabilities that the speed will have halved or doubled, for example?

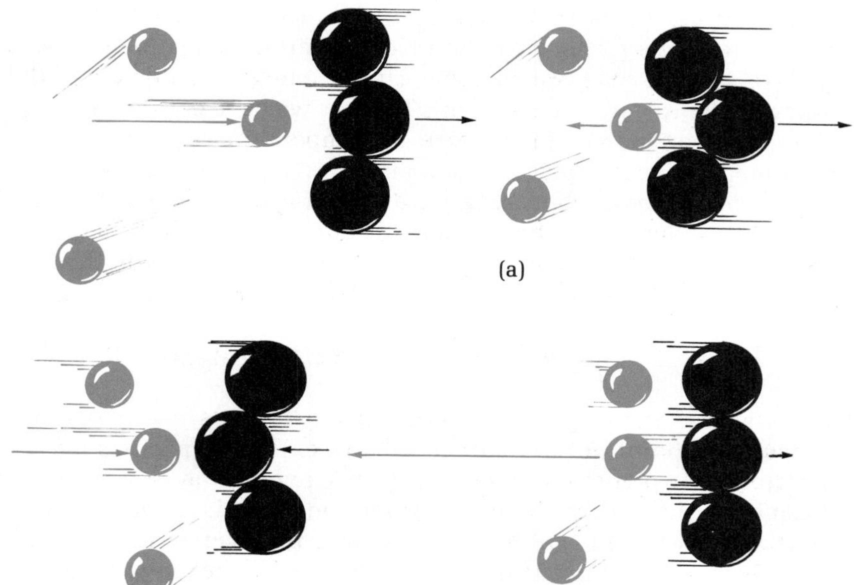

Figure 2–18
Depictions of collisions between gas molecules (*small spheres*) and atoms of a solid wall (*large spheres*). In (a) the wall atom is moving away from the gas molecule at the moment of contact; in (b) the wall atom is moving toward the gas molecule.

The problem posed is a purely statistical one and can be answered by reference to any statistical model in which successive, independent events occur, with each event having two possible outcomes of equal probability. This is exactly the situation with the throwing of coins, for which the outcomes are heads (*H*) and tails (*T*). Suppose that we assign numbers, +1 to heads and −1 to tails, and add the results of successive coin flips in analogy with the gain and loss of speed in molecular collisions with the wall. Consequently, the number zero is the analog of an unchanged molecular speed of 5×10^4 cm/sec. To see qualitatively what must happen with 100 coin flips, let us consider two coin flips. There are four possible results and they are *H–H*, *H–T*, *T–H*, and *T–T*, or the numbers 2, 0, 0, −2. The most probable value is zero, with a probability of 0.5. As we increase the number of coin flips, zero remains the most probable value and there is a sharp drop-off in probability in going away from zero in either direction. For the molecules, this means that 100 collisions are likely to be divided into approximately 50 speedups and 50 slowdowns. The likelihood of 100 successive speedups, to yield an extremely energetic molecule, or 100 successive slowdowns is extremely small. Thus, Figures 2–16 and 2–17 must contain a most probable speed near the average speed and the probability must fall off sharply at speeds less or greater than the most probable value.

2–4. Heat Capacities of Gases

Equation 2–19 gives the average molecular kinetic energy of **translation,** the process of moving from place to place within the container of gas. For one mole of gas, the total translational kinetic energy is $N(3kT/2) = 3RT/2$. This energy is temperature-dependent. Thus, as we have pointed out, energy must be added to the gas to increase its temperature, and this energy is usually added by heat input. The amount of energy required to raise the temperature of one mole of any substance by 1 °K is termed the **heat capacity** of that substance. For a gas, the predicted heat capacity is the kinetic energy of a mole of gas molecules at a temperature $(T + 1)$ °K minus the kinetic energy at temperature T.

$$\frac{3}{2}R(T + 1) - \frac{3}{2}RT = \frac{3R}{2} = \frac{3(1.987)}{2} = 2.981 \text{ cal/mole-°K}$$

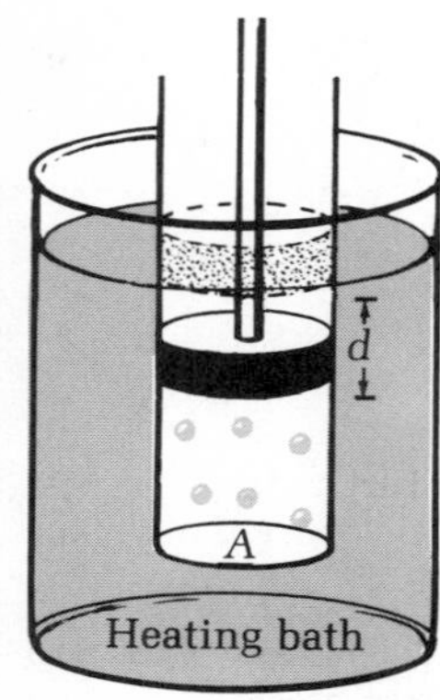

Figure 2–19
Illustration of the performance of work by a gas at constant pressure as a consequence of a temperature increase.

Before we compare the predicted heat capacity of gases with experimental values, we must point out a distinction between heat capacities measured and reported in two common ways. The first involves heating the gas in a closed container of fixed volume, for which the result is reported as C_v, the heat capacity at constant volume. The second involves heating the gas at constant pressure, as depicted in Figure 2–19. Before heating, the gas pressure is equal to the pressure exerted by the piston and no motion of the piston occurs. According to equation 2–10, the pressure of gas rises as it is heated at constant volume, and the resulting pressure imbalance must force up the piston in Figure 2–19 as the gas temperature increases. The gas molecules perform work as they push the piston up. Molecular kinetic energy is used to perform work.

How much work is done by raising the temperature of one mole of gas by one degree when the pressure is held constant? The work is given by the force the gas exerts on the piston (PA) times the distance (shown in Figure 2–19 as d) that the piston is displaced. But $A(d)$ is equal to ΔV, the increase in the volume of the gas. According to the ideal gas equation,

$$P\Delta V = P(V_{\text{final}} - V_{\text{initial}}) = nRT_{\text{final}} - nRT_{\text{initial}} = nR\Delta T$$

so that a ΔT of 1 °K involves an amount of work done by one mole of gas of R, 1.987 cal.

To increase the temperature of a gas at constant pressure, heat has been added to increase the molecular kinetic energies and to compensate for the extra kinetic energy required to do the work required for expansion. Thus, the **heat capacity at constant pressure** C_p for a gas is given by

$$C_p = C_v + R$$

and C_p is predicted to be $5R/2 = 4.968$ cal/mole-°K for all gases.

Table 2–2 shows the heat capacities of several gases, all at the initial temperature and pressure of 25 °C and 1 atm, respectively. The predictions of $C_v = 3R/2$ and $C_p = 5R/2$ are close for the monatomic gases helium and argon, and for any other gas whose fundamental units are individual atoms. All of the **diatomic** and **polyatomic** molecules possess heat capacities significantly greater than the predicted values. This discrepancy reveals that for such molecules energy can be stored in forms in addition to translational kinetic energy.

Gas	C_p	C_v
Theoretical prediction	4.97	2.98
He	4.98	2.99
Ar	4.98	2.99
H_2	6.89	4.86
HCl	6.96	5.07
O_2	7.02	4.97
Cl_2	8.11	6.15
I_2	8.81	6.80
CO_2	8.87	6.74
SO_2	9.51	7.70

Table 2–2
Comparison of Predicted and Measured Heat Capacities of Gases

In what other ways can molecules store energy? Suppose that molecules can rotate about their centers of gravity or that the atoms that compose a molecule can vibrate relative to each other. Each of these motions requires energy, and both should contribute to the heat capacity. At this point in our development of models, we have said nothing definite about molecular structure. Thus, we are in no position yet to consider molecular rotation and vibration. However these subjects are treated in detail in Chapter 12.

2–5. Transport of Information by Gases

The kinetic model of gases developed to this point considers the size of molecules to be negligible compared with the average distance between them in the gaseous state. All of the properties we have so far considered are insensitive to molecular sizes (as long as the molecules are sufficiently small). What are some properties of gases that are sensitive to molecular size, no matter how small the molecules may be? One answer is any property that is affected by the collision of gas molecules with each other. There are a number of properties that are related to the transport of information across matter. One is **thermal conductivity,** the transport of heat across a substance. A second property is **diffusion,** which is the passage of molecules of one substance between the molecules of a second substance. Molecular collisions impede such information transport, just as we find it more difficult to transport ourselves along a crowded sidewalk than along an empty one.

A Model of Gas Viscosity

One property by which molecular collisions and sizes in gases can be studied is **viscosity.** You are aware that matter resists the passage of objects through it. Thus, a weight that is more dense than water falls more slowly in water than in air and more slowly still in syrup. We commonly say that syrup is more viscous than water, which is more viscous than air. What, then, is the molecular source of these viscosity differences?

When any object passes through matter, collisions must occur at the surface of the object with the molecules of the medium through which the object is moving. Collision with gas molecules is shown in Figure 2–20. In these collisions, some of the momentum of the falling object is transferred to the gas molecules, as indicated in the figure by the increased length of the momentum vector of the molecule shown colliding with the falling object. According to Newton, momentum changes are manifested in forces. The object exerts a force on the gas and the gas exerts a force on the object. This force due to collisions is the basis of viscosity.

The model of gas viscosity is different from the model of liquid viscosity (which nevertheless is based on collisions). The following model applies

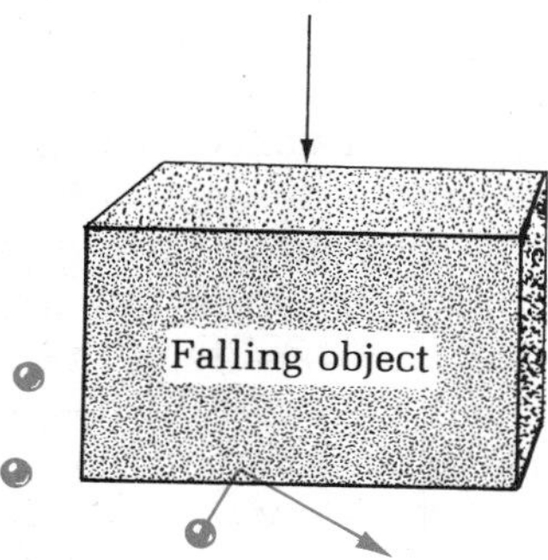

Figure 2–20
A gas molecule gaining momentum when struck by an object falling through the gas.

only to gases. Figure 2–21 shows a side view of two plates of area A separated by a distance d, and the top plate is moving parallel to the bottom plate at a velocity v. Molecules that collide with this moving plate acquire additional momentum in the direction in which the plate is moving. If

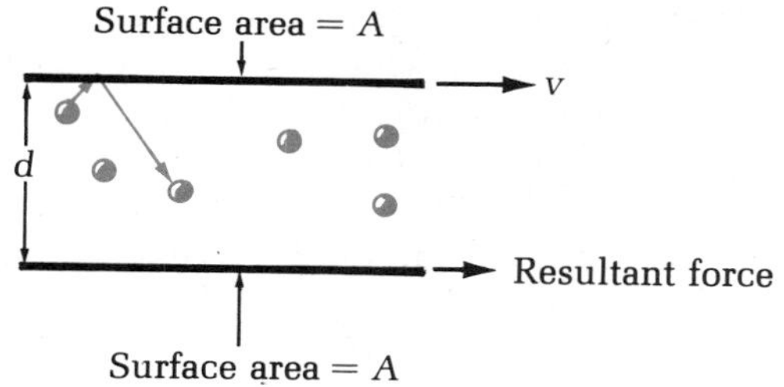

Figure 2–21
Transfer of momentum from a moving plate to a stationary plate by a gas molecule.

these molecules travel to the bottom plate, they can transfer this extra momentum to it. Thus, the top plate exerts a force on the bottom plate through the transport of momentum by gas molecules, and the force increases with the gas viscosity. You can simulate the experiment by placing your hands parallel to each other and about a half-inch apart. Move one hand rapidly parallel to the other and keep the other stationary. The "breeze" that the stationary hand feels is a consequence of the viscosity of air.

Experimentally, the force on the lower plate is given by

$$F = \frac{\eta A v}{d} \qquad \textbf{2–23}$$

where η (Greek lowercase *eta*) is the **coefficient of viscosity.** The units of η can be derived from equation 2–23 because the units of $\eta Av/d$ must be those of force. If η possesses units of dynes-sec/cm^2, we obtain this equality of units:

$$\text{Dynes} = \frac{(\text{dynes-sec/cm}^2)(\text{cm}^2)(\text{cm/sec})}{\text{cm}}$$

The combination of units dynes-sec/cm^2 is termed a **poise**, in honor of the French scientist Poiseuille.

Equation 2–23 holds for all substances, but the value of η changes from substance to substance. What are the factors that enter into η? Since gas molecule momentum is given by the molecular mass times the molecular velocity, the momentum picked up at the top plate increases with the molecular mass as well as with the velocity of the plate. The plate velocity v appears explicitly in equation 2–23 and the molecular mass m appears implicitly as a factor in η. The molecules gain momentum as a result of collisions at the moving plate. Thus, the viscosity must increase with the collision frequency at the plate, and equation 2–14 gives this collision frequency. This accounts for the appearance of A in equation 2–23. Two other terms that appear in equation 2–14 are (nN/V) and c_{rms}, and they must be factors in η. The final factor in η relates to d in equation 2–23. Figure 2–22(a) shows a very effective transport of momentum by a molecule traveling in a straight line between the moving and stationary plates. Figure 2–22(b) shows an ineffective transport. The molecule collides with several other molecules and loses the additional momentum gained at the moving plate. Clearly the greater d is, the more likely the molecule is to encounter another gas molecule, and this is the source of d in equation 2–23. The factor in η that relates to d is the **mean free path** (λ, Greek lowercase *lambda*), the average distance traveled by a gas molecule between collisions. Figure 2–22(a) shows a large mean free path and Figure 2–22(b) shows a small one.

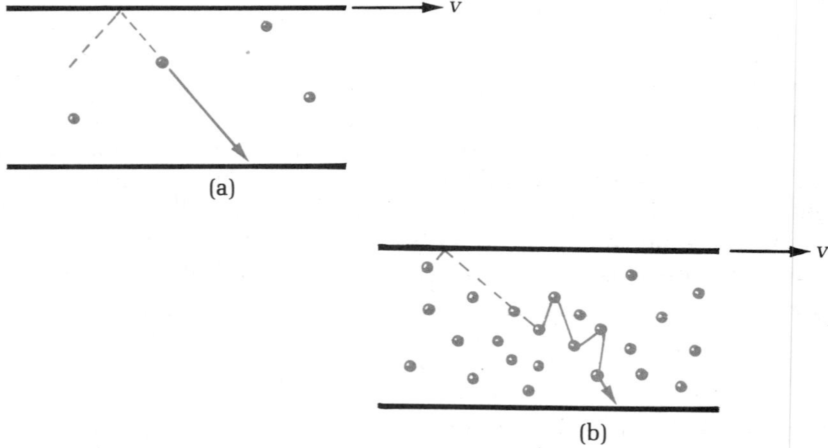

Figure 2–22
(a) Effective and (b) ineffective transfer of momentum from a moving plate to a stationary plate by a gas molecule.

All of the factors in η combine to yield

$$\eta = \frac{m\left(\frac{nN}{V}\right)c_{rms}\lambda}{3} \qquad \textbf{2–24}$$

The combined units on the right side of equation 2–24 are

$$\left(\frac{\text{g}}{\text{molecule}}\right)\left(\frac{\text{molecules}}{\text{cm}^3}\right)\left(\frac{\text{cm}}{\text{sec}}\right)(\text{cm}) = \frac{\text{g}}{\text{cm-sec}}$$

$$= \left(\frac{\text{g-cm}}{\text{sec}^2}\right)\left(\frac{\text{sec}}{\text{cm}^2}\right)$$

$$= \text{dynes-sec/cm}^2$$

the units of η from equation 2–23.

The Mean Free Path and the Molecular Diameter

At the beginning of this section, it was stated that the viscosity is somehow related to molecular size; but molecular size does not appear explicitly, either in equation 2–23 or equation 2–24. It appears implicitly in λ, as we can show by a consideration of Figure 2–23. The figure shows the path of a helium atom through other helium atoms in gaseous helium. The atoms are shown as hard spheres with a diameter D. This hard sphere picture is yet another model, which we shall subsequently modify. The path is a broken one because of numerous collisions, and the average length of path traveled in one second is $c_{\text{rms(He)}}$.

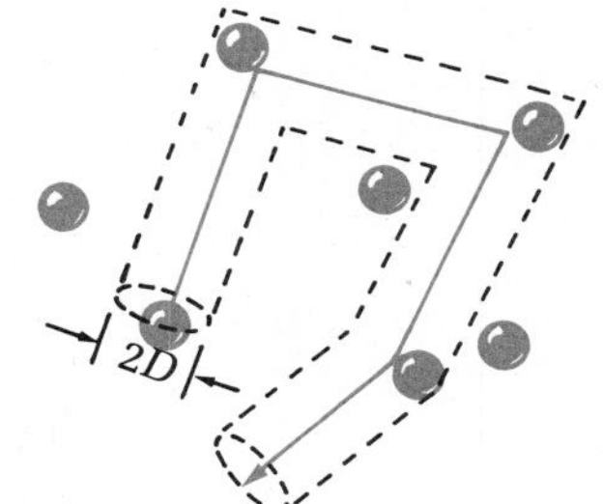

Figure 2–23 Typical path of a gas molecule through a collection of identical molecules.

How many other molecules will be encountered along the path? As shown in the figure, we can construct a bent cylinder of diameter $2D$ around the path as the axis. Any molecule having its center inside this cylinder represents a collision with the given molecule. Thus, the number of collisions in one second is the volume of the cylinder times the number of molecules per unit volume in the gas or

$$\frac{\text{collisions}}{\text{sec}} = (\pi D^2 c_{\text{rms}})\left(\frac{nN}{V}\right) \qquad \textbf{2–25}$$

The mean free path is given by the length traveled in one second divided by the number of collisions the molecule makes in one second:

$$\lambda = \frac{(\text{cm/sec})}{(\text{collisions/sec})}$$

$$\lambda = \frac{c_{rms}}{\pi D^2 c_{rms} nN/V} \qquad \textbf{2–26}$$

$$= \frac{V}{\pi D^2 nN}$$

representing cm/collision. Thus, the mean free path is large in gases at low pressures and in gases containing small molecules or small atoms.

A Calculation of a Molecular Diameter

Substitution of λ from equation 2–26 into equation 2–24 yields

$$\eta = \frac{mc_{rms}}{3\pi D^2} \qquad \textbf{2–27}$$

The measured η for He at 25 °C is 1.96×10^{-4} poise. The mass of the helium atom is $4.00/(6.023 \times 10^{23}) = 6.64 \times 10^{-24}$ g/atom. The value of $c_{rms(He)}$ can be calculated from the value of N_2 and Graham's law:

$$c_{rms(He)} = (5.15 \times 10^4)\left(\frac{28.0}{4.00}\right)^{1/2} = 13.6 \times 10^4 \text{ cm/sec}$$

With these values, D for He is calculated:

$$D_{He} = \left\{\frac{(6.64 \times 10^{-24})(13.6 \times 10^4)}{(3)(3.14)(1.96 \times 10^{-4})}\right\}^{1/2}$$

$$= 2.21 \times 10^{-8} \text{ cm}$$

The unit 10^{-8} cm occurs very commonly in the measurement of atomic and molecular sizes. As a consequence the unit is given its own name, the **angstrom,** and symbol (Å). The diameter of the He atom is therefore reported as 2.21 Å. In standard international units this would be reported in nanometres (nm). One nanometre is 10^{-9} metre (this is the spelling of *meter* in standard international units), and the helium diameter is 0.221 nm.

What volume is occupied by one mole of helium atoms? The volume of each helium atom is

$$\frac{4\pi(D/2)^3}{3} = 5.6 \times 10^{-24} \text{ cm}^3$$

so that a mole of them fills a volume of $(5.6 \times 10^{-24})(6.023 \times 10^{23}) = 3.37 \text{ cm}^3$. This is approximately three-fourths of the measured molar volumes of solid or liquid helium. Solids and liquids contain atoms and molecules in close contact while gases involve large intermolecular distances. The approximation comes largely from the fact that the packing together of spheres does not fill all of the available space.

2–6. Intermolecular Forces in Gases

One of the features of our kinetic model of gases is a lack of any repulsive or attractive force between gas molecules. A little thought convinces us that the idea of a lack of such forces cannot be strictly true. The molecules in liquids and solids must attract each other or these states of low molar volume would never exist. Furthermore, as molecules are forced closer together at short distances, they must begin to repel each other or otherwise liquids and solids would be easily compressible.

Does the pressure of a gas depend on intermolecular forces? Suppose that we have one mole of He atoms and the force of attraction between them is so strong that they exist as He_2 molecules. The number of moles of fundamental units is no longer one, but rather one-half. Since $PV = nRT$, the pressure of the gas is half what it would be if the atoms existed as separate atoms. Helium atoms do not actually form He_2 molecules, but the point just made is suited to the purpose. **Intermolecular attractions in gases yield lowered gas pressure.** The opposite conclusion is reached for intermolecular repulsions, which lead to increased pressure.

Real Gases Compared with the Ideal Gas

Since real gases exhibit intermolecular forces, the ideal gas law is not an exact law; it is only an approximation (although an extremely useful approximation). Intermolecular forces must increase as the molecules are brought closer together. Thus, the ideal gas law is found to be an excellent approximation at pressures of one atm and below, for which the average distance between molecules is relatively large. As the pressure is increased and this distance decreases, molecular attractions should become important and the pressure should be less than that given by equation 2–10. At very high pressures the molecules will all be in contact, and repulsion must predominate, with P considerably larger than equation 2–10 predicts. Furthermore, repulsion should begin to predominate at a molar volume that is simply related to the molecular sizes, and this offers another method for approximating molecular sizes. Equation 2–10 is termed the ideal gas equation because it applies exactly to only ideal gases, that is, gases in which intermolecular forces are zero.

Figure 2–24 shows plots of PV/RT for one mole of the gases N_2 and CH_4 at 25 °C plotted against pressure in atm. An ideal gas must have a value of PV/RT equal to unity at all pressures. For N_2, CH_4, and all other gases, PV/RT approaches unity as P approaches zero. The figure also shows that N_2 and CH_4, like most gases, are very nearly ideal at pressures of one atmosphere and below, and for our concerns we shall use the ideal gas law for gases at low to moderate pressure. Note that the abscissa scale in Figure 2–24 is in hundreds of atmospheres.

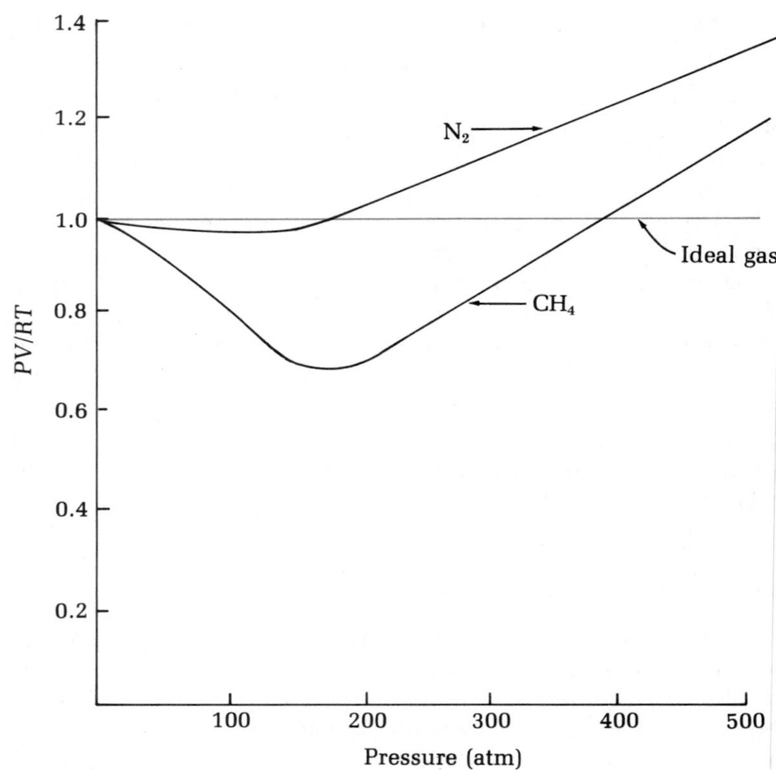

Figure 2–24
Plots of PV/RT for one mole N_2 and CH_4 as a function of pressure at 25 °C.

The deviations from ideality are as the attraction-repulsion model predicts. At low pressures (large average internuclear distance), the pressure is lower than equation 2–10 predicts and PV/RT is less than one. At high pressures, intermolecular repulsion predominates and PV/RT rises steeply.

Temperature Dependence of Deviations from Ideality

We have accounted for the qualitative features of Figure 2–24 by adding attractions and repulsions to our kinetic model of gases. What effect can we predict that an increase or decrease of temperature will have on the deviations from ideality shown in Figure 2–24? As the temperature increases, so do c_{rms} and the average molecular momentum. As this momentum increases, the push or pull exerted by the other molecules becomes less important. Consequently, the behavior of all gases at all temperatures should approach equation 2–10 as the temperature is increased. Figure 2–25 shows plots of PV/RT as a function of P for methane, CH_4, at three temperatures: 25 °C, 250 °C, and 750 °C. As the temperature increases, the curve flattens out and approaches the horizontal line at unity, as predicted.

As a final point, let us return to Figure 2–24. It is clear from the figure that

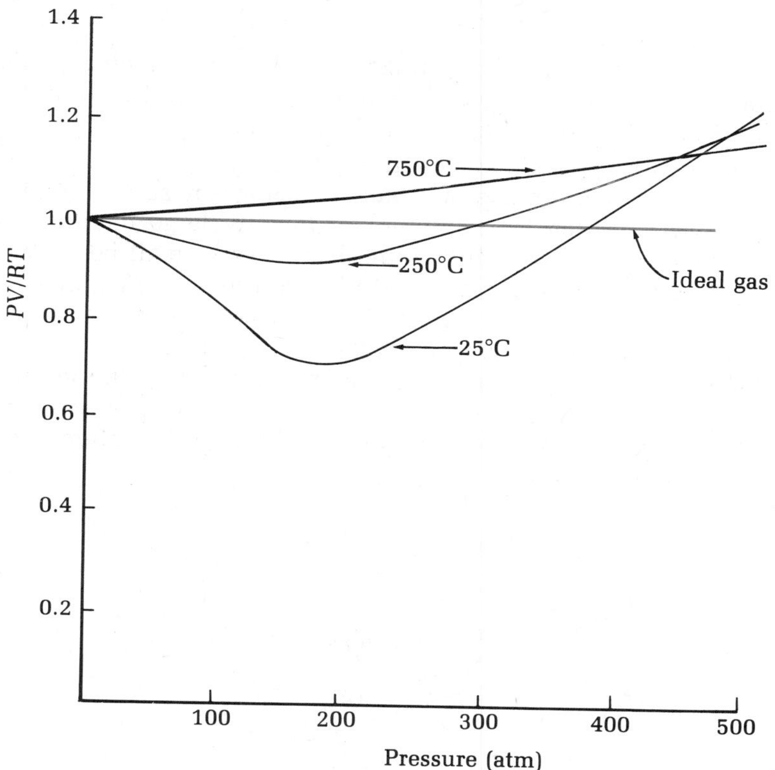

Figure 2–25
Plots of PV/RT for one mole of CH_4 as a function of pressure at three different temperatures.

at a given temperature and pressure, intermolecular attraction is considerably stronger in CH_4 than in N_2. The attraction should tend to pull molecules on the periphery of the gas in towards the center of the gas. Figure 2–25 shows that the attractive forces become increasingly more important as the temperature is lowered. If the temperature is lowered sufficiently, the attractive forces become so important that the gas collapses in on itself and a liquid results. With this view in mind, we should expect gaseous CH_4 at a given pressure to liquefy at a higher temperature than N_2 does. (The temperature at which liquefaction occurs at a pressure of 1 atm is the **normal boiling point** of the substance.) Nitrogen boils at 77 °K while methane boils at 112 °K.

SUGGESTIONS FOR ADDITIONAL READING

Porterfield, W. W., *Concepts of Chemistry*, Norton, New York, 1972.

Kauzmann, W., *Kinetic Theory of Gases*, Benjamin, New York, 1966.

Hildebrand, J. H., *An Introduction to Molecular Kinetic Theory*, Van Nostrand Reinhold, New York, 1963.

PROBLEMS

1. Pressure readings on home barometers are given in inches of mercury. On a given cold, clear day, the barometer reading is 30.42 in. Express this pressure in torr and in standard atmospheres.
2. Small pressures of residual gas in high vacuum systems are commonly measured through the use of a McLeod (pronounced McCloud) gauge. The gauge operates by trapping a relatively large quantity of low-pressure gas and then compressing it in a very small bore glass tube, using a column of mercury to supply the pressure. A drawing of the gauge, showing the gas compressed in the tube, is given in Figure 2–26.

   Suppose that the volume above the short dashed line in Figure 2–26 is 250 ml and that the volume of gas (shown in color) in the small bore glass tube is 0.051 ml. The gas pressure in the tube is due to a mercury height of 2.60 cm. What is the pressure of gas in the volume indicated by the scattered red dots?

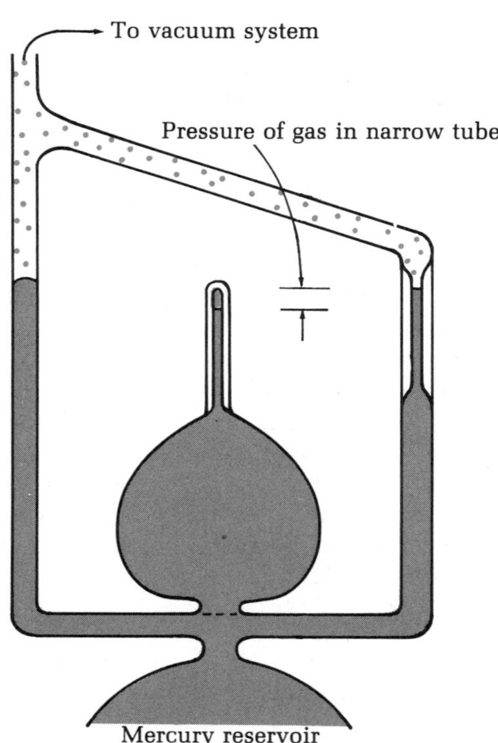

Figure 2–26
A McLeod gauge for measuring very low pressures.

3. Refer to Figure 2–3 and the measurement of gas compressibilities. On a day on which the atmospheric pressure is 752 torr, the following data

were recorded for the volume of trapped air as a function of the mercury height h:

Volume (ml)	10.00	9.00	8.00
h (cm)	4.35	13.19	24.23

Show that these data are in agreement with Boyle's law.

4. Indicate which of the following properties of a substance are extensive and which are intensive:
   a. Its energy.
   b. Its specific heat.
   c. Its molar volume.
   d. Its number of molecules.
5. The Celsius and Fahrenheit temperature scales cross at only one temperature. What is the value of this temperature that has the same numerical value on both scales?
6. The specific heat of liquid water is 1.00 cal/g-°C. What is the specific heat in cal/g-°F?
7. Charles's contribution to equation 2–4 was to show that all gases at a given pressure increase their molar volumes by a factor of 1.366 over a rise in temperature from the freezing point of water to the boiling point of water. Why is this relation insufficient to establish the validity of equation 2–4; that is, what must Gay-Lussac's contribution have been? Use the factor of 1.366 to calculate α of equation 2–4.
8. What is the molar volume of N_2 gas at its boiling point at a pressure of 1 atm if we assume ideal gas behavior? The boiling point of N_2 is 77 °K.
9. The ideal air pressure in a bicycle tire is 50 pounds per square inch (psi). Suppose that you are inflating the tires in a heated garage at 70 °F, and that the temperature outside is 20 °F. To what pressure should you inflate the tires in the garage to yield a pressure of 50 psi at the outside temperature?
10. Joseph Priestley first prepared pure oxygen gas by heating mercuric oxide (HgO) according to the following reaction:

$$2\,HgO(s) \rightarrow 2\,Hg(l) + O_2(g)$$

What volume of oxygen is produced at 25 °C and 1 atm pressure by the decomposition of 5.00 g of HgO?

11. The fizzing that occurs when certain antacid tablets are dropped into water occurs from the reaction

sodium bicarbonate + citric acid → sodium citrate + carbon dioxide + water

One mole of sodium bicarbonate ($NaHCO_3$) yields one mole of gaseous carbon dioxide (CO_2). The CO_2 can be collected and measured to determine the weight percentage of $NaHCO_3$ in a tablet. A 0.5133-g sample of tablet yielded 106.52 ml of CO_2 at 298 °K, and 35.54 cm of Hg. What is the percentage of $NaHCO_3$ in the tablet?

12. A compound with the empirical formula C_3H_6O is observed to boil at 56.2 °C and to be characterized by a vapor density of 2.14 g/liter. What are the molecular weight and molecular formula of this compound?

13. Air contains 0.27 mole of O_2 to each mole of N_2. What is the average molecular weight of this oxygen-nitrogen mixture?

14. The partial pressure of water vapor in air that is saturated with water at 25 °C is 24 mm. The relative humidity of air at any temperature is the partial pressure of water in that air divided by the saturation partial pressure at that temperature. What is the ratio of moles of H_2O to moles of O_2 in air at 25 °C when the relative humidity is 75 percent and the pressure is 740 torr?

15. Does air become more dense or less dense as the humidity increases at a given pressure and temperature?

16. As the result of the force of the earth's gravity, a 100-g mass falling toward the earth is observed to have a velocity that is steadily increasing at the rate of 980.7 (cm/sec)/sec = 980.7 cm/sec^2. What is the force of gravity in dynes on the object?

    A 50-g mass falls with exactly the same rate of velocity increase. What do you conclude about the relation between gravitational force and the mass of the falling object?

17. Derive the conversion factor 1 joule = 10^7 ergs.

18. What is the total kinetic energy of one mole of helium atoms at 25 °C? This is a relatively large amount of energy and it indicates that the energy of molecular motion in the atmosphere or the oceans is enormous. Yet this energy is not used even in a time of global shortage. You may speculate why this apparent neglect of an obvious free resource occurs. (We shall deal with the question in detail in Chapter 16.)

19. Most elements are known to be mixtures of atoms having different atomic weights. The different atoms of the same element are termed **isotopes** and are designated by the atomic weight. Two examples are ^{235}U and ^{238}U, two isotopes of the element uranium. During the Second World War, it became essential for the construction of the atomic bomb to obtain uranium enriched in ^{235}U from the mixture of isotopes that occurs in nature. The successful procedure was carried out as follows. The uranium is converted to gaseous UF_6, uranium hexafluoride, which is then passed through a long tube packed with a porous solid. The first fraction of gas that emerges is increased in the atom fraction of ^{235}U

compared with the gas that entered the tube. This enriched gas can then be passed through the tube again and another enriched fraction collected. Repetition of this process many times finally yields a sufficient enrichment of $^{235}UF_6$.

What is the principle behind the separation? Why is the separation so inefficient that the gas must be passed through the tube repeatedly?

20. The escape velocity from the moon's gravitational attraction is 1.8×10^5 cm/sec, or approximately 4000 miles an hour. The moon is bombarded with hydrogen atoms from the sun; yet the moon does not hold any significant amount of H_2 in an atmosphere. Use Figure 2–16 and equation 2–21 to explain why this is the case.

21. Suppose that the average speedup or slowdown in Figure 2–18 due to a wall collision is 10 percent of the average molecular speed. How many successive speedups are required to result in a molecular speed that is ten times the average speed? If the probability of a speedup in any given collision equals the probability of a slowdown, what is the probability of this many successive speedups? This indicates that the drop-off in the fraction at high speeds in Figure 2–16 is really quite precipitous.

22. Refer to Table 1–5 and calculate the heat capacities of copper, iron, and silver. The specific heats were determined by heating samples that were exposed to the atmosphere during heating. Is your calculated heat capacity a C_p or a C_v value? In the heating of gases, there is a sizable difference between C_p and C_v. Is this also true for the heating of solids? Explain.

23. Does the mean free path increase, decrease, or remain unchanged as a gas is compressed at constant temperature? Does λ increase, decrease, or remain the same if the gas is heated at constant volume? Explain.

24. By combining equations 2–24 and 2–26, you can make some predictions about the behavior of η. Does η increase, decrease, or remain unchanged as the gas is compressed at constant temperature? Does η increase, decrease, or remain unchanged as the gas is heated at constant volume? How does your prediction of the temperature dependence compare with your experience concerning the viscosity of liquids? Surprising as it may seem, gases and liquids do indeed show opposite temperature dependences of their viscosities, and this result represents one of the chief triumphs of the kinetic model.

25. The measured viscosity of gaseous N_2 is 1.77×10^{-4} poise at 25 °C. Calculate the value of D for N_2. Why are the viscosities of N_2 and He so similar, even though N_2 is so much more massive than He?

26. As mentioned in the text, thermal conductivity is, like viscosity, the transport of information across a substance. The rate of heat transport (in cal/sec) between two parallel plates of area A separated by a distance

d is

$$\text{Rate} = \frac{\kappa A \Delta T}{d} \tag{2–28}$$

where *kappa*, κ, is termed the **coefficient of thermal conductivity.** What are the units of kappa? Four possible factors contained in kappa are C_v, n/V, c_{rms}, and λ. Which of these terms should be in an expression for κ analogous to equation 2–24 for η? Arrange these terms so that the combined units are those of κ as deduced from equation 2–28.

27. The velocity of sound in air is 1129 ft/sec at 20 °C. Convert this number to cm/sec. Does this reveal the controlling factor in the propagation of sound through a gas? The velocity of sound in air at 1000 °C is 2297 ft/sec. Does this agree with your selection of the controlling factor? Is the speed of sound greater in helium as compared with air or less? Does this explain high-pitched voices in helium-filled rooms? Explain.

28. The heat capacity at constant volume of a monatomic ideal gas is $3R/2$. Suppose that we compress the gas until PV/RT is significantly greater than 1. Will C_v increase, decrease, or remain $3R/2$ in this compressed state? Will C_p be greater than, less than, or equal to $5R/2$ in this compressed state? Explain.

3/ Subatomic Particles

There are attractive and repulsive forces between atoms and molecules and the attractive forces are responsible for the liquefaction of gases. These are the same forces that cause liquids to solidify and atoms to join to form molecules. To discover the source of these forces, we must investigate the structure of the atom. This will give us a basis for predicting the chemical reactivity of the elements, and the sources of molecular bonding and structure. We are particularly interested in answering the question, If we could see inside an atom, what should we see?

3–1. Electrical Nature of Matter

Elemental sodium can be prepared by passing an electrical current through molten NaCl:

$$2\,NaCl(l) + \text{electricity} \rightarrow 2\,Na(l) + Cl_2(g) \qquad \textbf{3–1}$$

Electricity serves as a source of energy, but this is not its only role in reactions like 3–1 since merely input of energy does not bring about the reaction. For example, simply heating NaCl will not yield sodium and chlorine.

The Particulate Nature of Electricity

Michael Faraday, as a young assistant in Davy's laboratory, pursued the question of the special nature of electricity that allows it to bring about chemical transformations like reaction 3–1. His key measurements were the relative amounts of electricity needed to transform given amounts of chemical species. Electrolyses can be performed with many species, and two additional examples are

$$2\,KCl(l) + \text{electricity} \rightarrow 2\,K(l) + Cl_2(g) \qquad \textbf{3–2}$$

$$ZnCl_2(l) + \text{electricity} \rightarrow Zn(l) + Cl_2(g) \qquad \textbf{3–3}$$

Faraday found that one mole of NaCl and one mole of KCl require exactly the same amount of electricity apiece to be completely converted to constituent elements. Furthermore, every other chloride of the formula MCl, where M is the companion element, requires this same amount of electricity. In the case of one mole of $ZnCl_2$ or any chloride with the formula MCl_2, exactly twice as much electricity per mole is needed for complete reaction as is needed for one mole of NaCl. These results indicate to us that electricity is composed of fundamental units and that these units are added to and subtracted from atoms or molecules or both to produce chemical transformations. This means that atoms themselves must contain the elementary particles of electricity, and atoms must provide the basis of electricity. In 1874, G. J. Stoney, an English physicist, coined the word *electron* for this particle of electricity.

Coulomb's Law

By Faraday's time, it was known that electricity was of two types, termed positive and negative, and that objects with like types repelled each other and those with unlike types attracted each other, when the objects were located near each other. Electricity as it appears in such objects is termed **static electricity.** If a piece of metal is made to connect two objects of equal and opposite charge, an electrical current flows through the metal and both objects lose their charge. Under normal circumstances, matter is electrically neutral, but all matter can accept either positive or negative charges of electricity. Electricity as it is delivered through a wire consists of electrons; one of the two charges consists of an electron excess in matter and the other is an electron deficiency. Normal matter has no net charge and if neutral matter contains particles of one charge (namely electrons), it must also contain counterbalancing charges of the opposite kind. Removal of electrons from an atom or molecule is termed **ionization** and the resultant charged species is termed an **ion.**

The force that is exerted between charges was studied in the late eighteenth century by the French scientist Charles Coulomb. If charges q_1 and q_2 of the same type (that is, positive or negative) are placed on two suspended spheres as shown in Figure 3–1(a), the spheres repel each other against the force of gravity, and the repulsive force is easily measured. Figure 3–1(b) shows attraction between opposite charges, and this force is also easily

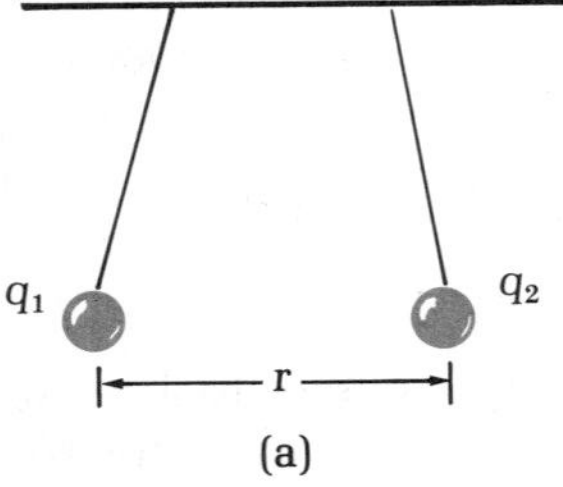

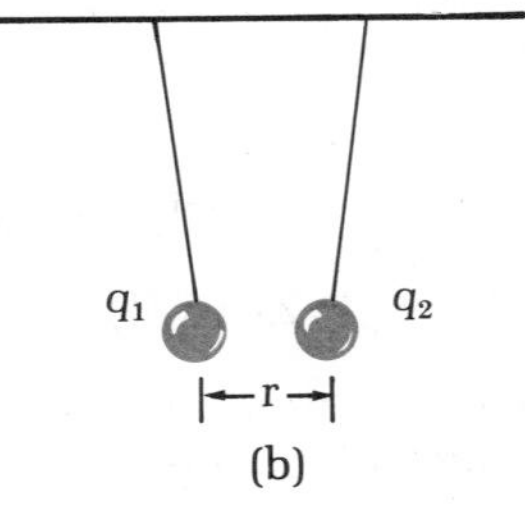

Figure 3–1 (a) Repulsion and (b) attraction between like-charged and oppositely charged spheres.

measured. Coulomb found that the relation between the force, the charges, and the distance r between the objects was

$$F = \frac{q_1 q_2}{r^2} \quad \textbf{3–4}$$

This result is known as **Coulomb's law** and it is one of the most important relations in chemistry.

If the force is given in dynes and r is given in cm, the units of q must be $dynes^{1/2}$-cm, and this is termed a **statcoulomb** (symbolized by esu for electrostatic unit) in honor of Charles Coulomb.

3–2. Distribution of Mass and Charge in the Atom

We have not yet said if electrons are units of positive or negative electricity. Nor have we reported the mass of the electron or the masses of the species having charges opposite to the charge of the electron. The diameter of a helium atom is calculated to be 2.21Å. How much of the volume of the helium atom is occupied by electrons and how much by oppositely charged species?

The Negative Charge of the Electron

Clearly any electrons whose properties we wish to determine are best studied isolated from all other matter. They are not isolated in a metal wire, but they are in the case of electrical conduction through a vacuum. Physicists in the late nineteenth century discovered that an electrical current could be made to flow through gases enclosed in glass tubes, as shown in Figure 3–2. The glass tube contains two metal electrodes and these are connected to a high-voltage source. What is depicted in the figure is what we commonly term a neon tube, though almost any gas can be used. The gas conducts electricity at the high voltage and also emits light.

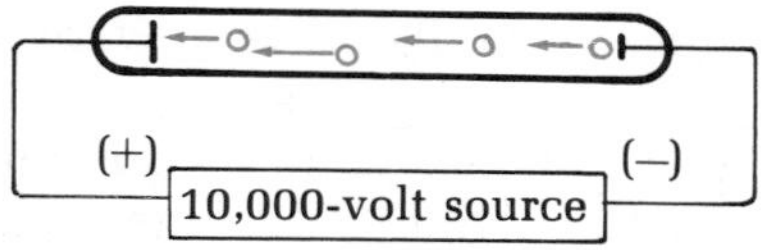

Figure 3–2
Electrons passing through a glass tube containing a low-pressure gas and two electrodes connected to a high-voltage source.

If the tube is connected to a vacuum pump and the gas is removed, the light emission from the gas stops and a greenish glow emanates from the glass walls as the electrical conduction through the tube continues. This glow is apparently caused by the collision with the glass wall of the electrical particles that traverse the tube. To identify the electrode that is the electron source, a solid object is placed in the tube. This object glows on the side exposed to the electron source and casts a shadow on the wall on the opposite side. Such an experiment shows that electrons emanate from the negatively charged electrode. Electrons are negatively charged particles, because they are attracted to the positively charged electrode. This flow of electrons is called an electrical current.

The Charge-to-Mass Ratio of the Electron

Another way to demonstrate that electrons are negatively charged particles is to make a narrow beam of them in an evacuated tube and then test for deflection of this beam in the presence of known charges. An apparatus for this purpose is depicted in Figure 3–3. The evacuated tube of Figure 3–2 is extended and the positively charged electrodes have small holes in them to allow the passage of a narrow electron beam in a horizontal direction. The beam then passes between the charged plates as shown in the figure. Because

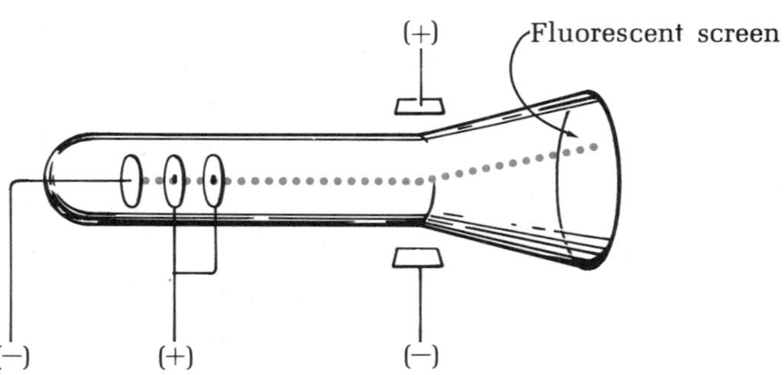

Figure 3–3 Apparatus for testing the deflection of an electron beam in an electric field.

of electrostatic attraction and repulsion, a negatively charged beam must be deflected upward and a positively charged beam downward. All substances bombarded with electrons of high kinetic energy will glow. However, some materials glow particularly brightly when so bombarded. The right end of the tube in the figure is coated with such a **fluorescent** material, and the spot at which the electron beam strikes is easily detected from the glow. The beam deflection in the figure is upward and thus electrons are shown to be negatively charged particles. The electrostatic force that the electrons experience, as shown in Figure 3–3, is proportional to the electronic charge. According to equation 2–12, this force produces an upward velocity and a displacement that are inversely proportional to the electronic mass. Thus, the observed upward deflection is directly proportional to e/m of the electron, where e is the electronic charge and m is the electronic mass.

A beam of charged particles is also deflected in moving through a magnetic field and again the deflection is proportional to e/m. In the late nineteenth century, the English physicist J. J. Thomson used the deflections by the electric and magnetic fields to obtain the ratio e/m for the electron. In succeeding years, more precise measurements have been made and the currently accepted value is 5.273×10^{17} esu/g.

Measurement of the Electron Charge

The American physicist R. A. Millikan measured the electronic charge in 1908. Millikan reasoned that it might be possible to put on an observable object an electrical charge of one or a few excess electrons or to remove one or a few electrons. This charge could be measured by the force exerted on the object due to a known electric charge. The experimental arrangement is shown in Figure 3-4. The objects chosen to be charged were tiny oil droplets produced by an atomizer and the charges were produced by exposing the droplets to radioactivity (the phenomenon of radioactivity had very recently been discovered). In the absence of an electric field, the oil droplets fall in air owing to gravity. This rate of fall can be slowed down by applying charges to two metal plates, and from the rate of fall of the droplet (as observed with a telescope), the charge on the droplet is calculable. The results were as Millikan hoped they might be. The smallest charge ever observed on any droplet was 4.8×10^{-10} esu, and all other droplets were found to carry an integral multiple of this charge. Therefore, 4.8×10^{10} esu must be the electronic charge.

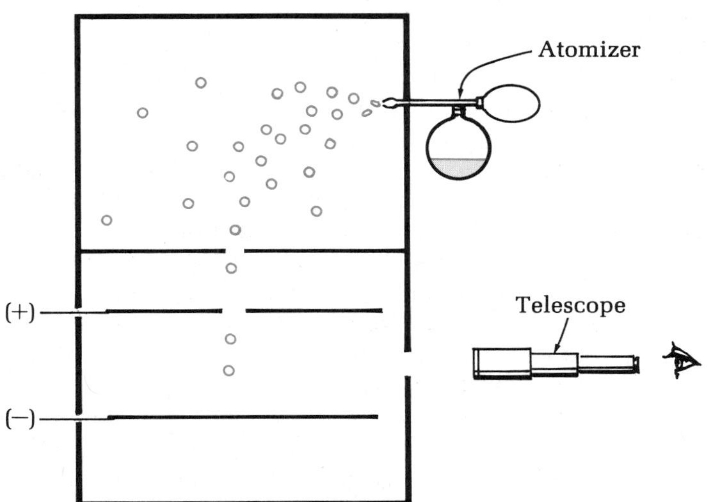

Figure 3-4
Schematic of the apparatus with which Millikan measured the electron charge.

With this value and Thomson's value of e/m, the electronic mass can be calculated.

$$m = \frac{4.8 \times 10^{-10}}{5.273 \times 10^{17}} = 9.1 \times 10^{-28} \text{ g}$$

The mass of the lightest atom known, hydrogen, is $1.008/6.023 \times 10^{23} = 1.674 \times 10^{-24}$ g. The electron is much lighter than any atom.

The Nucleus

With the work of Thomson and Millikan, it was known that most of the atomic mass resides in the positive charges. How, then, are these charges distributed in the atom? The simplest assumption is that electrons and positive charges possess the same densities and that the positive charges, which possess most of the atomic mass, occupy most of the atomic volume. Thus, in such a model, we picture the atom as a large, solid ball with tiny electron specks embedded in it. A test of this model can be provided by bombarding a collection of atoms (a target) with a beam of particles and observing the resulting scattering pattern. The English physicist Ernest Rutherford carried out this test in his laboratory, and the surprising result provided the first picture of the structure of the atom.

The collection of atoms Rutherford used as the target was very thin gold foil, and the bombarding particles were the products of radioactivity known as alpha particles (discovered to be the same as helium atoms that have lost their electrons). Alpha particles have very large kinetic energies as they emanate from radioactive sources, and Rutherford expected them to blast through the collection of charges in the gold atoms like a bowling ball through bowling pins. The results of the experiment showed that indeed the overwhelming majority of the alpha particles passed through the foil with little or no deflection. However, a very small number of alpha particles experienced very large deflections and some were deflected almost straight back from the foil. This result was totally inconsistent with the "ball-speck" model. In this model the mass and charge weren't concentrated enough in any one place to cause the highly energetic alpha particles to be deflected.

Rutherford reasoned that these large deflections could only arise from a large concentration of mass and charge in a very small volume (much smaller than the volume of the atom itself). In analogy to the bowling pins, a much greater deflection of the ball will occur with one steel pin than with ten wooden ones having a total mass equal to that of the steel pin. It is from the Rutherford experiment that we derive our current view of the atom as an extremely dense positively charged **nucleus** of approximate radius 10^{-12} cm surrounded by much lighter electrons. We thus discard the ball-speck atom model, and replace it with a nuclear atom model.

Atomic Number

How do we know the number of electrons in any given atom? We can make the simplest assumption possible and then test it to see if experiment bears out the assumption. Hydrogen is the lightest atom known and we shall assume that it comprises one electron and one unit of positive charge (the hydrogen nucleus is termed a **proton**). Helium is the next lightest element, and we might expect it to contain two electrons and a nucleus consisting of two protons. If such is the case, the helium atom should have a mass twice that of the hydrogen atom. However, the measured atomic weight of He is 4.00. Either the helium atom contains four electrons and four protons, or nuclei also contain uncharged particles that contribute to the nuclear mass but not to the charge. These uncharged particles, termed **neutrons,** were isolated and identified in 1932 by the English physicist D. N. Chadwick. An isolated neutron has a mass of 1.00893 amu, nearly identical to the 1.00757-amu mass of a proton. The helium atom atomic mass of 4.00 arises from two neutrons and two protons. The number of protons in a neutral atom (a number that equals the number of electrons) is termed the **atomic number** of the atom. The atomic numbers of hydrogen and helium are 1 and 2, respectively. The table of the elements on the inside front cover of this text lists the atomic numbers of all the known elements.

Isotopes

If any element is vaporized, ionized, and passed as a beam through a magnetic field, the beam will be deflected in the same manner as depicted in Figure 3–3 for electrons in an electric field. As with electrons, the degree of deflection is proportional to e/m. Since ionic charges are equal to the known electronic charge or a multiple of it, beam deflections may be used to measure atomic masses (or molecular masses if the ion is polyatomic like O_2^+). The instrument with which this mass measurement is accomplished is termed a **mass spectrometer,** and it is a very important asset in any modern research laboratory. The mass spectrometer provides a rapid and highly precise measurement of the mass of any molecule that can be obtained in the gaseous state and also the masses of molecular fragments produced in the ionization of the molecular beam. This information on the masses of the parent molecule and the fragments derived from it is invaluable in identifying the molecule or determining the molecular structure, or both.

If metallic magnesium is vaporized and the vapor analyzed in a mass spectrometer, the magnesium sample is found to contain three different kinds of atoms, termed **isotopes,** with masses of 23.993, 24.994, and 25.991 amu, respectively. Each of the three atoms contains 12 protons and the remaining mass arises from 12, 13, and 14 neutrons, respectively. The mass spectrometer also records the percentages of each of the three isotopes, and they are 78.60, 10.11, and 11.29, for the isotopes that are designated ^{24}Mg, ^{25}Mg, and ^{26}Mg, respectively. The measured atomic weight of Mg is the abundance-weighted average of the masses of the three isotopes, or $(23.993)(0.7860) + (24.994)(0.1011) + (25.991)(0.1129) = 24.32$. Most elements are composed of isotopic atoms. Another example is carbon, which consists of 99 percent ^{12}C, 1 percent ^{13}C, and a small percentage of radioactive ^{14}C. The atomic weights (listed inside the cover of this text) are calculated relative to the standard mass of ^{12}C, which is defined to be exactly 12. This is the reason, as we pointed out in Chapter 1, that H has an atomic weight of 1.008 instead of exactly 1, which Cannizzaro had previously designated.

Isotopes, particularly radioactive ones, have proved invaluable in following a given element through a physical or chemical process. For example, the American scientist Melvin Calvin used ^{14}C to map the sequential steps through which the carbon of CO_2 proceeds to sugar molecules in photosynthesis.

3-3. Atomic Spectra and Electronic Motion

How are we to learn about the motions of electrons within atoms? The obvious answer is that we should look at the electrons. How do we see objects? We see them by the light they scatter, reflect, or emit. The light carries information about the objects that our eyes can convert to an electrical signal; then our brains can convert the signal to images of the objects. What we know as visible light is just one of several forms of **radiant energy;** and to characterize the interaction of electrons and radiant energy, we must first characterize radiant energy. This interaction provides us with most of the detailed information we have on the structures of atoms and molecules.

Source of Radiant Energy

You are aware of radiant energy in the form of sunlight and radiant heat from substances at high temperatures. You also know about other forms of radiant energy such as radio signals, microwaves in a microwave oven, ultraviolet radiation from a "black light," and X-rays in the dentist's office. All of these are forms of energy because they raise the temperature of matter when absorbed by matter. What is the source of radiant energy and how does it interact with matter? We shall develop a picture of radiation that the English physicist James Maxwell first drew in the late nineteenth century. Each of the types of radiant energy emanates from matter and must be a consequence of the structures or motions of matter, or both. One motion that is known to give rise to radiation is the movement of electrons along a wire, and this motion is the basis of a radio broadcast antenna.

Consider the representation of the grounded antenna shown in Figure 3-5. The top of the antenna is connected to a radio-frequency oscillator that alternately attracts electrons from the ground along the antenna and repels them to the ground as a function of time. This periodic electron motion is indicated by the two-headed arrow alongside the antenna. Suppose that we have a single electron located at a fixed distance from the wire, as indicated

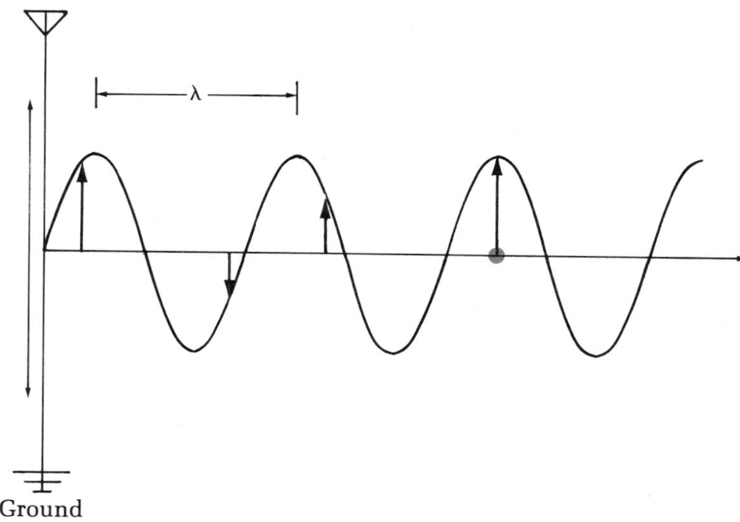

Figure 3-5
The electric field at a given time resulting from the periodic electron motion in a radio antenna.

by the colored sphere in the figure. Since electrical forces are transmitted through space according to Coulomb's law, the single electron, as well as the electrons in the wire, will move in an up-and-down direction in response to radio-frequency oscillation, although the force on it is less than the force on the electrons in the antenna wire because of the greater distance from the oscillator. There is also a small lag in the response of an electron at a distance from the antenna compared with the motion of the electrons in the antenna. Information about the motion of electrons in the antenna travels away from the antenna at a speed of 2.998×10^{10} cm/sec. This information, to which the outside electron responds, is a force that alternately pushes the electron towards the ground and pulls it away from the ground. This force is termed an electric field. The field travels as a wave from the antenna at the speed of 2.998×10^{10} cm/sec. This wave may be plotted for any instant of time, as shown in the figure. The electric field is a vector quantity, and the wave outlines the tips of the vector arrows at various distances from the antenna. A vector arrow pointing up indicates that an electron placed at that point experiences a force upward at the time shown and a downward arrow indicates a downward force. The electron in the figure, at that time, is experiencing the maximum upward force.

In addition to the electric field shown, a second consequence of the motion of electrons in a wire is a magnetic field. An electromagnet is produced by sending an electrical current through a coil of wire. Radiant energy is characterized by a periodic variation of both an electric field and a magnetic field. Thus, radiant energy is more commonly known as **electromagnetic radiation.**

Wavelengths and Frequencies

Maxwell deduced that all of the forms of radiant energy, as listed at the beginning of this section, are basically the same. They arise from the motions of charged particles, they all consist of periodic electric and magnetic fields, and they all travel at the same speed of 2.998×10^{10} cm/sec.

The characteristic differences between the forms of electromagnetic radiation occur in the **wavelengths** and **frequencies** of the waves. Figure 3–5 shows the wavelength λ as the distance from crest to adjacent crest, and all waves are characterized by such a distance. The frequency ν (Greek, lowercase *nu*) is defined as the number of crests passing a given point in one second. The frequency is related to the wavelength by

$$\nu = \frac{\text{no. of crests}}{\text{sec}} = \left(\frac{\text{cm}}{\text{sec}}\right)\left(\frac{\text{no. of crests}}{\text{cm}}\right) = \frac{2.998 \times 10^{10}}{\lambda} \qquad \textbf{3–5}$$

For any wave, the product of the wavelength and the frequency is the speed of the wave. The unit, crests/sec, is termed the hertz (symbolized Hz). A typical AM radio frequency is 100 kilohertz (100 kHz), or 1.00×10^5 crests passing a given point in one second. The wavelength corresponding to this frequency is calculated as follows:

$$\lambda = \frac{2.998 \times 10^{10}}{1.00 \times 10^5} = 3.00 \times 10^5 \text{ cm} = 3.00 \text{ km}$$

We see that AM radio wavelengths are extremely long. Figure 3–6 shows the division of electromagnetic radiation into forms based on wavelength and frequency ranges. Because the range covers many powers of ten, the ranges are shown by the **logarithms** of λ and ν. If you are unfamiliar with logarithms, refer to Appendix B for a short treatment of the subject. You will note that the region of visible light covers only a small range of wavelengths between approximately 3.5×10^{-5} cm and 7.5×10^{-5} cm.

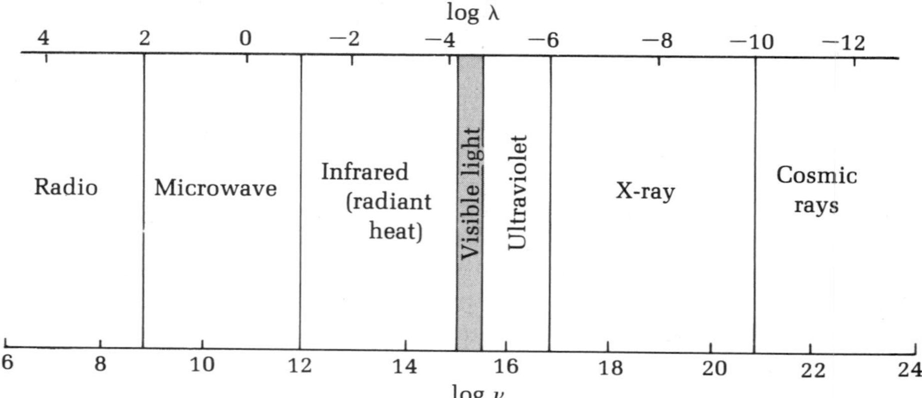

Figure 3–6 Wavelengths and frequencies of the different forms of electromagnetic radiation.

Emission Spectrum of the Hydrogen Atom

If the motions of electrons within atoms give rise to electromagnetic radiation, the electromagnetic radiation those atoms emit should provide details of these motions. Alternatively, electromagnetic radiation striking an atom excites electron motion, and thus electron motion can be studied by either the **emission** or the **absorption** of radiation.

The gas in the tube shown in Figure 3–2 emits light of a color characteristic of the gas. The electromagnetic radiation emitted can be spread into its component wavelengths by passing it through a glass plate on which a number of very closely spaced black lines are placed. The plate is called a **diffraction grating** (see Chapter 13, for the principle of its functioning in detail).

Figure 3–7 shows the photographic recording of the wavelengths of radiation that hydrogen gas emits. The diffraction grating spreads the radiation

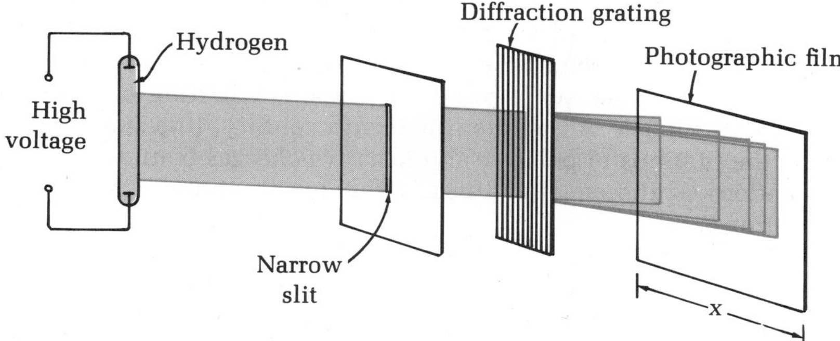

Figure 3–7
Schematic of apparatus for obtaining the emission spectrum of gaseous hydrogen.

so that each wavelength strikes the photographic film at a different value of x along the film. The resultant photographic pattern represents the intensity of emitted radiation as a function of wavelength, and it is termed the **emission spectrum** of gaseous hydrogen. As shown in the figure, the emission spectrum of hydrogen, like the emission spectrum of each of the other gases, consists of many well-defined, narrow lines. The existence of line spectra came as no particular surprise to the physicists who first measured these spectra, since there is a well-understood analog—the emission of sound from objects, for example from violin strings. Since the violin string is anchored at its ends, it can vibrate only with wavelengths that are simply related to its length. Two examples of allowed vibrations are shown in Figure 3–8. The stationary string is shown by the solid line of length a, and the vibrating string at some given time by the dashed line. The longest allowed λ is $2a$, as shown in Figure 3–8(a), and this yields the lowest possible frequency, the so-called fundamental frequency. The next-longest allowed λ is a, as shown in Figure 3–8(b), yielding the so-called first harmonic frequency of twice the fundamental. By the continuation of Figure 3–8, we see that the sound

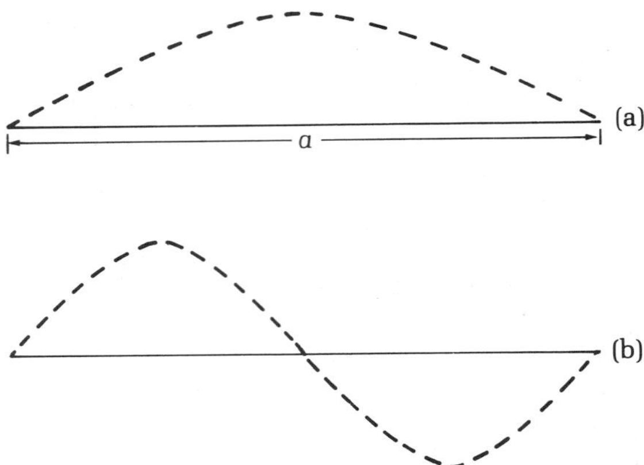

Figure 3–8
Two allowed vibrations of a violin string of length a.

spectrum consists of a family of discrete wavelengths (or frequencies) and that the wavelengths are all simply related through a. **All bounded vibrations yield only a discrete set of wavelengths,** and the pattern of wavelengths depends on the shape of the bounded, vibrating object. Surely, reasoned the physicists of the late nineteenth century, line spectra of gases arise from the motions of positive and negative charges bound inside atoms and the pattern of the spectral lines is the key to atomic shape and subatomic motion.

The spectrum of hydrogen gas became the focus of this attempt to unravel the pattern of lines since the hydrogen atom contains the smallest number of moving particles and yields the simplest spectrum of any gaseous atom. The visible portion of the hydrogen atom spectrum is shown in Figure 3–9. The figure shows a series of lines that appear to be related because they group together towards a limit at 3646.00 Å. When other regions besides the visible were investigated, a second family of lines that also group towards a limit was discovered in the ultraviolet region, and several such families were discovered in the infrared spectral region.

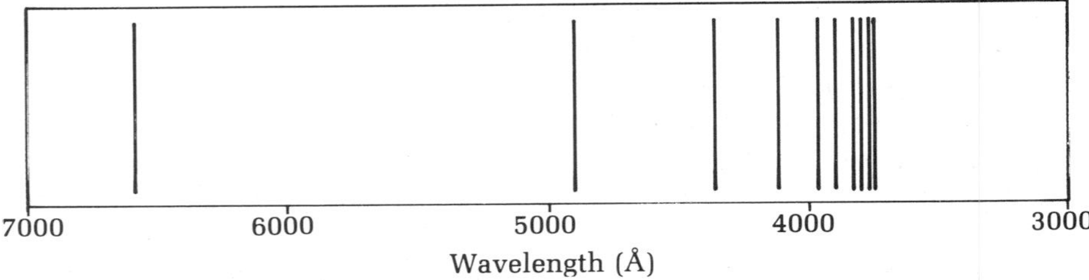

Figure 3–9
A drawing showing the wavelengths of the emission spectrum of gaseous hydrogen in the visible light range.

In 1885, a mathematician named Balmer discovered that the visible lines of the hydrogen spectrum are related through an equation; we write it here in the following form:

$$\frac{1}{\lambda} = 109{,}709.2\left(\frac{1}{4} - \frac{1}{n^2}\right)\ \text{cm}^{-1} \qquad \textbf{3–6}$$

where n takes the integral values 3, 4, 5, ... Rydberg related the Balmer family of lines to each of the other families; he showed that *all* of the hydrogen atom lines are related through

$$\frac{1}{\lambda} = 109{,}709.2\left(\frac{1}{n_1^2} - \frac{1}{n_2^2}\right)\ \text{cm}^{-1} \qquad \textbf{3–7}$$

For the Balmer lines, $n_1 = 2$ and $n_2 = 3, 4, 5$, etc.; while for the ultraviolet family of lines (termed the Lyman series), $n_1 = 1$ and $n_2 = 2, 3, 4$, etc. The infrared families are derived from sets based on $n_1 = 3$, $n_1 = 4$, and higher, with n_2 always greater than n_1.

The task of the physicist was to develop a model for bounded motion that would yield equation 3–7. In spite of a large amount of effort, no such model has ever been derived using the laws of physics as they were known in the year 1900. However, a perfect match is obtained between calculation and experiment if we accept some new physical concepts that began to be formulated shortly after 1900. After we see how these new concepts were developed, we shall then return to their application to the hydrogen atom spectrum.

The Quantum Hypothesis

The first radical break with pre-1900 physics came from the attempt by the German physicist Max Planck to derive a model that would successfully predict the emission spectra of heated solids. You know that as the temperature of any solid is increased, it first emits only radiant heat; then it successively glows red, yellow, and finally white when it has become very hot. The radiation can be split into component wavelengths, as in Figure 3–7, and the spectrum may be recorded on a photographic film. Unlike the spectra of gases, which consist of discrete lines, the spectra of heated solids show a continuous variation in intensity across the film (a common example of a continuous spectrum is a rainbow).

The intensity pattern on the film representing the emission spectrum of the heated solid is very similar to the pattern obtained in measuring the speed distribution for gas molecules (Figures 2–16 and 2–17), an observation that was not lost on the young Max Planck. Other physicists in the late nineteenth century had been trying for some time to devise a model for the motion of charges in solids that would quantitatively yield the observed emission spectrum. The first obvious question is why gases yield line spectra and solids yield continuous spectra. This is easily explained in terms of pre-1900 physics by an analogy to Figure 3–8. A gas is a collection of isolated moving charges that can be compared with a collection of isolated violin strings. The notes heard from the isolated strings are the individual fundamentals and harmonics from each string—a line spectrum, if you will. In the solid, the moving charges are in close proximity, analogous to violin strings in contact. Vibrating objects in contact yield extra frequencies in addition to the fundamentals and harmonics. These extra frequencies are all possible sums and differences of the frequencies from one string with the frequencies from other strings. When a huge number of strings are in contact (analogous to a solid made up of a huge number of atoms), the number of possible frequencies is also huge and a continuum of frequencies is detected by even the most sensitive measuring device.

The second question concerns the fact that the intensity on the film shows a single maximum, and the intensity falls off steeply to either side (higher and lower frequencies), analogous to the steep drop-offs in Figures 2–16 and 2–17. Pre-1900 physics could explain the drop-off at low frequencies, but not the high-frequency drop-off. The drop-off at high speed in Figures 2–16 and 2–17 is due to the very low probability that collisions can result in molecular speeds greatly in excess of the average speed. If the analogy between the situations as shown in Figures 2–16 and 2–17 and the frequency distribution from a heated solid were to be carried along by Planck, he had to propose that the energies of the motions of charged particles in solids are directly proportional to the frequencies of those motions. Thus, according to his proposal, high frequencies are present in the emission only in small intensities because high frequencies arise from high-energy motion in the solid and such high-energy motion is seldom encountered.

Planck proceeded to construct a model of the oscillating charges in solids that was able to agree quantitatively with the experimental frequency distribution from a heated solid. The elements of this model are as follows:

1. A huge collection of vibrators in mutual contact is characterized by a continuum of frequencies ν.
2. The allowed energies E of a vibration of frequency ν are given by the equation

$$E = nh\nu \qquad \textbf{3–8}$$

where h, termed Planck's constant, is 6.625×10^{-27} erg/Hz and $n = 1, 2, 3, 4, \ldots$.

Thus, a high-frequency vibration is a high-energy vibration and it is unlikely to be found, just as very high molecular speeds are unlikely to be found in a gas.

The Photoelectric Effect

The Planck proposal, equation 3–8, is complete heresy in terms of pre-1900 physics. The energy of vibration of a violin string, for example, is totally independent of the frequency of vibration. It is no wonder that the scientists of 1900 met this Planck proposal with almost universal incredulity. A notable exception was the young Albert Einstein, who saw in the proposal a way to explain another phenomenon, the details of which had puzzled physicists for several decades, namely the **photoelectric effect.** Figure 3–10 shows a photoelectric, or "electric eye," cell. Two electrodes are enclosed

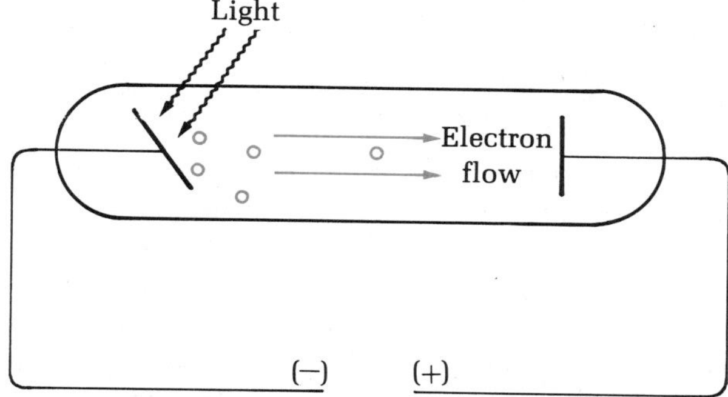

Figure 3–10
Schematic of the apparatus used for demonstrating the photoelectric effect.

in an evacuated glass tube. The negatively charged electrode is coated with a highly reactive metal, such as cesium. The voltage between the electrodes is made sufficiently low that no current flows if the cell is kept in the dark. However, if the cell is exposed to light, an electron flow occurs through the vacuum from the cesium electrode to the positive electrode. This light-induced electron flow is termed the photoelectric effect and it is extremely useful to the scientist in converting radiant energy directly to electrical energy so that, for example, the intensity distributions in spectra can be obtained rapidly, much more so than from photographic film, as in Figure 3–7.

The existence of the photoelectric effect is not surprising in terms of pre-1900 physics. The absorption of light produces an increase in energy in the absorber (the temperatures of most substances increase upon illumination), and if a sufficient amount of energy is absorbed by the cesium it is reasonable that some electrons could be freed from their atoms. Furthermore the electrical current is directly proportional to the light intensity, as pre-1900 physics would predict.

The difficulty came with the dependence of the electrical current on the frequency of the light used. According to pre-1900 physics, there should be no frequency dependence. Yet experiments showed clearly that every reactive metal possesses a characteristic threshold frequency below which the electrical current is zero, regardless of the light intensity. If a further experiment is set up to measure the kinetic energies of the electrons freed by light of a given frequency ν, which is above the threshold, the following relation is found:

$$\text{Kinetic energy} = \alpha\nu - \beta \tag{3–9}$$

where α is the same for all metals and β (Greek lowercase *beta*) varies from metal to metal.

Einstein recognized that the unexplained details of the photoelectric effect could be clarified by an extension of the Planck proposals. Planck's proposed energies of vibration are of the form $E = nh\nu$. Suppose that the emission of radiation occurs by a transition from the energy $nh\nu$ to the next lowest allowed energy $(n - 1)h\nu$. The energy difference is $h\nu$ and it should be released in a "burst," or **quantum,** of energy $E = h\nu$, as the downward transition in vibrational energy takes place. Thus, electromagnetic radiation should consist of bursts, or quanta, of waves. These quanta are commonly termed **photons.**

Einstein further speculated that if electromagnetic radiation is generated in bursts, perhaps it is absorbed in the same fashion; perhaps electrons and nuclei in any state are characterized by discrete energy levels and only those frequencies of radiation whose photon energies $h\nu$ match energy level differences in the absorber can be absorbed.

With these proposals in mind, we depict in Figure 3–11 Einstein's view of the photoelectric effect. The electrons of the metal were proposed to

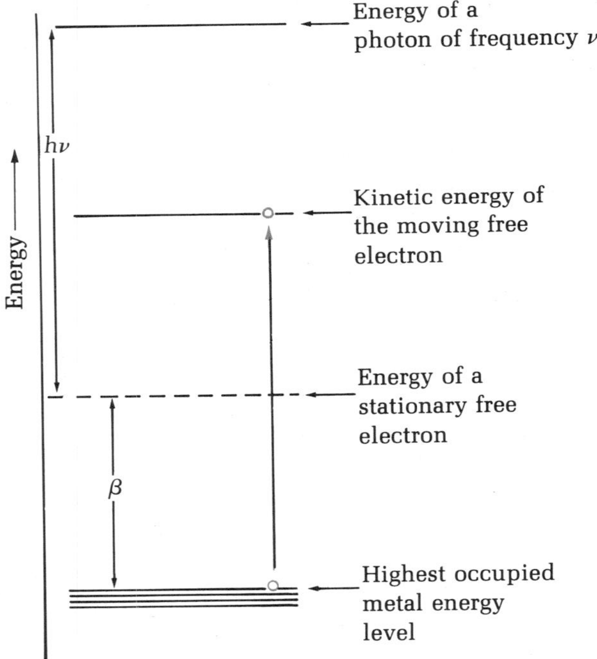

Figure 3–11 Energies involved in the freeing of an electron (*colored circle*) from the highest occupied energy level of a metal by a photon of energy ($h\nu$) greater than the threshold value (β).

occupy a series of energy levels of $E = nh\nu$ up to the highest level shown in the figure. There is a minimum energy β necessary to produce a stationary free electron by removing an electron from the highest level. Any photon of energy $E = h\nu$ less than β cannot produce a free electron, and hence there is a threshold frequency. The figure indicates the energy of a photon that is significantly greater than β and that does free an electron, indicated in color. The excess energy of the photon is converted to the kinetic energy of the freed electron, so that the kinetic energy is given by the equation

$$\text{Kinetic energy} = h\nu - \beta \tag{3–10}$$

If this picture is correct, the constant α in equation 3–9 should be equal to Planck's constant; experimentally it is.

The Planetary Atomic Model

In 1913, Niels Bohr, a Danish physicist, provided a very convincing argument for the existence of photons and quantized energy levels and also for the Rutherford model of the atom by his successful, precise calculation of the emission spectrum of the hydrogen atom. Bohr felt that certain integers, as shown in equation 3–7, could be related to the integers Planck had introduced, represented in equation 3–8. Equation 3–7 appears to represent a difference between energies, and with Einstein's proposals we should have

$$h\nu = \frac{h(2.998 \times 10^{10})}{\lambda} = E_2 - E_1 \tag{3–11}$$

or

$$\frac{1}{\lambda} = \frac{1}{h(2.998 \times 10^{10})}(E_2 - E_1) \tag{3–12}$$

where E_1 and E_2 are two electron energy levels in the hydrogen atom. By comparison of equations 3–7 and 3–12, we see that a successful model of the H atom must yield energy levels that are given, in general, by

$$E_i = \frac{-E_1}{n_i^2} \tag{3–13}$$

where E_i is the energy of the ith energy level and the constant E_1 must be $(109{,}709.2)(6.625 \times 10^{-27})(2.998 \times 10^{10}) = 2.179 \times 10^{-11}$ erg/electron.

Bohr assumed that the single electron of any given H atom revolves around the proton in a circular orbit, as shown in Figure 3–12, and that the

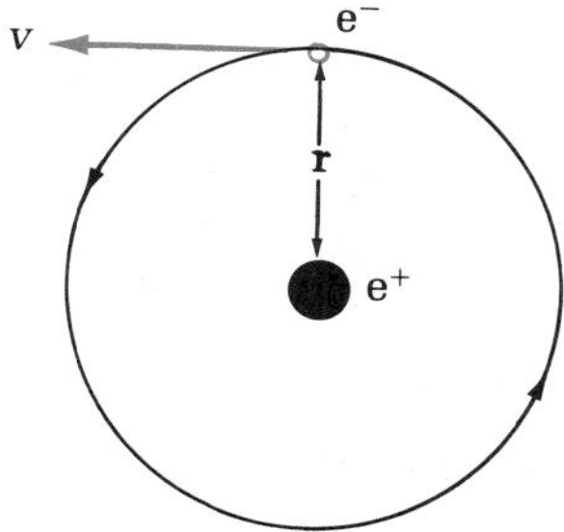

Figure 3–12
The planetary model of the hydrogen atom.

orbit is a stable one for the same basic reason that the planets move in stable orbits around the sun. In the case of planetary motion, it is known that the gravitational attraction between the planet and the sun is balanced by the centrifugal force of the motion. Bohr proposed that in the case of the H atom, the centrifugal force balances the force of electrostatic attraction. The quantitative expression for this force balance is

$$\frac{-e^2}{r^2} + \frac{mv^2}{r} = 0 \qquad \textbf{3–14}$$

where $-e$ is the electron charge, $+e$ is the proton charge, m is the electron mass, v is the magnitude of the electron velocity vector, and r is the electron-proton separation distance.

The energy of the electron in Figure 3–12 involves more than the kinetic energy of $mv^2/2$. Because of the attraction between the electron and the proton, force is required to pull the two particles apart and work must be performed. This indicates that the electron energy not only depends on v but also depends on r as well. The total energy of the electron is a sum of its kinetic energy, which depends on v, and its **potential energy,** which depends on r. In fact, the energy of any object under any conditions is given by the sum of its kinetic energy and its potential energy. To measure and report a value for the potential energy of an object, we must define where the zero of potential energy occurs. In the case of the electron in the hydrogen atom, the potential energy is defined as zero when the electrostatic force it experiences from the proton is zero. This occurs at r equal to infinity. The potential energy is negative when work must be done on the electron to remove it completely from the force exerted on it by the proton. By definition, **the potential energies of all electrons in all atoms are negative** because of the force of attraction between electrons and nuclei. The highest-energy bound electrons in Figure 3–11 are those that possess the least negative potential energy.

The value of the electron potential energy in Figure 3–12 is

$$\text{Potential energy} = \frac{-e^2}{r} \qquad \textbf{3–15}$$

so that the total electron energy is

$$E = \frac{mv^2}{2} - \frac{e^2}{r} \qquad \textbf{3–16}$$

However, from equation 3–14 we easily obtain

$$\frac{mv^2}{2} = \frac{e^2}{2r} \qquad \textbf{3–17}$$

or the electron kinetic energy is exactly minus one-half the potential energy. This is one example of a very useful relation known as the **virial theorem.**

Any situation in which the motions are all the result of electrostatic attractions or repulsions or both yields a combined kinetic energy for all of the moving objects that is minus one-half the combined potential energy. Such is the case in the motions of electrons and nuclei relative to each other in atoms and molecules; and the virial theorem is extremely important in chemical bonding (Chapter 5).

Equations 3–16 and 3–17 may be combined to yield

$$E = \frac{-e^2}{2r} = \frac{-mv^2}{2} \tag{3–18}$$

Bohr deduced that the integers shown as n_i in equation 3–13 could be introduced if the electron is forced to occupy only orbits i, in which

$$mv_i r_i = \frac{n_i h}{2\pi} \tag{3–19}$$

with $n_i = 1, 2, 3, \ldots$. Equating v between equations 3–14 and 3–19, we obtain

$$r_i = \frac{n_i^2 h^2}{4\pi^2 me^2} \tag{3–20}$$

and substituting r into equation 3–18 yields

$$E_i = \frac{-2\pi^2 me^4}{n_i^2 h^2} \tag{3–21}$$

Comparing equations 3–13 and 3–21 shows that they are of the same form. Of course, they were forced to be by the arbitrary insertion of equation 3–19 into the calculation. The stringent test comes in that requirement that

$$E_1 = -2.179 \times 10^{-11} \text{ erg} = \frac{-2\pi^2 me^4}{h^2} \tag{3–22}$$

Substituting the known values of each of these constants yields

$$\frac{2\pi^2 me^4}{h^2} = \frac{2(3.1416)^2(9.108 \times 10^{-28})(4.8029 \times 10^{-10})^4}{(6.625 \times 10^{-27})^2}$$

$$= 2.179 \times 10^{-11}$$

Thus, in spite of the arbitrariness of equation 3–19, the constant in equation 3–7 that applies to all of the hydrogen spectral lines was shown in the Bohr model to be a simple collection of known physical constants, one of which is Planck's constant. At this point, the concepts of photons and discrete energy levels seemed firmly established.

The first obvious extension of the Bohr model is clearly the helium atom, since it is the next simplest atom after hydrogen. All attempts to calculate the helium gas emission line spectrum based on the Bohr model of planetary orbits were dismal failures. By the early 1920s, many scientists had become convinced that while the concepts of photons and energy levels were correct, the view of well-defined electrons revolving around nuclei in well-defined orbits was incorrect. Therefore, leaving the Bohr model, we shall develop instead those concepts that have led to the current picture of the atom, a picture that permits quantitative predictions of all atomic emission spectra. These predictions are in excellent agreement with experiment.

The Uncertainty Principle

The German physicist Werner Heisenberg showed that the Bohr model of the hydrogen atom, which does not apply to atoms in general, contained a flaw that penetrated the very heart of the philosophy of model construction. Models exist to allow us to correlate and explain known facts and to predict new ones. Models should not contain features that cannot be observed under any circumstances. Implicit in the Bohr model is the assumption that we can look into an atom and see a sharply defined electron orbiting around a sharply defined nucleus in an orbit with a precisely measurable radius r at a precisely measurable velocity v. If such observations could never be made, the model is not applicable to the observations we can make, such as atomic spectra.

We have stated that if we are to see electrons, we must have light reflect from them or have them emit light. Let us imagine using light to look inside an atom. How do we go about looking into an atom? We see a thing by having light reflect from it to our eyes. In other words, nothing can be seen unless something interacts with it. In the case of large objects, the reflection of light produces very little disturbance of the object. Our eyes receive information that is a function of the object and not of the observation, and we get a faithful picture.

In the case of an atom of radius $\cong 10^{-8}$ cm, a problem arises in using reflected light. The wavelength range of visible light is from approximately 3.5×10^{-5} cm to 7.5×10^{-5} cm. Experimentally it is found that distances smaller than the wavelength of the radiation used for the observation cannot be distinguished; that is, if we wish to see inside an atom of radius $\cong 10^{-8}$ cm, we must use radiation having a wavelength less than 10^{-8} cm. By reference to Figure 3–6, we see that this limits the applicable range of electromagnetic radiation to X-rays and cosmic rays.

In 1923, the American physicist A. H. Compton investigated the encounter between a beam of electrons and a beam of X-rays. The X-rays and electrons were observed to scatter as though they were colliding particles. Furthermore, the X-ray wavelengths were changed as a result of collisions of the X-rays with the electrons. The X-ray photons appeared to be characterized by a mass calculable from Einstein's proposal that all particles possess a total energy given by

$$E = mc^2 \tag{3–23}$$

where m is the mass of the particle and c is the speed of electromagnetic radiation (2.998×10^{10} cm/sec). Since the energy of a photon is given by $E = h\nu$, Compton equated mc^2 and $h\nu$ to obtain the mass of a photon:

$$m = \frac{h\nu}{c^2} = \frac{h}{\lambda c} \qquad \textbf{3–24}$$

For an X-ray of a wavelength of 1.0×10^{-9} cm, the photon mass from equation 3–24 is 2.2×10^{-28} g, a value near that of the electron mass. If this photon mass is correct, there is no hope of our ever being able to observe the motions of electrons in atoms. For an observer to see an electron in an atom, λ must be significantly less than 10^{-8} cm. But then the photon is comparable to the electron in momentum, and the encounter of the two produces a very large disturbance of the electron. Any attempt at locating the electron would grossly disturb its motion. From this experiment and many others, Heisenberg became convinced that it is indeed impossible to specify simultaneously the exact position and motion of any particle. In our case, the Heisenberg uncertainty principle says that we can never see electrons moving in atoms; and a proper model of the atom must incorporate this uncertainty.

Wave Nature of Electrons

The French theoretician Louis de Broglie suggested the key to the current model of the atom. Equation 3–24 may be rearranged to

$$\lambda = \frac{h}{mc} \qquad \textbf{3–25}$$

where mc is the photon momentum. If photons have mass, reasoned de Broglie, perhaps electrons (and any other particles) have wavelengths given by Planck's constant divided by the momentum of the electron. Dramatic verification of this proposal came in 1927 from an experiment C. J. Davisson and L. H. Germer performed at the Bell Telephone Laboratories. They observed a beam of electrons reflected from the surface of metallic nickel to be diffracted just as the radiation is diffracted in Figure 3–7. Diffraction is a characteristic of waves and it can be used to measure the wavelength. Suppose that the electron momentum is produced as shown in Figure 3–2 and that the voltage difference is 100 volts. The energy the electron gains is commonly reported in **electron volts** (ev), where

$$1 \text{ ev} = 1.6 \times 10^{-12} \text{ erg}$$

Thus, the kinetic energy gained by the electron is 1.6×10^{-10} erg. Since the electron energy $= mv^2/2$ and the momentum $= mv$, the equation for the momentum of the 100-volt electron is

$$\text{Momentum} = [2m(\text{kinetic energy})]^{1/2}$$

$$= [(2)(9.1 \times 10^{-28})(1.6 \times 10^{-10})]^{1/2} \qquad \textbf{3–26}$$

$$= 5.4 \times 10^{-19} \text{ g-cm/sec}$$

With this result and the de Broglie relationship, the predicted wavelength of the 100-volt electrons is 1.2×10^{-8} cm. This is an ideal wavelength with which to obtain diffraction by almost any solid, and experiments show that the predicted wavelength is indeed correct.

SUGGESTIONS FOR ADDITIONAL READING

Gamow, G., *The Atom and its Nucleus*, Prentice-Hall, Englewood Cliffs, N.J., 1961.

Feinberg, G., "Light," *Sci. Amer.*, **219**, 50(1968).

Garrett, A. B., *The Flash of Genius*, Van Nostrand Reinhold, New York, N.Y., 1962.

PROBLEMS

1. A Faraday-type experiment was performed on the reaction of electricity with molten $MgCl_2$ and then with molten $ZnCl_2$. The same amount of electricity was passed through both salts, and then the metals produced were isolated and weighed to give the amounts of Mg and Zn. In such an experiment 0.124 g of Mg and 0.333 g of Zn were produced. Are these numbers consistent with the concept that electricity consists of discrete particles? What weight of sodium would have been produced if the same amount of electricity had been passed through molten NaCl?
2. Suppose that we remove an electron from each of two N_2 molecules, creating from each the ion N_2^+. What is the force between these two ions at a separation of 6.5 Å? According to Newton's gravitational law, the force of gravity between two identical objects of mass m is represented as follows:

$$F = \frac{(6.7 \times 10^{-8})\, m^2}{d^2} \text{ dyne}$$

where d is the distance between them. Calculate the gravitational force between the two N_2^+ ions. Does gravity make an important contribution to interionic attractions and repulsions? Explain.

3. Is the fact that an electron beam is deflected in passing through a magnetic field related to the operation of an electric motor? If so, how? Why cannot the charge-to-mass ratio of the electron be determined from an ordinary electric motor?
4. What fraction of the weight of a hydrogen atom is the weight of its single electron?
5. Naturally occurring chlorine is made up of two isotopes, ^{35}Cl and ^{37}Cl.

Refer to the measured atomic weight of chlorine and calculate the percentages of these two isotopes as they exist in naturally occurring chlorine.

6. A **light-year** is the distance that electromagnetic radiation travels in one year. How great a distance is this in cm? in miles?

7. At any given point in Figure 3–5, the magnetic-field vector is perpendicular to the electric-field vector. Does this agree with your knowledge of the magnetic "lines of force" around a conducting wire? Explain.

8. The velocity of sound in air is 1129 ft/sec at 20 °C. The musical note A is a sound frequency of 440 Hz. What is the wavelength of A in air? Is it apparent from this number why organ pipes are as long as they are?

9. The emission spectrum of the hydrogen atom shows a series of lines in the ultraviolet region. One of these lines occurs at 1025.7 Å. Show, by obtaining the values of n_1 and n_2 that yield this wavelength, that equation 3–7 accounts for it.

10. The light from the sun shows a continuous variation in intensity as a function of wavelength except for a few lines of very low intensity. One of these lines occurs at 6562 Å. Show, by obtaining n_1 and n_2 for it, that it corresponds to an emission line of hydrogen. What is the source of this absence in the light from the sun?

11. In Chapter 2 we showed that the most probable speed for gas molecules increases with increasing temperature. The same is true of the frequency appearing with highest intensity in the emission from a heated solid. Does this suggest a way to measure the temperature of a star? How?

12. The minimum energy required to remove an electron from an iron atom is 1.26×10^{-11} erg. Can electrons be removed from iron atoms by visible light if the shortest wavelength of visible light is taken to be 3500 Å?

13. What is the potential energy of the pair of N_2^+ species in problem 2? Is this energy significant compared with the average kinetic energy of a pair of N_2 molecules at 25 °C? What can you conclude about the importance of electrical forces between ions at 25 °C?

14. Calculate the energy required to remove an electron from the $n_1 = 1$ Bohr orbit of the hydrogen atom. Compare this energy with the average kinetic energy of H atoms at 25 °C. Are electrons likely to be freed from atoms as a result of the collisions of gaseous atoms with container walls at 25 °C?

15. Calculate the radii of the Bohr hydrogen atom orbits for $n_i = 1$ and $n_i = 2$. Are these approximately the sizes that you would expect for atomic orbits from the results of Chapter 2?

16. Heisenberg stated his principle of uncertainty in a semiquantitative fashion in the following way. If we try to follow the motion of any object along a direction x, we shall find an uncertainty in its momentum

(Δ momentum) and an uncertainty in its position x such that

$$(\Delta x)(\Delta \text{ momentum}) \cong h$$

Does equation 3–24 illustrate this relation? How?

17. An important technique for determining the structures of molecules is neutron diffraction, a method in which a beam of neutrons is diffracted by a solid consisting of the molecules in question. A good neutron wavelength for such experiments is 1.50 Å. According to the de Broglie relation, what velocity must the neutrons possess to give them a wavelength of 1.50 Å? The neutrons used in diffraction experiments are called **thermal neutrons.** Can you suggest the reason for this adjective? Hint: What is c_{rms} for neutrons in a gas at 25 °C?

4/ The Wave Model of Atoms

Different elements are composed of different kinds of atoms. Different elements also show different chemical and physical properties. The bases for different properties lie in the structure of the different atoms. Constructing a model for atoms is the first step in revealing the source of the different properties of different elements.

The fact that electrons may be pictured as waves of calculable wavelengths, represented in equation 3–25, is all that is required to construct a detailed and very useful model of atoms and molecules. All of the details of atomic and molecular structure are a consequence of electrostatic attractions and repulsions and the physics of waves.

4–1. The Wave Function

A wave such as a ripple on a pond or a sound wave in air represents the periodic variation of some physical property in space and time. With the ripple, this property is the surface height; and with the sound wave, it is the density of the air. What property of the electron varies periodically in an electron beam? The property that reveals itself in the diffraction experiment is the **detectability** of the electron. How can the detectability of a particle vary? We might argue that if a particle is in a particular spot it should be detectable and if it is not in that spot it should not be detectable. The uncertainty principle shows that this is an oversimplified view. An electron is in a given hydrogen atom but we can't detect exactly where. Can we say anything about where it is likely to be detected or not detected? The answer is yes, and the detectability function reports these likelihoods. The electron possesses a detectability function ψ (Greek lowercase *psi*), which is defined at all points at which the electron can be found; ψ^2 at a given point is the probability per unit volume of detecting the electron at that point. Since the electron detectability is wavelike (for example, it has a wavelength), ψ is commonly called the **wave function.** Electron detectability is also commonly referred to as **electron density.**

The Electron Confined to a Line

With this wave model, we should be able to predict how the electron would appear to us in a given situation. Suppose that we confine the electron to a line of length a, as shown in Figure 4–1. As with the violin string of Figure 3–8, we are confining a wave to a given length, and the only allowed wavelengths are $2a$, a, $2a/3$, . . . , or in general, $2a/n$, where n is any positive integer. If we define an x axis as shown in Figure 4–1, the wave function ψ must be

$$\psi = \left(\frac{2}{a}\right)^{1/2} \sin\left(\frac{n\pi x}{a}\right) \qquad \textbf{4–1}$$

This sine wave yields the wavelength $2a/n$, as required. The wave function is plotted in Figure 4–1(a) for $n = 1$. In Figure 4–1(b), the line is divided into ten equal segments. To a good approximation, the electron density in each

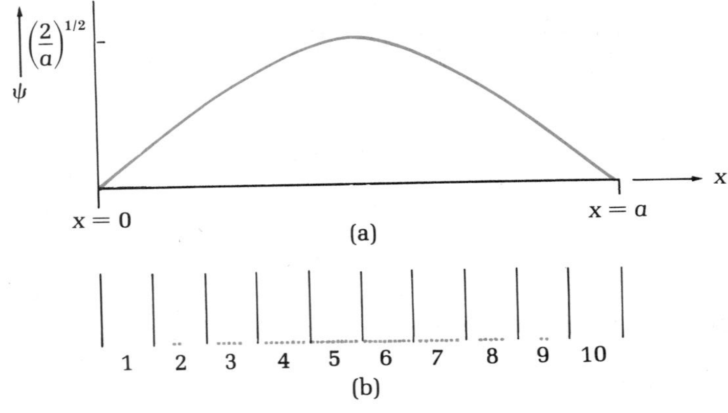

Figure 4–1
(a) Plot of ψ versus x for an electron restricted to a line having length a, when $n = 1$. (b) Dot plot of ψ^2 when $n = 1$.

segment is given by the value of ψ^2 at the middle of the segment times the length of the segment. The sum of the electron densities in all of the segments must be unity, because the electron is known to be somewhere along the line. The fact that the sum must be unity is called the condition of **normalization,** and all wave functions must be normalized. It is the normalization requirement that yields the term $(2/a)^{1/2}$ in equation 4–1.

The approximate electron density in the fifth segment from the origin when $n = 1$ is 0.20 electron. Let us define a unit of electron density as 0.020 electron and place a dot in each segment of Figure 4–1(b) for each unit of density calculated for that segment. Thus, for example, we place ten dots in segment 5. The resulting dot plot is a good representation of what we can expect to find if we attempt to observe the electron when it is characterized by the wave function for which $n = 1$. It is a picture (of sorts) of the electron.

Since λ is fixed at $2a$ in the state $n = 1$, the electron momentum is also fixed by the de Broglie relationship at $h/2a$. But if the momentum is fixed, so is the kinetic energy, through equation 3–26:

$$\text{Kinetic energy} = \frac{(\text{momentum})^2}{2m} = \frac{h^2}{8ma^2}$$

The wave function for $n = 1$ is just one of a family of functions characterized by various values of the number n. Each of the n values yields a different kinetic energy, given by

$$\text{Kinetic energy} = \frac{n^2h^2}{8ma^2} \qquad \textbf{4–2}$$

Thus the concept that elementary particles such as electrons can be represented as waves yields the conclusion that there must be discrete energy levels that are related to each other through integral **quantum numbers.** This is the conceptual basis that now justifies Planck's radical proposal that energy levels and quantum numbers exist. They are a logical consequence of the de Broglie relationship and the wave nature of electrons. **Any time an electron or any other particle is confined, energy levels and quantum numbers arise.**

The Electron Confined to a Box

What if the electron in Figure 4–1 is confined to a box of side lengths a, b, and c, instead of to the line in Figure 4–1? Now the electron possesses three independent directional motions: translation along x, y, and z directions. Motion in each of the three directions is bounded as shown in Figure 4–2(a). The electron must be characterized by three wavelengths of magnitudes $2a/n_x$, $2b/n_y$, and $2c/n_z$, corresponding to the three lengths between the boundaries. Thus three quantum numbers, n_x, n_y, and n_z, arise. By extend-

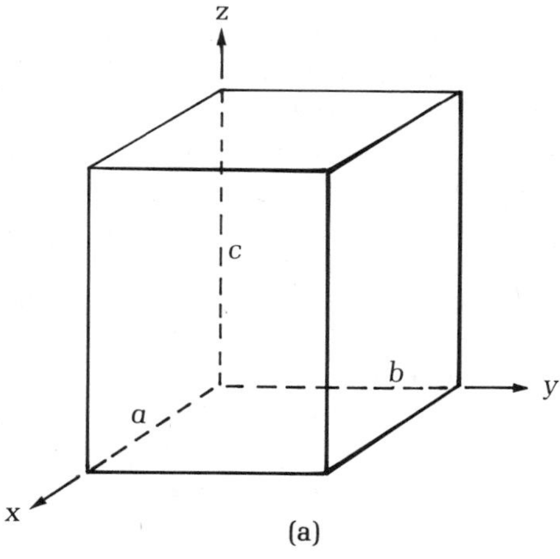

Figure 4–2
(a) Box for confining an electron.

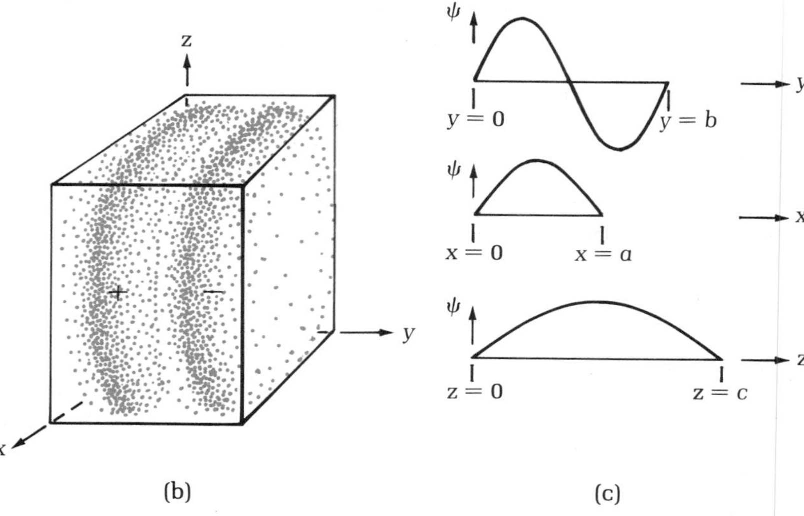

Figure 4–2 (continued) (b) Dot contour of the electron density in the state $n_x = 1$, $n_y = 2$, $n_z = 1$. (c) Plots of ψ along lines parallel to the y, x, and z axes, respectively.

ing this reasoning, we arrive at the conclusion that **there are as many quantum numbers as there are independent, confined motions of the particle.** This is a very useful conclusion, which will apply to atoms and molecules. The three quantized wavelengths yield three quantized kinetic energies from the de Broglie relationship and the total kinetic energy is the sum of these energies for motion along x, y, and z:

$$\text{Kinetic energy} = \frac{n_x{}^2h^2}{8ma^2} + \frac{n_y{}^2h^2}{8mb^2} + \frac{n_z{}^2h^2}{8mc^2} \qquad \textbf{4–3}$$

We should like to be able to plot the electron density in the box just as we did in the one-dimensional situation in Figure 4–1. The dot plot is a common method of plotting a function in any three-dimensional situation. We divide the box into small cubes, extending the use of the line segments of Figure 4–1. A very good approximation to the electron density in any given segment is the volume of the small cube times the value of ψ^2 calculated at the center of the cube. We can then define a unit of density and place a number of dots in the cube corresponding to the calculated number of density units. If we do this for the entire box, the result is a "cloud" of dots, which represents our best attempt to locate the electron in the box.

The cloud, or three-dimensional dot plot, is often too cluttered with dots to allow us to distinguish easily the shape of the electron density. This shape is more easily depicted by showing only those dots contained in cubes possessing some arbitrarily selected exact number of dots. Thus we plot a dot contour instead of all of the dots. Figure 4–2(b) shows such a dot contour for the electron energy state in which $n_x = 1$, $n_y = 2$, and $n_z = 1$. The contour consists of two separate lobes separated by a plane that divides the box in two. This plane in which the electron density is zero is termed a **node,** and **nodal surfaces** are an important characteristic of particles bound in three dimensions. We shall encounter them again in atoms. The source of the nodal plane is better seen in Figure 4–2(c), in which the bounded waves are shown along the x, y, and z directions.

The plot of ψ along the y direction in Figure 4–2(c) reveals that the electron wave possesses different signs in the two lobes of Figure 4–2(b), and we have indicated these signs in the figure. These signs are not related to any physical property of the electron (for example, they do not indicate its charge). However, the signs become very important if we introduce a second electron into the box, since the relative signs of the waves for the two electrons determine how the two electron waves interact. This becomes an extremely important point when considering bonding in molecules.

Energy Levels for Large Objects

Before we proceed further towards getting wave functions for electrons bound in atoms, we might ask why discrete, well-separated energy levels of the form of equation 4–3 are not detected for large objects that are bound in three dimensions. An example is a 100-g bird in a cubic cage of side length 1 m. As far as we can tell by observation, the bird can possess any kinetic energy between zero and an upper limit, determined by how fast it can fly. Surely if equation 4–3 is correct, it must apply to large objects as well as small ones. However, it appears to fail with large objects. Let us calculate the lowest kinetic energy for the bird from equation 4–3, with $n_x = n_y = n_z = 1$ and $a = b = c = 1$ m $= 10^2$ cm:

$$\text{Kinetic energy} = \frac{3(6.62 \times 10^{-27})^2}{8(100)(10^2)^2} = 1.6 \times 10^{-59} \text{ erg}$$

This is an impossibly small number to detect since it corresponds to a speed of the bird as given by the equation

$$\text{Speed} = \left(\frac{2 \text{ K.E.}}{m}\right)^{1/2} = \left[\frac{(2)(1.6 \times 10^{-59})}{100}\right]^{1/2}$$
$$= 6 \times 10^{-31} \text{ cm/sec}$$

The next lowest energy level allowed for the bird is the level for any one of the three combinations—($n_x = 1$, $n_y = 2$, $n_z = 1$); ($n_x = 2$, $n_y = 1$, $n_z = 1$); and ($n_x = 1$, $n_y = 1$, $n_z = 2$). The energy is 3.2×10^{-59} erg. Thus since the gap between the two adjacent levels is the nondetectably small 1.6×10^{-59} erg, equation 4–3 is quite consistent with a continuous range of energies (within the limits of our measurements) for large objects moving over a large distance.

The Electron in a Spherical Container

Our wave picture of the H atom is a nucleus surrounded by an electron density wave. A little thought leads us to conclude that there can be no preferred direction along which the electron density is concentrated since the potential energy of electron-nucleus attraction is given by

$$\text{Potential energy} = -e^2/r$$

and this is direction-independent. The wave functions of the H atom must be like the wave functions of an electron in a spherical enclosure. Thus, we get our first approximation to a picture of the hydrogen atom by examining the pictures of an electron in a spherical "box". As with Figure 4–2, the key condition that determines which waves are allowed is that the wavelength must be simply related to the distances between boundaries. In the case of the sphere, this is a single distance, the diameter. This diameter must be an integral number of half-wavelengths. Some allowed waves for a spherical box are depicted in Figure 4–3. These are circular cross sections taken through the center of the spherical dot plot in each case. These plots allow us to depict well the way in which electron density varies along the radius of the sphere. Contoured colored dashed lines are also drawn in each dot plot to indicate the shape of the complete contours in three dimensions. Thus, for example, the complete contours in Figure 4–3(a) are spherical surfaces. The electron wavelength in Figure 4–3(a) is the largest allowed wavelength, twice the diameter of the sphere. According to the de Broglie relationship, the state of largest λ possesses the smallest momentum, and thus the lowest energy. Thus, Figure 4–3(a) depicts the lowest energy level, or **ground state** as it is termed. Figure 4–3(b) depicts a state of $\lambda =$ two-thirds the spherical diameter. The complete contours in (b) are also spherical shells; and (b) contains a spherical nodal surface inside the sphere (shown by the black dashed line) in addition to the node at the spherical wall. The wave function changes sign at this interior node and this is indicated in the figure. Underneath figures (a) and (b), ψ is plotted as a function of distance along the diameter.

In addition to spherical nodes there can be planar nodes, as shown in Figure 4–3(c). Here we have added x, y, and z coordinate axes, and the node shown by the black dashed line is in the yz plane (the z axis is located perpendicular to the paper). Underneath Figure 4–3(c), ψ is plotted as a function of distance along the x axis, so that we see that one wavelength is contained in the spherical diameter. Thus, (c) represents an energy state intermediate between (a) and (b). The contours in (c) are two lobed surfaces analogous to the contours in Figure 4–2(b). There are two other wave functions that yield dot plots of the same shape as (c), but they are rotated relative to (c). With these other two functions, the node is located in the xz and xy planes, respectively. Since all directions in a sphere are equivalent, no electron pic-

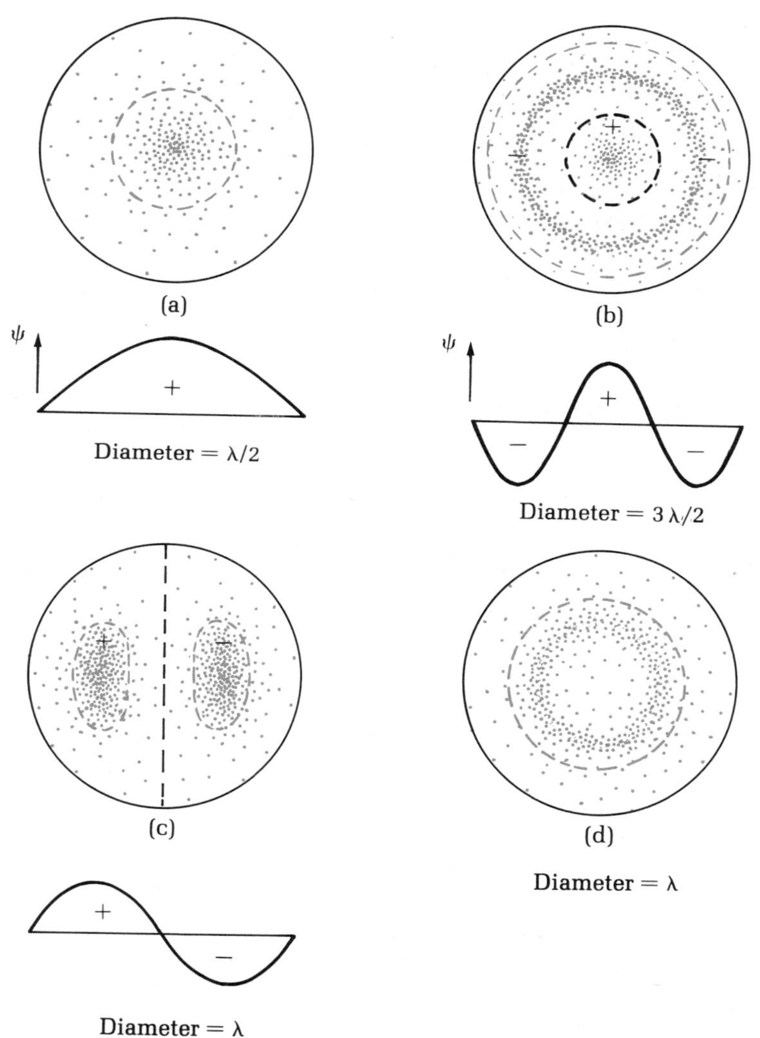

Figure 4–3
Cross sections of dot plots of the wave functions for an electron in a spherical container.

ture of the form of (c) would be observable. Instead the observed picture is an equal mixture of (c) and the two companion functions rotated with respect to (c). This combined observable picture is shown in (d). It contains a nodal point at the center of the sphere instead of a nodal plane since this is the only point in common between the three original nodal planes. Figure (d) contains one wavelength along any spherical diameter. The dashed line in (d) is a cross section of a spherical shell contour surface. It is very important to keep in mind that, while (d) very much resembles (a) and (b), (a) and (b) represent each a single wave function and (d) represents three wave functions. The situation as in (d), in which one energy level is composed of more than one wave function, is termed **degeneracy;** (d) represents a **triply degenerate** state. The concept of degeneracy is extremely important in considering atomic structure.

Three quantum numbers arise with the wave functions of Figure 4–3 because the situation involves one particle bound in three dimensions, just as in Figure 4–2. The only difference between Figures 4–2 and 4–3 is the shape of the three-dimensional container. In the box of Figure 4–2, the quantum number n_x minus 1 gives the number of interior planar nodes (nodes besides the walls) perpendicular to the x direction, and analogous statements can be made for n_y and n_z. Thus, the state $n_x = 1$, $n_y = 2$, $n_z = 1$ possesses no interior nodal planes perpendicular to the x and z axes and one plane perpendicular to the y axis. In the case of the spherical box, one of the three quantum numbers minus 1 lists the total number of interior nodal surfaces of all shapes. Thus, in (a) this quantum number is 1 because there are no interior nodes, and in (b) and (c) it is 2 because there is one interior node. A second quantum number lists the number of these internal nodes that are planar. This number is zero in (a) and (b) and 1 in (c). The third quantum number allows us to distinguish (c) from its two companion functions by revealing which of the three planes—xz, yz, or xy—contains the planar node.

Wave Model of the Hydrogen Atom

How does the hydrogen atom differ from a spherical box? The only difference in the two problems is in the nature of the electron binding. In the spherical box, the wall forces the electron density to become zero at the wall. In the hydrogen atom, the electron is bound by electrostatic attraction to the proton and there is no fixed boundary at which the electron density must be zero. Instead, it should show a continuous decrease as r increases at large r values. In all other aspects, the hydrogen atom and spherical box problems remain the same. Thus, the hydrogen atom wave functions will display spherical and planar nodes, and three quantum numbers appear that are related to those nodes.

The hydrogen wave functions are each given symbols that are commonly used, which are derived from the nodes displayed by each. The total number of nodes is $n - 1$, where n is termed the **principal quantum number.** The total number of planar nodes is l, and l is termed the **azimuthal quantum number.** The value of l is symbolized by a letter from the following set: $l = 0$, symbol = s; $l = 1$, symbol = p; $l = 2$, symbol = d; $l = 3$, symbol = f; $l = 4$, symbol = g; and on to h, i, etc. The third designation, as with Figure 4–3(c), indicates the orientation of the planar node(s). Let us consider some examples. The hydrogen ψ analogous to Figure 4–3(a) is designated 1s because it contains no nodes. The function analogous to Figure 4–3(b) is termed 2s because it contains one spherical node. That ψ analogous to Figure

4–3(c) is termed $2p_x$ because it possesses one planar node, and ψ^2 is at a maximum on the x axis (the node is in the yz plane). The two companion functions to $2p_x$ are termed $2p_y$ and $2p_z$. There are no $2d$ functions, since d symbolizes two planar nodes and this is impossible with a principal quantum number of 2.

With an n of 3, we have the functions 3s, $3p_x$, $3p_y$, and $3p_z$ together with five functions of the set $3d$. (The number of members in a set of given n and l is always $2l + 1$. Thus there is only one 3s function; there are three $3p$ functions, five $3d$ functions, seven $4f$ functions, and so forth.) The directional designations of the $3d$ functions are $3d_{xy}$, $3d_{xz}$, $3d_{yz}$, $3d_{x^2-y^2}$, and $3d_{z^2}$. These designations will be clearer when we consider the dot representations of the electron densities. Table 4–1 contains a summary of the relations between quantum numbers, numbers and kinds of nodes, and symbols for the wave functions. The locations of the nodes will be shown in dot plots, which follow shortly.

Table 4–1 Designations, Quantum Numbers, and Nodes for Hydrogen Atom Wave Functions

Symbol	n	l	Direction of Maximum Electron Density	Nodes
1s	1	0	All equal	None
2s	2	0	All equal	One spherical
$2p_x$	2	1	x axis	One planar
$2p_y$	2	1	y axis	
$2p_z$	2	1	z axis	
3s	3	0	All equal	Two spherical
$3p_x$	3	1	x axis	One spherical One planar
$3p_y$	3	1	y axis	
$3p_z$	3	1	z axis	
$3d_{xy}$	3	2	Between x and y axes	Two planar
$3d_{xz}$	3	2	Between x and z axes	
$3d_{yz}$	3	2	Between y and z axes	
$3d_{x^2-y^2}$	3	2	x and y axes	
$3d_{z^2}$	3	2	z axis	

We have concluded that ψ^2 for an electron in a hydrogen atom must decrease to zero as the electron-proton separation distance r increases to infinity. How steeply does this decrease occur? To answer that question, we need a procedure to obtain the mathematical expression for ψ for an electron bound in any fashion. The German theoretician Erwin Schrödinger published this procedure in 1926. The subsequent success of this procedure in allowing scientists to construct useful atomic and molecular models has made it the basis of all theories of atomic and molecular structure and motion. While the procedure itself is beyond our scope, we shall concentrate on the electron densities it predicts.

Figure 4–4 shows dot plots for several of the hydrogen atom wave functions. The plots are drawn to scale and the distance a_0 shown in the figure is the radius of the Bohr orbit with $n_i = 1$ in equation 3–20. This distance is termed the **Bohr radius,** and it is a convenient unit for reporting distances inside atoms. The value of a_0 is 0.529 Å. The axes at the top of the figure show the orientations of the wave functions shown. The black dot at the center of each plot represents the nucleus.

It is particularly easy to locate the nodes for each wave function in Figure 4–4. For example, the single spherical nodes for 2s and 3p are readily apparent as black dashed circles in the figure. The plot for $2p_z$ shows two lobes

Figure 4–4
Dot representation of some hydrogen atom wave functions.

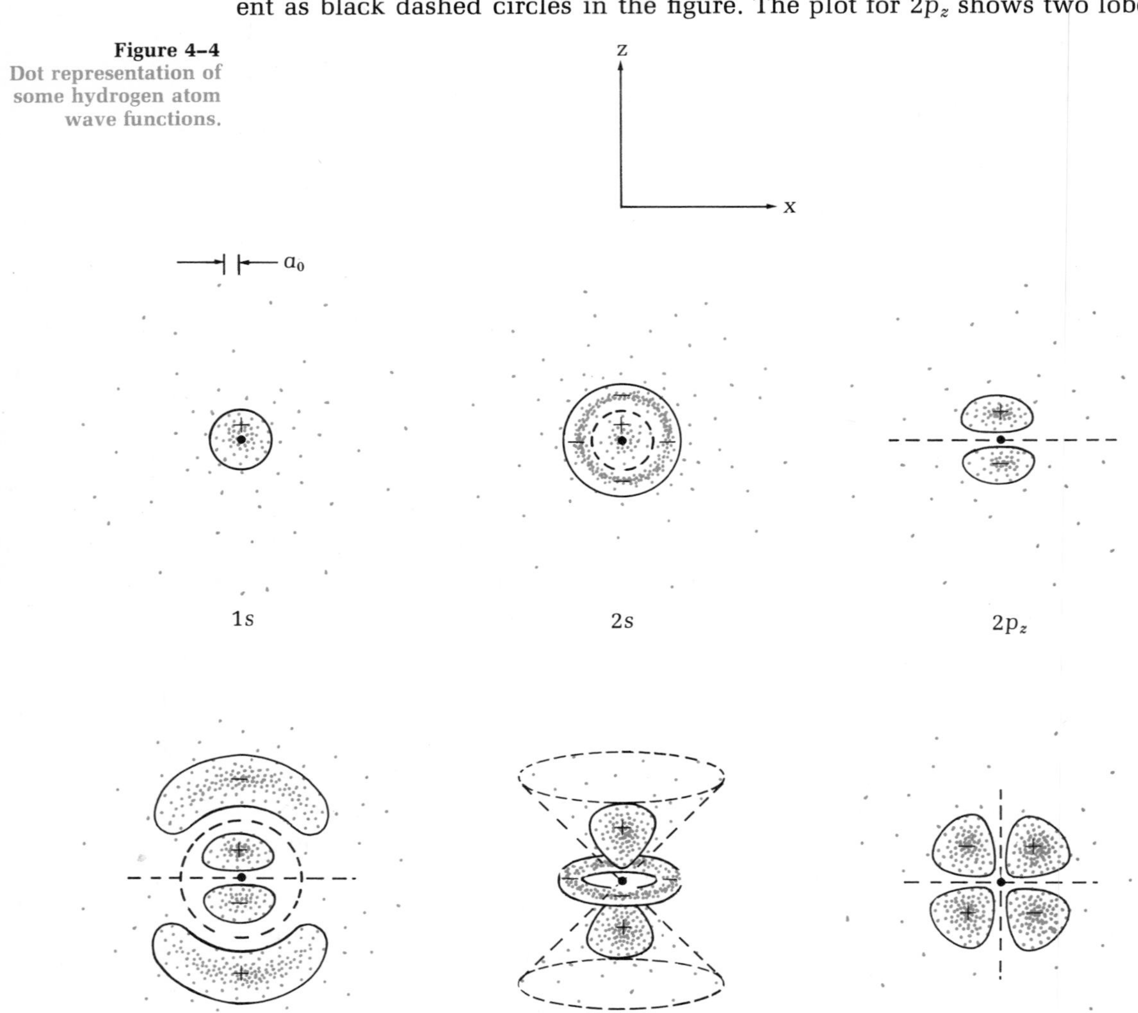

separated by a node in the xy plane. The dot plots of $2p_x$ and $2p_y$ are not shown. They look exactly like the plot of $2p_z$ save that the planar nodes are in the yz and xz planes, respectively. The planar node is also apparent in the plot for $3p_z$. With $3d_{xz}$, the two planar nodes are in the yz and xy planes, yielding the four-lobed figure shown. The functions $3d_{yz}$, $3d_{xy}$, and $3d_{x^2-y^2}$ have exactly the same shape as $3d_{xz}$ and they differ only in the locations of the nodal planes. The nodal planes for the $3d_{yz}$ function occur in the xz and xy planes and those for $3d_{xy}$ are the xz and yz planes. The four lobes of the $3d_{x^2-y^2}$ function point along the x and y axes. The function $3d_{z^2}$ differs in shape from the other functions of the 3d set as shown in the figure. The two planar nodes are bent to form two cones whose points meet at the nucleus. The electron density is thereby isolated into two identical lobes on either side of the xy plane and a ring around the z axis, with maximum detectability in the xy plane. The $3d_{z^2}$ plot in Figure 4–4 has been tilted to show the conical nodes and the ring in the xy plane.

A final feature that is apparent from Figure 4–4 is that for a given l value the regions of high electron density extend further from the nucleus as n increases. Thus, 2s is visibly "larger" than 1s, and $3p_z$ is larger than $2p_z$. It is impossible to emphasize too strongly the necessity of forming indelible mental pictures of the electron density in the hydrogen atom states plotted in Figure 4–4. All of the subsequent discussions of bonding in molecules, and also the structures and reactivities of the chemical substances to be presented, are based on the sizes and shapes of these wave pictures and the signs of ψ associated with the lobes. We shall often refer to these sizes, shapes, and signs, and so need some accurate representations of these features that are easier to draw than the detailed dot representation of Figure 4–4. The most common simplified representation of a given plot is a contour line enclosing 90 percent of the dots. These 90 percent contour lines (actually slices of 90 percent contour surfaces) are shown by the solid black lines in Figure 4–4. Thus, for example, the contour for 1s and 2s is a circle in each case, and the 1s circle is smaller than the circle for 2s.

The key test of the wave model of the hydrogen atom is its ability to yield the Bohr energies. Let us consider how we might calculate the energy of the electron in one of the states shown in Figure 4–4. In the Bohr model, r was fixed in a given energy level and E was given by $(-e^2/2r)$. In the wave model, r is variable and E is given by

$$E = \frac{-e^2}{2}\left\langle\frac{1}{r}\right\rangle \tag{4–4}$$

where $\langle 1/r\rangle$ is the average of the reciprocal r. The average reciprocal of a set of numbers is not equal to the reciprocal of the average number. Take the numbers 1, 2, and 4. The average reciprocal is given by

$$\left\langle\frac{1}{n}\right\rangle = \frac{1}{3}\left(\frac{1}{1}+\frac{1}{2}+\frac{1}{4}\right) = 0.58$$

and the reciprocal of the average is given by

$$\frac{1}{\langle n \rangle} = \frac{1}{\frac{1}{3}(1 + 2 + 4)} = 0.43$$

The wave functions derived from the Schrödinger procedure and plotted in Figure 4–4 have been found to yield electron energies that are identical to the Bohr energies. It is found that E depends only on the principal quantum number n, and consequently this is the n of equations 3–7 and 3–19. This triumph of the wave model was achieved by Schrödinger's calculations in 1926 and it touched off a five-year period of furious activity, in which most of our present concepts of atomic structure were developed.

4–2. The Wave Model of Multielectron Atoms

Atomic Orbitals

The second critical test of the wave model of atoms is its ability to yield correct predictions for atoms containing more than one electron. The Bohr orbit model failed this test, a result that led to discarding that model. Any atom that contains more than one electron represents a complicated situation in which electron-electron repulsion and electron wave–electron wave interaction in addition to electron-nucleus attraction are all considerations. Many models containing various degrees of complexity have been formulated for multielectron atoms. Our goal is to develop a model that allows us to make sound predictions based on our ability to form a visual image of atoms. It is to our advantage to sacrifice some sophistication in the model for the sake of forming a picture of atoms that allows us to correlate easily the properties of different atoms and also to predict which molecules should exist and what structures they should have. Such benefits indeed derive, as we shall see, from the visualizable model of the average electron densities, over a period of time, of the individual electrons in atoms. Since there are no unique directions in an atom, these average densities must be spherical just like the densities for the electron in the spherical box and for the hydrogen atom. Each electron has a wave function, the shape of which is similar to those of the electron in the hydrogen atom. In fact we can and do give the same symbols to these one-electron wave functions in many-electron atoms as to the corresponding hydrogen atom wave function. Thus we refer to 1s, 2s, $2p_x$, . . . electrons in many-electron atoms and we mean electrons whose dot plots are shaped similarly to those for electrons characterized by the 1s, 2s, $2p_x$, . . . functions of hydrogen. These one-electron wave functions are termed **atomic orbitals,** and our model of individual electrons in individual orbitals of visualizable shapes and sizes is termed the **atomic orbital model** of atoms. This is the model that chemists use almost universally, and the one that we shall use.

The Electronic Energy in the Helium Atom

What can we deduce about the electron energies in multielectron atoms? Let us begin with the helium atom. At first thought, we might say that each electron in the helium atom should "see" a charge of $+e$ from the nucleus and the other electron since the $-e$ charge of the electron should cancel one of the $+2e$ charges of the helium nucleus. However, this does not properly account for the average distribution of an electron about the nucleus nor for the most probable arrangement at any given time of electrons relative to each other. This most probable arrangement places the two electrons on opposite sides of the nucleus and minimizes the repulsion between them. Each electron is much closer, on the average, to the nucleus than to the other electron, so that each electron sees an effective positive charge greater than +1. Thus, in a comparison of the helium atom with the hydrogen atom, each electron is more strongly attracted to the nucleus in the helium atom. The dot plots for helium atomic orbitals are predicted to be smaller than the plots for the hydrogen orbitals. As a consequence of this size difference, the electron energy is considerably more negative in the helium 1*s* orbital than in the hydrogen 1*s* orbital.

Ionization Potentials

A key difference between atoms is a different energy for each given orbital, such as 1*s*, in each different atom. Our atomic orbital model appears capable of yielding well-defined predictions about the qualitative trends in these energies and we should like to have a method of checking these predictions. Such a method is available through the photoelectric effect (discussed in Chapter 3). Visible light can remove an electron from certain metal atoms. However, for visible light, $h\nu$ is not large enough to remove a 1*s* electron from any atom or to remove any electron from most atoms. X-rays have a much larger ν, however, and the most energetic X-ray from a molybdenum target, for example, can remove any electron from any atom with an atomic number significantly less than that of molybdenum. Thus, if molybdenum X-rays strike carbon atoms, electrons are expelled from each of the occupied carbon orbitals. In a **photoelectron spectrometer**, these expelled electrons are detected and their kinetic energies are measured. The energy necessary to expel a given electron from its atomic orbital is termed the **atomic orbital ionization potential,** and it is the negative of the total energy of the electron in that orbital. The ionization potential is calculable as follows:

$$h\nu = \text{ionization potential} + \text{kinetic energy} \qquad \textbf{4–5}$$

where ν is the photon frequency, and the kinetic energy is that of the emitted electron.

Let us use equation 4–5 to calculate an ionization potential for the helium atom from some photoelectron data. In the case of He, ultraviolet photons are sufficiently energetic to produce photo-ionization, and radiation of $\lambda = 3.00 \times 10^{-6}$ cm yields a large number of electrons with a kinetic energy of 2.70×10^{-11} erg. The photon energy is $h\nu = hc/\lambda = 6.62 \times 10^{-11}$ erg, so that the ionization potential is given as follows:

$$\text{Ionization potential} = h\nu - \text{kinetic energy}$$
$$= 3.92 \times 10^{-11} \text{ erg}$$

Electron energies and ionization potentials are usually reported in electron volts, with one electron volt equal to 1.60×10^{-12} erg. Thus the measured ionization potential for one of the helium orbitals is $3.92 \times 10^{-11}/1.60 \times 10^{-12} = 24.5$ ev. Further investigation shows that this is the largest ionization potential measurable for He, and therefore -24.5 ev is the energy of an electron in the most strongly bound helium state. In analogy with the hydrogen atom, this state of lowest total energy (ground state) is the 1*s*. The energy in the hydrogen 1*s* state is calculable from equation 3–21 with $n_i = 1$ as -2.18×10^{-11} erg $= -13.6$ ev. Thus our prediction for the helium electron energy is correct and a helium 1*s* electron is much more tightly bound than a hydrogen 1*s* electron.

Atomic Orbital Energies in Multielectron Atoms

The change in electron energy for the 1*s* orbital between H and He is predicted to continue to the lithium atom, where the least repulsive arrangement of electrons around the nucleus is a triangle, and two electron charges do not cancel two nuclear charges. Experimentally, the lithium 1*s* energy is shown to be -55.0 ev. This trend continues through all of the other elements. RULE 1: **The energy of any given atomic orbital decreases as the atomic number increases.**

We have demonstrated this for 1*s*, but it applies also to 2*s*, $2p_x$, or any other orbital.

The second energy difference we must investigate is between different orbitals in the same multielectron atom. We saw the effect of repulsion on the helium 1*s* orbital. Is the effect the same on the 2*s*? Figure 4–5 shows the most probable arrangement in a helium atom with one electron in a 1*s* orbital and the second electron in the larger 2*s* orbital. Again the electrons are on opposite sides of the nucleus. However, they are much farther apart than when both electrons are in a 1*s* orbital, and the repulsion is less in Figure 4–5. Thus, it is much more likely that the two electrons will be on the same side of the nucleus when one of them is in the 2*s* orbital than when both

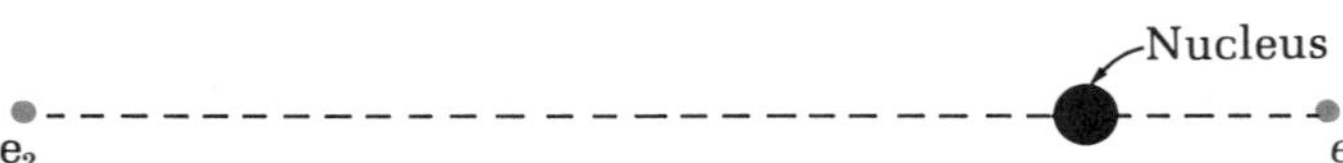

Figure 4–5 The most probable arrangement of electrons in a helium atom with one electron in a 1*s* orbital and one electron in a 2*s* orbital.

are in the 1*s*, and electron 1 should much more effectively cancel one unit of nuclear charge in Figure 4–5 than when both electrons reside in the 1*s* orbital. Our prediction is that as the principal quantum number increases between orbitals in many-electron atoms, the difference in energy between the orbital and the same orbital in hydrogen decreases. Thus, for example, the helium 2*s* energy should be nearer to the hydrogen 2*s* than the helium 1*s* is to the hydrogen 1*s*. The hydrogen 2*s* energy is $(-13.6/4) = -3.4$ ev and the helium 2*s* energy is -3.9 ev. Thus, the ordering of energies observed with the hydrogen atom is preserved in other atoms.

RULE 2: For a given *l* value in a given atom, the energy increases as *n* increases.

We have demonstrated this with *s* orbitals ($l = 0$), but it applies also to *p* orbitals (for example, a carbon 3*p* electron possesses a higher energy than a carbon 2*p* electron) and to all other orbitals.

A very important energy difference in many-electron atoms is that between orbitals of different *l* and the same *n*. In the hydrogen atom, this difference is zero, and for example, the energy of a 2*s* electron is identical to that of a 2*p* electron. In multielectron atoms, the presence of electron-electron repulsion produces an energy difference between these orbitals, and the energy of a helium 2*p* electron is -3.3 ev as compared with -3.9 ev for a 2*s* electron.

How does repulsion produce this 2*s*-2*p* energy difference, and what general rule can we state for energy trends within a given set of orbitals characterized by the same principal quantum number? We can answer this question by reference to Figure 4–6, which shows the electron density on spherical surfaces of radius *r*, for the hydrogen atom 2*s* and 2*p* wave functions. Since these two hydrogen orbitals have the same energy, they must possess the same value of $\langle 1/r \rangle$, according to equation 4–4. However, the 2*s* orbital is significantly larger than the 2*p*, as shown by the higher density for

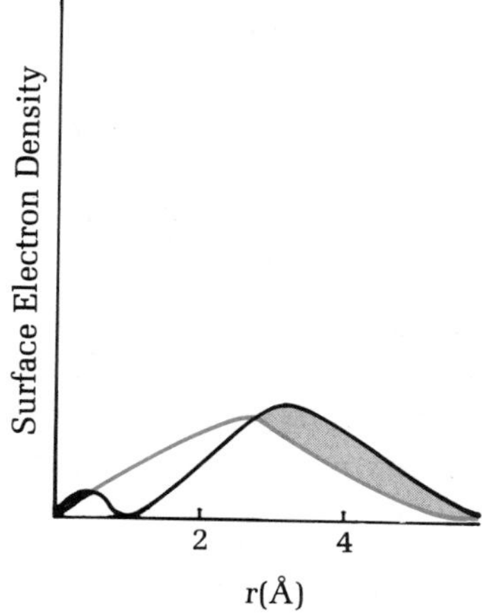

Figure 4–6
The electron density on spherical surfaces in the 2*s* (*black curve*) and 2*p* (*colored curve*) orbitals of the hydrogen atom, plotted as a function of the sphere's radius.

2s at large r (colored area of the figure). This is a general result, which is displayed in the contours of Figure 4–4; as l increases with a given n, the maximum distance from the nucleus to the 90 percent contour surface decreases. This is shown in the figure by the 2s and $2p_z$ contours and also by the $3p_z$ and the two 3d contours. From these contours, it would appear that $\langle 1/r \rangle$ must surely be smaller for 2s than for 2p. However, $\langle 1/r \rangle$ is very strongly influenced by electron density at small r values (which make a large contribution to 1/r). The black area in Figure 4–6 shows the additional density in 2s over 2p at small r. This additional density at small r is sufficient to counterbalance the larger 90 percent contour and make $\langle 1/r \rangle$ for 2s and 2p be the same in the hydrogen atom. We say that the 2s orbital is larger than 2p, but it also *penetrates* close to the nucleus better than 2p does.

Now let us add a second electron and make the helium atom. As an electron penetrates close to the nucleus, the effective positive charge it sees increases. Thus, the energy of the smaller helium 1s orbital is lowered more than the energy of the helium 2s orbital relative to the corresponding hydrogen orbitals. In the case of Figure 4–6, this effect must favor the orbital of greater penetration and this is 2s. Thus, the key factor that lowers 2s relative to 2p is the small additional detectability in 2s relative to 2p at small r.

RULE 3: The electron energy increases as *l* increases at constant *n* in any given atom possessing more than one electron.

The exact orbital energies change from atom to atom; the rules we have deduced for energy trends among these orbitals allow us, however, to construct a pattern of orbital energies for a typical many-electron atom, and this pattern is exemplified by the orbital energies for the chlorine atom shown in Figure 4–7. The energies are plotted on a logarithmic scale because they vary over a large range (from −2.2 ev for 4p to −3542 ev for 1s). This figure shows the increase in energy with increasing n within a given l (Rule 2) and the increase in energy with increasing l within a given n (Rule 3). The energy of the chlorine 1s (−3542 ev) is far below the hydrogen 1s (−13.6 ev) because of the high penetration of the 1s orbital into the region close to a nucleus containing 17 positive units of charge.

Determination of Atomic Numbers

The extremely large gap between 1s and all of the other orbitals in an atom of sizable atomic number provides the source of X-rays. The ionization potential of a chlorine 1s electron is 3542 electron volts. Electrons accelerated by a voltage greater than 3542 volts can eject a chlorine 1s electron by collision, producing a vacancy in that orbital. This vacancy is quickly filled by the transition of an electron from a higher-energy orbital. The highest-energy occupied level of Cl is 3p (as we shall see shortly) and its energy is −13.0 ev. Thus, the transition energy between 3p and 1s is $(3542 - 13) = 3529$ ev = 5.653×10^{-9} erg. The transition results in the emission of a photon of

$$\lambda = \frac{hc}{E} = \frac{(6.62 \times 10^{-27})(3 \times 10^{10})}{(5.653 \times 10^{-9})} = 3.5 \times 10^{-8} \text{ cm}$$

and this is a typical X-ray wavelength.

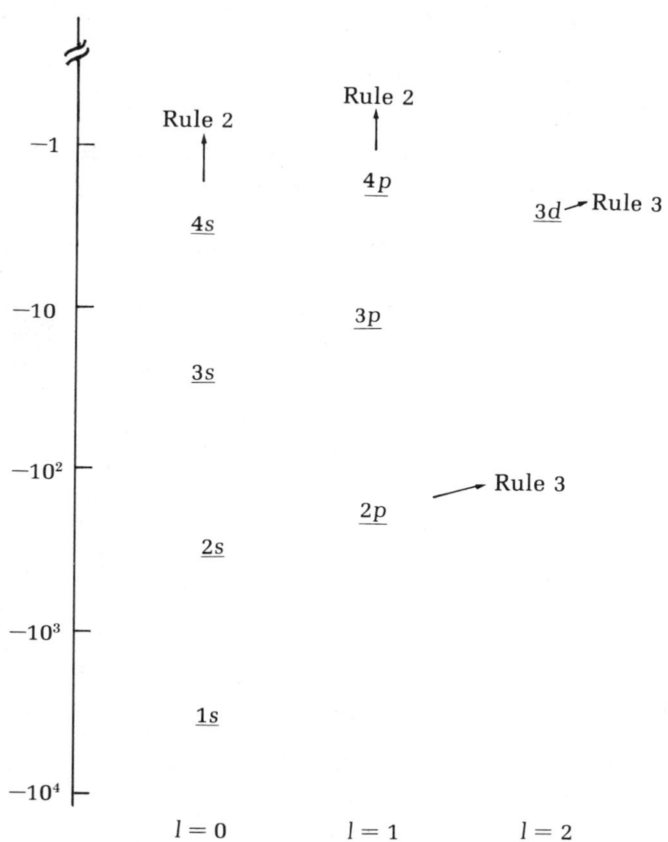

Figure 4–7 The energies of some of the atomic orbitals of the chlorine atom.

As the atomic number increases, the gap between the energy of the 1s orbital and the energy of the highest occupied orbital increases steeply and progressively. In 1913, Henry Moseley, a student of Ernest Rutherford, noticed that the wavelengths of the shortest-wavelength X-rays emitted by atoms shorten in almost all cases as the weight of the atom increases. However, a plot of $1/\lambda$ for this X-ray of shortest λ versus atomic weight of the emitting element revealed no simple quantitative relation between the two. In fact, in the case of argon and potassium, among other examples, $1/\lambda$ increases in going from Ar to K even though the atomic weight decreases. Thus they are exceptions to even the qualitative trend noted by Moseley. If, however, Moseley plotted $1/\lambda$ versus the atomic numbers for the elements (as listed in the inside front cover of this text), he could draw a smooth curve through the points. That curve has as its equation

$$\frac{1}{\lambda} = A(Z - 1)^2 \qquad \textbf{4–6}$$

where A is a constant and Z is the atomic number. This is the source of the atomic numbers as we now know them.

Electronic Configurations of Atoms

Now that we have pictures and energies for atomic orbitals, our next task is to assign the occupancies of these orbitals in various atoms. For example, does the helium atom exist with both electrons in the 1*s* orbital or does one electron occupy a 1*s* while the second occupies a 2*s*? The answer to this question of orbital assignments comes from the measurement of ionization potentials as in Figure 4–8, which shows the smallest ionization potentials actually measured for various atoms at room temperature. These smallest

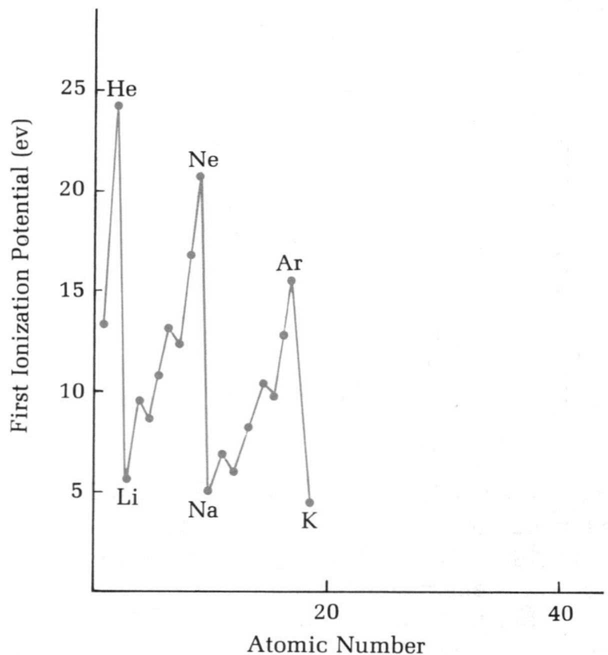

Figure 4–8 The first ionization potentials of the first nineteen elements, in order of increasing atomic numbers.

ionization potentials correspond to the negatives of the energies of electrons in the highest-energy atomic orbitals that are occupied to a significant degree at room temperature. They are commonly termed the **first ionization potentials,** meaning the ionization potential of the most easily ionized electron in each atom.

From Figure 4–7 we can make some predictions concerning first ionization potentials. First, because of the large energy gap between the 1s and 2s levels as shown in Figure 4–7, there should be a large decrease in the first ionization potential with the first atom containing a 2s electron as compared with those atoms that only contain 1s electrons. A similar sharp decrease should occur with the first 3s electron and the first 4s electron, because the 2p-3s energy gap is large and the 3p-4s gap is also. Figure 4–8 shows that these sharp decreases in the first ionization potential occur at lithium (3 electrons), sodium (11 electrons), and potassium (19 electrons). Thus it appears that an atom can contain a maximum of two 1s electrons and the third electron in Li must occupy a 2s orbital. If this limitation of 2 electrons per orbital holds for all orbitals, the $(n = 2)$ set of orbitals can hold a maximum of 8 electrons (2 for the 2s and 6 for the three 2p orbitals). The next sharp drop in the first ionization potential is then predicted to occur with the eleventh electron, and indeed it does. The set (3s and 3p) can also hold a maximum of 8 electrons and the next sharp drop in the first ionization potential should occur with the nineteenth electron, as indeed it does.

This type of analysis of atomic structure in terms of the occupation of atomic orbitals was first performed by the German physicist Wolfgang Pàuli, shortly after the Schrödinger wave model was first proposed. To obtain the electronic arrangement in any atom, all we have to do is to locate electrons in the orbitals, as displayed in Figure 4–7, starting from the bottom (1s) up and limiting the number of electrons in each orbital to two. This assignment of electrons to orbitals is termed the **electronic configuration** of the atom. Figure 4–9 shows a common representation of the electronic configurations of the five elements of smallest atomic number. The orbitals are represented by boxes and the electrons by circles. A shorter version of

Atom	1s	2s	2p			Configuration
H	•					1s
He	••					$1s^2$
Li	••	•				$1s^22s$
Be	••	••				$1s^22s^2$
B	••	••	•			$1s^22s^22p$

Figure 4–9 Electronic configurations of the elements with atomic numbers 1–5.

the configurations is shown in the column at the right of the figure, where the superscript indicates the number of electrons in the given orbital.

Table 4–2 lists the electron configurations of all of the atoms of known configuration. These configurations are the basis for our discussions of bonding and structure in molecules. The notations in brackets, such as [He], indicate the electron configuration of that element, and the dash indicates the configuration of the bracketed element above the dash. Thus, the electron configuration of Mg is $1s^2 2s^2 2p^6 3s^2$.

Table 4–2
Electronic Configurations of the Elements

Atomic Number	Element	Configuration	Atomic Number	Element	Configuration
1	H	$1s^1$	34	Se	—$3d^{10}4s^2 4p^4$
2	He	$1s^2$	35	Br	—$3d^{10}4s^2 4p^5$
3	Li	[He]$2s^1$	36	Kr	—$3d^{10}4s^2 4p^6$
4	Be	—$2s^2$	37	Rb	[Kr]$5s^1$
5	B	—$2s^2 2p^1$	38	Sr	—$5s^2$
6	C	—$2s^2 2p^2$	39	Y	—$4d^1 5s^2$
7	N	—$2s^2 2p^3$	40	Zr	—$4d^2 5s^2$
8	O	—$2s^2 2p^4$	41	Nb	—$4d^4 5s^1$
9	F	—$2s^2 2p^5$	42	Mo	—$4d^5 5s^1$
10	Ne	—$2s^2 2p^6$	43	Tc	—$4d^5 5s^2$
11	Na	[Ne]$3s^1$	44	Ru	—$4d^7 5s^1$
12	Mg	—$3s^2$	45	Rh	—$4d^8 5s^1$
13	Al	—$3s^2 3p^1$	46	Pd	—$4d^{10}$
14	Si	—$3s^2 3p^2$	47	Ag	—$4d^{10}5s^1$
15	P	—$3s^2 3p^3$	48	Cd	—$4d^{10}5s^2$
16	S	—$3s^2 3p^4$	49	In	—$4d^{10}5s^2 5p^1$
17	Cl	—$3s^2 3p^5$	50	Sn	—$4d^{10}5s^2 5p^2$
18	Ar	—$3s^2 3p^6$	51	Sb	—$4d^{10}5s^2 5p^3$
19	K	[Ar]$4s^1$	52	Te	—$4d^{10}5s^2 5p^4$
20	Ca	—$4s^2$	53	I	—$4d^{10}5s^2 5p^5$
21	Sc	—$3d^1 4s^2$	54	Xe	—$4d^{10}5s^2 5p^6$
22	Ti	—$3d^2 4s^2$	55	Cs	[Xe]$6s^1$
23	V	—$3d^3 4s^2$	56	Ba	—$6s^2$
24	Cr	—$3d^5 4s^1$	57	La	—$5d^1 6s^2$
25	Mn	—$3d^5 4s^2$	58	Ce	—$4f^2 6s^2$
26	Fe	—$3d^6 4s^2$	59	Pr	—$4f^3 6s^2$
27	Co	—$3d^7 4s^2$	60	Nd	—$4f^4 6s^2$
28	Ni	—$3d^8 4s^2$	61	Pm	—$4f^5 6s^2$
29	Cu	—$3d^{10}4s^1$	62	Sm	—$4f^6 6s^2$
30	Zn	—$3d^{10}4s^2$	63	Eu	—$4f^7 6s^2$
31	Ga	—$3d^{10}4s^2 4p^1$	64	Gd	—$4f^7 5d^1 6s^2$
32	Ge	—$3d^{10}4s^2 4p^2$	65	Tb	—$4f^9 6s^2$
33	As	—$3d^{10}4s^2 4p^3$	66	Dy	—$4f^{10}6s^2$

Atomic Number	Element	Configuration	Atomic Number	Element	Configuration
67	Ho	$—4f^{11}6s^2$	86	Rn	$—4f^{14}5d^{10}6s^26p^6$
68	Er	$—4f^{12}6s^2$	87	Fr	$[Rn]7s^1$
69	Tm	$—4f^{13}6s^2$	88	Ra	$—7s^2$
70	Yb	$—4f^{14}6s^2$	89	Ac	$—6d^17s^2$
71	Lu	$—4f^{14}5d^16s^2$	90	Th	$—6d^27s^2$
72	Hf	$—4f^{14}5d^26s^2$	91	Pa	$—5f^26d^17s^2$
73	Ta	$—4f^{14}5d^36s^2$	92	U	$—5f^36d^17s^2$
74	W	$—4f^{14}5d^46s^2$	93	Np	$—5f^46d^17s^2$
75	Re	$—4f^{14}5d^56s^2$	94	Pu	$—5f^67s^2$
76	Os	$—4f^{14}5d^66s^2$	95	Am	$—5f^77s^2$
77	Ir	$—4f^{14}5d^76s^2$	96	Cm	$—5f^76d^17s^2$
78	Pt	$—4f^{14}5d^96s^1$	97	Bk	$—5f^97s^2$
79	Au	$—4f^{14}5d^{10}6s^1$	98	Cf	$—5f^{10}7s^2$
80	Hg	$—4f^{14}5d^{10}6s^2$	99	Es	$—5f^{11}7s^2$
81	Tl	$—4f^{14}5d^{10}6s^26p^1$	100	Fm	$—5f^{12}7s^2$
82	Pb	$—4f^{14}5d^{10}6s^26p^2$	101	Md	$—5f^{13}7s^2$
83	Bi	$—4f^{14}5d^{10}6s^26p^3$	102	No	$—5f^{14}7s^2$
84	Po	$—4f^{14}5d^{10}6s^26p^4$	103	Lw	$—5f^{14}6d^17s^2$
85	At	$—4f^{14}5d^{10}6s^26p^5$			

Table 4–2 (continued)

Successive Ionization Potentials

The picture of electrons in discrete orbitals finds particularly strong support in the values of the successive ionization potentials of atoms. We have already considered that first ionization potentials can be obtained from photoelectron data. However, these ionization potentials have been obtained in practice by applying the mass spectrometer (discussed in Chapter 3). The mass spectrometer provides a measurement of (ne/m) for an ion, where n is the number of electrons added to or removed from the neutral atom or molecule and m is the atomic or molecular mass. An ion beam is produced in a mass spectrometer by electron bombardment, as shown in Figure 4–10. The uncharged beam of atoms or molecules leaves the sample chamber and passes through a beam of electrons whose kinetic energy is controlled by the variable positive voltage indicated in the figure. Suppose that the beam is sodium atoms. According to Table 4–2, the highest-energy electron in a sodium atom is a 3s and its ionization potential is 5.1 electron volts. Thus the voltage difference that accelerates the electrons in the wide beam must be at least 5.1 volts so that a free electron can transfer its energy to a sodium 3s electron to produce the ion Na^+. This measured voltage at which sodium ions first appear at the detector gives us a straightforward measurement of the first ionization potential.

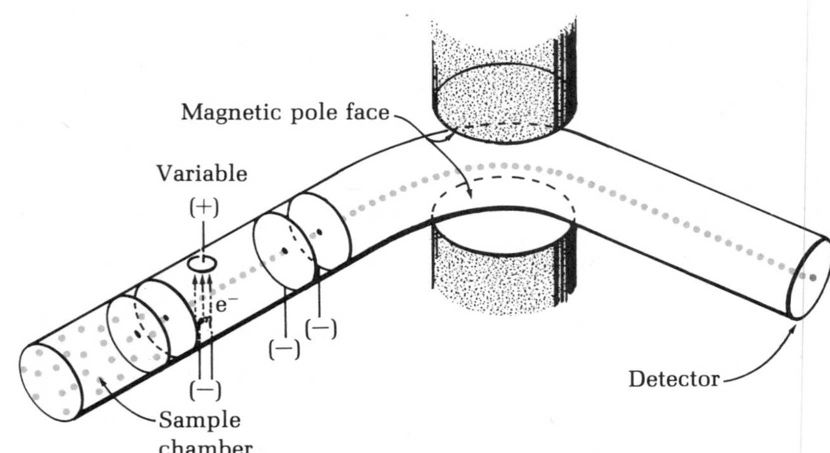

Figure 4–10
Schematic of apparatus for measuring successive ionization potentials.

In an electron beam, the Na^+ ions are capable of collision with a second electron to produce Na^{2+} ions. The energy required for this second ionization is termed the **second ionization potential.** The production of Na^{2+} requires a value of 47.3 ev, as shown in Table 4–3. The third ionization potential corresponds to the production of Na^{3+} from Na^{2+} and it is 71.6 ev. Table 4–3 lists first, second, and third ionization potentials for the first thirty elements in the order of increasing atomic number.

Table 4–3 shows much clearer evidence than Figure 4–8 in favor of the Pauli scheme of two-electron assignment to atomic orbitals in the order of increasing orbital energies. For example, with lithium the IP_1 is much lower than either IP_2 or IP_3 because the first electron removed is a 2*s* and the next two are much more tightly bound 1*s* electrons. In the case of Be, there are two easily removed 2*s* electrons, but there is a sharp increase at IP_3 because the third electron removed is a 1*s*.

4–3. Electron Spin

Electrons as Magnets

What is the source of the limitation deduced by Pauli that a maximum of two electrons can occupy any atomic orbital? Is there some property of the electron that can assume only two values and that leads to this limitation? The affirmative answer to this second question was supplied by the physicists Stern and Gerlach, who performed the experiment depicted in Figure 4–11. A beam of un-ionized silver atoms was produced by vaporizing silver in an oven and allowing the vapor to escape through a pinhole. This atomic beam was then passed through a magnetic field, as shown. Because the atoms have no net charge, there should be no deflection of the beam as there

Element	Atomic Number	(IP_1)	(IP_2)	(IP_3)
H	1	13.6		
He	2	24.6	54.4	
Li	3	5.4	75.6	122.4
Be	4	9.3	18.2	153.8
B	5	8.3	25.2	37.9
C	6	11.3	24.4	47.9
N	7	14.5	29.6	47.4
O	8	13.6	35.2	54.9
F	9	17.4	35.0	62.7
Ne	10	21.6	41.1	65.0
Na	11	5.1	47.3	71.6
Mg	12	7.6	15.0	80.1
Al	13	6.0	18.8	28.4
Si	14	8.1	16.3	33.5
P	15	11.0	19.6	30.2
S	16	10.4	23.4	35.0
Cl	17	13.0	23.8	39.9
Ar	18	15.8	27.6	40.9
K	19	4.3	31.8	47.7
Ca	20	6.1	11.9	51.2
Sc	21	6.6	12.9	24.8
Ti	22	6.9	13.6	28.1
V	23	6.7	14.2	29.7
Cr	24	6.8	16.5	31.0
Mn	25	7.4	15.6	33.7
Fe	26	7.9	16.2	30.6
Co	27	7.8	17.1	33.5
Ni	28	7.6	18.2	35.2
Cu	29	7.7	20.3	36.8
Zn	30	9.4	18.0	39.7

Table 4–3
First (IP_1), Second (IP_2), and Third (IP_3) Ionization Potentials of Isolated Atoms

is in a mass spectrometer. However, Stern and Gerlach saw that the beam was split by the magnetic field into two beams of equal intensity. Furthermore, the deflections shown in the figure are not in and out of the plane of the paper, as they would be for ions, but rather towards the respective magnetic pole faces, as they would be for a collection of magnets. Thus, the silver atoms are acting as magnets.

Silver atoms contain an odd number (47) of electrons and experiments show that all atoms with odd atomic numbers act as magnets, while many (but not all) elements with even atomic numbers are nonmagnetic. The magnetic atoms are termed **paramagnetic** and the nonmagnetic atoms are termed **diamagnetic.** The two deflections, z_1 and z_2 in Figure 4–11, are ex-

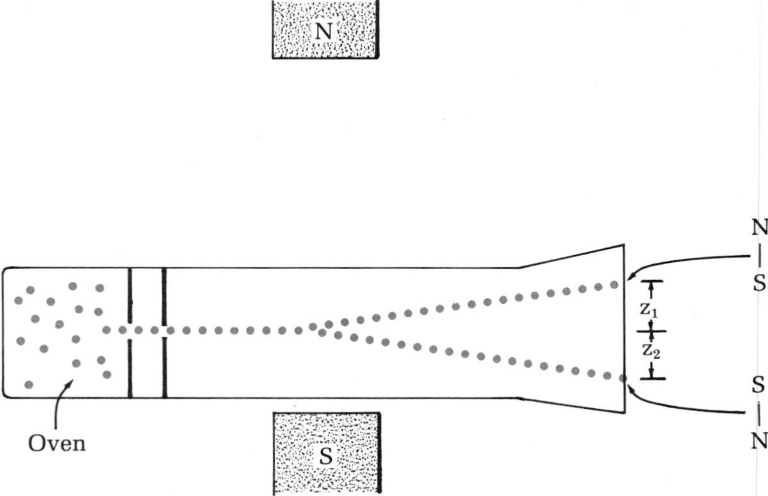

Figure 4–11
Schematic of the apparatus used in the Stern-Gerlach experiment.

actly equal. Thus, it appears that the magnetic strengths of all silver atoms are the same but the orientations of these magnets in a magnetic field are limited to two. The deflection z_1 is due to the $\begin{pmatrix} N \\ | \\ S \end{pmatrix}$ orientation along the magnetic field and z_2 is due to the $\begin{pmatrix} S \\ | \\ N \end{pmatrix}$ orientation. The north-south designation is rather cumbersome. From this point on we shall use arrows to indicate the two directions.

The fact that paramagnetism always occurs with atoms of an odd atomic number strongly indicates that paramagnetism resides in electrons. In each silver atom, for example, there must be an excess of one $\uparrow$ or $\downarrow$ orientation, while the remaining 46 electrons are equally paired between $\uparrow$ and $\downarrow$. Those atoms showing diamagnetism contain equal numbers of electrons with the two orientations.

What kind of motion of the electron could yield its intrinsic magnetism? We know that an electromagnet is produced by sending electrons through a wire wrapped in a circular fashion. Furthermore, if we reverse the direction of electron flow, we reverse the poles of the electromagnet. The magnetism of an electromagnet is not a consequence of the inherent magnetism of the electron. Rather it is a consequence of the electron charge and the circular motion. A similar model is commonly used to account for the intrinsic magnetism of the electron. The electron can be viewed as spinning about an internal axis, analogous to the spinning of the earth on its axis. This spin is a circular motion of a charged particle and it should yield magnetism. The two orientations to the magnetism arise from a clockwise and a counterclockwise sense to the spin motion. With this view, we say that every atomic orbital can contain two electrons, each with a spin sense opposite to the other.

Hund's Rules

One additional important feature of electron configurations is prefatory to considering the bonding of atoms in molecules. This is the assignment of electrons to orbitals within a given set, such as $2p$ or $3d$. For example, the configuration of carbon is $1s^2 2s^2 2p^2$. Are these two $2p$ electrons in the same $2p$ orbital, for example $2p_z$, or are they in different orbitals? If they are in different orbitals, are the spins paired or unpaired? Our intuition dictates the answer to the first question. Since electrons repel each other, they should not occupy the same orbital if another orbital of comparable energy is available. The second question concerning spin pairing in different orbitals cannot be treated intuitively, but must be answered by reference to experimental data. In the case of the carbon atom, isolated atoms are found to be paramagnetic and thus the spins are unpaired.

The answers to our two questions were first obtained in a general fashion by the German physicist Friedrich Hund, and they are known as **Hund's rules.** The first of these rules explains the decrease in the first ionization potential between nitrogen and oxygen and again between phosphorus and sulfur in Figure 4–8. The electronic configuration of N is $1s^2 2s^2 2p^3$, and according to Hund's first rule, the three $2p$ electrons must each be in different orbitals to minimize the electron-electron repulsion. With the oxygen atom, a fourth $2p$ electron is added and it must pair with one of the original three, giving rise to an extra electron-electron repulsion. This results in the decrease in the ionization potential shown in Figure 4–8. The same reasoning applies to P and S since the electron configurations of these atoms are $1s^2 2s^2 2p^6 3s^2 3p^3$ and $1s^2 2s^2 2p^6 3s^2 3p^4$.

SUGGESTIONS FOR ADDITIONAL READING

Adamson, A. W., "Domain Representations of Orbitals," *J. Chem. Educ.*, **42,** 141 (1965).

Hochstrasser, R. M., *The Behavior of Electrons in Atoms*, Benjamin, Menlo Park, Calif., 1964.

Sisler, H. H., *Electronic Structure, Properties, and the Periodic Law*, Reinhold, New York, 1963.

Pimentel, G. C., and Spratley, R. D., *Understanding Chemistry*, Chap. 12, Holden-Day, San Francisco, 1971.

Barrow, G. M., *General Chemistry*, Chap. 6, Wadsworth, Belmont, Calif., 1972.

PROBLEMS

1. Plot equation 4–1 for $n = 1$, $n = 2$, and $n = 3$ at a sufficient number of x values to verify that the wavelength is $2a/n$.
2. Verify the calculation that led to placing five dots in line segments 3 and 8 of Figure 4–1.
3. Consider the box in Figure 4–2. What are the nodal planes for the function characterized by the quantum numbers $n_x = 2$, $n_y = 2$, $n_z = 1$? How many lobes does any given dot contour possess? Assign a plus or a minus sign to ψ in each lobe.
4. What are the values of the principal quantum number and the azimuthal quantum number for a $4f$ wave function of the hydrogen atom? How many spherical and planar nodes does it possess? How many functions belong to the $4f$ set?
5. The energy of the 1s electrons of the helium atom is -24.6 ev. Electron-electron repulsion makes a large contribution to this energy. Modify equation 3–21 to make it apply to an imaginary helium atom in which there is no repulsion. (It will probably be necessary for you to retrace the derivation of equation with the electron potential energy that applies to the repulsionless He atom.) Calculate the energy of an electron in the 1s orbital of the repulsionless helium atom. Compare this energy with the measured electron energy of -24.6 ev. The difference is a measure of the importance of electron-electron repulsion in the He atom.
6. What is the sum of the kinetic energies of the two electrons in the helium atom? *Hint:* Use the virial theorem.
7. In discussing Figure 4–6, we made the point that electron density at small r values is very important in the calculation of $\langle 1/r \rangle$. Consider the following set of numbers: 0.100, 1.00, 1.50, 2.00, 3.00. Calculate the average of the reciprocal of these numbers. Now exclude the number 0.100, and recalculate the average reciprocal. Notice the importance of the number 0.100 in the first calculation.
8. As derived in the text, the energy of the highest-energy X-ray from a chlorine atom is 3529 ev. Use this number to calculate A in equation 4–6. Compare this A value with the constant in equation 3–7. Does this suggest the source of Moseley's A term? *Hint:* Refer to your derivation in problem 5.
9. Why do the first ionization potentials of He, Ne, and Ar in Figure 4–8 decrease in the order from He to Ne to Ar?
10. Without referring to Table 4–2, give the electron configurations of the atoms C, Al, P, Cl. The respective atomic numbers are 6, 13, 15, and 17.
11. The chemical properties of lithium, sodium, and potassium are similar. Refer to Table 4–2 and name two elements that you predict will have properties resembling those of Li, Na, and K.

12. Nitrogen and phosphorus have similar chemical properties. Refer to Table 4–2 and locate two elements that you predict will have properties resembling those of N and P.
13. Without referring to Table 4–3, predict which is larger, the second ionization potential of sodium or the first ionization potential of neon.
14. The ions K^+ and Cl^- are isoelectronic; that is, they contain the same number of electrons. Experimentally it is found that the species Cl^- is considerably larger than K^+. Can you explain this large difference in size?
15. Using Figure 4–8, predict which atom of these sets is larger, fluorine or carbon? helium or hydrogen? sodium or lithium?
16. Note that IP_2 for helium in Table 4–3 is the negative of the 1s orbital energy for the imaginary repulsionless helium atom. How does that relation come about? Without referring to Table 4–3, predict the value of IP_3 for the lithium atom.
17. Why isn't the silver atom beam in Figure 4–11 placed exactly between the two magnetic poles?
18. Which of the following atoms or ions can be predicted to be paramagnetic: Li, N, Ne, S, Fe, Al^{3+}, Cr^{2+}, Cu^{2+}?
19. Without referring to Table 4–3, predict whether the second ionization potential of fluorine is greater or less than the second ionization potential of oxygen.

5/ The Chemical Bond

A molecule like H_2 consists of two strongly bonded hydrogen atoms. What is the source of the "glue" that holds atoms together? On the other hand, gaseous helium consists of separate He atoms. Why doesn't helium form He_2 in analogy with H_2? The atomic orbital wave model of the atom accounts very nicely for chemical bonding. Also from this model, we can explain the relative strengths of many bonds. For example, the bond in the molecule N_2 is more than twice as strong as the bond in H_2.

5–1. Physical Basis of Chemical Bonding

Let us consider the formation of a chemical bond in a simple case

$$2\,H \rightleftharpoons H_2 \qquad \textbf{5–1}$$

The double half-arrows connecting $2\,H$ and H_2 indicate that H_2 can be formed from two H atoms, and that two H atoms can be formed from an H_2 molecule. In which state, H_2 or $2\,H$, do four electrons and two protons usually exist? Experiments show that H_2 is the prevalent species at room temperature and below, and experiments also show that energy must be given to H_2 molecules to dissociate them to H atoms. This energy is readily available at high temperatures and H atoms are the prevalent species at high temperatures. **In any situation in which chemical bonds exist, they owe their existence to the lower energy of the bonded state as compared with the separated atoms.** Our task is to discover the source of this energy difference.

Lowered Potential Energy as the Source of Bonding

When we consider the relative energies of H_2 molecules and H atoms, we are concerned with the energies of the stationary atoms since atoms and molecules may possess a wide range of translational energies. We are interested in the energy that is the sum of electrostatic potential energy and electron and nuclear kinetic energy within atoms and molecules. The virial theorem applies to all such cases. Figure 5–1 shows the relation between the total energy *E*, the potential energy *V*, and the kinetic energy *T* in any stationary atom or molecule. The potential energy *V* and total energy *E* are

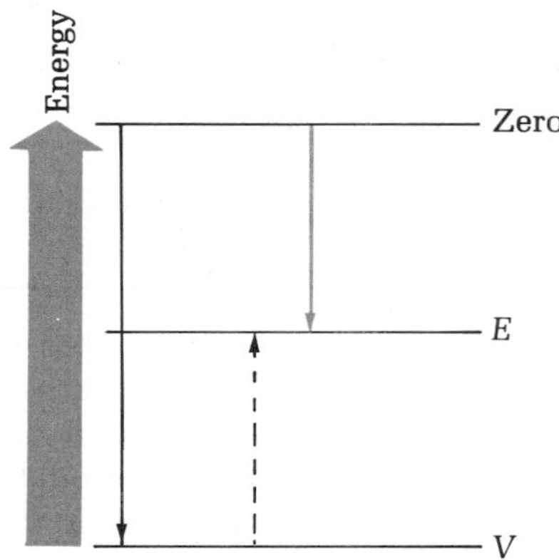

Figure 5–1
Relation between total energy (*colored arrow*), potential energy (*black arrow*), and kinetic energy (*dashed arrow*) for the particles in any stationary atom or molecule.

both negative, and $V = 2\,E$. The kinetic energy is positive and $T = -E$.

For reaction 5–1, we have for the energy difference

$$\Delta E = E_{H_2} - 2\,E_H \qquad \textbf{5–2}$$

where E_{H_2} is the total energy of a stationary H_2 molecule and E_H is the total energy of a stationary H atom. The energy E_H is −13.6 ev, the energy of an electron in a hydrogen atom 1*s* orbital. We can express equation 5–2 in terms of potential energies by using the virial theorem:

$$\Delta E = \frac{V_{H_2}}{2} - \frac{2\,V_H}{2} = \left(\frac{1}{2}\right)\Delta V \qquad \textbf{5–3}$$

This result shows that **the bond in H_2 (and any molecular bond) forms because the potential energy is lower in the molecule than it is in the isolated atoms.** We could also have substituted the kinetic energy into equation 5–2 to obtain

$$\Delta E = -T_{H_2} - 2(-T_H) = -\Delta T \qquad \textbf{5–4}$$

The kinetic energy actually increases in the formation of the molecular bond.

The Hydrogen Molecule Ion (H_2^+)

Equation 5–3 must mean that the electrostatic attraction between electrons and two nuclei in H_2 is sufficient to outweigh the electron-electron and nucleus-nucleus repulsions involved in having two H atoms approach at a short internuclear distance. Let us test this conclusion by calculation with the hydrogen molecule ion, H_2^+. This ion is produced by electron bombardment of H_2 in a mass spectrometer. The energy of H_2^+ has been found to be 2.65 ev lower than the energy of a separated proton and hydrogen atom. Thus, H_2^+ represents the simplest example of a chemical bond.

The measured internuclear distance in H_2^+ is 1.06 Å. Figure 5–2 shows the single electron of H_2^+ located at the electrostatically most favorable position, exactly halfway between the two protons. The equation for the potential energy in this arrangement is

$$V_{H_2^+} = e^2\left(\frac{-1}{r_{Ae}} - \frac{1}{r_{Be}} + \frac{1}{r_{AB}}\right)$$

$$= (4.8 \times 10^{-10})^2\left(\frac{-2}{0.53 \times 10^{-8}} + \frac{1}{1.06 \times 10^{-8}}\right)$$

$$= -6.5 \times 10^{-11} \text{ erg}$$

and by the virial theorem, the energy in this arrangement is $-6.5 \times 10^{-11}/2$ erg or -3.3×10^{-11} erg, or -20 ev. This energy is considerably lower than the energy of an electron in a hydrogen atom 1s orbital (-13.6 ev) and we can easily see that an electron intermediate between two protons has a lower energy than an isolated proton and an isolated hydrogen atom. Electrons will always be found associated with as many nuclei as possible. The possibilities are affected by several factors.

5–2. Molecular Orbitals

The difference in energy between H_2^+ and $H + H^+$ (calculated in the preceding section) is 6 ev, a considerably larger value than the experimental value of 2.65 ev. This discrepancy must arise because the electron density in Figure 5–2 is not actually concentrated completely at the energetically most favorable site, halfway between the two nuclei. This inability of the electron to reside exclusively at the energetically most favorable point is perfectly consistent with a wave model of the electron, since the wave is not confined to any one point.

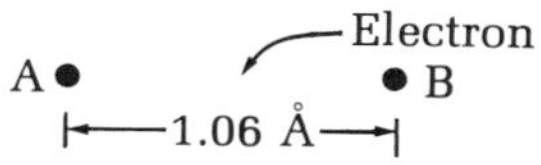

A • ⟵Electron • B

|⟵1.06 Å⟶|

Figure 5–2 Energetically most favorable location of the electron in H_2^+, relative to the two protons (A and B).

Addition of Atomic Orbitals

We now must construct a wave model of electrons in molecules that accounts for bonding and that can apply to a wide variety of molecules. Again, it should be a model that is easy to visualize and easy to use. Figure

5–3(a) shows two separated H atoms represented by 90 percent 1*s* contour spheres. In Figure 5–3(b), the two atoms have approached each other within a short distance and the two contours have overlapped.

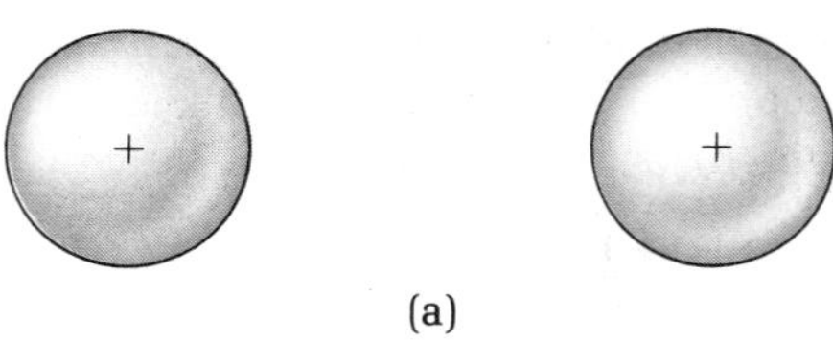

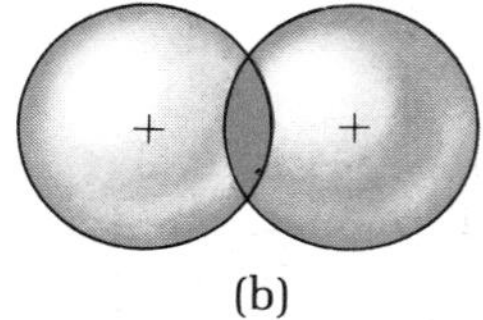

Figure 5–3
(a) Separate hydrogen 1*s* contours and (b) overlapping contours.

What happens to waves when they meet? Figure 5–4(a) shows two waves meeting in a **reinforcement** of each other. The maxima and minima of both waves occur at the same points along the direction in which the waves are

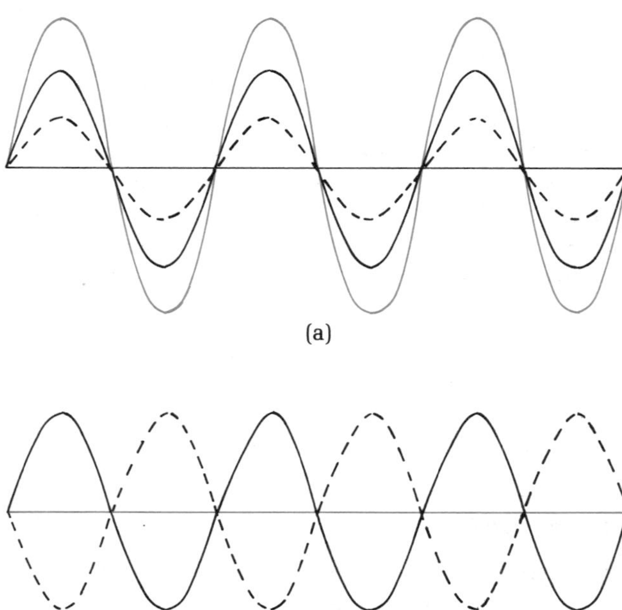

Figure 5–4
(a) Waves meeting in a reinforcing and (b) cancelling fashion. The waves represented by the solid and dashed black lines add, to give the wave represented by the colored line.

moving; this yields reinforcement, as shown by the sum of the two waves. In Figure 5-4(b), the waves meet in a **cancellation** of each other. Every maximum in one wave occurs at a point where the other wave has a minimum, and the sum of the two waves is zero at all points.

If the two spheres of Figure 5-3 meet in a reinforcing fashion, the two atomic wave functions simply add and the electron density increases in the electrostatically most favorable region of space, namely the overlap region in Figure 5-3(b). For this reinforcement to occur, the two wave functions must possess the same sign. In Figure 5-3(b), both signs are positive.

Figure 5-3(b) shows a **bonding molecular orbital** formed by the simple addition of atomic orbitals. Better models can be constructed of the electron density in H_2, but this simple addition model is very useful, while improved models are rather complex. The simple addition model is the one we shall use.

Antibonding

Figure 5-3 shows only one of the two ways in which hydrogen 1s waves can meet. The second way is the cancelling fashion shown in Figure 5-5. The molecular orbital this cancelling encounter forms has a depleted electron density in the overlap region, and the electron energy is actually higher than it is in either separated atom. Thus, Figure 5-5(b) depicts an **antibonding molecular orbital.**

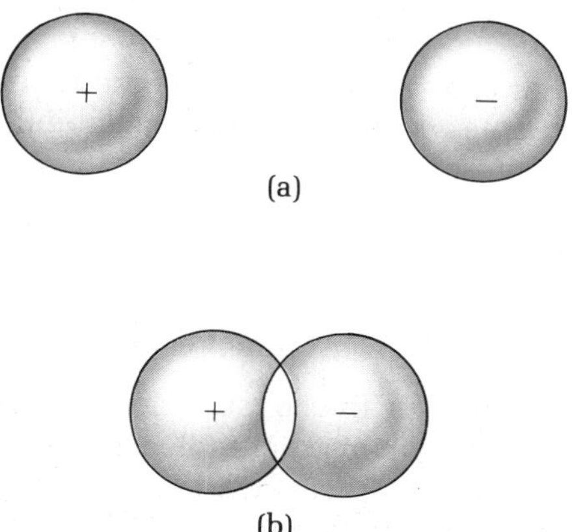

Figure 5-5
The encounter of hydrogen 1*s* orbitals in a cancelling fashion.

The encounter of any two atomic orbitals yields two molecular orbitals, one bonding and the other antibonding. Since there are atomic orbitals besides 1s, there are many molecular orbitals. (We shall consider a number of these shortly.) For reference we must devise a naming system for molecular orbitals and name the two orbitals shown in Figures 5-3 and 5-5. Both orbitals that are derived from the 1s atomic orbitals are completely symmetric about the internuclear axis. Any molecular orbital that is completely symmetric about the internuclear axis is designated a σ orbital (Greek lowercase *sigma*). The bonding orbital in Figure 5–3 is designated $1s\sigma$ and the antibonding orbital of Figure 5–5 is designated $1s\sigma^*$ (antibonding orbitals are always designated with an asterisk).

Electron Configurations

Figure 5–6 is an energy-level diagram showing the relative positions of the hydrogen 1*s* atomic orbitals and the two molecular orbitals derived from them. What rules do we have for assigning electrons to molecular orbitals? The rules are essentially the same as the combination of the Pauli and Hund rules for the electron configurations of atoms.

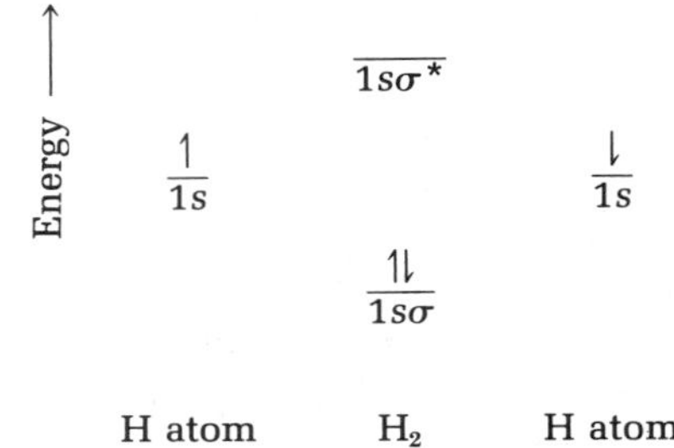

Figure 5–6 Energies of the molecular orbitals derived from hydrogen 1*s* atomic orbitals, relative to those atomic orbitals. Orbital occupations (number of electrons) are shown by the half arrows.

1. Add the electrons to the molecular orbitals in the order of increasing energy.
2. Any given molecular orbital can hold a maximum of two electrons.
3. Two electrons will enter different molecular orbitals of the same energy rather than pair in the same orbital.

According to these rules, the electron configuration of H_2 is $1s\sigma^2$ and this is shown by the arrows in Figure 5–6.

Molecular Orbitals Formed from Atomic Orbitals with $n = 2$

With our overlap procedure for the construction of molecular orbitals from atomic orbitals, we can easily construct molecular orbitals from the atomic orbitals 2*s*, $2p_x$, $2p_y$, and $2p_z$. These overlap pictures are shown in Figure 5–7. The 2*p* orbitals can meet in two ways in addition to a bonding and an antibonding fashion. The $2p_z$ orbitals (the internuclear axis is designated z) form sigma molecular orbitals; but $2p_x$ and $2p_y$ form molecular orbitals that are not completely symmetric about the internuclear axis. In fact, these molecular orbitals have a nodal plane that includes the internuclear axis. The molecular orbitals formed from $2p_x$ and $2p_y$ atomic orbitals are designated π orbitals (Greek lowercase *pi*). The orbitals $2p\pi_x$ and $2p\pi_x^*$ are shown in the figure. The orbitals $2p\pi_y$ and $2p\pi_y^*$ have the same shapes and

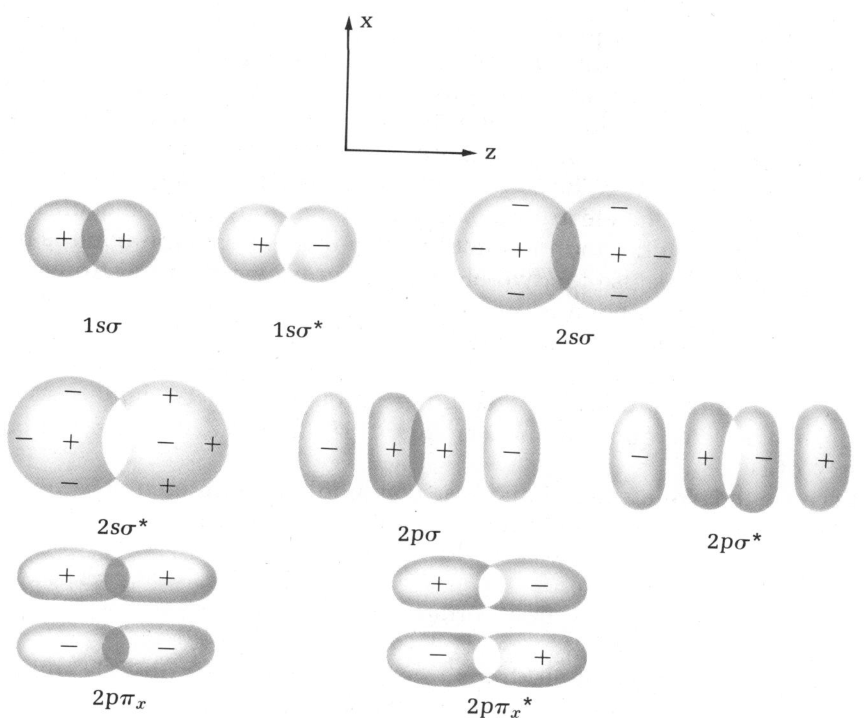

Figure 5–7 The formation of molecular orbitals by the reinforcing and cancelling overlap of atomic orbitals.

sizes as $2p\pi_x$ and $2p\pi_x{}^*$, respectively; but they are rotated by 90° about the internuclear axis.

5–3. Homonuclear Diatomic Molecules

Now that we have a set of molecular orbitals, we can consider the bonding in such molecules as N_2 and O_2. These molecules are homonuclear (both nuclei are the same) diatomic molecules. A systematic consideration of such molecules follows.

The Hydrogen Molecule (H_2)

The hydrogen molecule ion, $H_2{}^+$, has an internuclear separation of 1.06 Å, and its energy is 2.65 ev lower than the energy of a separated proton and hydrogen atom. The internuclear separation between bonded atoms is termed the **bond length** and the energy gap between the bonded and nonbonded species is termed the **bond energy.** Bond energies are most commonly reported in units of kilocalories per mole, and the conversion factor is

$$1 \text{ ev} = 23.06 \text{ kcal/mole}$$

Thus, the H_2^+ bond energy is 61.1 kcal/mole.

The electronic configuration of H_2^+ is $1s\sigma$. When a second electron is added to produce H_2, it also occupies the bonding $1s\sigma$ orbital. Hence, we expect the bond length in H_2 to be less than 1.06 Å and the bond energy to be greater than 61.1 kcal/mole because H_2 contains two bonding electrons instead of one as in H_2^+. The bond length in H_2 is 0.76 Å and the bond energy is 104 kcal/mole. One-electron bonds are seldom found in molecules. The reason they are seldom found is not because the one-electron bond is not a strong bond but because the two-electron bond is significantly stronger.

If the two-electron bond is preferable to the one-electron bond, shouldn't a three-electron bond be even better? The problem with H_2^- is that the third electron must occupy the next highest energy orbital above $1s\sigma$; this orbital is $1s\sigma^*$. The antibonding character of $1s\sigma^*$ subtracts from the bonding by the two $1s\sigma$ electrons, and the bond energy of H_2^- is even less than the bond energy of H_2^+.

Because the two-electron bond is so commonly encountered, it is the basis of the definition of the **bond order** of a molecule. The bond order is defined as follows:

$$\text{Bond order} = \frac{n - n^*}{2} \qquad \textbf{5–5}$$

where n is the number of electrons in bonding molecular orbitals and n^* is the number of electrons in antibonding molecular orbitals. Thus, the **bond order is the number of two-electron bonds,** and the bond order is one in H_2 and one-half in both H_2^+ and H_2^-.

An example of a four-electron molecule would be He_2, but such a molecule would possess the electron configuration $1s\sigma^2 1s\sigma^{*2}$, with bond order zero. Actually, as two helium atoms approach each other within a distance corresponding to substantial bonding in H_2 or H_2^+ (an internuclear distance of approximately 1 Å), electron-electron repulsion and nucleus-nucleus repulsion strongly dominate electron-nucleus attraction, and the He—He interaction is strongly repulsive. This is the reason for the very low boiling point of helium (4 °K), and for the existence of helium as a monatomic gas, He.

The Lithium Molecule (Li_2)

Most of the molecules in lithium vapor are the diatomic species Li_2. The bond energy of Li_2 is 25 kcal/mole, and the equilibrium internuclear separation is 2.68 Å. Since Li_2 contains six electrons, two electrons are contained in each of the two molecular orbitals $1s\sigma$ and $1s\sigma^*$. These four electrons do not produce any net bonding and they are commonly termed the **nonbonding core electrons.** The electron density in the core is very similar to the density in the separated atoms, namely two spherical orbitals each containing two electrons. Two electrons are located in a bonding molecular orbital derived from atomic orbitals with a principal quantum number of 2. These two bonding electrons are termed **valence electrons** and the molecular orbital that contains them is a **valence molecular orbital.**

Which of the four bonding orbitals—$2s\sigma$, $2p\sigma$, $2p\pi_x$, or $2p\pi_y$—is occupied in Li_2? The lithium 2s atomic energy level is 45 kcal/mole lower than the 2p atomic level and this results in a lower energy for the $2s\sigma$ molecular orbital in Li_2 than for $2p\sigma$ or either of the $2p\pi$ orbitals. Thus, the electronic configuration of Li_2 is $1s\sigma^2 1s\sigma^{*2} 2s\sigma^2$ and the bond order is $(4 - 2)/2 =$ one two-electron **sigma bond** (a sigma bond is defined as one in which the bonding valence electrons occupy a σ molecular orbital). A two-electron bond is commonly indicated in molecular-structure representations by a single line connecting the two atoms. Thus, Li_2 is represented Li—Li. The bond energy of Li_2 is considerably smaller than the bond energy of H_2, and the bond length in Li_2 compared with H_2 is approximately four times as great. This is to be expected, since the larger 2s atomic orbitals form the bonding molecular orbital in Li_2 and **as the size of the bonding molecular orbital increases, the bond strength decreases and the bond length increases.**

The Hypothetical Beryllium Molecule (Be_2)

Let us now move along to the eight-electron species Be_2. There is now some ambiguity in our thinking about the ordering of energy levels in the molecule. The sequence $1s\sigma$, $1s\sigma^*$, $2s\sigma$ (in order of increasing energy) is preserved through all of the homonuclear diatomic molecules from Li_2 up to the heaviest such molecules known. This is the case because the separation in energy between the atomic 1s and 2s and the atomic 2s and 2p increases progressively with increasing atomic number. The seventh and eighth electrons occupy either the $2s\sigma^*$, the $2p\sigma$, or one of the $2p\pi$ molecular orbitals, depending on which of these orbitals offers the lowest energy. In the case of Be_2, which of these orbitals is occupied hinges on the fact that beryllium vapor contains only Be atoms. This must mean that the $2s\sigma^*$ orbital has the lowest energy of the three possibilities; and the electronic configuration of Be_2 is $1s\sigma^2 1s\sigma^{*2} 2s\sigma^2 2s\sigma^{*2}$. The bond order is $(4 - 4)/2 =$ zero, and Be_2 does not exist.

The Boron Molecule (B_2)

The second part of the ambiguity encountered with Be_2 must be resolved in B_2. Since the 2s-2p separation increases with increasing atomic number, we can confidently assign the eight electrons of lowest energy in B_2 to the configuration $1s\sigma^2 1s\sigma^{*2} 2s\sigma^2 2s\sigma^{*2}$, but the ninth and tenth electrons have a choice between $2p\sigma$ or $2p\pi$. Reference to Figure 5-7 shows $2p\sigma$ involves overlap in the electrostatically most favorable region. The orbitals $2p\pi_x$ and $2p\pi_y$ do not involve overlap on the internuclear axis. However, if two electrons are placed in $2p\sigma$, there is a sizable amount of electron-electron repulsion, as shown in Figure 5–8(a). The figure shows favorable locations of the ninth and tenth electrons of B_2 relative to the two nuclei of B_2 (labeled A and B). Figure 5–8(a) shows the two electrons along the internuclear axis where the $2p\sigma$ electron density is concentrated. Figure 5–8(b) shows the two electrons on a line midway between the two nuclei and perpendicular to the

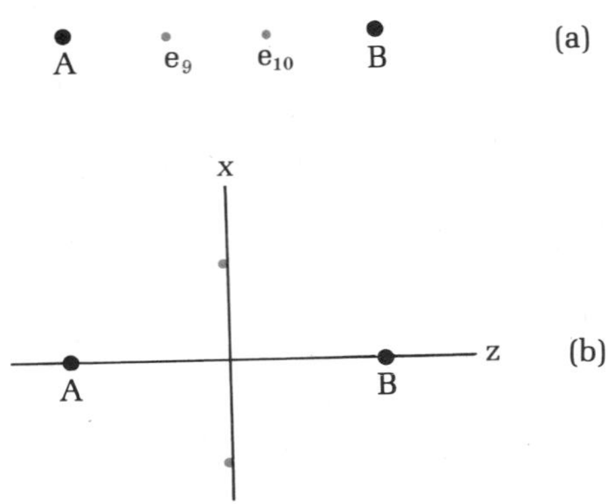

Figure 5–8 Favorable arrangements of the two valence electrons (e_9 and e_{10}) of B_2, relative to the nuclei (A and B) in the gaseous boron molecule, (a) sigma bonding and (b) pi bonding.

internuclear axis. In Figure 5–8(b) the electrons are in the $2p\pi_x$ orbital. The electron placement in Figure 5–8(b) results in a poorer electron-nucleus interaction than for placement as shown in Figure 5–8(a), but the electron-electron repulsion is considerably higher for (a) than for (b); hence the situation leads to ambiguity. Again, as with Be_2, the ambiguity is resolved by reference to experiment. In this case, it is found that the B_2 molecule is paramagnetic. The two electrons in Figure 5–8(a) must have paired spins, but Hund's second rule leads us to conclude that the two electrons in Figure 5–8(b) should really be in the two different orbitals, $2p\pi_x$ and $2p\pi_y$, and that the spins should be unpaired. Thus, the paramagnetism shows that the electronic configuration of B_2 is $1s\sigma^2 1s\sigma^{*2} 2s\sigma^2 2s\sigma^{*2} 2p\pi_x^{\,1} 2p\pi_y^{\,1}$ and the bond order is $(6 - 4)/2 =$ one two-electron **pi bond** (a pi bond is defined as one in which the bonding valence electrons occupy π molecular orbitals).

The Carbon Molecule (C_2)

The molecule C_2 is encountered in carbon vapor and its visible emission is responsible for the bluish color of a bunsen burner flame. The fact that C_2 gas does not dissociate appreciably at flame temperature is a reflection of its high bond energy of 144 kcal/mole. The bond order of C_2 must be two, since the two possible electronic configurations are $1s\sigma^2 1s\sigma^{*2} 2s\sigma^2 2s\sigma^{*2} 2p\pi^4$ and $1s\sigma^2 1s\sigma^{*2} 2s\sigma^2 2s\sigma^{*2} 2p\sigma^2 2p\pi_x^{\,1} 2p\pi_y^{\,1}$, and the bond order in both cases is $(8 - 4)/2 =$ two. In the first case, C_2 has two pi bonds and in the second case it has one sigma bond and one pi bond. Again, the question of the configuration is resolved by the magnetism of the compound. The substance is diamagnetic and this means that a C_2 molecule possesses two pi bonds. A

double bond is commonly represented by double lines between the atoms, and C_2 is represented by C=C.

The Molecules N_2, O_2, F_2, and the Hypothetical Molecule Ne_2

In the case of N_2, there is no ambiguity, since the configuration must be $1s\sigma^2 1s\sigma^{*2} 2s\sigma^2 2s\sigma^{*2} 2p\pi^4 2p\sigma^2$ and the bond order is $(10 - 4)/2 = 3$. There are two pi bonds and a sigma bond. The **triple bond** is represented by N≡N, and we should predict that N_2 has a large bond energy. The bond energy is found to be 225 kcal/mole and the N_2 bond is one of the strongest bonds found in any molecule.

The ambiguity in the ordering of levels arises again with O_2, since the fifteenth and sixteenth electrons can occupy either the $2p\sigma^*$ or $2p\pi^*$ molecular orbitals. Again, experiments reveal the answer by showing that O_2 is paramagnetic, and paramagnetism can only arise in this molecule with the electronic configuration $1s\sigma^2 1s\sigma^{*2} 2s\sigma^2 2s\sigma^{*2} 2p\sigma^2 2p\pi^4 2p\pi_x^{*1} 2p\pi_y^{*1}$. The bond order is $(10 - 6)/2 = 2$, and the double bond in O=O consists of one sigma bond and one pi bond. The bond energy of O_2 is 118 kcal/mole. This double bond is somewhat weaker than the double bond of C_2 because of the extra electron-electron repulsion introduced by antibonding pi electrons; these are present in O_2 but not in C_2.

The repulsive character of $2p\pi^*$ occupation is even more dramatically shown in F_2. This molecule is diamagnetic, so its electronic configuration must be $1s\sigma^2 1s\sigma^{*2} 2s\sigma^2 2s\sigma^{*2} 2p\sigma^2 2p\pi^4 2p\pi^{*4}$. The bond order is $(10 - 8)/2 =$ one sigma bond. The bond energy is only 36 kcal/mole, contrasting with 69 kcal/mole for the single bond in B_2. The low value of the bond energy for F_2 is largely a result of the extra electron-electron repulsion introduced with the antibonding pi electrons.

If the compound Ne_2 were to exist, its electronic configuration would have to be $1s\sigma^2 1s\sigma^{*2} 2s\sigma^2 2s\sigma^{*2} 2p\sigma^2 2p\pi^4 2p\pi^{*4} 2p\sigma^{*2}$, with a bond order $(10 - 10)/2$ = zero. As with helium, the encounter of the two neon atoms is highly repulsive.

Figure 5–9 summarizes the electron configurations of the real and hypothetical homonuclear diatomic molecules from H_2 to Ne_2. The energy levels are indicated in the proper order; they are not, however, drawn to scale. Thus all $1s\sigma$ orbitals are indicated at the same level, though in reality they markedly decrease in energy from hydrogen to neon. You will note that the relative energies of the $2p\sigma$ and $2p\pi$ (both bonding and antibonding) invert between N_2 and O_2.

Table 5–1 summarizes the bond orders, bond lengths, magnetism, and bond energies for those species in Figure 5–9 that actually form. The table reveals a very distinct correlation between bond energy and bond order. The ratios of bond energies from B_2 through O_2 are $\frac{69}{69} : \frac{144}{69} : \frac{225}{69} : \frac{118}{69} =$ 1 : 2.1 : 3.3 : 1.7, or approximately the ratio of the bond orders, 1 : 2 : 3 : 2. The

H_2	He_2	Li_2	Be_2	B_2	C_2	N_2	O_2	F_2	Ne_2	
— —	— —	— —	— —	— —	— —	— —	—	—	↑↓	$2p\pi^*, 2p\sigma^*$
—	—	—	—	—	—	—	↑ ↑	↑↓ ↑↓	↑↓ ↑↓	
—	—	—	—	—	—	↑↓	↑↓ ↑↓	↑↓ ↑↓	↑↓ ↑↓	$2p\pi, 2p\sigma$
— —	— —	— —	— —	↑ ↑	↑↓ ↑↓	↑↓ ↑↓	↑↓	↑↓	↑↓	
—	—	—	↑↓	↑↓	↑↓	↑↓	↑↓	↑↓	↑↓	$2s\sigma^*$
—	—	↑↓	↑↓	↑↓	↑↓	↑↓	↑↓	↑↓	↑↓	$2s\sigma$
—	↑↓	↑↓	↑↓	↑↓	↑↓	↑↓	↑↓	↑↓	↑↓	$1s\sigma^*$
↑↓	↑↓	↑↓	↑↓	↑↓	↑↓	↑↓	↑↓	↑↓	↑↓	$1s\sigma$

Figure 5–9
Electron configurations of real and hypothetical homonuclear diatomic molecules from H_2 through Ne_2, in the order of their increasing atomic numbers.

bond order is also reflected qualitatively in the bond length. As we go from B_2 to F_2, the bond length first decreases from B_2 to N_2 as the bond order increases from 1 to 3, and then the bond length increases from N_2 to F_2 as the bond order decreases from 3 to 1.

Table 5–1
Properties and Parameters of Several Homonuclear Diatomic Molecules

Molecule	Bond Energy (kcal/mole)	Bond Order	Bond Length(Å)	Magnetism
H_2	104	1	0.76	Diamagnetic
Li_2	25	1	2.68	Diamagnetic
B_2	69	1	1.59	Paramagnetic
C_2	144	2	1.24	Diamagnetic
N_2	225	3	1.10	Diamagnetic
O_2	118	2	1.21	Paramagnetic
F_2	36	1	1.44	Diamagnetic

5–4. Heteronuclear Diatomic Molecules

The study of bonding between identical atoms leads to consideration of bonding between different atoms. The simplest example of bonding between different atoms can lead to conclusions and procedures to be carried over to more complicated situations.

The Lithium Hydride Molecule (LiH)

The simple molecule we focus on is lithium hydride, LiH, for which the bond energy is 58 kcal/mole and the bond length is 1.61 Å. Both the bond energy and the bond length are intermediate between the values listed in Table 5–1 for H_2 and Li_2, and we should like to account for that fact.

The fundamental reason that molecules form is the lowered potential energy of electrons associated simultaneously with two nuclei compared with electrons associated with single nuclei in isolated atoms. However, the simultaneous association is not always favorable. For example, the molecule He_2 does not form. In that case, the electrons prefer to remain associated with isolated helium atoms. We saw one view of He_2 in the occupation of molecular orbitals; another very instructive view is based on the atomic orbital picture of the helium atoms approaching each other. Each helium atom possesses a filled 1*s* atomic orbital, and that orbital cannot accept electrons from any other atom. If He atoms are to accept electrons, they must do so in the 2*s* orbital or an orbital of even higher energy. But a molecular orbital that is made up equally of a helium 1*s* and a helium 2*s* has a much higher energy than that of the helium atom 1*s* alone. Thus, the helium electrons would gain energy in forming the species He_2, and the encounter is highly repulsive. This is an illustration of an exceedingly important principle of chemical bonding. **Electrons from one atom will not move into the vicinity of another atom unless the second atom possesses an unfilled atomic orbital with an energy less than the energy of the electron in the first atom or comparable with it.** In the case of H_2, in which strong bonding occurs, the H atom 1*s* orbitals are unfilled and an electron from one hydrogen atom can be accepted by the other, resulting in a lowered electron energy. The resultant association of electrons with two nuclei is commonly termed the **covalent bond.**

From this principle, it is clear that the key to understanding the bonding between different atoms lies in the relative energies and the occupations of the atomic orbitals of the separated atoms. Figure 5–10 shows the energy levels and electron occupations of the lithium and hydrogen atoms. The lithium 1*s* level is so far below the hydrogen 1*s* level that there is little or no advantage for the two lithium 1*s* electrons in moving into the vicinity of the hydrogen nucleus. These electrons will remain in the vicinity of the lithium nucleus and will be **nonbonding. Such a situation invariably occurs with electrons below the valence level.** These electrons cannot move into valence orbitals on a second atom because the energy gap is too large. Likewise the nonvalence (core) electrons cannot move into nonvalence orbitals on a second atom because these orbitals are fully occupied. Thus, for ex-

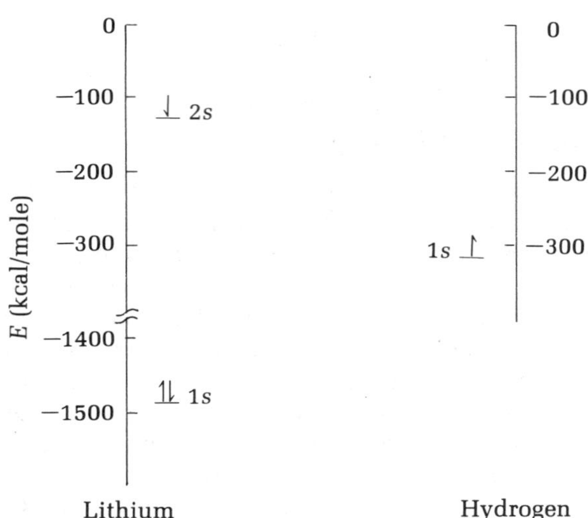

Figure 5–10 Energy levels and electron occupation of orbitals for hydrogen and lithium atoms.

ample, in Li_2 there can be no bonding involving the 1*s* electrons because both lithium atom 1*s* orbitals are filled.

As we can see in Figure 5–10, the lithium 2*s* and hydrogen 1*s* levels are much closer in energy than the lithium 1*s* and hydrogen 1*s* levels, and neither is fully occupied in the atom. The lithium 2*s* and hydrogen 1*s* orbitals can form a bonding molecular orbital by reinforcing overlap, as shown in Figure 5–11. In molecular orbitals for homonuclear diatomic molecules, the electron density must be equally divided by a plane located perpendicular to the internuclear axis and midway between the nuclei. This equal division will, in general, not occur in a heteronuclear diatomic molecule because the two atomic orbitals have different energies. Thus we see in Figure 5–10 that the lithium 2*s* electron has much more of an advantage in moving into the vicinity of the hydrogen nucleus than the hydrogen 1*s* electron has in moving into the vicinity of the lithium nucleus. There is a considerably greater electron density to the hydrogen side of the bonding molecular orbital than to the lithium side.

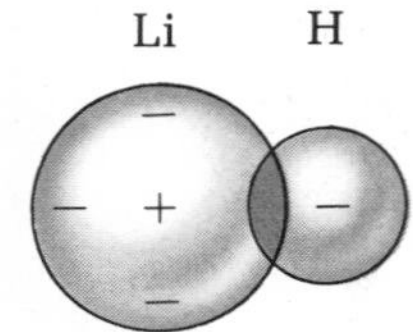

Figure 5–11 Bonding atomic overlap between a lithium 2*s* orbital and a hydrogen 1*s* orbital in lithium hydride. The signs shown are those of the atomic wave functions.

The Dipole Moment

If we count electrical charges on the Li and H portions of LiH, we conclude that the lithium portion is positively charged (3*e* for the nucleus minus 2*e* for the nonbonding 1*s* electrons minus less than one electron in the bonding molecular orbital), and the hydrogen portion is negatively charged (*e* for the nucleus minus more than one electron in the bonding molecular orbital). The LiH bond is a **polar bond** and the LiH molecule is a **polar molecule.** There is a net charge of $+\delta$ (Greek lowercase *delta*) in the lithium portion and a net charge of $-\delta$ in the hydrogen portion. The molecule as a whole possesses no net charge, since the sum of the charges on the two portions is $+\delta - \delta = 0$. This situation of two separated charges equal and opposite in sign is termed an **electric dipole** (as contrasted with a single charge, which is an electric pole).

How large is δ in the LiH molecule? This can be determined by measuring the **dipole moment** of the molecule, where the dipole moment is defined as the product of δ and the Li—H bond length. The bond length is measurable independent of the dipole moment and is 1.61 Å. The dipole moment is measured by placing LiH vapor between electrically charged plates, as shown in Figure 5–12. (Lithium hydride is a solid at room temperature and it melts at 680 °C. Lithium hydride diatomic molecules do not exist in the

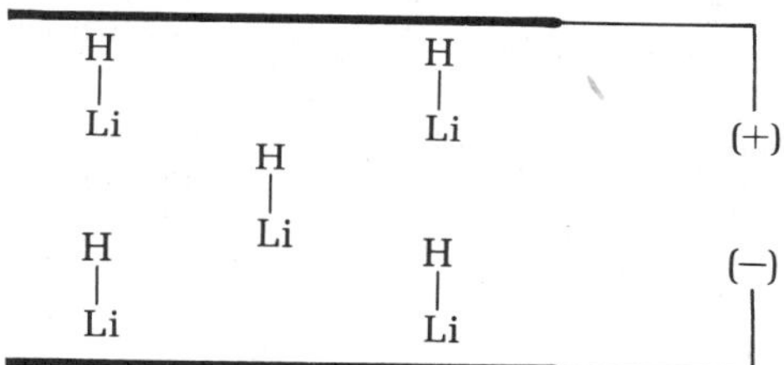

Figure 5–12 Electrostatically most favorable orientation of LiH molecules between charged plates.

solid. The bonding in solids that yield highly polar vapor species such as LiH is considered in the next section.) The LiH molecules orient themselves in the electric field, as shown in Figure 5–12, and the presence of the charges $+\delta$ and $-\delta$ changes the force between the two charged plates. The gas between the plates is termed a **dielectric medium** and its effect on the force between the plates is measurable and reported as the **dielectric constant** of the gas. The dielectric constant, in turn, is simply related to the dipole moment of the LiH molecule.

The dipole moment (symbolized by μ, Greek lowercase *mu*) of LiH is found to be 5.9×10^{-18} esu-cm = 5.9 Debye. The Debye is a unit named in honor of Peter Debye, an American chemist who made large contributions in this century to our understanding of the behavior of dielectric media. Thus, δ for LiH is represented:

$$\delta = \frac{\mu}{\text{(electronic charge)(bond length)}}$$

$$= \frac{(5.9 \times 10^{-18})}{(4.8 \times 10^{-10})(1.61 \times 10^{-8})} = 0.76 \text{ electron}$$

Atomic Orbital Ionization Potentials and Bond Polarity

As Figure 5–10 shows, the energies that determine bond polarity are the valence electron energies of the constituent atoms. Since the valence electrons are the highest-energy electrons (they have the least negative energies) in an atom, their energies are simply related to the ionization potentials, as given in Figure 4–8 and Table 4–3. For example, Figure 4–8 shows that the highest-energy electron of Li is easier to remove by approximately 8 ev than the single electron of H. This translates to the approximately 180 kcal/mole gap between the lithium 2*s* and hydrogen 1*s* levels in Figure 5–10. Thus, Figure 4–8 is extremely useful in allowing us to make qualitative predictions about bond polarities. **A large gap in the first ionization potentials usually indicates a highly polar bond; the atom with the higher ionization potential usually possesses the $-\delta$ charge in the bond.** Some examples of highly polar vapor species that follow this principle are LiF, LiCl, NaF, NaCl, KF, KCl, and many others that are solids at room temperature.

An exception to the principle occurs with such proposed species as LiHe and LiNe. With helium and neon, the valence orbitals are filled and they cannot accept electrons at a low-energy as F and Cl can. Thus He, Ne, Ar, Kr, and Xe, all of which have filled valence orbitals, can function, if at all, in chemical bonding only as the $+\delta$ atom, like Li, Na, and K.

5–5. Bonding in Ionic Solids

The substances LiH, NaCl, KCl, and so forth are solids at room temperature. In investigating why these compounds are solids, we shall devote our attention to sodium chloride since this solid is so common (for example, in salt shakers on the dining table). The sodium 3*s* orbital energy is obtainable from Table 4–3 as $-IP_1 = -5.1$ ev/molecule $= -118.4$ kcal/mole. The chlorine atom 3*p* orbital energy is given in the same table as -13.0 ev $= -300$ kcal/mole. Thus the gaseous NaCl molecule is highly polar, as we have predicted, with the sodium portion positively charged relative to the chlorine portion.

The Ion Pair Model

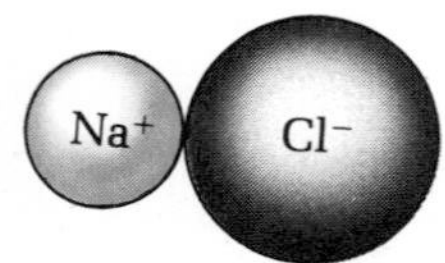

Figure 5–13 The ion pair model of the gaseous NaCl molecule.

Sodium chloride boils at 1413 °C, and the vapor consists of highly polar NaCl molecules with a bond length of 2.38 Å. There is a model of such a highly polar species that is commonly used, and we shall develop it here. In the model we view the NaCl molecule as a Na^+ ion and a Cl^- ion, both of which are spherical and in contact, as shown in Figure 5–13. Positive ions are termed **cations** and negative ions are **anions,** so the model is of a cation-anion pair. The ionic model corresponds to a δ of unity for the molecule, and δ is actually close to unity in many highly polar bonds.

Let us test the ionic model by using it to calculate ΔE for the following reaction:

$$Na(g) + Cl(g) \rightarrow NaCl(g) \qquad \textbf{5–6}$$

The measured value of ΔE for reaction 5–6 is -98 kcal/mole, or 98 kcal/mole is the NaCl bond energy. Reaction 5–6 may be considered as the sum of the following three reaction equations:

$$Na(g) \rightarrow Na^+(g) + e^- \qquad \textbf{5–7}$$

$$e^- + Cl(g) \rightarrow Cl^-(g) \qquad \textbf{5–8}$$

$$Na^+(g) + Cl^-(g) \rightarrow NaCl(g) \qquad \textbf{5–9}$$

The ΔE for equation 5–7 is the first ionization potential of sodium = 118.4 kcal/mole. The ΔE for equation 5–8 is the negative of the **electron affinity** of Cl^-. The electron affinity (symbolized by A) of any species is defined as the energy input required to separate an electron from the negative ion of that species to generate the species. In the case of the chlorine atom, the measured electron affinity is 83.4 kcal/mole, meaning that the electronic energy is −83.4 kcal/mole for each of the six electrons in the three $3p$ orbitals of the chloride ion. This energy is higher than the −300 kcal/mole for the $3p$ electrons in the chlorine atom because of the extra electron-electron repulsion introduced in Cl^-.

The ΔE for equation 5–9 is given in our ion pair model by the potential energy of two charges, $+e$ and $-e$, separated by 2.38 Å, the equilibrium internuclear separation of the gaseous NaCl pair. This ΔE is written:

$$\Delta E = \frac{-(4.8 \times 10^{-10})^2}{(2.38 \times 10^{-8})}$$

$$= -9.7 \times 10^{-12} \text{ erg/molecule} = -140 \text{ kcal/mole}$$

The ΔE for reaction 5–6 is the sum of the ΔE's for the three constituent steps = 118.4 − 83.4 − 140 = −105 kcal/mole. This is a good approximation to the actual ΔE of −98 kcal/mole, and the ionic model is a good one for NaCl. Because of the success of this model, the bonding in NaCl is commonly termed **ionic bonding.**

Solid Sodium Chloride

If an electron in NaCl is at a lower energy in the neighborhood of two nuclei rather than one, would more than two nuclei be even better? In Chapter 13 we shall consider how arrangements of atoms, ions, and molecules in solids are determined experimentally. Figure 5–14 shows the arrangement of Na^+ and Cl^- ions in solid NaCl. Each chloride ion is surrounded by six sodium ions as nearest neighbors, and the chloride ion electrons experience attraction from all seven nuclei simultaneously. For example, the Cl^- ion labeled 1 in Figure 5–14 is in contact with Na^+ ions labeled 2, 3, 4, 5, 6, and 7. Each of these Na^+ ions lies in the center of a cubic face. The Na—Cl internuclear distance in the crystal is 2.83 Å, a considerably larger distance than for the NaCl molecule in the vapor. This difference is due to the existence of more, but weaker, bonds in the solid as compared with the gas

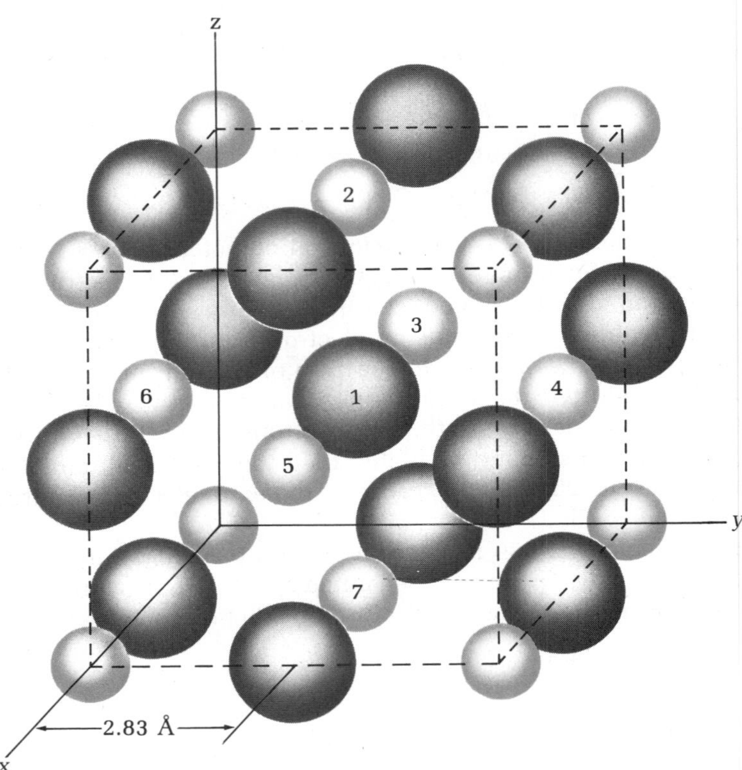

Figure 5–14
The placement of ions in solid NaCl. Large black spheres represent Cl^-; small colored spheres represent Na^+.

phase. The weaker bonds are reflected in a longer bond distance. (In the solid the spheres are in contact, as in Figure 5–13. They are drawn apart in Figure 5–14 so that the spheres can be more clearly distinguished.)

In addition to the six nearest neighbor sodium ions located 2.83 Å from every Cl^-, there are twelve other chloride ions located at a distance of $(2.83)(2)^{1/2} = 4.00$ Å away from each Cl^-, and eight sodium ions at a distance of $(2.83)(3)^{1/2} = 4.90$ Å. If we examine a larger model of the NaCl structure, we can continue to add numbers of ions and separation distances of ions around a given ion.

Using the ion pair model, we can combine these ion numbers, charges, and distances to calculate ΔE for the following reaction:

$$Na^+(g) + Cl^-(g) \rightarrow NaCl(s) \qquad \textbf{5–10}$$

This ΔE is the potential energy of charges $+e$ and $-e$, located at the nuclear positions in the NaCl crystal. If we indicate the NaCl distance of 2.83 Å by r_{AB}, this potential energy can be written:

$$\text{Potential energy} = \frac{-e^2}{r_{AB}}\left(6 - \frac{12}{\sqrt{2}} + \frac{8}{\sqrt{3}} - \frac{6}{\sqrt{4}} + \frac{24}{\sqrt{5}} - \frac{24}{\sqrt{6}} + \cdots\right) \qquad \textbf{5–11}$$

The terms within the brackets form an infinite series of alternating attractions and repulsions; the series converges to the value 1.7476. All ionic crystals are characterized by such a series of terms, and all salts that adopt the NaCl ionic arrangement are characterized by the value 1.7476. This important constant is termed the **Madelung constant.**

The potential energy in equation 5–11 is evaluated:

$$\text{Potential energy} = \frac{-(4.8 \times 10^{-10})^2(1.7476)}{(2.83 \times 10^{-8})}$$

$$= 1.42 \times 10^{-11} \text{ erg/molecule}$$

$$= -204 \text{ kcal/mole}$$

This value of -204 kcal/mole is much lower than the value of -140 kcal/mole for formation of gaseous molecules from the separated ions (equation 5–9), and it reveals why highly polar substances are typically solids at room temperature. It also shows that the many-nucleus bonding in the solid is stronger than the two-nucleus bonding in the vapor, just as we should predict.

SUGGESTIONS FOR ADDITIONAL READING

Gray, H. B., *Chemical Bonds*, Benjamin, Menlo Park, Calif., 1973.

Companion, A., *Chemical Bonding*, McGraw-Hill, New York, 1964.

Pimentel, G. C., and Spratley, R. D., *Understanding Chemistry*, Chaps. 13, 14, and 15, Holden-Day, San Francisco, 1971.

PROBLEMS

1. What is the average kinetic energy of an H_2^+ ion at 25 °C in a gas containing H_2^+ species? Compare this energy with the bond energy of H_2^+. Do you expect to find many of the H_2^+ species dissociated into hydrogen atoms and protons at 25 °C?
2. If the electron is at a lower energy for H_2^+ than for the H atom, perhaps it would be at an even lower energy for H_3^{2+}, where it can simultaneously associate with three nuclei. Place three protons on a straight line with 1.06 Å between each pair. Put an electron density corresponding to one-half electron exactly halfway between each nuclear pair. What potential energy do you calculate for this arrangement? Is a linear H_3^{2+} species likely to exist?
3. If the waves in Figure 5–4(b) are moving in opposite directions, the situation is that of the vibrating violin string, as shown in Figure 3–8. One result is permanent nodes (points at which there is no vertical motion at any time), such as the center of the string in Figure 3–8(b).

Locate some permanent nodes in Figure 5–4(b). Locate some points at which the vertical displacement will maximize, as the center of the string in Figure 3–8(a).

*4. The simple addition of atomic orbitals provides a model for molecular orbitals. However, it fails to account for the effect of a second nucleus on atomic orbitals. This effect leads to an electron density between the two nuclei greater than the density for the simple addition. What is this effect?

5. The regions of a diatomic molecule are sometimes referred to in terms of **bonding space** and **antibonding space.** Describe where these spaces are located.

6. Construct an arrangement of the four particles in H_2 analogous to the locations of the three particles in H_2^+ in Figure 5–2. Make your arrangement with an $r_{AB} = 0.76$ Å and place the two electrons in positions that you believe should lead to a low potential energy. Calculate the potential energy in this two-electron bond. Does this calculation indicate that the two-electron bond is stronger than the one-electron bond of H_2^+?

7. The text reports the bond energy of Li_2 as 25 kcal/mole. The bond energies for the diatomic molecules Na_2, K_2, Rb_2, and Cs_2 are -17, -12, -11, and -10 kcal/mole, respectively. Explain this trend.

*8. Why doesn't Figure 5–7 show molecular orbitals derived from such overlaps as $2p_z - 2p_x$, and $1s - 2s$?

9. Draw a figure comparable with those in Figure 5–8, showing favorable locations of the four bonding electrons of C_2. Try to come as close to the most favorable arrangement as you can without performing any calculations.

10. An intermediate species that exists in the conversion of O_2 to water in living systems is superoxide, O_2^-. What is the electronic configuration of this species? What is its bond order?

11. Consider the validity of the following statement: Since the bond energy is much greater for N_2 than for F_2, the first ionization potential must be much greater for N_2 than for F_2.

12. The three-electron species He_2^+ is stable with respect to splitting into He and He^+. However the three-electron species HeH is not stable. Why does one of these species form and not the other?

13. The dipole moment of HCl is 1.08 Debye and the bond length is 1.27 Å. What is δ for this molecule? Refer to the ionization potentials listed in Table 4–3 and predict whether the $+\delta$ resides on the H or Cl portion of HCl. Be prepared to be surprised in problem 18.

14. Which atomic orbitals are involved in the formation of the bonding molecular orbital in a molecule of LiF? Using Table 4–3, predict whether δ for LiF is larger or smaller than δ for LiH?

15. Why doesn't NaCl solid consist of Na^{2+} ions and Cl^{2-} ions? How do we know from Faraday's experimental work that it doesn't?

16. What ions are present in the following ionic solids: Na_2O, Na_2S, $MgCl_2$, MgO, Al_2O_3? Give the electronic configurations of each of these ions.

17. Using the data in Table 4–3, we might predict that the electron affinity of fluorine is greater than the electron affinity of chlorine. However, A is 70.5 kcal/mole for fluorine, compared with 83.4 kcal/mole for Cl. What factor did we neglect in making our faulty prediction? Does the larger A value for Cl mean that the NaCl bond is stronger than the NaF bond in the two gaseous molecules? Explain.

18. In the calculation of ΔE for reaction 5–6, we showed that it is reasonable to view NaCl(g) as an Na^+ ion and a Cl^- ion in contact with each other at an internuclear separation of 2.38 Å. Repeat the calculation for

$$H(g) + Cl(g) \rightarrow HCl(g)$$

where the internuclear separation in HCl(g) is 1.27 Å. Ionization potentials are given in Table 4–3 and the electron affinities of H and Cl are 17.3 and 83.4 kcal/mole, respectively. Try both the H^+—Cl^- pair and the Cl^+—H^- pair. The measured bond energy of HCl is 103 kcal/mole. Is the ionic picture a good one for HCl?

19. Examine Figure 5–14 and verify that if this arrangement is extended in three dimensions, every sodium ion has six chloride ions as nearest neighbors, and each chloride ion has six sodium ions as nearest neighbors. Verify that each ion has twelve next-nearest neighbors, and that the separation distance is $(2.83)(2)^{1/2}$ Å.

20. Corundum (Al_2O_3) is one of the most stable ionic solids known. An enormous amount of energy is liberated when it is formed from Al(s) and O_2(g). In the text we considered the factors that cause NaCl to be a stable ionic solid. Which of these factors is responsible for the great stability of Al_2O_3(s)?

6/ Molecular Structure

We are now in a position to consider molecules that have more than two nuclei. Some important questions to answer are the following:

1. When two given elements form a compound can we make a correct prediction of the formula of that compound?
2. Can we make a correct prediction of the geometric arrangement of nuclei in molecules of that compound?
3. Can we deduce the polarities of the bonds in those molecules?

The answer to each of these questions, we shall see, is in the affirmative.

6–1. Molecular Formulas

Atoms such as hydrogen and lithium, when bonded to other atoms in the gas phase, are invariably found to bond to only one other atom at a time. Thus we find H_2, Li_2, and LiH, and not H_3, Li_4, LiH_2, or Li_2H, for example. The reason for the restriction to a single bond between two nuclei is that both H and Li possess a single valence orbital. In LiH, the hydrogen atom can accept only a single electron in the low-energy region characterized by the 1*s* orbital, and this is the reason Li_2H does not form. Likewise the fluorine atom typically participates in only a single two-nucleus bond in the gaseous state, since the only low-energy, unfilled orbital is the 2*p* and it can accept only a single electron.

All of the elements between Li and F have either two or more valence electrons that can bond to atoms possessing low-energy acceptor orbitals, or they contain two or more vacancies in such acceptor orbitals. For example, the electronic configuration of the carbon atom is $1s^2 2s^2 2p^2$, and the electronic energies of the valence electrons are -450 kcal/mole for the 2*s* electrons and -247 kcal/mole for the 2*p* electrons. The energy of the fluorine 2*p* level is -431 kcal/mole. Thus, each of the four carbon valence electrons provides the basis for bonding to a single fluorine atom, and we should predict the existence of the molecule CF_4. Indeed, CF_4 is a very stable molecule and the only stable molecule known containing a single carbon atom and only fluorine atoms bonded to it.

Table 6–1 lists the valence electron energies and electronic configurations of the atoms from hydrogen through neon. Let us use this table to examine the abilities of the atoms in it to combine with fluorine. The molecules HF and LiF have already been discussed. In the case of helium, the filled He 1*s* level lies somewhat below the F 2*p* level. Bonding between He and F would

Table 6–1
Energies (kcal/mole) of Electrons in the Valence Orbitals together with Electronic Configurations of the Atoms H through Ne

Atom	Configuration	E_{1s}	E_{2s}	E_{2p}
H	$1s^1$	−313		
He	$1s^2$	−565		
Li	$1s^2 2s^1$		−124	
Be	$1s^2 2s^2$		−214	
B	$1s^2 2s^2 2p^1$		−333	−116
C	$1s^2 2s^2 2p^2$		−450	−247
N	$1s^2 2s^2 2p^3$		−588	−302
O	$1s^2 2s^2 2p^4$		−745	−367
F	$1s^2 2s^2 2p^5$		−1070	−431
Ne	$1s^2 2s^2 2p^6$		−1116	−491

involve electron density from He moving into the relatively high energy fluorine region. No species containing He—F bonds has ever been detected. Helium does not function as an electron donor in chemical bonding. Helium also does not serve as an electron acceptor in molecular bonding, since the helium 2*s* orbital is the lowest-energy acceptor orbital, and the addition of an electron to He to produce He^- has a positive ΔE (helium has a negative electron affinity because the repulsion between the three electrons of He^- is greater than the electron-nuclear attraction). **No molecule has ever been found that contains helium, since helium can serve neither as an electron donor nor as an electron acceptor in bond formation.**

The 2*s* level of Be is considerably higher in energy than the fluorine 2*p* level and the compound BeF_2 is predicted to form with beryllium donating two electrons to bond formation. For boron, there are three electrons at energies higher than the fluorine 2*p*, and the predicted compound is BF_3. The prediction of the molecule CF_4 for the combination of carbon and fluorine has already been made.

With nitrogen, there are five electrons in the valence levels and we might predict that NF_5 is the molecule that forms. However, such is not the case and NF_3 contains the largest number of fluorines to which a nitrogen atom can bond. The reason for this limitation is that the F 2*p* orbital that accepts electron density from other atoms already contains an electron, and in bond formation this electron is shared by the atom bonded to fluorine. In NF_5, there would be five N—F bonding orbitals, each of which would contain two electrons. The total of ten electrons could not be accommodated in the sum of the single 2*s* and the three 2*p* orbitals. A maximum of eight electrons can be accommodated in the nitrogen valence orbitals or in the valence orbitals of any atom from lithium through neon. The limitation at eight electrons applies exclusively to those eight elements, and it is termed the **octet rule.** The rule limits the number of F atoms bound to an N atom to three, which yields eight valence electrons around N (five from the N atom and one each from the three fluorines). Application of the octet rule to oxygen results in the prediction that the molecule that forms is OF_2, and such is the case. With neon, the octet is complete in the atom itself, and like helium, neon has never been found in molecular form.

In our subsequent consideration of the combinations of atoms in many molecules, we shall see that in most cases a table of atomic electron configurations and orbital energies is quite sufficient to allow us to predict molecular formulas.

6–2. Molecular Geometry

In our study of molecular geometry, we shall center on the compounds formed by hydrogen with the elements from beryllium to oxygen, because several very important compounds are found in this group. The compounds are termed the **hydrides** of the five elements. The molecular formulas of the molecules we shall consider are all simply related to the formulas of the corresponding fluorides, since hydrogen, like fluorine, can both accept and donate a single electron in the formation of a single bond. The formulas of the hydrides to be considered are BeH_2, BH_3, CH_4, NH_3, and H_2O.

Beryllium Hydride—A Linear Molecule

The species BeH_2 contains two sigma bonds just as BeF_2 does. Such a triatomic molecule can have a bent structure in which the angle between the two Be—F internuclear axes (this angle is termed the **bond angle**) is less than 180°, or the molecule can be **linear** with a bond angle of 180°. The fundamental principle of bonding and molecular structures that allows us to decide between these two possibilities is that **molecules exist in the structure of lowest electrostatic potential energy available to the constituent particles.** The most favorable regions for the valence electrons in BeH_2 are the bonding regions between the Be and each of the two hydrogen atoms. Each of the bonding molecular orbitals concentrated in these regions can accommodate a maximum of two electrons, and the beryllium valence orbitals contain a total of four electrons (two from beryllium and one each from two hydrogens).

The two bonding electron pairs strongly repel each other and they are predicted to be located as far from each other as possible in the BeH_2 molecule. This occurs in the linear geometry with a bond angle of 180°. Figure 6–1 shows two common representations of the structure and bonding in BeH_2. Figure 6–1(a) is a simple line representation, in which a line between two atoms denotes a single two-electron bond. Figure 6–1(b) is a **Lewis dot representation** (first devised by the American chemist G. N. Lewis), in which each dot represents a valence electron. The two dots between each pair of atoms denote a two-electron bond in each case.

Figure 6–1 Two representations of the structure and bonding in the beryllium hydride molecule (BeH_2), (a) line representation and (b) dot representation.

H—Be—H (a)   H:Be:H (b)

Beryllium hydride is not a common species, but there are a large number of common triatomic molecules that have the linear geometry. We shall subsequently develop a general procedure for predicting when this geometry occurs.

Borane—A Triangular Molecule

In the same way that BF_3 is formed, boron forms a hydride with the formula BH_3, and the species is called **borane.** Borane is not stable under normal conditions and it forms **diborane,** B_2H_6. Nevertheless, this does not prevent us from considering the geometry of BH_3.

The number of valence electrons around the B in BH_3 is six, divided into three sigma bonding pairs. The geometry that offers the minimum repulsion between these three bonded pairs places the boron nucleus in the center of an equilateral triangle with hydrogen nuclei defining the corners of the triangle. All four nuclei are in the same plane and the bond angle is 120°. The line and dot representations are given in Figure 6–2. The triangular geometry is found in several important species and we shall see shortly how to predict its occurrence.

Figure 6–2 Two representations of the structure and bonding in the borane molecule (BH_3).

Methane—A Tetrahedral Molecule

The molecule CH_4 is one of the most common molecules. The species is called **methane** and it is the principal component of the fuel natural gas. The molecule contains four sigma bonding electron pairs and the geometry that minimizes the repulsion between these pairs is the tetrahedral arrangement shown in Figure 6–3(a). A convenient way to represent a tetrahedron is to locate its four corners at alternating corners of a cube, as shown in the figure. This is particularly convenient for calculating the bond angle θ,

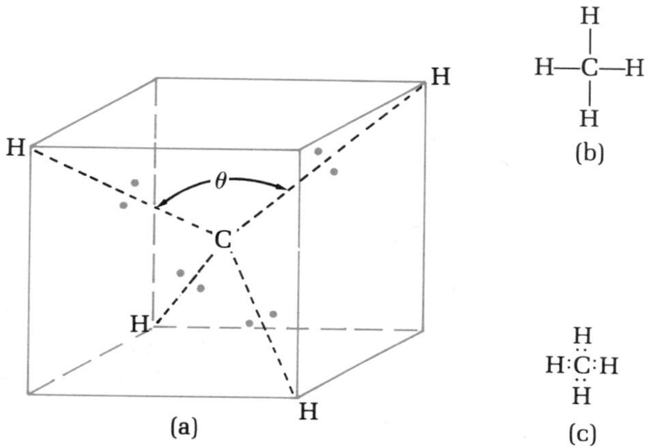

Figure 6–3 Representations of the structure and bonding in the methane molecule (CH_4).

shown in the figure. The bond angle is calculated from the cosine law and the fact that the H—H distance, r_{H-H} in Figure 6–3(a), is exactly $(2\sqrt{2}/\sqrt{3})$ times the C—H distance, r_{C-H}.

$$r^2_{H-H} = 2r^2_{C-H} - 2r^2_{C-H}\cos\theta$$

$$\tfrac{8}{3} = 2 - 2\cos\theta$$

$$\theta = 109.45°$$

The line and dot representations of CH_4 are shown in (b) and (c) of Figure 6–3. No attempt has been made in these representations to show the tetrahedral geometry; such representations do show the bonding, however, and are commonly used.

Ammonia—A Pyramidal Molecule

The **ammonia** molecule, NH_3, has eight valence electrons surrounding the nitrogen. However, only six of these are bonding; the other two are a **nonbonded pair** (sometimes called a **lone pair**). The molecular geometry depends on the relative repulsions nonbonding valence electron pairs exert as compared with bonding pairs. That is, the geometry is determined by the relative sizes of the electron dot density 90 percent contour surfaces for nonbonded pairs and for bonded pairs. Since bonded pairs are simultaneously bound by two nuclei, we should expect the electron density to be highly concentrated near the internuclear axis, and the 90 percent contour surface should be smaller than the contour surface for nonbonded electrons. Thus, we predict, the three bonded pairs are forced closer together in NH_3

by repulsion between them and the nonbonded pair, and the bond angle is less than 109.45°. The experimentally determined bond angle for NH_3 in gaseous NH_3 is 107.3°, and our prediction is borne out. The pyramidal structure along with line and dot representations are shown in Figure 6–4.

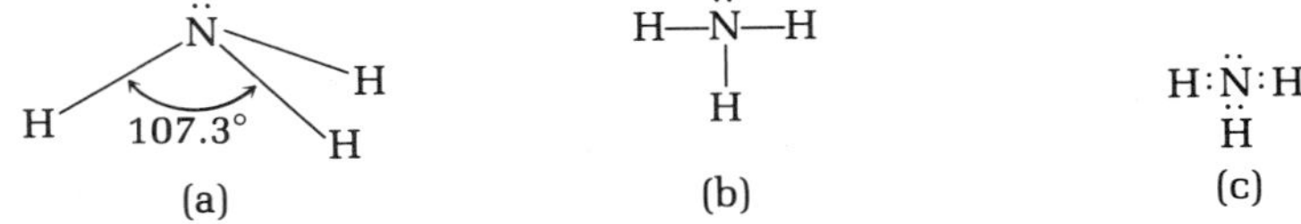

Figure 6–4 Three representations of the structure and bonding in the ammonia molecule (NH_3).

Water—A Bent Triatomic Molecule

For the water molecule, the octet rule dictates that the molecular formula be H_2O, and the eight valence electrons consist of two bonding pairs and two nonbonding pairs. In an extension of the reasoning employed with NH_3, we predict that two nonbonded pairs repel each other more than the bonded pairs do. As a consequence, the H—O—H bond angle in isolated water molecules in water vapor should be less than 109.45°. The bond angle is found experimentally to be 104.5°. Representations of the water molecule are given in Figure 6–5.

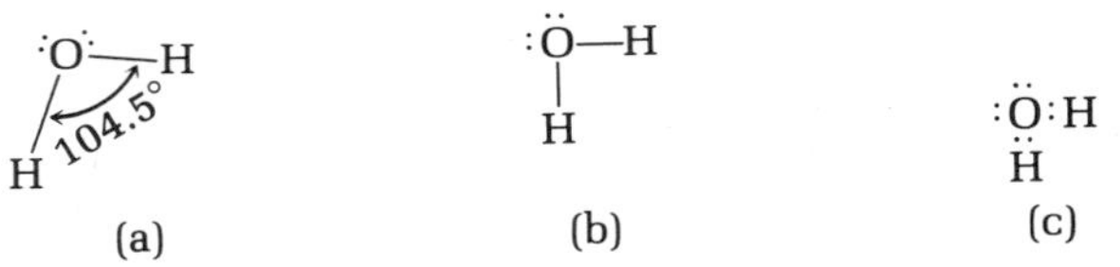

Figure 6–5 Representations of the structure and bonding in the water molecule (H_2O).

O—H
H
104.5°
(a)

:O—H
H
(b)

:O:H
H
(c)

6–3. Prediction of Molecular Geometry and Bonding

Geometry in a molecule, then, is determined by the repulsion between bonded (and in cases where they exist, nonbonded) electron pairs. We have not yet considered a molecule containing more than two nuclei in which pi bonding occurs. However, **pi-bonding electrons play no role in determining the geometry of a molecule.** Since only sigma-bonded electrons and nonbonded electrons determine the geometry, we must devise a general procedure for counting the numbers of each. To do this we shall *mentally* destroy all bonding in the species under investigation and systematically reintroduce only sigma bonding. The procedure for the prediction of molecular structure and bonding proceeds through well-defined steps. These steps will be applied to the **carbonate ion,** CO_3^{2-}. This ion is commonly found in ionic solids, and it is one of the principal constituents of limestone.

Step 1:

Strip as many valence electrons from the central atom as possible and add them to the valence orbitals of the atoms bonded to it (these surrounding atoms are termed **ligands**). The central atom can usually be identified as the one that has the smallest subscript in the formula. In the case of CO_3^{2-}, the central atom is carbon. The carbon atom contains four valence electrons. Each oxygen atom can accept two electrons because there are two vacancies in the oxygen valence orbitals. Thus, it is possible to give the four carbon electrons and the two extra electrons of the doubly minus ion to oxygen to mentally decompose CO_3^{2-} to C^{4+} and $3O^{2-}$. In these separated species, the valence electrons of carbon have been completely removed and the valence orbitals of oxygen are completely filled. It must be emphasized again that electron stripping and separation into ions is a mental process only and bears no necessary resemblance to any actual process.

Electrons that are nonbonding will remain on the central atom after the stripping process, and their number is easily obtained. In the case of CO_3^{2-}, there are no nonbonding carbon valence electrons.

Step 2:

Reintroduce only sigma bonding by having each ligand supply two valence electrons to sigma-bond formation with the central atom. In the case of CO_3^{2-}, this means that three sigma bonds are formed.

Step 3:

Count the number of valence electrons surrounding the central atom after reintroducing the sigma bonding. This is the number of nonbonding electrons plus twice the number of ligands. The mutual repulsion between sigma-bonded and nonbonded electron pairs determines the molecular geometry. In the case of CO_3^{2-}, the C^{4+} ion possesses no nonbonded valence electron pairs, and six valence electrons are reintroduced by sigma bonding, so that the sum obtained in this step of the procedure is six.

Step 4:

The conclusions reached in this step depend on the numbers of nonbonded and identical bonded electron pairs. If all of the valence electrons counted in Step 3 are contained in bonded pairs involving identical ligands, precise geometries and bond angles can be predicted; these are given in Table 6–2. We are familiar with the linear, triangular, and tetrahedral

Pair Number	Geometry	Bond Angle
2	Linear	180°
3	Triangular	120°
4	Tetrahedral	109.45°
5	Trigonal-bipyramidal	120° and 90°
6	Octahedral	90°

Table 6–2
Geometries and Bond Angles as a Function of the Number of Identical Sigma-Bonded Electron Pairs

geometries of BeH_2, BH_3, and CH_4, respectively. The trigonal-bipyramidal and octahedral geometries are shown in Figure 6–6. In the trigonal-bipyramidal arrangement, three of the ligands form the corners of an equilateral

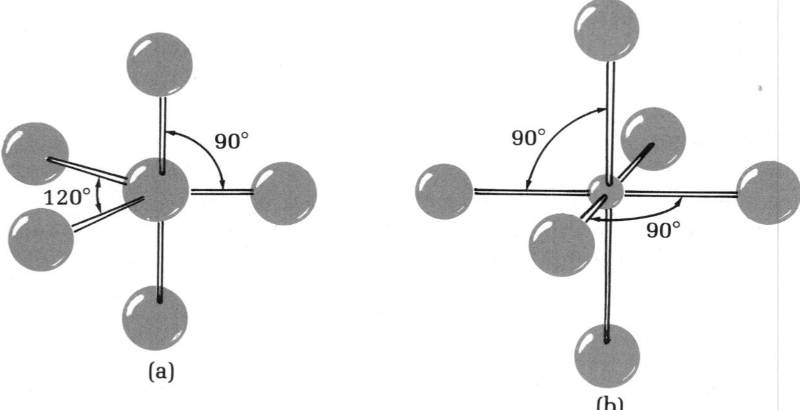

Figure 6–6
(a) Trigonal-bipyramidal and (b) octahedral geometries of the atomic arrangements found in some molecules.

triangle, and the bond angle in this plane is 120°. The remaining two ligands form a linear arrangement about the central atom, and the bond angle between them and the triangular plane is 90°. In the octahedron, the six ligands are identically situated relative to the central atom and all bond angles are 90°. In the case of CO_3^{2-}, there are three identical bonded pairs and the geometry is that previously met with BH_3. All nuclei are in one plane and the oxygens define the corners of an equilateral triangle.

If the valence electrons consist of both bonded pairs and nonbonded pairs, the geometry is approximately that given by Table 6–2, where the pair number refers to the sum of bonded and nonbonded pairs. However, the geometry will be somewhat distorted from the precise geometries given in the table because nonbonded pairs are, in general, larger than bonded pairs. We shall illustrate such a case at the conclusion of this complete consideration of CO_3^{2-}.

Step 5:

Once the geometry has been determined, the pi bonding consistent with that geometry can be reintroduced. Pi bonding invariably occurs if the sum of the valence electrons as given in Step 3 is less than eight. With CO_3^{2-}, this valence electron sum is six, and pi bonding occurs. The number of pi bonds equals one-half the difference between eight and the electron sum as given in Step 3. In the case of CO_3^{2-}, this means one pi bond. The pi bonding must occur because a carbon atom will never exist in a molecule with only six electrons in its valence orbitals. These orbitals represent a low-energy region for electrons and they can and will accept electron density from the ligands.

Resonance

While the procedure that we have just carried out gives a clear picture of the structure and bonding of CO_3^{2-}, there is a problem in depicting these bonds in dot or line representations. Examples of representations we might draw are given in (a) and (b) of Figure 6–7. These show one pi bond between the carbon and the oxygen numbered (1) and full negative charges on oxygens

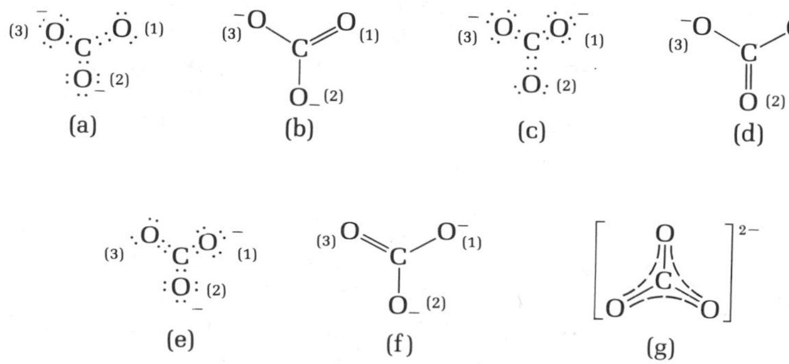

Figure 6–7 Representations of the structure and bonding of the carbonate ion (CO_3^{2-}).

numbered (2) and (3). These representations cannot be correct, since all three oxygens are identical; we could just as well have placed the pi bond between the carbon and oxygens (2) or (3), as depicted in Figure 6–7, in (c) and (d), and (e) and (f), respectively. The true bonding description lies midway between the three representations in Figure 6–7. Each C—O pi bond contains electron density amounting to $\frac{2}{3}$ of an electron and thus each C—O bond order is $\frac{4}{3}$ arising from [($\frac{1}{2}$)(two sigma-bonded electrons plus $\frac{2}{3}$ pi-bonded electrons)]. The true representation is commonly termed a **resonance hybrid,** in this case of the three **canonical resonance forms** depicted in Figure 6–7. (The term *resonance* is unfortunate, since students often erroneously interpret it as indicating an oscillation between canonical resonance forms.) The true bonding and structure are often indicated by the dashed lines, as in Figure 6–7(g), indicating equal pi bonding between the central carbon and all three oxygens. This situation of pi bonding involving more than two nuclei at a time is common in chemistry.

Examples of Molecular Geometry Predictions

Now that we have illustrated the procedure with CO_3^{2-}, let us consider an example involving nonbonded electrons. As an example we shall use ammonia, a molecule that we have already considered by a less systematic procedure.

The first step splits NH_3 into N^{3+} and $3H^-$. Each hydrogen atom can accept only one valence electron; only three of the five nitrogen valence electrons can be removed, leaving two nonbonding electrons. In Step 2, we reintroduce three sigma bonds (three bonding electron pairs) for a total of eight valence electrons in Step 3. The geometry is close to tetrahedral, but the repulsion by the lone pair results in the bond angle of 107.3° instead of the tetrahedral angle of 109.45°. The number of valence electrons in Step 3 is eight and hence NH_3 contains no pi bonding. The total bond order is three, from three sigma bonds.

As a second example, we pick the **azide ion,** N_3^-. Step 1 splits N_3^- into $N^{5+} + 2N^{3-}$, in which the valence electrons have been completely stripped from the central atom and the valence orbitals of the ligands have been filled. In Step 2, we reintroduce two sigma bonds and two bonding electron pairs so that the number of valence electrons in Step 3 is four. The geometry must be linear with a 180°-bond angle, as depicted for BeH_2 in Figure 6–1. The number of electrons in Step 3 is less than eight, and there are two pi bonds in N_3^-. A problem arises again in the line and dot representations of this ion. Figures 6–8(a) and (b) show a triple bond between atoms (1) and (2) and the negative charge on atom (3). Figures 6–8(c) and (d) show the reverse of this situation, and Figure 6–8(e) shows the resonance hybrid representation. There are two half pi bonds between each N—N pair and the negative charge is equally divided between the right and left halves of the molecule.

[N≡N≡N]⁻

(e)

Figure 6–8
Representations of the structure and bonding in the azide ion (N_3^-).

Limitation of the Octet Rule

Why are the numbers 5 and 6 included in Table 6–2 if the octet rule limits the number of valence electron pairs to four? As mentioned previously, the octet rule holds for the atoms lithium through neon because a maximum of eight electrons can occupy the sum of the 2*s* and 2*p* orbitals. This restriction does not hold for any of the elements with atomic numbers larger than 10. For example, the sum of the 3*s*, 3*p*, and 3*d* orbitals can hold a maximum of 18 electrons. Thus, while NF_5 does not exist, the molecule PF_5 does exist, and it possesses the trigonal-bipyramidal geometry. (We do not venture into an investigation of molecules involving atoms with atomic numbers greater than ten in this chapter. However, Chapters 8 and 9 are heavily devoted to such investigations and there we shall systematically investigate the structure and bonding in PF_5, and also in many other molecules.)

6–4. Hybridization

Orbital Hybridization in the Tetrahedral Geometry

Suppose that we wish to deduce the polarity of the C—H bonds in the methane molecule, CH_4. The first thing we must know is which atomic orbitals are participating in bonding molecular orbital formation in this molecule. Table 6–1 reveals that the hydrogen 1*s* and the carbon 2*s* and 2*p* orbitals are of comparable energies. The geometry of CH_4 is tetrahedral and consequently we must examine the overlap between hydrogen 1*s* contours placed at the corners of a tetrahedron with carbon orbital contours centered at the center of the tetrahedron. All carbon 2*s* and 2*p* orbitals showing positive net overlap with the four hydrogen orbitals participate in the bonding. Figure 6–9 shows the bonding combinations of each of the four carbon orbitals with the four hydrogen 1*s* orbitals. The hydrogen 1*s* spheres are represented by either plus or minus signs located at alternate corners of a cube that contains a carbon nucleus at the center. The carbon 2*s* orbital is represented in Figure 6–9(a) by colored plus and minus signs, and Figures 6–9(b), (c), and (d) show the carbon $2p_x$, $2p_y$, and $2p_z$ orbital lobes with colored plus and minus signs. For instance, in Figure 6–9(b), the carbon 2*p* wave function in the front half of the cube is positive and in the rear half negative.

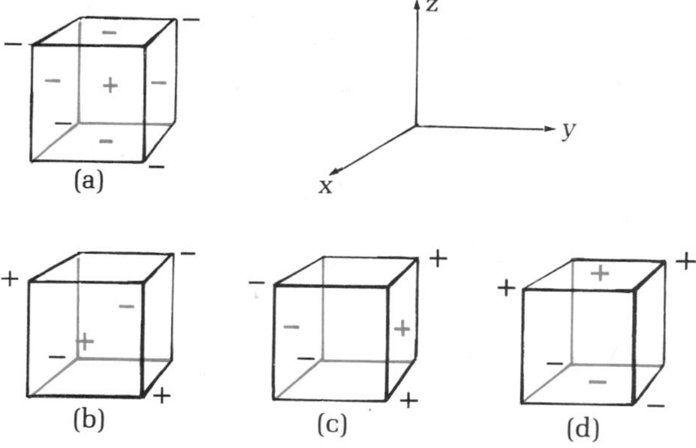

Figure 6–9
Signs and positions of orbitals involved in bonding in the methane molecule (CH_4). The signs in black indicate hydrogen 1*s* orbitals and the signs in color the carbon orbitals. The carbon orbitals depicted are (a) 2*s*, (b) $2p_x$, (c) $2p_y$, (d) $2p_z$.

Figure 6–9 shows that all four carbon atomic orbitals participate in the sigma bonding in methane. Sigma bonding that involves more than one orbital on the same atom defines the **hybridization** of the atom. In the case of methane (or any other molecule characterized by the tetrahedral geometry), the hybridization is termed **sp³**, indicating that the carbon 2*s* and each of the three 2*p* orbitals contribute to the bonding.

With this picture of the bonding in methane, we are in a position to consider the polarity of the C—H bonds. Figure 6–10 shows the hydrogen and carbon atomic valence orbital energies and electron occupations for forming

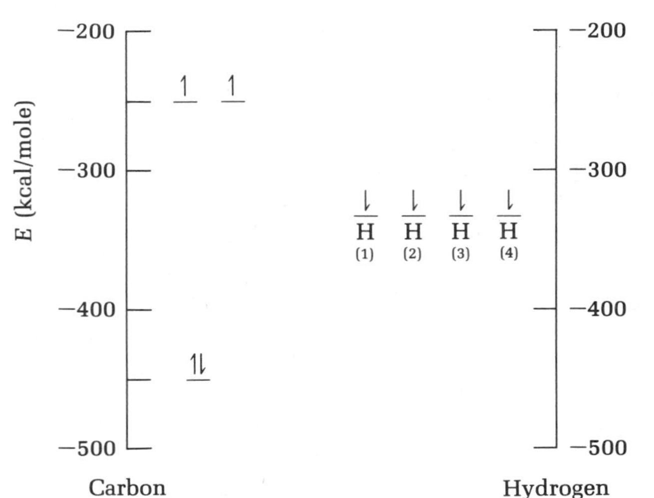

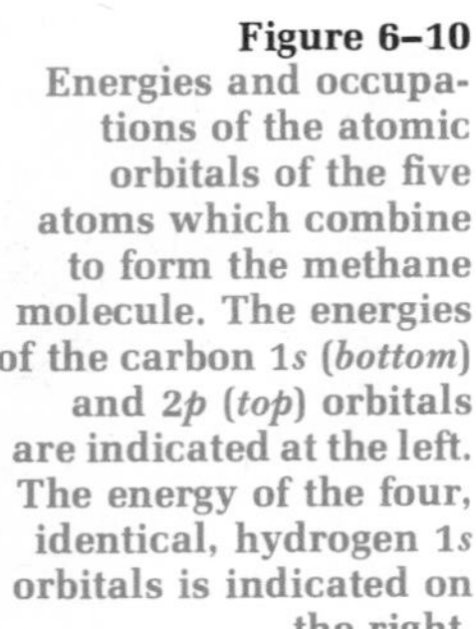
Figure 6–10 Energies and occupations of the atomic orbitals of the five atoms which combine to form the methane molecule. The energies of the carbon 1*s* (*bottom*) and 2*p* (*top*) orbitals are indicated at the left. The energy of the four, identical, hydrogen 1*s* orbitals is indicated on the right.

methane. The carbon 2*s* orbital is considerably lower in energy than the hydrogen 1*s* is, and the electron density in the 2*s* bonding orbital depicted in Figure 6–9(a) is strongly polarized in the direction of the carbon. On the other hand, the hydrogen 1*s* energy level is lower than the carbon 2*p* level, and the electron density in the three molecular orbitals depicted in Figures 6–9(b), (c), and (d) is polarized toward the four hydrogens. Thus each C—H bond involves opposite polarizations in the molecular orbitals that contribute to it. The net effect is the same as though the four carbon atomic orbitals possess a common energy that is their average. This average energy is $[3(-247)-450]/4 = -297$ kcal/mole. This energy is very close to the energy of the hydrogen 1*s* orbital, and the C—H bonds in methane have very little polarity. This lack of polarity is characteristic of C—H bonds in most compounds, and it results in relatively low chemical reactivities of compounds that contain only carbon and hydrogen (hydrocarbons).

Orbital Hybridization in the Triangular Geometry

The tetrahedral geometry is common in chemistry, but so are the triangular and linear geometries of Table 6–2. The tetrahedral geometry is characterized by sp^3 hybridization of the orbitals of the central atom. What hybridization is characteristic of the triangular geometry? The molecule BF_3 possesses the triangular geometry, and Figures 6–11(a), (b), and (c) show the atomic orbital overlaps involved in the sigma bond formation. The participating orbitals are a single fluorine 2*p* orbital from each fluorine atom and the $2p_x$, $2p_y$, and 2*s* boron orbitals. All three orbitals show net bonding overlap with the three fluorine orbitals. Figure 6–11 shows 90 percent contour surfaces for each of the orbitals. Since the boron 2*s* and two boron 2*p* orbitals participate in the sigma bonding, this molecule (and any other

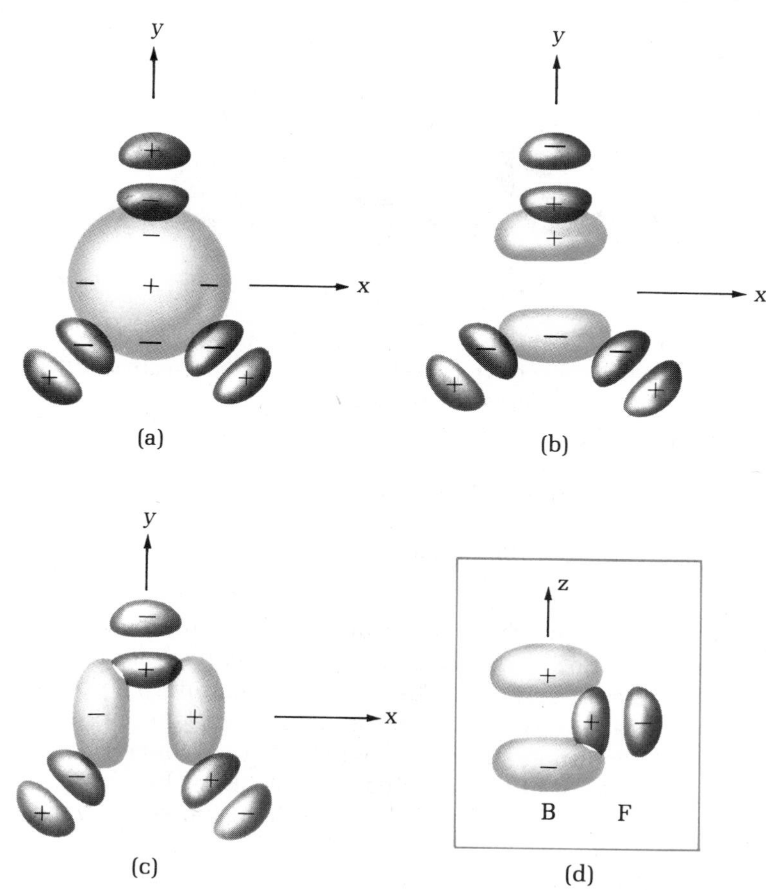

Figure 6–11
(a,b,c) Atomic orbital overlaps involved in sigma bond formation in BF_3. (d) Nonbonded overlap of the boron $2p_z$ orbital and a sigma bonding fluorine orbital.

molecule possessing the triangular geometry of BF_3) is characterized by **sp² hybridization.** The boron $2p_z$ orbital is perpendicular to all three of the fluorine orbitals that participate in the sigma bonding. Figure 6–11(d) shows the overlap between this boron $2p_z$ orbital and one of those three fluorine orbitals. The region of bonding overlap is exactly canceled by a region of antibonding overlap and the interaction is nonbonding. However, there are fluorine $2p_z$ orbitals and they contain two electrons apiece. Since the boron $2p_z$ orbital is empty, it represents a low-energy region for electrons, and B—F pi bonding occurs. The boron $2p_z$ orbital and the three fluorine $2p_z$ orbitals are shown in Figure 6–12. The orbitals are not drawn in an overlapping fashion and they may therefore be clearly seen.

The BF_3 molecule provides another example of resonance. The molecule has the same number of electrons and nuclei as CO_3^{2-} (it is said to be **isoelectronic** with CO_3^{2-}) and the structure and bonding in BF_3 can be obtained by replacing C by B and O by F in Figure 6–7.

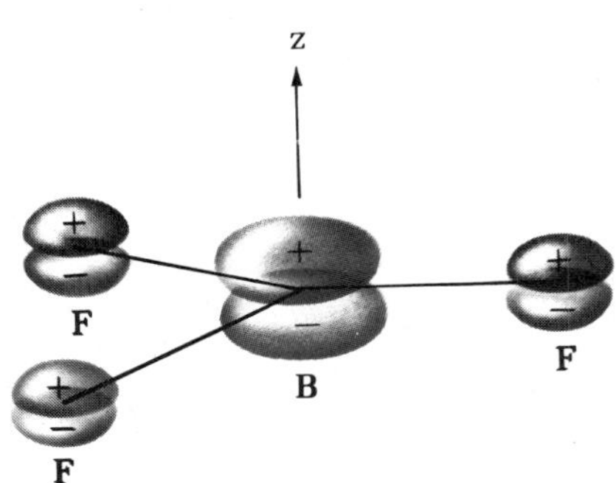

Figure 6–12
Contours of the boron and fluorine $2p$ orbitals whose overlap results in pi bonding in boron trifluoride.

The qualitative nature of the B—F bond polarity can be judged by reference to Table 6–1. The average of the boron $2s$, $2p_x$, and $2p_y$ orbital energies is $[-333 + 2(-116)]/3 = -188$ kcal/mole. The fluorine $2p$ energy is -431 kcal/mole, and thus the B—F sigma-bond density is highly polarized toward the fluorines, like the pi electron density.

Orbital Hybridization in the Linear Geometry

The species BeH_2, BeF_2, and N_3^-, among others, show linear geometry. Figures 6–13(a) and (b) show the orbital overlap involved in the sigma-bond formation in BeF_2. The atomic orbitals involved are the fluorine $2p_z$ and the beryllium $2s$ and $2p_z$. The fluorine $2p_x$ and $2p_y$ orbitals are both perpendicular to the axis of the BeF_2 molecule, and these fluorine orbitals participate in pi bonding rather than sigma bonding. Since an s and a single p orbital on the central atom participate in the sigma bonding in BeF_2 (and in any other species with the linear geometry), the molecule is characterized by **sp hybridization.**

Figure 6–14 shows one of the two sets of p orbitals in BeF_2 that overlap to

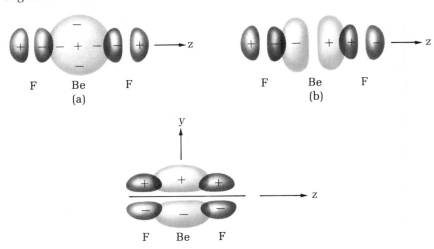

Figure 6–13
Atomic orbital overlaps involved in the formation of the sigma bonds in the BeF_2 molecule.

Figure 6–14
The overlap of $2p_y$ orbitals to yield pi bonding in BeF_2.

form pi bonding. The set shown comprises the $2p_y$ orbitals. The $2p_x$ orbitals in the other set form an identical pi-bonding arrangement, but they are rotated by 90° about the internuclear axis.

The molecule BeF_2 is isoelectronic with N_3^-, and its structure and bonding can be obtained by substituting Be for the central N and F for each ligand N in Figure 6–8.

Hybrid Atomic Orbitals

If we were now to consider the polarity of the Be—F sigma bonds in BeF_2, we should proceed as with CH_4 and BF_3 to obtain the average energy of the Be orbitals involved in the sigma bonding. Since it is the average orbital that concerns us in BeF_2, BF_3, and CH_4 (or in any other case in which hybridization occurs), why don't we construct pictures of such average atomic orbitals and use them to describe the bonding? These average orbitals are called **hybrid atomic orbitals.**

The two *sp* hybrid orbitals are the average of an *s* and a *p* orbital and they are obtained by adding the two orbitals, as shown in Figure 6–15(a), and subtracting one from the other, as shown in Figure 6–15(c). The *sp* hybrid orbital that results from the addition is shown in Figure 6–15(b); and (d) shows the result of subtraction. Figure 6–15(e) shows the arrangement of the negative lobes of the hybrid orbital contours about the nucleus. The linear

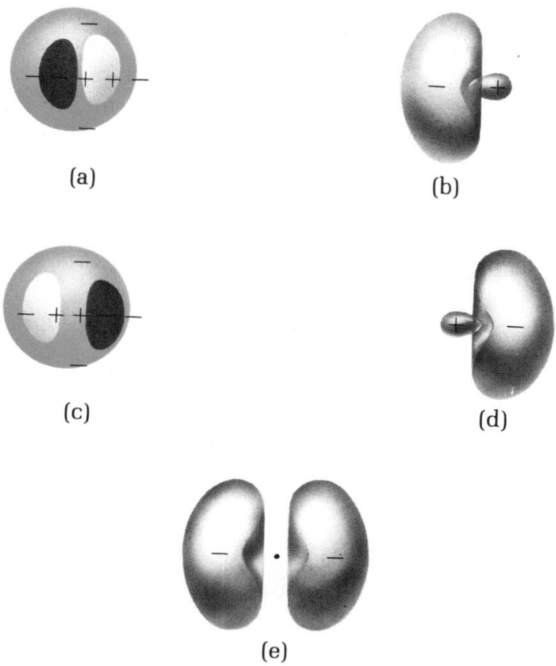

Figure 6–15
Formation of two *sp* hybrid atomic orbitals by the (a,b) addition and (c,d) subtraction of an *s* and a *p* orbital centered on the same nucleus. Figure (e) represents the two negative lobes of the two *sp* hybrid orbitals centered about the nucleus.

geometry associated with the bonding is quite apparent in Figure 6–15(e). This ready appearance of the geometry is the principal reason for the common reference to hybridized atomic orbitals. Figure 6–16 shows the negative lobes of the hybrid orbital contours of the three sp^2 hybrid orbitals and the four sp^3 orbitals. The triangular and tetrahedral geometries, respectively, are readily apparent.

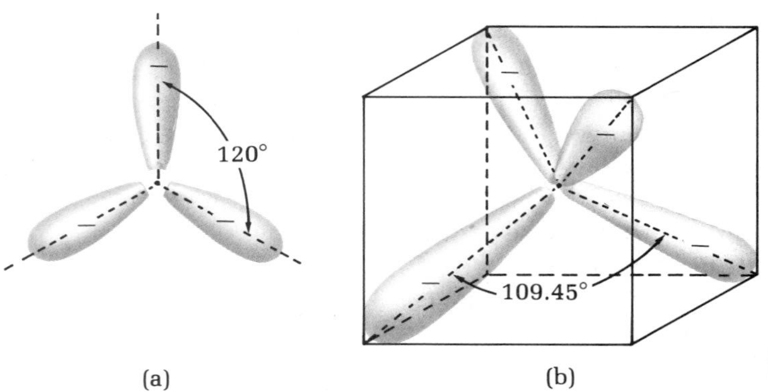

Figure 6–16 The negative lobes of orbital contours for (a) sp^2 hybridization and (b) sp^3 hybridization.

Polarity of the Water Molecule

The most important single compound on earth is probably water, because it is the most abundant constituent of living systems. Much of the important chemistry of water arises from the polarity of its O—H bonds. Let us approximate the H_2O geometry as a tetrahedral arrangement of two bonding and two nonbonding electron pairs around the central oxygen, so that the approximate oxygen orbital hybridization is sp^3. From Table 6–1 we obtain the oxygen atomic orbital energies as −745 kcal/mole for 2*s* and −367 kcal/mole for 2*p*. The average oxygen atomic orbital energy in the combination of orbitals involved in the bonding in water is

$$\langle E \rangle = 0.25(-745) + 0.75(-367) = -461 \text{ kcal/mole}$$

This average energy is much lower than the −313 kcal/mole for the hydrogen 1*s* orbital. As a consequence, the water molecule is highly polarized as $H^{\delta+}—O^{2\delta-}—H^{\delta+}$, and the value of δ is substantial.

6–5. Bond Energies and Electronegativity

Energy Changes in Chemical Reaction

The energy that H_2 molecules absorb when they split to form H atoms is easily understood in terms of the breaking of the chemical bond. Chemical reactions, in general, involve the making and breaking of bonds between atoms. For example, reaction 6–1 involves the breaking of two H—H bonds and an O═O bond with the formation of four O—H bonds:

$$2\ H_2(g) + O_2(g) \rightarrow 2\ H_2O(g) \qquad \textbf{6–1}$$

It would aid our thinking about this reaction if we could consider the O—H bonds to be independent of each other in water. Since the bonding electron pairs in H_2O, and also in a great many other molecules, maintain the maximum possible separation from each other, this model of independent two-atom bonds is widely valid. It is called the **valence bond** model of molecules. It is a very useful model for all molecules that do not show resonance. Resonance necessarily involves electrons that are simultaneously bound to three or more nuclei.

If the O—H bonds in water are independent of each other, this O—H bond energy in H_2O should be equal to the O—H bond energy in any other molecule and we can list an O—H bond energy in a table. Table 6–3 lists this energy along with a number of other energies for commonly encountered bonds. The combustion of hydrogen results in release of a large quantity of

Bond	Energy	Bond	Energy	Bond	Energy
H—H	104	C≡C	200	N═N	100
H—Li	58	C—N	73	N≡N	225
H—C	99	C═N	147	N—O	48
H—N	93	C≡N	213	N—F	65
H—O	111	C—O	86	O—O	51
H—F	134	C═O	176	O═O	118
H—Cl	103	C—F	106	F—F	37
H—Br	87	C—Cl	80	Cl—Cl	58
H—I	71	C—Br	66	Br—Br	46
H—S	82	C—I	55	I—I	36
Li—Li	25	C—S	62	F—Cl	36
C—C	83	C═S	128	S—S	54
C═C	146	N—N	60		

Table 6–3
Bond Energies (kcal/mole)

energy. Let us use Table 6–3 to see why. **The ΔE for reaction 6–1 and for all other reactions is the negative of the sum of the bond energies of the products minus the sum of the bond energies of the reactants.** From Table 6–3 we obtain values in kcal/mole of 111, 104, and 118 for the O—H, H—H, and O═O bond energies respectively. Thus, ΔE for reaction 6–1 is

$$\Delta E = -[4(111) - 2(104) - 118] = -118 \text{ kcal/mole}$$

and 118 kcal of energy is released when two moles of H_2 are burned. Energy is liberated in this reaction because of the O—H bond. The O—H bond is weaker than the O═O bond, but the O═O bond is really two bonds (one sigma and one pi). The average of the two bonds of O_2 is 118/2 = 59 kcal/mole. In the two water molecules formed in reaction 6–1, each of the four single bonds has a bond energy of 111 kcal/mole. Thus reaction 6–1 is largely the substitution of strong O—H bonds for weak O—O bonds. This substitution of strong bonds for weak bonds is the principal factor in most chemical reactions.

Electronegativity

Let us use the bond energies of Table 6–3 to calculate the ΔE of the combustion of methane.

$$CH_4(g) + 2\ O_2(g) \rightarrow CO_2(g) + 2\ H_2O(g) \qquad \textbf{6–2}$$

Carbon dioxide (CO_2) is isoelectronic with N_3^- and it may be considered to have two C═O double bonds. The ΔE for reaction 6–2 is

$$\begin{aligned}\Delta E &= -[2(\text{Bond energy})_{C=O} + 4(\text{Bond energy})_{O-H} \\ &\quad -2(\text{Bond energy})_{O=O} - 4(\text{Bond energy})_{C-H}] \\ &= -[2(176) + 4(111) - 2(118) - 4(99)] \\ &= -164 \text{ kcal}\end{aligned}$$

In reaction 6–2, four C—H single bonds are converted to four O—H single bonds, and two O═O double bonds are converted to two C═O double bonds. The C—H and O═O bonds are nonpolar, while the C═O and O—H bonds are distinctly polar. The ΔE of reaction 6–2 is an example of a very important point. **As the polarity of a bond increases, the strength of the bond increases.**

Bond polarities arise from gaps between the energies of the atomic orbitals involved in the bonding, as we showed in the case of LiH. It would be very convenient if we had some way to translate atomic orbital energy differences or their equivalent into bond energies. The American chemist Linus Pauling did this effectively in the 1930s when he defined and reported the relative **electronegativities** of the atoms. The electronegativity of an atom is a measure of its tendency to serve as the electron acceptor in bond formation. A large electronegativity correlates with a high first ionization potential coupled with a large, positive electron affinity. Table 6–4 shows the electronegativity values of some elements Pauling deduced by considering specific bond energies.

By reflecting on a large number of known bond energies, Pauling concluded that the bond energy for a heteronuclear, two-center bond between elements A and B is given to a good approximation by equation 6–3, where X_A and X_B are the electronegativities of atoms A and B, respectively:

$$\begin{aligned}&(\text{Bond energy})_{AB} \\ &\quad = \{[(\text{Bond energy})_{A_2}][(\text{Bond energy})_{B_2}]\}^{1/2} + 23.06\ (X_A - X_B)^2 \qquad \textbf{6–3}\end{aligned}$$

H						
(2.1)						
Li	Be	B	C	N	O	F
(1.0)	(1.5)	(2.0)	(2.5)	(3.0)	(3.5)	(4.0)
Na	Mg	Al	Si	P	S	Cl
(0.9)	(1.2)	(1.5)	(1.8)	(2.1)	(2.5)	(3.0)
K	Ca	Ga	Ge	As	Se	Br
(0.8)	(1.0)	(1.6)	(1.8)	(2.0)	(2.4)	(2.8)
Rb	Sr	In	Sn	Sb	Te	I
(0.8)	(1.0)	(1.7)	(1.8)	(1.9)	(2.1)	(2.5)

Table 6–4 Electronegativity Values of Some of the Elements

The number 23.06 is the conversion factor from electron volts/molecule to kcal/mole; thus the units of electronegativity are $(\text{ev/molecule})^{1/2}$. According to equation 6–3, the bond energy increases with the difference between electronegativities, just as it does with the difference between atomic orbital energies. From this point on, we shall use Table 6–4 in predicting the polarities of chemical bonds.

Let us illustrate the use of Table 6–4 and also check how well the electronegativity values yield a given bond energy. Our illustration will be the C—F bond, which Table 6–3 lists with a bond energy of 106 kcal/mole. The C—C energy is 83 kcal/mole and F—F is 37 kcal/mole. The electronegativity values are 2.5 for C and 4.0 for F, so that the C—F bond energy is calculated to be $[(83)(37)]^{1/2} + (23.06)(4.0 - 2.5)^2 = 107$ kcal/mole, in excellent agreement with the value from Table 6–3.

6–6. Review and Outlook

We are now familiar with the basic principle that atoms bond to form molecules in order to minimize the electrostatic potential energies of electrons. We have elaborated this principle by considering covalent bonds in such species as H_2 and O_2, and have shown how the geometries of such molecules as CF_4 and NH_3 can be predicted. We investigated bond polarity and showed why ionic solids such as NaCl exist.

We have by no means exhausted the important examples of chemical bonding. For example, we have said nothing about the bonding in metallic solids. A bar of iron is obviously different in its properties from a block of ice or a lump of sodium chloride, and the properties are a reflection of the unique bonding in a metal. We have said nothing about bonding between molecules such as that which causes all gases to liquefy at a sufficiently low temperature. As far as possible in this text, models will be developed to explain the chemical and physical properties of matter. These further bonding considerations and other related matters are treated subsequently (Chapters 10 and 11) in their proper context.

SUGGESTIONS FOR ADDITIONAL READING

Ryschkewitz, G. E., *Chemical Bonding and the Geometry of Molecules,* Van Nostrand Reinhold, New York, 1972.

Linnett, J. W., *The Electronic Structure of Molecules,* Wiley, New York, 1964.

Coulson, C. A., *Valence,* Clarendon Press, Oxford, 1961.

Pimentel, G. C., and Spratley, R. D., *Understanding Chemistry,* Chap. 16, Holden-Day, San Francisco, 1971.

PROBLEMS

1. The species C_2 is very stable, not readily separated into carbon atoms, since the bond energy of C_2 is 144 kcal/mole. However, C_2 is only found in unusual circumstances (as in flames) and is highly reactive. Why?

2. Count the number of valence electrons around each of the atoms in the following compounds and show that the octet rule is obeyed in each case: (a) NH_3; (b) N_2; (c) O_2; (d) PCl_3; (e) SiF_4.

3. In which of the atoms in problem 2 *must* the octet rule be obeyed?

4. Give line and Lewis dot representations for the bonding in each of the following species: (a) CCl_4; (b) OF_2; (c) HCl; (d) PH_3.

5. Use the geometry procedure given in the text to obtain the structures of the following species: (a) BH_4^-; (b) NH_4^+; (c) CO_2; (d) NO_3^-; (e) OF_2; (f) N_2O (nitrogen is the central atom).

6. Each of the following species is represented by canonical resonance forms. Give the Lewis dot and line representations of these resonance forms: (a) NO_3^-; (b) O_3; (c) NO_2^-.

7. What is the total bond order (the sum of the orders of each of the bonds) in the following species: (a) CO_2; (b) NO_2^-; (c) H_2O?

8. Give a Lewis dot representation of CO. Show that each atom obeys the octet rule.

9. What orbital hybridization of the carbon atom(s) is involved in the bonding in each of the following species:

(a) CO_2; (b) CO_3^{2-}; (c) $H_2C{=}CH_2$ (each C bonded to two H); (d) CO; (e) H—C≡C—H.

10. What is the approximate hybridization of the nitrogen orbitals in ammonia?

11. What is the orbital hybridization of the nitrogen atom in NO_3^-?

12. What is the energy of an electron in an sp^3 orbital of a nitrogen atom?

***13.** Do the lone pair electrons in NH_3 have a higher or lower energy than that calculated in problem 12? Explain.

14. The δ value is larger in LiF than in LiH, but not much larger (0.84 electron as compared with 0.76 for LiH). However, the LiF bond is found to be much stronger than the LiH bond (the LiF bond energy is 137 kcal/mole compared with 58 kcal/mole for LiH). Do Figures 6–12 and 6–14 indicate the source of the extra bond energy and surprisingly small δ of LiF? Explain.

15. The dipole moment of a molecule is a vector quantity with its direction along the axis connecting δ^+ and δ^-. Thus in the case of a diatomic molecule, the dipole moment is directed along the internuclear axis. The dipole moment of a multinuclear molecule such as water may be considered to result from the vector addition of **bond dipole moments,** where the direction of each bond moment is along the axis connecting a bonded nuclear pair and each pair is characterized by a δ^+ and a δ^-. The water molecule in the gas phase possesses a dipole moment of 1.86 Debye and δ^- resides on the oxygen. The structure and charge separation of gaseous water are

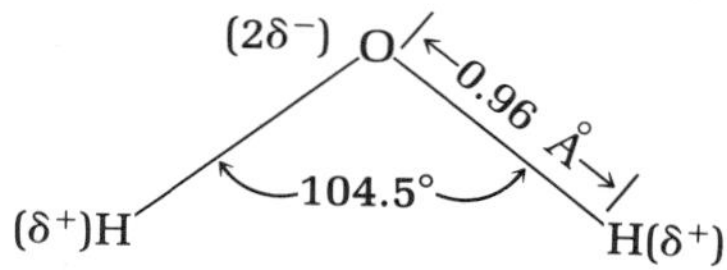

From the information given, obtain a value for δ^+. Is a good model for water that of ions 2 H^+ and O^{2-} in contact with each other?

16. The molecules BeF_2, BF_3, and CF_4 possess no dipole moment, in spite of the fact that the bonds in each of the three molecules are highly polar. Can you explain why? *Hint:* Consider the geometries of these molecules.

17. The C—H bonds in acetylene, H—C≡C—H, are much more polar than the same bonds in methane. Consider the carbon orbital hybridization and explain the polarity difference. Is the hydrogen δ^- or δ^+ in acetylene?

18. Use the bond energies of Table 6–3 to calculate the ΔE for each of the following reactions:

a. $2\ H_2(g) + H{-}C{\equiv}C{-}H(g) \rightarrow C_2H_6(g)$.
b. $3\ H_2(g) + N_2(g) \rightarrow 2\ NH_3(g)$.

19. In the text, we showed that reaction 6–1 liberates a large quantity of energy because the O—H bond is significantly stronger than the average energy of the two bonds of O_2. What is the average energy of the three bonds of N_2? Compare this with the N—H single-bond energy. Explain why reaction (b) in problem 18 liberates energy even though the extremely strong N≡N bond is broken.

20. From equation 6–3 and data in Tables 6–3 and 6–4, calculate the N—F bond energy. Compare your calculated value with the N—F bond energy listed in Table 6–3.

7/ Chemical Periodicity and Chemical Properties

The bases we have used for model building so far are physical properties of matter. Little has been said in this development about chemical transformations of matter. The number of known chemical species is very great, and so is the number of reactions between them. Each year chemical research increases the number of known chemical compounds and their reactions, so that the proportions become awesome. Our models enable us to understand this wealth of chemical facts, and even to predict those facts in many cases. Our principal model is the wave model of the atom. A second model involves viewing molecules as containing bonded atoms. Molecules really do not contain bonded atoms; molecules contain only electrons and nuclei. However, a model involving bonds between pairs of atoms approximates the arrangement of electrons and nuclei in many molecules. With this model we can account for the existence, structures, and reactions of many molecules by considering their "constituent" atoms.

7–1. Chemical Periodicity

Figure 4–8 shows the variation of the first ionization potential with atomic number for some elements, and Table 4–3 lists ionization potentials for the first thirty elements in the order of increasing atomic number. In these ionization potentials we have already seen one striking instance of recurrence, and these are the sharp drops between He and Li, Ne and Na, and Ar and K, as shown in Figure 4–8. These sharp drops also occur between Kr and Rb, and between Xe and Cs. The drops provide us with the primary basis for the electronic configurations given in Table 4–2.

Long before ionization potentials were measurable, scientists noted that periodic recurrences were a function of elemental atomic weight. The Cannizzaro procedure and related methods were used in the mid-nineteenth century to determine many molecular formulas. It was noted, for example, that the metallic elements Li, Na, K, Rb, and Cs all form solid oxides with the formula M_2O, where M stands for the metal. Furthermore, each of these oxides reacts in a similar fashion with the same substance. For example, all of them react with HCl according to the equation

$$M_2O(s) + 2\,HCl(g) \rightarrow 2\,MCl(s) + H_2O(l) \qquad \textbf{7–1}$$

The solids M_2O are ionic, like the chlorides that are products of reaction 7–1. Each of the elements with an atomic number one greater than M in equation 7–1 forms an oxide of the formula MO. The oxides are BeO, MgO, CaO, SrO, and BaO. The elements with atomic numbers one greater than those elements yield the set of oxides B_2O_3, Al_2O_3, Ga_2O_3, In_2O_3, and Tl_2O_3.

For the elements lithium through neon, we have already considered in some detail how molecular formulas can be explained in terms of the electron configuration. Lithium has an electron configuration of $1s^22s^1$; the single relatively high-energy valence electron causes lithium to form such molecules as LiF (Chapter 5). The bond polarity of LiF involves a substantial positive charge (δ^+) on the lithium portion of the molecule. Like fluorine, oxygen has valence electrons that lie at relatively low energy; also like fluorine, oxygen has valence orbitals that serve largely as electron acceptors in bond formation rather than electron donors. The two electron vacancies in the oxygen valence shell limit oxygen to forming compounds like H_2O and Li_2O. The electronic configuration of Na is $1s^22s^22p^63s^1$. The single high-energy valence electron should lead to sodium's having a chemistry quite similar to that of lithium; this is indeed the case, as the identical formulas and reactivities of the two oxides show. For K, Rb, and Cs, the single valence electron is 4s, 5s, and 6s, respectively, and these elements are very similar in chemistry to Li and Na.

For Be, Mg, Ca, Sr, and Ba, there are two relatively high-energy valence electrons; compounds should be found of the formula MF_2, where a positive charge in excess of 1 resides on the metal atom portion. Since oxygen can accept two electrons, compounds should also be found with the formula MO, and again M should be characterized by a significant positive charge in the bonding. The elements—Be, Mg, Ca, Sr, and Ba—are found most commonly in ionic solids that contain the M^{2+} ion.

For B, Al, Ga, In, and Tl, the number of high-energy valence electrons has risen to three, and we should predict the existence of compounds such as BF_3 and AlF_3, and also such oxides as B_2O_3 and Al_2O_3. Continuing to reason from the number of high-energy valence electrons, we further predict the existence of the series of oxides CO_2, SiO_2, GeO_2, SnO_2, and PbO_2, and these oxides are indeed found.

The Periodic Table

With a distinct correlation between chemical properties of an atom and its electronic configuration, we can construct a **periodic table** of the elements, which groups in columns those elements possessing the same numbers and kinds (*s*, *p*, *d*, *f*) of valence electrons. Such a table is presented in Figure 7–1.

s | d | p

I A	II A	III B	IV B	V B	VI B	VII B	VIII B			I B	II B	III A	IV A	V A	VI A	VII A	VIII A
1 H																	2 He
3 Li	4 Be											5 B	6 C	7 N	8 O	9 F	10 Ne
11 Na	12 Mg											13 Al	14 Si	15 P	16 S	17 Cl	18 Ar
19 K	20 Ca	21 Sc	22 Ti	23 V	24 Cr	25 Mn	26 Fe	27 Co	28 Ni	29 Cu	30 Zn	31 Ga	32 Ge	33 As	34 Se	35 Br	36 Kr
37 Rb	38 Sr	39 Y	40 Zr	41 Nb	42 Mo	43 Tc	44 Ru	45 Rh	46 Pd	47 Ag	48 Cd	49 In	50 Sn	51 Sb	52 Te	53 I	54 Xe
55 Cs	56 Ba	57–	72 Hf	73 Ta	74 W	75 Re	76 Os	77 Ir	78 Pt	79 Au	80 Hg	81 Tl	82 Pb	83 Bi	84 Po	85 At	86 Rn
87 Fr	88 Ra	89–	104 Rf	105 Ha													

f

57 La	58 Ce	59 Pr	60 Nd	61 Pm	62 Sm	63 Eu	64 Gd	65 Td	66 Dy	67 Ho	68 Er	69 Tm	70 Yb	71 Lu
89 Ac	90 Th	91 Pa	92 U	93 Np	94 Pu	95 Am	96 Cm	97 Bk	98 Cf	99 Es	100 Fm	101 Md	102 No	103 Lw

Figure 7–1
A periodic table of the elements.

The number of columns in the table is 18 and this is the sum of two *s* electrons, six *p* electrons, and ten *d* electrons. In two series of elements, Ce through Lu, and Th through Lw, *f* orbitals are being filled. Since a 32-column table is awkward to display, these two sets of 14 elements are typically displayed below the other elements, as in the figure, and their locations in the table are indicated by including with them the element that precedes them in atomic number. The series Ce through Lu immediately follows the element lanthanum, and thus the series is termed the **lanthanide series.** The series Th through Lw immediately follows actinium and is termed the **actinide series.**

Each of the columns of the table is given a roman numeral and a letter (A or B). In most cases, the roman numeral indicates the maximum number of valence electrons that reflect themselves in the bonding characteristic of the members of a given column (group) of elements. Thus, for example, both carbon and titanium possess four valence electrons and form the compounds CF_4 and TiF_4 as a consequence. The distinction A and B is drawn in most cases depending on whether or not *d* electrons are included in the number to which the roman numeral refers. The number and directional characteristics of *d* orbitals differ markedly from the similar characteristics of *p* orbitals, and there are marked differences in the characteristic chemistry between A and B groups with the same roman numeral.

We have based Figure 7–1 on our knowledge of atomic orbitals and electron occupations. However, long before the birth of the modern wave model of the atom, the Russian chemist Dimitri Mendeleev had arrived at a very similar table, using his knowledge of the chemical and physical properties of the elements—information such as the formulas of their oxides. From this table, Mendeleev was able to predict the existence and properties of several elements that were as yet undiscovered. These elements were predictable from the gaps that Mendeleev found when he tried to arrange the elements known to him at that time in an order reflecting their similar properties. Moreover, he could predict the properties of the missing elements from the known properties of the elements in the column containing the gaps.

Metals and Nonmetals

Knowing about chemical periodicity greatly simplifies our task of predicting the characteristic chemistry of each of the elements. The problem is largely reduced to predictions for the separate groups. We shall also distinguish between **metallic** and **nonmetallic** elements, since the characteristic chemistries of these two groups differ markedly. Everyone is familiar with the lustrous appearance of metals, and the fact that metals are excellent conductors of heat and electricity. These properties are in great contrast to the properties of ionic solids such as NaCl.* Every metallic element has a

* We do not discuss the details of bonding in metals in this chapter; this topic is treated at the beginning of Chapter 11.

relatively low first ionization potential. Li, Na, K, Rb, and Cs are all metallic in their elemental states, and each has a low first ionization potential.

Those elements whose first ionization potentials are large, such as O, F, P, S, Cl, form solids that do not show any of the characteristic properties of metals. In Figure 7–1, the "stair steps" to the left of and below elements B, Si, Ge, Sb, and Po have been added as a dividing line between the metallic and nonmetallic elements. You will recognize that there must be some arbitrariness in drawing such a line. As we move along a row of the table from left to right, the properties change in smooth fashion from characteristically metallic at the left edge of the table to characteristically nonmetallic at the right. Elements near the dividing steps, such as boron, silicon, germanium, antimony, and tin, behave neither as characteristic metals nor as nonmetals. In spite of this degree of arbitrariness, the division of the periodic table into metals and nonmetals proves very useful.

Our immediate concern is the nonmetallic elements with the exception of carbon. Carbon is the most important element in living (organic) systems, and although a nonmetallic element, its chemistry is considered separately.

7–2. Acids and Bases

Donor-Acceptor Complexes

Atoms bond together to form molecules if electrons from one atom can transfer into low-energy regions around other atoms. Atoms that bond well possess either relatively high-energy electrons or low-energy sites for electrons. The same applies to molecules, many of which also possess relatively high-energy electrons or low-energy sites, or both. For example, the nonbonded valence electrons of NH_3, shown in Figure 6–4, are at a relatively high energy because they are associated with only one nucleus, and because of repulsion between them and the three bonded electron pairs. The boron portion of BF_3 represents a low-energy site because of electron depletion around the boron nucleus due to the high electronegativity of fluorine.

In view of this characterization of NH_3 and BF_3, it is not surprising that the two molecules react to form the molecule whose structure is shown in Figure 7–2. The high-energy electrons of NH_3 are shared with the low-energy region of the boron. The orbital hybridization around both N and B in Figure 7–2 is very nearly sp^3 and this hybridization is revealed in the figure by the

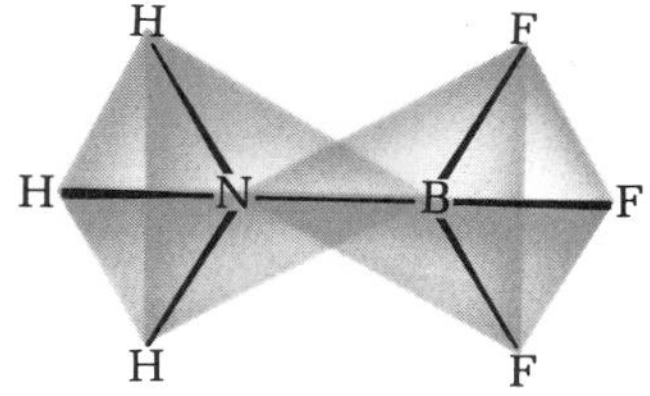

Figure 7–2
The structure of the donor-acceptor complex of ammonia and boron trifluoride.

tetrahedral geometry around each atom. The B—N bond formed is commonly termed a **coordinate covalent bond** to indicate that both electrons come from the same atom (N in this case). This contrasts with the normal covalent bond (as in H_2), in which one electron comes from each atom.

The NH_3 serves as an electron donor in the formation of the coordinate covalent bond. There is a commonly used term for such an electron donor. It is **nucleophile,** meaning a "nucleus-loving" species. The BF_3 serves as an electron acceptor and it is termed an **electrophile.** The species formed from the combining of a nucleophile and an electrophile is a **donor-acceptor complex.**

The Hydrogen Bond

Water is a polar molecule that can be both a nucleophile and an electrophile. Water has two nonbonded electron pairs, which form the nucleophilic parts of the molecule. The nucleophilic character of the water molecule is significant in the chemistry of water. A sizable positive charge characterizes the hydrogen regions of H_2O, and electrostatic attraction occurs between these regions and negatively charged regions in other species. One such species is another water molecule. A model for the water-water interaction is ionic bonding, that is, a purely electrostatic attraction (although neither δ^+ nor δ^- is as large as one electron).

Figure 7–3 shows a portion of the structure of ice. Each oxygen is located at the center of a tetrahedron and participates in four bonds. The black lines indicate strong covalent bonds with two hydrogens. The colored lines are

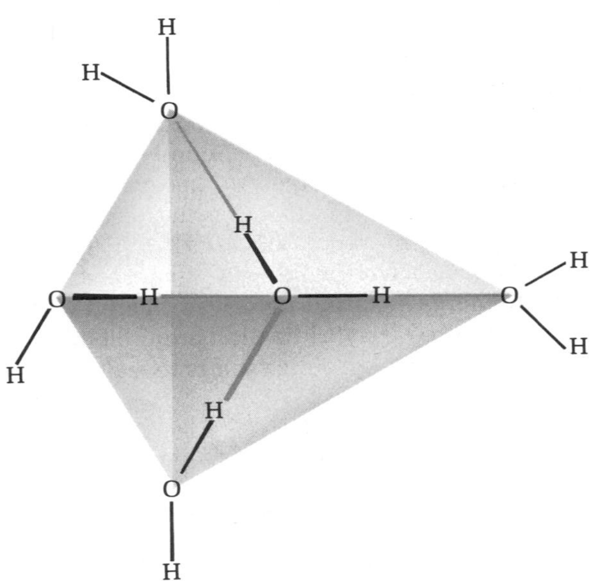

Figure 7–3
A portion of the structure of ice.

electrostatic bonds between water molecules. Ionic-like bonds that involve H as the positively charged species occur very commonly and the bond is termed the **hydrogen bond,** regardless of the nature of the negatively charged species involved in the bond. Thus, each oxygen in ice participates in two hydrogen bonds.

Acids in Aqueous Solution

Liquid HCl is not a good conductor of electricity nor is liquid water. Both have a high electrical resistance. However, if we dissolve HCl in water, the resultant solution is an excellent electrical conductor due to formation of ions according to the reaction

$$HCl(aq) \rightarrow H^+(aq) + Cl^-(aq) \qquad \textbf{7–2}$$

Why should HCl dissociate into ions in aqueous solution when it will not do so as pure HCl? The answer comes from the nature of H^+ in solution. The proton is a strong electrophile and it bonds to a lone pair of electrons from a water molecule to form the cation H_3O^+, which contains three strong O—H bonds.

Reaction 7–2 reflects a donor-acceptor competition. The HCl may be viewed as a donor-acceptor complex formed from H^+ and Cl^- (even though HCl is not an ionic compound). When HCl is exposed to water, the H^+ has a choice between Cl^- and H_2O as electron donors, and Cl^- has a choice between forming a bond with H^+ and forming hydrogen bonds with some unknown number of water molecules. Since water can associate with protons and anions, such a competition is set up when any of a large number of hydrogen-containing molecules are added to water.

In the case of HCl in water, the donor-donor and acceptor-acceptor competitions are won overwhelmingly by water and one mole of HCl yields one mole of H_3O^+ in aqueous solution. By definition, any species whose addition to water increases the concentration of H_3O^+ in water is termed an **acid.** Species such as HCl that produce H_3O^+ quantitatively in water are termed **strong acids.** A strong acid is strong because water overwhelmingly wins the struggle for the proton. There are many important species that are **weak acids** in water and they dissociate only partially to produce H_3O^+ in most solutions. (See Chapter 19 for a quantitative definition of acid strengths.)

Bases in Aqueous Solution

Liquid ammonia is a poor electrical conductor. When ammonia is added to water, electrical conductivity for the solution is much higher than for either pure NH_3 or pure H_2O. The conductivity increase arises from a partial ionization according to the reaction

$$NH_3(l) + H_2O(l) \rightarrow NH_4^+(aq) + OH^-(aq) \qquad \textbf{7–3}$$

This is a second example of donor-acceptor competition in aqueous solution. The H_2O is now the complex whose proton is being competed for by the good nucleophile NH_3. To a large extent, the H_2O wins and retains its proton, but NH_3 extracts a small fraction of protons to form a complex, the **ammonium ion** (NH_4^+), leaving the **hydroxide ion** (OH^-). By definition, any species whose addition to water increases the concentration of OH^- in water is a **base,** and ammonia is a **weak base.** An example of a strong base is the ionic solid **sodium hydroxide,** which consists of Na^+ and OH^- ions. When added to water, sodium hydroxide dissociates almost entirely into Na^+ and OH^-:

$$NaOH(s) + H_2O(l) \rightarrow Na^+(aq) + OH^-(aq) + H_2O(l) \qquad \textbf{7–4}$$

7–3. Oxidation and Reduction

Redox Reactions

Many reactions are encountered in which the electron energy difference between an electron donor and an electron acceptor is so large that instead of a donor-acceptor complex being formed, actual electron transfer occurs between the two species. Such is the case, for example, when electron transfer from sodium atoms to chlorine atoms causes formation of an ionic solid. Such is also the case when solid zinc is added to an aqueous solution containing Cu^{2+} ions:

$$Cu^{2+}(aq) + Zn(s) \rightarrow Cu(s) + Zn^{2+}(aq) \qquad \textbf{7–5}$$

The solid Zn dissolves and solid copper comes out of the solution because the electron affinity of Cu^{2+} is much greater than the electron affinity of Zn^{2+}. Two electrons transfer from each zinc atom to each Cu^{2+} ion.

Electron loss is termed **oxidation,** and electron gain is termed **reduction.** The species that gains electrons is the **oxidizing agent** and the one that loses electrons is the **reducing agent.** In reaction 7–5, Zn(s) is the reducing agent and Cu^{2+} is the oxidizing agent. The reaction itself is termed a **redox reaction** (a name derived from reduction-oxidation). Redox reactions are very common in both organic and inorganic chemistry.

Oxidation Numbers

Reaction 7–5 is easily recognized as a redox reaction because the electrical charges on the copper and zinc species change. Consider reaction 7–6:

$$CH_4(g) + 2\,O_2(g) \rightarrow CO_2(g) + 2\,H_2O(l) \qquad \textbf{7–6}$$

Is it a redox reaction? There are no charges involved and thus it is hard to tell at a glance. We can easily recognize redox reactions if we set up and carry through a system of **oxidation numbers** for atoms in molecules and polyatomic ions. In most cases, these numbers do not correspond to real charges on atoms. Nevertheless, they are very useful. We shall endeavor to make these numbers agree with real charges on atoms in those species, such as Zn^{2+} and Cu^{2+}, in which real charges appear. The complete set of rules is rather complex, but most cases can be handled by the few rules that follow:

1. The sum of the oxidation numbers of the atoms in any given species equals the electrical charge on the species.
2. Atoms in their elemental states are assigned oxidation number zero, regardless of the bonding in that state. Thus, Al(s) is Al(0), H_2(g) is H(0), and Ar(g) is Ar(0). This agrees with a total charge of zero on the species.
3. Hydrogen bound directly to any atom other than itself or a metal has an oxidation number of +1; it is H(I). (Wherever possible, oxidation numbers are given by roman numerals to distinguish them from real charges.)
4. Oxygen bound to any element other than itself has oxidation number −2; it is O(−II). Water is considered to be composed of two H(I) atoms and one O(−II) atom for a combination of oxidation numbers and actual charge of zero.
5. The atoms of Group VIIA of the periodic table have an oxidation number of −1 except when bonded to each other or to oxygen. Thus, HCl contains H(I) and Cl(−I); and **perchlorate ion,** ClO_4^-, contains Cl(VII) and four O(−II) for a total oxidation number and charge of −1.

Let us now assign oxidation numbers to the species in reaction 7–6. Methane (CH_4) must involve H(I) [Rule 3], and according to Rule 1 this means that methane also contains C(−IV). Carbon dioxide involves O(−II) [Rule 4], and C(IV) [Rule 1]. Thus, the conversion of CH_4 to CO_2 involves an increase in oxidation number by carbon of 8. In O_2, we have two O(zero), so the conversion of O_2 to 2 H_2O is a total change in oxidation number of 4. Hydrogen is H(I) in both methane and water.

Let us now see how these oxidation numbers can be used to give us the number of electrons lost and gained by given species. Methane involves a total change in oxidation number of 8 in reaction 7–6 (a change of 8 for one carbon and zero for four hydrogens). When the total oxidation number increases as it does for carbon between CH_4 and CO_2, electron loss is signalled for the molecule containing that atom. Thus, each methane molecule loses eight electrons in the conversion to 2 H_2O and CO_2. Methane is the reducing agent in reaction 7–6.

Each of the four oxygen atoms of O_2 molecules in reaction 7–6 has its oxidation number reduced by 2 in going to H_2O. Reduction of the oxidation number by an atom corresponds to electron gain by the species containing the atom. Thus, 2 O_2 gain the eight electrons lost by CH_4.

The oxidation number procedure just outlined is an arbitrary one (we could construct other rules); but the results in terms of electrons transferred are real. Each methane molecule does lose eight electrons in reaction 7–6 and these electrons are gained by oxygens. Thus, the introduction of the concept is justified by its usefulness.

7–4. Formal Charges

The Lewis dot representations of CO_2, CO, and one canonical resonance form of N_2O are given in Figure 7–4. Each bonded pair of electrons is shared by the nuclei to either side of the pair. If both atoms that make a two-atom bond have the same electronegativity, these electrons are equally shared by the two nuclei. Let us ignore electronegativity differences for the moment and simply count electrical charges in the bonded atoms of CO_2. Each oxygen has two core (1*s*) electrons, four nonbonding valence electrons, and half of four bonding valence electrons for a total of eight electrons. Since the oxygen nuclear charge is 8, the oxygen atoms have no charge in the absence of electronegativity differences in CO_2. A similar analysis can be performed for C in CO_2 to show that it would possess no charge in the absence of electronegativity differences.

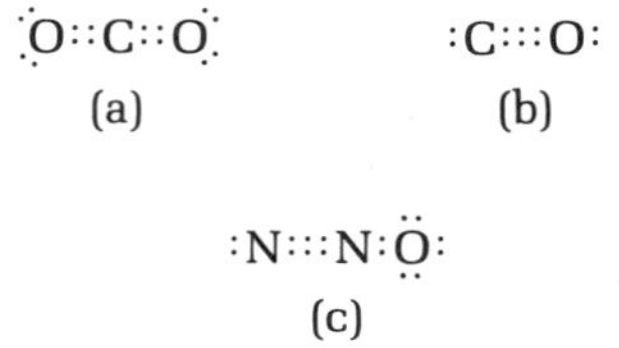

:Ö::C::Ö: (a)

:C:::O: (b)

:N:::N:Ö: (c)

Figure 7–4 Lewis dot representations of (a) CO_2, (b) CO, and one canonical resonance form of N_2O.

The situation is quite different in CO. The oxygen contains two core electrons, two nonbonding valence electrons, and three bonded valence electrons, for a total of seven. This results in a +1 charge on the oxygen and a −1 charge on the carbon. These charges do not actually exist in CO, since the oxygen atom is more electronegative than the carbon atom. However, the formal charge is a useful concept in correlating chemical properties.

SUGGESTIONS FOR ADDITIONAL READING

Rich, R. L., *Periodic Correlations*, Benjamin, Menlo Park, Calif., 1965.

Sanderson, R. T., *Chemical Periodicity*, Reinhold, New York, 1960.

PROBLEMS

1. Given the formulas CF_4, Al_2O_3, Mg_3N_2, and SiO_2, refer to Figure 7–1 and predict the formulas of the compounds formed by combining
   a. Scandium and oxygen.
   b. Calcium and phosphorus.
   c. Silicon and chlorine.
   d. Carbon and sulfur.

2. Boron trifluoride is a good electrophile. Do you predict that CF_4 is a good electrophile also?
3. Water is a poorer nucleophile than ammonia. Can you explain why?
4. Water molecules associate by hydrogen bonding. The strong water-water interaction causes water to boil at a relatively high temperature (100 °C). Both NH_3 and CH_4 have much lower boiling points (−33 °C and −161 °C respectively). Can you account for these boiling points that are lower relative to water?
5. Propose a structure for the H_3O^+ ion. Draw an electron dot representation of the species.
6. The solid $NaNH_2$ is a stronger base than NaOH, and the NH_2^- ion extracts protons from H_2O to form NH_3 and OH^-. Can you explain why NH_2^- is a better proton acceptor than OH^-? Is CH_3^- an even better proton acceptor than NH_2^-? Explain.
7. Cider vinegar is made by the oxidation of ethyl alcohol to acetic acid by natural organisms. The structures of ethyl alcohol and acetic acid are

```
H         H                 H          O
 \       /                   \        //
H—C—C—H                    H—C—C
 /       \                   /        \
H         O                 H          O—H
         /
        H

  Ethyl alcohol                 Acetic acid
```

Assign the oxidation numbers of each of the atoms in ethyl alcohol and acetic acid and show from these that ethyl alcohol is indeed oxidized in the process of conversion to acetic acid. How many electrons does each ethyl alcohol molecule lose? The oxidizing agent is O_2 and it is reduced to H_2O.
8. Assign oxidation numbers to all of the atoms in each of the following species: (a) H_3CCl; (b) Al_2O_3; (c) ClO^-; (d) $S_2O_3^{2-}$; (e) H_3AsO_4; (f) NOCl.
9. How many electrons are lost or gained in the following conversions?
   a. H_3PO_3 to H_3PO_4 and H_2O.
   b. 2 NOCl to N_2, HCl, and H_2O.
   c. $S_2O_3^{2-}$ to 2 H_2S and H_2O.
10. Is the following reaction a redox reaction?

$$H_3CCl(g) + H_2O(l) \rightarrow H_3COH(l) + HCl(g)$$

11. Assign formal charges to each of the atoms in Figure 7–4(c).
12. Which atom in CO do you predict to be more nucleophilic?

8/ The Chemistry of Groups VA and VIA

The molecular formulas, structures, and reactivities show characteristic differences between any two columns of the periodic table. We should like to be able to explain these differences (and even predict them) from our model of bonding between wavelike atoms. Within each column there is usually a progressive change in chemical properties from the element at the top of the column to the element at the bottom. This progressive change is also understandable in terms of the model of bonded atoms. Group IVA elements form the largest wealth and variety of compounds of all the nonmetallic elements, largely because for each element an unreacted atom possesses four valence electrons and four vacancies in the valence electron level. This offers the maximum opportunity for each element to bond to atoms of its own kind in a nonpolar fashion, or to elements to its right and left in the periodic table in a polar fashion. The next most extensive chemistry is displayed by the Group VA elements, N, P, As, Sb, and Bi. Carbon is the element of Group IVA that is of greatest interest to us, so much so that an entire chapter (Chapter 10) is devoted to its chemistry. Consequently we shall not begin this discussion with the Group IVA elements. Instead, the chemistry displayed by the Group VA elements N, P, As, Sb, and Bi will be taken up, to be followed by that of the elements O, S, Se, and Te of Group VIA.

8–1. The Chemistry of the Group VA Elements

There are three significant factors in the chemistries of each of the nonmetallic groups of elements. The first is the ability of the atoms of an element to bind to other atoms of that element to yield the various physical forms of the elements. Elemental carbon, for example, is a solid at room temperature and elemental nitrogen is a gas at room temperature.

Second is the ability of the atoms of the nonmetallic group to be the electronegative atom in polar or ionic bond formation with elements to their

III A	IV A	V A	VI A	VII A	VIII A
					He
B	C	N	O	F	Ne
	Si	P	S	Cl	Ar
	Ge	As	Se	Br	Kr
	Sn	Sb	Te	I	Xe
	Pb	Bi			

Figure 8–1 The groups of nonmetallic elements whose chemistries are discussed in Chapters 8, 9, and 10.

left in the periodic table. Representative elements for this type of bonding are those of Groups IA and IIA; these atoms possess the smallest electronegativities in the periodic table.

Third is the ability of the atoms of the nonmetallic group to be electron donors in polar bond formation with elements to their right in the periodic table. Representative electronegative elements for such binding are those of Group VIIA (termed the **halogens**) because they are the most electronegative group of elements in the periodic table. The bonding of the elements of the nonmetallic group with oxygen is also significant because many of the most important compounds in nature contain bonds between oxygen and other kinds of atoms.

Elemental Forms

Nitrogen can form N—N single bonds, N═N double bonds, or N≡N triple bonds. Elemental nitrogen is found only as gaseous N_2. Let us see if we can understand why N_2 is the only elemental form. Table 6–3 lists the N—N, N═N, and N≡N bond energies as 60, 100, and 225 kcal/mole, respectively. Since the triple bond is more than three times as strong as the single bond and more than 50 percent stronger than the double bond, the bond model leads to the prediction that N_2 is the predominant state of elemental nitrogen.

The great strength of the N_2 bond presents a problem to us because nitrogen atoms form a key part of such important molecules as proteins. The growing of food requires large amounts of fertilizers and nitrogen-containing compounds are an important component of fertilizers. Gaseous nitrogen must be converted to nitrogen atoms in other molecules.

Phosphorus is another important atom in the molecules that make up fertilizer. Many important phosphorus-containing molecules are required in the chemistry of living systems. Since phosphorus atoms contain the same number of valence electrons as nitrogen atoms, we might expect phosphorus atoms to form P_2 molecules as a gas at room temperature, but such is not the case. Apparently 3*p* orbitals (and *p* orbitals larger than 3*p*) are simply too large to form good pi-bonding molecular orbitals. The electron density in the plane midway between the two nuclei is concentrated too far from both nuclei, and pi bonding is quite weak compared with sigma bonding. This weakness of pi bonding involving *p* orbitals with *n* greater than 2 is quite general, and a factor in attempting to predict molecular structures.

In the case of NH_3, each nitrogen atom can provide a low-energy region of space that can accommodate a maximum of three electrons. Thus, we should not be surprised to find that in elemental phosphorus each P atom is bonded to three other P atoms. Elemental phosphorus, like many other elements, can exist in more than one structural form; such forms for any element are termed **allotropes.** The two common forms of phosphorus are termed **red phosphorus** and **white phosphorus.** Red phosphorus is related to a relatively rare allotrope, **black phosphorus,** which has the corrugated sheet structure depicted in Figure 8–2. The colored and black spheres represent P atoms.

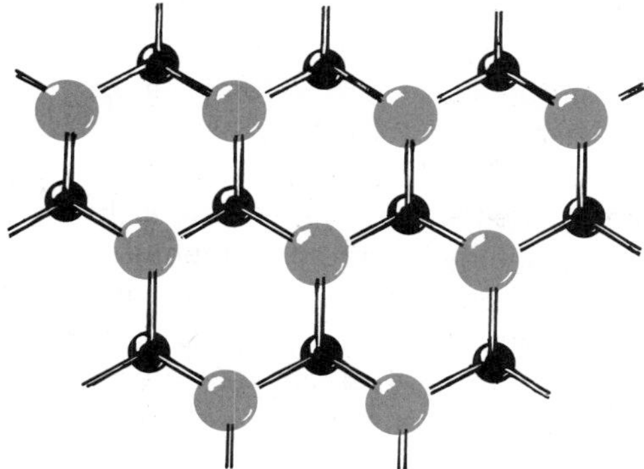

Figure 8–2
The corrugated sheet structure of black phosphorus. The colored spheres, which represent phosphorus atoms, are in the plane of the paper, with the black spheres, which also represent phosphorus atoms, lying in a parallel plane below the paper.

Each of the colored spheres lies in the plane of the paper and the black spheres lie in a parallel plane below the paper. The geometry around each phosphorus atom is the pyramidal geometry of NH_3, just as we might expect it to be.

Black phosphorus is a **crystalline compound,** meaning that all of the P atoms belong to an ordered array such as that shown in Figure 8–2. Red phosphorus is an **amorphous compound.** In an amorphous compound, the high order of a crystalline compound is absent. Red phosphorus involves the same pyramidal geometry around each phosphorus atom as black phosphorus does, but the ordered sheet array of Figure 8–2 does not exist in red phosphorus.

The most reactive of the allotropic forms of phosphorus is white phosphorus, which consists of P_4 molecules with the structure shown in the Lewis dot representation in Figure 8–3. As in the other two allotropic forms, white phosphorus involves a pyramidal geometry around each phosphorus.

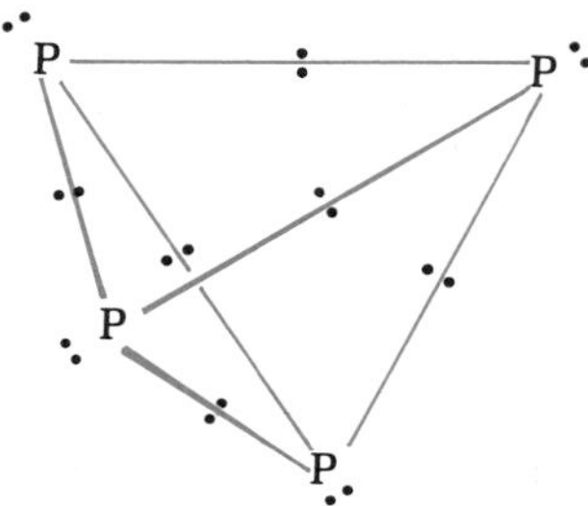

Figure 8–3
Lewis dot representation of the P_4 molecule. The tetrahedral geometry of the molecule is shown in color.

However, in P_4 the nuclei are situated at the four corners of a tetrahedron. The resultant P—P—P bond angle is 60° instead of the 107° H—N—H angle in NH_3, and the bonded electron pairs in P_4 experience sizable mutual repulsion in this structure. The structure is said to be **strained,** and it is not surprising that white phosphorus is highly reactive. For example, it spontaneously ignites when exposed to air to produce phosphorus oxides.

Phosphorus vapor consists largely of P_4 molecules. At very high temperatures this molecule is largely split into P_2 molecules. Most molecules will dissociate into smaller species at higher temperatures (Chapters 15 and 16). In P_4 the P—P bond distance is 2.21 Å, while in P_2 it is 1.88 Å. The shorter bond distance in P_2 reflects P—P pi bonding, which is, however, not strong enough to allow the P_2 species to predominate at lower temperatures.

Arsenic and antimony exist in their elemental forms in structures analogous to those for phosphorus. The more stable form in each case, like black phosphorus, has a sheet structure. These forms are generally termed metallic arsenic and metallic antimony. The fact that these forms are metallic is an example of the ambivalent character of the elements near the dividing line between metals and nonmetals in Figure 7–1. The more reactive forms of As and Sb consist of As_4 and Sb_4 molecules, as we might have predicted.

Combination with the Elements of Groups IA and IIA

The ability of the elements of Group VA to accept electrons in ionic bonding and polar covalent bonding varies within the group. The electronegativities presented in Table 6–4 give a good indication of this variation in electron-accepting ability. Nitrogen equals chlorine in electronegativity (the value is 3.0). If chlorine forms ionic chlorides with such elements as sodium, lithium, and calcium, it is not surprising that nitrogen does also.

The **nitride ion,** N^{3-}, exists in ionic solids with cations of Groups IA and IIA metals. For example, lithium nitride is formed by reaction of molten lithium with nitrogen gas according to the following reaction:

$$6\,Li(l) + N_2(g) \rightarrow 2\,Li_3N(s) \qquad \textbf{8–1}$$

You might ask why N^{3-} forms instead of N^- or N^{2-}. In N^{3-} the nitrogen valence orbitals are completely filled. Monatomic anions typically contain filled valence orbitals. We shall show why this is the case in Chapter 11, but it is a fact to be kept in mind as we examine the chemistries of other elements.

The electronegativity typically decreases from top to bottom within any given column of the periodic table, as shown in Table 6–4. Thus, we would predict that the tendency towards monatomic anion formation decreases from N to P to As to Sb to Bi. White phosphorus will react with reactive metals, forming **phosphides.** An example is the formation of calcium phosphide according to the reaction

$$6\,Ca(s) + P_4(s) \rightarrow 2\,Ca_3P_2(s) \qquad \textbf{8–2}$$

The ions As^{3-}, Sb^{3-}, and Bi^{3-} do not exist in ionic solids, and this is consistent with our prediction.

The nitride ion is a very strong base and reacts with water to produce ammonia according to the reaction

$$Li_3N(s) + 3\,H_2O(l) \rightarrow NH_3(aq) + 3\,OH^-(aq) + 3\,Li^+(aq) \qquad \textbf{8–3}$$

It is not surprising that N^{3-} is a strong base. Ammonia serves as a base because of its relatively high-energy nonbonded electron pair. The N^{3-} ion has four such pairs, each of which has a considerably higher energy than the NH_3 pair since N^{3-} has three fewer hydrogen nuclei than NH_3.

Phosphides are also strong bases and react with water to yield **phosphine,** the phosphorus analog of ammonia.

$$Ca_3P_2(s) + 6\,H_2O(l) \rightarrow 2\,PH_3(g) + 3\,Ca(OH)_2(s) \qquad \textbf{8–4}$$

Despite being an analog of NH_3, phosphine is quite a poor base. This is somewhat surprising in view of the fact that the 3*p* electrons of phosphorus are at a higher energy than the 2*p* electrons of nitrogen, and thus should be more easily donated in the formation of donor-acceptor complexes. In a later section of this chapter, and in Chapter 9, we shall discuss the relative strengths of acids and bases.

Ammonia is probably the most important nitrogen-containing compound and it, or the ammonium ion (NH_4^+), is an important constituent of fertilizer. Ammonia for use in fertilizer is obtained from the direct reaction between H_2 and N_2. The reaction is slow at room temperature, however, and is carried out at elevated temperatures. The production and maintenance of elevated temperatures is expensive, and thus ammonia is a relatively expensive fertilizer. What is needed is an inexpensive way to produce nitrogen-containing compounds from N_2 on a huge scale at room temperature. At present no such process exists.

Ammonia burns in oxygen at flame temperatures according to the following reaction:

$$4\,NH_3(g) + 3\,O_2(g) \rightarrow 2\,N_2(g) + 6\,H_2O(g) \qquad \textbf{8–5}$$

Since neither N_2 nor H_2O is an atmospheric pollutant, ammonia is under study as a possible replacement for gasoline in internal combustion engines.

Combination of Nitrogen with the Halogens

Table 6–3 shows that the bonds between halogen atoms in the species F_2, Cl_2, Br_2, and I_2 are relatively weak. The four bond energies are 37, 58, 46, and 36 kcal/mole, respectively. The table also shows that the N—N single bond is weak. According to equation 6–3, nitrogen-halogen bonds should also be weak unless there is a large electronegativity difference between the N atom and the halogen atom. Thus, we predict that N and Cl, for example, should form weak bonds with each other. The N—Cl bond energy is 48 kcal/mole, and the reaction

$$N_2(g) + 3\,Cl_2(g) \rightarrow 2\,NCl_3(g) \qquad \textbf{8–6}$$

has a calculated ΔE from bond energies of 111 kcal. Two NCl_3 molecules represent a much higher energy situation than an N_2 molecule and three Cl_2 molecules. According to our chemical bond model, NCl_3 is predicted to be unstable, readily decomposing to its elements. Indeed, NCl_3, like NBr_3 and NI_3, decomposes explosively.

The N—F bond is stronger than the N—Cl, N—Br, or N—I bonds, consistent with the larger difference in electronegativities of N and F. Thus, the N—F bond energy is 65 kcal/mole, and ΔE for the reaction

$$N_2(g) + 3\,F_2(g) \rightarrow 2\,NF_3(g) \qquad \textbf{8–7}$$

is $6(-65) - (-225) - 3(-37) = -54$ kcal; NF_3 is predicted to be a stable gas at room temperature and this prediction is correct.

The N—N single bond is weak. Some compounds do contain the N—N single bond, however, and one example is **hydrazine,** N_2H_4. This substance, which is a liquid at room temperature, has a molecular structure that may be considered to be the fusion of two ammonia molecules with an N—N bond replacing two N—H bonds. Since each N—H bond possesses a bond energy of 93 kcal/mole and an N—N single bond possesses an energy of 60 kcal/mole, hydrazine is not as stable as ammonia. Hydrazine has been used as a rocket propellant.

The N═N double bond is not much more stable than the N—N single bond. The N═N bond is contained in **dinitrogen difluoride** (N_2F_2). The nitrogen orbital hybridization is sp^2 and all four nuclei lie in one plane. This gives rise to the existence of two **geometrical isomers** (see Figure 8–4) termed **cis** (the two fluorines on the same side of the N═N axis) and **trans** (the fluorines on opposite sides of the N═N axis). The ΔE for the formation of N_2F_2 from its elements is $(-100) - 2(65) - (-225) - (-37) = 32$ kcal/mole and N_2F_2 is an extremely reactive molecule.

F–N═N–F (a) F–N═N–F (b)

Figure 8–4 Structures of the geometrical isomers of dinitrogen difluoride: (a) *trans*, (b) *cis*.

Combination of P, As, Sb, and Bi with the Halogens

The electronegativities of P, As, Sb, and Bi are considerably lower than the electronegativity of nitrogen and they form much more stable compounds with the halogen atoms. Phosphorus forms stable compounds with all of the halogens. In addition to compounds of formula PX_3, where X is any halogen element, phosphorus also forms compounds with the formula PX_5 with X = F, Cl, and Br. Such a formula is a violation of the octet rule, as we can see by applying the procedure for the consideration of molecular structure and bonding (Chapter 6). Let us go through that procedure with PX_5 in steps. Step 1 splits PX_5 into P^{5+} and $5X^-$, thus emptying the valence level of phosphorus and filling each X valence level. In Step 2, five sigma bonds are introduced, and in Step 3, the valence electron count around the P nucleus is found to be 10, yielding a trigonal-bipyramidal structure.

Which orbitals on the P atom participate in the bonding in PX_5? Nitrogen is limited to the formation of NX_3 because the only orbitals available for bonding are the 2*s* and 2*p*. However, the phosphorus valence level includes the 3*d* orbitals; thus bonding that involves phosphorus is not limited by the octet rule.

The triangular belt of the trigonal bipyramid of PX_5 clearly involves sp^2 hybridization of the phosphorus orbitals; but which phosphorus orbitals are involved in the other two bonds? Clearly one of them is the third 3*p* orbital (we shall designate it the $3p_z$). The other is a *d* orbital. (In one of the problems at the end of this chapter you are asked to show that the only 3*d* orbital that yields bonding overlap with the two ligand p_z orbitals is the $3d_{z^2}$ orbital.) Thus, five phosphorus orbitals (3*s*, $3p_x$, $3p_y$, $3p_z$, and $3d_{z^2}$) participate in the bonding in PX_5, and the trigonal-bipyramidal geometry corresponds to **dsp^3 hybridization.**

Both PF_5 and PCl_5 are strong electrophiles since the phosphorus atom contains additional empty, low-energy 3*d* orbitals, and the species PF_6^- and PCl_6^- are found in ionic solids (solid PCl_5 is an ionic substance containing

PCl_4^+ ions and PCl_6^- ions). What is the geometry and orbital hybridization in PX_6^-? Step 1 of our procedure for geometry predictions splits PX_6^- into P^{5+} and six X^-. Six sigma bonds are introduced for a total of twelve valence electrons around the P nucleus. This yields a prediction of the octahedral geometry and this prediction is verified experimentally. (In a problem at the end of this chapter you are asked to show that the phosphorus orbital hybridization in PX_6^- is $\mathbf{d^2sp^3}$.)

The tendency toward formation of non–octet rule compounds is even higher in As and Sb than with phosphorus, since the empty *d* orbitals are closer in energy to the occupied *p* valence orbitals in those atoms than in phosphorus, and the As—X and Sb—X bonds are more polar than the P—X bond. In fact, SbF_5 is such a strong electrophile that it will react with many compounds containing a bonded fluorine atom to extract a fluoride ion and form the octahedral SbF_6^- ion.

The Nitrogen Oxides

Nitrogen compounds such as NH_3 are important to living systems. Other nitrogen-containing compounds represent some of the most troublesome atmospheric pollutants. Among these are certain of the nitrogen oxides. The simplest of these oxides is **nitrogen oxide,** NO (commonly called **nitric oxide**). This molecule contains an odd number of electrons, and thus it must be paramagnetic. Since the nitrogen and oxygen atomic orbitals have comparable energies, the molecular orbitals carry the same designations as in N_2 and O_2. The assignment of the 15 electrons in NO is $1s\sigma^2 1s\sigma^{*2} 2s\sigma^2 2s\sigma^{*2} 2p\sigma^2 2p\pi^4 2p\pi^{*1}$. The bond order in NO is $(10 - 5)/2 = 2.5$. Thus, we expect the bond energy to be intermediate between the 225 kcal/mole for N≡N and the 118 kcal/mole for O═O. The experimental value is 150 kcal/mole. The single antibonding valence electron can be removed by an oxidizing agent and the NO^+ ion is found in several crystalline compounds. Consistent with the removal of an antibonding electron in its preparation, the bond energy of NO^+ is 244 kcal/mole, an energy characteristic of a triple bond.

Nitric oxide is produced in internal combustion engines from the combination of nitrogen and oxygen from the air. The reaction is also brought about by lightning in the atmosphere. When NO is released in automobile exhaust, it is rapidly oxidized to produce **nitrogen dioxide,** NO_2, a noxious red brown gas that is one of the principal components of urban air pollution. Let us use our procedure for obtaining the geometry and bonding in NO_2. Step 1 separates NO_2 into N^{4+} and two O^{2-}, and Step 2 introduces two sigma bonds for a total of five valence electrons in Step 3. We do not have any geometry designations for odd numbers of valence electrons, so we shall have to reason from a neighboring structure involving an even number of electrons. The two neighboring species are the **nitronium ion,**

NO_2^+, and the **nitrite ion,** NO_2^-. We shall use the nitrite ion as the basis for reasoning to the structure of NO_2. For NO_2^-, our geometry procedure yields six valence electrons in Step 3. The structure of NO_2^- is predicted to be based on the triangle. The six valence electrons consist of one nonbonding pair and two sigma-bonding pairs, so that the O—N—O bond angle should be somewhat smaller than 120°; this is indeed the case as determined by experiments. The number of electrons in Step 3 of the geometry procedure is two fewer than eight, so NO_2^- contains a pi bond. Figures 8–5(a) and (b) show the two canonical resonance forms of NO_2^-, an ion in which the total bond order is 3. Figure 8–5(c) shows the NO_2^- resonance hybrid.

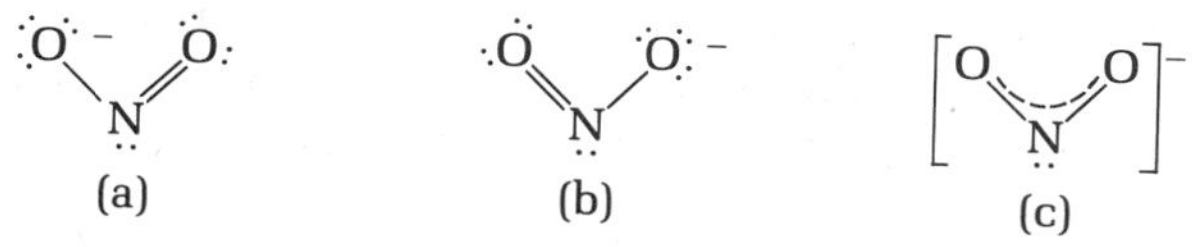

(a) (b) (c)

Figure 8–5 (a, b) Canonical resonance forms of NO_2^- and (c) resonance hybrid representation.

Nitrogen dioxide can be produced by removing one electron from NO_2^-. The easiest way to do this is to add a strong acid to an aqueous solution containing nitrite ions. The nitrite ion is a base and forms nitrous acid, HNO_2, by reacting with strong acids. Nitrous acid is unstable and undergoes self–oxidation-reduction, a process termed **disproportionation:**

$$2\,HNO_2(aq) \rightarrow H_2O(l) + NO(g) + NO_2(g) \qquad \textbf{8–8}$$

The nitrogen oxidation numbers in reaction 8–8 are III, II, and IV for HNO_2, NO, and NO_2, respectively. Thus, reaction 8–8 involves the transfer of one electron from one nitrous acid molecule to another with an accompanying decomposition to NO and NO_2.

We can arrive at the structure of NO_2 by removing the highest-energy electron from NO_2^-. Clearly this will not be a bonding electron, but rather will be a nonbonding electron from either N or O. Since the valence electrons are at a higher energy in the N atom than in the O atom, we expect the electron removed from NO_2^- to be an electron of the nitrogen nonbonded pair shown in Figure 8–5. This should maintain the basic structure and bonding of Figure 8–5 in NO_2, but the O—N—O bond angle should enlarge because of the lesser repulsion from the single nonbonded nitrogen electron. The actual bond angle is 134°. (In a problem at the end of the chapter you are asked to deduce the bond angle in NO_2^+.) The color of NO_2 is due to the single nonbonded electron (although not all species containing single nonbonded electrons have color). The vacancy in this nonbonding orbital can be filled by the **dimerization** of NO_2 to form **dinitrogen tetroxide,** N_2O_4, which is colorless.

A fourth commonly encountered nitrogen oxide is **dinitrogen oxide,** N_2O (commonly termed **nitrous oxide**). One canonical resonance form of this molecule is shown in Figure 7–4(c). The compound is quite a stable gas at room temperature. It is commonly termed "laughing gas" and was the first compound to be used as a general anesthetic.

The Oxyacids of Nitrogen

Two additional nitrogen oxides are formed based on bonding involving the odd electron in NO_2 and NO. Just as two NO_2 molecules can join in N_2O_4 through an N—N bond, they can be joined by an O atom to make **dinitrogen pentoxide,** N_2O_5, whose structure in the gas phase is shown in Figure 8–6 in one canonical resonance form. At room temperature, dinitrogen pentoxide is a white ionic solid composed of NO_2^+ and **nitrate ions,** NO_3^-. The last of the nitrogen oxides is N_2O_3, which is stable only as a pale blue solid at low temperatures and involves the linking of an NO molecule and an NO_2 molecule through an N—N bond. As the temperature is raised, N_2O_3 largely decomposes to NO and NO_2.

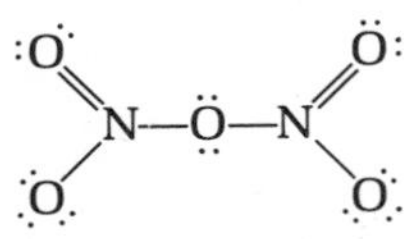

Figure 8–6 One canonical resonance form of the gaseous N_2O_5 molecule.

Both N_2O_3 and N_2O_5 are acids because both produce H_3O^+ when added to water. The oxides of most of the nonmetallic elements are acids. The reactions produced by the addition of N_2O_3 and N_2O_5 to water are

$$N_2O_3(s) + 3\,H_2O(l) \rightarrow 2\,NO_2^-(aq) + 2\,H_3O^+(aq) \qquad \textbf{8–9}$$

$$N_2O_5(s) + 3\,H_2O(l) \rightarrow 2\,NO_3^-(aq) + 2\,H_3O^+(aq) \qquad \textbf{8–10}$$

Dinitrogen pentoxide is a strongly acidic oxide, which produces H_3O^+ quantitatively in water. Dinitrogen trioxide (N_2O_3) produces nitrous acid (HNO_2) by reaction with water, and HNO_2 ionizes only slightly to yield H_3O^+ and NO_2^- ions. Thus, N_2O_3 is a considerably weaker acid than N_2O_5.

We have introduced the concept of an acid in the molecule HCl, which can produce H_3O^+ by donating a proton to H_2O. Such a proton donor is termed a **Lowry-Brønsted** acid, after the two scientists who recognized as the common feature of such acids that they donate protons. Dinitrogen pentoxide is also an acid. However, N_2O_5 does not donate protons to water, because it contains no protons to donate. Instead it accepts electron pairs from water and protons are produced as a consequence. Thus, strong electrophiles are also acids and they are termed **Lewis acids** (after G. N. Lewis of the Lewis dot representation). The names *Lewis acid* and *electrophile* are often used interchangeably.

If one mole of N_2O_5 is added to one mole of water, two moles of **nitric acid** (HNO_3) are produced. Nitric acid is one of an important class of compounds termed **strong mineral acids** because it can be produced from nitrate-containing minerals. Pure HNO_3 is a liquid at room temperature and boils at 84 °C. The acid is commonly obtained commercially as a concentrated aqueous solution (68 percent HNO_3 by weight).

Why is HNO_3 a strong acid while HNO_2 is a weak acid? An explanation lies in the formal charges on the nitrogen atom in each molecule. Figure 8–7 shows Lewis dot representations of HNO_3 and HNO_2. With HNO_2, the

(a) (b)

Figure 8–7 Lewis dot representations of (a) nitric acid and (b) nitrous acid.

formal charge on N is zero and in HNO_3 it is +1. This model describes an electron deficiency at the nitrogen in HNO_3, and this provides an explanation for the strong acidity. The deficiency can be satisfied to some extent by the expulsion of the proton with the production of the nitrate anion with its excess of one electron. Nitrous and nitric acids are termed **oxyacids** of nitrogen. We should predict that **in the series of oxyacids of any element the strength of the oxyacid increases as the formal charge on the element increases.**

The electron affinity of nitrogen with a formal charge of +1 is sufficiently high that HNO_3 and NO_3^- can remove electrons completely from other species; thus nitric acid and NO_3^- are predicted to be very strong oxidizing agents. The reduction of NO_3^- offers a convenient method for preparing nitrogen oxides. Two such reactions are

$$3\,Cu(s) + 8\,H^+(aq) + 2\,NO_3^-(aq) \rightarrow 3\,Cu^{2+}(aq) + 2\,NO(g) + 4\,H_2O(l) \qquad \textbf{8–11}$$

$$NH_4NO_3(s) \rightarrow N_2O(g) + 2\,H_2O(l) \qquad \textbf{8–12}$$

Reaction 8–12 illustrates the oxidizing ability of NO_3^- even in the solid state. This reaction can be carried out in the laboratory by gently heating solid ammonium nitrate. If the temperature of solid NH_4NO_3 is made too high, the decomposition occurs explosively; NH_4NO_3 has long been used as an inexpensive explosive. Perhaps the oldest application of the oxidizing ability of NO_3^- is as the oxidizing agent in gunpowder.

The Oxides and Oxyacids of Phosphorus

The oxidation numbers of nitrogen in the nitrogen oxides range from I in N_2O to V in N_2O_5. This broad range results from the similar electronegativities of the nitrogen atom and the oxygen atom together with the existence of strong pi bonding. When the electronegativity gap is considerably enlarged, as it is between phosphorus and oxygen, and the strong pi-bonding possibility is removed, only high oxidation states are observed. Thus phosphorus forms oxides in which phosphorus has oxidation numbers III and V.

If white phosphorus is burned in a limited supply of air, the product is P_4O_6, which has the structure shown in Figure 8–8. This phosphorus oxide involves each P atom bonded to three oxygen atoms in a pyramidal fashion. The four P nuclei define the corners of a tetrahedron as in P_4 and the six oxygens are associated with the six edges of the tetrahedron. If white

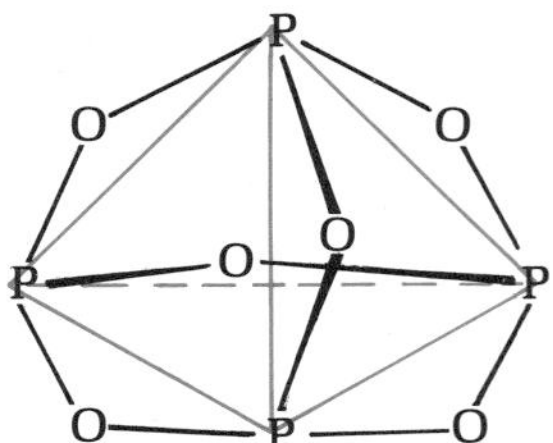

Figure 8–8
The structure of P_4O_6.

phosphorus is burned in an excess of oxygen, the oxide obtained is P_4O_{10}. This oxide has essentially the same structure as P_4O_6 save that each phosphorus is bonded to an additional oxygen atom so that the arrangement of O atoms around each P atom is approximately tetrahedral.

Both P_4O_6 and P_4O_{10} are acidic oxides. If one mole of P_4O_6 is added to six moles of water, four moles of **phosphorous acid,** H_3PO_3, are obtained. The oxide P_4O_{10} yields **orthophosphoric acid,** H_3PO_4. Possible Lewis representations of H_3PO_3 and H_3PO_4 are shown in Figures 8–9(a) and (b), respectively.

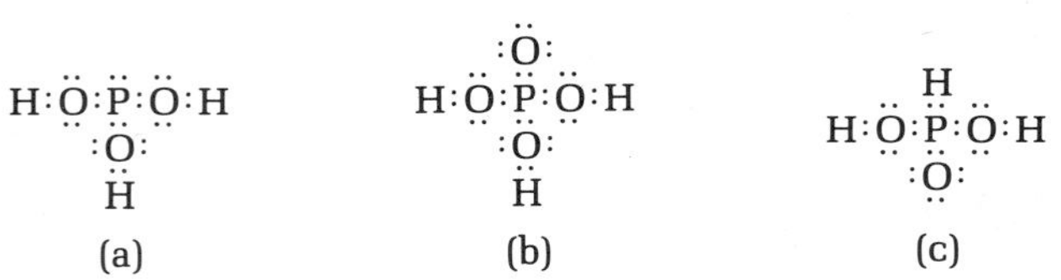

Figure 8–9
Lewis representations of (a) a hypothetical and (c) an actual structure of the phosphorous acid molecule, H_3PO_3. (b) Lewis representation of the orthophosphoric acid molecule, H_3PO_4.

```
                        :Ö:                   
 H:Ö:P̈:Ö:H          H:Ö:P̈:Ö:H             H
    :Ö:                 :Ö:             H:Ö:P̈:Ö:H
     H                   H                 :Ö:
    (a)                 (b)                (c)
```

The formal charges on phosphorus in H_3PO_3 and H_3PO_4 are zero and +1. Thus, we predict that H_3PO_4 is the stronger acid. Actually H_3PO_3 and H_3PO_4 are very nearly equal in acid strength. The explanation is that Figure 8–9(a) is an incorrect representation of H_3PO_3 and that the correct picture of the bonding is shown in Figure 8–9(c). The formal charge on the phosphorus atom in Figure 8–9(c) is +1, just as it is in H_3PO_4.

Both H_3PO_3 and H_3PO_4 are much weaker acids than HNO_3 in spite of the fact that the central atom has a formal charge of +1 in all three molecules. This is consistent with the greater electronegativity of N as compared with P.

Orthophosphoric acid can donate one, two, or three protons to water. The anions produced by this proton transfer are **dihydrogen orthophosphate** ($H_2PO_4^-$), **hydrogen orthophosphate** (HPO_4^{2-}), and **orthophosphate** (PO_4^{3-}). The structure of PO_4^{3-} is tetrahedral, the structure our geometry procedure would yield. It contains four identical P—O bonds.

Phosphates are very important in the growth of foodstuffs because they are incorporated into several very important biochemical compounds. Much of their biological importance derives from their ability to form chains through P—O—P linkages. The structure of one of these species, **pyrophosphate ion,** is shown in Figure 8–10. Pyrophosphate ion ($P_2O_7^{4-}$) derives its name from its mode of formation. The prefix *pyro* alludes to fire, and the pyrophosphate ion is prepared by heating a solid hydrogen orthophosphate.

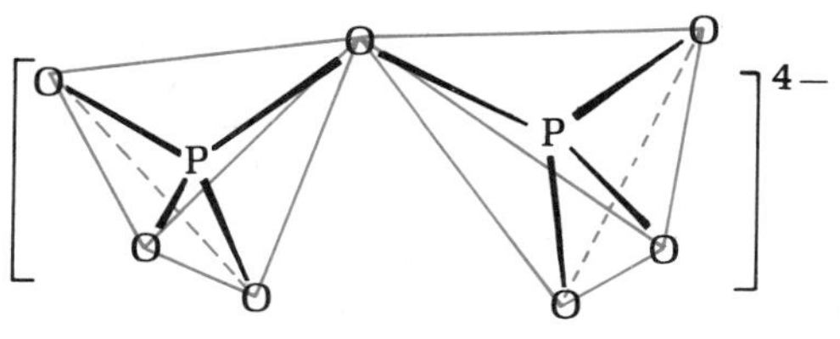

Figure 8–10 The structure of pyrophosphate ion, $P_2O_7^{4-}$.

$$2\ Na_2HPO_4(s) \rightarrow Na_4P_2O_7(s) + H_2O(g) \qquad \textbf{8–13}$$

A longer chain, **tripolyphosphate,** is produced by heating a mixture of hydrogen orthophosphate and dihydrogen orthophosphate:

$$2\ Na_2HPO_4(s) + NaH_2PO_4(s) \rightarrow Na_5P_3O_{10}(s) + 2\ H_2O(g) \qquad \textbf{8–14}$$

The geometry around each phosphorus in the chainlike phosphates is tetrahedral, as shown in Figure 8–10, and the linking occurs through oxygen atom bridges ($P_3O_{10}^{5-}$ contains two such bridges). These bridging bonds are extremely important biologically because it is the splitting of such bonds by the addition of water (the reverse of reactions 8–13 and 8–14) that provides the energy for many biological processes.

8–2. The Chemistry of the Group VIA Elements

Elemental Forms

The O═O double-bond energy is 118 kcal/mole and this is more than double the O—O single-bond energy. This explains why the most stable form of elemental oxygen is gaseous O_2. Oxygen also exists in the less stable form **ozone** (O_3) largely in the upper atmosphere, where it is formed from O_2 by the absorption of high-energy radiation from the sun. This reaction is very important in that it screens this radiation from the earth's surface; if the radiation were allowed to reach this planet, it would destroy life on it. The structure and bonding in O_3 are obtained by our procedure for molecular structure and bonding. Step 1 splits the molecule into O^{4+} and two O^{2-}; Step 2 introduces two sigma bonds, and the valence electron sum in Step 3 is six (two bonding pairs and one nonbonding pair). The bond angle is predicted to be approximately 120° (it is 117°). The molecule contains one three-atom pi bond, and Figure 8–11 shows two canonical resonance forms of the molecule. Ozone is a very strong oxidizing agent, even stronger than O_2.

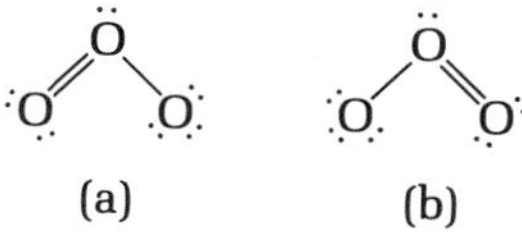

Figure 8–11 Canonical resonance forms of ozone.

Like P, As, and Sb, the other elemental forms in Group VIA have their structures based exclusively on sigma bonding, and the maximum number of sigma bonds in which each atom participates is two. The most stable molecular form of elemental sulfur is the S_8 ring shown in Figure 8–12. All of the bond distances (2.12 Å) and bond angles (105°) in the S_8 ring are the same.

Figure 8–12 The puckered ring structure of the S_8 molecule.

Two crystalline forms contain the S_8 molecule. **Rhombic sulfur** is the more stable form at room temperature and **monoclinic sulfur** is the stable form above 96 °C.

The S—S bond is relatively weak (54 kcal/mole is the bond energy) and the S_8 rings are easily broken at moderately high temperatures. Thus, rhombic sulfur melts to form a liquid of normal viscosity. If this liquid is heated above about 160 °C, it becomes very viscous because the S_8 rings break and reform into long, tangled chains. This chain form can be "trapped" by sudden cooling and that form is amorphous **plastic sulfur.** If plastic sulfur is left to stand at room temperature, it slowly converts to the more stable rhombic sulfur. Two other unstable forms can be prepared also. **Rho sulfur** consists of S_6 rings, which are somewhat more strained than the S_8 rings and consequently less stable. Sulfur vapor at very high temperatures consists largely of S_2 molecules; if this vapor is condensed on a cold surface, **purple sulfur** is formed.

Just as As and Sb show tendencies toward metallic structures, so do Se and Te. Selenium exists in a form containing Se_8 molecules and in a metal-like form, while tellurium exists only in the metallic structure.

Combination with the Elements of Groups IA and IIA

The electronegativities of the elements in Group VIA are higher than the electronegativities of the neighboring elements in Group VA, and O, S, Se, and Te are predicted to form stronger, more polar bonds with Groups IA and IIA elements than N, P, As, and Sb do; indeed this is the case.

Because of this greater electronegativity the anionic form of the Group VIA elements, as compared with Group VA, should be more commonly encountered. The **oxide ion** is quite commonly found in crystalline solids involving Group IA and Group IIA elements, and these elements are much more reactive towards gaseous oxygen than towards gaseous nitrogen. An example is

$$2\ Ca(s) + O_2(g) \rightarrow 2\ CaO(s) \qquad \textbf{8–15}$$

Like the nitride ion, the oxide ion is a strong base. However, the electron energy in oxide ions is considerably more negative than in nitride ions. As a consequence, the electrons of O^{2-} have a lesser tendency to associate with protons than the electrons of N^{3-} do. The molecules that result from the protonation of O^{2-} and N^{3-} are H_2O and NH_3. If O^{2-} has a lesser tendency than N^{3-} to accept protons, water must have a greater tendency than ammonia to lose protons. Thus, water is predicted to be a stronger acid than ammonia, and this agrees with fact (although both H_2O and NH_3 are very weak acids). Water is a stronger acid than ammonia and this correlates with the greater electron affinity of oxygen as compared with nitrogen. In any given row of the periodic table, the acidities of the hydrides show a strong correlation with the electron affinities of the elements combined with the hydrogen. The electron affinity of Li is quite small (18 kcal/mole), and LiH reacts as a base with water, as all of the hydrides of groups IA and IIA do except for the hydride of hydrogen itself (H_2):

$$LiH(s) + H_2O(l) \rightarrow Li^+(aq) + OH^-(aq) + H_2(g) \qquad \textbf{8–16}$$

The highest electron affinity in the same row of the periodic table as Li, N, and O occurs with fluorine, and HF is predicted to be the strongest acid of the hydrides of the elements of this row. Such is indeed the case.

The principal compound formed between oxygen and hydrogen is water; but a second compound is also formed, **hydrogen peroxide** (H_2O_2). This molecule is the oxygen analog of hydrazine and Lewis representations of N_2H_4 and H_2O_2 are shown in Figure 8–13. No attempt has been made to depict the correct bond angles in either case.

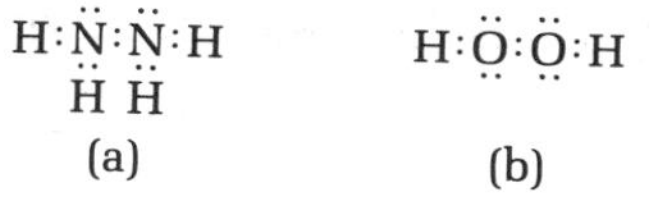

$$\underset{\text{(a)}}{\mathrm{H\!:\!\overset{\cdot\cdot}{\underset{\cdot\cdot}{N}}\!:\!\overset{\cdot\cdot}{\underset{\cdot\cdot}{N}}\!:\!H} \atop \mathrm{H\ H}} \qquad \underset{\text{(b)}}{\mathrm{H\!:\!\overset{\cdot\cdot}{\underset{\cdot\cdot}{O}}\!:\!\overset{\cdot\cdot}{\underset{\cdot\cdot}{O}}\!:\!H}}$$

Figure 8–13
Lewis representations of (a) hydrazine and (b) hydrogen peroxide.

Hydrogen peroxide is one of the few compounds in which oxygen is not assigned an oxidation number of –II. Instead it is –I. The molecule is easily oxidized to O_2 and reduced to H_2O, and it is both a good oxidizing agent and a good reducing agent. The weakness of the O—O single bond results in the disproportionation of H_2O_2 according to the following reaction:

$$2\,H_2O_2(l) \rightarrow 2\,H_2O(l) + O_2(g) \qquad \textbf{8–17}$$

The ΔE for reaction 8–17 as calculated from bond energies is $(-118) - 2(-51) = -16$ kcal/mole, and the instability of H_2O_2 derives from the fact that the O—O single bond is less than half as strong as the O═O double bond. (This is the same reason, as we have seen, for the instability of ozone.) Reaction 8–17 accounts for the fact that H_2O_2 is not a commonly used solvent. In many other ways it closely resembles water. For example, the O—H bonds are highly polar and H_2O_2 molecules strongly interact with each other by hydrogen bonding. This strong interaction results in a boiling point of 150 °C.

Sulfur possesses a considerably higher electron affinity than phosphorus does, and as a consequence, the **sulfide ion** (S^{2-}) is much more commonly encountered than the phosphide ion. Many metallic elements occur naturally in the earth as sulfide salts. The sulfide ion is also a base, but it is a weaker base than O^{2-}. As a consequence, H_2S is a stronger acid than H_2O. This increase in the acidity of hydrides from top to bottom in a given column of the periodic table is a general phenomenon. The compound H_2Se is a stronger acid than H_2S, and H_2Te is stronger than H_2Se. All of the compounds, H_2S, H_2Se, and H_2Te, are poisonous, extremely malodorous gases. The odors of many Se and Te compounds are so offensive that the chemistries of these elements have not been extensively developed. While the ions As^{3-} and Sb^{3-} do not occur, the ions Se^{2-} and Te^{2-} are commonly found in small amounts in sulfide ores.

Combination with the Halogens

The elements of Group VIA should, like the Group VA elements, display a wide range of reactivities toward an element such as fluorine; they also have a wide range of electronegativities. Oxygen forms many compounds with the elements of Group VIIA. Like the combinations of nitrogen with these elements, all of these oxygen-containing compounds are, with the lone exception of OF_2, explosively unstable, readily decomposing to the elements. Just as NF_3 is stable because of the strength and polarity of the N—F bond, OF_2, which contains two O—F bonds, is also stable. The species OF_2 is also a strong oxidizing agent, as the following reaction indicates:

$$OF_2(g) + H_2O(l) \rightarrow O_2(g) + 2\,HF(aq) \qquad \textbf{8–18}$$

Like phosphorus, sulfur forms many stable compounds with the halogens. The difference between the S and F electronegativities is large and the sulfur appears with high oxidation numbers. The three stable fluorides are SF_4, SF_6, and S_2F_{10}. The SF_6 molecule possesses an octahedral arrangement of fluorines around a central sulfur, as our geometry and bonding procedure (Chapter 6) would yield. The S_2F_{10} molecule involves two such octahedra with an S—S bond replacing two S—F bonds. Let us apply the geometry procedure to SF_4, because its structure is less obvious. Step 1 splits it to S^{4+} and four F^- and Step 2 introduces four sigma bonds for a total of ten valence electrons in Step 3. Thus, the structure is based on the trigonal bipyramid. The nonbonding electron pair has a choice between the triangular belt or the direction perpendicular to the belt. The belt is chosen because there are two fluorines located ~90° from the direction of the pair rather than the three fluorines the other structure calls for. The structure of SF_4 is shown in Figure 8–14. Selenium and tellurium form fluorides that are similar in formula and structure to the sulfur fluorides.

Figure 8–14 The structure of sulfur tetrafluoride.

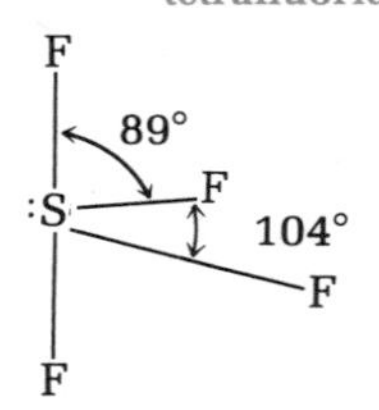

The electronegativities are much more nearly equal for S and Cl than for

S and F, and thus less stable compounds are formed between S and Cl, and the S oxidation numbers are lower. The compounds are SCl_2, S_2Cl_2, and SCl_4. Both SCl_2 and S_2Cl_2 are liquids at room temperature. The molecule SCl_2 is the chloride analog of H_2S, and S_2Cl_2 is the fusion of two SCl_2 molecules with an S—S bond replacing two S—Cl bonds. Sulfur also forms the bromide SBr_2, which is also a liquid at room temperature. Sulfur tetrachloride is the least stable of the sulfur chlorides and must be stored at low temperatures to avoid decomposition. Selenium and tellurium form tetrachlorides in addition to the species Se_2Cl_2, Te_2Cl_2, $SeCl_2$, and $TeCl_2$.

The Oxides and Oxyacids of Sulfur

The oxides of sulfur, SO_2 and SO_3, are both dangerous atmospheric pollutants. Sulfur dioxide (SO_2) is the principal product from the combustion of sulfur. Unfortunately, sulfur is contained in relatively high concentration (as sulfides) in much of the coal that is available as a natural resource in large supply in the United States. This is the principal reason that petroleum has supplanted coal as a fuel in the United States. However, the high cost and short supply of petroleum have led to considerable research recently on removing sulfur from coal to give a "clean" fuel.

Sulfur dioxide has the same number of valence electrons as O_3 and SO_2 is predicted to be similar in structure to O_3. The O—S—O bond angle in SO_2 is 120°.

Sulfur trioxide is a more dangerous pollutant than SO_2. It is produced in the atmosphere in the following reaction.

$$2\ SO_2 + O_2 \rightarrow 2\ SO_3 \qquad \textbf{8–19}$$

Reaction 8–19 is normally slow; but the rate is increased by solids that are suspended in smoke from the burning of coal. Our geometry procedure yields the correct structure of SO_3, namely the sulfur at the center of an equilateral triangle of oxygens.

Figure 8–15 shows Lewis representations of SO_2 and SO_3. The formal charges on sulfur in these representations of SO_2 and SO_3 are +2 and +3,

Figure 8–15
Lewis representations of (a) SO_2 and (b) SO_3.

respectively. Both oxides are Lewis acids and strong oxidizing agents consistent with the formal charges, and SO_3 is the stronger acid and oxidizing agent of the two.

When SO_2 is added to water it forms the donor-acceptor complex, **hydrogen sulfite ion** (HSO_3^-), according to the reaction

$$SO_2(aq) + H_2O(l) \rightarrow HSO_3^-(aq) + H^+(aq) \qquad \textbf{8–20}$$

The hydrogen sulfite ion is a weak Lowry-Brønsted acid that can ionize in water to produce $H^+(aq)$ and the **sulfite ion** (SO_3^{2-}). There is no evidence for the existence of H_2SO_3, the molecule that we might predict would result from the protonation of HSO_3^-. Protonation of HSO_3^- leads to the formation of SO_2 and water.

Sulfur trioxide reacts slowly with water to produce **sulfuric acid,** H_2SO_4, one of the most important strong acids known. In aqueous solution, H_2SO_4 ionizes essentially quantitatively to **hydrogen sulfate ion** (HSO_4^-), which ionizes more weakly to **sulfate ion** (SO_4^{2-}). Sulfate ion is a common anion in ionic crystals. Our geometry procedure yields the correct structure of SO_4^{2-} with the sulfur in the center of a tetrahedron of oxygens; all four S—O bonds are identical.

Sulfuric acid is the most widely produced industrial chemical per year by weight. This is because it is a strong acid, a good oxidizing agent when heated, and a good dehydrating agent (it forms strong hydrogen bonds with water). It is relatively inexpensive to obtain from sulfur or sulfides. An inexpensive compound with these three properties finds many applications in industrial processes. The key steps in the production of H_2SO_4 include reaction of the sulfur or sulfide with oxygen to produce SO_2 followed by conversion to SO_3 by further reaction with O_2. Pure sulfuric acid is a liquid at room temperature and it is significantly self-ionized according to the reactions

$$2\,H_2SO_4 \rightleftharpoons H_3SO_4^+ + HSO_4^- \qquad \textbf{8–21}$$

When SO_3 is added to pure H_2SO_4, a donor-acceptor complex is formed involving SO_3 as the electrophile and hydrogen sulfate ion as the nucleophile. The resultant ion ($HS_2O_7^-$) is termed **hydrogen pyrosulfate ion,** and protonation yields $H_2S_2O_7$, termed **pyrosulfuric acid.** These two species contain the S—O—S bridging link, which is analogous to the P—O—P link in the pyrophosphate ion. As with pyrophosphate ion, the S—O—S link is split by the addition of water and the pyrosulfuric acid produced from SO_3 and H_2SO_4 is reconverted to H_2SO_4 by the addition of water.

The dehydrating ability of sulfuric acid is so strong that it can remove the elements of water from other compounds. The nitronium ion (NO_2^+), a very useful species in synthesizing certain commercially important explosives, is produced in solution by mixing concentrated solutions of nitric and sulfuric acids:

$$H_2SO_4(aq) + HNO_3(aq) \rightarrow NO_2^+(aq) + HSO_4^-(aq) + H_2O(l) \qquad \textbf{8–22}$$

The strong acid and dehydrating properties of H_2SO_4 can be employed at room temperature with little interference from the oxidizing ability of the molecule, since H_2SO_4 is a relatively poor oxidizing agent at room temperature. However, the oxidizing ability of H_2SO_4 increases with increasing temperature, and the most common way to produce SO_2 in the laboratory is by reducing hot concentrated H_2SO_4 by metallic copper:

$$Cu(s) + 2\ H_2SO_4(aq) \rightarrow SO_2(g) + Cu^{2+}(aq) + SO_4^{2-}(aq) + 2\ H_2O(l) \quad \textbf{8–23}$$

The dangerous character of SO_3 as an atmospheric pollutant derives from its reaction with water to produce H_2SO_4.

Two other sulfur-oxygen species are commonly used in redox (reduction-oxidation) reactions. **Peroxydisulfuric acid,** a powerful oxidizing agent, involves the fusion of two H_2SO_4 molecules and the production of a S—O—O—S link. The strong oxidizing power of the molecules derives from the combination of a positive formal charge on sulfur together with the presence of the weak O—O bond. Peroxydisulfate salts in acidic aqueous solution are strong enough oxidizing agents to oxidize many elements to their highest observable oxidation states.

If one of the four oxygens of SO_4^{2-} is replaced with a sulfur atom, we obtain the **thiosulfate ion** ($S_2O_3^{2-}$). Thiosulfate is easily and rapidly converted to **tetrathionate ion** ($S_4O_6^{2-}$) by weak oxidizing agents such as elemental iodine, and $S_2O_3^{2-}$ is commonly used in quantitative analyses for iodine.

8–3. Summary

We can correlate a large number of chemical facts not only through models, but also by recognizing and remembering trends in chemical properties within groups and between groups of the periodic table. It helps enormously if those trends make sense to us, and models are important in making sense out of these trends.

Some trends shown by groups in the periodic table are the following:

1. The tendency of an element to form monatomic anions increases from left to right along any given row of the periodic table (excluding Group VIIIA).
2. The tendency toward monatomic anion formation decreases from top to bottom in a given column of the periodic table.
3. The hydrides of the nonmetallic elements are acids. The acid strengths increase from left to right along a given row of the periodic table, and they also increase from top to bottom in a given column of the table.
4. The tendency toward higher positive oxidation numbers increases from top to bottom within a given column of the table.
5. The oxides of the nonmetallic elements are acidic. The acid strength increases with an increase in formal charge on the nonmetallic element in the oxide. With a given formal charge, the acid strength decreases from top to bottom in a column of the table. With a given formal charge, the acid strength increases from left to right along a given row of the periodic table.

6. The oxides of the nonmetallic elements react with water to form oxyacids. The strengths of these oxyacids show the same trends as the oxides that yield them.
7. The oxidized forms of the nonmetallic elements are typically oxidizing agents. For a given oxidation number, the oxidizing strength typically increases from left to right along a row of the periodic table and the oxidizing strength decreases from top to bottom in a given column of the table.

SUGGESTIONS FOR ADDITIONAL READING

Jolly, W. L., *The Chemistry of the Non-Metals,* Prentice-Hall, Englewood Cliffs, N.J., 1966.

Sanderson, R. T., *Inorganic Chemistry,* Reinhold, New York, 1967.

PROBLEMS

1. Suppose that somehow nitrogen could be converted to a solid with a structure analogous to that of black phosphorus. Calculate the ΔE for the conversion of one mole of this solid to N_2 gas.
2. Approximately 1 percent of hydrogen nuclei contain a neutron as well as a proton. This heavy hydrogen is called deuterium (symbolized by D). Suppose that you had a sample of D_2O and you wished to convert it to ND_3. How would you accomplish the conversion?
3. Using bond energy data, calculate ΔE for reaction 8–5. Compare this value with ΔE for the combustion of one mole of methane. Is it reasonable to consider using ammonia as a fuel? Explain.
4. Do you predict that NF_3 is a stronger base or a weaker base than NH_3?
5. Elemental chlorine, Cl_2, is a fairly strong oxidizing agent in the process of going to Cl^-. From the ΔE for reaction 8–6, do you predict that NCl_3 is a stronger or weaker oxidizing agent than Cl_2 in the process of going to N_2 and Cl^-? Explain.
6. In the trigonal-bipyramidal structure of PF_5, the phosphorus orbital hybridization is dsp^3. Show by investigating the overlap that if the triangular plane is the xy plane, the $3d_{z^2}$ orbital of phosphorus can bond to fluorines above and below this plane but the other four $3d$ orbitals of phosphorus cannot.
7. The three P—F bonds in the triangular plane of the PF_5 molecule are shorter than the two bonds directed perpendicular to this plane. Explain why.
8. What is the geometry and orbital hybridization in PCl_4^+?
9. Show that if the six halogens in PX_6^- are located along the x, y, and z

axes with the phosphorus nucleus at the origin, only the phosphorus d_{z^2} and $d_{x^2-y^2}$ orbitals of the five d orbitals can participate in the bonding, so that the phosphorus hybridization is d^2sp^3.

10. The NO^+ ion possesses a very strong NO bond, stronger even than N_2. However, N_2 is a very unreactive species and NO^+ is extremely reactive. Why?

11. What is the O—N—O bond angle in NO_2^+?

***12.** Nitric acid dimerizes to a small extent to produce N_2O_2. Propose a structure for N_2O_2 and draw a Lewis dot representation of the bonding. Explain why NO shows a much smaller tendency to dimerize than NO_2.

13. Draw a Lewis representation of N_2O_3.

14. What are the formal charges on nitrogen in N_2O_3 and N_2O_5?

15. Why doesn't phosphorus form the oxides PO and PO_2 in analogy with NO and NO_2?

16. Use the geometry procedure to show that the structure of PO_4^{3-} is a tetrahedral arrangement of four oxygens around a central phosphorus. On which of the four oxygens do the negative charges reside?

17. Besides forming H_3PO_3 and H_3PO_4, phosphorus also forms an acid with the formula H_3PO_2, and it is approximately equal in acid strength to H_3PO_3 and H_3PO_4. Draw a Lewis representation of H_3PO_2.

18. Hydrogen peroxide is a somewhat stronger acid than water. Explain why.

19. What dipole moment value is predicted for SF_6? Use the result of this prediction to account for the fact that SF_6 is a gas at room temperature, while SCl_2 and S_2Cl_2 are liquids.

20. Draw a Lewis representation of fluorosulfuric acid, FSO_3H. Would you predict FSO_3H to be a stronger acid than H_2SO_4 or a weaker one?

21. Which do you predict to be the stronger oxidizing agent, H_3PO_4 or H_2SO_4? Explain.

9/ The Chemistry of Groups IIIA, IVA, VIIA, and VIIIA

Our emphasis so far has been on relatively common compounds like nitric acid, sulfuric acid, and sulfur dioxide. We used those species in examining common chemical properties such as acidity, and in establishing systematic trends in chemical properties. Models have been helpful in explaining and in some cases predicting the trends.

The properties of the remaining nonmetallic elements can be established from the most important compounds that the elements of Groups IIIA, IVA, VIIA, and VIIIA form.

9–1. The Chemistry of the Halogens

Elemental Forms

Much of the chemistry of the halogens is pertinent to the chemistries of other elements. The fluorine atom has the highest electronegativity of any atom, and thus it becomes an electron acceptor to more different kinds of atoms than any other element does. The extreme reactivity of elemental fluorine as an oxidizing agent makes it the most difficult of the halogens to prepare in elemental form. By definition, no chemical agent can remove electrons from the strongest oxidizing agent, nor can any chemical agent add electrons to the strongest reducing agent. The preparation of extremely strong oxidizing agents and extremely strong reducing agents is normally accomplished electrolytically. Such is the case with fluorine, which is prepared from molten ionic fluorides. Chlorine (Cl_2), which at room temperature is a green yellow, poisonous gas, is also produced electrolytically—from molten ionic chlorides.

The valence electrons of bromide ion are of sufficiently high energy that they can be removed by elemental chlorine. Seawater is a good source of bromide ion, which by reaction with Cl_2 produces bromine:

$$2\,Br^-(aq) + Cl_2(g) \rightarrow 2\,Cl^-(aq) + Br_2(l) \qquad \textbf{9–1}$$

Elemental bromine is a reddish liquid with a boiling point of 59 °C. It is the only nonmetallic element that is a liquid at room temperature. Iodine is a dark-colored solid at room temperature, and this solid contains I_2 molecules.

The bond energies of F_2, Cl_2, Br_2, and I_2 are 37, 58, 46, and 36 kcal/mole. This is a rather unusual trend. Most properties show a progressive change from top to bottom in a given column of the periodic table. These progressive changes reflect the progressive changes in electronegativities.

The electron affinities of the halogen atoms increase from F to Cl and then decrease again from Cl to Br to I. Thus, the bond strengths of the homonuclear diatomic halogen molecules reflect the electron affinities. When identical atoms share electrons, a key factor must be the ability of one atom to accept electrons from the other atom. This tendency of an atom to accept additional electron density is measured by the electron affinity.

Combination with Elements of Groups IA and IIA

All of the halogens possess relatively high electronegativities and they combine with metallic elements to form ionic salts containing the ions F^-, Cl^-, Br^-, and I^-.

Consistent with the low valence electron energies in the atoms, the halide ions are relatively poor bases and the acids obtained by protonating the halide ions are relatively strong acids. The hydrogen halides HF, HCl, HBr, and HI are all gases at room temperature and can be prepared by heating a mixture of the halide salt and orthophosphoric acid (chosen because it has a high boiling point and because it is a poor oxidizing agent). An example of such a preparative reaction is

$$NaBr(s) + H_3PO_4(l) \rightarrow HBr(g) + NaH_2PO_4(s) \qquad \textbf{9–2}$$

The acid strengths of the hydrogen halides increase with increasing atomic number of the halogen. Hydrogen fluoride is a relatively weak acid while HCl, HBr, and HI are all strong acids. We have seen that the acid strengths of the hydrides in a given row of the periodic table reflect the electron affinities of the elements bound to hydrogen. Thus, for example, HF is a stronger acid than H_2O, which is a stronger acid than NH_3. This correlation between acid strength and electron affinity does not hold within any given column of the periodic table. For instance, with the halogens, the electron affinities of the halogen atoms do not increase progressively from F to I as the strengths of the acids HF to HI do. The acid dissociation of a hydrogen halide occurs according to the reaction

$$HX(aq) + H_2O(l) \rightarrow H^+(aq) + Cl^-(aq) + H_2O(l) \qquad \textbf{9–3}$$

As we can see, the HX bond is broken in the dissociation process. Thus, in addition to the electron affinity of X, the HX bond energy is a second factor that influences the acid strength of HX. (As the bond strength decreases, the acid strength should increase.) This bond strength markedly decreases from top to bottom in any given column of the periodic table, accounting for the trend in acid strength.

Combination of the Halogens with Each Other

Because the electronegativities decrease markedly in going from F to I, the different halogens can form polar bonds with each other, and relatively high oxidation numbers can appear, just as in the binding of the Group VIA elements with the halogens. The lowest oxidation numbers should occur in compounds between neighboring halogens; the highest numbers should appear in compounds involving iodine and fluorine, since the difference in electronegativities is greatest between those two elements. The interhalogen species are characterized by some unusual molecular geometries and they offer us an opportunity to expand a bit on our geometry procedure.

Chlorine and fluorine are neighboring halogens and they form ClF and ClF_3. The structure of ClF_3 can be deduced by our geometry procedure. The valence electron sum in Step 3 is ten, consisting of three bonding pairs and two nonbonding pairs. The structure is based on the trigonal bipyramid, and the resulting T-shaped molecule is shown in Figure 9–1. Resembling SF_4 (Chapter 8), the lone pairs lie in the triangular belt. Repulsion by the lone pairs lowers the F—Cl—F bond angle from 90° to 87.5°. Since the elemental halogens are moderate (I_2) to very strong (F_2) oxidizing agents, the interhalogen molecules are also moderate to strong oxidizing agents. Both ClF and ClF_3 are strong oxidizing agents.

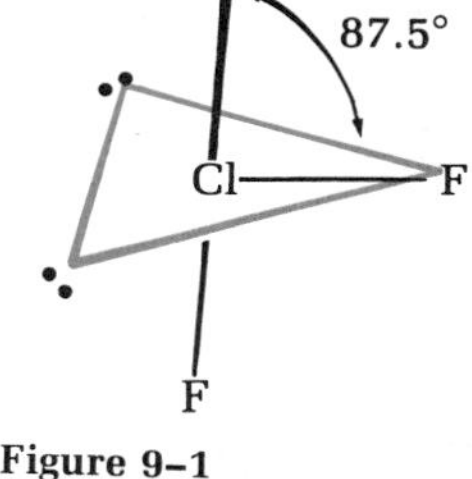

Figure 9–1 The structure of the chlorine trifluoride molecule.

As expected, bromine appears in a higher oxidation state with fluorine than chlorine does. Besides forming the compounds BrF and BrF_3 (BrF_3 has essentially the same structure as ClF_3), these two elements also form BrF_5. Our geometry procedure for BrF_5 yields twelve valence electrons in Step 3, and the structure is based on the octahedron. Two of the twelve valence electrons are a nonbonding pair; the structure is shown in Figure 9–2. Because of repulsion by the lone pair, all of the fluorine nuclei in the figure are above the bromine nucleus, so that the molecule has an "inverted umbrella" structure.

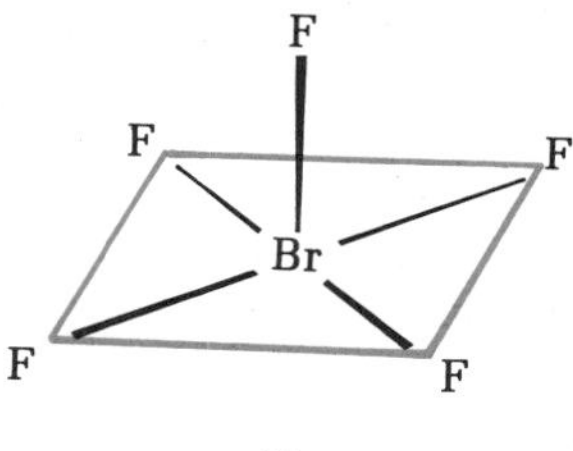

Figure 9–2 The structure of bromine pentafluoride.

There is only one compound formed between the neighboring elements Br and Cl and that is BrCl. With iodine and fluorine the compounds formed contain iodine in high oxidation states, as we should predict; these compounds are IF_5 and IF_7. Iodine pentafluoride is essentially like BrF_5 in structure. According to our geometry procedure, iodine heptafluoride possesses fourteen valence electrons around the iodine nucleus. The structure is believed to involve the I in the center of a pentagon defined by five fluorines, and there are two additional fluorines, one above and one below the pentagonal plane. Iodine and chlorine form ICl and ICl_3, whereas iodine and bromine form only IBr.

Because of the high electron affinities of the halogens, anions can be formed from halide ions and halogen or interhalogen molecules. The most common of these is the reddish brown **triiodide ion,** I_3^-, formed by dissolving iodine (I_2) in an aqueous solution of an iodide salt. Our geometry procedure would split I_3^- into $I^+ + 2\ I^-$, and the valence electron sum in Step 3 is ten, with two bonding pairs and three nonbonding pairs. Again, as with SF_4 and ClF_3, the lone pairs are at a lower energy in the triangular belt of the trigonal bipyramid and the I—I—I bond angle is 180°, as shown in Figure 9–3. Other anions with this linear structure include ICl_2^- and IBr_2^-.

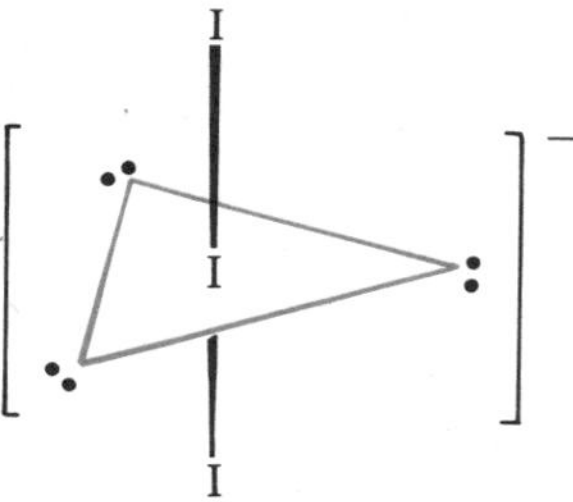

Figure 9–3 The structure of triiodide ion.

Compounds Containing Halogen-Oxygen Bonds

The halogens form a number of species containing halogen-oxygen bonds. The electronegativities of all of the halogens except fluorine are lower than the electronegativity of oxygen. Thus, Cl, Br, and I can be expected to have relatively high oxidation numbers when bound to oxygen. The known chlorine oxides are Cl_2O (analogous to OF_2), ClO_2, Cl_2O_6, and Cl_2O_7. Lewis representations of these molecules are given in Figure 9–4. Each of these compounds is a strong oxidizing agent, and each is capable of explosive

Figure 9–4 Lewis representations of (a) OCl_2, (b) ClO_2, (c) Cl_2O_6, and (d) Cl_2O_7.

(a)	(b)	(c)	(d)
Cl:O:Cl	:O:Cl:O:	:O::O: :O:Cl:Cl:O: :O::O:	:O: :O: :O:Cl:O:Cl:O: :O: :O:

decomposition to Cl_2 and O_2. The molecule ClO_2 contains an odd number of electrons; yet it does not dimerize to form Cl_2O_4. By way of contrast, Cl_2O_6 shows little tendency to split into ClO_3 molecules. (In a problem at the end of the chapter you are asked to speculate why ClO_3 dimerizes and ClO_2 does not.)

The chlorine oxides are strong Lewis acids and they yield a family of Lowry-Brønsted acids with a wide range of oxidation numbers. Just like the oxides themselves, these oxyacids are strong oxidizing agents because they preserve the chlorine-oxygen bonding. The strengths of the oxyacids of a given element increase as the formal charge on the element increases, as we know from HNO_2 and HNO_3. The four chlorine oxyacids are **hypochlorous acid** (HOCl), **chlorous acid** (HOClO), **chloric acid** ($HOClO_2$), and **perchloric acid** ($HOClO_3$). The Lewis representations of these four acids are given in Figure 9–5. The formal charges on chlorine, as shown in Figure 9–5, are zero, +1, +2, and +3 in (a), (b), (c), and (d), respectively. The acid strengths increase progressively from hypochlorous acid, which is a relatively weak acid, to perchloric acid, which is one of the strongest acids known.

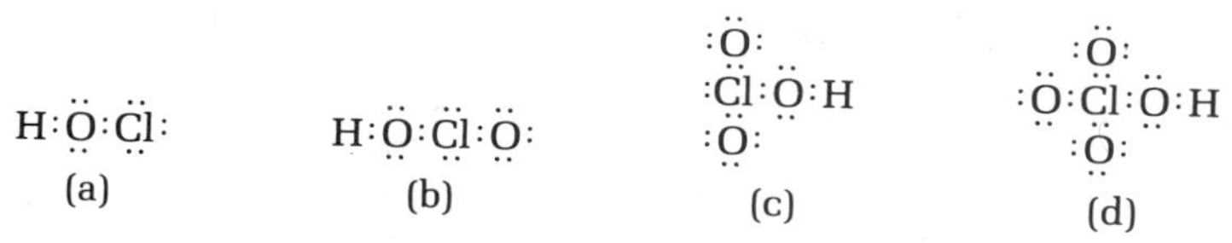

H:Ö:C̈l: (a) H:Ö:C̈l:Ö: (b) :Ö: / :Cl:Ö:H / :Ö: (c) :Ö: / :Ö:C̈l:Ö:H / :Ö: (d)

Figure 9–5 Lewis representations of the four oxyacids of chlorine.

The **hypochlorite anion, ClO^-**, is the active ingredient in laundry bleach. Hypochlorite ion is a relatively strong oxidizing agent and that is the source of its bleaching ability. Liquid laundry bleach is obtained by the addition of Cl_2 to an aqueous solution of NaOH:

$$Cl_2(g) + 2\,OH^-(aq) \rightarrow Cl^-(aq) + ClO^-(aq) + H_2O(l) \quad \textbf{9–4}$$

Solid laundry bleach is formed from Cl_2 and solid calcium hydroxide:

$$Cl_2(g) + Ca(OH)_2(s) \rightarrow CaCl(ClO)(s) + H_2O(l) \quad \textbf{9–5}$$

Bromine and iodine also form oxyacids. (Some of these are the subjects of a problem at the end of this chapter.)

9–2. The Chemistry of the Group VIIIA Elements

Until 1962, one sentence could have described the compounds of He, Ne, Ar, Kr, and Xe—there were no compounds known for these elements. However, in 1962 it was discovered that xenon bonds to fluorine, a result that has strongly influenced our views of the sources of chemical bonding. The ionization potentials of the Group VIIIA elements are quite high, but the first ionization potential of Xe is 12.1 ev and this is lower than the first ionization potentials of Cl, O, and N. Since chlorine, oxygen, and nitrogen bond to fluorine because of the low-energy valence-shell vacancy in the F atom, there is no reason why Xe should not also bond to fluorine; in 1962 it was discovered that it does.

The first xenon fluoride prepared was XeF_4, a colorless solid at room temperature. Our geometry procedure splits it into Xe^{4+} and four F^-. The Xe valence electrons are considered to be two 5*s* and six 5*p* electrons, and thus Xe^{4+} possesses four valence electrons. The valence electron sum in Step 3 is twelve and the structure of XeF_4 is based on the octahedron. There are two nonbonding electron pairs in XeF_4 and they are located on either side of a square-planar arrangement of the four bonding pairs, as shown in Figure 9–6. Other known fluorides are XeF_2 and XeF_6, both of which are also solids at room temperature.

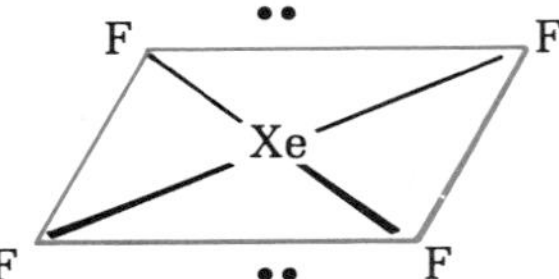

Figure 9–6
The structure of xenon tetrafluoride.

Since the valence electron energy of Xe as compared with oxygen is also higher, species should also be formed containing xenon-oxygen bonds. Known examples are **xenon trioxide** (XeO_3), the **xenate ion** (XeO_4^{4-}), and the **perxenate ion** (XeO_6^{4-}). (In a problem at the end of this chapter, you are asked to deduce the structures of XeO_3 and XeO_4^{4-}.)

Since the halogen fluorides are strong oxidizing agents, we are not surprised at this characteristic in the xenon fluorides. Just as fluorine can oxidize water to O_2, XeF_4 can also, and the process offers a route to XeO_3 in solution:

$$6\,XeF_4(s) + 12\,H_2O(l) \rightarrow 4\,Xe(g) + 3\,O_2(g) + 24\,HF(aq) + 2\,XeO_3(aq) \qquad \mathbf{9\text{–}6}$$

The element krypton possesses a first ionization potential of 14.0 ev. Since this is still considerably higher than IP_1 for fluorine, Kr—F bonds should form. The two known compounds are KrF_2 and KrF_4. The valence electron energy is also less for argon than for fluorine, but the gap is relatively small and there are no known compounds containing argon (or neon or helium).

9–3. The Chemistry of the Group IVA Elements

The C—C single bond energy is more than one-half the C═C double bond energy and more than one-third the C≡C triple bond energy. **Diamond** is the allotropic form of carbon that incorporates C—C single bonds. A portion of the diamond structure is shown in Figure 9–7. Each carbon atom in diamond is bonded to four other carbon atoms, and the carbon orbital hybridization is sp^3. In the figure, atoms 2, 3, 4, and 5 define the corners of a tetrahedron and they are bonded to atom 1 at the center of the tetrahedron. A second elemental form of carbon is **graphite** and graphite has the struc-

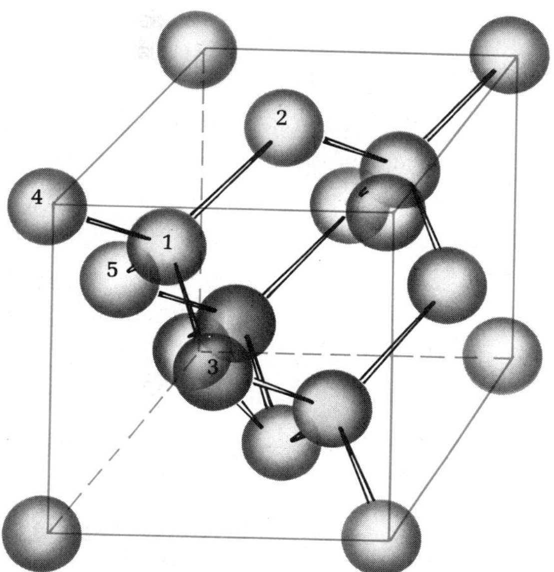

Figure 9–7
The arrangement of carbon atoms in diamond.

ture shown in Figure 9–8. It consists of stacked sheets of fused hexagonal rings of carbon atoms. The bonding within a given sheet involves sp^2 hybridization of the carbon atoms as contrasted with the sp^3 hybridization of diamond. The bonding between sheets is relatively weak. As a consequence the sheets are easily separated, a fact that finds its widest application in the ordinary graphite pencil. In contrast to diamond, which does not conduct electricity, graphite is a good electrical conductor.

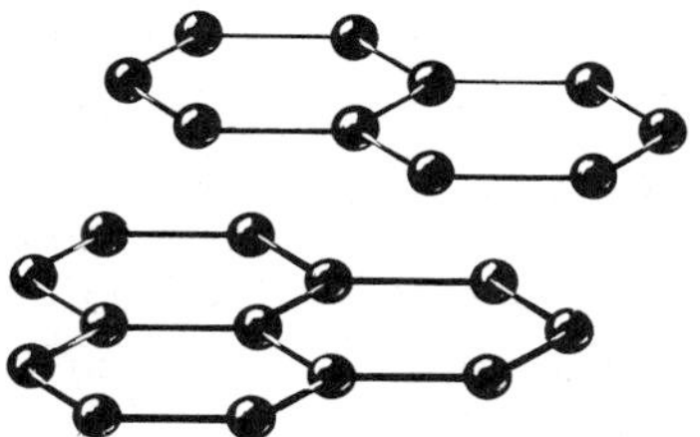

Figure 9–8
A portion of the structure of graphite. The spheres represent carbon atoms.

Silicon is the second most abundant element on earth (oxygen is the most abundant) and Si is a principal constituent of a large variety of rocks, particularly those of volcanic origin. Just as with phosphorus and sulfur, the structure of elemental silicon is based entirely on sigma bonding. Thus, there is no silicon analog of graphite, and silicon exists only in the diamond structure. Silicon carbide (SiC) also exists in the diamond structure. This substance, called **carborundum,** is nearly as hard as diamond. Germanium also exists in this structure, while elemental tin and lead possess structures characteristic of metals.

Combination with the Elements of Groups IA and IIA

The electronegativities in the members of Group IVA are not so high as for the members of Group VA, and anions analogous to N^{3-} and P^{3-} can be found only with carbon. **Calcium carbide,** CaC_2, can be produced by heating CaO (lime) and elemental carbon to a high temperature in a furnace. The reaction is

$$CaO(s) + 3\,C(s) \rightarrow CaC_2(s) + CO(g) \qquad \textbf{9–7}$$

The anion in CaC_2 is C_2^{2-}, a species that is isoelectronic with N_2 and that therefore contains a carbon-carbon triple bond. Like N^{3-} and P^{3-}, C_2^{2-} is a strong base and the addition of CaC_2 to water generates **acetylene** gas, C_2H_2:

$$CaC_2(s) + 2\,H_2O(l) \rightarrow Ca(OH)_2(s) + C_2H_2(g) \qquad \textbf{9–8}$$

The triple bond remains, and acetylene is usually written H—C≡C—H. The ion C^{4-} is found in aluminum carbide, Al_4C_3, and the addition of Al_4C_3 to water produces methane, CH_4.

The reactivity of the hydrides of the Group VA elements increases markedly with increasing atomic number. Thus, NH_3 does not react with oxygen at room temperature, but PH_3 spontaneously ignites in air. The Group IVA hydrides show an analogous trend, and methane will not react with oxygen at room temperature, but **silane** (SiH_4) spontaneously combusts when exposed to oxygen. We shall see the reason for this trend in Chapter 21 when we consider the factors which influence the rates at which chemical reactions proceed.

The chemistry of carbon is very extensive, largely because the carbon-carbon bond is relatively strong, the bond can occur as a single, double, or triple bond, and most carbon compounds are stable with respect to reaction with O_2 under normal conditions at room temperature. The Si—Si bond is also strong, and the compound **disilane,** Si_2H_6, can be prepared, as well as silanes containing longer chains of bonded silicon atoms. Unfortunately, all of the silanes react rapidly with oxygen at room temperature. This high reactivity of silanes can be lowered by replacing the Si—H bonds with Si—C bonds. Thus, **tetramethylsilane** [$Si(CH_3)_4$], a compound containing a tetrahedral arrangement of carbons around a central silicon, is quite stable at room temperature in the presence of oxygen.

Combination with the Halogens

Just as Group IVA elements are less reactive towards Group IA elements than the members of groups of higher roman numerals are, the Group IVA elements should be more reactive towards the halogens. While NCl_3, NBr_3, and NI_3 are explosively unstable, the corresponding carbon tetrahalides are all quite stable compounds, though the stability decreases in going from CCl_4 to CI_4, as expected.

Silicon-fluorine bonds are quite strong and this fact causes problems in handling hydrogen fluoride. Glass contains a high percentage of SiO_2, normally a quite unreactive substance. However, HF reacts with SiO_2 to etch glass according to the reaction

$$SiO_2(s) + 4\,HF(g) \rightarrow SiF_4(g) + 2\,H_2O(l) \qquad \textbf{9–9}$$

Even though HCl is a stronger acid than HF, HCl will not etch glass because the Si—Cl bond is not strong enough. In fact, $SiCl_4$ reacts with water in a reaction that is analogous to the reverse of reaction 9–9:

$$SiCl_4(l) + 2\,H_2O(l) \rightarrow SiO_2(s) + 4\,H^+(aq) + 4\,Cl^-(aq) \qquad \textbf{9–10}$$

As the atomic number increases in Group IVA, the tendency towards metallic behavior increases. Metals typically react with the halogens to form ionic salts, and the cations typically display small charges such as +1 or +2. Thus, in germanium, tin, and lead, the oxidation state II appears as well as state IV. The relative stability of the state II increases in going from Ge to Sn to Pb. The strong oxidizing power of PbO_2, which contains lead as Pb(IV), is employed in the lead storage battery; the product of its reduction in the lead storage battery is $PbSO_4$, an ionic salt containing Pb^{2+}. Electron removal is more difficult from tin atoms than from lead atoms. Thus, the **stannous ion,** Sn^{2+}, exists in ionic crystals, but Sn^{2+} is easily oxidized to non-ionic species that contain Sn in the oxidation state IV.

The Oxides of Carbon and Silicon

Carbon forms three oxides, carbon monoxide (CO), carbon dioxide (CO_2), and carbon suboxide (C_3O_2). Of these three, C_3O_2 is by far the least stable and the least commonly encountered.

Carbon monoxide is formed from the combustion of carbon or of carbon-containing compounds under conditions in which the supply of oxygen is limited. The limited supply of oxygen often occurs in the burning of gasoline in internal combustion engines, and thus CO is a significant component of urban air pollution. Carbon monoxide is isoelectronic with N_2 and possesses a very strong carbon-oxygen triple bond. The carbon orbital hybridization is *sp* and there is a nonbonding electron pair in the carbon region with a relatively high energy (higher for instance than in N_2). This nonbonding pair results in the formation of donor-acceptor complexes between CO and certain Lewis acids. One of these complexes forms between hemoglobin and CO; the resultant inactivation of hemoglobin as an oxygen carrier in human blood makes CO a lethal poison. The carbon lone pair of electrons in CO can also bond to a second oxygen atom to yield CO_2, as shown in Figure 9–9. The representation of CO_2 shown on the right of the arrow in Figure 9–9(a)

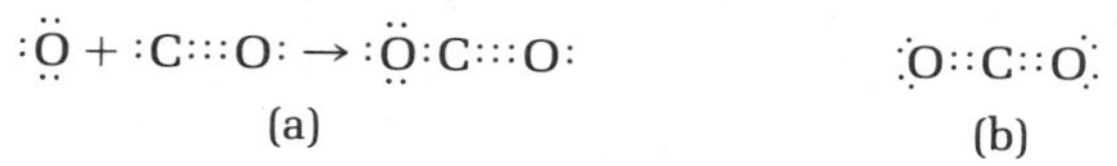

:Ö + :C:::O: → :Ö:C:::O:

(a)

:O::C::O:

(b)

Figure 9–9
(a) Lewis representation of the formation of CO_2 from CO and an oxygen atom. (b) Lewis representation of CO_2, showing two equivalent oxygens.

is one canonical resonance form. The two oxygens in CO_2 are equivalent, and a more proper representation is that shown in Figure 9–9(b).

The C═O bonds of CO_2 are quite polar and there is a significant positive charge on the carbon region (even though CO_2 has no dipole moment because of its linear structure). Thus CO_2 can serve as a weak Lewis acid and it does so in the formation of the **hydrogen carbonate ion** in solution:

$$CO_2(aq) + H_2O(l) \rightarrow HCO_3^-(aq) + H^+(aq) \qquad \textbf{9–11}$$

The hydrogen carbonate ion, HCO_3^-, serves as either a proton donor or a proton acceptor. It can accept a proton to form **carbonic acid,** H_2CO_3, or lose a proton to form the **carbonate ion,** CO_3^{2-}.

While carbon and oxygen can form both sigma and pi bonds to yield the gases CO and CO_2, silicon and oxygen form only Si—O single bonds, and SiO_2 is a solid as a consequence. Each silicon atom forms single bonds to four oxygens, as shown in Figure 9–10. Each oxygen is bonded to two silicon atoms, and this is also shown in the figure. Because the Si—O—Si bond

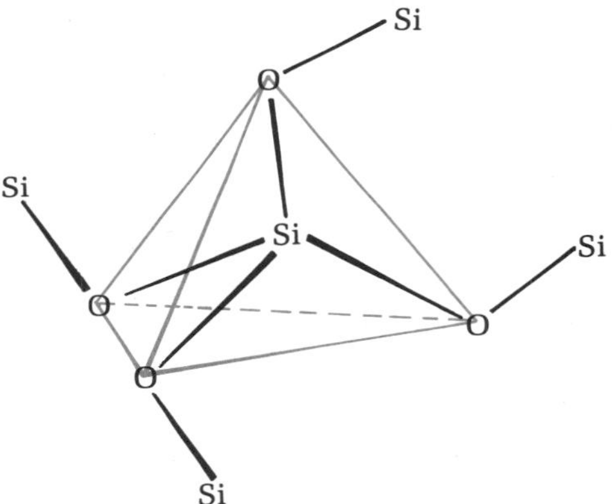

Figure 9–10
A portion of the structure of silicon dioxide.

angle is not 180°, there are a huge number of possibilities for the arrangement of the four silicons shown in the figure about the central silicon. These arrangements are generated by rotation about the Si—O bonds. In **quartz,** the atoms are located in an orderly array and quartz is a crystalline solid. When quartz is melted, many Si—O bonds must be broken (thus quartz has a high melting point). When the melted quartz is cooled below its melting point, the bonds would have to re-form in the correct, ordered way out of a huge number of ways in order to re-form in the quartz structure. The likelihood that this will occur is small, and instead, an amorphous substance forms. This substance is glass. Glass is not normally obtained commercially from quartz because quartz is expensive. Sand is used instead, because sand consists largely of SiO_2 and sand is inexpensive.

Silicon dioxide is a weakly acidic oxide and salts containing the SiO_4^{4-} anion can be prepared. Silicate anions also exist in linked forms analogous to pyrophosphate and pyrosulfate, and the Si—O—Si link is far more resistant to splitting by the addition of water than either the P—O—P or S—O—S links are. Extremely large silicate anions are found in many kinds of rocks and the strength of the Si—O—Si links is responsible for the hardness of those rocks.

The Si—O—Si link can also be incorporated into very nonreactive liquids. The **silicones** are such liquids, and the sizes and flexibilities of the Si—O—Si networks are controlled by introducing **methyl groups** (CH_3). For example, in equation 9–10 we saw that $SiCl_4$ reacts with water to give the large Si—O—Si network of SiO_2. The compound $(CH_3)_2SiCl_2$ reacts with water to yield the chain silicone structure shown in Figure 9–11. Just as SiO_2 is nearly chemically inert, so are the silicones; they find applicability, for instance, as lubricants in situations where low chemical reactivity is essential.

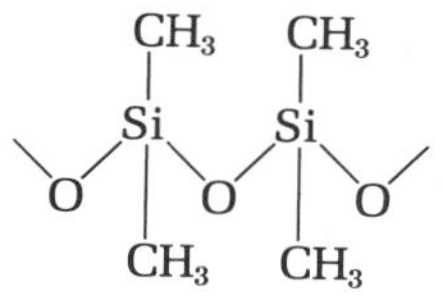

Figure 9–11
A portion of the structure of a silicone.

9–4. The Chemistry of the Group IIIA Elements

Elemental Forms

Aluminum, gallium, indium, and thallium are typical metals and the structures of these elements are those of typical metals.* Boron is the only nonmetal in this group. Elemental boron exists in a giant, covalently bonded network analogous to the diamond structure of elemental carbon. As a consequence, the melting point of elemental boron is quite high.

Combination of Boron and Hydrogen

The set of compounds containing B—H bonds is particularly interesting and unusual. Since boron forms the compound BF_3 with fluorine, we should expect also to find the analogous BH_3, since carbon forms CF_4 and CH_4, nitrogen forms NF_3 and NH_3, and oxygen forms F_2O and H_2O. Instead, the simplest boron hydride is **diborane,** B_2H_6, a compound that possesses the structure shown in Figure 9–12. The two boron atoms and four of the hydrogen atoms lie in a plane, and the remaining two hydrogen atoms are

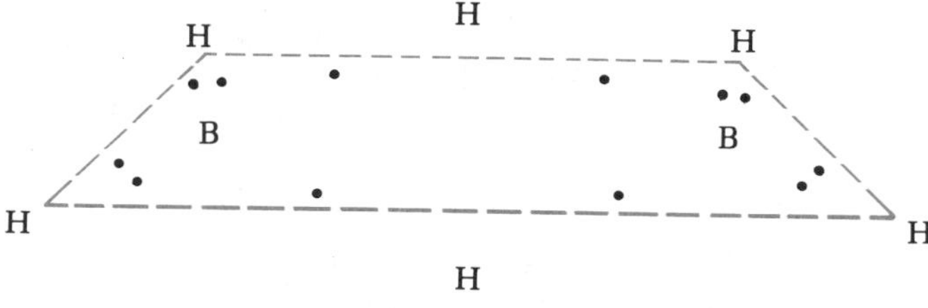

Figure 9–12
Lewis representation of diborane. The dashed lines outline a plane containing four hydrogen nuclei and two boron nuclei.

* We shall consider the structures of metals in Chapter 13.

located directly above and below the center of mass of the B_2H_4 unit. The four B—H internuclear distances in the plane are shorter (1.19 Å) than the four B—H distances (length = 1.37 Å) above and below the plane.

The BH_3 molecule does not exist, since it would contain one completely empty boron $2p$ orbital. This is the orbital that participates in pi bonding in BF_3. However, such pi bonding cannot occur in BH_3 because the hydrogen atom contains no p electrons. In B_2H_6, the multinuclear pi bonding of BF_3 finds its analog in multinuclear sigma bonding involving the two boron atoms and the two bridging hydrogen atoms. The total number of valence electrons in B_2H_6 is twelve, three from each boron and one from each hydrogen. The four B—H bonds in the plane are normal two-electron sigma bonds and account for eight electrons. The other two electron pairs are each contained in molecular orbitals extending over three centers (both borons and a hydrogen). These three-center, two-electron bonds are weaker than the four normal B—H bonds, and this explains the longer B—H distances above and below the plane in Figure 9–12. Boron hydrides containing more than two borons can also be prepared (B_4H_{10} is an example) and they also involve both normal and multicenter B—H bonding.

Diborane and the higher boranes are strong Lewis acids because they possess the same valence electron vacancy that would be present in BH_3. Thus they form anions such as BH_4^- (**hydroborate ion**), a tetrahedral ion that may be viewed as a donor-acceptor complex between BH_3 and H^-. The electronegativity of the boron atom is lower than that of the carbon atom, and B—H bonds are polarized as $B^{\delta+}$—$H^{\delta-}$. This is the same type of polarity found in Li—H, and a species such as BH_4^- is a strong reducing agent. The strong reducing character of boron-hydrogen compounds is also reflected in a very high reactivity of boranes towards atmospheric oxygen. Like phosphine and the silanes, most boranes spontaneously combust in air.

Combination with the Halogens

Like carbon, boron forms strong bonds with the halogens and the resulting compounds decrease in stability in going from BF_3 to BI_3. The boron-halogen bond is polarized as $B^{\delta+}$—$X^{\delta-}$, but the boron halides all have low melting points (the highest is BI_3 at 43 °C) and thus they are far from being ionic in the solid state. This nonionic character of the halides is partially preserved in the aluminum halides. Aluminum fluoride (AlF_3) is an ionic species, but $AlCl_3$, $AlBr_3$, and AlI_3 all have relatively low melting points and are nonionic solids. Like BF_3, aluminum chloride is a strong Lewis acid and forms such ions as $AlCl_4^-$, in analogy with BF_4^-.

Combination with Oxygen

Boron trioxide (B_2O_3) is one of the products of the combustion of boranes. Just as oxides such as N_2O_5 and SO_3 are related to the acids HNO_3 and H_2SO_4

by the addition of water, so B_2O_3 is related to **boric acid,** H_3BO_3. Boric acid is a compound similar in structure to BF_3 and contains three boron-oxygen bonds and three O—H bonds. Like BF_3, boric acid is a Lewis acid; it reacts with water to form the **borate anion,** which contains a tetrahedral arrangement of oxygens around a central boron. The reaction is

$$H_3BO_3(aq) + H_2O(l) \rightarrow H^+(aq) + B(OH)_4^- \qquad \textbf{9–12}$$

Just as silicates are usually found in nature as large anions containing Si—O—Si bridges, borates are commonly found as ions containing B—O—B bridges. The most common mineral containing boron is **borax** ($Na_2B_4O_7 \cdot 10 H_2O$) found in large deposits in the Mojave Desert of California. The commercial laundry agent known as Borax is derived from this mineral and its formula is $Na_2H_4B_4O_9$. The anion $B(OH)_4^-$ may be viewed as a donor-acceptor complex of H_3BO_3 and OH^-, and it is a weak base. The ion $H_4B_4O_9^{2-}$ is a donor-acceptor complex of $H_2B_4O_7$, the protonated form of $B_4O_7^{2-}$, and two hydroxide ions. Borax is also a weak base, and weak bases serve as good cleaning agents because they react with fats and oils; another weak base commonly used as a cleaning agent is aqueous ammonia.

Aluminum oxide is one of the most stable compounds found in nature, and Al_2O_3 is the form in which aluminum is found in the ore **bauxite.** Since Al_2O_3 is so stable, the pure metal can be obtained only by electrolysis. In practice, the Al_2O_3 is dissolved in fused cryolite, Na_3AlF_6, at an elevated temperature and the electrolytic product is molten aluminum.

SUGGESTIONS FOR ADDITIONAL READING

Gould, E. S., *Inorganic Reactions and Structure,* Holt, Rinehart, and Winston, New York, 1962.

Cotton, F. A., and Wilkinson, G., *Advanced Inorganic Chemistry,* Wiley-Interscience, New York, 1972.

PROBLEMS

1. From the fact that the S—S bond energy is greater than the energy of the O—O single bond, what do you conclude about the relative electron affinities of the oxygen and sulfur atoms?
2. Why isn't sulfuric acid used as the high-boiling acid in reaction 9–2?
3. Why is phosphine a poorer base than ammonia?
4. Can you suggest an explanation for the fact that ClO_3 dimerizes whereas ClO_2 does not?
5. Why is perchloric acid a stronger acid than either nitric acid or sulfuric acid?

6. Predict the order of the acid strengths of the acids HOCl, HOBr, and HOI. Give the basis for your prediction.
7. If a strong acid is added to liquid laundry bleach, Cl_2 gas is formed. Write a balanced equation for the reaction that produces Cl_2.
8. Why are there no oxyacids of fluorine?
9. Deduce the structures of XeO_3 and XeO_4^{4-}.
10. The ion C_2^{2-} is isoelectronic with N_2. Why is C_2^{2-} a strong base and N_2 an extremely weak base?
11. Why is there no molecule with the structure O=C=C=O?
12. Draw Lewis representations of the hydrogen carbonate ion and of carbonic acid.
13. Aqueous solutions of strong bases such as NaOH etch glass. Explain why this occurs.
14. Propose a structure for the molecule Al_2Cl_6.
15. Propose a structure for $H_2B_4O_7$. Show how this molecule can accept two hydroxide ions to form $H_4B_4O_9^{2-}$.

10/ The Chemistry of Carbon

The chemistry of carbon is particularly valuable in showing how bond models of molecules may be used to predict and correlate chemical properties. The chemistry of living, or organic, systems is largely the chemistry of carbon compounds, which therefore has been termed organic chemistry. Carbon is also represented extensively in many nonliving systems, including important substances like plastics. Important biological molecules and man-made carbon compounds are included in our examination.

10–1. The Saturated Hydrocarbons

Almost all carbon compounds can be considered to be derived from chains or rings containing only carbon-carbon and carbon-hydrogen single bonds. These **saturated hydrocarbons** form a key part of the system of naming organic molecules.

Nomenclature

The simplest of the saturated hydrocarbons is methane (CH_4) (a molecule previously considered in Chapter 6). The next member of the series is **ethane** (C_2H_6), whose structure is shown in Figure 10–1. Ethane may be viewed as two fused methane molecules with a C—C bond replacing two C—H bonds [Figure 10–1(a)]. The molecule can also be envisioned along the

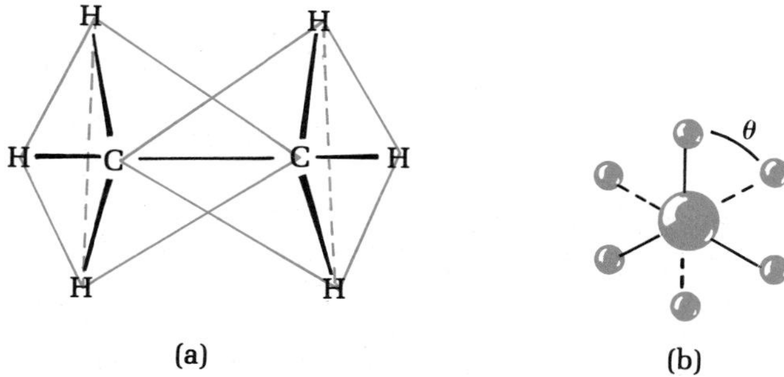

Figure 10–1
Two views of the structure of ethane: (a) perpendicular to the C—C bond; (b) along the C—C bond.

C—C axis, as in Figure 10–1(b); three C—H bonds (the solid lines) extend up from the plane of the paper and three others (the dashed lines) extend below the plane of the paper. This "staggered" configuration minimizes the repulsion between the solid and dashed C—H bonds, and it is the most energetically favorable configuration. However, the energy difference between the staggered and "eclipsed" (when the dashed lines are directly beneath the solid lines) configurations is not large and the staggered and eclipsed forms rapidly interconvert by rotation of the CH_3 groups around the C—C bond (changing the angle θ shown in the figure).

The next member in the series is **propane** (C_3H_8). As the size of the hydrocarbon increases, precise representation of the geometry, as in Figure 10–1, proves to be increasingly inconvenient. Furthermore it is unnecessary since we know that, in the non-ring saturated hydrocarbons, the atoms around each carbon are arranged in a tetrahedral fashion. We can represent the bonding in the three-dimensional molecule C_3H_8 by the two-dimensional picture

```
    H  H  H
    |  |  |
H—C—C—C—H
    |  |  |
    H  H  H
```

The compound C_4H_{10} yields an example of a phenomenon commonly found in organic chemistry, namely isomerism. Compounds are isomers of each other when their molecular formulas are identical but the arrangement of atoms within the molecules is different. Several distinct types of isomerism occur in organic compounds. The type that C_4H_{10} exemplifies is structural isomerism. Two or more compounds are **structural isomers** if they have the same molecular formula but differ in the bonding of at least two atoms. For C_4H_{10}, the structural isomers are shown in Figure 10–2. The **straight-chain hydrocarbon** (a) contains two carbons that are bonded to three hydro-

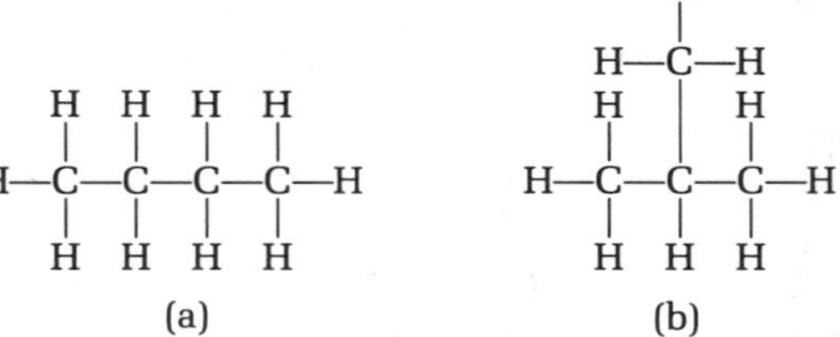

Figure 10–2 The two isomers of molecular formula C_4H_{10}: (a) normal butane; (b) isobutane.

gens and a carbon, and two carbons that are bonded to two hydrogens and two carbons. The **branched hydrocarbon** (b) contains three carbons that are bonded to three hydrogens and a carbon, and one carbon that is bonded to one hydrogen and three carbons. Thus, two carbons in (a) are bonded differently from two carbons in (b) and the two molecules are structural isomers.

The molecule depicted in Figure 10–2(a) is termed **normal butane** (*n*-butane) and the molecule in Figure 10–2(b) is termed **isobutane.** The system of the International Union of Pure and Applied Chemistry (IUPAC) names each hydrocarbon as a derivative of its longest continuous chain. Thus, for example, isobutane is considered a derivative of propane in this system. The carbons in the continuous chain are given numbers, starting with 1 for an end carbon. Each substituent hydrocarbon fragment bonded to the chain is named as follows, using isobutane as our example. The stem of the name (*meth-*) is taken from the hydrocarbon having the same number of carbons (methane, 1 carbon), and *-yl* replaces the *-ane* of the hydrocarbon (methyl, from methane). This is combined with the name of the compound of which the hydrocarbon is a derivative (propane). The location of the substituent is indicated by the number of the carbon atom to which it is bonded. Thus, in the IUPAC system, isobutane is called 2-methylpropane.

To apply this system to more complicated hydrocarbons, we require names for the nonbranched hydrocarbons that form the basis of the system. We have already seen the names *methane, ethane, propane,* and *butane* for the chains C_1 to C_4. The names for C_5 to C_{10} use Greek prefixes, and thus become *pentane, hexane, heptane, octane, nonane,* and *decane*. The names also extend beyond ten carbons, but C_1 to C_{10} will suffice for our purposes. Consider the structure shown below (for the sake of space lines have been eliminated for C—H bonds, and a methyl group is shown simply as H_3C— or —CH_3, for example):

```
       CH3  CH2—CH3
        |    |
H3C—CH—CH—CH2—CH2—CH3
```

The longest continuous chain in the molecule is six carbons long and the molecule is named as a derivative of hexane. The substituents are a methyl group and an ethyl group and they are bonded to carbons 2 and 3 respectively (the numbering always starts at the carbon nearest to the substituents); so the molecule is 2-methyl-3-ethylhexane. If both substituents had been methyl groups, the name would have been 2,3-dimethylhexane. Table 10–1 lists the formulas and names of some saturated hydrocarbons, along with the melting points and boiling points of each.

Each of the three isomers of C_5H_{12} is listed in the table, but from C_6H_{14} through the higher-carbon substances, only the nonbranched hydrocarbon is listed. The two branched C_5H_{12} compounds are also commonly named by the trivial names *isopentane* and *neopentane,* and these names are listed in the table along with the IUPAC names.

Boiling Points and Molecular Structures

Table 10–1 displays a phenomenon that we may have previously observed, the trend in a family of similar compounds for the boiling point to increase

Formula	Name	Melting Point (°C)	Boiling Point (°C)
CH_4	Methane	−183	−162
C_2H_6	Ethane	−172	− 89
C_3H_8	Propane	−187	− 42
C_4H_{10}	*n*-Butane	−135	0
C_4H_{10}	2-Methylpropane	−145	− 10
C_5H_{12}	*n*-Pentane	−130	36
C_5H_{12}	2-Methylbutane (Isopentane)	−160	28
C_5H_{12}	2,2-Dimethylpropane (Neopentane)	− 20	10
C_6H_{14}	*n*-Hexane	− 94	69
C_7H_{16}	*n*-Heptane	− 90	98
C_8H_{18}	*n*-Octane	− 57	126
C_9H_{20}	*n*-Nonane	− 54	151
$C_{10}H_{22}$	*n*-Decane	− 29	174

Table 10–1
Melting and Boiling Points of Some Saturated Hydrocarbons

as the molecular weight increases. We know that the stronger the attraction between molecules, the higher the boiling point of the substance they compose. Thus, a liquid such as water, in which there is strong hydrogen bonding, has a much higher boiling point than a nonpolar substance such as methane. However, all of the hydrocarbons listed in Table 10–1 are nonpolar; why then should there be a higher boiling point and concomitantly greater molecular attraction in these substances along with increasing molecular weight? The answer lies not in the weight of the molecules as such, but rather in their size and shape, and their effect on intermolecular interactions.

It is easy to picture the interaction between water molecules as the attraction between positively charged hydrogen atoms and negatively charged oxygen atoms; it is more difficult to explain why "perfect" nonpolar species such as helium atoms attract each other. Helium displays such attraction because it liquefies at 4 °K and 1 atm pressure. We need a model for intermolecular attraction that is appropriate for such nonpolar substances as helium.

The atomic orbital model depicts the He atom (or any other atom) as a sphere of electron density having a dipole moment that is exactly zero. However, this is only a time-average view, and at any instant in any atom or molecule the particles are usually arranged in such a way that a dipole moment results. Such an arrangement in He is the following:

$$\bullet^{-} \qquad\qquad \bullet^{-}$$
$$\bullet^{2+}$$

The upper portion of the atom depicted is negatively charged, and the bottom portion is positively charged. As the electrons move, the dipole moment fluctuates; the time-average dipole moment of the He atom is zero. All atoms and molecules have fluctuating dipole moments, and the time-average value is the permanent dipole moment. Because of the fluctuating dipoles all atoms and molecules attract each other, but the attraction is weaker between so-called nonpolar molecules than between polar molecules. Thus nonpolar liquids have lower boiling points than polar liquids of the same molecular weight. This is why methane has a much lower boiling point than water.

This analysis shows that the interaction of a nonpolar molecule with its neighbors is proportional to the number of electrons near its "surface" (valence electrons)—their motion gives rise to fluctuating dipoles, and they are close enough to the neighboring molecules to exert a sizable force on them. In general, the greater the surface area of the molecule (which we might take as the 90 percent contour surface), the greater the combined interaction with the neighbors. Ethane (with a larger surface area than methane) boils at a higher temperature than methane, because ethane interacts more with neighbors than methane does.

One would predict from such reasoning that a branched hydrocarbon such as neopentane should possess a lower boiling point than the normal hydrocarbon would. The reason is that the central carbon in neopentane is "buried" inside four methyl groups and the molecular surface is less than in normal pentane. This prediction is verified in the data of Table 10–1. The effect is general: branched hydrocarbons have lower boiling points than their nonbranched isomers. This reasoning also gives us an explanation for the low boiling points of certain inorganic species. For example, uranium hexafluoride (UF_6) has a molecular weight of 352 and yet boils at 56 °C. Most of the mass and electron density in UF_6 is concentrated in the uranium and this central atom is shielded from neighboring UF_6 molecules by the six fluorines; thus the number of surface electrons is much smaller than in a hydrocarbon of comparable molecular weight.

Reactivity

The chemical reactivity of the compounds of Table 10–1 is largely restricted to reaction with very electronegative species such as oxygen or chlorine. The principal driving force for these reactions derives from the larger bond energies of polar bonds as compared with the nonpolar bonds in the hydrocarbons.

$$CH_4(g) + 2\,O_2(g) \rightarrow CO_2(g) + 2\,H_2O(g) \qquad \textbf{10–1}$$

$$CH_4(g) + 4\,Cl_2(g) \rightarrow CCl_4(l) + 4\,HCl(g) \qquad \textbf{10–2}$$

There are other saturated hydrocarbons that contain rings instead of (or in addition to) chains. The simplest of these molecules is **cyclopropane** (C_3H_6), in which the three carbon atoms define the corners of an equilateral triangle. The C—C—C bond angle in cyclopropane is 60°, a much smaller angle than the 109.5° angle in propane, and the C—C bond is much weaker than in a non-ring hydrocarbon. Whenever there is strain in rings, for example, in cyclopropane and cyclobutane, the compounds are considerably more reactive than the non-ring hydrocarbons and unstrained ring hydrocarbons.

The names of the ring hydrocarbons are derived from the corresponding non-ring hydrocarbon by adding the prefix *cyclo-*. Thus C_5H_{10} is cyclopentane and C_6H_{12} is cyclohexane. Saturated hydrocarbons can also contain more than one ring, as shown in Figure 10–3 for **bicyclobutane** (C_4H_6).

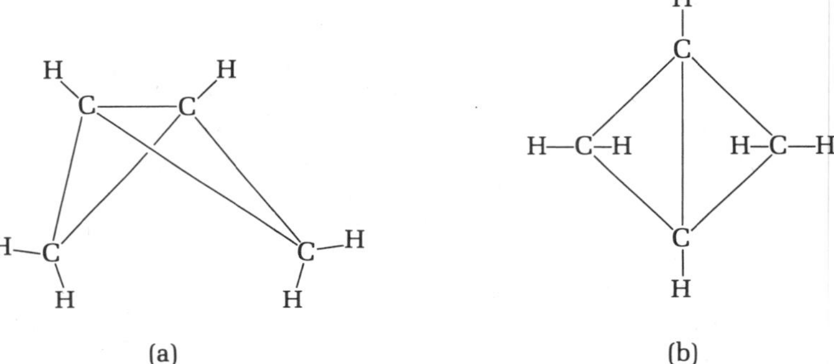

Figure 10–3
The structure of bicyclobutane: (a) Side view; (b) top view.

10–2. Alkenes and Alkynes

Chemical modifications of saturated hydrocarbons usually convert them to considerably more reactive molecules, and the chemical reactivity is centered in the added or altered atoms or groups of atoms, which are termed **functional groups.** The functional group most closely related to the parent saturated hydrocarbon is the carbon-carbon double bond. Saturated hydrocarbons are commonly termed **alkanes;** and hydrocarbons containing one or more C═C bonds are termed **alkenes,** or **olefins.** Because the olefins do not contain the maximum possible number of sigma bonds (as the saturated hydrocarbons do), they belong to the class of compounds termed **unsaturated hydrocarbons.**

Nomenclature

The simplest alkene is **ethene** (commonly called **ethylene**), C_2H_4, a molecule containing one C═C double bond. The IUPAC system names alkenes by the characteristic suffix *-ene*. Thus, the two simplest alkenes are ethene and propene. The presence of a C═C bond yields the possibility of two additional kinds of isomerism. The first is based on the location of the double bond or bonds, and the second is based on the arrangement of atoms directly bonded to the double-bonded carbons. Examples of both kinds of isomerism are shown in Figure 10–4 for C_4H_8, whose IUPAC name is butene. Figure 10–4(a) shows **1-butene,** a molecule in which the double bond is between carbons 1 and 2. Figures 10–4(b) and (c) show the two isomers of **2-butene.** These two isomers represent different species because they have different physical properties. Rotation about the C═C bond requires that the pi bond be broken, a process that normally does not occur at room temperature, and each isomer can be obtained in essentially pure form. We have experienced this type of isomerism (termed **geometrical isomerism**) with N_2F_2; it is commonplace in organic chemistry. The fourth butene isomer is the structural isomer **2-methylpropene,** shown in Figure 10–4(d).

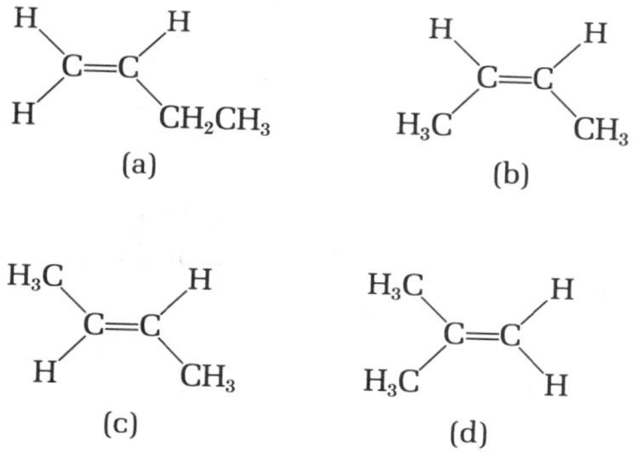

H, H, C═C, H, CH_2CH_3
(a)

H, H, C═C, H_3C, CH_3
(b)

H_3C, H, C═C, H, CH_3
(c)

H_3C, H, C═C, H_3C, H
(d)

Figure 10–4
Four butene isomers: (a) 1-butene; (b) *cis* 2-butene; (c) *trans*-2-butene; (d) 2-methylpropene.

Reactivity

In general, the C═C bond in alkenes is reactive with respect to forming two additional sigma bonds. Thus, the most characteristic reaction of alkenes is **addition,** to yield saturated compounds.

$$H_2C{=}CH_2(g) + H_2(g) \rightarrow H_3C{-}CH_3(g) \qquad \textbf{10–3}$$

$$H_3C{-}CH{=}CH_2(g) + Br_2(l) \rightarrow \underset{\text{(1,2-dibromopropane)}}{H_3C{-}CHBr{-}CH_2Br} \qquad \textbf{10–4}$$

The H_2 and Br_2 bonds must be broken for the two C—H (equation 10–3) or C—Br (equation 10–4) bonds to form. Since the bonds must be broken, and this does not happen readily at room temperature, a catalyst must be used. Hydrogenation (equation 10–3) of double bonds is a procedure that is common; the reaction can be catalyzed by adding finely divided metallic nickel.

The double bond in ethylene (ethene) can be replaced by C—C sigma bonds by the **polymerization** of ethylene to form **polyethylene.** This familiar plastic substance contains very long molecular chains (in excess of one thousand carbons) formed by repeated additions of the type shown:

$$H_2C{=}CH_2 + H_2C{=}CH_2 \rightarrow -CH_2-CH_2-CH_2-CH_2-$$
$$-CH_2-CH_2-CH_2-CH_2- + H_2C{=}CH_2 \rightarrow -CH_2-CH_2-CH_2-CH_2-CH_2-CH_2- \quad (10\text{–}5)$$

The short lines at the ends of these chains indicate incomplete bonding at the carbons and thus the capacity for further chain lengthening.

A variety of plastics are produced by polymerizing derivatives of ethylene, and their properties depend on the nature of the substituent groups. Some of these plastics (**polymers**) and the ethylene derivatives (**monomers**) that yield them are listed in Table 10–2.

Table 10–2
Some Common Polymers from Ethylene Derivatives

Monomer	Polymer	Common Name
$H_2C{=}CHCH_3$	Polypropylene	Polypropylene
$H_2C{=}CHCl$	Polyvinyl chloride	PVC
$H_2C{=}CHC_6H_5$	Polystyrene	Polystyrene
$H_2C{=}CHC{\equiv}N$	Polyacrylonitrile	Acrilan, Orlon
$F_2C{=}CF_2$	Polytetrafluoroethylene	Teflon
$H_2C{=}CCl_2$	Polyvinylidene chloride	Saran
$H_2C{=}C(CH_3)C({=}O)OCH_3$	Polymethylmethacrylate	Lucite, Plexiglas

Natural rubber is a polymer of the molecule **isoprene**, shown in Figure 10–5(a). This molecule is termed a **diene** (pronounced die-een), and polymerization occurs when carbons 1 and 4 add to the corresponding carbons of another monomer (the carbons are numbered in the figure). This yields a double bond between carbons 2 and 3, and a portion of the polymeric structure of natural rubber is shown in Figure 10–5(b). This polymer has a high degree of flexibility and very little structural strength. The strength is markedly increased by heating the rubber with elemental sulfur. In this process of **vulcanization**, sulfur atoms add to the double bonds and serve as bridges between the chains, as shown in Figure 10–5(c). The hardness of the vulcanized rubber can be controlled through the percentage of double bonds to which sulfur has been added.

Alkynes

A second type of unsaturation is produced by introducing a triple bond, and the simplest compound of this type is **ethyne**, C_2H_2 (commonly called **acet-**

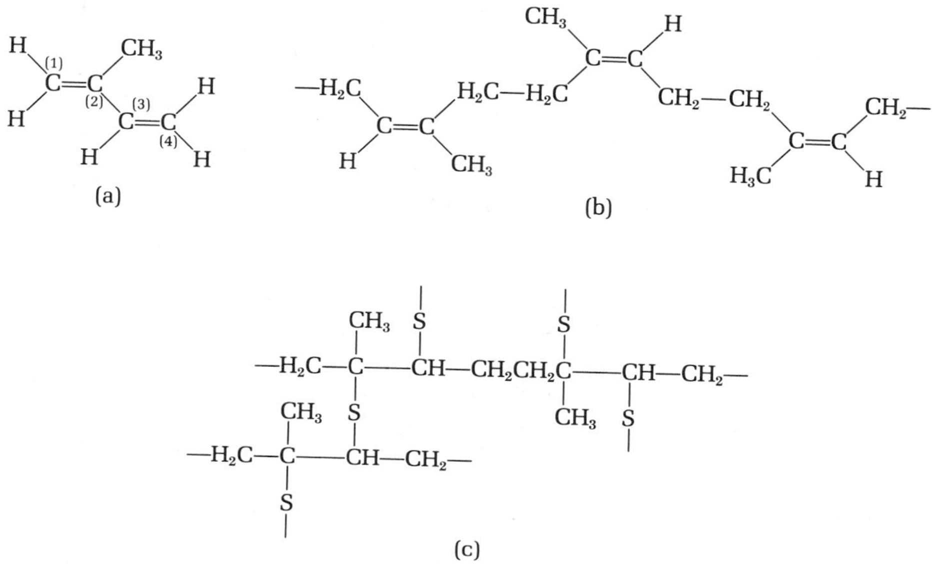

Figure 10-5
(a) Isoprene, and two of its polymers; (b) natural rubber and (c) vulcanized rubber.

ylene), a linear molecule involving *sp* hybridization of the carbon atomic orbitals. The naming of these compounds (termed the **alkynes**) is identical to the naming of the alkenes, save that the ending *-yne* is used. The compound having the formula

$$H_3C—C{\equiv}C—CH_3$$

is called 2-butyne. As we might expect, the most characteristic reaction of alkynes is addition, just as with alkenes.

10-3. Aromatic Hydrocarbons

The hydrocarbon **benzene** (C_6H_6) displays a chemistry that is markedly different from that of ethylene because it contains multicenter bonding. Benzene and related compounds are of the general class termed **aromatic hydrocarbons.**

Multicenter Pi Bonding in Benzene

A fundamental principle of chemical bonding is that electrons tend to be found simultaneously associated with as many nuclei as possible. This tendency toward multicenter bonding is strongly displayed in ionic solids. Thus sodium ions are associated with six chloride ions at a distance of 2.83 Å in the solid state at room temperature rather than one Cl^- at 2.38 Å in the gaseous state. Multicenter bonding (or **delocalized bonding,** as it is commonly called) is also found in the pi bonds of BF_3, BeF_2, and CO_3^{2-}. In these cases, molecular orbitals are formed from 4, 3, and 4 atomic p orbitals, respectively.

Benzene is perhaps the most extensively studied example of delocalized bonding. The six carbon nuclei in benzene define the corners of a regular hexagon, and the orbital hybridization in the sigma bonding is sp^2. Thus, all twelve nuclei lie in one plane and all bond angles are 120°. Let us define x as the direction perpendicular to the hexagon. The six carbon $2p_x$ orbitals combine to form the molecular orbitals whose occupation results in pi bonding.

Up to this point we have largely been concerned with two-center bonding, and we have constructed molecular orbitals simply by reinforcing or cancelling overlap of atomic orbitals. We have seen that the overlap of two atomic orbitals yields two molecular orbitals, one bonding and the other antibonding. The six $2p_x$ carbon orbitals in benzene yield six pi molecular orbitals, three of which are bonding and three of which are antibonding. Figure 10–6 shows the six combinations of atomic orbital contours that yield the six molecular orbitals by bonding overlap or antibonding overlap, or both. (The contours are drawn very small so that each can be observed separately.)

The best bonding molecular orbital is depicted in Figure 10–6(a). There are six bonding overlaps between adjacent orbitals (though they have not been drawn in an overlapping fashion). Both (b) and (c) have two net bond-

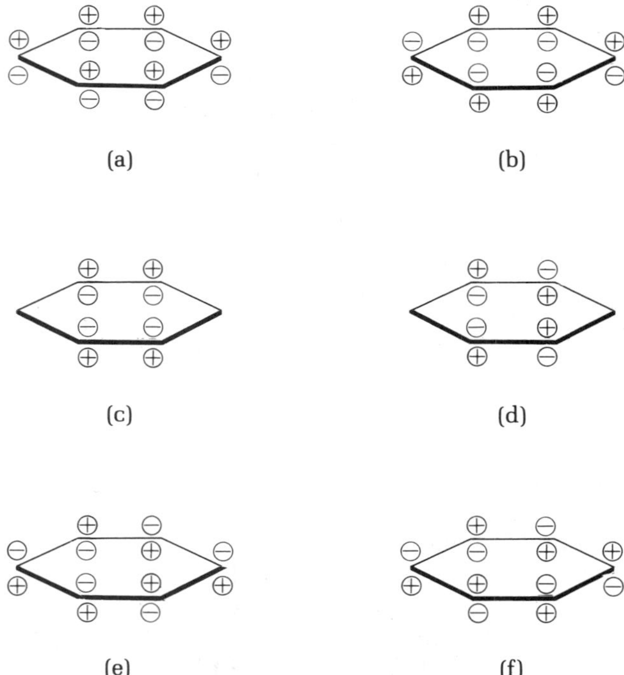

Figure 10–6 Atomic orbital combinations in the six pi molecular orbitals of benzene.

ing overlaps; (b) has four bonding overlaps and two antibonding overlaps while (c) has two bonding overlaps and four nonbonding overlaps. Both (c) and (b) are bonding molecular orbitals and they lie at the same energy (but higher than (a)). The orbitals (d) and (e) are both antibonding and each contains two net antibonding overlaps. The highest-energy molecular orbital is (f), since it is formed from six antibonding overlaps. Figure 10–7 is an energy-level diagram showing the level of the six carbon $2p_x$ orbitals and the six molecular orbitals derived from them along with the orbital occupations. The six electrons all occupy bonding molecular orbitals and the pi orbital system of benzene is characterized by three bonds.

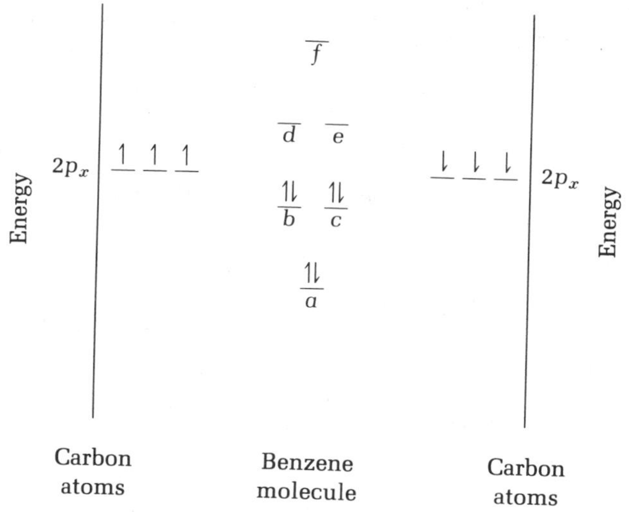

Figure 10–7 Energy levels and orbital occupation of the pi molecular orbitals of benzene, as compared with the carbon atomic orbitals from which they are derived.

As with any species that contains delocalized bonding, we have in benzene a difficulty in giving a simple line or dot representation of the bonding. We have represented CO_3^{2-} by a combination of canonical resonance structures or by the dashed resonance hybrid representation (Chapter 6). Figure 10–8 shows line and dot representations of canonical resonance forms of benzene and also the single dashed representation. Figures 10–8(a) and (b) depict localized two-center carbon-carbon double bonds between carbons numbered 2–3, 4–5, and 6–1, while in Figures 10–8(c) and (d) these double

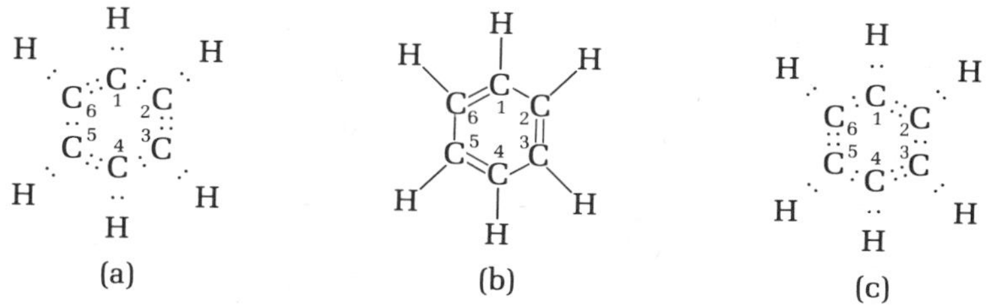

Figure 10–8 Representations of the benzene molecule.

Fig. 10–8 (cont.) (d) (e)

bonds are between carbons 1–2, 3–4, and 5–6. Figure 10–8(e) shows the hybrid form involving six equivalent carbon-carbon bonds of bond order $1\frac{1}{2}$. This delocalized pi bonding is quite common, and it occurs any time the canonical resonance forms contain an alternation of single and double bonds. This alteration is termed **conjugation** and benzene is a **conjugated molecule.** The isoprene molecule shown in Figure 10–5 is also conjugated.

Reactivity

Because the pi bonding is stronger in benzene than in nonconjugated systems, the characteristic reaction of benzene is not addition. Rather, the characteristic reaction is substitution, as illustrated by the reaction

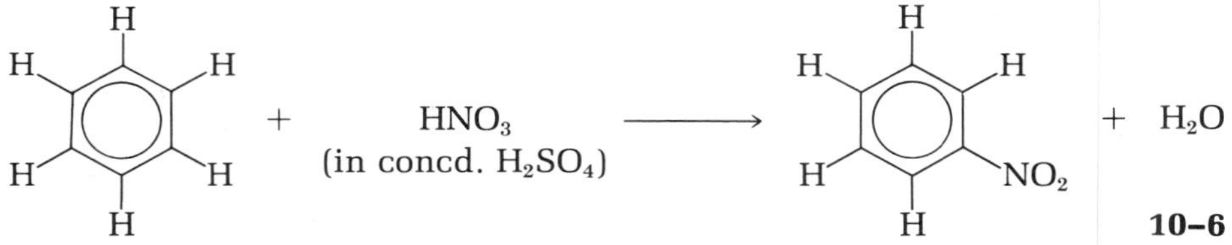

$$C_6H_6 + HNO_3 \text{ (in concd. } H_2SO_4) \longrightarrow C_6H_5NO_2 + H_2O \qquad \textbf{10–6}$$

In reaction 10–6, the **nitro group** (—NO_2) has been substituted for a benzene hydrogen atom.

The product of reaction 10–6 is called **nitrobenzene.** The nitric acid dissociates in concentrated sulfuric acid according to the reaction

$$HNO_3 + 2\,H_2SO_4 \rightarrow NO_2^+ + H^+ + H_2O + 2\,HSO_4^- \qquad \textbf{10–7}$$

and the reaction is between benzene and the nitronium ion (NO_2^+). The ion NO_2^+ is isoelectronic with CO_2 and thus is linear. The N atom portion of NO_2^+ may be considered to possess most of the electron deficiency, and NO_2^+ is a very strong electrophile. Two pi electrons of benzene can be donated to this electrophile in the formation of a donor-acceptor complex of the following structure:

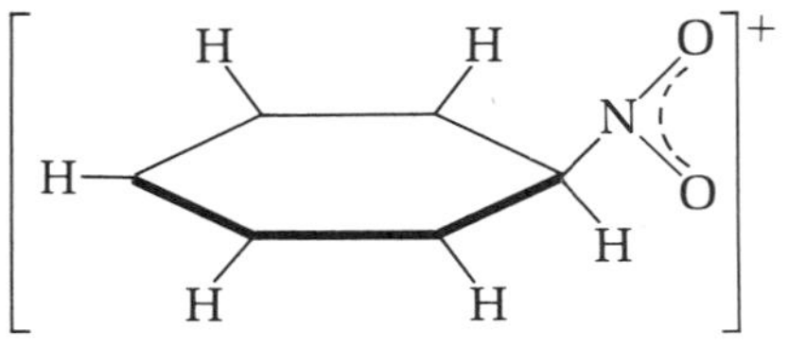

In this complex, a two-electron sigma bond is formed between the nitrogen and one of the six benzene carbons. The positive charge of the NO_2^+ is effectively shifted into the benzene ring portion of the complex and three canonical resonance forms can be drawn for the location of the charge, as shown in Figure 10–9. This shows that two of the original six pi electrons of benzene have been shifted to make the N—C sigma bond, leaving one of three carbon atoms with a positive charge. This positive charge can be relieved by the expulsion of a proton from the carbon bonded to the nitrogen; the product of the expulsion is nitrobenzene.

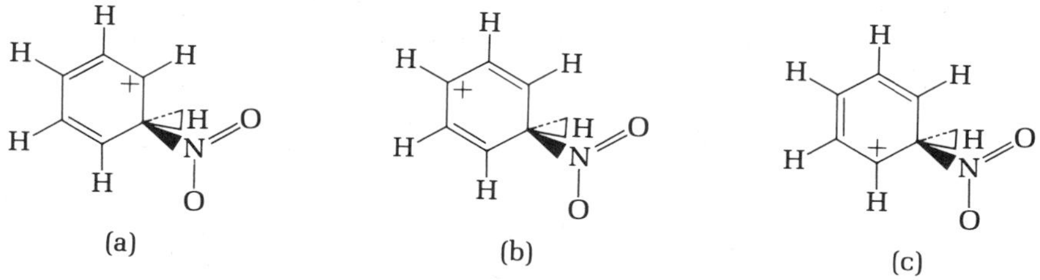

Figure 10–9 Canonical resonance structures of the nitronium ion-benzene complex.

In the nitration of benzene, we have our first example of a **reaction mechanism.** A mechanism is a model for the sequence of reactions and structures through which a chemical change proceeds from reactants to products. All reactions, both organic and inorganic, proceed through such sequences and can be assigned mechanisms. We shall be investigating several mechanisms in the remainder of the text. A very common species formed as an **intermediate** in many chemical reactions is a donor-acceptor complex, and the $[C_6H_6—NO_2]^+$ species is only one example. One characteristic that we will be seeking in any given molecule is a good electron-acceptor site or a good electron-donor site.

This donor-acceptor factor can be illustrated further in the nitration of nitrobenzene. Substitution of a second nitro group onto the benzene ring is considerably more difficult than substituting the first; the principal product is the molecule shown in Figure 10–10(a). The three possible **positional isomers** of dinitrobenzene are indicated by the names (a) **meta,** (b) **para,** and (c) **ortho.** Why does the presence of the first nitro substituent make the benzene ring less reactive toward NO_2^+, and why is *meta*-dinitrobenzene the predominant product? The NO_2 group contains delocalized pi bonding (compare analysis of structure and bonding of NO_2 and NO_2^- in Chapter 8). The nitrobenzene molecule contains all fourteen nuclei in one plane and

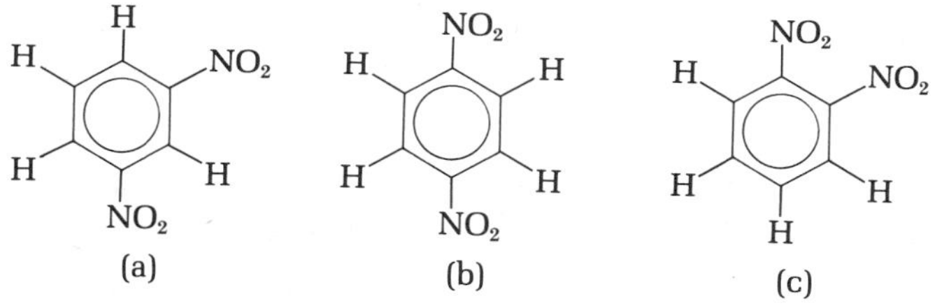

Figure 10–10 The three positional isomers of dinitrobenzene: (a) *meta*; (b) *para*; (c) *ortho*. The circle within the hexagon denotes delocalized pi bonding, as in Figure 10–8(e).

the delocalized pi bonding extends over six carbons, the nitrogen, and two oxygens. Since the nitrogen and oxygens are more electronegative than the carbons, the pi electron density is polarized toward the nitro group. The resultant electron deficiency in the ring renders the molecule a poorer nucleophile than benzene and less reactive towards NO_2^+.

The pi electron deficiency in the ring of nitrobenzene is not equally distributed among the six carbons. Figure 10–11 shows the three canonical resonance forms of the dipolar species that would result if one electron

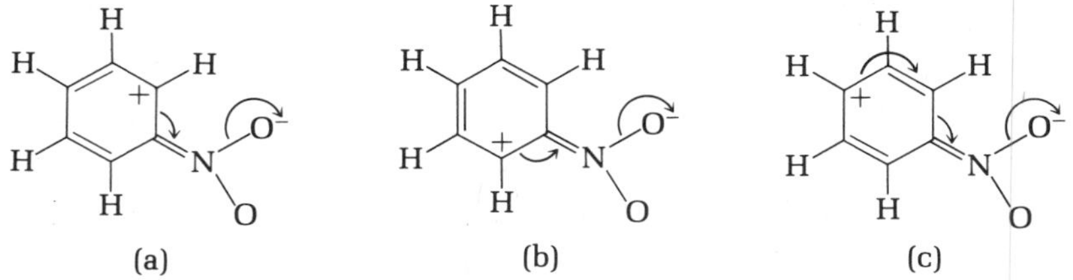

Figure 10–11 Canonical resonance forms of the dipolar nitrobenzene molecule.

(a) (b) (c)

were to transfer from the benzene portion of the pi system to the nitro group portion. The shift of electron pairs is indicated by the arrows. The pi electron deficiency is concentrated at the carbons *ortho* and *para* to the nitro group, and these positions are much less nucleophilic than the two *meta* positions. This explains why the product of reaction with the electrophilic NO_2^+ is *meta*-dinitrobenzene.

The effect that substituents (such as $—NO_2$) have on chemical reactivity is a common and important one in organic chemistry. The effect occurs in a wide variety of molecules with a wide variety of substituents. What would happen if we tried to nitrate a benzene derivative containing a substituent that is much less electronegative than the nitro group? Such a substituent serves as a supplier of electron density to the carbons *ortho* and *para* to it by the reverse of the electron shifts shown in Figure 10–11, and substitution occurs at the *ortho* and *para* positions. Examples of such substituents are OH, Cl, and CH_3.

The benzene ring is susceptible to attack by any strong Lewis acid, and a large variety of substitution reactions can be carried out with benzene. For example, hydrocarbon substituents can be introduced by complexing a halohydrocarbon with the strong Lewis acid $AlCl_3$. A typical reaction is

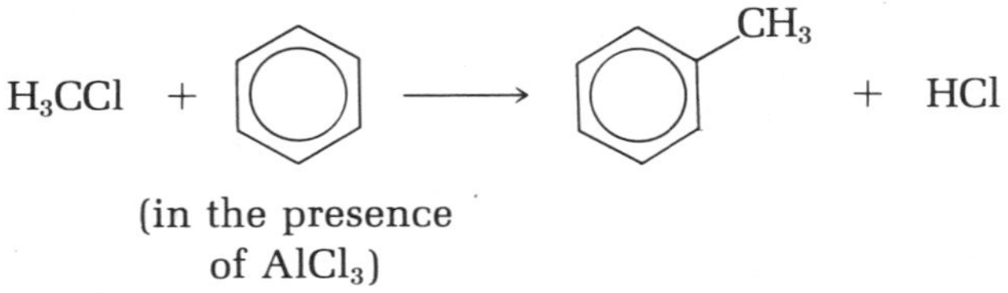

$$H_3CCl + C_6H_6 \xrightarrow{\text{(in the presence of } AlCl_3)} C_6H_5CH_3 + HCl \qquad \textbf{10–8}$$

and the product is called **methylbenzene** (or more commonly **toluene**). (The benzene ring is drawn without the H atoms in reaction 10–8, and we shall continue this practice in the remainder of the text.) The Lewis acid that attacks the benzene ring in equation 10–8 is the carbon portion of a complex formed between H_3CCl (chloromethane) and $AlCl_3$, which is polarized as

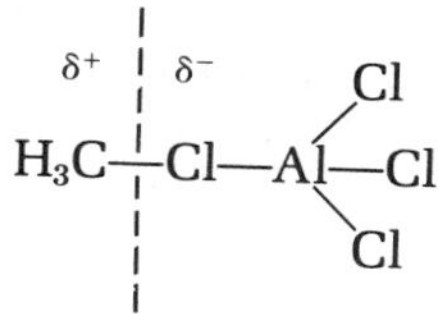

$$\overset{\delta+}{H_3C} \,\vdots\, \overset{\delta-}{Cl{-}}AlCl_3$$

The $AlCl_3$ is a catalyst in this reaction because it facilitates the reaction without being consumed in it, and that is the definition of a catalyst. From this point on we shall indicate the catalyst in reactions by placing it over the arrow connecting reactants and products.

10–4. Functional Groups That Contain Oxygen

Nomenclature of the Alcohols

While nonpolar species such as H_2 and Br_2 can be added to carbon-carbon double bonds, so can polar species. One example of addition of a polar species is the industrially important reaction in which ethene from petroleum is hydrated to form a compound identical to grain alcohol (which is distilled from fermented mash):

$$H_2C{=}CH_2 + H_2O \xrightarrow{H_2SO_4} H_3CCH_2OH \qquad \textbf{10–9}$$

The product contains the functional group —OH, and is thus a member of the class of compounds termed **alcohols.** In the IUPAC system, alcohols are named by changing the *e* on the end of the hydrocarbon name to *ol*. Thus, the product of reaction 10–9 is **ethanol** and the simplest alcohol, H_3COH, is **methanol.** With alcohols of higher molecular weight, we encounter positional isomerism, since the —OH group can be located on different carbons. The simplest examples are

$$H_3CCH_2CH_2OH \qquad\qquad H_3C{-}\underset{\displaystyle OH}{\underset{|}{CH}}{-}CH_3$$

(a) (b)

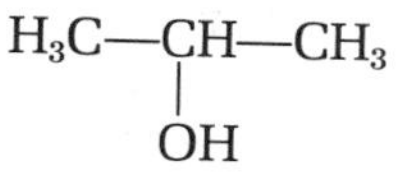

The IUPAC system uses numbers to denote positions; thus (a) is **1-propanol** and (b) is **2-propanol.**

Many simple organic compounds had been named and were in common use long before the IUPAC system was formulated. Thus, chemists continue to use several nonsystematic, or common, names for compounds. Some alcohols are referred to by the old names still. These names are based on names given to the hydrocarbon species attached to the —OH group as well as the point of attachment to that group, and these names of **alkyl radicals** are given in Table 10–3. In each case the open-ended dash indicates the carbon to which the functional group is attached. The prefixes *sec*- and

Table 10–3
Names of Some Common Alkyl Radicals

Name	Structure	Name	Structure
Methyl	$H_3C—$		
Ethyl	$H_3CCH_2—$	Isobutyl	$H_3C—C(H)(CH_3)—CH_2—$
n-Propyl	$H_3CCH_2CH_2—$		
Isopropyl	H_3CCHCH_3 (attached at the central C)		
n-Butyl	$H_3CCH_2CH_2CH_2—$	*Tert*-butyl	$H_3C—C(CH_3)(CH_3)—$
Sec-butyl	$H_3CCH_2CHCH_3$ (attached at the CH)		

tert- are shortened forms of secondary and tertiary. A carbon that is bonded to two other carbons is termed a **secondary carbon,** and a carbon that is bonded to three other carbons is a **tertiary carbon.** The alcohols can be named by giving the appropriate word from Table 10–3 followed by alcohol. Thus, H_3CCH_2OH is often called **ethyl alcohol.**

Properties of Alcohols

Alcohols of low molecular weight, such as methanol and ethanol, may be considered to be derivatives of water since they have the —OH group in common with water. They are waterlike in many ways. Like water, they can serve as both acids and bases, and they are good solvents for ionic substances and polar substances in general. They interact with each other by hydrogen-bonding in the liquid state, although since methanol and ethanol have only one hydroxyl group (—OH) per molecule and water has two, hydrogen bonding is less extensive in the alcohols. Consequently ethanol boils at 78 °C in spite of having a molecular weight approximately three times that of water.

Both methanol and ethanol are infinitely soluble (**miscible**) in water. As the molecular weight of an alcohol containing one —OH group increases, the hydrocarbon portion begins to dominate in the determination of physical properties. Thus 1-hexanol (*n*-hexyl alcohol) is only slightly soluble in water. Its boiling point is 156 °C, which is only 58° higher than the temperature at which *n*-heptane boils, whereas methanol has a boiling point 154° higher than that of ethane.

Methanol and ethanol are somewhat weaker acids in water than water itself; and the corresponding bases, **methoxide ion** (H_3CO^-) and **ethoxide ion** ($H_3CCH_2O^-$), react almost quantitatively with water to produce the alcohols and hydroxide ion. However, hydroxyl derivatives of the aromatic hydrocarbons are much stronger acids than water. The simplest of these aromatic species is **phenol**, whose structure is

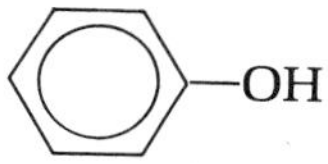

—OH

The origin of this enhanced acidity lies in delocalized bonding in the corresponding base, the **phenolate ion** ($C_6H_5O^-$). Figure 10–12 shows several canonical resonance forms of this ion. Delocalized bonding is stronger than

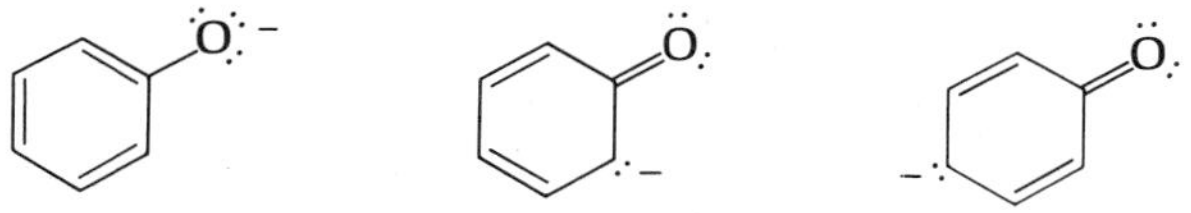

Figure 10–12 Some canonical resonance forms of the phenolate ion.

localized bonding, and consequently the phenolate ion is more stable than the methoxide ion (or the hydroxide ion), and phenol is a stronger acid than either methanol or water.

Aldehydes and Ketones

Just as water can be oxidized to molecular oxygen, alcohols can be oxidized to give **aldehydes, ketones,** and **carboxylic acids.** Aldehydes result from the two-electron oxidation of alcohols in which the oxygen atom is bonded to the terminal carbon atom (primary alcohols). An example is the oxidation of ethanol according to the reaction

$$H_3CCH_2OH + \text{Oxidizing agent} \rightarrow H_3C\text{—}C(=O)\text{—}H \qquad \textbf{10–10}$$

The other product of reaction 10–10 is the reduced form of the oxidizing agent. Organic reactions are usually written showing as product or products only the principal organic substance or substances produced.

The IUPAC names of the aldehydes are obtained from the parent hydrocarbon by replacing the terminal *e* with *-al*. Thus the product of reaction 10–10 is **ethanal.** As with the alcohols, a number of names for aldehydes were in common use before the adoption of the IUPAC system and these names are still used for the simplest aldehydes. The names are based on prefixes indicating the number of carbon atoms and in some cases the arrangement of the carbon atoms. Table 10–4 lists several of these prefixes.

Table 10–4
Some Common Aldehyde Prefixes and Their Structures

Name	Structure	Name	Structure
form-	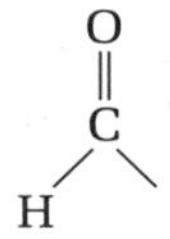 $H-C(=O)-$	isobutyr-	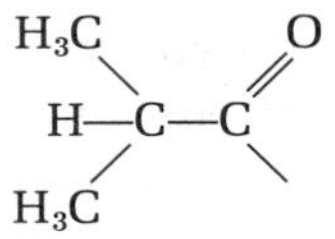 $(H_3C)_2CH-C(=O)-$
acet-	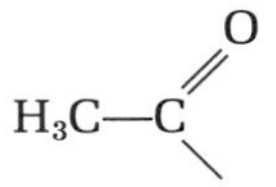 $H_3C-C(=O)-$		
propion-	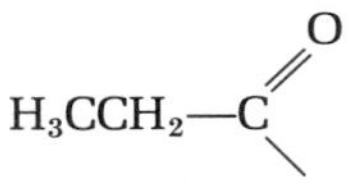 $H_3CCH_2-C(=O)-$		
n-butyr-	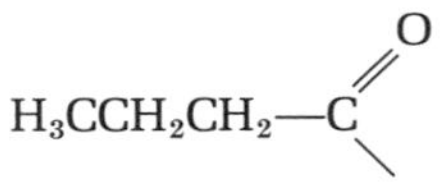 $H_3CCH_2CH_2-C(=O)-$	*n*-valer-	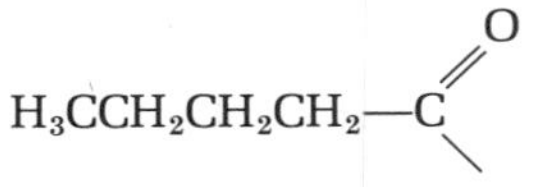 $H_3CCH_2CH_2CH_2-C(=O)-$

The suffix that is added to the prefixes of Table 10–4 in the case of aldehydes is *-aldehyde,* and the two simplest aldehydes are called formaldehyde and acetaldehyde.

Ketones are the products of 2-electron oxidation of secondary alcohols. The simplest ketone is produced by the oxidation of 2-propanol (isopropyl alcohol) according to the reaction

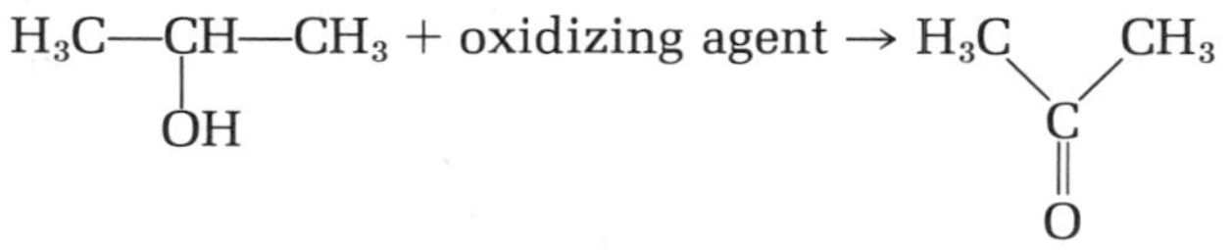

$$H_3C-CH(OH)-CH_3 + \text{oxidizing agent} \rightarrow H_3C-C(=O)-CH_3 \qquad \textbf{10–11}$$

The IUPAC name for the ketones involves the substitution of *-one* for the terminal *e* in the name of the parent hydrocarbon. Thus, the product of reaction 10–11 is **propanone.** The common naming system for ketones identifies the two alkyl radicals bonded to the **carbonyl group**

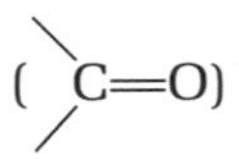

$$(>C=O)$$

Thus, propanone is also known as **dimethyl ketone** and **butanone** is **methyl ethyl ketone.** Propanone is most widely known by the trivial name **acetone.**

The carbonyl bond (C=O) of aldehydes and ketones is polar, with the charge distribution $C^{\delta+}=O^{\delta-}$, and the molecules associate relatively strongly in the liquid state. The boiling point of acetone is 56.5 °C, while 2-propanol boils at 82.3 °C. Table 10–5 contains a comparison of the boiling points of three compounds of approximately equal molecular weights.

Compound	Molecular Weight	Boiling Point (°C)	Association
n-Butane	58	0	Weak
Acetone	58	56.5	Relatively strong
2-Propanol	60	82.3	Strong

Table 10–5 Comparison of Boiling Points of a Hydrocarbon, a Ketone, and an Alcohol of Approximately Equal Molecular Weights

Because of the charge separation, the carbon atom in the carbonyl group displays an electrophilic character and the oxygen a nucleophilic character. Thus, aldehydes and ketones in water exist to some extent in a hydrated form shown in the following structure for acetaldehyde, in which OH^- has added to the carbonyl carbon and H^+ to the carbonyl oxygen

$$H_3C\text{—}C(OH)_2\text{—}H$$

The polarity of the carbonyl group is the principal source of the reactivity of aldehydes and ketones.

Carboxylic Acids

Carboxylic acids result from the two-electron oxidation of aldehydes; an example is

$$H_3C\text{—}C(=O)\text{—}H + \text{Oxidizing agent} \rightarrow H_3C\text{—}C(=O)\text{—}OH \qquad \textbf{10–12}$$

The IUPAC names of the carboxylic acids are derived from the parent hydrocarbons by substituting *-oic acid* for the terminal *e*. Thus, the product of reaction 10–12 is **ethanoic acid.** The prefixes of Table 10–4 apply to the common names of the carboxylic acids, and ethanoic acid is more commonly known as **acetic acid.** Perhaps the most obvious characteristic of the lower-molecular-weight carboxylic acids is their odor. Both acetic and formic acids have the sharp odor characteristic of vinegar (the "distilled white vinegar" in the grocery store is a dilute solution of acetic acid in water), while ***n*-butyric acid** is responsible for the unpleasant odor of rancid butter and some cheeses.

The acidity of carboxylic acids results from delocalized bonding in their corresponding bases. The carbonyl carbon is always characterized by sp^2 hybridization and thus **carboxylate anions** are similar in structure and bonding to the carbonate ion. Figure 10–13 shows the two canonical resonance forms of the **acetate ion.**

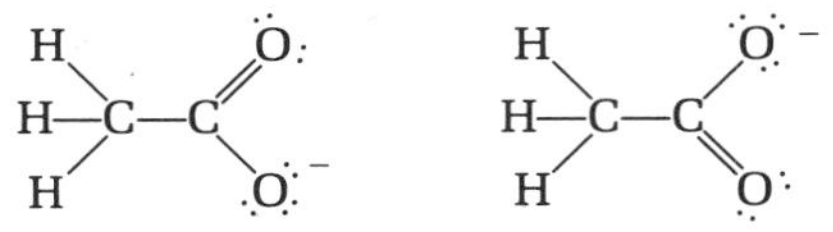

Figure 10–13
The two canonical resonance forms of the acetate ion.

The carboxylic acids offer another good example of the effect of substituent groups on chemical reactivity. A typical carboxylic acid is a much stronger acid than phenol in spite of the fact that delocalized bonding occurs in the phenolate ion as well as in carboxylate ions. We have seen that the acid strengths of oxyacids increase with the formal charge of the central atom (Chapters 8 and 9). For example, the acidities of the oxyacids of chlorine increase by many orders of magnitude between hypochlorous acid (HOCl) and perchloric acid ($HOClO_3$). The effect is explained in terms of a postulated progressive withdrawal of electron density from the highly electronegative Cl in going from HOCl to $HOClO_3$. Electron withdrawal is also proposed as the explanation why fluorosulfuric acid (FSO_3H) is a stronger acid than H_2SO_4; in this case the electrons are withdrawn from the sulfur. The carbonyl oxygen serves in an electron-withdrawing capacity in carboxylic acids and this is why carboxylic acids are stronger acids than phenol. The acidity can be enhanced by the substitution of more electronegative atoms in other places in the molecule. Thus we should predict that trifluoroacetic acid, F_3CCOOH, is a stronger acid than acetic acid. Trifluoroacetic acid is a strong acid that dissociates completely in water.

Esters

Carboxylic acids, in addition to their acid character, are quite reactive because of the polarity of the carbonyl group. If acetic acid is dissolved in ethanol, the acid can add a molecule of ethanol in the same way that water is added to acetaldehyde, and the resulting structure is represented by

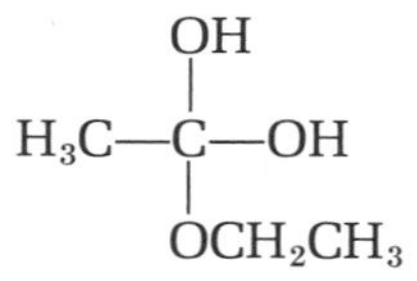

$$H_3C—C(OH)(OH)—OCH_2CH_3$$

In the case of hydrated acetaldehyde, re-formation of the carbonyl bond by dehydration gives acetaldehyde back again. However, the addition product of acetic acid and ethanol can yield the C=O bond in two ways: by the elimination of ethanol to re-form the acid, or by the elimination of water to form the structure

$$H_3C-\overset{\overset{\displaystyle O}{\|}}{C}-O-CH_2CH_3$$

Such compounds formed from acids and alcohols are termed **esters.** The IUPAC name derives from the name of the alcohol and that of the acid and the characteristic suffix is *-oate.* Thus, the ester above is **ethyl ethanoate.** The common names are based on the names of the alcohols and prefixes of Table 10–4, and the characteristic suffix is *-ate*. The ester formed from acetic acid and ethanol is named **ethyl acetate.** Unlike the acids from which they are derived, esters generally have very pleasant odors and they are responsible for the odors of many fruits and flowers. For example, *n*-butyl acetate has the odor of bananas.

Esters of certain high-molecular-weight carboxylic acids are important in living systems because they form the class of compounds termed **fats.** Fats are esters of the trihydric (three —OH groups) alcohol, **glycerol,** whose structure is

$$\underset{\displaystyle HO}{\overset{}{H_2C}}-\underset{\displaystyle OH}{CH}-\underset{\displaystyle OH}{CH_2}$$

The acids (termed fatty acids) commonly found esterified with glycerol in fats are given in Table 10–6.

Name	Structure
Palmitic acid	$H_3C(CH_2)_{14}COOH$
Stearic acid	$H_3C(CH_2)_{16}COOH$
Oleic acid	$H_3C(CH_2)_7CH{=}CH(CH_2)_7COOH$
Linoleic acid	$H_3C(CH_2)_4CH{=}CHCH_2CH{=}CH(CH_2)_7COOH$

Table 10–6
Names and Formulas of the Common Fatty Acids

Each fat molecule is a triester made from one glycerol molecule and three fatty acid molecules with the elimination of three water molecules. The common fatty acids are unbranched. The saturated fatty acids, palmitic acid and stearic acid, yield fats that are solids at room temperature (saturated fats), whereas oleic and linoleic acids, both unsaturated compounds, yield fats that are liquids at room temperature (unsaturated fats). Hydrogen can be added to the C=C bonds of unsaturated fats to convert them to solid saturated fats. This hydrogenation process is used to convert vegetable oils to margarine (though margarine still contains a high percentage of unsaturated fats).

Soaps and Detergents

Just as an alcohol can be added to the C═O bond of an acid to yield the intermediate in the formation of an ester, water can be added to an ester to yield this same intermediate in the **hydrolysis** of an ester to an acid and an alcohol. This is shown in the following reactions:

$$R\text{—}C(=O)\text{—}OR' + H_2O \rightarrow R\text{—}C(OH)(OH)\text{—}OR'$$

$$R\text{—}C(OH)(OH)\text{—}OR' \rightarrow R\text{—}C(=O)\text{—}OH + HOR'$$

Since the addition of water to the C═O bond involves OH^- added to the carbon and H^+ to the oxygen, the hydrolysis is catalyzed by both acids and bases. The base-catalyzed hydrolysis of a fat is termed **saponification**, and the product is glycerol and the salt of the fatty acid.

Fatty acid salts are termed **soaps**, and they serve as mediators between nonpolar materials and water. This mediator ability arises from the highly polar character of the carboxylate anion end of the species and the decidedly nonpolar character of the hydrocarbon portion. The ionic end possesses a strong affinity for water (it is said to be **hydrophilic**) and the other end avoids water (or more properly water avoids it, because hydrogen bonding cannot occur with it), and the hydrocarbon portion is said to be **hydrophobic.** As a consequence of this "split personality," soaps form **micelles** in water, of the structure shown in Figure 10–14(a). The hydrocarbon portions associate to form a "greasy ball" surrounded by charges. Such a structure can easily suspend a droplet of oil by surrounding it as shown in Figure 10–14(b), and this gives soap its ability to serve as a laundry agent.

Clearly any water-soluble molecule possessing an ionic "head" and a long hydrocarbon "tail" will serve as a laundry agent. Magnesium and calcium ions are commonly found in significant concentrations in "hard" water and these ions form precipitates (water-insoluble substances) with soaps (bath-

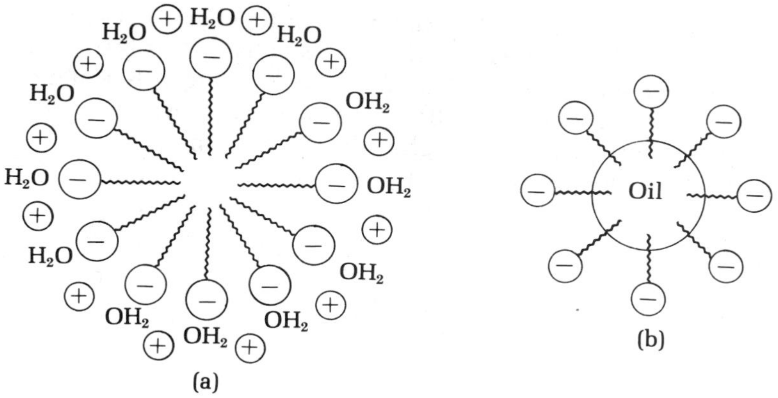

Figure 10–14
(a) A soap micelle.
(b) The structure of an oil droplet and soap anions which allow suspension in water. The hydrocarbon portion of the soap anion is represented by a wavy line.

tub ring). This precipitation is avoided when molecules derived from such anions as sulfate and phosphate are involved. These species belong to the general class of substances termed **detergents,** and a typical example is **sodium cetyl sulfate,** whose structure is

$$[H_3C(CH_2)_{14}CH_2—O—\overset{\displaystyle O}{\overset{|}{\underset{\underset{\displaystyle O}{|}}{S}}}—O]^-Na^+$$

In highly concentrated soap solutions a layered structure is formed, as shown in Figure 10–15. This structure is known as a **liquid crystal** because

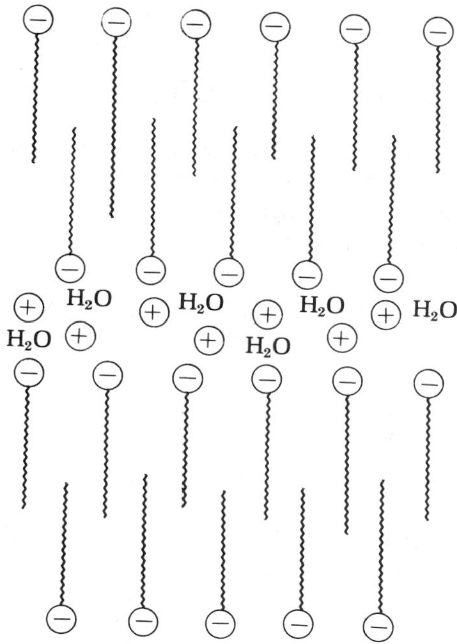

Figure 10–15
Structure of a liquid crystalline phase of soap in water.

it flows like a liquid yet has the long-range order characteristic of a crystal. This type of **bilayer** phase is also mainly responsible for the structure and properties of **cell membranes** in animals. The detergentlike molecules in membranes are **phospholipids**, fatty acid molecules, in which one fatty acid is replaced by a derivative of orthophosphoric acid. One of the most common phospholipids is **lecithin**, whose structure is

$$\begin{array}{l} H_2C-O-\overset{\overset{O}{\|}}{C}-R \\ \;\;| \\ HC-O-\overset{\overset{O}{\|}}{C}-R \\ \;\;| \\ H_2C-O-\underset{\underset{O^-}{|}}{\overset{\overset{O}{|}}{P}}-O-CH_2CH_2N^+(CH_3)_3 \end{array}$$

where R symbolizes the long hydrocarbon chain of a fatty acid. In animal cell membranes the phospholipids form a bilayer "sandwich" (two parallel bilayer sandwiches are shown in Figure 10–15), with the phosphate portion exposed to body fluid at the inside and outside membrane surfaces.

Ethers

The final class of oxygenated hydrocarbons that we shall consider comprises the **ethers.** Ethers are made by the dehydration of alcohols; an example is

$$2\ H_3CCH_2OH \xrightarrow{H_2SO_4} H_3CCH_2OCH_2CH_3 + H_2O \qquad \textbf{10–13}$$

The IUPAC system names ethers as alkoxy derivatives of hydrocarbons, and the ether produced in reaction 10–13 is thus **ethoxy ethane.** The ethers are more commonly named by identifying the two alkyl radicals bonded to oxygen and then adding the name *ether*. In this way, the product of reaction 10–13 is **diethyl ether.** The ethers are relatively nonpolar molecules that boil at almost the same temperature as the corresponding hydrocarbon. For example, diethyl ether boils at 35 °C and *n*-pentane boils at 36 °C.

10–5. Functional Groups Containing Nitrogen

Amines

Just as the nitrogen analog of water is ammonia, the nitrogen analog of an alcohol is a **primary amine** of the general formula RNH_2. The N—H bond is less polar than the O—H bond. Consequently ammonia possesses a much lower boiling point than water and primary amines boil at a considerably lower temperature than the corresponding alcohols. For example, **ethylamine** boils at 17 °C, and ethyl alcohol boils at 78 °C. Just as ethanol is a stronger base (and weaker acid) than water, most primary amines are stronger bases than ammonia. For example, methylamine is more than ten times as strong a base as ammonia. An exception to this increased basicity occurs when R is an aromatic radical. Thus, **aniline** ($C_6H_5NH_2$) is less than 10^{-4} times as strong a base as ammonia. This is the analog of the increased acidity of phenol as compared with methanol. (You are asked to explain the effect in a problem at the end of the chapter.) **Secondary amines** contain two alkyl groups and a single N—H bond, while **tertiary amines** contain three alkyl groups. Examples of each are **dimethyl amine,** $(H_3C)_2NH$, and **trimethyl amine,** $(H_3C)_3N$.

The amines can be prepared by the substitution of a C—N bond for the weaker C—I bond of an iodoalkane:

$$H_3CI + NH_3 \rightarrow H_3CNH_2 + HI \qquad \textbf{10–14}$$

The resultant methylamine can react with excess iodomethane (also commonly called methyl iodide) to produce dimethylamine and trimethylamine. The reaction can even proceed one step further:

$$(H_3C)_3N + H_3CI \rightarrow [(H_3C)_4N]^+I^- \qquad \textbf{10–15}$$

wherein the product is a salt containing the **tetramethyl ammonium ion,** a highly stable ion having a tetrahedral arrangement of carbons around a central nitrogen.

Amides

Perhaps the most important reaction of amines is their reaction with carboxylic acids, in the amine analog of ester formation. Amines can add to the carbonyl group of carboxylic acids to give the structure

```
      OH
      |
  R—C—OH
      |
     HNR'
```

where R and R′ are two alkyl groups. The addition is shown for a primary amine, but secondary amines can add as well. As in ester formation, this intermediate can decompose by the loss of a water molecule to yield an **amide:**

```
        O
       //
  R—C
       \
        NHR'
```

This reaction provides the basis for the synthesis of a large number of important long-chain molecules (polymers). Carboxylic acids that have two carboxyl groups, one at each end of a hydrocarbon chain, can be obtained. One such dicarboxylic acid is **adipic acid,** $HOOC(CH_2)_4COOH$. Also available are diamines having two amino groups on the two terminal carbons of a hydrocarbon chain. Some of these diamines are produced in natural processes, and one example is **1,5-diaminopentane,** which is known by the common name **cadaverine,** because it is the source of the putrid odor of rotting flesh.

When 1,5-diaminohexane is caused to react with adipic acid, a polymer with the following partial structure forms.

$$\text{—NH—(CH}_2)_6\text{—NH—}\overset{\overset{\displaystyle O}{\|}}{C}\text{—(CH}_2)_4\text{—}\overset{\overset{\displaystyle O}{\|}}{C}\text{—NH—(CH}_2)_6\text{—NH—}$$

The polyamide polymer is known as **nylon-66** and it was the first of a series of nylons, all of which are based on the linkage of molecules through the formation of amides. Another such polymer is **nylon-6,** in which the monomer contains an amino group and a carboxyl group at opposite ends in a six-carbon molecule, $H_2N(CH_2)_5COOH$.

If polymers can be made by amide formation, surely they can also be made by ester formation. The most common of such **polyester** substances is **Dacron,** which is synthesized from **terephthalic acid,**

$$\text{HOOC—C}_6\text{H}_4\text{—COOH}$$

and **ethylene glycol,** $HOCH_2CH_2OH$, to yield the structure

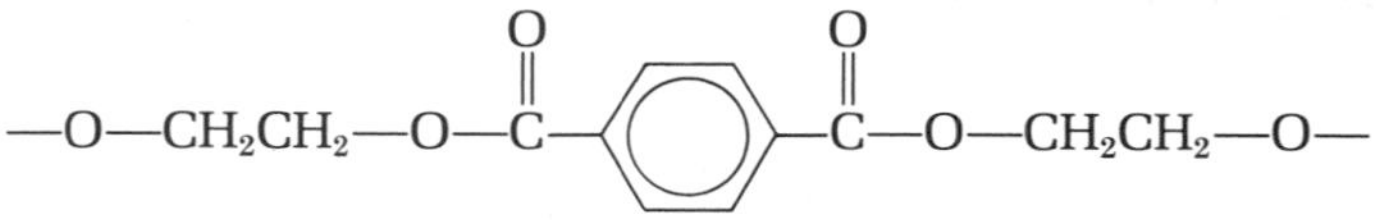

$$\text{—O—CH}_2\text{CH}_2\text{—O—}\overset{\overset{\displaystyle O}{\|}}{C}\text{—C}_6\text{H}_4\text{—}\overset{\overset{\displaystyle O}{\|}}{C}\text{—O—CH}_2\text{CH}_2\text{—O—}$$

Polyurethanes

We have considered polymers that are formed from the addition to carbon-carbon double bonds and to carbon-oxygen double bonds (in the formation of the intermediates that lead to esters and amides). The carbon-nitrogen double bond also provides a basis for polymerization; it occurs in molecules whose functional group is derived from the **cyanate ion,** which has the structure and bonding

$$[\text{N=C=O}]^-$$

The organic molecules termed **isocyanates** may be considered to result from the combining (at the nitrogen) of a cyanate ion and a **carbonium ion** (an organic cation). An example is **methyl isocyanate,** $H_3C—N=C=O$. The C—N double bond is polarized as $C^{\delta+}=N^{\delta-}$ and both alcohols and amines will add to it. A typical **polyurethane** plastic is obtained from the addition of a dihydric alcohol to a diisocyanate:

$$HO—(CH_2)_4—OH + O=C=N—(CH_2)_5—N=C=O \rightarrow$$

$$—\underset{}{\overset{H}{\overset{|}{N}}}—\overset{O}{\overset{\|}{C}}—O—(CH_2)_4—O—\overset{O}{\overset{\|}{C}}—NH—(CH_2)_5— \qquad \textbf{10–16}$$

The urethanes derive their name from their structural similarity to the animal waste product **urea,** which may be formed by the addition of ammonia to the N=C bond of cyanate ion:

$$NCO^- + NH_3 + H_2O \rightarrow H_2N—\overset{O}{\overset{\|}{C}}—NH_2 + OH^- \qquad \textbf{10–17}$$

10–6. Optical Isomerism

Mirror Images of Molecules

The subject of isomerism would not be complete without looking into a type of isomerism that is extremely important in biological processes, namely **optical isomerism.** We have previously encountered three types of isomerism. The first is structural isomerism (also called functional when functional groups are involved), illustrated by butane and isobutane and by propionaldehyde and acetone. The latter pair are an example of the functional version of structural isomerism; both have the molecular formula C_3H_6O, but propionaldehyde is an aldehyde and acetone is a ketone. The second type of isomerism is positional isomerism, involving the same functional group or groups and the same hydrocarbon structure but different locations of the functional group or groups. An example occurs with 1-propanol and 2-propanol. The third type of isomerism is geometrical isomerism, which we illustrated with the positioning of groups in a *trans* or *cis* fashion about double bonds.

Optical isomerism is illustrated in Figure 10–16 with the biologically important molecule **lactic acid.** The carbon drawn as the central atom in the figure has four different substituents bonded to it (a hydrogen, a methyl group, a hydroxyl group, and a carboxyl group). Figures 10–16(a) and (b) are mirror images of each other, and the mirror that relates the two isomers is shown midway between them. The two molecules in the figure are differ-

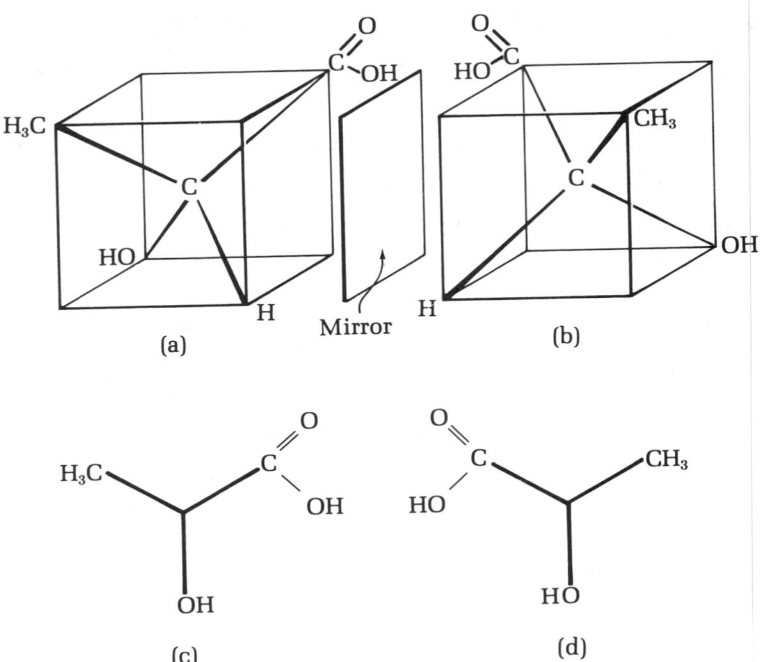

Figure 10–16
(a, b) The two optical isomers of lactic acid. (c, d) Views of (a) and (b) from H to C along the H—C axis.

ent, as an attempt to superimpose them shows. In Figures 10–16(c) and (d), we are looking along the H—C axis and we see again the mirror image relation between the two isomers. Figure 10–16(c) is the same molecule as (a), and (b) and (d) are also identical. There is no possible orientation of (d) that makes it identical with (c), just as there is no orientation of a right hand that makes it indistinguishable from a left hand. Optical isomerism occurs in all cases in which a carbon atom is bonded to four different substituents. (Such carbon atoms are called **asymmetric** carbon atoms.) This is not the only situation that gives rise to nonidentical mirror image forms (witness your right and left hands), but it is the one most commonly found in organic chemistry.

Why is the existence of optical isomers important? At first glance the two forms of lactic acid, for example, would appear to possess the same reactivity because the same functional groups are attached to the carbon atom. The important difference, however, can be seen by the analogy with right and left hands. Right and left hands can grasp a hammer with equal ease (there are no right-handed or left-handed hammers) because a hammer possesses an identical mirror image. However, a hand cannot clasp another hand in a way that is independent of the handedness (people normally clasp right hands). Thus, optical isomerism becomes important when two **optically active** molecules (the designation used for molecules that possess optical isomers) react with each other. Most biologically important molecules are optically active, and thus optical activity is extremely important in biological chemistry.

Optical Activity

What kind of activity is denoted in the phrase *optical activity?* Electron motion in a radio antenna generates radio waves. The electromagnetic radiation produced by the electron motion restricted to one direction (that of the antenna) is said to be **plane-polarized** because the electric field vectors point only in that direction (vertical in Figure 3–5). There is no force in the horizontal direction on any electrical charges. Almost any form of electromagnetic radiation can be obtained in a plane-polarized form. Any single optical isomer of any optically active substance produces a rotation of the plane of plane-polarized visible light, and the sense of this rotation is opposite for the two enantiomers. (Two optical isomers such as the lactic acid isomers are commonly termed **enantiomers.**) As shown in Figure 10–17, one of the isomers produces a clockwise rotation or rotation to the right. It

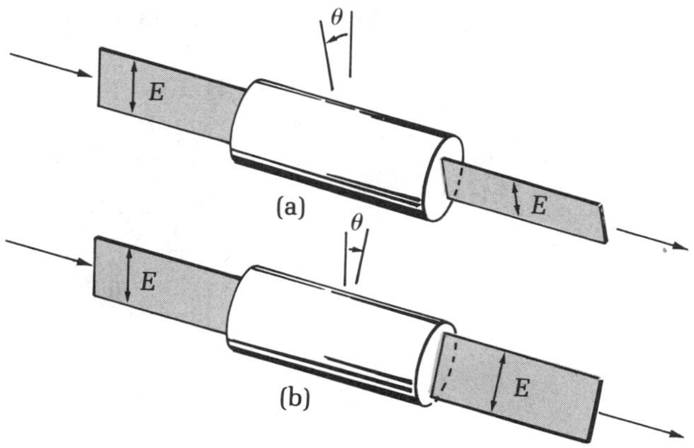

Figure 10–17
(a) Dextro and (b) levo rotation of the plane of plane polarized light by passage through tubes, each of which contains one of the two enantiomers of an optically active substance.

is the **dextrorotatory** isomer and is indicated by *d*- preceding the name of the compound. Thus, one of the lactic acid enantiomers is ***d*-lactic acid.** The other enantiomer produces an equal counterclockwise rotation and it is said to be **levorotatory.** Thus, the second lactic acid isomer is ***l*-lactic acid.** An equimolar mixture of the *d*- and *l*- forms of any molecule is known as a **racemic mixture** and it produces no rotation of the plane of polarized light.

SUGGESTIONS FOR ADDITIONAL READING

Allinger, N. L. and Allinger, J., *Structures of Organic Molecules*. Prentice-Hall, Englewood Cliffs, N.J., 1965.

Herz, W., *The Shape of Carbon Compounds*, Benjamin, Menlo Park, Calif., 1963.

PROBLEMS

1. Draw the structures and give the names of all of the isomers of C_6H_{14}.
2. Most fluorides that have a zero dipole moment have low boiling points. For example, CF_4 boils at $-128\,°C$. This is much lower than neopentane, which has a comparable molecular weight. Can you explain this phenomenon? *Hint:* Consider the polarity of the C—F bond.
*3. The discussion in the text reveals that the boiling points listed in Table 10–1 can be neatly correlated with molecular structures. However, the melting points do not correlate with the boiling points. For example, the melting point of neopentane is much higher than the melting points of the two hydrocarbons immediately above and below it in the table. Explain this lack of correlation.
4. From bond energy data given in Table 6–3, draw a conclusion about the bonding that is principally responsible for the ease with which reaction 10–2 proceeds. Almost any hydrogen-containing compound reacts well with Cl_2. Explain why.
5. Figure 10–18 shows two arrangements of the atoms in the puckered ring molecule cyclohexane. These two arrangements are termed the **chair**

(a)

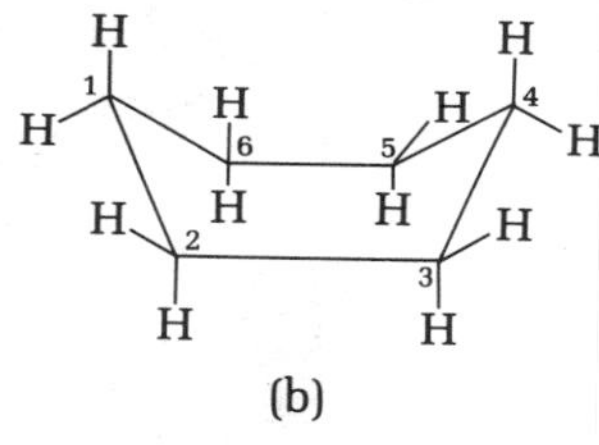

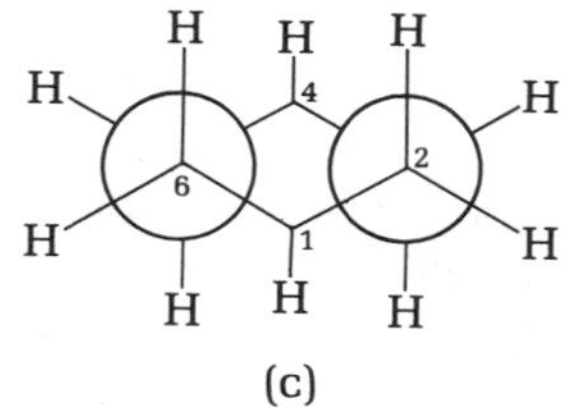

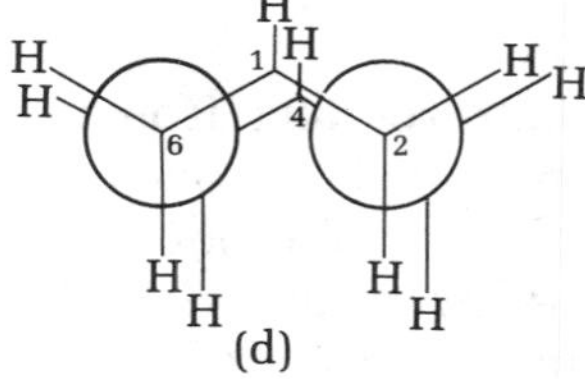

Figure 10–18
(a, c) The chair and (b, d) boat arrangements of cyclohexane (C_6H_{12}). Structures (a) and (b) are in perspective and (c) and (d) are end views, as in Figure 10–1(b). The six carbon atoms are not explicitly indicated, but they lie at the intersections of the straight lines. Only ten of the twelve hydrogens are shown in (c) and (d).

form (a) and (c), and the **boat** form (b) and (d). Predict which form predominates in liquid cyclohexane at room temperature. *Hint:* Refer to Figure 10–1.
6. Draw all of the possible isomers of C_5H_{10} and name as many as you can.
7. From the bond energy data of Table 6–3, calculate ΔE for reaction 10–4,

and for reaction 10–18:

$$H_2C{=}CH_2(g) + O_2(g) \rightarrow 2\,H_2C{=}O(g) \qquad \textbf{10–18}$$

*8. Reaction 10–4 proceeds well, but reaction 10–18 is not observed. We shall deal later with such anomalies, but you have enough information now to speculate on the key difference between the two reactions.

9. Show three propylene molecules joining as in equation 10–5 to form polypropylene. In the case of equation 10–5, only one structure can be formed. How many different structures can be formed in the polypropylene segment? Draw these different structures.

10. What is the IUPAC name for isoprene?

11. Over a long period of time, rubber can become hard and brittle by reacting with molecular oxygen. What reaction with oxygen could lead to this hardening?

12. Draw all of the possible isomers of C_4H_6. Name as many of them as you can.

13. The energy released by reaction 10–19 can be calculated quite satisfactorily by using the C—H, C═C, O═O, C═O, and O—H bond energies listed in Table 6–3:

$$C_2H_4(g) + 3\,O_2(g) \rightarrow 2\,CO_2(g) + 2\,H_2O(g) \qquad \textbf{10–19}$$

If this calculation is repeated with benzene, with the benzene structure taken as one of the canonical resonance forms in Figure 10–8, the calculated energy released is 42 kcal/mole higher than the energy actually observed. Why is less energy liberated than that calculated?

14. If benzene is heated carefully in the presence of concentrated sulfuric acid, the product is **benzenesulfonic acid,** whose structure is

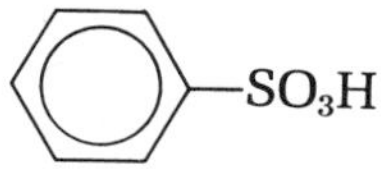

$$C_6H_5{-}SO_3H$$

In a fashion analogous to that in the text for the nitration of benzene, propose a series of steps for the sulfonation of benzene. *Hint:* Remember that H_2SO_4 contains HSO_4^- and SO_3H^+.

15. Naphthalene, $C_{10}H_8$, is also an aromatic hydrocarbon whose canonical resonance structures show an alternation of C—C and C═C bonds. Consider the nitronapthalene molecule shown:

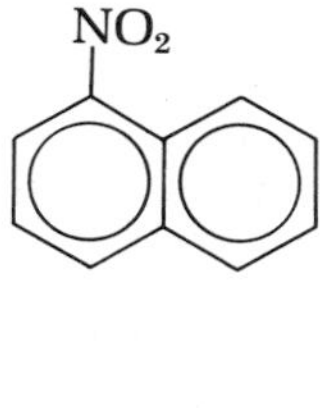

NO_2

Suppose that you further nitrate this molecule by adding a second nitro group. Through canonical resonance forms analogous to those in Figure 10–11, predict which dinitronaphthalene structures will *not* be observed in the products of the nitration.

16. Draw the three canonical resonance structures of fluorobenzene that reveal why fluorobenzene nitrates to yield predominantly *ortho*- and *para*-nitrofluorobenzene.

17. The structure of the explosive TNT (trinitrotoluene) is

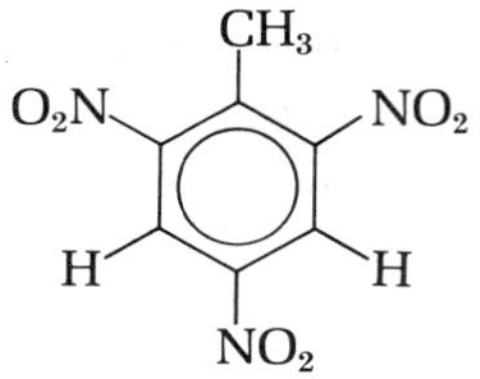

CH_3
O_2N NO_2
H H
NO_2

Explain why this molecule is explosive.

18. *Sec*-butyl alcohol is more soluble in water than *n*-butyl alcohol, and *tert*-butyl alcohol is miscible with water. Explain this trend in solubilities.

19. Show that the conversion of a primary alcohol to an aldehyde is a two-electron oxidation.

20. Draw the structures of all of the non-ring isomers of C_4H_8O that contain the carbonyl group. Give two names for each of the compounds.

21. The lower-molecular-weight aldehydes and ketones are very soluble in water. For example, both acetaldehyde and acetone are miscible in water. What is the source of this high solubility?

22. **Cyanohydrins** result from the reaction of the weak Lowry-Brønsted acid HCN with aldehydes and ketones. From the explanation in the text about the principal source of reactivity of aldehydes and ketones, predict the structure of the product of the reaction between acetone and HCN.

23. Strong oxidizing agents can split the C=C bond with the introduction of oxygen according to the following sample reaction:

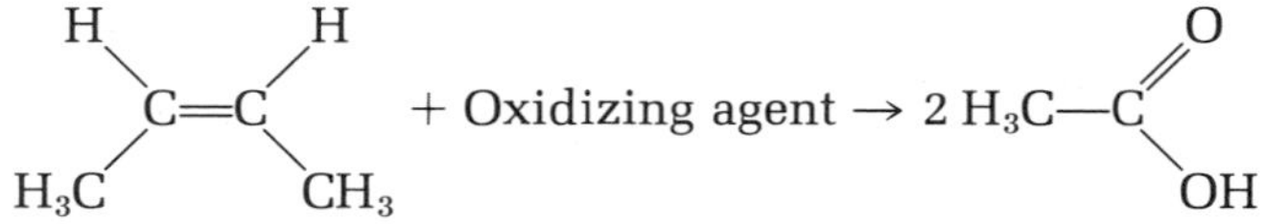

$$(H)(H_3C)C{=}C(H)(CH_3) + \text{Oxidizing agent} \rightarrow 2\ H_3C{-}C({=}O)OH$$

How many electrons are lost by each *cis*-2-butene molecule in this reaction?

***24.** The structure of fumaric acid is

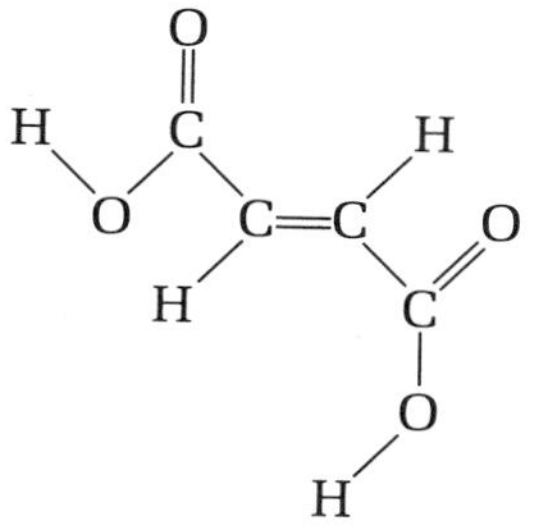

$$HO-C(=O)-CH=CH-C(=O)-OH \quad (trans)$$

Maleic acid is the *cis* form of fumaric acid. It is over ten times stronger an acid than fumaric acid. Propose a reason why.

25. Dichlorodiphenyltrichlorethane (DDT) is highly soluble in cell membranes, where it interferes with the normal passage of chemical species across those membranes. The DDT molecule is nonpolar and only slightly water soluble. Refer to Figure 10–15 and indicate the location of the DDT molecules in cell membranes.

26. Account for the unusually low basicity of aniline in an explanation analogous to that for the unusually high acidity of phenol.

27. Amides are quite poor bases as compared to amines. Explain why.

28. The boiling points of the low-molecular-weight amides are unusually high. For example acetamide boils at 222 °C, while acetic acid boils at 118 °C. Explain why acetamide has a higher boiling point than acetic acid. The structure of acetamide is

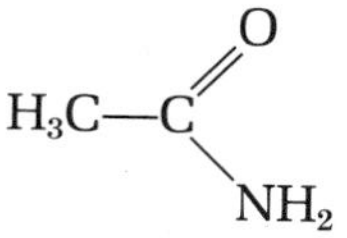

$$H_3C-C(=O)-NH_2$$

29. Each of the following structures represents one optical isomer of the species shown. Identify the asymmetric carbon atoms in each case.

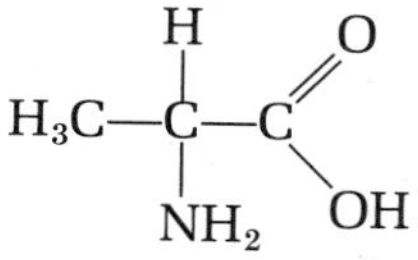

$$H_3C-CH(NH_2)-C(=O)-OH$$

(a)

$$HO-C(=O)-CH(OH)-CH(OH)-C(=O)-OH$$

(b)

$$\begin{array}{ccc} & CH_2-CH_2 & \\ HO-CH & & CH_2 \\ & CH=CH & \end{array}$$

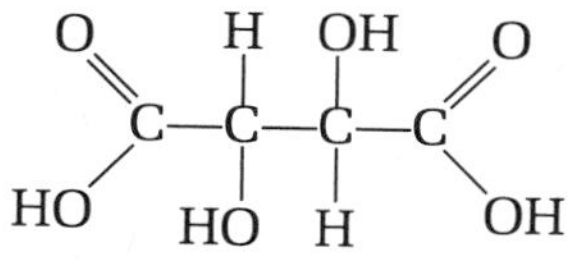

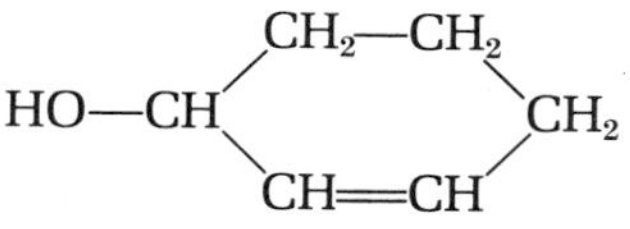

(c)

11/ The Chemistry of the Metallic Elements

The elements whose chemistry we shall consider in this chapter are those shown in Figure 11–1. Each of these elements (except mercury) is a solid at room temperature and each possesses metallic properties.

I A	II A	III B	IV B	V B	VI B	VII B	VIII B			I B	II B
Li	Be										
Na	Mg										
K	Ca	Sc	Ti	V	Cr	Mn	Fe	Co	Ni	Cu	Zn
Rb	Sr	Y	Zr	Nb	Mo	Tc	Ru	Rh	Pd	Ag	Cd
Cs	Ba	La	Hf	Ta	W	Re	Os	Ir	Pt	Au	Hg

Figure 11–1
The groups of metallic elements considered in Chapter 11.

The metallic elements are characterized by relatively low electronegativities as compared with the nonmetallic elements. Thus, cation formation and the structures and reactions of cations are the dominant features of the chemistries of the metallic elements. This contrasts markedly with the nonmetallic elements for which anion formation predominates.

11–1. Bonding in Metals

Delocalized Bonding in Graphite

One of the most remarkable differences in physical properties between closely related substances occurs between graphite and diamond. Graphite is black and an excellent conductor of electricity, while diamond is colorless and an electrical insulator; yet both graphite and diamond are pure carbon. Surely this difference in properties must arise from the different structures that reflect the different bonding.

Graphite, like benzene, involves carbon 2*p* delocalized pi bonding, but in graphite, in any commonly encountered sample of the substance, the bonding extends over a huge number of atoms. In benzene, six carbon 2*p* orbitals yield six pi molecular orbitals. In graphite, with a much larger number of orbitals involved, there is a huge number of pi molecular orbitals between the most bonding orbital and the most antibonding orbital. These orbitals compose a **band** of very closely spaced orbitals. Figure 11–2 shows this band on an energy-level diagram analogous to that for benzene (Figure 10–7). Half of the orbitals (the bonding half) are filled, and this is indicated by the shading of the lower half of the band in Figure 11–2.

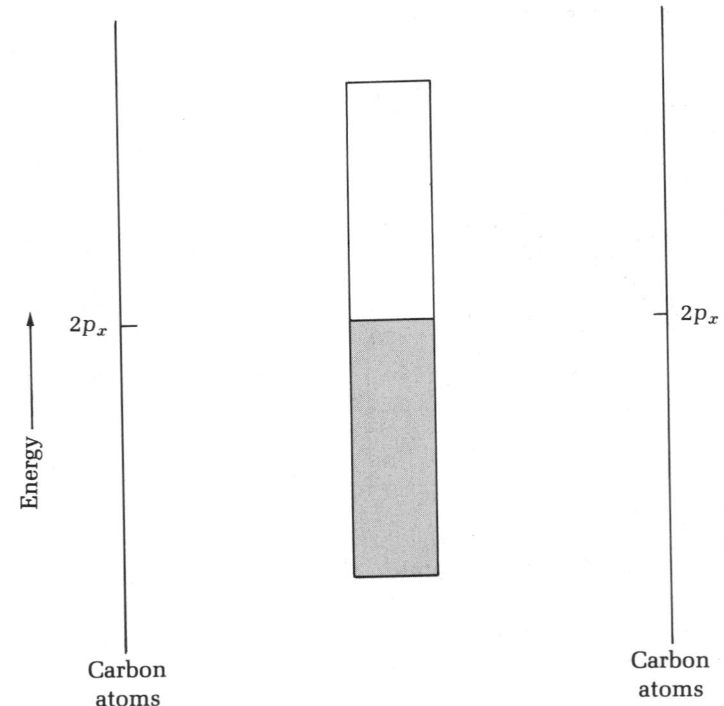

Figure 11–2
Pi molecular orbital band of graphite. The shaded area represents filled bonding molecular orbitals.

How does this band picture of graphite yield a black color and good electrical conductivity? For conduction to occur, the electrons that carry the current must be able to acquire extra kinetic energy from voltage applied across the conducting substance. In benzene or diamond or any other electrical insulator, the energy gap between the highest occupied orbital and the lowest unoccupied orbital is very large and a very high voltage (typically tens of thousands of volts) must be applied to supply the extra energy to obtain conduction. In graphite, there are empty antibonding pi orbitals immediately above the highest-energy, filled orbitals, and even very small voltages can supply the extra energy and yield electrical conduction.

Graphite is black because it (and any other black substance) absorbs visible light strongly and uniformly throughout the visible range. To absorb a frequency ν, a substance must possess an energy difference, ΔE, between an occupied level and an unoccupied level such that

$$\Delta E = h\nu$$

Graphite has a virtual continuum of such gaps between the pi bonding levels and the pi antibonding levels, and graphite absorbs almost all frequencies of electromagnetic radiation.

The Band Model of Metals

In examining the models of beryllium, we found that Be_2 does not exist. Yet beryllium exists at room temperature as a metallic solid. How do the beryllium atoms bond in the solid when they do not bond in Be_2? In any solid element, the atomic orbitals of any given atom overlap those of its neighboring atoms, which overlap those of their neighbors, and so on. Thus, the possibility of highly delocalized bonding exists, as it does in graphite. We can construct molecular orbital bands for each of the atomic orbitals of Be, as we did for the carbon $2p_x$ orbital band of graphite in Figure 11–2. The lower part of Figure 11–3 shows the 2*s* band. Since the Be atom has two 2*s* electrons, this band should be full and the antibonding should exactly cancel the bonding, as in the hypothetical Be_2 molecule. However, a band also forms from the beryllium 2*p* orbitals and this band extends below the top of the 2*s* band, as shown in the figure. The 2*s* bonding levels are filled, and the 2*s* antibonding and 2*p* bonding levels are only partially filled, as shown in the figure. More bonding levels than antibonding levels are occupied, and as a consequence, there is net bonding in solid beryllium. There are also empty levels immediately above filled levels in both the 2*s* and 2*p* bands. Thus, beryllium is an electrical conductor and has other properties characteristic of a metal.

The overlap of bands as in beryllium is the key reason that **the majority of the elements in the periodic table form metallic solids.** As the atomic numbers of the elements increase, the gap between atomic orbital energies decreases, and band overlap becomes even more likely than with Be. Thus, all of the solid elements of Group IIA are metals because of the overlap of *s* and *p* bands. For example, with Mg the overlap is between the 3*s* and 3*p* bands.

Metallic behavior is also shown by all of the elements from scandium

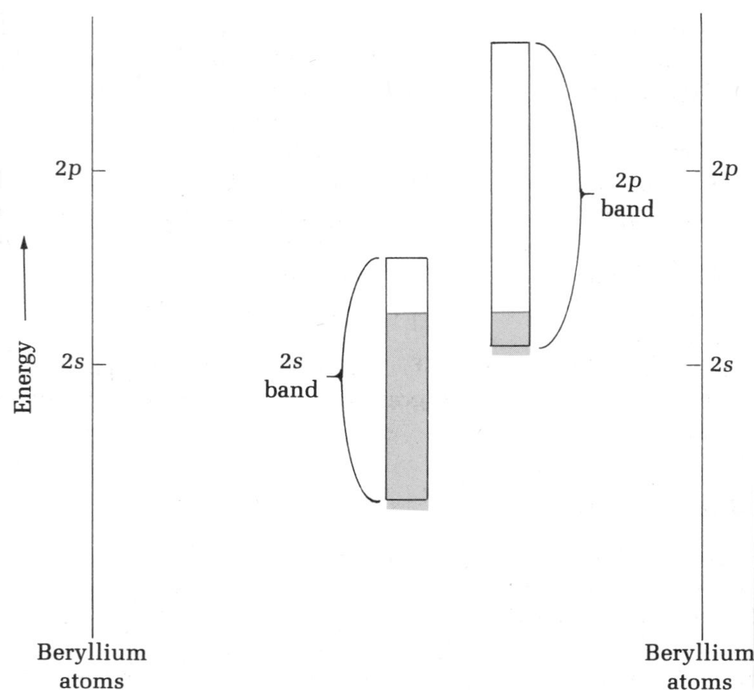

Figure 11–3
Molecular orbital bands of beryllium metal. The shaded areas represent filled molecular orbitals.

through zinc because of the 3*d*–4*p* overlap. These ten elements are called the first series of **transition metals,** because of their metallic behavior and the fact that they represent a transition in chemical behavior between groups IIA and IIIA. The elements yttrium through cadmium and lanthanum through mercury (except for the lanthanide elements) are termed the second and third transition metal series, respectively. The lanthanide elements are also metals because of overlap between the 4*f* band and the 5*d* and 6*p* bands.

It is not surprising that a great majority of the solid elements are metals because this is consistent with our fundamental bonding principle. Wherever possible, delocalized bonding should be found, since delocalization maximizes the number of nuclei associated with any given electron. Only when the number of bonding orbitals of low energy becomes limited does localized bonding occur. They are limited for the elements near the right edge of the periodic table in which there is a large gap between the low-energy valence electrons and the next empty atomic orbital. This gap, exemplified by that between the valence electrons of fluorine and sodium, is so large that band overlap cannot occur. Thus, those elements to the right of the periodic table are nonmetals.

There must be at least a few elements near the dividing line between metals and nonmetals for which the gap between bands is neither large (as in fluorine) nor small (as in sodium). The electrical conductivity of such a substance (silicon and germanium are examples) is a sensitive function of the applied voltage, and such a substance is a **semiconductor** rather than either a conductor (metal) or an insulator (nonconductor).

11-2. The Chemistry of the Group IA Elements

We have already encountered much of the characteristic chemistry of the elements of Group IA in conjunction with that of the nonmetallic elements. Each of the elements Li, Na, K, Rb, and Cs possesses a single valence electron, and the energy of this electron is higher on the average than for any other group in the periodic table. As a consequence, these elements react vigorously with most of the nonmetallic elements to form ionic solids; these metals are very strong reducing agents. Each of these elements M reacts vigorously with water to produce hydrogen gas:

$$2\ M(s) + 2\ H_2O(l) \rightarrow 2\ M^+(aq) + 2\ OH^-(aq) + H_2(g) \qquad \textbf{11-1}$$

The oxides of the Group IA metals are strong bases because they are ionic compounds containing the oxide ion, which is a strong base. The addition of one of these oxides to water produces hydroxide ions. For example,

$$Na_2O(s) + H_2O(l) \rightarrow 2\ Na^+(aq) + 2\ OH^-(aq) \qquad \textbf{11-2}$$

Because the Group IA metals and their oxides yield alkaline solutions (solutions containing an excess of OH^- over H^+) when they are added to water, they are commonly termed the **alkali metals.**

Since the alkali metals are the least electronegative elements in the periodic table, they are usually found as the ions M^+ in combination with elements that are more electronegative. These cations, like all cations, are Lewis acids, but compared with other atomic and ionic species, their electron affinities are relatively small. Thus, Na_2O consists of Na^+ and O^{2-} ions, and as reaction 11-2 shows, the strong base O^{2-} instead of the weak Lewis acid Na^+ largely determines the chemistry of this oxide. A salt like NaCl, which contains the extremely weak acid Na^+ and the extremely weak base Cl^-, produces only an imperceptible change in the amount of OH^- or H^+ when it is added to water. Alkali metal salts of strong acids such as HNO_3, HCl, $HClO_4$, HBr, and HI are neither acids nor bases but neutral species.

One of the outstanding characteristics of most alkali metal salts is their high solubility in water (100 ml of water will dissolve 37 g of NaCl at 25 °C). This solubility results from the combination of their low positive charge (+1 compared with +2 or +3 for cations in some other groups), their weak Lewis acid character, and the polarity of the water molecule. In solid NaCl, each sodium ion is bonded to six chloride ions arranged in an octahedral geometry. In aqueous solution, polar water molecules can surround each sodium ion. One possible arrangement is the octahedral one shown in

Figure 11–4. The nucleophilic oxygen portions of H_2O molecules associate with the electrophilic M^+. Chloride ions are also **aquated** in aqueous solution, with association between the H portions of H_2O molecules and the chloride ion.

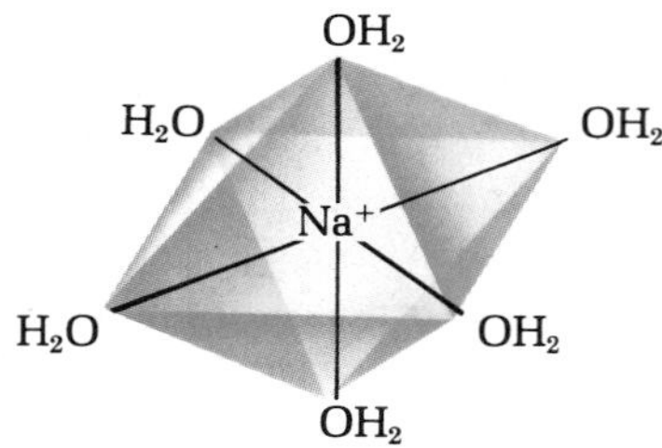

Figure 11–4
The structure of an aquated sodium ion. The black lines indicate Na—O bonds; the octahedral geometry is shown in color.

Because of the aquation of Na^+ and Cl^-, sodium chloride possesses very nearly the same energy dissolved in water as it does as NaCl(s). Table salt dissolves in water with almost no heat either absorbed or liberated. Any dissolution process proceeds extremely well when little or no heat is involved in the dissolution.

That carbon can form ionic carbides with the very metallic elements suggests that perhaps negatively charged organic species can be prepared in combination with ions such as Li^+ and Na^+. Indeed such compounds, termed **organometallic compounds,** do form, and they are very important both in the laboratory and industrial synthesis of organic molecules. Typical alkali-metal examples are ***n*-butyl sodium** (containing Na^+ and $H_3CCH_2CH_2CH_2^-$) and **phenyllithium** (containing Li^+ and $C_6H_5^-$).

11–3. The Chemistry of the Group IIA Elements

Ionic Bonding

Each of the elements Be, Mg, Ca, Sr, and Ba possesses two relatively high-energy valence electrons. Hence the chemistry of these elements, commonly termed the **alkaline earth elements,** is very similar to the chemistry of the alkali metals. The alkaline earth elements characteristically react vigorously with the nonmetallic elements to yield ionic salts containing the M^{2+} ion. The monovalent ions M^+ are never observed as stable species either in the solid state or in solution, and it is of some interest to discover why this is so. To do this, we shall center on the species CaO. Gaseous calcium and oxygen atoms are our reference point for calculation. Consider the relative energies of ionic solids that contain the species Ca^+O^- and $Ca^{2+}O^{2-}$. In the first case the solid is formed in the following three steps:

$$Ca(g) \rightarrow Ca^+(g) + e^-$$

$$e^- + O(g) \rightarrow O^-(g)$$

$$O^-(g) + Ca^+(g) \rightarrow CaO(s)$$

The energy required in the first reaction is the first ionization potential of Ca, or 141 kcal/mole, while the energy of the second reaction is the negative of the electron affinity of oxygen, or −34 kcal/mole. The energy of the third reaction is given by

$$\Delta E = \frac{-1.76e^2}{r_{AB}} \text{ erg/molecule} \qquad \textbf{11–3}$$

where 1.76 is the Madelung constant for the NaCl ionic arrangement adopted by CaO. The experimental value of r_{AB} in CaO is 2.40 Å and ΔE in equation 11–3 is -1.69×10^{-11} erg/molecule = −244 kcal/mole. The overall ΔE for the formation of $Ca^+O^-(s)$ from the separated atoms is the sum of these three energies: 141 − 34 − 244 = −137 kcal/mole.

Now let us repeat the calculation of ΔE for $Ca^{2+}O^{2-}(s)$ as the product. Five steps are involved in the process:

$$Ca(g) \rightarrow Ca^+(g) + e^-$$

$$Ca^+(g) \rightarrow Ca^{2+}(g) + e^-$$

$$e^- + O(g) \rightarrow O^-(g)$$

$$e^- + O^-(g) \rightarrow O^{2-}(g)$$

$$Ca^{2+}(g) + O^{2-}(g) \rightarrow CaO(s)$$

Steps 1 and 3 have already been discussed. The energy change in Step 2 is the second ionization potential of Ca, or 273 kcal/mole. The energy change in Step 4 is the negative of the electron affinity of O^-, or 210 kcal/mole. Steps 2 and 4 are very unfavorable, but Step 5 is an extremely favorable one. The ΔE for Step 5 is

$$\Delta E = \frac{-(1.76)(2e)^2}{r_{AB}} \text{ erg/molecule} = -976 \text{ kcal/mole} \qquad \textbf{11–4}$$

and the overall ΔE for all five steps is 141 + 273 − 34 + 210 − 976 = −386 kcal/mole. Thus, the formation of $Ca^{2+}O^{2-}$ is favored over Ca^+O^- because of the more favorable electrostatic interaction between the larger charges and in spite of the considerable energy that must be expended to produce those charges.

A very similar calculation can be carried out for the production of Ca^{2+} and Ca^+ in solution. The ΔE for aquation of Ca^{2+} in water is −395 kcal/mole while that for a typical singly charged cation such as Na^+ is −104 kcal/mole. This difference of nearly 300 kcal/mole is more than enough to offset the ΔE of 273 kcal/mole for the ionization of Ca^+ to produce Ca^{2+}.

Reactivity

The alkaline earth elements have higher ionization potentials, in general, than their alkali metal neighbors. The Group IIA elements are very strong reducing agents, but they are not as strong as the alkali metals. For example, Mg will not react at any significant rate with water at room temperature to evolve hydrogen, though it will yield H_2 when added to an aqueous solution of an acid such as HCl.

The alkaline earth oxides are basic, but much more weakly so than the alkali metal oxides. This is consistent with the higher electronegativities of the alkaline earth atoms and the increased tendency of the M—O bond to be preserved because of the increased Lewis acid character of the metal center. This dependence of basicity on the electronegativity of the metal is also shown within the group where the electronegativity decreases with increasing atomic number. Consistent with this trend, BaO and SrO are stronger bases than MgO and CaO. All four of these species react with water to form the hydroxides $M(OH)_2$. The solubilities of these hydroxides are considerably smaller than the alkali-metal hydroxides, increasing in the order $Mg(OH)_2 < Ca(OH)_2 < Sr(OH)_2 < Ba(OH)_2$.

In the case of BeO, the Lewis acid character of the metal center is sufficiently great so that BeO is said to be **amphoteric,** meaning that it can act as either an acid or a base. It reacts with strong acids according to the reaction

$$BeO(s) + 2\ HNO_3(aq) \rightarrow Be^{2+}(aq) + 2\ NO_3^-(aq) + H_2O(l) \qquad \textbf{11–5}$$

and with strong bases according to the reaction

$$BeO(s) + H_2O(l) + 2\ NaOH(aq) \rightarrow Be(OH)_4^{2-} + 2\ Na^+(aq) \qquad \textbf{11–6}$$

Amphoteric behavior is common for oxides of elements near the dividing line between metals and nonmetals.

The relatively low solubilities of the Group IIA hydroxides result from the stronger Lewis acid character of the M^{2+} ions as compared with the M^+ alkali metal cations. The M^{2+}-hydroxide interaction in the solid is considerably stronger than the M^{2+}-water interaction in solution. Regardless of the metal, hydroxides of the formula $M(OH)_n$ typically show low solubilities in water when n is two or higher.

Calculating the energy of formation of CaO(s) showed that solids containing ions with charges greater than 1 have much lower energies than solids in which the charges are unity. Thus, we should predict that such salts as the alkaline earth carbonates and sulfates, in which all of the ions possess either +2 or −2 charges, would be much less soluble than an alkali halide such as NaCl. Such is indeed the case. For example, the solubility of $CaCO_3$ in water is 1.5×10^{-3} g/100 ml of H_2O at 25 °C. **With few exceptions, ionic**

solids in which both cation and anion charges are 2 or greater show only a low solubility in water.

Grignard Reagent

Both the alkali metals and the alkaline earth metals form organometallic compounds. The most commonly used compound of this type is the **Grignard reagent,** named for the French chemist who first prepared it. The reagent possesses the general formula RMgX, where R is an organic radical and X is a halogen atom. The species are commonly prepared in an ether solvent by causing metallic magnesium to react with an ether solution of an alkyl halide. A typical example is

$$H_3CCH_2CH_2Cl + Mg \xrightarrow{\text{Ether}} H_3CCH_2CH_2MgCl \qquad \mathbf{11\text{–}7}$$

The C—Mg and Mg—Cl bonds are highly polar, with the C and Cl atoms negatively charged with respect to the magnesium. This is an unusual polarity for carbon and it leads to facile reactions with other molecules that have positively charged centers. One such positively charged center is the carbon of the carbonyl group. Thus, for example, *n*-propyl magnesium chloride reacts with acetaldehyde in the following fashion:

$$H_3CCH_2CH_2MgCl + H_3C\overset{\displaystyle O}{\overset{\|}{C}}\text{—}H \xrightarrow{\text{Ether}} H_3CCH_2CH_2\overset{\displaystyle CH_3}{\underset{\displaystyle H}{C}}\text{—}OMgCl \qquad \mathbf{11\text{–}8}$$

in which addition has occurred to the C═O bond, with the negatively charged carbon of the Grignard reagent adding to the positively charged carbon of the carbonyl group. If water is subsequently added to the product of reaction 11–8, the reaction is

$$2\ H_3CCH_2CH_2\overset{\displaystyle CH_3}{\underset{\displaystyle H}{C}}\text{—}OMgCl + 2\ H_2O \xrightarrow{\text{Ether}} 2\ H_3CCH_2CH_2\underset{\displaystyle OH}{\underset{|}{C}H}CH_3 + Mg(OH)_2(s) + MgCl_2(s) \qquad \mathbf{11\text{–}9}$$

The combination of reactions 11–8 and 11–9 converts an aldehyde to a secondary alcohol. In an analogous fashion, formaldehyde can be converted to a primary alcohol of any desired chain length, and ketones can be converted to tertiary alcohols. Carboxylic acids can be prepared by adding the Grignard reagent to CO_2.

$$H_3CMgBr + CO_2 \xrightarrow{\text{Ether}} H_3C\overset{\displaystyle O}{\overset{\|}{C}}\text{—}OMgBr \qquad \mathbf{11\text{–}10}$$

Water is then added:

$$2\ H_3C\overset{\overset{O}{\|}}{C}\text{—}OMgBr + 2\ H_2O \xrightarrow{\text{Ether}} 2\ H_3C\overset{\overset{O}{\|}}{C}\text{—}OH + Mg(OH)_2(s) + MgBr_2(s) \qquad \textbf{11–11}$$

11–4 The Chemistry of the Group IB and IIB Elements

Oxidation States

An element such as Cu in the IB group is similar to the element potassium in Group IA, since the highest-energy electron in each atom is in a 4*s* orbital. The elements differ, in many ways, however, because Cu contains ten 3*d* electrons whose energies are comparable to those of the 4*s*, while the 4*s* electron is the sole high-energy electron in potassium. Thus, while K displays only the oxidation state I in its compounds, Cu, Ag, and Au display higher oxidation states also. The common oxidized states of Cu are Cu(I) and Cu(II), though the ion Cu^{2+} is more commonly encountered than Cu^+ in ionic crystals. Silver occurs in the oxidation states Ag(I) and Ag(II), though Ag(II) is relatively rare. Gold exhibits the oxidation states Au(I) and Au(III).

With few exceptions ionization potentials increase with atomic number in atoms within a given row of the periodic table. This effect is displayed for a number of elements in Figure 4–8. This same trend is also displayed in Table 4–3 with the elements K through Zn, and it also occurs between Rb and Cd and between Cs and Hg. Table 11–1 lists the first two ionization potentials for nine elements of Groups IA, IB, and IIB.

Table 11–1
First (IP_1) and Second (IP_2) Ionization Potential for Nine Elements of Groups IA, IB, and IIB

Element	Group	IP_1 (in ev)	IP_2 (in ev)
K	IA	4.3	31.8
Cu	IB	7.7	20.3
Zn	IIB	9.4	18.0
Rb	IA	4.2	27.5
Ag	IB	7.6	21.5
Cd	IIB	9.0	16.9
Cs	IA	3.9	25.1
Au	IB	9.2	20.5
Hg	IIB	10.4	18.8

The higher oxidation potentials for the Group IB and IIB elements are reflected in lower reductive capability for these B elements as compared with their A group counterparts. Thus, while potassium, for example, reacts vigorously with water to evolve hydrogen, metallic copper will not evolve hydrogen from even a concentrated solution of hydrochloric acid in water.

Reactivity of the Cations

Consistent with the higher oxidation potentials of the IB and IIB atoms, the IB and IIB cations are much stronger Lewis acids than their counterparts in Groups IA and IIA. We should predict that this increased Lewis acid character would be reflected in lower solubilities for salts of the B cations as compared with salts of the A cations, and this is indeed the case. For example, AgCl is only slightly soluble in water, as many other silver salts and also salts of Cu(I) are. The elements Zn and Cd are most commonly encountered as Zn^{2+} and Cd^{2+}, and the ions Zn^{+} and Cd^{+} are never encountered. The reason for this exclusion of the monovalent cations is essentially the same as the reason for the exclusion of Ca^{+} (see previous section). With mercury, the oxidation state Hg(I) occurs in addition to Hg(II), but the Hg(I) ion still has a double positive charge: the ion is Hg_2^{2+}, and it contains a rather strong mercury-mercury bond. Consistent with the stronger Lewis acid character of the B cations, the salt Hg_2Cl_2 is relatively insoluble in water.

Since the B group cations are more easily reduced than the A group cations, the pure B elements are more easily obtained from their naturally occurring salts than the A elements are. The sulfide is generally the form in which B elements naturally occur; the sulfides can be converted to oxides by heating in air:

$$2\ ZnS(s) + 3\ O_2(g) \rightarrow 2\ ZnO(s) + 2\ SO_2(g) \qquad \textbf{11–12}$$

The oxide can then be reduced chemically by using an inexpensive reducing agent. In the case of ZnO, this agent is carbon, which is oxidized to carbon monoxide as Zn^{2+} is reduced to Zn(s). In contrast, the Group IA and IIA elements must be obtained electrolytically because of the difficulty of chemical reduction.

The Group IB and IIB cations form strong donor-acceptor complexes in solution with a number of nucleophiles stronger than water. One such nucleophile is ammonia. While AgCl(s) does not dissolve to any significant extent in water, it is highly soluble in an aqueous ammonia solution because of the reaction

$$AgCl(s) + 2\ NH_3(aq) \rightarrow Ag(NH_3)_2^{+}(aq) + Cl^{-}(aq) \qquad \textbf{11–13}$$

The bonding in the silver-ammonia complex is between the silver ion and the nucleophilic nitrogens of the two ammonia molecules. The cations of Groups IA and IIA show little tendency to form complexes with ammonia in aqueous solution, in contrast to reaction 11–13.

What is the geometry of the species $Ag(NH_3)_2^{+}$? In predicting the structures of such complexes containing a complete ten *d* electrons in the valence level, we do not count the *d* electrons formally as valence electrons. Thus, in our geometry procedure, the complex is split into Ag^+ and 2 NH_3, two sigma bonds are then introduced, and the valence electron sum is 4. Thus, the N—Ag—N bond angle is 180° and the two silver orbitals involved in the bonding are the 5*s* and one of the 5*p*. In a similar fashion, we conclude that the complex $HgCl_2$ is linear, while $ZnCl_4^{2-}$ and $HgCl_4^{2-}$ are tetrahedral.

The geometries of complexes of Cu^{2+} are not easy to predict, because the central ion contains only nine *d* electrons. With ten *d* electrons as in Zn^{2+}, we may assume a spherical distribution of these electrons with no resulting influence on the geometry. With nine *d* electrons this cannot be true, and the *d* electron distribution has a large effect on the geometry. Most of the transition metal cations show this effect.

11–5. The Chemistry of the Transition Metals

Oxidation States

The ionization potentials in Table 4–3 reveal that the transition metals should show a progressive variation between the behavior characteristic of the A and B groups. For example, scandium shows ionization potentials that are very similar in magnitude to those of aluminum, and we should predict that Group IIIB displays a chemistry very similar to that of Group IIIA. As the periodic table shows, this includes Sc, Y, La, and all of the lanthanide elements. The only oxidized form encountered with Al is Al^{3+} and scandium, yttrium, and lanthanum are like aluminum in having only one such oxidized form, M(III). The lanthanide elements differ from Sc, Y, and La in the addition of 4*f* electrons. However, the *f* electrons are located quite deeply within the atom and they are not very important in the chemistry of these elements. Thus, the predominant oxidation state for all of the lanthanide elements is M(III).

Titanium has ionization potentials that are between the potentials for Si and Ge, and this indicates that the predominant oxidation state is Ti(IV), analogous to the behavior of Si and Ge. This is borne out in fact, since Ti forms the volatile, nonionic $TiCl_4$ in analogy with $SiCl_4$ formation, and the nonionic solid TiO_2 in analogy with SiO_2 formation. The ions Ti^{2+} and Ti^{3+} can be prepared, but they are extremely strong reducing agents consistent with Ti(IV) as the most stable oxidation state. The oxidation state M(IV) is also commonly displayed by zirconium and hafnium, and the best-known compounds of these elements are ZrO_2 and HfO_2.

The first member of Group VA (nitrogen) is characterized by a wide range of oxidation states up to N(V) as the maximum. For the other elements of the group, the range is considerably more narrow and V is the most stable state. The Group VB species also show this behavior. Vanadium exists in a wide range of oxidation states, while niobium and tantalum are most commonly encountered in oxidation state V.

As we have noted, the ionization potentials of the transition metal atoms increase in the main with atomic number, just as they do between Li and Ne. However, the slope of this increase is much greater for the light elements, Li to Ne, than for the transition elements. As a result, nitrogen has considerably more negative atomic orbital energies than vanadium does. Thus, the compound V_2O_5 is a less acidic oxide than N_2O_5 or P_4O_{10}; in fact, it is amphoteric. Vanadium pentoxide dissolves in basic solution to produce the **vanadate ion,** which is analogous to the orthophosphate ion:

$$V_2O_5(s) + 6\ OH^-(aq) \rightarrow 2\ VO_4^{3-}(aq) + 3\ H_2O(l) \qquad \textbf{11–14}$$

The vanadate ion can form polymeric ions containing the V—O—V bridge (recall similar behavior of the phosphate ion), and it does so readily in solution. The amphoteric character of V_2O_5 is reflected in its dissolution in acidic as well as basic solution; in acidic solution, the resulting species is the **vanadyl(V) ion,** VO_2^+:

$$V_2O_5(s) + 2\ H^+(aq) \rightarrow 2\ VO_2^+(aq) + H_2O(l) \qquad \textbf{11–15}$$

The vanadyl(V) ion is analogous to NO_2^+, though it is less reactive than NO_2^+. The oxidation state V(IV) exists as VO_2, and very commonly as the vanadyl(IV) ion VO^{2+}. The states V(II) and V(III) exist as the ions V^{2+} and V^{3+}; V^{3+} is the more stable.

Just as there is some parallel between N and V, there is a parallel between chromium and sulfur. Both exist in a number of oxidation states up to a maximum of VI. This highest oxidation state is most commonly represented by SO_4^{2-}, **chromate ion** (CrO_4^{2-}) and **dichromate** ion ($Cr_2O_7^{2-}$).

Because of the lower ionization potentials of the transition metals as compared with their A group analogs, the lower-oxidation-state ionic species have substantial stabilities. Dichromate ion is an extremely strong oxidizing agent; in oxidizing something else, it is reduced to the **chromium(III) ion** (Cr^{3+}), the most stable form of the element. The stability of the oxidation state Cr(III) is reflected in the fact that chromium is found most commonly in nature as Cr_2O_3 in the ore termed **chromite.** The **chromium(II) ion** Cr^{2+} also exists in aqueous solution, but it is an extremely strong reducing agent. Just as Nb and Ta (Group VB) normally occur in the V oxidation state, molybdenum and tungsten (Group VIB) most commonly occur in the VI oxidation state.

The element analogous to manganese is chlorine, and both display oxidation states up to VII as a maximum. The common species with this highest oxidation number are perchlorate ion, ClO_4^-, and **permanganate ion,** MnO_4^-. Both perchlorate and permanganate are extremely strong oxidizing agents, and the two most stable oxidation states of manganese are Mn(IV) (manganese is found in nature as MnO_2) and Mn(II) (as the ion Mn^{2+}). It is interesting that the adjacent elements Cr and Mn display a dramatic reversal in the relative stabilities of the ions M^{2+} and M^{3+}. With Cr, the ion Cr^{3+} is much more stable than Cr^{2+}, and with manganese, Mn^{2+} is much more stable than Mn^{3+}. (The reason for this reversal will become apparent in the next section of this chapter.) Technetium and rhenium are relatively rare elements that also display a number of oxidation states up to a maximum of VII.

The next three columns of the periodic table (headed by iron, cobalt, and nickel) are linked together as Group VIIIB since they display similar chemistries. For Fe, Co, and Ni, the predominant oxidation states are the ions M^{2+} and M^{3+}. In the case of iron, both Fe^{2+} and Fe^{3+} are commonly found (iron is found in nature as Fe_2O_3). Cobalt also exists commonly as Co^{2+} and Co^{3+}, although the ion Ni^{3+} does not exist. The heavier members of Group VIIIB display higher oxidation numbers than Fe, Co, and Ni, as expected. Osmium and platinum, for example, form the fluorides OsF_6 and PtF_6.

Ligand-Field Stabilization

We have encountered many examples of donor-acceptor complexes involving a large number of elements. The transition metal cations, however, especially stand out in their ability to form extremely stable complexes of this type. The source of this stability lies in their incomplete *d* electron configurations. Figure 11–5 shows this extra stability as a plot of the aquation energies of the divalent ions from Ca^{2+} to Zn^{2+}. The aquation energy of an ion is defined as the energy released when one mole of ions transfer from the gaseous state to aqueous solution. The aquation of many of these ions

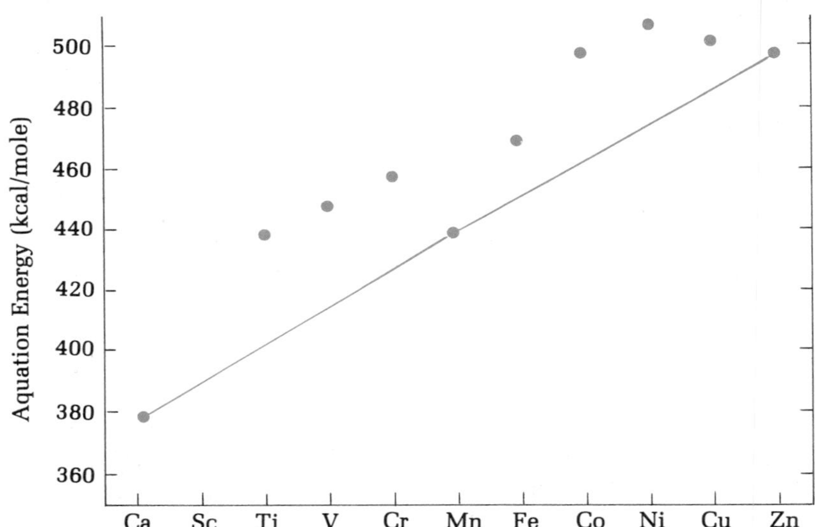

Figure 11–5 The aquation energies of the divalent cations of the elements calcium through zinc in the order of increasing atomic number.

involves the formation of octahedral cation-water complexes (the structure depicted in Figure 11–4).

In general, we expect the aquation energy to increase with the Lewis acid strength of the cation, and this should increase with the ionization potentials of the atom forming the cation. Thus, the larger value of the Zn^{2+} aquation energy compared with the value for Ca^{2+} reflects the higher ionization potentials of Zn. However, Figure 11–5 shows an additional feature that is not explained by ionization potentials, namely the two maxima in the aquation energies at Cr^{2+} and Ni^{2+}. The reason for these maxima lies in the ability of an incomplete set of *d* electrons to adapt to the presence of the ligands (the nucleophilic electron donors) so as to maximize the electrostatic attraction and subsequent bonding between the metal nucleus and the ligand electrons. Let us take Ni^{2+} as an example. The gaseous ion contains eight *d* electrons and these are contained with equal probability in each of the five *d* orbitals, yielding a spherical distribution. Now let us bring in six water molecules to aquate the ion, as shown in Figure 11–6. The *d* electrons distribute themselves into six electrons in the combined $3d_{xy}$, $3d_{xz}$, and $3d_{yz}$ orbitals, which are directed between the ligands (these orbitals not shown in figure), and one electron each in the $3d_{z^2}$ and $3d_{x^2-y^2}$ orbitals, which point at the ligands (these orbitals are shown). This arrangement maximizes the metal ion-ligand interaction and also minimizes the *d* electron–*d* electron repulsion consistent with Hund's rules. Thus, $Ni(H_2O)_6{}^{2+}$ contains two unpaired electrons, shown by the arrows in Figure 11–6, and is paramagnetic.

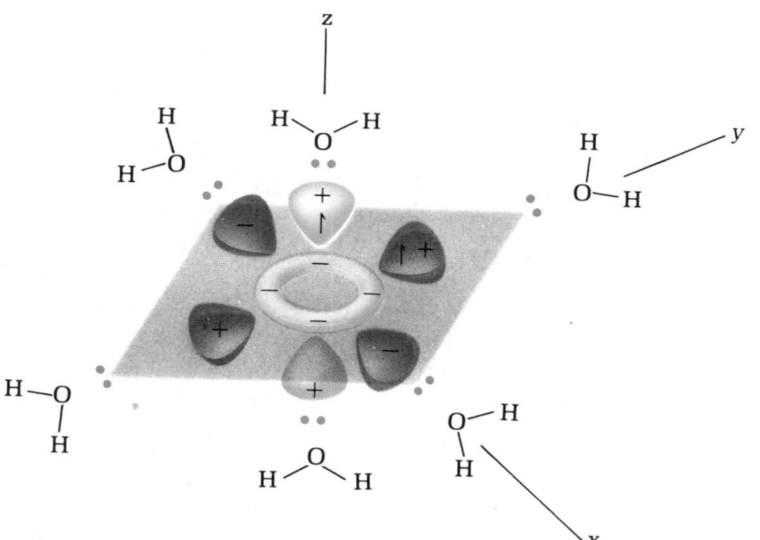

Figure 11–6
Contours and electron occupations of the 3*d* orbitals that are directed toward the ligands in $Ni(H_2O)_6{}^{2+}$. The 3*d* electrons are indicated by half arrows. The colored contours are those of the $3d_{z^2}$ orbital and the black contours are those of the $3d_{x^2-y^2}$ orbital.

The aquation energy of Ni^{2+} arises from the sum of two terms. The first is the cation-water interaction if the *d* electrons were spherically distributed, and the second is the increase in the cation-water interaction due to the redistribution of the *d* electrons. The first term can be estimated for each ion by drawing a solid line between Ca, Mn, and Zn as shown in Figure 11–5. The second term for Ni^{2+} is given to a good approximation by the energy between the Ni^{2+} dot in Figure 11–5 and the solid line connecting the Ca, Mn, and Zn dots; this difference is 29 kcal/mole. This energy, which depends on the relative arrangement of the ligands and the *d* electrons, is termed the **ligand-field stabilization energy** (LFSE). It is largely because of the LFSE that transition metal cations form a particularly wide and varied array of very stable coordination complexes.

The LFSE is seen to be zero for $Mn(H_2O)_6^{2+}$. This arises because Mn^{2+} contains five *d* electrons and these are all unpaired in the aquated ions. The resultant spherical *d* electron distribution results in a zero LFSE, and Mn^{2+} complexes are much less stable in general than Ni^{2+} complexes.

With species that are stronger nucleophiles than water, the possibilities for metal ion–ligand binding may be so strong that metal ion–electron pairing occurs in order to vacate completely the *d* orbital which points at the ligands. A common example of such a strong nucleophile is cyanide ion, CN^-. The ion $Fe(H_2O)_6^{3+}$ possesses a zero LFSE (it is isoelectronic with $Mn(H_2O)_6^{2+}$). The ion $Fe(CN)_6^{3-}$ is extremely stable and possesses only one unpaired electron, compared with five in $Fe(H_2O)_6^{3+}$. The bonding between Fe^{3+} and CN^- is so strong that Hund's rules are overcome and electrons vacate the $3d_{z^2}$ and $3d_{x^2-y^2}$ orbitals to pair with other electrons in the set $3d_{xz}$, $3d_{yz}$, and $3d_{xy}$. Transition metal complexes are characterized as **high-spin complexes** if they obey Hund's rules and **low-spin complexes** if they do not.

One immediate application of the ligand-field model is in explaining the instability of Cr^{2+} compared with Cr^{3+}. The ion Cr^{3+} in aqueous solution contains three *d* electrons that are unpaired in the $3d_{xz}$, $3d_{yz}$, $3d_{xy}$ set in $Cr(H_2O)_6^{3+}$. This provides the ideal situation to maximize the metal ion–ligand interaction and to minimize *d* electron–*d* electron repulsion. Introducing a fourth *d* electron in Cr^{2+} must either decrease the metal ion–ligand interaction or increase the *d* electron–*d* electron repulsion.

In an ion such as $Ni(H_2O)_6^{2+}$, which possesses a specific orientation of the ligands relative to the *d* orbitals, the *d*-orbital energies are not all equal as they are in the gaseous Ni^{2+}. This splitting of *d*-orbital energies depends on the ligand arrangement. The pattern for an octahedral arrangement like that in $Ni(H_2O)_6^{2+}$ is shown in Figure 11–7. The pattern shown incorporates into

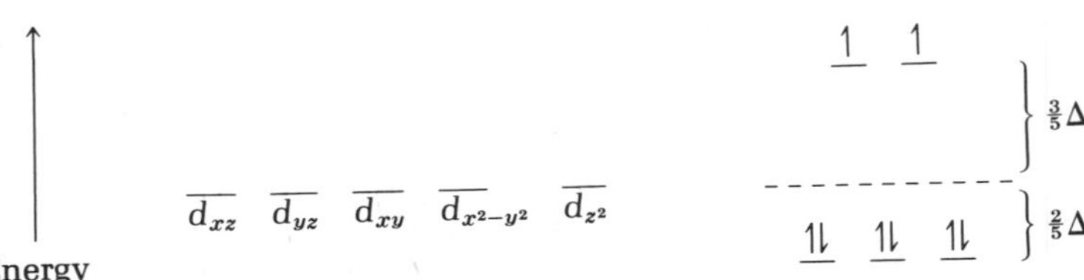

Figure 11–7
Splitting of the energies of the 3*d* orbitals in $Ni(H_2O)_6^{2+}$.

the d-orbital energies the favorable bonding of the ligands to the ion, so that the splitting parameter Δ shown in the figure is closely related to the LFSE. Thus, the lowered energies of d_{xy}, d_{xz}, and d_{yz} in the complex largely reflect the enhanced ligand–metal ion interaction that occurs when the electrons occupy those orbitals instead of the $d_{x^2-y^2}$ or d_{z^2} orbitals.

In an octahedral environment, the orbitals $d_{x^2-y^2}$ and d_{z^2} are each raised in energy by $\frac{3}{5}\Delta$ with respect to the gaseous ion, while the other three orbitals are each lowered by $\frac{2}{5}\Delta$. The fractions $\frac{3}{5}$ and $\frac{2}{5}$ simply reflect the relative numbers of orbitals in the two sets. The relation between LFSE and Δ is therefore written:

$$\text{LFSE} = (\tfrac{2}{5}\Delta)(\text{number of electrons in low-energy set}) - (\tfrac{3}{5}\Delta)(\text{number of electrons in high-energy set}) \qquad \textbf{11–16}$$

From equation 11–16 and Figure 11–7, using an LFSE of 29 kcal/mole for $Ni(H_2O)_6^{2+}$, we can calculate Δ as $\frac{5}{6}$(LFSE) = 24 kcal/mole. With this value, we can obtain an approximate energy for the excited state of $Ni(H_2O)_6^{2+}$, in which one electron has been promoted from the low-energy set to the high-energy set. This state lies at an energy Δ above the ground state. This energy is easily converted by calculation to the wavelength of electromagnetic radiation that produces the transition:

$$\lambda = \frac{hc}{\Delta} = \frac{(6.62 \times 10^{-27})(3 \times 10^{10})}{(24)(6.9 \times 10^{-14})}$$

$$= 1.2 \times 10^{-4}\ \text{cm}$$

This value is within a factor of 2 of the red limit of the visible region. The hydrated Ni^{2+} absorbs red light and therefore is green both in the solid state and in solution. Most transition metal complexes are colored and the color arises from transitions between d orbitals, as in Figure 11–7. Since the transitions depend on the splitting pattern of the d orbitals, which in turn depends on the ligand geometry, color and geometry are often closely related.

Nomenclature

In naming salts that contain complex cations, complex anions, or both, a set of rules has been devised. These rules are as follows:

1. The cation is always expressed before the anion in the name, just as in the simple salt sodium chloride.
2. In naming a complex or complexes, use the order—first, the anionic ligands, then the neutral ligands, then any cationic ligands, and finally the metal.
3. The numbers of ligands of any particular kind are indicated by the prefixes *di-*, *tri-*, . . . , (or *bis-*, *tris-*, *tetrakis-*, . . . , when the former prefixes are also contained in the ligands).

4. Anionic ligands take on the suffix *-o*; neutral ligands keep their names; and cationic ligands take on the suffix *-ium*. Exceptions are water, which becomes **aquo,** and ammonia, which becomes **ammine.**
5. The metal ion is named by naming the element if the complex is uncharged or positively charged. If the complex is negatively charged, the suffix *-ate* is added. In some cases the name of the element is changed to its Latin name for phonetic reasons. Examples are *ferrate* for "iron" and *cuprate* for "copper."
6. The oxidation state of the metal is indicated by a roman numeral after its name. Some examples are:
   $K_3[Fe(CN)_6]$, potassium hexacyanoferrate(III)
   $[Co(NH_3)_3(H_2O)_3]Cl_3$, triaquotriammine cobalt(III) chloride.
   $[Co(en)_3]Cl_3$, tris-ethylenediamine cobalt(III) chloride.
   Note: Ethylenediamine has the structure $H_2N—CH_2—CH_2—NH_2$, and it is commonly symbolized by (en).
   $Pt(NH_3)_2Cl_2$, dichlorodiammine platinum(II).

Structure

Although a number of geometries are encountered in transition metal complexes, the octahedral and tetrahedral arrangements of atoms bound to the metal ions are the most common. Geometrical and optical isomers can occur with octahedral complexes. Two geometrical isomers of the dichloro-*bis*-ethylenediamine cobalt(III) cation are shown in Figure 11–8. In analogy with geometrical isomers of organic molecules, the ion with the chlorides on *opposite* sides of the plane containing the Co^{3+} and the four nitrogens is called the *trans* isomer; the other ion is the *cis* isomer.

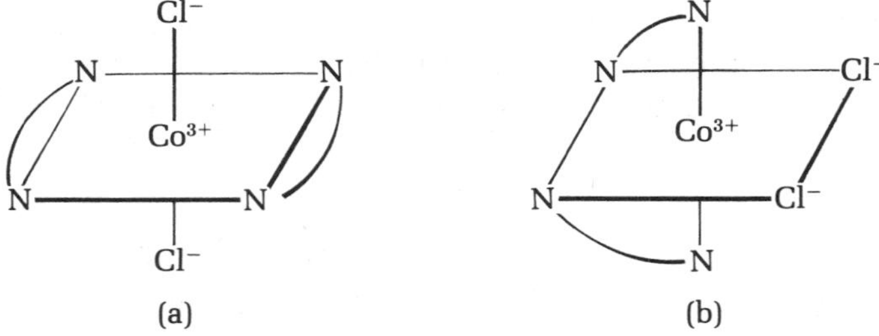

Figure 11–8 The geometrical isomers of dichloro-*bis*-ethylenediamine cobalt(III) ion: (a) *trans*; (b) *cis*. The ethylenediamine ligand is symbolized by N—N.

There are two enantiomeric (mirror-image) *cis* forms, and these are shown in Figure 11–9. The fact that these mirror images are not identical is illustrated in the (c) and (d) portions of the figure, where the view is perpendicular to one of the eight triangular faces of the octahedron. There are two such faces in the view and they are parallel. The face above the plane of the paper is indicated by solid lines and the face below the paper by dashed lines.

Tetrahedral complexes do not display geometrical isomerism, since all of the ligands have the same relation to each other. These complexes do

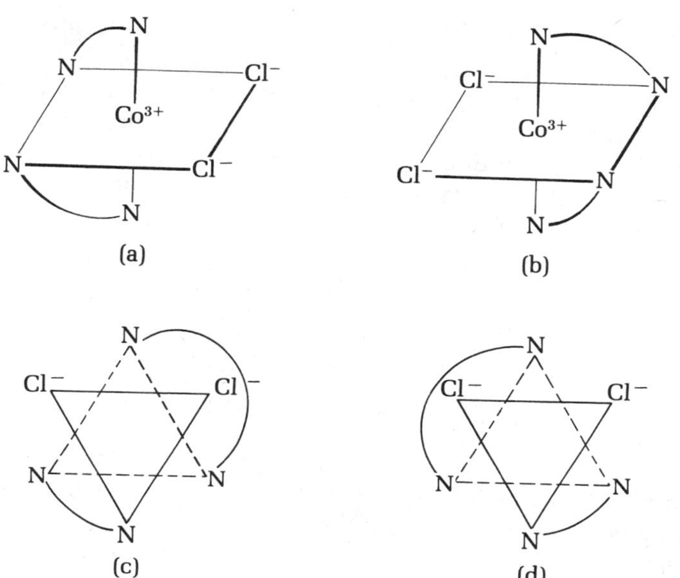

Figure 11-9 The two optical isomers of *cis*-dichloro-*bis*-ethylenediamine cobalt(III) ion. Drawings (a) and (c) are two views of one isomer and (b) and (d) are two views of the other isomer.

display optical isomerism, however, if all four ligands are different. A third type of geometry often displayed by transition metal complexes is square-planar, in which the metal ion is directly bonded to four atoms, and all five nuclei lie in the same plane. An example is dichlorodiammine platinum(II); the two geometrical isomers are shown in Figure 11-10.

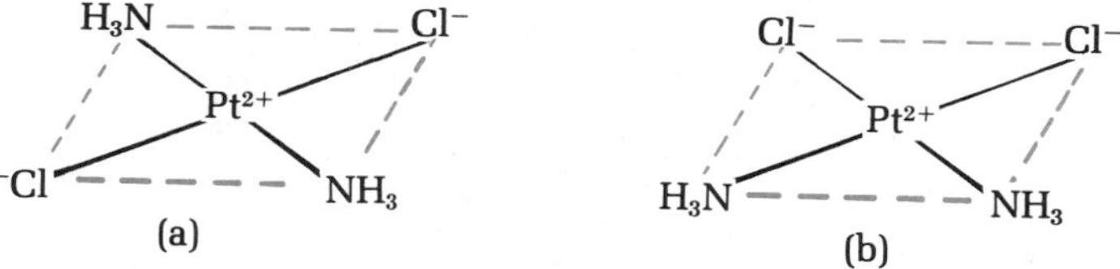

Figure 11-10 The geometrical isomers of dichlorodiammine platinum(II): (a) *trans*, (b) *cis*.

Reactivity

Our emphasis for unifying chemical reactions has been the relation between structure and reactivity. Two common types of reactions of transition metal complexes are a good illustration of this relation. The first type is **ligand substitution,** exemplified by the reaction

$$Cr(H_2O)_6{}^{3+}(aq) + Cl^-(aq) \rightarrow Cr(H_2O)_5Cl^{2+}(aq) + H_2O(l) \qquad \textbf{11-17}$$

Such ligand substitution reactions involving octahedral complexes of Cr^{3+} are quite slow at room temperature. However, this is not typical of other transition metal ion complexes formed from M^{3+} ions. For example, if the hexaquo complexes in reaction 11-17 involve Ti^{3+}, V^{3+}, or Fe^{3+}, the reaction proceeds essentially to completion in less than one second. The contrast between Cr^{3+} and the ions V^{3+} and Fe^{3+} is particularly striking because these two elements are on opposite sides of Cr in the periodic table. Clearly, the unusually slow reactivity of Cr^{3+} complexes does not reflect any property that changes progressively along the first transition metal series.

The key to understanding these rates of substitution lies in the mechanism of the reaction. Two types of mechanisms commonly encountered in ligand substitution reactions are

$$\begin{aligned} Cr(H_2O)_6{}^{3+}(aq) &\rightleftharpoons Cr(H_2O)_5{}^{3+}(aq) + H_2O(l) \\ Cr(H_2O)_5{}^{3+}(aq) + Cl^-(aq) &\rightarrow Cr(H_2O)_5Cl^{2+}(aq) \end{aligned} \qquad \text{11–18}$$

$$\begin{aligned} Cr(H_2O)_6{}^{3+}(aq) + Cl^-(aq) &\rightleftharpoons Cr(H_2O)_6Cl^{2+}(aq) \\ Cr(H_2O)_6Cl^{2+}(aq) &\rightarrow Cr(H_2O)_5Cl^{2+}(aq) + H_2O(l) \end{aligned} \qquad \text{11–19}$$

The first mechanism is termed a **dissociative** one, since the Cr—O bond breaks completely before the Cr—Cl bond forms. Since the metal ion–oxygen bond energy is substantial, this mechanism will yield to the mechanism of reactions 11–19 when the latter mechanism can occur. In this second mechanism, the chloride ion reacts with the octahedral complex to form a seven-ligand species that rapidly expels a water molecule to form the substitution product. Thus, mechanism 11–19 is analogous to the mechanism for ester formation from a carboxylic acid and an alcohol (Chapter 10) and is termed an **associative** mechanism.

In $Cr(H_2O)_6{}^{3+}$ there are three *d* electrons, one each in the d_{xy}, d_{xz}, and d_{yz} orbitals, and the two *d* orbitals pointing toward the six ligands are vacant. The seventh ligand must enter the complex along a direction containing a *d* electron, and thus the Cr—Cl interaction is considerably weaker in the seven-ligand complex than the Cr—O interaction is. In the subsequent rapid decomposition of the seven-ligand complex, it is much more likely that the Cl^- will be lost than an H_2O. Thus, permanent substitution of a chloride by this mechanism is unlikely.

This problem does not occur with Ti^{3+}, V^{3+}, or Fe^{3+}. The complex $Ti(H_2O)_6{}^{3+}$ contains a single *d* electron, so that there are two vacant orbitals of the set d_{xy}, d_{xz}, d_{yz} in which the incoming ligand can be accepted. The ion $V(H_2O)_6{}^{3+}$ has one such vacant orbital. In $Fe(H_2O)_6{}^{3+}$ there are no vacant *d* orbitals in the set d_{xy}, d_{xz}, and d_{yz}, but neither are there vacancies in the set $d_{x^2-y^2}$ and d_{z^2}. All five orbitals contain one electron, and the bonding of the seventh ligand is equivalent to the bonding of the departing ligand in the seven-ligand complex.

A second type of characteristic reaction of metal ion complexes is chemical alteration of the ligands. If the metal ion is unchanged in this process, the metal ion is a catalyst in the ligand reaction and the ability of the metal ion and ligand to form the complex is the basis for catalysis. This is a common role of metal ions in biological processes, and it is one of the principal reasons why so-called minerals are required for proper nutrition. One of the most important biological reactions is the hydrolysis of polyphosphate esters to yield inorganic phosphate and a smaller phosphate ester. Such reactions provide the immediate driving force for many biological reactions. Figure 11–11 shows the mechanism for the hydrolysis (a) in the absence and (b) in the presence of a divalent metal ion. (The organic portion of the phosphate ester is indicated by R, and it will be identified in Chapter 22.)

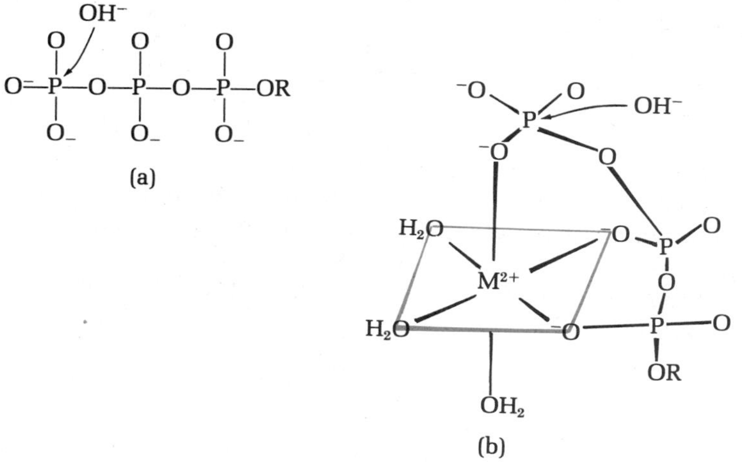

(a)

(b)

Figure 11–11 Attack points of the hydroxide ion in the hydrolysis of a polyphosphate ester, (a) in the absence of a metal ion and (b) in the presence of a metal ion.

The hydrolysis mechanism is similar to that for the hydrolysis of carboxylate esters. The P—O bonds are polarized as $P^{\delta+}$—$O^{\delta-}$ and the P atom is electrophilic. The nucleophile OH^- attaches itself to the terminal phosphate as shown in the figure, producing a trigonal-bipyramidal geometry about the terminal phosphorus. This intermediate may decompose to yield the starting materials or to split the P—O—P bond in the following overall reaction:

$$ROP_3O_9^{4-} + OH^- \rightarrow ROP_2O_6^{3-} + HPO_4^{2-} \qquad \textbf{11–20}$$

Anything that increases the electrophilic character of the terminal phosphorus atom promotes the hydrolysis reaction, and this is an important role of the metal ion in the complex depicted in Figure 11–11(b). The doubly charged metal ion depletes the electron density in all regions of the ligand, resulting in an increased electrophilic character of all ligand regions. Thus, the attack by OH^- on the terminal phosphorus is promoted.

A third characteristic of transition metal ions and complexes in solution is very facile oxidation or reduction in the presence of suitable oxidizing or reducing agents. One reason for this facile electron transfer in many cases is the *d*-electron arrangement that results in LFSE. A good example is provided by $Fe(H_2O)_6^{2+}$, which is rapidly oxidized to $Fe(H_2O)_6^{3+}$ by any suitably strong oxidizing agent. Four of the six *d* electrons of $Fe(H_2O)_6^{2+}$ are in the orbitals d_{xy}, d_{xz}, d_{yz}, which point between the ligand water molecules. Thus, one of them is easily extracted by an oxidizing agent without the necessity of the oxidizing agent's replacing a water molecule to form a metal ion–oxidizing agent complex.

Organometallic Compounds of the Transition Metals

The great ability of transition metal ions to serve as Lewis acids allows them to form a wide variety of organometallic species that other metals do not form. In a species such as *n*-butyllithium, the association is between a weak electrophile, Li^+, and a very strong nucleophile, $H_3C(CH_2)_2CH_2^-$. With many transition metal ions, the carbon compound can be a much weaker nucleophile, and complex formation still occurs. One common example occurs with pi electrons in organic molecules, which can delocalize by associating with a transition metal ion. Such a compound is trichloroethylene platinate(II) ion, whose structure is shown in Figure 11–12. The ion is square-planar and the platinum accepts eight electrons from the ligands, two each from the chlorides and the two pi electrons of the C=C bond of ethylene.

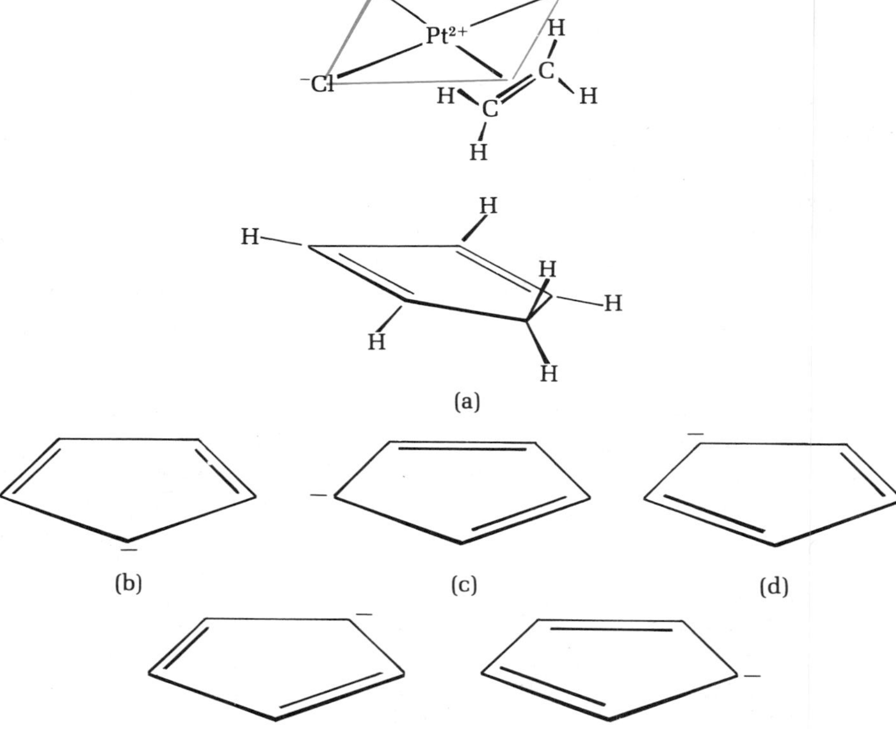

Figure 11–12 The structure of trichloroethylene platinate(II) ion.

Figure 11–13 (a) Cyclopentadiene and (b-f) the five canonical resonance forms of the anion derived from it.

Transition metal ions can even accept pi electrons from already delocalized organic pi systems. The first compound of this type that was ever characterized combines Fe^{2+} and the anion derived by proton loss from **cyclopentadiene.** Figure 11–13 shows cyclopentadiene and the five canonical resonance forms of the anion derived from it. The anion is perfectly pentagonal with all five carbons identical. The complex bis-cyclopentadieno iron(II) is known by the common name **ferrocene,** and its structure is shown in Figure 11–14. It is a "sandwich," in which the metal ion is between two parallel cyclopentadiene rings. The number of electrons donated to the metal ion is twelve (the six pi electrons in each ligand), just as twelve electrons are donated in $Fe(H_2O)_6^{2+}$.

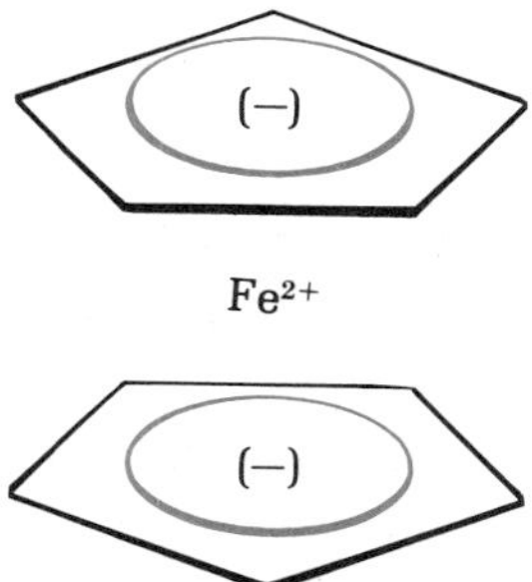

Figure 11–14
The structure of *bis*-cyclopentadieno iron(II), "ferrocene."

The ability of transition metal ions to bind molecules at C═C bonds and also molecules that can add to those C═C bonds causes those ions to be used as catalysts in many important reactions of organic molecules. One such reaction is the polymerization of ethylene in which titanium is the transition metal catalyst in the form of $TiCl_4$, and a source of alkyl radicals is a compound such as triethyl aluminum, $Al(C_2H_5)_3$.

It is proposed that by ligand substitution the structure shown in Figure 11–15(a) forms about the titanium. The ethylene is then inserted into the Ti—C bond to yield the structure shown in Figure 11–15(b). One of the ethylene carbons is given an asterisk in the figure so that its location is marked in the insertion product. Structure (b) adds a second ethylene molecule to yield structure (c), and so the polymerization proceeds indefinitely.

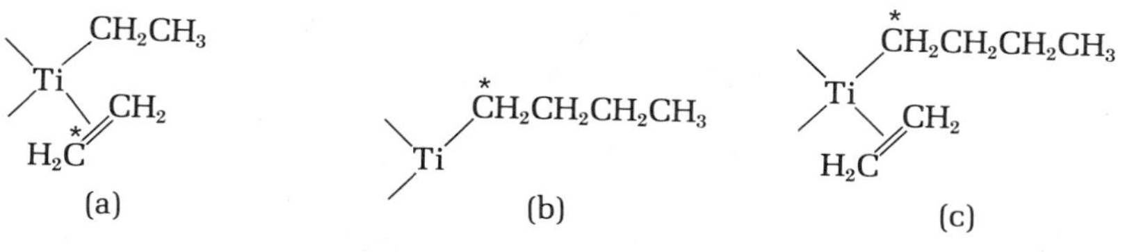

(a) Ti with CH_2CH_3 and $H_2\overset{*}{C}{=}CH_2$

(b) Ti—$\overset{*}{C}H_2CH_2CH_2CH_3$

(c) Ti with $\overset{*}{C}H_2CH_2CH_2CH_3$ and $H_2C{=}CH_2$

Figure 11–15
Intermediate structures in the proposed mechanism of polymerization of ethylene. The structures (a), (b), and (c) are formed successively.

Because of the wide variety of species that transition metals can accept as ligands, there are a large number of well-characterized reactions that are based on insertion as in the polymerization of ethylene.

Transition Metal Carbonyls

In the discussion of ligand-field stabilization, we showed the *d* electrons to be largely passive; that is, they position themselves so as to interfere as little as possible with the metal ion–ligand interaction. Are there situations in which these *d* electrons after being removed to the regions between the ligands can nevertheless actively participate in the bonding? Such a situation occurs with cyanide ion as the ligand, and it is the principal contributor to the low-spin electronic configuration of $Fe(CN)_6^{3-}$ and $Fe(CN)_6^{4-}$ as well as to the very high stability of those complex ions. Figure 11–16 shows a view of the doubly occupied d_{xz} orbital in $Fe(CN)_6^{4-}$ and two ligand atomic

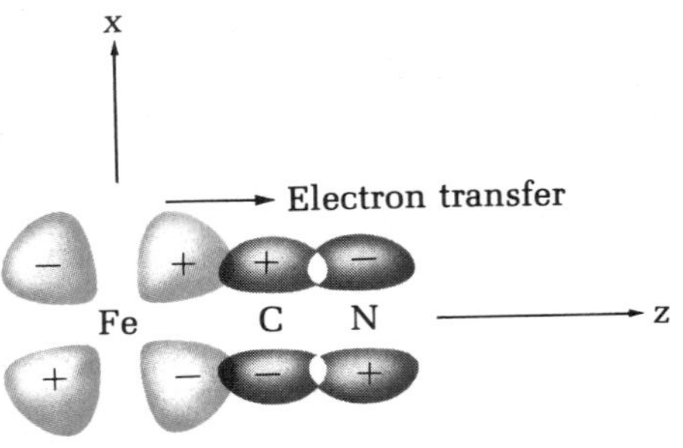

Figure 11–16 The bonding overlap of the doubly occupied metal ion $3d_{xz}$ orbital (colored contours) and the empty $2p\pi_x^*$ molecular orbital of cyanide ion in $Fe(CN)_6^{4-}$.

p orbitals in the combination that produces one of the two empty antibonding pi molecular orbitals of CN^-. The metal ion–ligand orbital overlap is clearly a bonding one, and considerable *d* electron density transfers onto the ligand. This model of the transfer of electron density to a ligand is termed **back bonding,** and it makes a very significant contribution to the strength of the metal-ion–ligand bond. Carbon monoxide (CO) is isoelectronic with CN^- and might also be expected to form complexes with transition metals that involve strong back bonding if our model of back bonding is correct. Such is indeed the case, and the fact that CO can associate with transition metals is due in large part to the back bonding since CO is a very poor nucleophile that does not appreciably associate with nontransition metal atoms or ions.

Unlike cyanide complexes, which are most commonly encountered with transition metal ions in their normal (II) and (III) oxidation states, CO complexes form most commonly with the un-ionized metals. This is explainable in terms of the relationship between sigma electron donation from CO to the metal and back donation in a pi fashion of electrons from the metal to CO. Ionization of the metal substantially lowers the energies of the remaining *d* electrons, so that they are far below the energy of the $2p\pi^*$ orbitals of CO. This prevents back bonding from occurring and leaves only the sigma donation of the carbon lone pair of CO to the metal ion. This sigma donation alone is insufficient for stable complex formation between CO and Fe(II) or Fe(III).

The compounds formed between metals and CO are termed **metal carbonyls** and their formulas lead us to conclude that the *d* electrons and *d* orbitals are involved in the bonding. Table 11–2 lists five of these compounds and the total number of metal atom valence electrons in each. This valence electron sum is composed, in those molecules containing one metal atom, of the valence electrons of the un-ionized metal and two electrons from each of the CO ligands. In the case of **nickel tetracarbonyl,** $Ni(CO)_4$, this sum is ten electrons for the Ni atom plus two electrons from each of four CO molecules for a total of eighteen electrons. With **iron pentacarbonyl** this is eight electrons from the Fe atom plus ten electrons from five CO molecules for a total of eighteen electrons again.

Compound	Valence Electrons			
	Metal	*Ligands*	*Other Metal*	*Sum*
$Cr(CO)_6$	6	12	0	18
$Mn_2(CO)_{10}$	7	10	1	18
$Fe(CO)_5$	8	10	0	18
$Co_2(CO)_8$	9	8	1	18
$Ni(CO)_4$	10	8	0	18

Table 11–2
Metal Valence Electron Sums in Several Transition Metal Carbonyls

The magic number 18 is the maximum allowed electron occupation of the sum of the five 3*d*, three 4*p*, and one 4*s* orbital. Clearly all of these orbitals are involved in the bonding. The atomic electron configuration $[Ar]3d^{10}4s^24p^6$ is that of the gas krypton, which possesses no significant electron affinity. The metal atom in $Ni(CO)_4$, for example, is said to possess an **effective atomic number** equal to that of krypton (36) because there are 36 electrons around the Ni nucleus. Eighteen of these 36 are in valence orbitals. Thus, $Ni(CO)_4$ would be predicted to possess no further electron affinity and no compound of the formula $Ni(CO)_5$ or $Ni(CO)_6$ should form. Such is indeed the case.

The existence of a magic number of eighteen valence electrons for metal carbonyls (and many other organometallic species) allows us to understand why **dimanganese decacarbonyl,** Mn_2CO_{10}, contains two Mn atoms. The molecule $Mn(CO)_5$ contains only seventeen electrons around the Mn atom and $Mn(CO)_6$ would contain nineteen. The number 18 can be achieved by having two $Mn(CO)_5$ molecules form a two-electron Mn—Mn bond in Mn_2CO_{10}. An analogous problem arises with the cobalt carbonyl, and it too contains a metal-metal bond.

An alternative to metal-metal bonding as a way to achieve the 18-electron configuration is reduction. Thus, for example, sodium metal reduces

$Co_2(CO)_8$ to produce the anion $Co(CO)_4^-$, according to the following reaction:

$$Co_2(CO)_8 + 2\ Na \rightarrow 2\ Na[Co(CO)_4] \quad \textbf{11–21}$$

The $Co(CO)_4^-$ anion also possess 18 metal valence electrons since the added electron may be viewed as a tenth cobalt 3*d* electron. The $Co(CO)_4^-$ ion is a nucleophile and reacts with acids to form a hydride containing an M—H bond or it can react with electrophilic centers in organic molecules to yield organometallic carbonyl compounds:

$$Co(CO)_4^- + H^+ \rightarrow HCo(CO)_4 \quad \textbf{11–22}$$

$$Co(CO)_4^- + CH_3I \rightarrow H_3CCo(CO)_4 + I^- \quad \textbf{11–23}$$

Compounds such as the product of reaction 11–23 are susceptible to insertion of the CO into the Co—C bond according to the following reaction:

$$(OC)_4CoCH_3 \rightarrow (OC)_3Co\text{—}\overset{\overset{\displaystyle O}{\|}}{C}CH_3 \quad \textbf{11–24}$$

and reaction 11–24 is the basis for the convenient synthesis of organic compounds containing the carbonyl group. For many years chemists have used transition metals and their compounds as catalysts in organic reactions. However, it has been only recently that we have come to recognize the organometallic structures and reactions that yield this catalytic activity. The predictable result of this new understanding has been the rational design of specific catalysts for specific reactions.

SUGGESTIONS FOR ADDITIONAL READING

Basolo, F. and Johnson, R. C., *Coordination Chemistry*, Benjamin, Menlo Park, Calif., 1964.

Rochow, E. G., *Organometallic Chemistry*, Reinhold, New York, 1964.

PROBLEMS

1. We have seen the overlap pictures of pi molecular orbital formation in benzene and the energies that result from bonding and antibonding overlap. The molecule **cyclobutadiene,** C_4H_4, in which all of the nuclei lie in the same plane, has never been synthesized nor does it occur in nature. Draw the atomic orbital overlap pictures for the four pi molecular orbitals of C_4H_4. From this overlap picture, construct an energy-level diagram analogous to that for the six pi orbitals of benzene. Assign the four pi electrons to these energy levels in the lowest-energy fashion. What is the total bond order of the carbon-carbon bonding in this molecule? Can you explain why cyclobutadiene does not exist?

2. The electronic band model of metals explains their high electrical conductivity. Use the band model to explain why metals are also excellent heat conductors.

3. If the lowest 2*p* bonding molecular orbital in Be metal is at a lower energy than the highest 2*s* antibonding orbital, why don't the $2p\pi$ or $2p\sigma$ levels of Be_2 lie below the $2s\sigma^*$ level? If they did (or if either one did), Be_2 would be strongly bonded.

4. The species Be_2 and He_2 do not exist for essentially the same reason. Yet Be is a metallic solid at room temperature. Why isn't helium a metal?

*5. Semiconductors, unlike metals, have electrical conductivities that increase with increasing temperature. Explain why.

6. Most aquated cations of known structure are observed to have the octahedral geometry shown for $Na(H_2O)_6^+$ in Figure 11–4. A notable exception is $Be(H_2O)_4^{2+}$, which possesses the tetrahedral geometry. Why does Be^{2+} associate with only four water molecules?

7. If $Ca^{2+}O^{2-}$ is preferable to Ca^+O^- in solid calcium oxide, why isn't $Ca^{3+}O^{3-}$ even better?

8. The anions nitrate and carbonate are isoelectronic. Yet there is a great contrast between the solubilities in water of nitrate and carbonate salts. With the exception of the alkali metal salts, most carbonates show only a low solubility in water. However, almost all metal ion nitrates are highly soluble in water. Explain this large difference between nitrates and carbonates. Do you predict the same sort of difference between sulfate and perchlorate? Explain.

9. The organic product of the addition of ethyl magnesium bromide to methyl acetate in ether solution is methyl ethyl ketone. Show from the structures of the intermediates how this product is formed.

10. Water must be scrupulously excluded from the system in the preparation of a Grignard reagent. Why?

11. The first ionization potentials typically decrease with increasing atomic number within a given group of the periodic table, and this is demonstrated with K, Rb, and Cs in Table 11–1. The same trend is shown between Cu and Ag, but Au is out of line. Furthermore, Hg is also out of line with Zn and Cd. Explain these two misplacements.

12. Metallic calcium very quickly reacts with oxygen to form a white oxide coating when exposed to air. Copper very slowly reacts with oxygen. Why is copper less reactive with O_2 than calcium is? Which is predicted to be the stronger base, CaO or CuO?

13. The substance $HgCl_2$ is a solid at room temperature but melts at 276 °C and boils at 302 °C. These are very low values for a compound that combines a metal with a halogen. For example, sodium chloride melts at

801 °C and boils at 1413 °C. Explain this unusual behavior of $HgCl_2$ by referring to the ionization potentials listed in Table 11–1.

14. The ions Zn^{2+} and Hg^{2+} commonly form complexes that contain four metal ion–nucleotide bonds and have a tetrahedral geometry. Why aren't there six bonds and an octahedral geometry as there are with many other metal ion complexes?
15. Cerium(IV) salts are very strong oxidizing agents that are commonly used in chemical analysis. Why should we predict this strong oxidizing ability of Ce(IV)? Which oxidation state is predicted to be the product of the reduction of Ce(IV)?
*16. Carbon tetrachloride, like all of the tetrachlorides of Groups IVA and IVB, is unstable with respect to reaction with water. In the case of CCl_4, the reaction is

$$CCl_4(l) + 2\ H_2O(l) \rightarrow CO_2(g) + 4\ HCl(g)$$

However, the reaction between CCl_4 and water is extremely slow under normal conditions. Such is not the case with $SiCl_4$ and $TiCl_4$, which react vigorously with water to form HCl and SiO_2 and TiO_2, respectively. Why is CCl_4 unique in this regard? *Hint:* Try to picture a mechanism for the reaction in each case.

17. What is the O—V—O bond angle in VO_2^+? The ion NO_2^+ is a very strong oxidizing agent and Lewis acid. The ion VO_2^+ is considerably weaker than NO_2^+ as an oxidizing agent and a Lewis acid. Why?
18. Propose a structure and draw a Lewis dot representation for the dichromate ion.
19. Propose a structure and draw a Lewis dot representation for the permanganate ion.
20. Why is there no point in Figure 11–5 that represents the aquation energy of Sc^{2+}?
21. High-spin complexes of Ni^{2+} are octahedral while low-spin Ni^{2+} complexes contain only four ligands and are square-planar. Explain this change in geometry.
22. Low-spin complexes are favored over high-spin complexes as the atomic number increases within a group. Thus, Pt(II) complexes, for example, are invariably low-spin, square-planar ones. Explain this trend.
23. Complexes of Co^{3+} are usually extremely stable and diamagnetic. Explain both the high stability and the diamagnetism.
24. What geometry do you predict for the ion $Cu(NH_3)_4^{2+}$? It is difficult to add more than four NH_3 molecules to Cu^{2+}. Explain this difficulty.
25. Assign the seven d electrons to the five d orbitals in high-spin octahedral Co^{2+} complexes. We showed in the text that for $Ni(H_2O)_6^{2+}$, LFSE $= 1.2\Delta$. Show for $Co(H_2O)_6^{2+}$ that LFSE $= 0.8\Delta$.

26. Name the following compounds:
    a. $[Co(NH_3)_5Cl][BF_4]_2$.
    b. $Na_2[CuCl_4]$.
    c. $K_3[Co(CN)_6]$.
    d. $Pt(en)Cl_2$.
27. Draw all of the possible isomers of the chloroammine-bis-ethylenediamine cobalt(III) ion.
28. The text explains why substitution reactions are slow with octahedral Cr^{3+} complexes. Deduce the identity of two other metal ions that you would predict to show such anomalous slow substitution rates for their octahedral complexes.
29. Metal ions are required in the synthesis of proteins from amino acids in living systems. Propose one important role for these metal ions in protein synthesis. *Hint:* Protein synthesis involves amide formation.
30. Deduce the geometries of $Ni(CO)_4$ and $Cr(CO)_6$. These carbonyls do not react with water, but they are rapidly oxidized by molecular oxygen. Explain this difference in reactivity.
31. Nickel tetracarbonyl has a relatively low boiling point (43 °C). Explain this low boiling point.
32. The CO molecules in $Ni(CO)_4$ can be replaced by other molecules. Do you predict that these ligand substitutions occur by dissociation, as in reaction 11–18, or by association, as in reaction 11–19? Explain.
33. The text introduces the carbonyl anion $Co(CO)_4^-$. Propose a structure for it.
34. What is the effective atomic number of iron in ferrocene?

12 / Spectroscopy

We have developed models for the prediction of molecular structures, and we have applied these structural models to the prediction and correlation of the chemical properties of the elements and their compounds. How do we know that the molecular structures we have deduced are correct? Much of the definitive information concerning molecular structure comes from analyses of the interaction between matter and electromagnetic radiation.

12–1. A Classical Model of the Interaction between Electromagnetic Radiation and Gaseous Molecules

Molecular Rotation

Figure 12–1 shows plane-polarized radiation encountering an isolated HF molecule in the gas phase. The double arrows indicate the direction of the electric field vector E and hence indicate the alternating directions along which charges are moved by the field. The signs on E indicate the instantaneous direction of the electric field. The minus sign at the top and the plus sign at the bottom indicate that plus charges are attracted up and minus charges down by the field at the time shown.

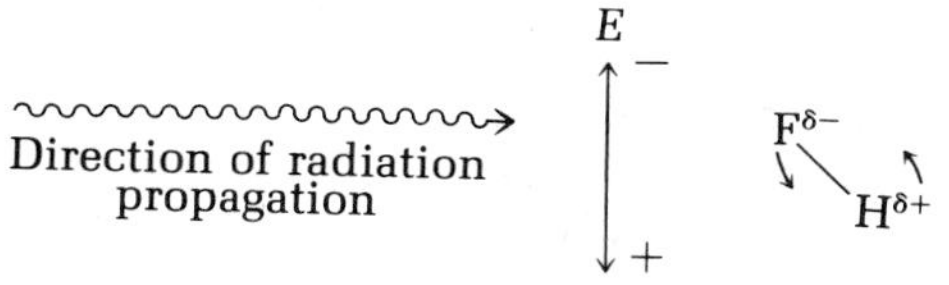

Direction of radiation propagation

E − +

$F^{\delta-}$ $H^{\delta+}$

Figure 12–1
Rotation of a HF molecule produced by the electric field of plane-polarized electromagnetic radiation.

The HF molecule is polar as shown in the figure and its charges, $\delta+$ and $\delta-$, undergo three changes in response to the field E. The first change is **molecular rotation.** In the orientation shown in the figure, the HF molecule is unfavorably oriented with respect to E since the most favored arrangement has $\delta+$ up and $\delta-$ down in a vertical molecular orientation at the time shown. Thus, the molecule rotates in the direction shown by the arrows. Since E is alternating, the signs of E switch at the frequency of the radiation and the HF molecule continuously rotates to follow it. The rotation of any molecule with a dipole moment can increase in the gas phase in response to electromagnetic radiation. In contrast, molecules such as O_2, N_2, CO_2, and CH_4 that have dipole moments equal to zero do not show a rotational response to radiation.

Molecular Vibration

The second response of the HF molecule is **vibration,** as shown in Figure 12–2. The molecule shown is in the most favored orientation, but further motion of the δ+ upward and δ− downward can occur by stretching the H—F bond as shown. The vibration of all polar molecules increases when subjected to electromagnetic radiation. Figure 12–3 shows the motions of the gaseous water molecule that radiant energy induces.

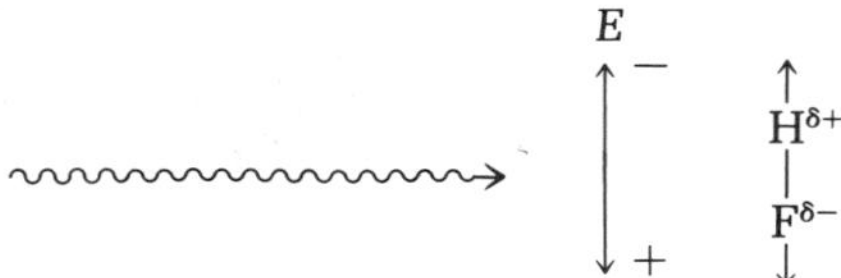

Figure 12–2 Vibration of a HF molecule produced by the electric field of plane-polarized electromagnetic radiation.

The motion of a gaseous water molecule can be resolved into nine independent components, termed the **degrees of motional freedom** of the molecule. For any gaseous molecule, three degrees of motional freedom are translation along x, y, and z axes. These motions yield a contribution to the two heat capacities, C_v and C_p, of $\frac{3}{2}R$ and $\frac{5}{2}R$, respectively (Chapter 2). One of our aims is to account for the C_v and C_p values in Table 2–2 that are larger than $\frac{3}{2}R$ and $\frac{5}{2}R$, respectively.

Nonlinear molecules have three rotational degrees of freedom, namely rotation about x, y, and z axes. These three motions are shown in Figures 12–3(c), (b), and (a), respectively. There are three independent vibrational motions of the water molecule and they are shown in Figures 12–3(d), (e), and (f). The motions are called **symmetric stretch** (both O—H bonds lengthen and shorten in unison), **asymmetric stretch** (one O—H bond length maximizes when the other minimizes), and **bend** (the bond lengths remain constant and the bond angle changes). The arrows in Figures 12–3(b), (c), (d), (e), and (f) show activation of each of the five motions by plane-polarized electromagnetic radiation. The rotation in Figure 12–3(a) does not occur in response to electromagnetic radiation. (In a problem at the end of the chapter you are asked to explain why no response occurs.)

The total number of degrees of freedom of the water molecule is nine, which is three times the number of nuclei in the molecule. **The total number of degrees of freedom for any molecule is three times the number of nuclei in the molecule.** For the gaseous HF molecule, there are six degrees of freedom (three translations, two rotations, and one vibration).

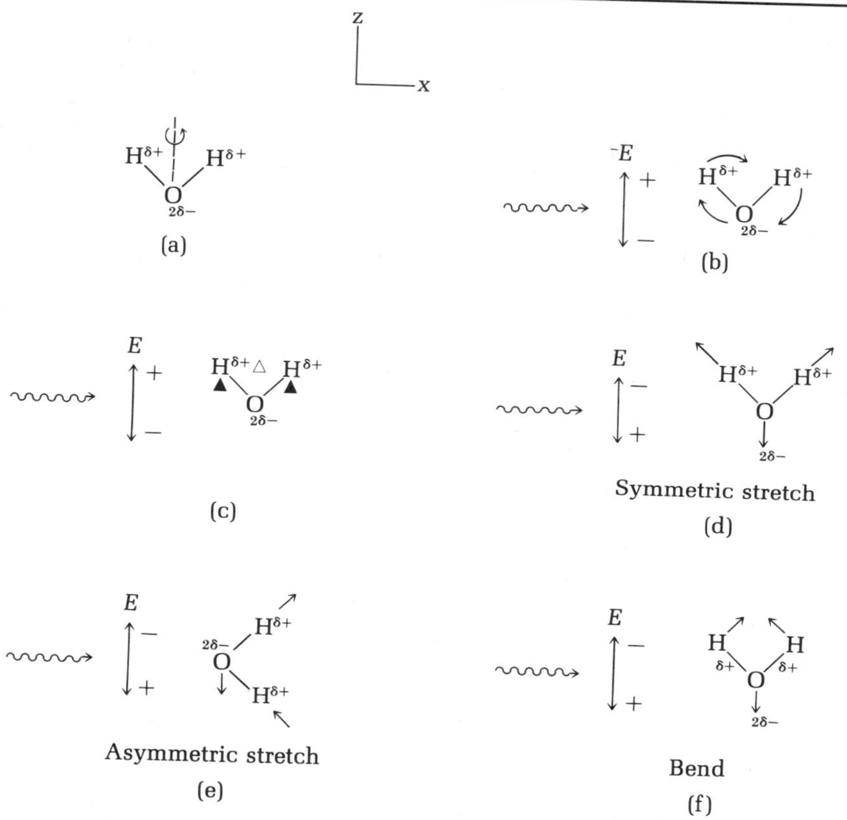

Figure 12–3
The rotational and vibrational motions of a gaseous water molecule. Parts b, c, d, e, f show responses to electromagnetic radiation. In (c), (△) and (▲) indicate motion directly in and out of the plane of the paper, respectively. (The coordinate axes at the top of the figure apply to all six motions.)

Homonuclear diatomic molecules such as H_2 and N_2 do not vibrate in response to radiant energy. However, molecules such as CO_2, which possess no dipole moment in the absence of molecular vibration, can and do respond vibrationally to radiant energy (even though they do not respond rotationally). Figure 12–4 shows the four independent vibrational motions of CO_2 and the response of three of them to electromagnetic radiation. The fourth, symmetric stretch, does not respond because the motion does not send negative charge in one direction and positive charge in the other. The independent vibrational motions of any molecule are termed its **normal modes of vibration.**

Both H_2O and CO_2 are triatomic molecules, but they respond quite differently to electromagnetic radiation; for example, the rotation of CO_2 does not increase, but the rotation of H_2O does. These different responses are a consequence of the different structures of H_2O and CO_2. The response of molecules to electromagnetic radiation gives us very detailed information about molecular structures.

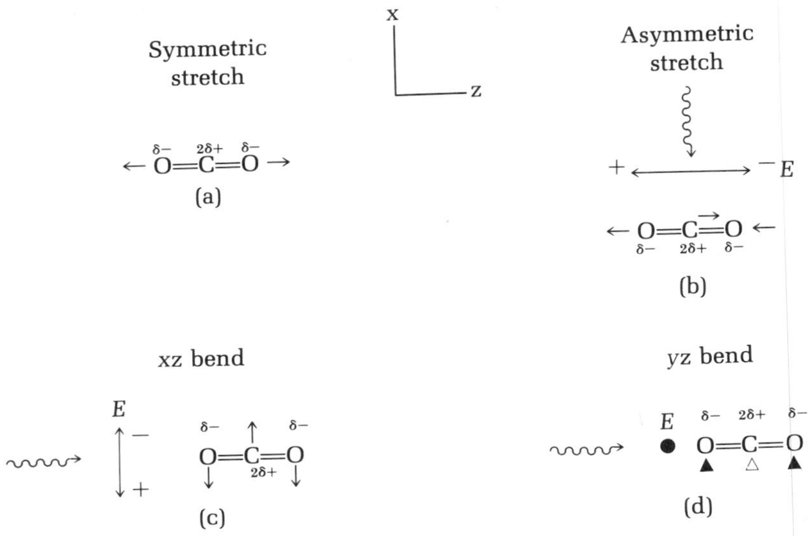

Figure 12–4
The four vibrational motions of the CO_2 molecule. Parts b, c, d show responses to electromagnetic radiation. In (d), the field E is directed perpendicular to the plane of the paper. Motion directly in and out of the plane of the paper is indicated by the symbols △ and ▲, respectively. (The coordinate axes at the top of the figure apply to all four motions.)

Electron Displacement within Molecules

Figure 12–4 shows that an electric field can produce a dipole moment in a molecule that otherwise has a zero dipole moment by displacing the nuclei, that is, by making the molecule vibrate. Another way to produce polarity is to leave the nuclei fixed and displace an electron from one region of the molecule to another. In Figure 12–2, the electric field not only can produce the HF vibrational motion shown but can increase δ+ and δ− by displacing an electron from the H atom region to the F atom region. Conversely, if the HF molecule is rotated by 180° with respect to the orientation in Figure 12–2, the electron displacement is from F to H; and if the rotation is 90°, the electron displacement is perpendicular to the H—F axis, as shown in Figure 12–5. **All molecules can respond to radiant energy by electron displacement.**

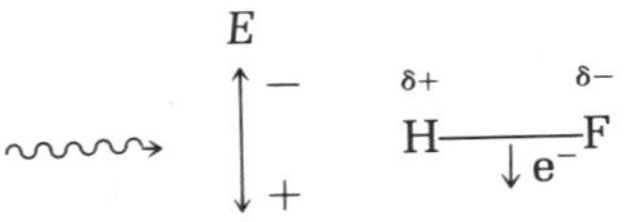

Figure 12–5
Electron displacement in HF in response to electromagnetic radiation.

Energy Absorption

The fact that molecules respond to electromagnetic radiation by the motion of charges does not, of itself, indicate how radiant energy is absorbed. The electron to the right of the antenna in Figure 3–5 responds to radiation by a vertical motion. However, in responding, it becomes itself an antenna and serves to reradiate the energy incident upon it. The electron does not absorb energy but rather scatters radiation.

Absorption of radiant energy by atoms and molecules occurs because the atoms and molecules interact. Figure 12-6 shows an HF molecule rotating in response to an alternating electric field. As it is rotating, it collides

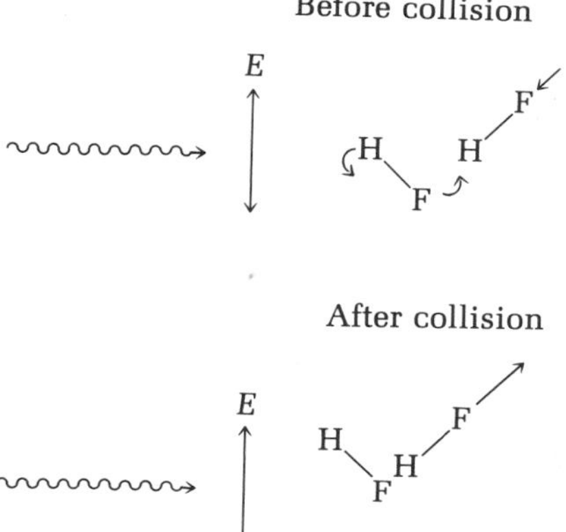

Figure 12-6 The indirect transfer of energy from electromagnetic radiation to molecular translation through a molecular collision.

with another molecule, as shown, and much of its rotational energy is passed to the second molecule as translational energy. Two steps are depicted in the figure:

$$\text{Radiation} + \text{nonrotating HF} \rightarrow \text{rotating HF} \qquad \textbf{12–1}$$

$$\text{Rotating HF} + \text{slowly translating HF} \rightarrow \text{nonrotating HF} + \text{rapidly translating HF} \qquad \textbf{12–2}$$

The overall process is the sum of these two steps, or

$$\text{Radiation} + \text{slowly translating HF} \rightarrow \text{rapidly translating HF} \qquad \textbf{12–3}$$

Equation 12-3 is merely the conversion of radiant energy to the energy of molecular translational motion—the absorption of radiant energy. To the extent that molecules collide with each other, absorption predominates over scattering. The molecules of liquids and solids are in almost constant collision with each other and absorption predominates over scattering in most liquids and solids.

12-2. Rotational Spectroscopy

Microwave Absorption

In our classical model in section 12-1, we made no reference to the frequency of plane-polarized radiation that excites molecular motion. This is because, in a classical picture, all frequencies of electromagnetic radiation are equally effective in producing rotation, vibration, and electronic displacement. However, this is not the case experimentally. Rotational absorption occurs almost exclusively in the microwave region, as shown in Figure 3–6, vibration is centered in the infrared region, and electronic displacement occurs almost exclusively in the visible, ultraviolet, and X-ray regions. The relative placement of these absorption regions could have been predicted in a very straightforward fashion from the wave model (Chapters 3 and 4). According to the wave model, the absorption occurs in discrete energy steps of energy $\Delta E = h\nu$, so that the order of step size must be electronic > vibration > rotation, since $\nu_{\text{visible}} > \nu_{\text{infrared}} > \nu_{\text{microwave}}$. As illustrated in equation 4–2 for a particle confined to a line, gaps between energy levels for any kind of bounded motion increase as the mass of the moving particle decreases and as the distance between boundaries decreases. Of the three motions—rotational, vibrational, and electronic—the smallest gap is predicted for rotation because the nuclei move over the circumference of circles whose center is at the center of gravity of the molecule. Next comes vibration, in which nuclei move short distances about and along the directions of chemical bonds. Finally, electronic displacement is the highest in energy because the moving particle is the very light electron.

We can use equation 4–2 to obtain an estimate of ν for the rotational energy absorption by a diatomic molecule such as HCl through a model in which the nuclei move on one-dimensional paths of constant potential energy, namely circular orbits. The energy of rotation of a diatomic molecule about its center of gravity can be shown to be equal to the energy of the **reduced mass** of the two nuclei moving in a circular orbit with a radius equal to the internuclear distance. The reduced mass μ of two nuclei of masses m_1 and m_2 is given as follows:

$$\mu = \frac{m_1 m_2}{m_1 + m_2} \qquad \textbf{12–4}$$

Let us calculate the reduced mass (in amu) of an HCl molecule that contains a proton and a ^{35}Cl nucleus:

$$\mu = \frac{(1.01)(35.0)}{1.01 + 35.0} = 0.98 \text{ amu}$$

In this case of HCl, μ very nearly equals the mass of the proton, so that the rotation of HCl is very nearly the rotation of a proton around a stationary chlorine nucleus.

Equation 4–2 for the energy levels of a particle of mass m confined to a line of length a is

$$E = \frac{n^2 h^2}{8ma^2}$$

For HCl rotation, the line is wrapped into a circle and a is the circumference of that circle ($2\pi r_{HCl}$). Thus, the rotational energies are predicted to be

$$E_{rotation} = \frac{n^2h^2}{8\mu(2\pi r_{HCl})^2} = \frac{n^2h^2}{32\pi^2\mu r_{HCl}^2} \qquad \text{12–5}$$

A typical transition is from the level $n = 1$ to the level $n = 2$, which occurs at a frequency given as

$$\nu_{1\to2} = \frac{\Delta E_{1\to2}}{h} \cong \frac{3h}{32\pi^2\mu r_{HCl}^2}$$

$$= \frac{(3)(6.62 \times 10^{-27})(35.0 + 1.01)(6.02 \times 10^{23})}{(32)(3.14)^2(35.0)(1.01)(1.27 \times 10^{-8})^2}$$

$$= 2.4 \times 10^{11} \text{ Hz}$$

where μ is expressed in g/molecule and r_{HCl} is 1.27×10^{-8} cm. According to Figure 3–6, a frequency of 2.4×10^{11} Hz is near the upper end of the microwave frequency range. Since ν is inversely proportional to μ, and since μ for HCl has very nearly the smallest value encountered for any diatomic molecule, ν is smaller than 10^{12} Hz for most molecules and typically falls in the range of 10^{10}–10^{12} Hz.

Equation 4–2 is not an exact one for the rotation of a diatomic molecule since the rotation of particles that manifest themselves as waves is not confined to a well-defined circle. However, we can use the Schrödinger procedure to obtain the correct energies. The actual frequency for the transition from the lowest rotational energy level of HCl to the next-lowest level is 6.3×10^{11} Hz.

When we considered the energy levels of the hydrogen atom (Chapter 3), we derived the line spectrum by considering transitions between all possible energy levels. In most cases, the transitions are more limited than in the H atom in that they are subject to **selection rules** derivable from the Schrödinger procedure. The selection rules dictate which transitions are allowed and which forbidden. The Schrödinger procedure yields the following expression for the rotational energies of a diatomic molecule:

$$E_{rotation} = \frac{h^2}{8\mu r_{AB}^2} J(J + 1) \qquad \text{12–6}$$

where μ is the reduced mass, r_{AB} is the bond length, and J is the **rotational quantum number.** Equation 12–6 is quite similar to equation 12–5, showing that equation 12–5 is not a bad approximation.

The rotational selection rule is $\Delta J = \pm 1$. Several of these allowed transitions are shown in Figure 12–7. The ordinate in the figure is in units of $h^2/8\mu r_{AB}^2$.

If a beam of microwave radiation is passed through a tube containing gaseous HCl for example, absorption occurs only if the frequency of the

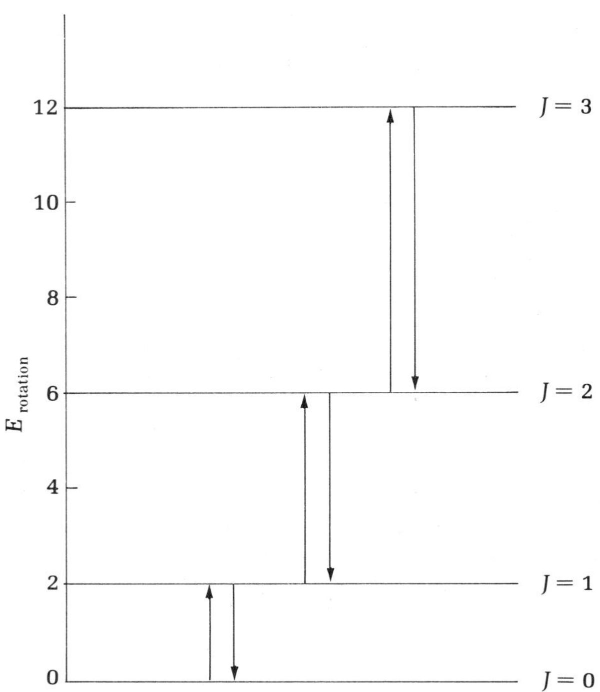

Figure 12–7 Allowed transitions between rotational energy levels of a diatomic molecule.

radiation matches $\Delta E/h$ for one of the allowed transitions. These frequencies can be found by varying the microwave frequency continuously from one end of the microwave range to the other with simultaneous monitoring of the degree of absorption by HCl. The result is a plot of energy absorbed as a function of frequency, and such a plot is termed an **absorption spectrum.** The absorption spectrum is a series of lines like those shown in Figure 3–9, though the pattern of lines is quite different from that in Figure 3–9.

Both equation 12–5 and the exact expression for the rotational energy levels of HCl (equation 12–6) contain the internuclear distance r_{HCl}. Thus, the values of microwave absorption frequencies can be used to calculate very precise values of internuclear distances.

The microwave absorption spectrum of a polyatomic molecule such as water is more complicated than for HCl because there is more than one type of rotation, as depicted in Figures 12–3(b) and (c). Nevertheless, the spectrum can be unraveled, and very precise values for the O—H bond distance and H—O—H bond angle have been obtained for gaseous water. In fact, the structures of a very large number of small molecules have been determined in this way.

Molecular Rotation and Heat Capacities

One of our aims is to show the contributions of molecular rotation and vibration to the heat capacities of gases. Table 2–2 shows that C_v and C_p for HCl are approximately $\frac{5}{2}R$ and $\frac{7}{2}R$, respectively. These values are higher by an amount R than the values for translational motion alone. Let us see if we can account for the extra R by investigating molecular rotation.

The energy gap between the lowest and second-lowest rotational energy levels of HCl is

$$\Delta E_{1\to 2} = h\nu_{1\to 2} = (6.62 \times 10^{-27})(6.3 \times 10^{11}) = 4.2 \times 10^{-15} \text{ erg/molecule}$$

The average translational energy of HCl molecules at 25 °C is 6×10^{-14} erg/molecule. There is more than enough translational energy available to promote HCl molecules to the second-lowest level (and several higher levels) as a result of molecular collisions. Thus, many rotational levels have sizable populations at room temperature.

As the temperature is raised, even more translational energy becomes available and even higher rotational levels become populated. The population of the higher levels requires an energy input, and this is the rotational contribution to C_v and C_p. The HCl molecule has two rotational degrees of freedom. Each rotational degree of freedom contributes $R/2$ to C_v just as each translational degree does.

Table 2-2 shows that C_v for O_2 is also $\frac{5}{2}R$, C_v for H_2 is slightly smaller than $\frac{5}{2}R$, and the other diatomic molecules yield a C_v larger than $\frac{5}{2}R$. Furthermore, in the cases of I_2 and Cl_2, the deviation from $\frac{5}{2}R$ increases with the weight of the molecule. (The next section reveals the source of this additional heat capacity.)

12-3. Vibrational Spectroscopy

Infrared Absorption

Let us use equation 4-2 to estimate ν for a transition between typical vibrational energy states. To make this estimation, we need to know the mass of the vibrating particle or particles and the distance within which the motion is confined. The vibration of a diatomic molecule may be treated by the model shown in Figure 12-8(a), in which the two nuclei of masses m_1 and m_2 are held together by a spring. The stretching and compression of a spring appears to be a crude model for molecular vibration; but it leads to predictions that are remarkably reliable, and the model is very commonly used.

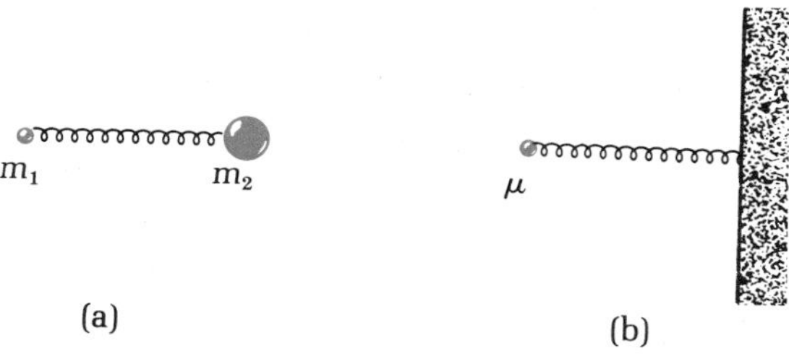

Figure 12-8
Ball and spring model of a diatomic molecule: (a) both nuclei free to move, (b) equivalent model involving the reduced mass.

Figure 12–8(b) shows an equivalent model for a diatomic molecule, namely one of a mass μ, which is the reduced mass of m_1 and m_2, tied to a wall by a spring of a length equal to the internuclear distance in Figure 12–8(a). The model in Figure 12–8(b) is the one that we shall use in our calculation.

Now let us see how the motion is confined in molecular vibration. Atoms attract each other at large distances and repel each other at very short distances. In terms of potential energy, the potential energy of the atom-atom interaction is negative at large distances and positive at very short distances. The curve of potential energy versus internuclear distance for a diatomic molecule typically has the shape shown in Figure 12–9. The minimum in the curve is the bond length r_{AB} that we have been using.

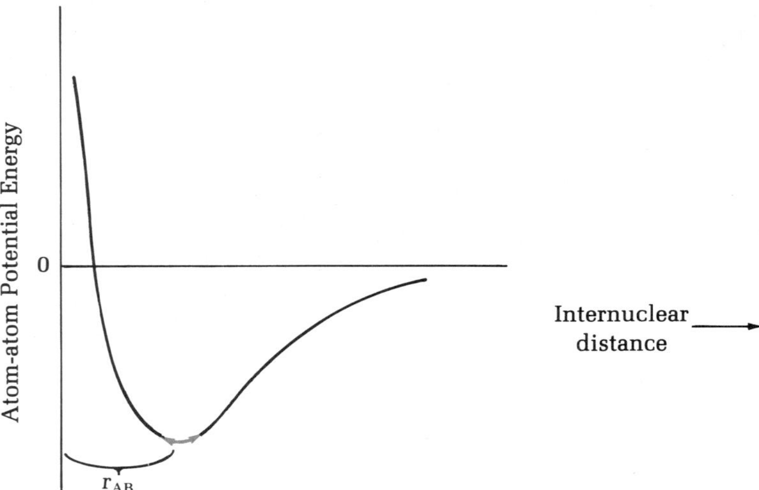

Figure 12–9 Typical shape of the curve for the potential energy of atom-atom interaction as a function of internuclear distance in a diatomic molecule.

The vibrational motion in the lowest-vibration energy state is shown by the two-headed colored arrows in Figure 12–9. We know the amplitude of the vibration cannot be large or the molecule will vibrate apart at room temperature. Most diatomic gases must be heated to over 1000 °C before their vibrational energies become so large that a sizable fraction of the molecules dissociate. In our calculation for vibration at room temperature, we shall assume that the vibrational motion occurs over a distance equal to 20 percent of the equilibrium bond distance in the HCl molecule. This distance d of motion is 2.54×10^{-9} cm. With our model, equation 4–2 becomes

$$E_{\text{vibration}} = \frac{n^2h^2}{8\mu d^2} \qquad \textbf{12–7}$$

and

$$\nu_{1\to 2} = \frac{\Delta E_{1\to 2}}{h} = \frac{3(6.62 \times 10^{-27})(6.02 \times 10^{23})}{(8)(0.98)(2.54 \times 10^{-9})^2}$$

$$= 2.4 \times 10^{14} \text{ Hz}$$

This frequency corresponds to a wavelength of 1.2×10^{-4} cm and this wavelength is an infrared wavelength quite close to the visible region of wavelengths. Experimentally, the wavelength for the transition between the lowest and second-lowest vibrational energy levels of HCl occurs at 3.5×10^{-4} cm.

Equation 12–7 shows that μ appears in the denominator, so that molecules that possess a higher μ than HCl (and that includes most other molecules) absorb energy vibrationally further into the infrared region (further away from the visible region) than HCl does.

Molecular Vibrations and Heat Capacities

The energy gap between the lowest and second-lowest energy levels for HCl vibration is $(3.0 \times 10^{10})(6.62 \times 10^{-27})/3.5 \times 10^{-4} = 5.7 \times 10^{-13}$ erg/molecule. This value is much larger than the average translational energy of HCl molecules at 25 °C. In this case and in the case of most molecular vibrations involving light atoms at room temperature, only the ground vibrational state is significantly populated. A small increase in temperature will not alter this situation appreciably; little energy must be absorbed vibrationally to accomplish the temperature increase and the vibrational contribution to C_v and C_p for molecules such as HCl is near zero. This is the reason that C_v for H_2, CO, and O_2 shows no contribution from vibration at 25 °C even though each possesses one vibrational degree of freedom. This contribution will begin to appear, however, if the temperature is sufficiently raised.

There are two reasons why I_2 shows a sizable contribution to C_v and C_p from vibration. The first is that μ is much greater for I_2 than for HCl, and the second is that the HCl bond energy is much larger than the bond energy of I_2. Figure 12–10 shows the effect of a change in bond energy on molecular vibration. The greater the bond energy, the steeper the potential energy curve and the more confined the vibration. According to equation 12–7, this results in a larger gap between vibrational energy levels for the stronger bond. In the case of I_2, the weak bond results in a small gap and a sizable vibrational contribution to C_v and C_p.

Equation 12–7 is also the basis of the rule of Dulong and Petit, since our model applies to vibrational levels in solids as well as in liquids and gases. For light atoms, there is little more than one vibrational level occupied for each atom at room temperature and the vibrational contribution to C_v and C_p is small. A heavier element possesses a larger number of populated vibrational levels per atom, and these heavier elements all show a $C_v \cong C_p \cong 3R$.

What is the source of the value of $3R$? In a solid, the only degrees of freedom are vibrational degrees, and each atom possesses three such degrees. Thus, each vibrational degree of freedom contributes the value R to the heat capacity; this is in contrast to the $R/2$ contribution of each rotational and

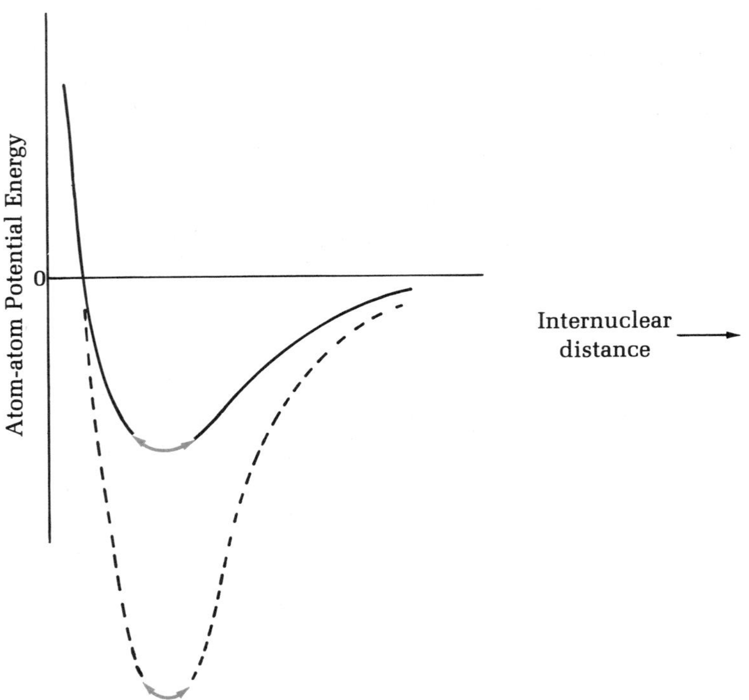

Figure 12–10
Molecular vibration (colored lines) for a diatomic molecule with a small bond energy (solid black line) and large bond energy (dashed black line).

translational degree for gas molecules. Vibration involves two kinds of energy, kinetic energy and potential energy (the potential energy is shown as the ordinate in Figures 12–9 and 12–10). Each vibrational degree of freedom "stores" energy in two forms, while translational energy and rotational energy involve only one form (kinetic energy). This explains why each of the vibrational degrees of freedom contributes R to the heat capacity.

Vibrational Spectra

The Schrödinger procedure also yields an expression for vibrational energy levels; this expression is

$$E_{\text{vibration}} = \frac{h}{2\pi}\left(\frac{k}{\mu}\right)^{1/2}\left(v + \frac{1}{2}\right) \qquad \textbf{12–8}$$

In equation 12–8, μ is the reduced mass and v is termed the **vibrational quantum number.** The constant k is called the **force constant** and it is a measure of the steepness of the potential energy curve, such as the curves shown in Figure 12–10. Figure 12–11 shows several vibrational energy levels superimposed on the potential energy curves of Figure 12–10.

A selection rule applies to vibrational transitions. This selection rule is $\Delta v = +1$ for energy absorption and $\Delta v = -1$ for energy emission. Since only

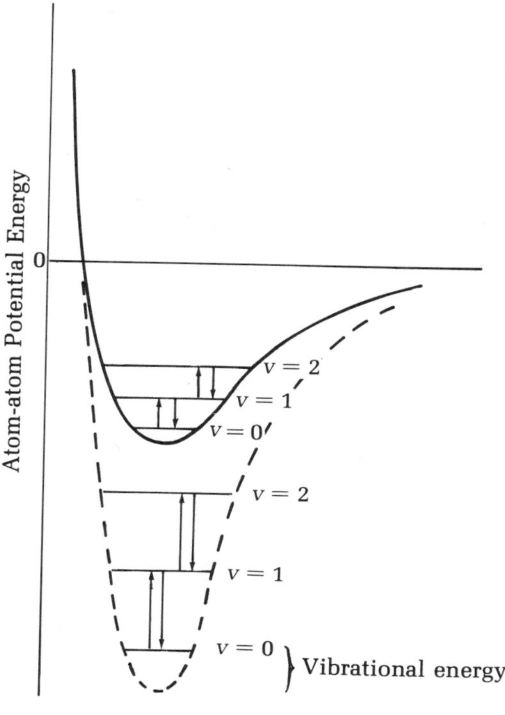

Figure 12-11 Vibrational energy levels and allowed transitions for two molecules having different potential energy curves.

the lowest vibrational level is significantly populated at room temperature for most molecules in either the gaseous or condensed states, the only prominent transition observed in infrared absorption spectroscopy is from the lowest level to the second-lowest level ($v = 0$ to $v = 1$). Since the second-lowest level is relatively low in population at room temperature, infrared emission is sparse from substances at room temperature. However, the intensity of this emission increases markedly with increasing temperature, producing the familiar "radiant heat" that is given off by substances at high temperatures. This is due to the increase in population of higher vibrational levels as the temperature is increased.

An **infrared absorption spectrum** of a substance is obtained with the apparatus depicted in Figure 12-12. Infrared radiation with a continuous distribution of frequencies is generated by electrically heating a solid, as shown on the left in the figure. A beam of this radiation passes through a narrow slit, then through the sample (which is either a gaseous sample of the substance under investigation or a sample of it dissolved in some appropriate solvent), and then through a second slit. After passing through the sample, the beam will show strong attenuation at those frequencies corresponding to $\Delta E/h$ for vibrational transitions in the sample. The beam is then split into its component frequencies by passage through a diffraction grating. A revolving mirror reflects the individual frequencies one at a time

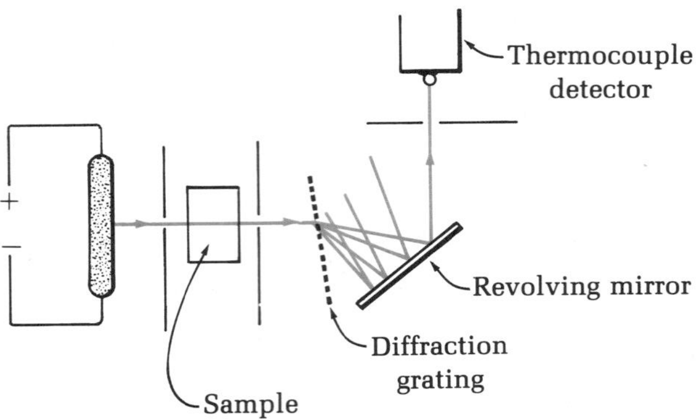

Figure 12–12
Simplified diagram of an infrared absorption spectrometer.

to the detector. The detector is a heat-sensing device such as a thermocouple or thermistor, which converts the radiant heat to an electrical signal. This signal is then sent to a recorder that plots it as the percentage of infrared radiation of a given frequency transmitted by the sample as a function of that frequency. This plot of percent transmittance as a function of frequency is the infrared absorption spectrum of the substance. An abscissa scale commonly used on absorption spectra is $\bar{\nu} = 1/\lambda = \nu/c$. This is a scaled frequency that is smaller and less cumbersome than ν itself.

Figure 12–13 shows the infrared absorption spectrum of gaseous HCl. The large number of inverted peaks arise because gaseous molecules absorb energy at infrared frequencies to change both vibrational and rotational energies simultaneously. All of the peaks arise from transitions from the lowest vibrational level to the second-lowest level and between adjacent rotational levels (combination of Figures 12–7 and 12–11). The rotational transitions are labeled in the figure by assigning the J quantum numbers: zero to the lowest rotational level, 1 to the next-lowest, and so forth. Thus, the transition $0 \rightarrow 1$ is from the lowest J level to the second-lowest J level. The closely spaced splitting of each inverted peak is due to the natural occurrence of two chlorine isotopes, ^{35}Cl and ^{37}Cl, in comparable amounts. The two isotopes yield slightly different μ values for $H^{35}Cl$ and $H^{37}Cl$, and

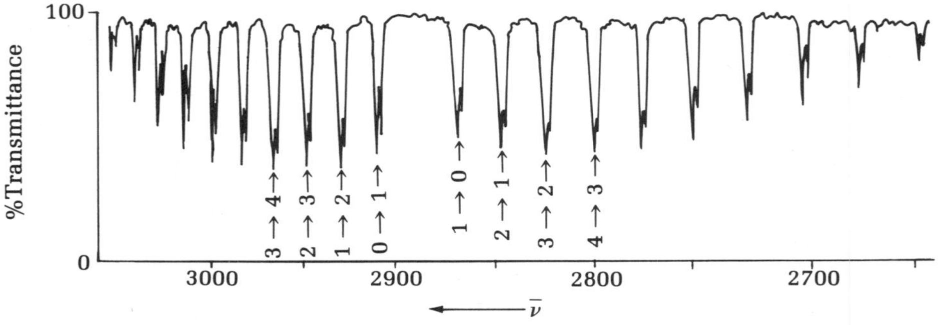

Figure 12–13
The infrared absorption spectrum of gaseous HCl. Each set of closely split inverted peaks is labeled with the initial and final rotational levels involved in the transition.

according to equation 12–8, this yields a different ΔE and $\bar{\nu}$ for $H^{37}Cl$ than for $H^{35}Cl$.

Since many molecules do not vaporize to any significant extent at room temperature, infrared spectra of these molecules are most commonly obtained when they are dissolved in some appropriate solvent. In solution, free rotation of the molecules under study cannot occur and the rotational splitting shown in Figure 12–13 is not observed. Furthermore, the molecules are in contact with solvent molecules and this complicates the picture of molecular vibration because the molecules do not vibrate independent of the solvent. The infrared absorption spectrum for HCl dissolved in a solvent such as CCl_4 is one relatively broad inverted peak with a maximum absorption at the frequency corresponding to the center of Figure 12–13 (at $\bar{\nu} = 2886$ cm^{-1}). With a molecule possessing more than one vibrational degree of freedom, a single broad absorption peak is found for each degree whose motion is excitable by electromagnetic radiation. Thus, for example, CO_2 in solution shows infrared absorptions corresponding to asymmetric stretch and bend, but not to symmetric stretch. The absorption for bond stretching appears to higher $\bar{\nu}$ than for bending, indicating the general conclusion that bending is easier to accomplish energetically than stretching.

Infrared Spectra in Determining Molecular Structure

For molecules of any significant complexity, the number of normal vibrational modes becomes very large and it appears hopeless to analyze the infrared spectrum of such a molecule in order to discern the motions that it reveals. Fortunately, this pessimistic outlook is not borne out completely in fact, and infrared spectroscopy has been used for many years to obtain valuable information about the identities and structures of molecules. Many modes can almost completely be described as the motion of one light atom with respect to a stationary collection of heavier atoms. Thus, for example, in a compound containing H—C bonds, one almost completely isolated motion is the movement of the H atoms in the stretching of this bond. Since H—C bond energies do not vary widely from compound to compound, all compounds containing H—C bonds are predicted to show infrared absorption for C—H stretching over a relatively narrow range of $\bar{\nu}$ values. Such is indeed the case, and depending on the compound, the absorptions occur anywhere between approximately 2850 and 3300 cm^{-1}. This range can be further subdivided since the C—H bond energy shows a small systematic variation depending on the bonding of the carbon to atoms other than the hydrogen in question. For example, CH_3 groups show absorptions at 2960 cm^{-1} for symmetric stretching and 2870 for asymmetric stretching, benzene rings show C—H stretching absorptions at 3030 cm^{-1}, and acetylenic (—C≡C—H) groups show C—H stretching absorption between 3270 and 3300 cm^{-1}, depending on the compound.

Another common atomic grouping that shows a characteristic absorption peak is the carbonyl group (C═O), which shows an absorption for stretching between 1580 and 1900 cm^{-1}. As with C—H stretch, the exact position of this peak depends on the additional bonding of the carbon, so that aldehydes, ketones, esters, and other compounds show their own small range for $\bar{\nu}$ for C═O absorption due to stretching. Thus, the infrared absorption spectrum is invaluable in identifying functional groups in molecules because these groups yield characteristic **group frequencies** in the infrared spectra of molecules. Table 12–1 lists a number of common group frequencies (in cm^{-1}).

Table 12–1
Some Common Group Infrared Absorption Stretching Frequencies

Group	Absorption Maximum (cm^{-1})
CH_3	2870, 2960
CH_2	2850, 2930
C═CH	3010–3090
Benzene ring	3030
C≡CH	3270–3300
OH (nonhydrogen-bonded)	3500–3700
OH (hydrogen-bonded)	3200–3700
NH (nonhydrogen-bonded)	3300–3500
C≡N	2250
C═O (aldehyde)	1695–1740
C═O (ketone)	1700–1730
C═O (carboxylic acid)	1680–1730
C═O (ester)	1720–1750
C—O	1000–1250
—C═C—	1600–1680
—C≡C—	2200–2260
CF	1000–1400
CCl	600–800

Figure 12–14 reveals how definitive infrared spectroscopy can be in revealing the presence or absence of particular functional groups in molecules. Figure 12–14(a) shows the infrared spectrum of 3-pentanone. The most prominent peak in this spectrum is the carbonyl stretching frequency at 1715 cm^{-1}. By contrast, the spectrum of diisopropyl ether in Figure 12–14(b) shows no peak of any significance in the carbonyl stretching region; it does show strong absorption in the C—O stretching region. If the molecular formulas of both compounds had been determined by analysis and a molecular weight measurement prior to the obtaining of their infrared spectra, there would be no doubt, from Figure 12–14(a), that the former contains its single oxygen as C═O, and, from Figure 12–14(b), that the latter contains its oxygen as C—O—.

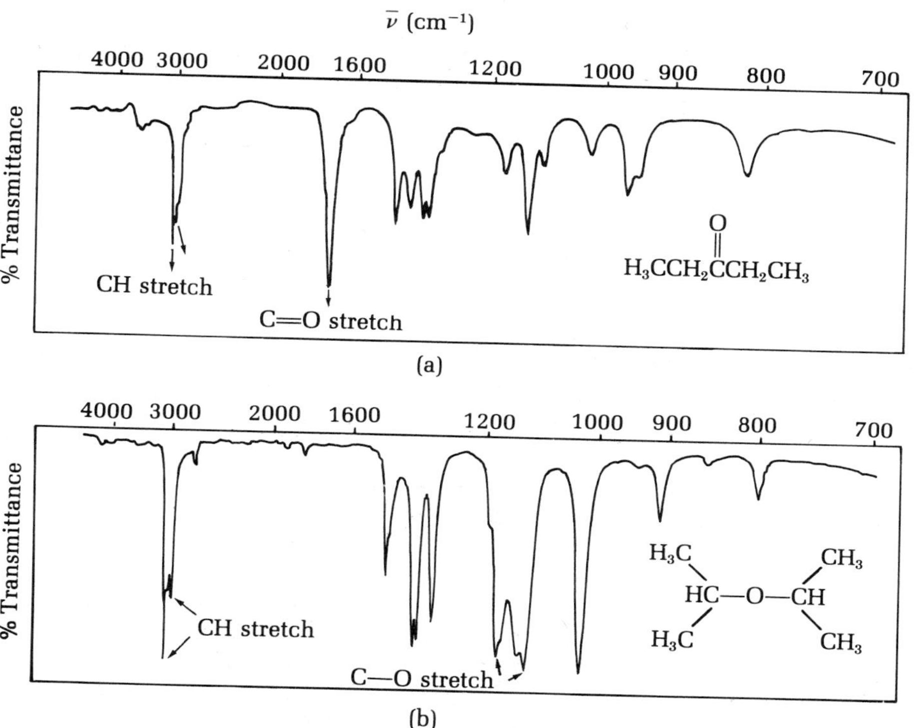

Figure 12–14 The infrared absorption spectra of liquid 3-pentanone and liquid diisopropyl ether. (Adapted from Silverstein, R. M., and Bassler, G. C., *Spectrometric Identification of Organic Compounds*, John Wiley and Sons, Inc., New York, 1964.)

12–4. Electronic Spectroscopy

Electronic Energy Levels

Absorption of electromagnetic radiation can occur by the displacement of an electron within a molecule. We can use equation 4–2 to obtain an estimate of a characteristic frequency for this type of absorption. The particle mass in this case is that of the electron and the distance a is of the order of an atomic diameter in most molecules. We will pick 2.0 Å as a in our calculation to yield the following value of $\bar{\nu}$:

$$\bar{\nu} = \frac{\Delta E_{1\to2}}{hc} = \frac{3h}{8ma^2c} = 2.1 \times 10^5 \text{ cm}^{-1}$$

The visible range extends from approximately 15,000–30,000 cm^{-1}, so that our characteristic frequency is in the ultraviolet region of electromagnetic radiation. The value of 2.0 Å is characteristic of localized bonding where the electrons are restricted to a relatively small region of space (small value of a). Molecules such as H_2, H_2O, CH_4, CCl_4, and acetone, which contain localized bonds, are transparent to visible light and they are colorless as a consequence. However, in delocalized systems, strong absorption in the visible region occurs if the number of atoms involved in the delocalized system is sufficiently large. A common example of such a colored, delocalized system is **vitamin A aldehyde,** whose structure is shown in Figure 12–15. This substance is reddish orange in color and it is the primary absorber of visible light in the rods of the retinal cells of the eye. The strong visible absorption arises from the extended conjugated system of the molecule. Many dyes used in the clothing industry are molecules containing extended conjugation, and consequently they have large effective a values so that $\bar{\nu}$ is in the visible range.

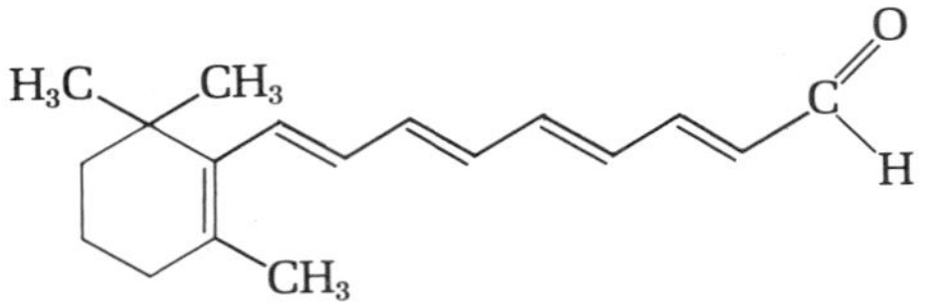

Figure 12–15
One canonical resonance structure of Vitamin A aldehyde.

We have already introduced to some extent the energy levels between which transitions occur in electronic spectroscopy (the energies of the molecular orbitals introduced in Chapter 5). Thus, for example, a possible absorptive transition in H_2 is from the $1s\sigma$ ground state molecular orbital to an excited state such as $1s\sigma^*$. This transition, which occurs in the ultraviolet region, leads to dissociation of H_2 into H atoms, since a molecule that contains one electron in a bonding orbital and one electron in an antibonding orbital contains no net bond. The absorption of ultraviolet light commonly produces bond dissociation in many molecules because of electronic transitions between bonding and antibonding molecular orbitals. For example, $H_2(g)$ and $Cl_2(g)$ do not react with each other at a measurable rate at room temperature if light is excluded. However, in the presence of sunlight they react explosively, to yield HCl(g). The role of the sunlight is to produce dissociation of $Cl_2(g)$ through a bonding-antibonding transition of one of the two bonding electrons; this results in a nonbonding species. The mechanism of the HCl formation consists of the following sequence of steps:

$$\begin{aligned} Cl_2(g) + h\nu &\rightarrow 2\ Cl(g) \\ Cl(g) + H_2(g) &\rightarrow HCl(g) + H(g) \\ H(g) + Cl_2(g) &\rightarrow HCl(g) + Cl(g) \end{aligned} \qquad \textbf{12–9}$$

This mechanism is one of a class termed **chain mechanisms,** which commonly occur in explosive reactions. In a chain mechanism, a highly reactive species (Cl in this case) is produced in an **initiation step.** This reactive species goes on to produce products, but it is regenerated in the process (in Step 3 in this case). This regeneration is the key to the chain, since a single Cl atom can result in the formation of a large number of HCl molecules.

A typical ultraviolet frequency is 60,000 cm^{-1}, which corresponds to a ΔE for the transition of 1.2×10^{-11} erg/molecule $= 1.7 \times 10^2$ kcal/mole, which is sufficient to dissociate all but the strongest bonds. The absorption of ultraviolet or visible light proves to be a convenient way to initiate a large number of chemical reactions, and the branch of chemistry concerned with such initiation is termed **photochemistry.**

Electronic Spectra

The selection rules we have examined for rotational and vibrational transitions are relatively simple. The rules are also relatively simple for electronic transitions in isolated atoms. There is no restriction on Δn, the change in the principal quantum number. The rule for Δl is $\Delta l = \pm 1$. This second rule was obscured in the H atom spectrum (Chapter 3) because all l states within a given n level possess the same energy in that atom. A third important selection rule in the case of all electronic transitions (atoms or molecules) is that the electron-spin pairing or unpairing must remain constant in the transition. Helium atoms, for example, contain two spin-paired electrons in the ground electronic configuration of $1s^2$. The allowed transition of lowest energy involves the promotion of one electron to a $2p$ orbital, so that the excited atom has the configuration $1s^1 2p^1$ but with the spins still paired.

The atomic electronic selection rules can be carried over to molecular orbitals to some extent; new ones must be added however, since atomic orbitals combine to form molecular orbitals that may be bonding, nonbonding, or antibonding. Table 12–2 lists a number of allowed molecular electronic transitions along with some transitions that are forbidden.

Table 12–2
Some Allowed and Forbidden Transitions between Molecular Electronic Orbitals

Allowed	Forbidden
$\sigma \rightarrow \sigma^*$	$\sigma \rightarrow \sigma$
$\sigma \rightarrow \pi$	$\sigma^* \rightarrow \sigma^*$
$\pi \rightarrow \pi^*$	$\pi \rightarrow \pi$
Nonbonding $\rightarrow \pi^*$	$\pi^* \rightarrow \pi^*$

The visible absorption of vitamin A aldehyde is due to a nonbonding $\rightarrow \pi^*$ (commonly symbolized as $n \rightarrow \pi^*$) transition involving one of the four nonbonding electrons of the aldehyde oxygen atom. Such $n \rightarrow \pi^*$ transitions are commonly the source of colors in dyes.

In transitions between electronic states, all of the vibrational selection rules are relaxed; that is, all normal modes can be observed and transitions can occur between a given vibrational level of the lowest electronic level and any of a large number of vibrational levels in the excited electronic level. Figure 12–16 shows potential energy curves for the lowest-energy level and one of the excited levels of the N_2 molecule. Since there are six bonding electrons in N_2, one electron can undergo a transition to an antibonding

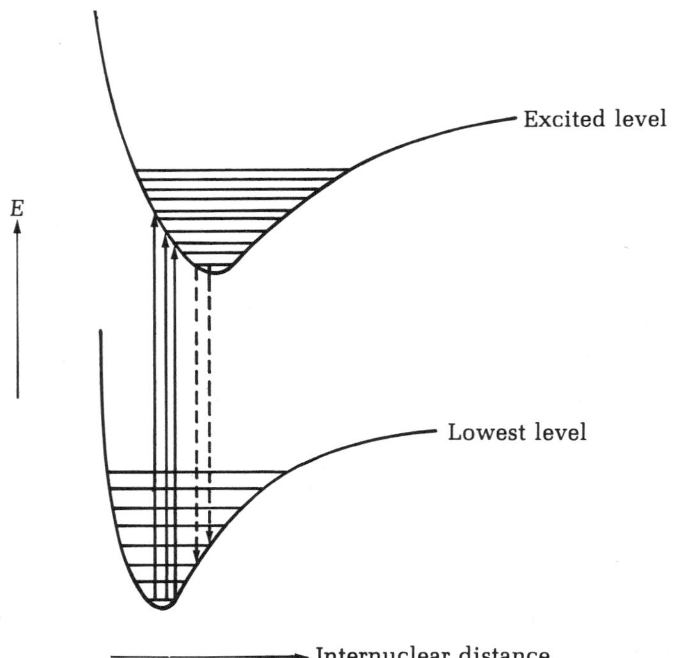

Figure 12–16
Some electronic transitions in the N_2 molecule. The solid vertical lines indicate energy absorption and the dashed lines indicate emission.

molecular orbital without bond dissociation. The horizontal lines inside each curve are vibrational energy levels.

An important rule that controls transitions as shown in Figure 12–16 is the **Franck-Condon** principle, which states that electronic transitions occur at fixed nuclear positions. This principle is a natural consequence of the relative masses of electrons and nuclei. During the time of the electronic transition, the nuclear motion is essentially zero. The transitions drawn in the figure must therefore be drawn vertically to reflect the lack of nuclear motion. Thus, transitions from the ground state occur with the greatest probability to excited vibrational states of the excited electronic state, and emission from the excited state occurs with high probability to excited vibrational levels of the ground electronic state, as shown.

Vibrational energy is very easily exchanged with neighboring molecules through collisions, so that very soon after electronic absorption (indicated by the solid vertical lines in Figure 12–16) the electronically excited molecule finds itself in the lowest vibrational level of that excited state. To return to the ground electronic state, it reemits radiation, as shown by the vertical dashed lines in the figure. The cycle of absorption, vibrational energy loss, and reemission describes the generation of **fluorescence,** such as the mechanism of a "black light." The black light emits invisible ultraviolet radiation, which is absorbed by the fluorescent substance, but as

Figure 12–16 shows the emission from the fluorescent substance is at a significantly smaller ΔE and $\bar{\nu}$ than the absorption. In many cases this emission is visible light when ultraviolet radiation is absorbed and this is the source of the glow produced by the black light.

If gaseous molecules are heated to a sufficiently high temperature, they emit radiation that can be analyzed to give a wealth of information about those molecules. An enormous number of lines are observed in such an emission spectrum since there are simultaneous electronic, vibrational, and rotational transitions. The complete emission spectrum of N_2 can be analyzed to obtain the N—N bond distance and bond energy, energies of the excited levels, and the detailed shape of the potential energy curves shown in Figure 12–16.

The vibrational and rotational detail is lost when the electronic absorption spectrum is obtained of a substance in solution. Instead, each electronic transition appears as an extremely broad absorption peak. A **visible-ultraviolet absorption spectrometer** is similar in many ways to the infrared spectrometer diagrammed in Figure 12–12. One of the key differences is that the visible-ultraviolet spectrometer uses the photoelectric effect (Chapter 3) in the detector to convert radiation intensity to electrical energy. Also, recorders on visible-ultraviolet spectrometers typically plot the absorption of radiation as a function of frequency (or wavelength) so that absorption peaks are not inverted as they are in infrared spectra. Figure 12–17 shows the predominant feature of the visible absorption spectrum of $Cu(H_2O)_4^{2+}$ in aqueous solution. The single broad peak that occurs at the visible-infrared boundary arises when a Cu^{2+} electron transfers from one 3d orbital to a 3d orbital of higher energy. The strong absorption in the red region of the visible range leads to the dominance of blue light in the visible light transmitted by this solution, and $Cu(H_2O)_4^{2+}$ yields sky-blue solutions.

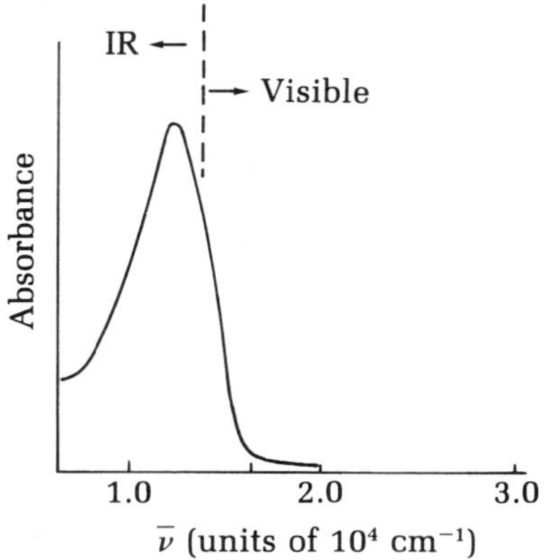

Figure 12–17 The most prominent portion of the visible absorption spectrum of $Cu(H_2O)_4^{2+}$ in aqueous solution.

12-5. Nuclear Magnetic Resonance

Magnetic Energy Levels

Up to this point, we have centered exclusively on the electric-field component of electromagnetic radiation and have ignored the magnetic field. Since electrons possess intrinsic magnetism because of their spin motion, atoms and molecules containing unpaired electrons interact with a magnetic field, as shown in Figure 4–11. The beam splitting occurs in the figure because of the two possible orientations of the electron magnet in the magnetic field, that is,

$$\begin{pmatrix} \mathrm{S} \\ | \\ \mathrm{N} \end{pmatrix} \quad \text{and} \quad \begin{pmatrix} \mathrm{N} \\ | \\ \mathrm{S} \end{pmatrix}.$$

If the electron beam passes slightly outside of the volume between the pole caps, those electrons with the former spin orientation will be attracted into the field while the latter will be repelled by the field. Thus, the two kinds of electrons have different potential energies due to the presence of the field: the former is lowered in energy through attraction by the field and the latter is raised in energy by repulsion. Figure 12–18 shows how these spin energies vary with increasing magnetic-field strength (symbolized by H and

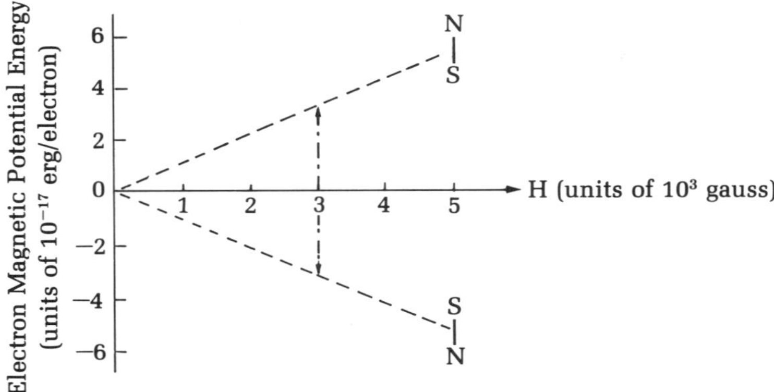

Figure 12–18 The variation in energies of the two electron spin states with applied magnetic field.

reported in **gauss**). The zero of electron magnetic potential energy has been defined to occur at a magnetic-field strength of zero gauss. It should be emphasized that H in Figure 12–18 is measured **at the position of the electron** since the field produced by a magnet is a function of the distance and direction away from the pole faces.

If an alternating magnetic field in the form of electromagnetic radiation is applied to electrons in a magnetic field, energy absorption occurs if the frequency matches $\Delta E/h$, where ΔE is the energy difference between the two spin states. The vertical line in the figure shows ΔE for a field of

3000 gauss, a field that is easily obtained with a small electromagnet. The ΔE is approximately 6×10^{-17} erg/electron and $\Delta E/h$ is approximately 9×10^9 Hz. This is a typical microwave frequency and any species containing unpaired electrons does indeed absorb microwave energy when placed in a strong magnetic field. The phenomenon of microwave absorption by unpaired electrons is termed **electron paramagnetic resonance** (EPR) and it is a powerful tool for studying the magnetic properties of matter.

We shall not dwell at length on EPR because it applies only to paramagnetic species, and most molecules do not contain unpaired electrons. However, all atoms and molecules do contain nuclei and these nuclei provide the basis for an extremely useful form of magnetic resonance. If electrons, as elementary particles, possess spin, there is no reason that nuclei cannot also, and indeed most nuclei are magnetic.

The most useful of the magnetic nuclei has been the proton because it appears in such a wide variety of compounds. Carbon and oxygen are also ubiquitous atoms, but neither the ^{12}C nucleus nor the ^{16}O nucleus acts as a magnet. The proton reveals two spin states in a magnetic field as the electron does, but the energy difference between the proton spin states is much smaller at a given H than with the electron. This is not surprising, in view of our discussion surrounding equation 4–2, since the proton is much more massive than the electron. At a magnetic field of 14,000 gauss (which is produced by a relatively large electromagnet), $\Delta E/h = 60 \times 10^6$ Hz = 60 MHz for **nuclear magnetic resonance** (NMR) energy absorption by protons. This is a typical radio frequency, and so are the characteristic NMR absorption frequencies for all other nuclei at this magnetic field. Table 12–3 lists these characteristic absorption frequencies for a number of nuclei along with the number of nuclear spin states encountered with each.

Nucleus	Spin States	Frequency (MHz)
^{1}H	2	60
^{2}H	3	9.2
^{13}C	2	15
^{14}N	3	4.3
^{17}O	6	8.1
^{19}F	2	56
^{31}P	2	24

Table 12–3
Characteristic NMR Absorption Frequencies for $H = 14{,}000$ Gauss and Numbers of Spin States for Several Nuclei

The Chemical Shift

The phenomenon of nuclear magnetic resonance would be little more than a laboratory curiosity were it not for the key fact that the NMR absorption frequency varies slightly for a given type of nucleus depending on its chemical environment in molecules. This arises because the NMR absorption frequency depends on the magnetic field **at the nucleus.** To a very small but very important extent, this field at the nucleus depends on the numbers and kinds of neighboring nuclei and on the numbers and motions of neighboring electrons. It is these neighbors that define the chemical environment, and hence there is a very good correlation between chemical environment and nuclear absorption frequency. Let us take protons as an example. Some chemically different types of protons are the following:

—C(=O)H   —O—H   >N—H   —C—H (with three bonds to C)

Protons in these four environments show four different radio absorption frequencies. This makes NMR an extremely powerful tool for the elucidation of molecular structure.

Figure 12–19 is a diagram of an NMR spectrometer. The electromagnetic radiation is generated by a radio-frequency oscillator and radiated to the

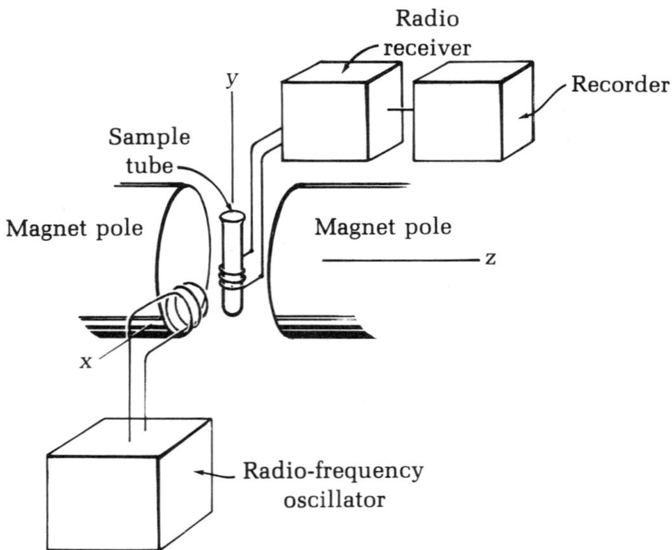

Figure 12–19
Diagram of a nuclear magnetic resonance spectrometer.

sample by a set of radio transmitter coils mounted on either side of the sample (only one of these coils is shown in the figure). The substance under investigation is studied as either a pure liquid or dissolved in some appropriate solvent (preferably a solvent that does not contain the nucleus under study). Wrapped around the sample (contained in a narrow glass tube) is a radio receiver coil. The absorption of energy by the nuclei results in their reorientation between

$$\begin{pmatrix} N \\ | \\ S \end{pmatrix} \text{ and } \begin{pmatrix} S \\ | \\ N \end{pmatrix}.$$

This reorientation is analogous to turning a magnet inside the coil of an electrical generator, and an electrical signal is induced in the receiver coil, shown in Figure 12–19. This signal is then amplified and passed to a recorder, where it is plotted as a function of the radio frequency.

As we have stated, the variation in NMR absorption frequency with chemical environment is small. For example, almost all proton absorptions for a sample in a field of 14,000 gauss occur in the range of frequencies 60 MHz ± 1,000 Hz. Thus, the absorptions of protons are well isolated from the absorption of the other types of nuclei listed in Table 12–3. If absorptions are observed near 60 MHz at 14,000 gauss, they can only arise from protons.

Figure 12–20 shows the proton NMR spectrum of toluene. The two peaks arise because there are two chemically distinct kinds of protons, the five

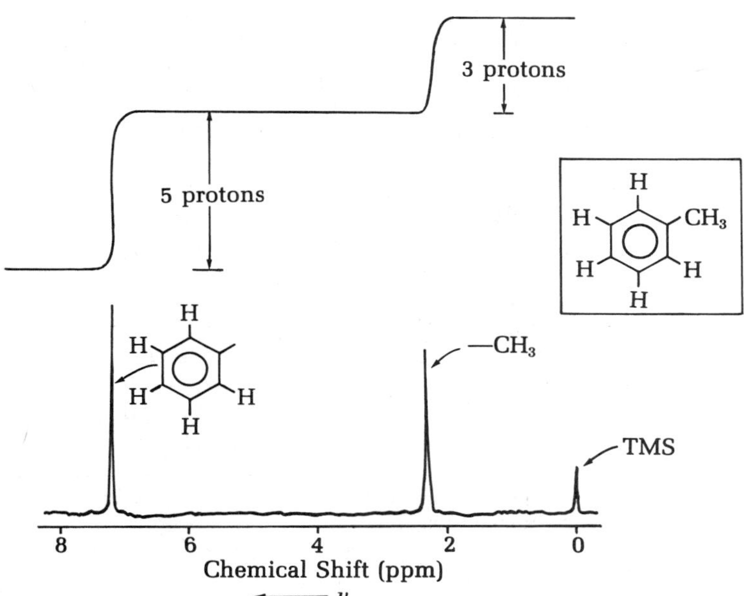

Figure 12–20 The proton NMR spectrum of toluene.

benzene ring protons and the three protons of the methyl group. Nuclear magnetic resonance spectra also yield in a simple manner the relative number of nuclei giving rise to the various peaks, since the peak areas are proportional to those numbers. Almost all NMR spectrometers are designed to measure peak areas in addition to recording the peaks; the steplike trace in Figure 12–20 is this graphic area measurement for toluene. The height of the step for the taller peak is five-thirds that for the smaller peak, and this results in the assignment of the two peaks as shown. The small peak at zero on the abscissa scale arises from the twelve chemically equivalent protons of tetramethyl silane (TMS), which is commonly used as a standard to which other peak positions are referred. The abscissa scale in the figure is termed the **chemical shift** and symbolized by δ; δ is defined as follows:

$$\delta = \frac{10^6(\nu_{\text{peak}} - \nu_{\text{TMS}})}{\nu_{\text{TMS}}} \tag{12-10}$$

or it is the displacement of the peak to higher frequency from the TMS peak as measured in parts per million (ppm). Thus, the toluene methyl peak occurs at a frequency 2.3 ppm = 140 Hz higher than the TMS peak. The use of a standard allows proper reporting of peak positions without the necessity of measuring and reporting radio frequencies to seven or eight significant figures. Tetramethyl silane was chosen as the standard because very few types of protons absorb at frequencies below it, and hence most chemical shift values are positive.

Table 12–4 shows the ranges of chemical shift values for various proton chemical environments. For example, the first entry shows that the chemical

Table 12–4
Proton Chemical Shifts (δ, ppm) for Some Common Chemical Environments

Environment	Shift
$H_nC—C—C$[a]	0.8– 1.0
$H_nC—C—OH$ (or —OR)	1.2– 1.4
$H_nC—\phi$[b]	2.3– 2.9
$N_nC—Cl$	3.0– 4.0
$H_nC—OH$ (or —OR)	3.4– 3.9
$H_nC{=}C$ (nonaromatic)	5.0– 7.0
$H—\phi$	7.0– 8.0
$H—C{\equiv}C$	2.3– 2.9
$H—C{=}O$	9.5–10.0
$HO—C{=}O$	10.4–12.0

[a] The subscript n stands for H_3C, H_2C, or HC.
[b] The symbol ϕ denotes an aromatic ring.

shift for most saturated hydrocarbon protons is between 0.8 and 1.0. The symbol n denotes H_3C, H_2C, and HC. The chemical shift values within a given range increase, in general, from H_3C to H_2C to HC. The toluene peaks in Figure 12–20 are at 2.3 and 7.2 ppm from TMS and these values fall into the ranges $H_nC—\phi$ and $H—\phi$, respectively. In the case of the methyl peak, it lies at the lower end of its range since it is H_3C instead of H_2C or HC.

Peak Splitting

Figure 12-21 shows the proton NMR spectrum of acetaldehyde, which reveals a common and immensely useful feature of NMR spectra in general.

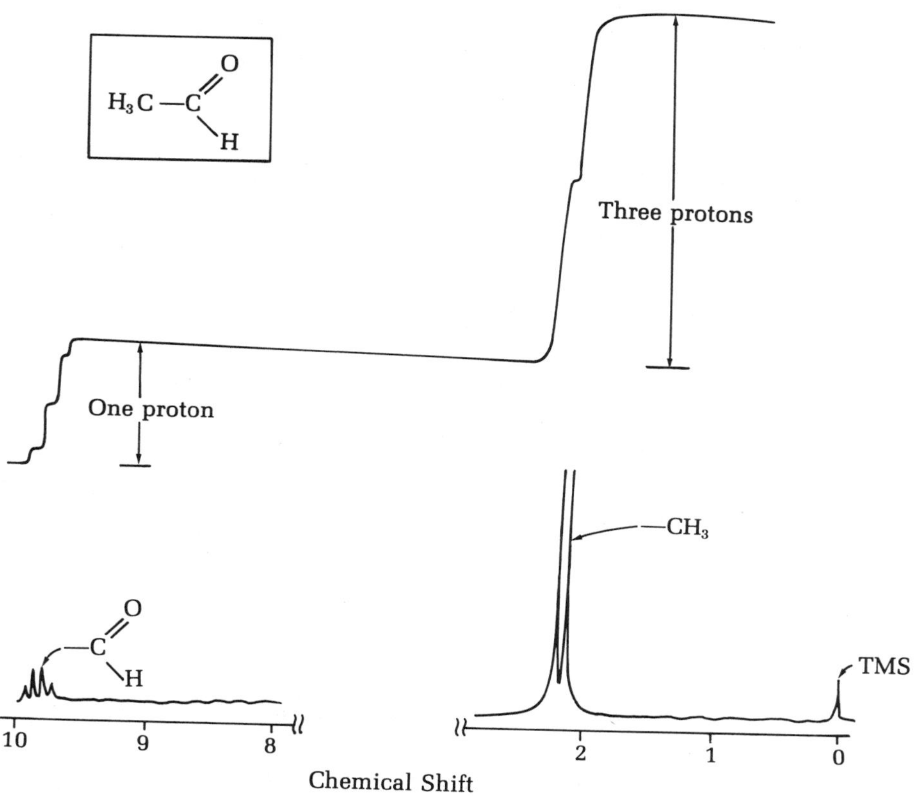

Figure 12-21
The proton NMR spectrum of acetaldehyde.

Acetaldehyde contains two chemically distinct kinds of protons, namely those of the methyl group and those of the aldehyde group. Thus, there should be two absorption peaks. However, the figure shows that each of these peaks is split, one into a **doublet** (two peaks) and the other into a **quartet** (four peaks). Furthermore, the distance between the two peaks of the doublet is exactly the same as the distance between adjacent peaks of the quartet, indicating that the peak splitting arises from an interaction between the methyl protons and the aldehyde protons. Actually, we should expect such an interaction because protons must experience the magnetic fields due to neighboring magnetic nuclei and these fields must be reflected in the absorption frequencies. Since protons can adopt only two orientations in a magnetic field, the number of possible fields and absorption frequencies due to the interaction is very limited and easily analyzed. Figure 12–22 shows these orientations by arrows for the three methyl protons and the single aldehyde proton. The aldehyde proton can have two orientations, and consequently the methyl group proton peak in Figure 12–21 is a doublet. The methyl group protons can yield four different fields at the aldehyde proton, and the ratio of the number of times each field occurs is the ratio

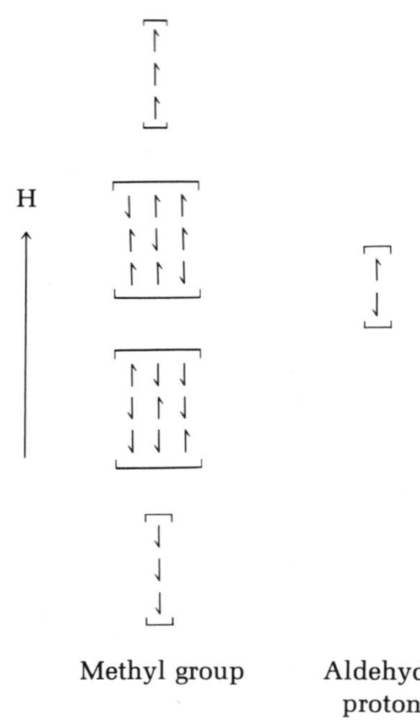

Figure 12–22 All possible spin orientations of the protons of acetaldehyde.

of the numbers of ways the field can be produced, as shown by the brackets in Figure 12–22. Thus, the aldehyde absorption consists of a quartet of peaks with peak areas in the ratio 1:3:3:1, as shown by the steplike trace in the upper portion of Figure 12–21.

If we follow the same sort of reasoning for two equivalent protons, we see that they will result in a triplet for a neighboring proton, and the area ratio for the triplet is 1:2:1.

At this point two questions may well have occurred to you. The first concerns the lack of splitting in the CH_3 peak of acetaldehyde due to interactions between the three identical protons. These protons do interact strongly with each other, but **chemically equivalent protons never split the NMR absorption peaks of each other.** The reason for this lack of splitting is well understood but beyond our scope. Nevertheless, it is a very important fact to be kept in mind when interpreting NMR spectra (as you are asked to do at the end of this chapter).

The second question concerns the definition of "neighboring" nuclei for purposes of the splitting of NMR absorption peaks. Except in rare instances, nuclei do not split the peaks of each other if more than three chemical bonds separate them, and with three bonds or less the peak separations decrease as the number of bonds increase.

A Molecular Structure Determination

Let us now apply all of these considerations to the elucidation of a molecular structure from a nuclear magnetic resonance spectrum. The compound in question possesses the molecular formula $C_4H_{10}O$, and its proton NMR spectrum is given in Figure 12–23. This formula corresponds to either an ether or an alcohol, and the possible structures are the following:

$H_3CCH_2CH_2CH_2OH$
(a)

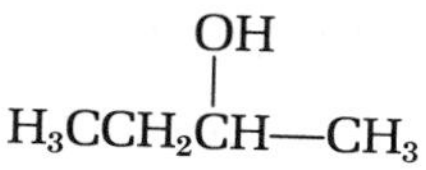

$H_3CCH_2CH(OH)—CH_3$
(b)

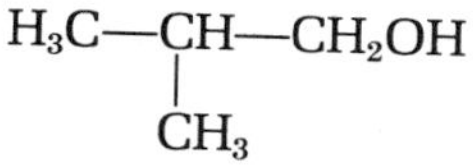

$H_3C—CH(CH_3)—CH_2OH$
(c)

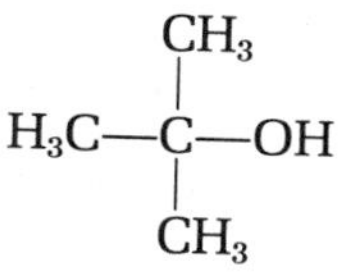

$H_3C—C(CH_3)_2—OH$
(d)

$H_3CCH_2CH_2OCH_3$
(e)

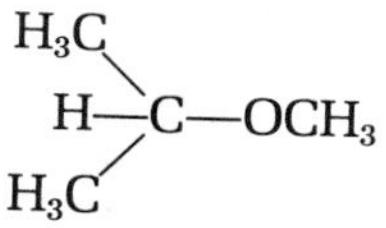

$(H_3C)_2CH—OCH_3$
(f)

$H_3CCH_2OCH_2CH_3$
(g)

Structure (a), 1-butanol, has five chemically nonequivalent types of protons (CH_3, OH, and three nonequivalent CH_2 groups). As a consequence, the spectrum of 1-butanol is much more complex than the relatively simple triplet and quartet observed in Figure 12–23. In fact, the only alcohol containing only two kinds of protons is 2-methyl-2-propanol, structure (d). The protons in (d) are in the number ratio 9:1, while the area ratio in Figure 12–23 is 3:2. Furthermore, there is no peak splitting in the 2-methyl-2-propanol spectrum since the CH_3 and OH protons are separated by four bonds. Ethers (e) and (f) contain four and three chemically distinct kinds of protons respectively, and both yield a large unsplit peak for the OCH_3 protons. This leaves diethyl ether as the only possible compound. Diethyl ether contains two kinds of protons, the H_2C and the H_3C, in the form of two equivalent ethyl groups. The triplet and quartet in Figure 12–23 are centered at 1.2 and 3.4 ppm, respectively. These positions are consistent with the assignments of the CH_3 groups of diethyl ether as the source of the triplet and the CH_2 groups as the source of the quartet according to the characteristic shifts listed in Table 12–4. The CH_3 and CH_2 protons of the same ethyl group are separated by three bonds, and consequently the CH_2 absorption is a quartet due to the interaction with three equivalent protons and the CH_3 absorption is a triplet because of the two neighboring protons. Finally, the peak area ratio of the triplet/quartet is predicted to be $\frac{6}{4} = 1.5$ from the relative numbers of CH_3 and CH_2 protons in diethyl ether, and this is the observed ratio in the figure.

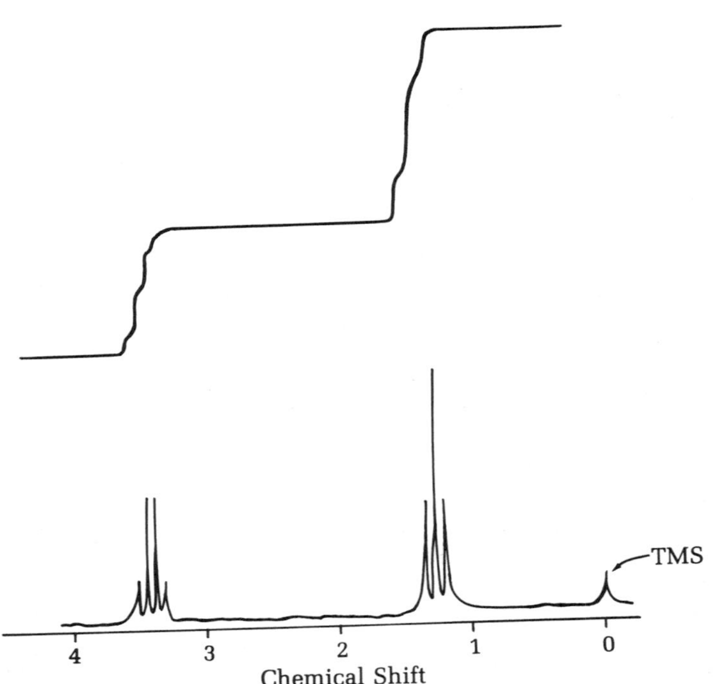

Figure 12–23 Proton NMR spectrum of one of the isomers of $C_4H_{10}O$. The ratio of peak areas for triplet : quartet is 3 : 2.

It is clear that for large, complex molecules, the NMR spectrum will also be complex since the number of chemically nonequivalent protons will be large. Nevertheless, NMR spectra have yielded very valuable structural information concerning molecules with molecular weights as high as tens of thousands of atomic mass units.

SUGGESTIONS FOR ADDITIONAL READING

Silverstein, R. M., and Bassler, G. C., *Spectrometric Identification of Organic Compounds*, 2nd ed., Wiley, New York, 1967.

Barrow, G. M., *The Structure of Molecules*, Benjamin, New York, 1963.

PROBLEMS

1. The rotation in Figure 12–3(a) does not respond to electromagnetic radiation. Explain why.
2. Gaseous methane does not respond rotationally to electromagnetic radiation, but gaseous H_2O does respond. This difference provides an explanation for a phenomenon commonly observed in the operation of a microwave oven. In a microwave oven, water is quickly heated to its boiling point; yet a piece of polyethylene is hardly heated at all in the same period of time. Explain this strange difference.

*3. The normal modes of vibration of a molecule must be independent of each other in order to be degrees of motional freedom. Show that none of the three normal modes of water can be obtained as a combination of the other two.

4. How many translational, rotational, and vibrational degrees of freedom does each of the following molecules possess in the gaseous state: (a) CF_4; (b) PF_5; (c) BeF_2?
5. The molecule BF_3 is planar and triangular, as we showed in Chapter 6. It possesses six vibrational degrees of freedom. By analogy with Figures 12–3 and 12–4, deduce and show with arrows a normal vibrational mode of this molecule that is not excitable by electromagnetic radiation. Show a vibrational mode that is excited and show the orientation of the electric field that excites it.
6. The ammonia molecule is pyramidal. Draw the symmetric stretching mode for this molecule. Is this motion excitable by electromagnetic radiation?
7. Figure 12–6 depicts the transfer of rotational energy to translational energy. What type of energy transfer is depicted in Figure 2–18?
8. Gamma rays arise from transitions between energy levels inside nuclei, and the frequencies of gamma rays are larger even than frequencies of X-rays. Explain the extremely large values of ν for gamma rays based on equation 4–2.
9. The C_v value of H_2 in Table 2–2 is less than the C_v value for any other diatomic molecule. Propose an explanation.

*10. The lowest vibrational energy of a molecule cannot be zero. Show that this is demanded by the Heisenberg uncertainty principle.

11. One of the naturally occurring isotopes of hydrogen is deuterium (symbolized as D), in which the nucleus contains a neutron as well as a proton so that its atomic weight is almost exactly twice that of H. Which molecule possesses the larger vibrational energy in its ground vibrational state, HCl or DCl? The curve of potential energy as a function of r_{AB} may be taken as identical for HCl and DCl. Which molecule has the larger bond energy?

12. The heat capacities of most gaseous molecules increase with increasing temperature. Propose an explanation.

13. The vibrational contribution to C_v and C_p for HCl exceeds that for N_2, CO, and O_2. How can that be when the reduced masses of all three molecules are significantly greater than the reduced mass of HCl?

14. The average atomic weight of chlorine is 35.45 amu. What are the relative percentages of ^{35}Cl and ^{37}Cl in a natural sample of HCl? In the narrowly separated peak pairs in Figure 12–13, the larger peak always appears to higher $\bar{\nu}$. Show that this is consistent with the relative percentages of ^{35}Cl and ^{37}Cl calculated above.

15. The absorption frequency for the CO stretching motion of carbon monoxide is shifted to a significantly lower value when the CO is bonded to a metal in a metal carbonyl. Refer to the discussion of the bonding in metal carbonyls in Chapter 11 and explain this shift.

16. Table 12–1 shows that typical OH stretching frequencies are higher than NH stretching frequencies, which are in turn higher than those of CH stretch. Why should this be so?

17. Table 12–1 shows that the stretching frequencies for hydrogen-bonded OH groups occur significantly below the stretching frequencies of free OH groups. Can you explain why?

18. The ion $Mn(H_2O)_6^{2+}$ is almost completely colorless in aqueous solution. It contains five *d* electrons and is a high-spin ion. Explain why this ion is colorless when other transition metal hexaquo complexes, such as $Cr(H_2O)_6^{3+}$ and $Ni(H_2O)_6^{2+}$, are highly colored.

19. Figure 12–16 shows an electronic transition of the N_2 molecule. Refer to Table 12–2 and suggest an assignment of electrons to molecular orbitals in the excited state in the figure. Make the excited state the lowest possible in energy.

20. If a pure *cis*- or *trans*-substituted ethylene is dissolved in some suitable solvent and exposed to sunlight for several days, it is converted to a mixture of the *cis* and *trans* isomers. Explain in terms of a transition between electronic energy levels how this *cis-trans* conversion takes place.

21. Benzene is colorless, but nitrobenzene is yellow. In fact, most aromatic compounds containing the nitro group (—NO_2) are colored. Explain

how the addition of a nitro group to benzene results in color. *Hint:* Consider the electronic transition in each case.

22. Refer to the discussion in Chapter 8 concerning the color of nitrogen dioxide. What electronic transition is responsible for the color of NO_2?
23. What is the characteristic proton NMR absorption frequency in a magnetic field at 23,000 gauss?
24. At a given magnetic field such as 3,000 gauss, the gap between the two electron-spin energy levels is of the order of 10^3 times as great as the gap between the proton-spin energy levels. Explain this in terms of equation 4–2.
25. Figure 12–22 shows the possible orientations of a set of three equivalent protons in a magnetic field. These orientations result in a 1:3:3:1 ratio of peak areas for a quartet as the NMR absorption pattern of a nucleus near these three protons. Construct a similar orientation diagram for four equivalent protons. What pattern do they yield for the NMR absorption of a neighboring nucleus?
26. The proton NMR spectrum in Figure 12–24 is that of a nitrated hydro-

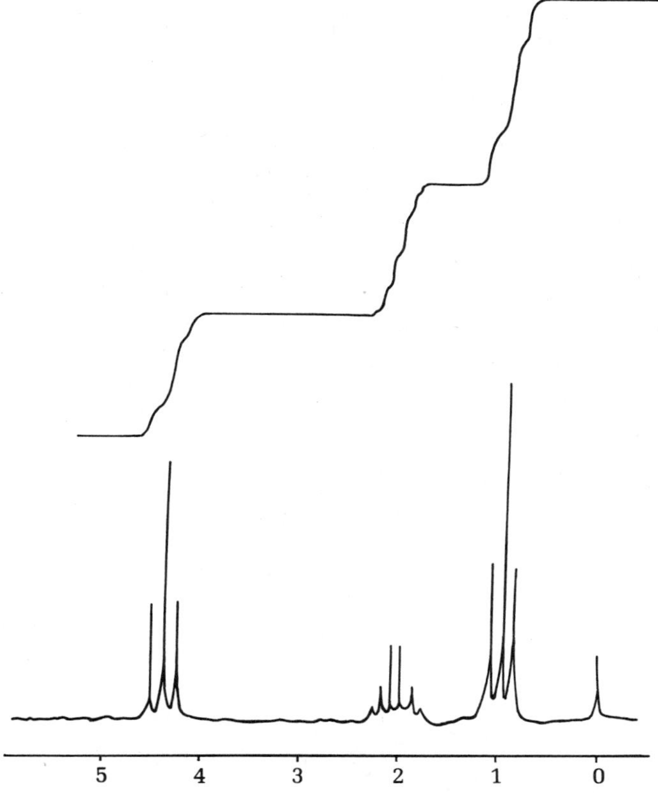

Figure 12–24
Proton NMR spectrum and area trace for an isomer of $C_3H_7NO_2$.

carbon with the molecular formula $C_3H_7NO_2$. Draw the structure that yields the proton spectrum in the figure. (The nitrogen nucleus does not cause peak splitting.)

27. The proton NMR spectrum in Figure 12–25 is that of an isomer of $C_4H_8O_2$. Draw the structure that yields this spectrum.

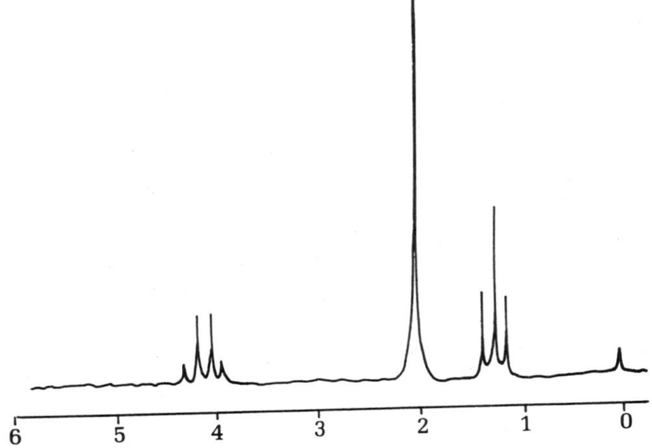

Figure 12–25
The proton NMR spectrum of an isomer of $C_4H_8O_2$.

13/ Models of Solid Substances

All crystalline solids are characterized by an orderly array of atoms, molecules, or ions. One such array is that of sodium chloride, shown in Figure 5–14. An appropriate question to ask is, "How do scientists discover the structures of the orderly arrays in crystals?" We have seen the bases for such arrays as those of diamond, graphite, and black phosphorus in the directed character of covalent bonding. Those arrays and others that are held together by covalent bonding have structures that can be predicted by applying principles of covalent bonding and molecular structure. There are additional factors that determine arrays in ionic and metallic bonding and also the bonding for solids whose units are discrete small molecules. The arrays and the bonding that produces them can furnish good explanations for the physical properties of solids.

13–1. The Structures of Solids

X-ray Diffraction

We can discover the array of atoms, molecules, or ions in any given solid by bombarding that solid with X-rays. In order to see, in principle, how this can be done, we must investigate the diffraction of electromagnetic radiation. Figure 3–5 shows the generation of electromagnetic radiation by a radio antenna. Figure 13–1 shows two such antennae in which the electron motions occur in phase; that is, the maximum and minimum electron displacements in each antenna are reached at the same time. The dashed line shows a wave emanating from the left antenna, and the black solid line is a wave from the right antenna. In Figure 13–1(a) the antennae are exactly one wavelength apart, so that an electric-field minimum from the left antenna arrives at the right antenna just as the right antenna emits a wave minimum. As a consequence, the two waves reinforce each other perfectly at all points as both waves travel further to the right. The combined wave is shown in color in the figure. In Figure 13–1(b), the right antenna has been moved to the right by an additional one-half wavelength. Now the two waves meet at the right an-

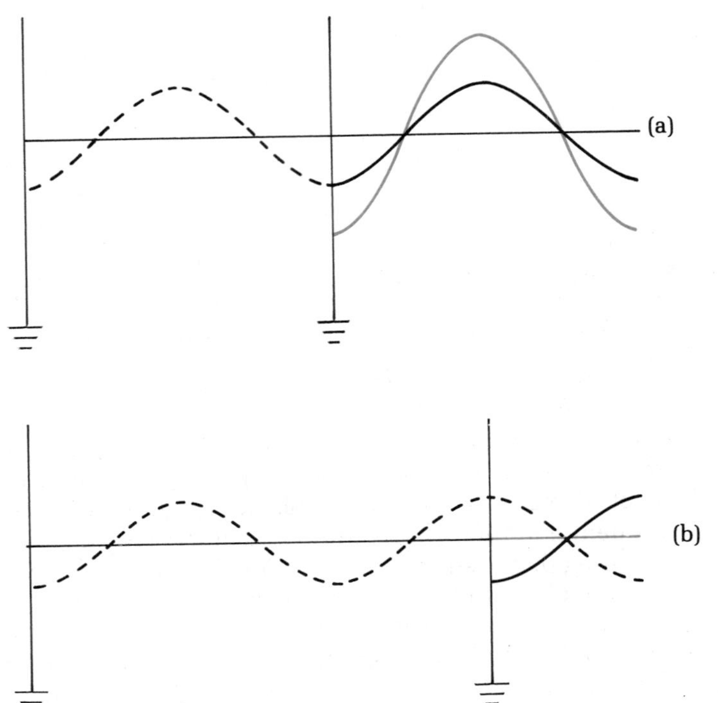

Figure 13–1
(a) Reinforcement and (b) cancellation of waves emitted by two antennae.

tenna with exactly opposite electric-field vectors. The resulting horizontal colored line to the right shows complete cancellation of the two waves.

With this picture of reinforcement and cancellation in mind, let us consider the interaction of electromagnetic radiation with the orderly array of a crystalline substance. We have already considered many of the ways in which matter absorbs radiation. The scattering of radiation by matter is another important phenomenon. In Figure 3–5 we showed that electromagnetic radiation results in a vertical motion of the electron to the right of the antenna. This electron itself becomes an antenna because of the vertical motion, and the result is the scattering of the radiation incident on the electron. This is better shown by the top view of Figure 3–5, shown in Figure 13–2. Radiation travels out in a circular pattern from the antenna, much as water waves on the surface of a pond into which a stone has been dropped.

Figure 13–2
Top view of the emission of radiation from a radio antenna and the secondary radiation from an electron to the right of the antenna. The concentric colored circles represent maxima in the electric field at a given time.

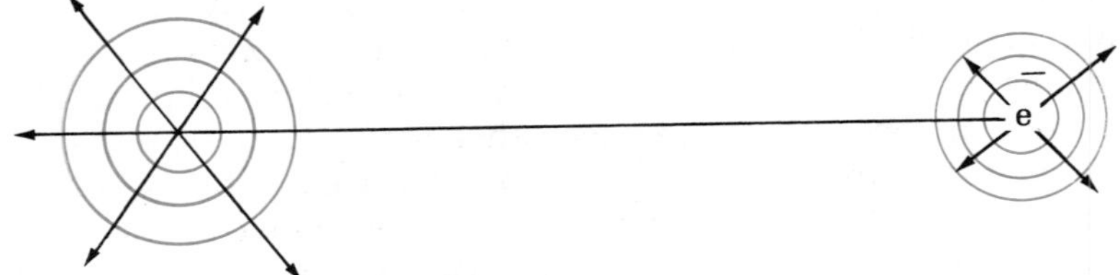

The concentric circles shown in the figure represent maxima in the electric field vector (wave crests in Figure 13–1) at some fixed time. To the right of the antenna the waves encounter the electron and set it in motion, and a second circular wave pattern comes from the electron. This is analogous to the water waves on a pond striking a protruding rock, which then serves as a new source of a circular wave pattern. All matter serves to scatter electromagnetic radiation by this mechanism. Any sizable sample of bulk matter contains a huge number of secondary sources of radiation, namely all of the electrons and nuclei. We should expect the pattern of scattered radiation to resemble the waves on a pond surface containing many protruding rocks; that is, it should be a completely chaotic pattern. There is an important exception to this expectation, however, if the scatterers are arranged in an orderly pattern relative to each other. Let us consider the array shown in Figure 13–3. Radiation of the wavelength λ shown is directed at the array at the angle θ between the direction of wave propagation and the surface of the

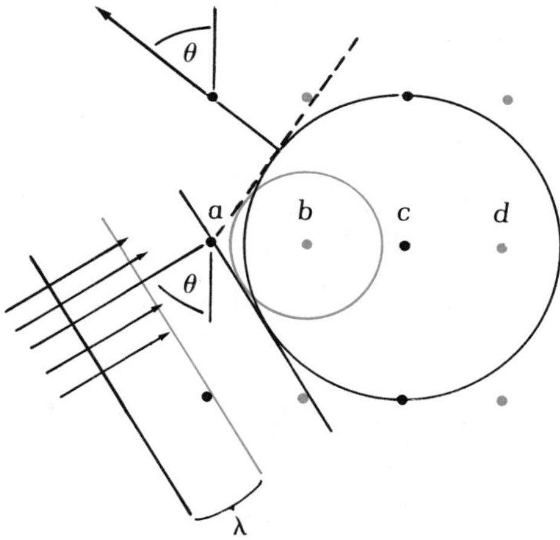

Figure 13–3
Reinforced scattering of radiation from an ordered molecular array.

array. In the figure, a crest of the wave (maximum electric field) has just arrived at the molecule labeled *a*, and a circular crest of scattered radiation immediately emanates from molecule *a*. The incident crest continues through the array, diminished somewhat in magnitude, and strikes molecule *b*, which also emits a circular crest. The figure shows circular crests around molecules *b* and *c* that have been generated by crests in advance of the one shown striking molecule *a*. We have indicated every other column of molecules in color and have placed a colored circle with a colored scatterer and a black circle with a black scatterer. This has been done only for clarification. All of the scattering molecules are intended to be identical. Every scatterer at any given time is surrounded by a large number of circles that have been generated by successive crests of the incident X-ray beam striking the scatterer. For purposes of clarity, most of these circles have been deleted from Figure 13–3.

The radius of the colored circle around b is exactly equal to λ and the radius of the circle around c is exactly 2λ. Under these circumstances, the circles have a common tangent, shown as the dashed line. This line is a front of reinforcing crests. This picture could be continued for d and so on, and we should find that scattering from all molecules in horizontal rows in the figure yields reinforced scattering that leaves the array in a direction perpendicular to the dashed front, as shown by the arrow in the figure. The orderly arrangement shown leads to a strong **reflection** of the incident radiation from the array at an angle of reflection θ equal to the incident angle θ.

Figure 13–3 shows only one of several angles θ for which strong reflection occurs. The critical condition for strong reflection is that the radius of the colored circle shown be $(n/2)\lambda$, where n is any integer. In Figure 13–3, n equals 2. Figure 13–4 shows that this radius equals $d \sin \theta$, where d is the

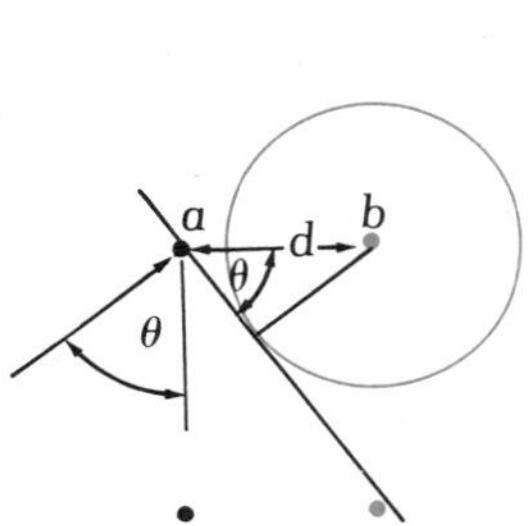

Figure 13–4 Trigonometric basis of the Bragg equation. The figure is a portion of Figure 13–3, and the radius of the colored circle is λ. Angles θ are equal because the lines which define them are mutually perpendicular.

spacing between the vertical columns of molecules in Figure 13–3. Thus, the condition for strong reflection is

$$n\lambda = 2d \sin \theta \tag{13–1}$$

Equation 13–1 was first formulated early in this century by English scientists William and Lawrence Bragg, father and son, and it is known as the **Bragg equation.** The general phenomenon in which strong reflection occurs only for a limited set of incidence angles is termed **diffraction.**

We are unaware of any diffraction phenomenon for the reflection of visible light from solids under normal circumstances. Clear images can be seen in a mirror no matter what the viewing angle is. For an explanation, we shall first center on a situation in which visible light is diffracted.

We have already referred to the diffraction of visible light in connection with the diffraction grating in Figure 3–7. The grating is a plate of transparent material containing a set of very closely ruled parallel lines. A view of the plate along these lines is given in high magnification in Figure 13–5. The lines are directed in and out of the paper, and the separation between the adjacent openings between lines is d.

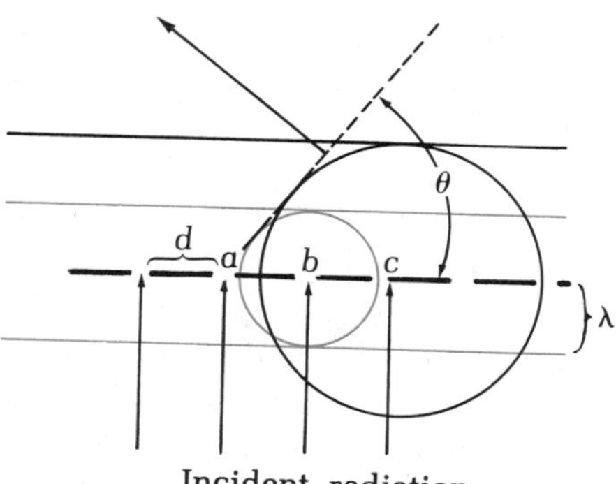

Figure 13–5
Diffraction of visible light by a ruled grating of lines.

Visible light (or infrared or ultraviolet radiation) is directed at the grating as shown. The narrow openings between dark lines serve as radiation scatterers, and circular wave patterns emanate from each as shown. The circles around openings b and c are crests (electric field maxima), as in Figure 13–3. Alternate crests of the incident light are shown as the alternating colored and black horizontal lines. A black crest has just struck the grating at the time shown. The colored circle around opening b was generated by the colored line above the grating and the black circle around c by the black line above the grating. The radius of the colored circle is λ and the radius of the black circle is 2λ. In this case, the circles have a common tangent that includes opening a, and the wave is strongly reinforced in the direction defined by θ in the figure.

In the case of the diffraction grating, the general requirement for reinforcement is that the radius of the circle around b must be an integral number of wavelengths, or

$$n\lambda = d \sin \theta \tag{13–2}$$

Since θ depends on λ, the passage of visible light through the grating results in a splitting of the light into its component wavelengths and colors. Violet light possesses the shortest λ (approximately 3500 Å), and θ is smallest for violet light compared with all visible light. Red light possesses the longest λ (approximately 7500 Å) and it yields the largest θ. Equation 13–2 reveals that diffraction cannot occur if λ is larger than d since n cannot be smaller than 1 nor can $\sin \theta$ be larger than 1. The distance between adjacent scatterers in solids is of the order of 10^{-8} cm, and λ for visible light is larger than 10^{-5} cm. Thus, diffraction is not observed in the reflection of visible light from a mirror.

The Bragg equation shows that λ must be smaller than $2d$ for diffraction to occur. Since $2d$ in crystals is of the order of 10^{-8} cm, crystalline diffraction is limited to X-rays and cosmic rays. Since X-rays may be conveniently generated in high intensities, this is the form of radiation used in practice. As a general rule with either equation 13–1 or equation 13–2, λ and d should be of the same order of magnitude to obtain the best diffraction. Since X-rays of $\lambda \cong 10^{-8}$ cm are easily generated, this provides another reason why X-rays are used for crystalline diffraction.

Crystalline Substances and Amorphous Substances

The diffraction of X-rays by a solid depends on the arrangement of the molecules in an orderly array. Consequently, the information obtained from diffraction experiments may be used to deduce the structure of a given array.

How do we identify a substance having an ordered array in the first place? Consider the array shown in Figure 13–6. Suppose that we take a measurable quantity of a substance with this array and try to compress it. Solids and

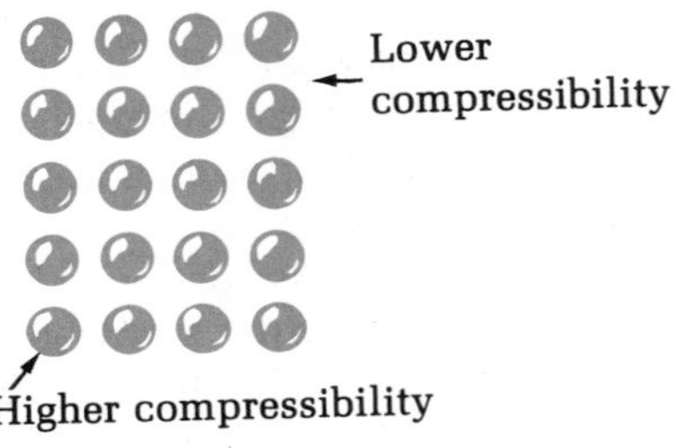

Figure 13–6 Illustration of the anisotropy in the compressibility of an orderly molecular array.

liquids have very low compressibilities because of high intermolecular repulsion at short intermolecular distances; but the compressibilities are measurable nevertheless. Figure 13–6 reveals that the substance should be more easily compressible if the pressure is applied diagonally than if it is applied horizontally, since the molecules that are forced together are initially further apart along the diagonal. This dependence of a physical property on direction is termed **anisotropy,** and it should occur with many properties of orderly arrays. In practice, the most convenient property is the absorption of plane-polarized light, which changes as the ordered substance is rotated. A key difference between solids and liquids is that liquids do not show anisotropy of their physical properties, and thus have no long-range order.

We tend to think of solids as substances that hold a given shape even when subjected to considerable stress. Thus, a piece of quartz and a piece of window glass would both be termed solids. However, the piece of quartz is anisotropic and the glass is not. One of the most dramatic demonstrations of this difference is trying to cleave each substance by placing a sharp knife on it and hitting the knife with a hammer. If this is done properly with the quartz, the substance will split neatly in two, showing two new, perfectly flat surfaces. This is what is expected of an array like that shown in Figure 13–6, where such cleavage can occur between any adjacent rows (or planes, in a three-dimensional figure).

Glass will shatter when cleavage is attempted and no flat surface can ever be so obtained (a flat window pane is obtained by cooling molten glass on a flat surface). Instead, curved surfaces are always obtained. Glass does not possess long-range molecular order; and it is not really a solid at all, but a very viscous liquid.

There is an even more obvious distinction between quartz and glass—the reproducibility of symmetry that is present in quartz and lacking in glass. Quartz as it occurs naturally in the earth commonly forms rods having six flat sides, as shown in Figure 13–7. These sides form a regular hexagon, which can be viewed from the top of the rod, with an angle θ between sides of 120°. The word *crystal* was first introduced to apply to substances that show such reproducible symmetry. Glass is an amorphous substance and does not possess such symmetry.

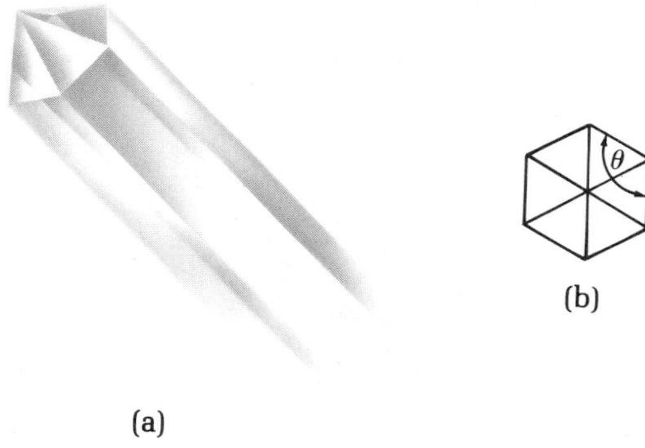

Figure 13–7
A portion of a quartz crystal: (a) side view, (b) top view.

Lattices and Unit Cells

Crystalline symmetry is a natural consequence of an orderly molecular array, and the details of the crystal symmetry should result from a similar symmetry on the molecular level. For example, if we continued the square array of Figure 13–6 into three dimensions the entire structure would consist of a repetition of identical cubes. The angle between adjacent cubical faces is 90°, and this angle will also appear between the large, flat faces of the crystal. Under no circumstances could the array of Figure 13–6 ever yield a reproducible crystal angle of 120°, because no such angle exists on the molecular level. Thus, the molecular arrangement in quartz must be something different from the arrangement shown in Figure 13–6.

How many possibilities are there for the orderly arrangement of identical units in space? First of all, some comment must be made about the nature of the identical units of crystals. In a crystal made up of identical atoms, these units are the atoms (examples are metallic iron and metallic copper). In a crystal made up of distinct molecules, these units are the molecules. In an ionic crystal, the units are collections of ions having the formula of the compound. Thus, in NaCl they are pairs of Na^+ and Cl^- ions, and in CaF_2 each unit consists of one Ca^{2+} ion and two F^- ions. This ion pairing is not a description of the bonding (there are no NaCl molecules in solid NaCl), but the arrangement of ions in space may be described in terms of ion pairs.

Fortunately, and perhaps surprisingly, there are only fourteen possible arrangements of identical units in space. The simplest of these fourteen is the extension of the square arrangement of Figure 13–6 into three dimensions to make a cubical arrangement. Such a regular three-dimensional array is termed a **lattice.** Each lattice consists of a collection of identical **unit cells,** and Figure 13–8 shows the unit cell for a simple cubic lattice. Each repeating unit is represented by a sphere. The sides of the unit cell cut each of the eight spheres at the corners of the cube into eighths, so that each unit cell contains $8(\frac{1}{8})$ = one sphere = one repeating unit. The three-dimensional lattice is produced by packing the unit cells side by side along the x, y, and z directions.

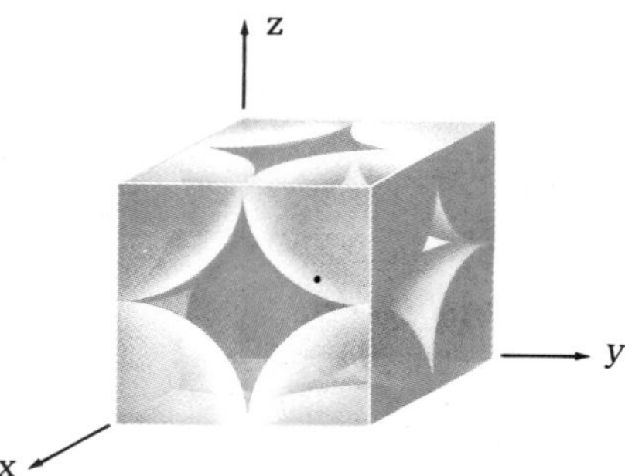

Figure 13–8
A simple cubic unit cell.

Each of the fourteen possible unit cells is characterized by three lengths—a, b, and c—and three angles—α, β, and γ—as shown for the cubic cell in Figure 13–9. The relationships of lengths and angles for the fourteen unit

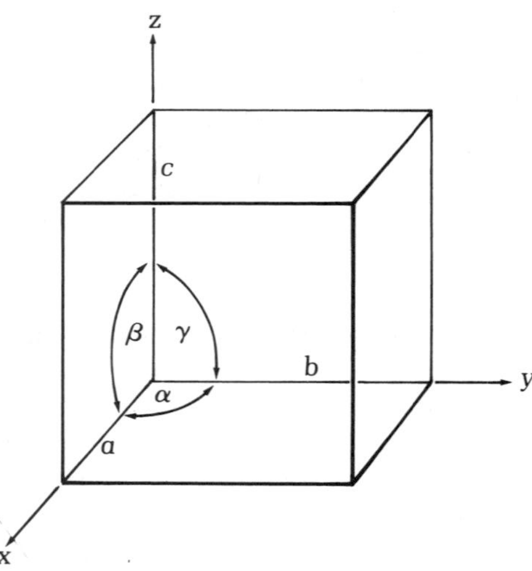

Figure 13–9
The lengths and angles which define a cubic unit cell.

cells are summarized in Table 13–1. In addition, a shape such as the cubic shape yields three unit cells, depending on how the repeating units are placed within the cell. The simple cubic cell has units at each corner, while

Shape	Unit Cell	Lengths	Angles
Cubic	Simple Body-centered Face-centered	$a = b = c$	$\alpha = \beta = \gamma = 90°$
Tetragonal	Simple Body-centered	$a = b \neq c$	$\alpha = \beta = \gamma = 90°$
Orthorhombic	Simple Body-centered Face-centered End-centered	$a \neq b \neq c$	$\alpha = \beta = \gamma = 90°$
Rhombohedral	Simple	$a = b = c$	$\alpha = \beta = \gamma \neq 90°$
Monoclinic	Simple End-centered	$a \neq b \neq c$	$\alpha = \gamma = 90° \neq \beta$
Triclinic	Simple	$a \neq b \neq c$	$\alpha \neq \beta \neq \gamma \neq 90°$
Hexagonal	Simple	$a = b \neq c$	$\alpha = \beta = 90°; \gamma = 120°$

Table 13–1
Unit Cell Length and Angle Relationships for Each of the Fourteen Three-Dimensional Unit Cells

body-centered cubic adds one in the exact center of the cube. Face-centered cubic has six additional units over simple cubic, one in the center of each of the six faces. Figure 13–10 shows the fourteen unit cells. The spheres in Figure 13–10 have not been clipped by the sides of the unit cells as shown in Figure 13–8, but you should keep in mind that units that are located at boundaries between unit cells are clipped.

Any crystalline substance, regardless of the size and complexity of the molecules, is composed of one of fourteen unit cells, as shown in Figure 13–10.

Structure Determination from X-ray Diffraction Data

X-ray diffraction information can be gathered for a given crystalline substance and this information then translated into information about the structure of the material. The area of chemistry and physics that involves the use of X-rays to probe crystal structure is termed **X-ray crystallography.**

Our principal calculation device is equation 13–1 and the distances d calculable from measured θ values (termed the Bragg angles) for reinforced scattering of X-rays. Each lattice possesses a huge number of sets of parallel planes, and each set possesses its own orientation in space and its own d

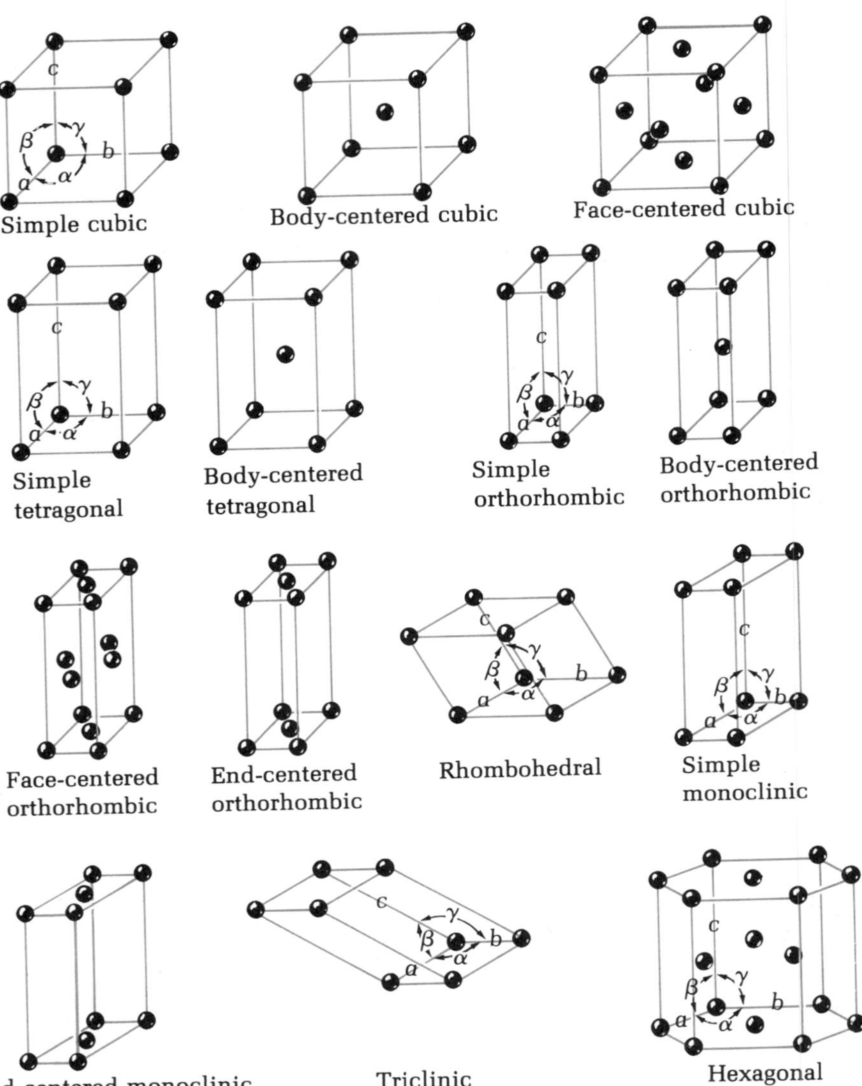

Figure 13–10
The fourteen three-dimensional unit cells.

value. Each of the sets gives rise to a number of Bragg angles in which reinforced X-ray scattering is observed. These correspond to the different values of the integer n of equation 13–1. Each of the fourteen unit cells yields a characteristic pattern of θ values, and this pattern can be used to determine which unit cell corresponds to a given crystal and to obtain the values of the unit cell parameters a, b, c, α, β, and γ.

Figure 13–11 shows an experimental arrangement for obtaining X-ray diffraction data. A beam of X-rays is passed through a small hole in a lead

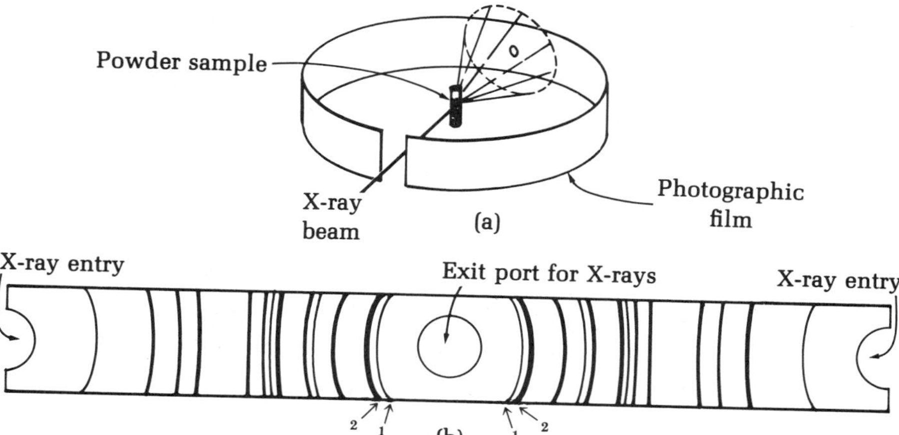

Figure 13-11
The powder method of X-ray diffraction. (a) Experimental arrangement, (b) film from diffraction by NaCl powder. (Adapted from Moore, W. C., *Physical Chemistry*, Prentice-Hall, Englewood Cliffs, N.J., 1973.)

plate (lead is an excellent absorber of X-rays) and directed at a powdered crystal sample contained in a glass tube, as shown in Figure 13-11(a). The purpose of the powder is to provide a huge range of crystal orientations about the X-ray beam. All possible Bragg angles for all possible sets of parallel planes are represented by one or more tiny crystals, and all possible strong reflections are observed at the same time. The resultant diffraction pattern is a series of cones, one of which is shown in Figure 13-11(a). The diffracted X-ray beam strikes the cylindrically wound photographic film, yielding a set of lines. Figure 13-11(b) shows the photographic film with the X-ray diffraction pattern of powdered NaCl. The pattern of lines shown is easily converted to a pattern of Bragg angles. It is beyond our scope to examine the relation of unit cell and pattern of θ values. Thus, we shall simply state that the pattern shown in Figure 13-11(b) is that of a face-centered cubic lattice and the unit cell length is 5.66 Å.

Figure 13-12 shows the placement of Na^+ and Cl^- ions in the NaCl structure. The sodium ions are located at each of the corners and at the centers of each of the faces of the cube shown. Thus, the Na^+ ions alone form a face-centered cubic array. The chloride ions are located at the centers of each edge of the cube and at the body center of the cube. Each chloride ion can be paired with each sodium ion, as shown in color in Figure 13-12. The array of these pairs is face-centered cubic. For example, the pairs numbered 1, 2, 3, 4, and 5 form the top face of a face-centered cubic array. The cell length is the Na^+—Na^+ distance shown in the figure.

How many Na^+—Cl^- pairs are contained in a unit cell? We can represent every NaCl pair by a sphere (regardless of what the actual shape is) and each corner of the cell contains one-eighth of a pair while each face center contains one-half of a pair. The number of pairs per unit cell is $8(\frac{1}{8}) + 6(\frac{1}{2}) = 4$.

We can use this value along with the value of the unit cell length to calculate Avogadro's number. First we must measure the density of crystalline

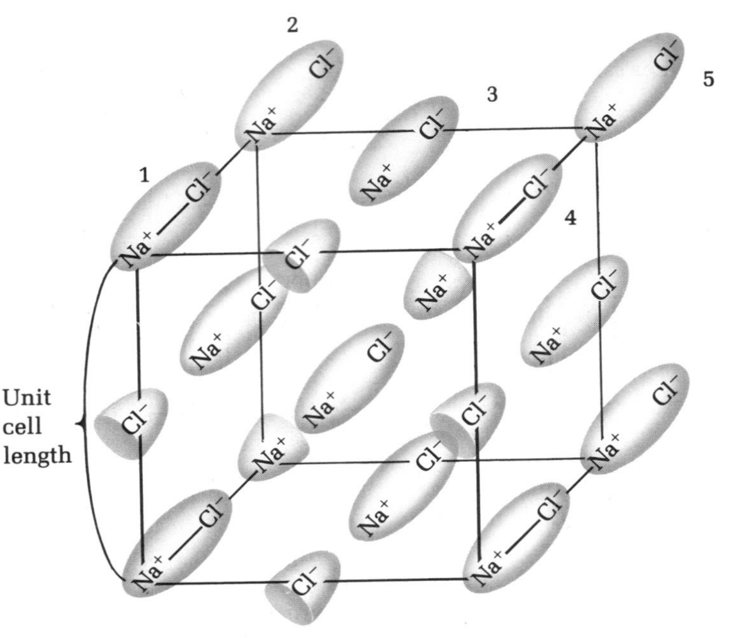

Figure 13–12
The face-centered cubic arrangement of sodium and chloride ion pairs in solid sodium chloride. Pairs 1, 2, 3, 4, and 5 form one face of a face-centered cube.

sodium chloride, which is found to be 2.165 g/cm^3. The volume of the unit cell is $(5.66 \times 10^{-8})^3 = 1.81 \times 10^{-22}$ cm^3. The molecular weight of NaCl is $23.00 + 35.45 = 58.45$ amu $= (58.45/N)$ g. The NaCl density is expressible as the mass of the unit cell divided by the volume of the unit cell:

$$2.165 = \frac{4(58.45)}{N(1.81 \times 10^{-22})}$$

which yields N as 6.0×10^{23}. With a precise value of N determined from any substance of known molecular weight, the preceding calculation can be reversed to obtain the molecular weight from the density of the solid, together with the information concerning the unit cell.

From the pattern of Figure 13–11(b), we know how the NaCl pairs are arranged relative to each other, but we do not know the locations of each of the ions. These are determined from relative intensities of the lines on the film. For example, it is highly significant that the two lines labeled 1 in the figures are faint while those labeled 2 are dark. In advanced study, precise ionic positions can be determined from the intensities of these lines. The intensities of diffraction lines are the keys that allow us to define the atomic positions in even very complex molecules. Within the past twenty years, even biological molecules with molecular weights as high as 20,000 have had their atomic positions determined in the crystalline state through X-ray crystallography.

13-2. Why Given Crystal Structures Occur

At this stage we know how any given crystal and molecular structure can be determined, but we do not know any of the factors that determine which of the possible structures any given species adopts. For example, why doesn't NaCl adopt the simple cubic lattice?

The Structures of Metals

Sodium chloride is an example of multicenter bonding just as all ionic solids and metals are. In such a case, the guiding principle to structure is the packing of as many atoms or ions around any given atom or ion as will maximize the extent of the bonding. In the case of the metallic elements, the structure should be that of the most efficient packing of identical spheres.

We are all aware of the most efficient way of packing together spheres in a single layer, that is, the triangular arrangement shown in Figure 13-13. A second layer is then added to the first, represented by the transparent spheres in the figure. These new spheres are located over the centers of equilateral triangles in the first layer, and the centers of the four spheres in

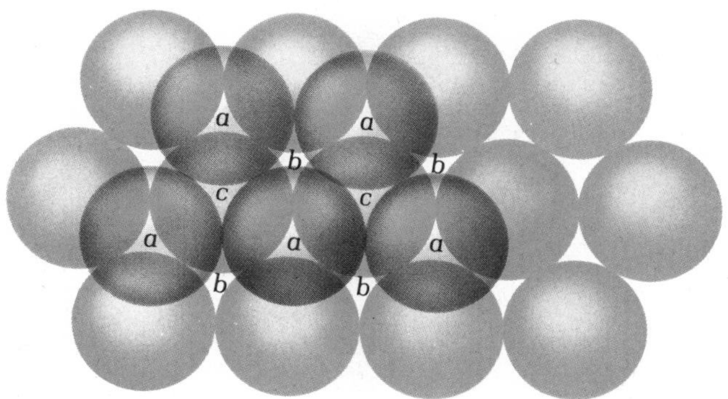

Figure 13-13 Layers of closest-packed spheres.

mutual contact form the four corners of a tetrahedron. From the figure we see that there are two identical sets of triangular centers in the first layer, those labeled *a* and those labeled *b*. The second layer of spheres can be located above only one of these sets of centers, and in the figure this is set *a*.

When a third set of spheres is added on top of the second, there are again two different possibilities. The third set of spheres can occupy sites *b* or the sites *c* directly over the centers of the first layer of spheres. These two possibilities for the third layer define two quite different lattices with different unit cells. If sites *c* are chosen, we have the arrangement whose side view is shown in Figure 13-14, where the three layers have been separated so that the spheres can be better displayed. The unit cell is the hexagonal cell and the entire arrangement of spheres is called **hexagonal closest-packing.**

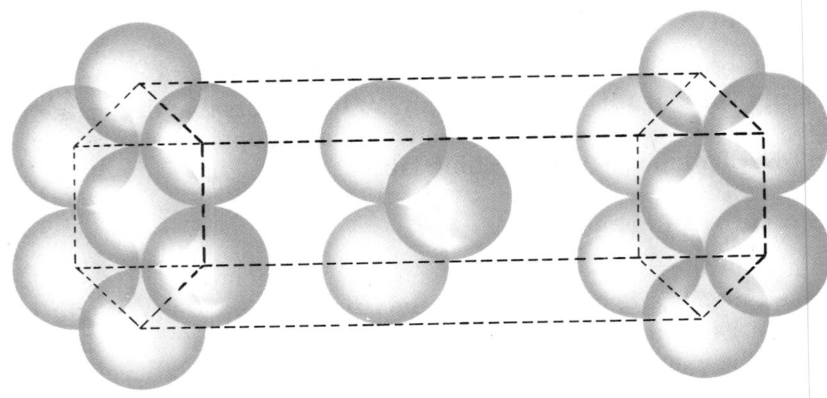

Figure 13–14 The unit cell for hexagonal closest-packing of spheres.

If the third layer of spheres occupies sites *b* of Figures 13–13, we have the arrangement shown in Figure 13–15, where the entire lattice has been expanded to display all of the spheres. This figure shows four layers of spheres.

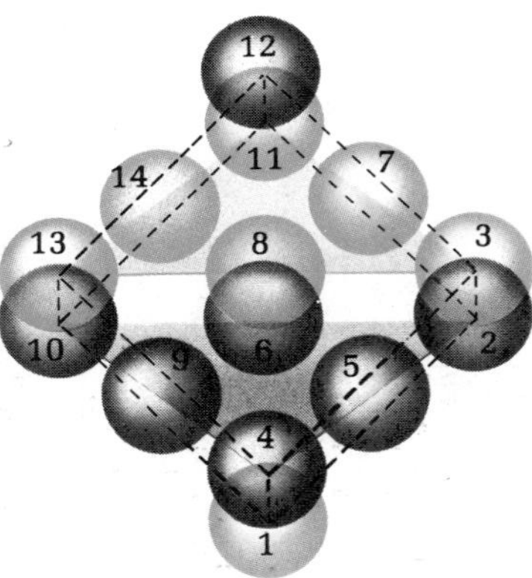

Figure 13–15 The unit cell for cubic closest-packing of spheres.

Sphere 1 belongs to the bottom layer; spheres 2, 4, 5, 6, 9, and 10 belong to the second layer; spheres 3, 7, 8, 11, 13, and 14 belong to the third layer; and sphere 12 belongs to the fourth layer. The pattern of site occupation is second layer *a*, third layer *b*, and fourth layer *c*, and this *abc* pattern then repeats through successive layers. The unit cell in this closest-packed lattice is face-centered cubic. Figure 13–16 shows a clearer view of the face-centered cubic unit cell, and the numbering of the atoms in Figures 13–15 and 13–16 is identical. The vertical direction in Figure 13–15 is the body diagonal connecting molecules 1 and 12 in Figure 13–16.

We predicted that solid metallic elements would adopt closest-packed structures. This is found to be the case with approximately two-thirds of the metallic elements. The remainder adopt a structure with a body-centered cubic unit cell.

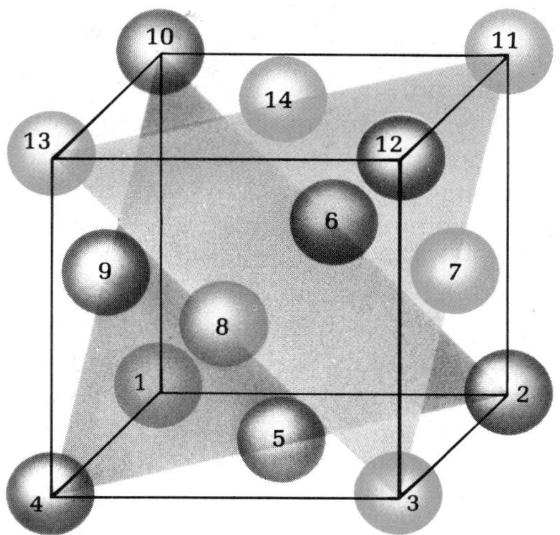

Figure 13-16
A face-centered cubic array. The numbering corresponds to that in Figure 13-15.

The Structures of Ionic Crystals

Structural considerations are more complicated with ionic crystals than with metals because the cations and anions, in general, possess different sizes. From theoretical considerations of the factors influencing ionic sizes, together with a large number of experimentally determined internuclear distances in ionic crystals, Linus Pauling constructed a table of ionic radii; this table is now in common use. Some of these radii are given in Table 13-2. The sum of radii for Na^+ and Cl^- from the table is $0.95 + 1.81 = 2.76$ Å, as compared with the experimental value of 2.83 Å. This is due to small

Ion	Radius (Å)	Ion	Radius (Å)
Li^+	0.60	Fe^{2+}	0.75
Be^{2+}	0.31	Co^{2+}	0.72
O^{2-}	1.40	Ni^{2+}	0.69
F^-	1.36	Cu^+	0.96
Na^+	0.95	Zn^{2+}	0.74
Mg^{2+}	0.65	Br^-	1.95
Al^{3+}	0.50	Rb^+	1.48
S^{2-}	1.84	Sr^{2+}	1.13
Cl^-	1.81	Ag^+	1.26
K^+	1.33	Cd^{2+}	0.97
Ca^{2+}	0.99	I^-	2.16
Sc^{3+}	0.81	Cs^+	1.69
Cr^{3+}	0.64	Ba^{2+}	1.35
Mn^{2+}	0.80	Hg^{2+}	1.10

Table 13-2
Pauling's Radii for Some Elemental Ions

variations in the effective ionic radii from one cation-anion pair to another, and the Pauling radii represent average values.

For a given charge, the radii in the table show the expected increase with increasing atomic number within any given column of the periodic table. Thus, for example, the radii increase in going from Li^+ to Na^+ to K^+ to Rb^+ to Cs^+. The effect of charge on ionic radii is very dramatic. Even an anion such as F^-, with a quite small atomic number, is very sizable, and larger than many cations of much greater atomic number.

Because of the attraction between anions and cations due to their opposite charges, the controlling principle in ionic crystal structures should be: **Cations should be found in contact with as many anions as possible; anions should be found in contact with as many cations as possible; cations should not be found in contact with other cations; and anions should not be found in contact with other anions.**

Figure 5–14 shows that each sodium ion is in contact with six chloride ions (the spheres have not been drawn in contact in the figure; it was desired thus to show the ionic positions more readily) and each chloride ion is in contact with six sodium ions. Furthermore, in each case the six neighbors define the six corners of an octahedron.

Sodium chloride is a close relative to the cubic closest-packed structure because both yield the face-centered cubic unit cell. The NaCl structure may be considered to be an expanded closest-packed array of chloride ions with sodium ions in **octahedral holes,** where an octahedral hole is a vacancy formed by six chloride ions that define the corners of an octahedron. Examination of Figure 13–16 shows that the centers of all twelve edges and the body center of the face-centered cubic unit cell are octahedral holes. In the NaCl crystal, all of these holes are filled with sodium ions.

What is the size of an octahedral hole relative to the spheres that define it? Figure 13–17 shows an octahedral site made by the closest-packing of spheres, and it also shows the largest sphere that can occupy this site with-

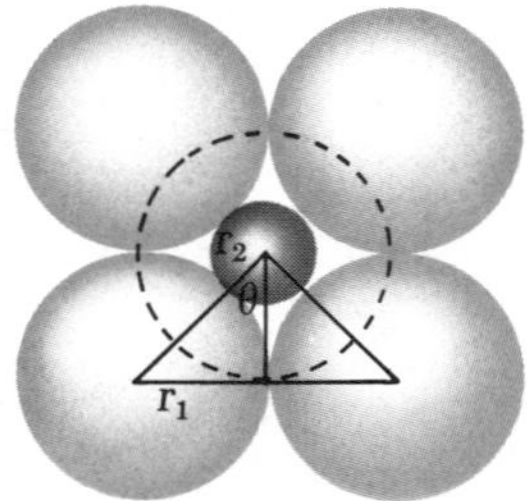

Figure 13–17
Small sphere in an octahedral site made by the closest-packing of larger spheres. The dashed sphere lies above the plane of the paper and there is another sphere below the plane of the paper.

out pushing the closest-packed spheres apart. The angle θ shown in Figure 13–17 is 45° and the sine of this angle is given by the equation

$$\sin \theta = \frac{r_1}{r_1 + r_2}$$

where r_1 and r_2 are the radii of the large spheres and small sphere respectively. Since $\sin 45° = 1/\sqrt{2}$, this yields

$$r_2 = (\sqrt{2} - 1)r_1 = 0.414\ r_1$$

The ionic radii of Na^+ and Cl^- are 0.95 Å and 1.81 Å, respectively, so that the radius ratio is $0.95/1.81 = 0.52$. Thus, the structure of NaCl does not contain closest-packed chloride ions. Of course there is no reason why it should have chloride ions in contact with each other. To the contrary, our principle says it should not.

This analysis shows that Na^+ can certainly accommodate six Cl^- ions around it such that all six chlorides are in contact with the sodium ion and none of the chlorides are in contact with each other. Why isn't each Na^+ surrounded by eight Cl^- ions in a cubic arrangement instead; and each Cl^- by a cube of Na^+ ions? This would appear to be more consistent with our principle of maximum cation-anion contact. In fact there may be an arrangement that permits even more than eight nearest neighbors. Figure 13–18 shows the atomic positions in the structure of crystalline cesium chloride, CsCl. The lattice has been expanded for clearer viewing of the ions. The small, darkened spheres are cesium atoms and the large colored spheres are chloride ions. Both the Cs^+ and Cl^- ions alone form a simple cubic array. Each cesium ion is in contact with eight chloride ions and each chloride ion is in contact with eight cesium ions. For example, the chloride ions numbered 2 to 9 are in contact with the cesium ion numbered 1. Each Cs^+ occupies a **cubic hole**, and there are eight such holes in the dashed cube in Figure 13–18. The number of chloride ions in the dashed cube shown is $8(\frac{1}{8})$ for the corners plus $12(\frac{1}{4})$ for the centers of edges plus $6(\frac{1}{2})$ for the centers of faces plus one for the body center for a total of eight. Thus, the formula of the compound is indeed CsCl.

How large is a cubic hole? That is, what is the relative radius of the largest sphere that can occupy a cubic site made by the contact of larger spheres? The large spheres meet along the edge of the cube that they define; so the edge length is $2r_1$. The large spheres are in contact with the small sphere along the body diagonal, which is $\sqrt{3}$ times as large as the edge length. Thus,

$$\sqrt{3}(2r_1) = 2r_1 + 2r_2$$

or

$$r_2 = (\sqrt{3} - 1)r_1 = 0.732r_1$$

If NaCl were to adopt the CsCl simple cubic structure, the chlorides would be in contact with each other and the sodium ions would not fill the cubic hole; that is, they would not be in contact with the chlorides. What is the radius ratio for Cs^+ to Cl^- that permits the occupation of cubic holes? From Table 13–2, it is $1.69/1.81 = 0.93$, and all eight chloride ions are in contact with the Cs^+ ion and no Cl^- is in contact with another Cl^-.

What structure should we expect for a radius ratio less than 0.414? At a

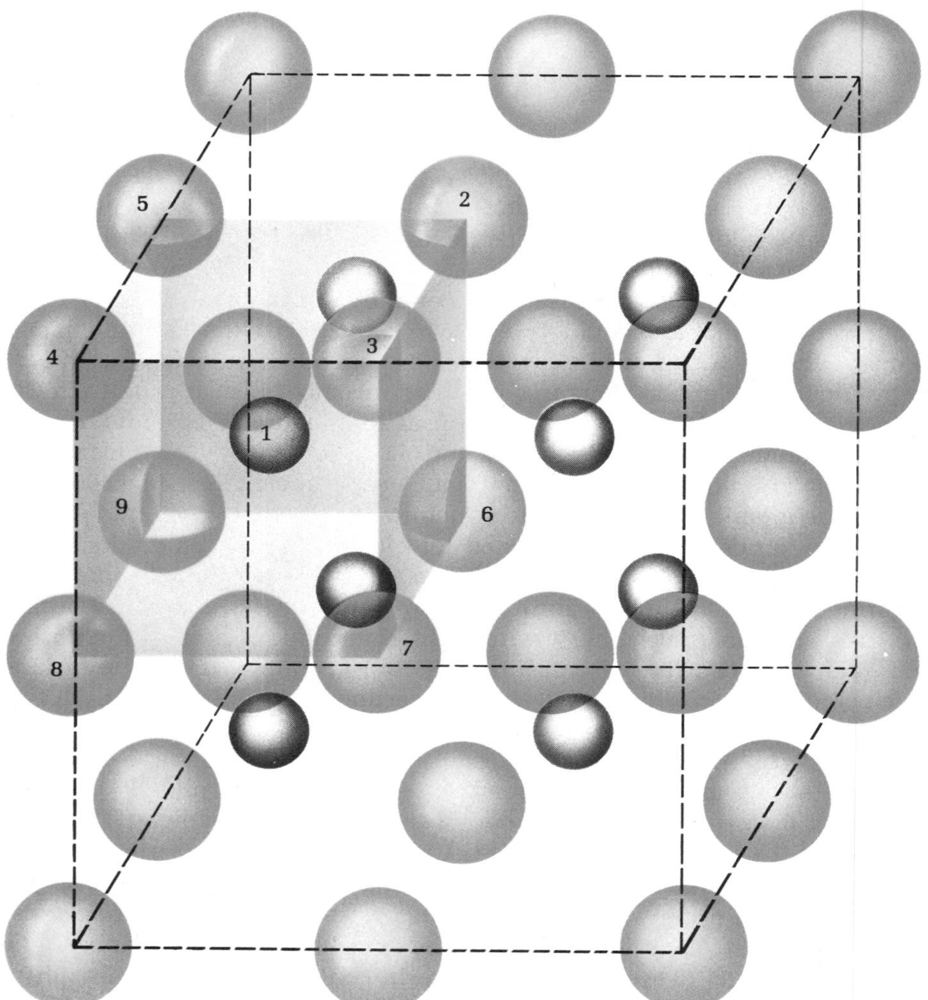

Figure 13–18
The placement of ions in the CsCl structure. Large spheres are Cl^-, small spheres are Cs^+.

ratio of 0.414, the larger ions would be in contact if the smaller ion occupies octahedral sites. Hence, the NaCl structure will not be adopted when the radius ratio is less than or equal to 0.414. (Actually it must be somewhat in excess of 0.414 before we can safely predict that it will adopt the NaCl structure.)

Figure 13–19 shows the structure of zinc sulfide, ZnS, in the crystalline form known as **zincblende.** The sulfide ions form a face-centered cubic array. Four sulfide ions, such as those numbered 2, 3, 4, and 5 in the figure, define a **tetrahedral hole** and there are eight such tetrahedral holes in the dashed cube in Figure 13–19. Half of these tetrahedral holes are filled with

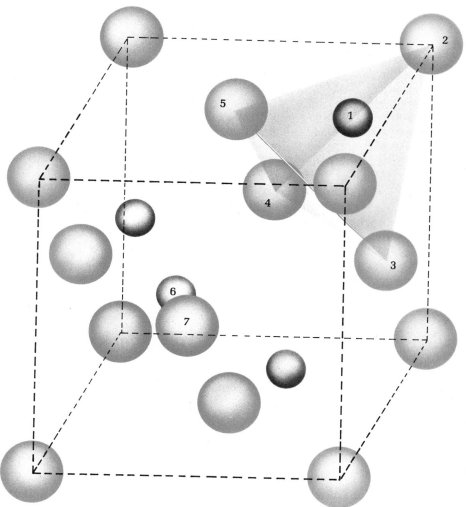

Figure 13–19
The placement of ions in the unit cell of ZnS (zincblende). Large spheres are S^{2-} and small spheres are Zn^{2+}.

Zn^{2+} ions. (The zinc ion numbered 6 would normally be hidden behind the S^{2-} numbered 7 in this view of the structure.

What is the size of a tetrahedral hole? Again we consider large spheres in contact defining the hole and a smaller sphere that just fits inside the hole. Figure 13–20 shows a sphere in a tetrahedral hole in contact with two of the spheres that define that hole. The angle θ is half the tetrahedral angle, or $109.54°/2 = 54.72°$. The sine of θ is 0.816, and this is related to r_1 and r_2 according to the equation

$$\sin \theta = \frac{r_1}{r_1 + r_2} = 0.816$$

Figure 13–20
A sphere in a tetrahedral hole in contact with two of the four larger spheres which define that hole.

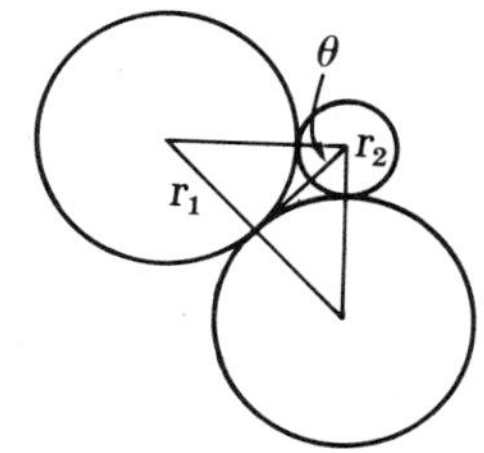

or

$$r_2 = 0.225r_1$$

From Table 13–2 we obtain the radius ratio for Zn^{2+} to S^{2-} as 0.74/1.84 = 0.40. This is too small a ratio for the occupation of octahedral holes; but it is perfectly consistent with the observed occupation of tetrahedral holes in the zincblende structure.

From these considerations of radius ratios, we should be able to predict the types of holes occupied by the cations in many crystalline solids, and also the possible lattice structures containing these holes. Some examples are given in Table 13–3. The table lists various cation-anion combinations and the cation/anion radius ratio (r_+/r_-) for each. Our prediction is that tetrahedral holes should be occupied when r_+/r_- is between approximately 0.225 and 0.414, octahedral holes should be occupied when the ratio is between approximately 0.414 and 0.732, and cubic holes should be filled for ratios between 0.732 and 1.0. The three crystal structures that result from occupation of these holes are the ZnS structure (tetrahedral), NaCl (octahedral), and CsCl (cubic). The table lists the known structure for each cation-anion pair as one of these, so that the actual and predicted structures can be compared.

Table 13–3
Predictions of Crystal Structures of Ionic Solids from Radius Ratios

Solid	r_+/r_-	Structures	
		Predicted	Observed
ZnSe	0.37	ZnS	ZnS
CuI	0.44	(ZnS-NaCl)?	ZnS
MgO	0.46	(ZnS-NaCl)?	NaCl
AgBr	0.65	NaCl	NaCl
CaO	0.71	NaCl	NaCl
BaS	0.73	NaCl	NaCl
TlCl	0.80	CsCl	CsCl
CsBr	0.87	CsCl	CsCl

The table reveals an excellent agreement between prediction and fact with the ionic solids listed. It would be unfair to claim, however, that maximum cation-anion contact in the absence of cation-cation and anion-anion contact is the only factor that influences ionic crystal structures, because we can quote many examples in which structure predictions based on our simple considerations of radius ratios are in error. For example, r_+/r_- for BeSe is 0.16 and yet it adopts the ZnS structure.

The Structures of Molecular Crystals

Most solid organic compounds and many solid inorganic species such as ice consist of discrete small molecules that are relatively weakly bonded to each other. Because the bonding is weak, it is often difficult to predict what the molecular array will be in any given case. One case in which a prediction can be made occurs when the intermolecular interaction occurs through hydrogen bonding. The directed character of hydrogen bonding would allow us to predict the portion of the structure of ice shown in Figure 7–3.

With nonpolar organic molecules such as the hydrocarbons, intermolecular interactions occur because of fluctuating dipoles (Chapter 10). The principal factor in the formation of the solid structure is the attaining of the lowest-energy arrangement by having the maximum possible contact between molecular surfaces. This is most easily achieved with highly symmetric molecules. There is a distinct correlation between the melting point of a hydrocarbon and its symmetry.

13–3. Physical Properties of Crystals

We began our treatment of crystals by pointing out a unique characteristic of the physical properties of crystals, namely their anisotropy. The crystal structures and the factors that lead to them can be used in making further predictions about the physical properties of crystals.

Mechanical Properties

We are all aware that different crystalline solids respond to forces in different fashions. If a block of ice is hit with a hammer, it breaks easily. The same is true of a block of NaCl. If a block of copper is struck with a hammer, the metal does not usually break. Rather, a permanent impression is left in the metallic surface. These differences arise from the differences between metals and other substances in their bonding characteristics. Any force applied to a solid moves atoms, ions, or molecules with respect to other atoms, ions, or molecules. Figure 13–21(a) shows a two-dimensional portion of a closest-packed array of a metal. In (b) one of the rows has been moved with respect to its neighboring row. The bonding in metals is largely nondirected and the bonding between the rows is preserved during the motion. Thus, metals are relatively easy to distort without breaking them. Metals are said to be **malleable.**

Figure 13–22(a) shows a two-dimensional ionic array from the structure of NaCl. In (b) one row has been moved with respect to its neighbor as in Figure 13–21. In the case of the ionic crystal, the bonding is highly directional and the arrangement shown in Figure 13–22(b) is highly repulsive between the two rows shown, and the crystal will split between those two rows. Ionic crystals do not distort as metals do, and ionic crystals are said to be **brittle.** All solids in which the bonding is highly directed are predicted to be brittle. The brittleness of ice derives from the highly directed hydrogen bonds.

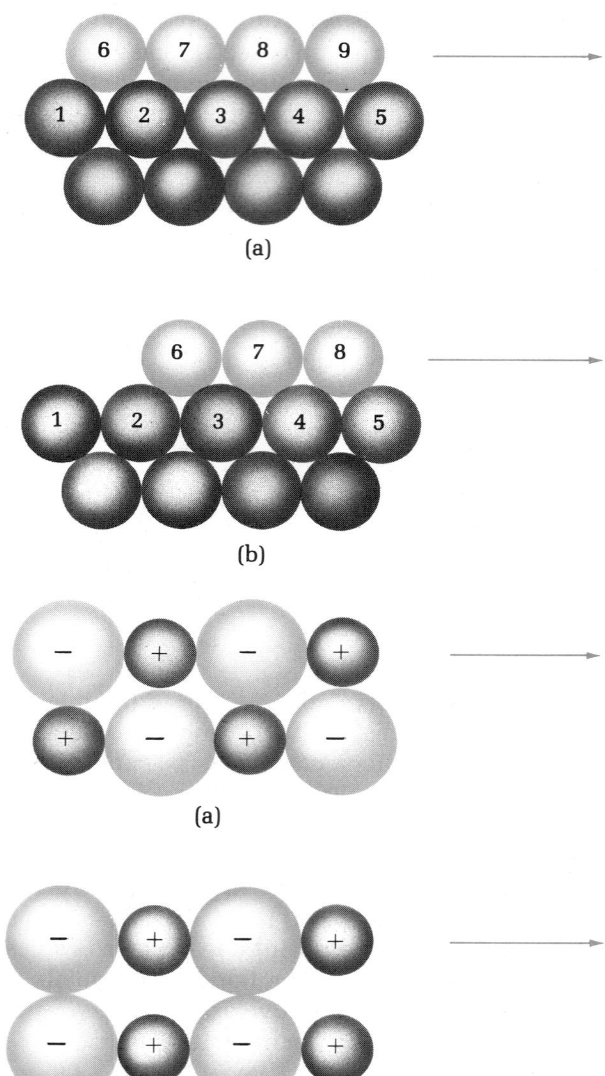

Figure 13–21 The motion of one row of atoms in a metal with respect to another. Part (a) before the motion and (b) after the motion.

Figure 13–22 The production of strong repulsion between adjacent planes of a sodium chloride crystal through the motion of one plane relative to the other: (a) before the motion and (b) after the motion.

A second important mechanical property of a solid is its **hardness,** which is a measure of the difficulty with which distortions, as shown in Figures 13–21 and 13–22, are made. The hardness increases with the strength and directed character of the bonds that must be strained as one row of a crystal is moved relative to another row. The hardest substances known are materials such as diamond in which all of the atoms are held together by strong, directed covalent bonds. Metals like sodium and potassium are relatively soft, and both Na(s) and K(s) can be cut by hand with an ordinary table knife. Other metals like copper, iron, and silver have stronger bonding than sodium and potassium have, and these metals are considerably harder than Na and K. The bonding is quite weak between the molecules of a molecular crystal and these crystals are very easily cut. Such a weak molecular interaction is that between the molecular sheets of graphite. Because of the weak interaction, graphite is a "soft" form of carbon in comparison with the extreme hardness of diamond.

Melting Points

Ionic crystals such as sodium chloride and certain nonionic crystals, such as diamond, are in reality giant molecules. While the formula of sodium chloride is NaCl, there are no NaCl molecules in solid sodium chloride. Rather, sodium ions are bound to chloride ions, which are bound to other sodium ions, and so forth, and this strong cation-anion binding extends throughout the entire crystal. The same extended bonding from carbon atom to carbon atom to carbon atom has been displayed for diamond (Chapter 9). Because of this extended bonding, both ionic crystals and giant molecular nonionic crystals would be predicted to possess very high melting points. Since liquids are isotropic and possess no long-range order, much of the extended bonding in the solid must be lost in the liquid. The melting point of NaCl is 801 °C, and that high value is indeed typical of ionic crystals. Diamond melts at a temperature in excess of 3500 °C; and another giant molecular, nonionic crystal, quartz, melts at 1610 °C.

Extended bonding also occurs in metals. Since bonds must be broken in the melting process, we should predict that soft metals have considerably lower melting points than hard metals. Such is indeed the case. Sodium melts at 98 °C and potassium melts at 62 °C, while copper melts at 1083 °C, and iron melts at 1535 °C.

The correlation between hardness and melting point is also shown by molecular crystals, which are predicted to have relatively low melting points as compared with quartz, sodium chloride, and iron. Table 10–1 lists the melting points of a number of saturated hydrocarbons, and they are indeed relatively low. Table 10–1 shows that the melting points of the unbranched hydrocarbons increase with increasing molecular weight just as the boiling points do. As with the boiling point, this increase comes from the increased number of molecules to which any given molecule can bond as the "surface area" of the molecule increases. In most cases, branching has the same effect on the melting point as it does on the boiling point, as Table 10–1 shows. A notable exception is neopentane, which has a much lower boiling point although a much higher melting point than *n*-pentane. The high melting point arises from the high symmetry of the molecule, which permits efficient packing in the crystal with a maximization of intermolecular contact.

A Model for the Melting of Crystals

What sort of motion do we associate with atoms, ions, and molecules in crystalline solids? The heat capacity expressed in terms of a mole of atoms in a crystalline solid is $3R$ for a large number of crystalline solids. The value $3R$ derives from three vibrational degrees of freedom for each atom of the solid. This indicates that atoms, ions, and molecules are largely fixed to their lattice sites in crystals. They can vibrate around those sites, but they are not normally free to move from site to site.

This picture of "a place for everything and everything in its place" that we have drawn for crystalline solids is certainly correct for most of the atoms or ions of any crystalline solid, but there is evidence that it is not true for every atom or ion. For example, hydrogen gas can diffuse through a number of crystalline solids. A perfect crystal in which all species are fixed to lattice sites would not permit such diffusional motion. Another indication that crystals contain small numbers of imperfections comes from the temperature dependence of the heat capacity. Most solids show a sharp rise in $C_v \cong C_p$ over the Dulong-Petit value as the solid nears its melting point. The additional energy is going into the production of structures or motions that are closely related to melting.

In a perfect crystal of NaCl, each Na^+ is surrounded by six Cl^- ions and each Cl^- is surrounded by six Na^+ ions, and this yields the long-range order of the crystal. Since liquids do not possess such long-range order and since the melting of most solids is accompanied by a volume increase (most liquids have a lower density than the corresponding solids at the melting temperature), liquids must be characterized by a large number of **vacancies.** For example, many Na^+ ions will be found with only four or five chloride ion neighbors in molten sodium chloride.

The fact that the heat capacity rises just below the melting point must mean that these vacancies are being created by ion migration in the solid, and the nearer the temperature comes to the melting point, the more the vacancies in the crystal. At the melting temperature, the vacancy concentration becomes so large that the long-range order can no longer be preserved and the entire structure "suddenly crumbles" and the crystal melts. The situation is entirely analogous to the random removal of bricks from a brick building. The removal of some critical number of bricks will result in the sudden and complete collapse of the building. This is why crystalline materials melt at a fixed temperature rather than over a range of temperatures.

SUGGESTIONS FOR ADDITIONAL READING

Addison, W. E., *Structural Principles of Inorganic Compounds*, Wiley, New York, 1961.

Hannay, N. B., *Solid State Chemistry*, Prentice-Hall, Englewood Cliffs, N.J., 1967.

Castellan, G. W., *Physical Chemistry,* 2nd ed., Addison-Wesley, Reading, Mass., 1971.

PROBLEMS

1. With the use of a compass, ruler, and graph paper draw a picture of reinforced scattering analogous to that in Figure 13–3, but for the case in which n in equation 13–1 is equal to 1. Must θ be greater or less than the value of θ shown in Figure 13–3 if λ is the same in both cases?
2. Figure 13–5 shows that a diffraction grating reflects light as well as transmits it. Is the reflected light diffracted also? A phonograph record viewed at certain angles shows a rainbow. Explain this phenomenon. If the light strikes the phonograph record in a perpendicular fashion, the rainbow can be seen by looking at the record along a direction almost perpendicular to the record's surface. Explain this by reference to Figure 13–5 or equation 13–2 or both.
3. A model of X-ray diffraction from an ordered molecular array can be obtained from an array of steel bearings imbedded in plastic blocks. For a distance between adjacent bearings of 3 cm, what form of electromagnetic radiation (ultraviolet, infrared, and so on) should be used to obtain the best diffraction from this array?
4. Many crystalline substances show **dichroism** when viewed with plane-polarized light. A dichroic substance possesses different colors when viewed along different directions. Explain this phenomenon by reference to Figure 13–6 and to the discussion of electronic spectroscopy in Chapter 12.
5. Figure 13–10 shows a face-centered cubic unit cell. Why doesn't it show a face-centered tetragonal cell?
6. If the sample in Figure 13–11 is not powdered finely enough, the lines on the photographic film are not obtained; instead we see linear arrays of spots. Explain why.
7. Calcium oxide forms a solid with a face-centered cubic unit cell. The density of CaO is 3.37 g/cm^3. What is the length of the CaO unit cell?
8. A certain metal of relatively low reactivity has a measured density of 19.3 g/cm^3. The metal forms with a face-centered cubic unit cell of cell length 4.08 Å. Identify the metal by calculating its atomic weight.
9. How many nearest neighbors does each sphere have in the cubic closest-packing arrangement of spheres? In hexagonal closest-packing?
10. Figures 13–15 and 13–16 are two views of the same cubic closest-packed arrangement. The spheres 2, 4, 5, 6, 9, and 10 belong to a layer of spheres oriented perpendicular to the body diagonal connecting spheres 1 and 12. Spheres 3, 7, 8, 11, 13, and 14 belong to another layer of spheres.

This planar layer arrangement occurs perpendicular to any of the body diagonals of the cube in Figure 13–16. Give the numbers of all of the spheres in each of the layers perpendicular to the line connecting spheres 3 and 10 in Figure 13–16.

11. In both cubic closest-packing and hexagonal closest-packing, 74 percent of all space is occupied by spheres. Prove that this is the case for cubic closest-packing by considering the unit cell for that structure and showing that 74 percent of the unit cell volume is occupied by hemispheres or one-eighth spheres.

12. Would you predict a correlation between the radius of an ion and the ionization potential or potentials of the atom that yields the ion? Refer to Tables 4–3 and 11–1 and predict which ion in each of the following pairs of ions has the smaller radius.
    a. K^+ and Cu^+.
    b. Ca^{2+} and Zn^{2+}.
    c. Li^+ and Na^+.
    d. Mg^{2+} and Ca^{2+}.
    e. Ni^2 and Mn^{2+}.

    Check your predictions with the values given in Table 13–2.

13. The text contains calculations of r_2/r_1 for octahedral, cubic, and tetrahedral holes. What is r_2/r_1 for a triangular hole?

14. The density of CsCl(s) is 3.97 g/cm^3. What is the length of the CsCl unit cell? This length is the smallest Cs^+—Cs^+ distance in Figure 13–18. Calculate from this distance the sum of the Cs^+ and Cl^- ionic radii. Compare this value with the value calculated from the data in Table 13–2.

15. The radius ratio of Cu^+ to Cl^- is 0.53, yet CuCl adopts the ZnS structure. Refer to the discussion of the Group IB elements in Chapter 11 and suggest a reason that this structure occurs with CuCl.

*16. It is much more difficult to cleave a sodium chloride crystal by separating planes oriented perpendicular to any body diagonal of the cube, as shown in Figure 5–14, than to cleave it by separating planes that are parallel to the sides of the cube. Explain why from a comparison of Figures 5–14 and 13–22.

17. The melting point of toluene (methylbenzene) is $-95\,°$ while that of benzene is 5.5 °C. Explain this large difference.

18. Fats derived from saturated fatty acids such as palmitic acid and stearic acid are typically solids at room temperature. Fats derived from unsaturated fatty acids are liquids (oils) at room temperature. Propose a reason for the different melting points of saturated and unsaturated fats.

14/ Models of Liquid Substances

The three principal states of matter are gases, solids, and liquids, and each state possesses its own characteristic physical properties. These properties can be accounted for in terms of atomic and molecular models. Models of the liquid state can help to explain the unique properties of liquids, as the models for gases and solids have so far done for those substances.

14–1. The Structure of the Liquid State

A model of the liquid state resembles the model of a solid, but contains so many vacancies that the long-range order of the solid is destroyed. One possible molecular model of a liquid is shown in Figure 14–1. Figure 14–1(b) shows a two-dimensional square lattice that represents a crystalline solid. In the (a) part of the figure this lattice has been broken into small pieces, each piece displaying the same square unit cell with the same cell length as the larger array in (b). This model of the liquid views it as an extremely fine powder, so fine that the tiny crystals can no longer be distinguished by eye or microscope.

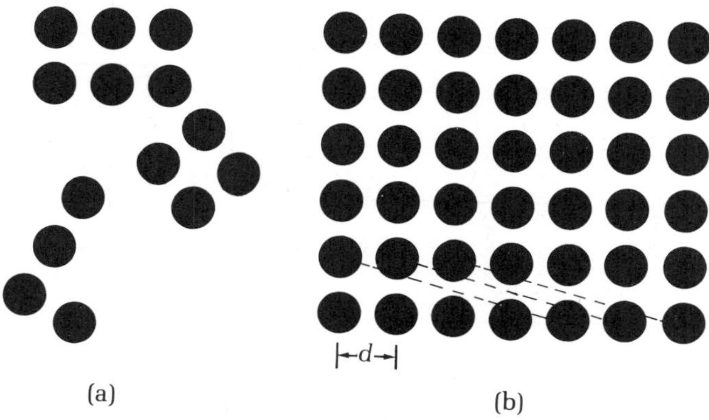

Figure 14–1
(b) A solid lattice and (a) a model of a liquid derived from it.

Probably the most obvious difference between solids and liquids is the extreme ease of distortion of most liquids; liquids flow easily and crystals do not. Crystalline solids have long-range directed bonding that strongly resists distortion. The model of Figure 14–1(a) displays no such long-range directionality. The tiny crystals can easily rotate and translate past each other to permit easy distortion.

Another obvious difference between liquids and solids is the relative rate of diffusion of some second substance through them. Copper and zinc can mix to form the alloy brass. Alloy formation occurs extremely slowly if solid Cu and solid Zn are mixed, but the two metals mix rapidly when molten. This is also perfectly consistent with the model of Figure 14–1(a), since there is ample room for microcrystals of one substance to penetrate microcrystals of the other.

An excellent test of Figure 14–1(a) is provided by X-ray crystallography. Figure 14–1(b) contains many sets of parallel rows such as the vertical rows with the spacing d shown. The dashed lines indicate another set of rows with a considerably smaller spacing. Equation 13–1 reveals that as the spacing between rows (planes in a crystal) decreases, $\sin\theta$ increases, and consequently θ increases. The rows indicated by the dashed lines in Figure 14–1(b) yield lines relatively far away from the circle in the center of the photographic film of Figure 13–11, since a large distance from the center indicates a large θ value. None of the three microcrystals shown in Figure 14–1(a) possesses rows with the separation between the dashed lines of Figure 14–1(b) because the microcrystals are not large enough. However, the microcrystals are large enough to yield parallel rows with relatively large d values, which yield small θ values in an X-ray diffraction experiment. Thus, the microcrystal model yields an X-ray pattern nearly identical to that of the corresponding solid at small θ, but the intensity should drop off very sharply at increased θ.

Figure 14–2 shows a typical comparison of the X-ray patterns of a liquid and the corresponding solid. As predicted for the liquid, the intensity of the lines at large θ drops off sharply. However, even for small θ, the lines are

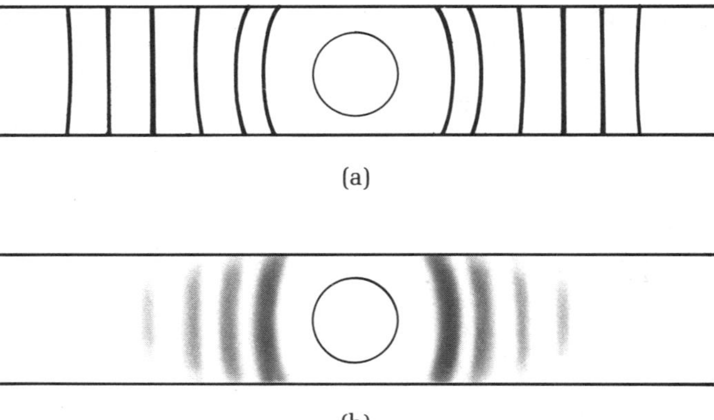

Figure 14–2 X-ray pattern (a) from a crystalline solid and (b) from the corresponding liquid.

relatively faint and diffuse compared with the sharp lines from the solid. This diffuseness reveals a relatively broad variation in even nearest-neighbor distances in the liquid, contrary to the microcrystal model of Figure 14–1(a).

The liquid state appears to be a compromise between the nearly perfect order of the solid state and the nearly perfect disorder of the gaseous state. It is not difficult to construct models of perfect order or perfect disorder according to our experience with gases and solids. However, it is extremely difficult to construct models for the compromise that have any general validity and that are also simple enough to be easily applicable. The simple models of liquids already developed do not agree well with experimental data, and those models that do yield good quantitative agreement with experimental data are extremely complex. However, we can understand many of the physical properties of liquids in a qualitative fashion, by developing simple models.

14–2. Physical Properties of Liquids

Viscosity

We have previously considered a model for transport by gases based on the kinetic model of the gaseous state. In that picture intermolecular collisions interfere with transport, and such properties as the thermal conductivity, diffusion rate, and viscosity increase as the mean free path increases. In the liquid state, molecules are almost constantly in a state of collision with their neighbors, and we might expect all three transport properties to be considerably smaller in liquids than in gases. This prediction is borne out only in the case of diffusion, that is, the transport of matter. Heat transport and momentum transport (viscosity) are much more efficient in liquids than in gases. This must be a consequence of an entirely different mechanism for heat and momentum transport in liquids as compared with gases.

Figure 14–3 shows a familiar mechanism for extremely efficient momentum transport. A row of identical spheres is suspended from a beam. Sphere 1 is then raised a distance h and dropped as shown in Figure 14–3(a). On collision of spheres 1 and 2, sphere 1 stops abruptly and sphere 5 swings out to a height h, as shown in Figure 14–3(b). Clearly this momentum transfer occurs **because** of collisions, not **in spite** of them; this is in contrast with the situation for gases, and it demonstrates the extremely high viscosity of solids.

The ordered spheres of Figure 14–3 provide a good model for momentum transport by solids. How can the figure be modified for liquids? We can lower the order to some extent by putting the spheres in motion in and out of the plane of the paper. Now the collision of spheres 1 and 2 will, in general, increase the motion of all of the spheres, and the efficiency of the transport

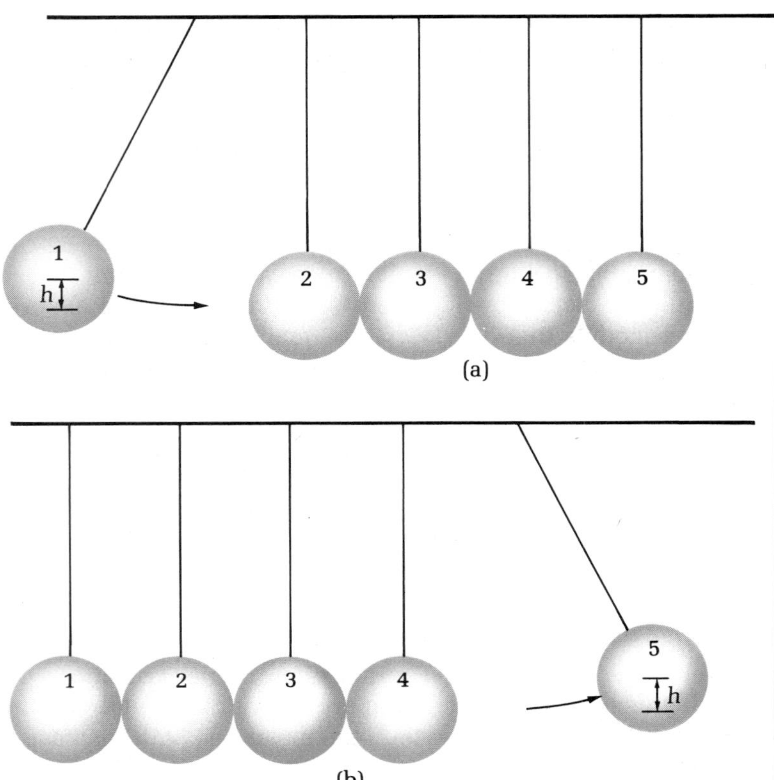

Figure 14–3 Momentum transfer through suspended spheres: (a) before transfer, (b) after transfer.

between 1 and 5 greatly diminishes. This model leads to the prediction that liquid viscosities should decrease with increasing temperature because the molecular motion is increased. This is indeed the case with almost all liquids.

One way to diminish the relative motion of liquid molecules is to have polar molecules in which intermolecular bonding is stronger than in nonpolar liquids. Thus, a second prediction is that polar liquids are more viscous than nonpolar liquids. This prediction is also borne out by experiment.

Surface Tension

Intermolecular attraction tends to make a gas pull in on itself; and if the temperature is lowered sufficiently, the gas will collapse into a liquid. The liquefaction, of course, does not remove the intermolecular attraction and the molecules at the liquid surface experience a strong force in the direction of the center of the liquid. It is this force that must be overcome when a molecule enters the vapor over the liquid.

The force we have just described on liquids manifests itself in one of the most outstanding properties of liquids, namely **surface tension.** One effect of the intermolecular attraction is to minimize the liquid surface area. This is the reason that falling droplets of liquids are spherical. Of all possible shapes for a given volume, the sphere possesses the minimum surface area.

Consider the circular surface of a liquid shown in Figure 14-4. This surface could be the top of a liquid in a beaker. The arrows shown indicate

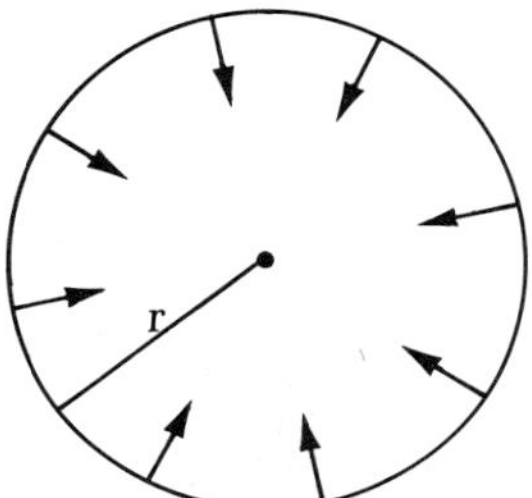

Figure 14-4
Surface tension on the circumference of a circular surface. The arrows indicate the inwardly directed forces.

the inward-directed surface tension, which attempts to minimize the surface area. This force F is exerted on the circumference, $2\pi r$, and the surface tension γ of the liquid is defined as follows:

$$\gamma = \frac{F}{2\pi r} \tag{14-1}$$

and its units are dynes/cm.

Perhaps the most outstanding consequence of surface tension in liquids is the spontaneous rising of many liquids, such as water, in narrow, hollow tubes made of a substance like glass. Figure 14-5 shows such a situation. When the tube is inserted in the beaker of water, the level slowly rises in the tube to the final height h, shown. The rising of the liquid in the capillary is a

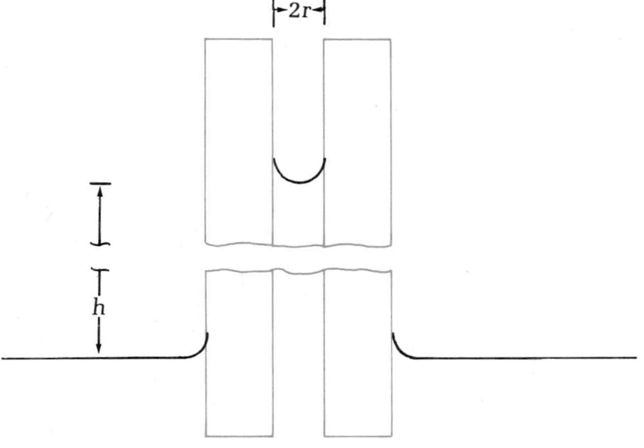

Figure 14-5
The demonstration and measurement of the surface tension of water by its rise in a glass capillary (magnified view) of radius r.

consequence of two factors and both are related to the properties of surfaces. Surfaces represent an unsatisfied potential for intermolecular association. This is particularly true of the surfaces of ionic solids, polar solids, and polar liquids. If an ionic surface is exposed to water vapor, water molecules are **adsorbed** to the solid surface, as shown in Figure 14–6. The surface of glass strongly attracts water molecules and we say that water "wets" glass. In the case of Figure 14–5 the attraction between glass and water is manifested in the **meniscus,** or curved surface at the top of the liquid.

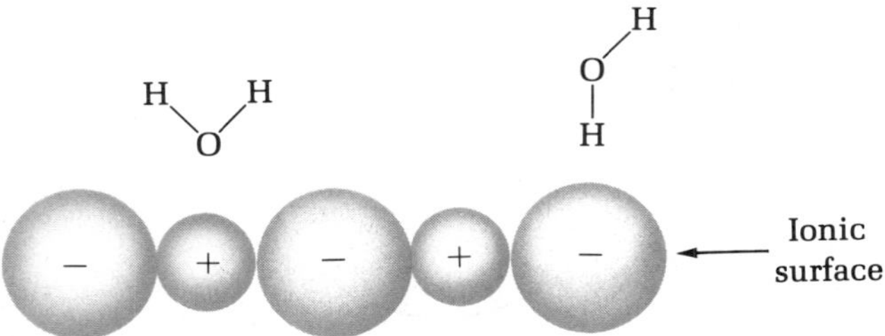

Figure 14–6
The adsorption of water molecules to an ionic surface.

The meniscus forms as soon as the capillary is inserted in water. Meniscus formation, however, is opposed by the water surface tension since the water surface area is increased in the curved meniscus over the area of a flat surface. The liquid can restore the flat surface by rising up the tube. The entire liquid rise to the height h may be considered to result from a large succession of meniscus formations and destructions. As the liquid column rises, meniscus flattening becomes more difficult since the weight of liquid to be raised increases. At the height h, the surface tension force exactly equals the force of gravity on the liquid column and no further flattening of the meniscus can occur. This force balance is given by

$$2\pi r\gamma = \pi r^2 \rho h g \qquad \textbf{14–2}$$

where ρ is the density of the liquid. Equation 14–2 may then be used to calculate γ from the measured h, r, and ρ.

Soap and detergent solutions usually lather easily. A lather consists of a large number of small bubbles. A bubble is characterized by a very large surface area compared with the total volume of the liquid in the bubble. Thus, bubble formation is strongly resisted by the surface tension of the liquid. Since soaps and detergents make it easier to form water bubbles, they must lower the surface tension of water. Figure 14–7 shows how this is done by a combination of micelle and layer structures.

Figure 14–7(a) shows a portion of the air-water interface and the incomplete hydrogen bonding that occurs there. The surface water molecules possess a higher energy than a completely hydrogen-bonded water molecule such as the one shown in color in the interior of the liquid. To produce the additional surface of a bubble, low-energy interior molecules must be raised

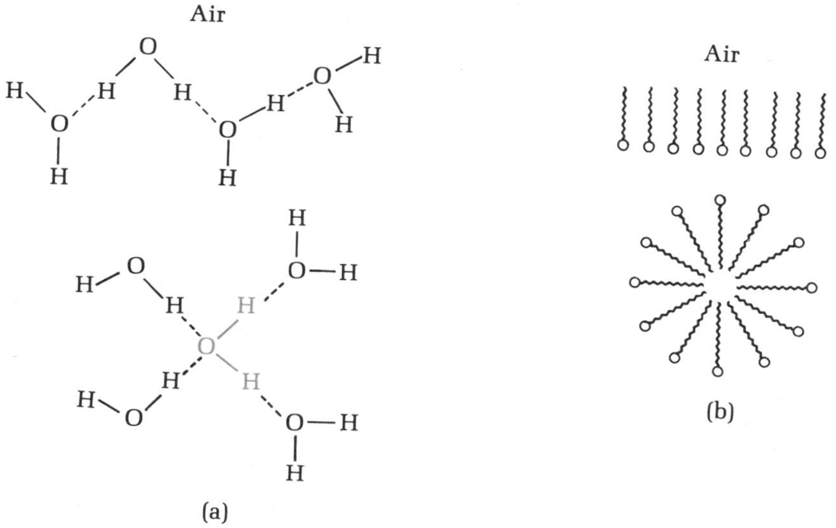

Figure 14–7 Structures which influence the surface tension of (a) pure water and (b) a soap or detergent solution.

in energy to that of the surface. The interior molecules resist such energy raising, and this is the source of surface tension.

In the soap or detergent solution, the surface contains the hydrophobic hydrocarbon tails of soap or detergent molecules, as shown in Figure 14–7(b), and the energy of the surface molecules is comparable to the energy of the molecules in the interior micelle because the two structures are similar. Additional surface is relatively easily produced by the transferral of interior soap or detergent molecules to the air-solution interface, and the surface tension of the solution is much lower than the surface tension of pure water.

14–3. The Freezing of Liquids

Our model for the melting of solids is based on producing a critical concentration of vacancies so that the long-range order in the crystal is destroyed. The result is a sharply defined melting temperature as measured in the manner shown in Figure 14–8. In the figure is shown a small glass tube containing a crystalline solid. The tube is immersed in a bath of some liquid with a high boiling point, e.g., mineral oil. The mineral oil is then slowly heated while the crystalline solid and the thermometer reading are closely observed. At the melting temperature, the solid typically will liquefy completely in a very short time, and the melting temperature can be obtained with a precision of a few tenths of a degree Celsius.

If a more precise measurement of the melting point is desired, it is accomplished by the reverse procedure, namely freezing the liquid. The experimental arrangement is shown in Figure 14–9(a) for the determination of the melting point of benzene, C_6H_6, which is 5.48 °C. A sizable quantity of pure liquid benzene is placed in a large glass tube along with a wire stirrer

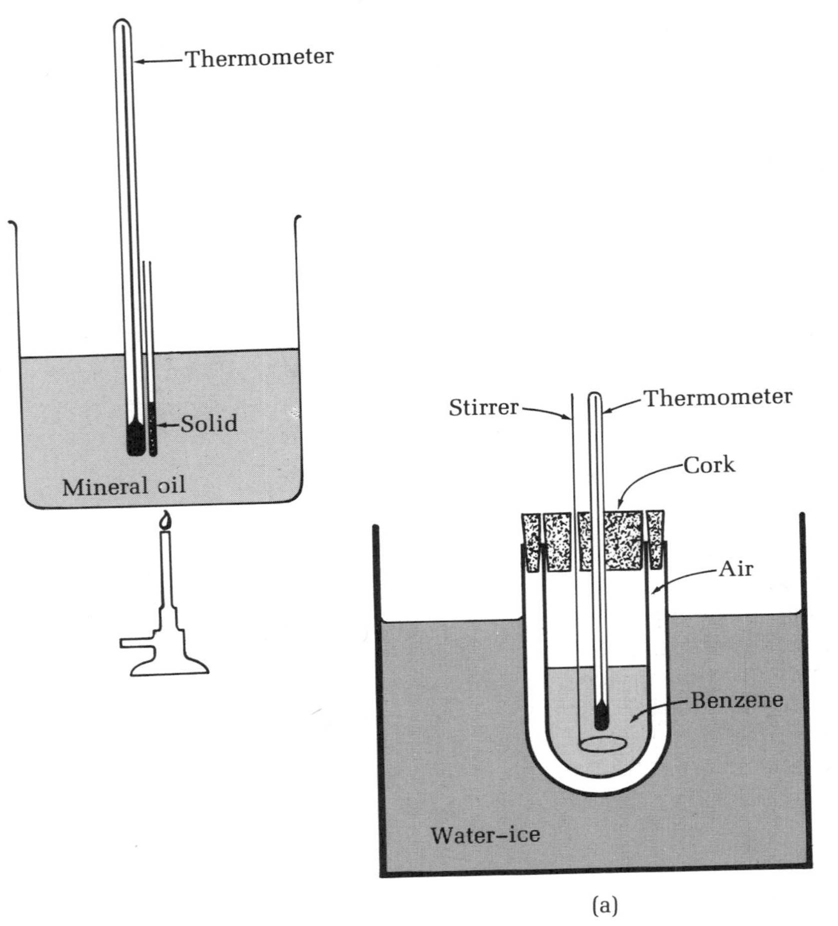

Figure 14–8
A simple arrangement for the measurement of the melting points of solids.

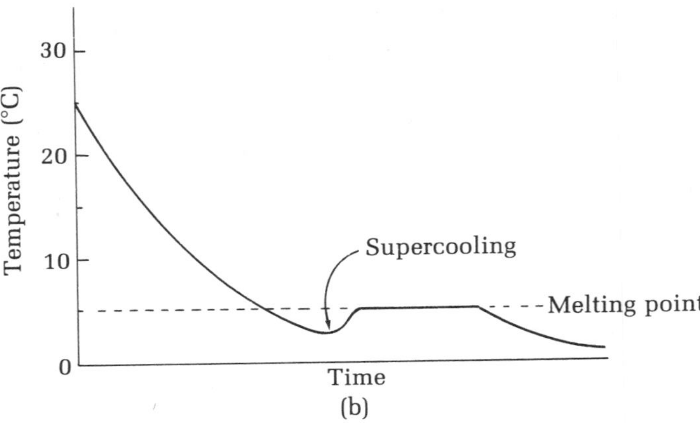

Figure 14–9
The determination of the melting point of benzene from the freezing of the liquid: (a) experimental arrangement, (b) time-temperature plot.

and a thermometer (the stirrer is used to maintain uniform temperature in the liquid). The glass tube is then placed in a second glass tube as shown, and both tubes are immersed in an ice-water bath at 0 °C. The purpose of the second tube is to provide an insulating air gap between the benzene and the ice-water bath so that heat transfer from the benzene is relatively slow.

Figure 14–9(b) shows a plot of the benzene temperature as a function of time as it cools. A surprising phenomenon, but one that is noted in the freezing of most liquids, is the **supercooling** of the liquid below its melting point, usually by several degrees. However, as soon as the first benzene crystals appear the temperature rapidly rises to the equilibrium melting point, where it remains until all of the liquid has solidified. Thus, the melting point can be precisely determined as that on the horizontal portion of the cooling curve.

Why do liquids supercool whereas solids do not superheat; that is, why does the solid not exist above its melting point? The answer to this question is provided by the result of an experiment that is commonly performed on supercooled liquids. Suppose that we have a sample of liquid benzene at 3 °C and we add a small crystal of solid benzene. Almost invariably this "seed crystal" will initiate crystallization of the entire liquid sample. The seed crystal is a **nucleus** for further crystallization, and the neighboring molecules of the liquid easily bond to it in the ordered crystalline pattern. The problem in the supercooled liquid is the spontaneous formation of tiny crystals that can be nuclei. A relatively disordered arrangement of liquid molecules must spontaneously form a very ordered arrangement in these tiny crystals, and this process must take place largely by chance as the molecules encounter each other at various distances and orientations in the liquid. Thus, even though the melting point of the solid may be reached in cooling the liquid, there will, in general, be some time before the crystal nuclei form by chance, and the liquid will supercool.

The solid does not superheat because the formation of vacancies can be accomplished by one energetic molecule at a time, quite independent of the other molecules. All that is required for the critical number of vacancies that causes melting is sufficient energy and this is reached at the melting temperature.

PROBLEMS

1. The "lines" in Figure 14–2(b) are relatively diffuse. However, the diffraction pattern from a solid does not yield perfectly sharp lines either, although they are considerably narrower than those from a liquid. Name two characteristics of a solid that cause broadening of diffraction lines. Which of these two characteristics is greatly exaggerated in the liquid, and consequently yields the much broader diffraction lines from the liquid?

2. The **compressibility** of any substance is defined as the fractional change in its volume caused by a pressure increase of 1 atm. The compressibilities of many liquids at room temperature show an abrupt decrease when the initial volume (at $P = 1$ atm) has been decreased by about 3 percent (at P equal to approximately 1000 atm). Propose an explanation for this marked change.
3. The heat capacities of most organic compounds increase with increasing temperature, and the steepness of this increase is particularly dramatic with polar compounds. What is the source of this increasing heat capacity?
4. The viscosities of organic compounds typically increase with increasing molecular weight. Thus, for example, the viscosity of ethanol is approximately twice the viscosity of methanol at room temperature. Propose an explanation.
5. A concentrated solution of sulfuric acid in water is very viscous. Propose an explanation.
6. Do you predict that surface tension of a liquid increases, decreases, or remains unchanged with increasing temperature? Explain.
7. There is a rough correlation between the surface tension of a liquid and its boiling point. For example, the surface tension of molten aluminum is 840 dynes/cm at 700 °C. Molten NaCl has a γ of 113.8 dynes/cm at 803 °C. The surface tension of water is 72.0 dynes/cm at 25 °C. Explain why this correlation exists.
8. The phenomenon of water rising in capillaries is perhaps most commonly illustrated in nature by water rising in tree trunks (although this is not the only reason water rises spontaneously in trees). From equation 14–2, what must be the radii of the small vertical tubes in tree trunks to permit water to rise to a height of 100 feet by the capillary effect alone?
9. Is the pressure inside a soap bubble greater than, equal to, or less than the air pressure outside the bubble? Explain.
*10. Does a rubber balloon have a surface tension? If so, what is the source? That is, give a molecular picture of what happens to the rubber in the balloon as it is inflated.
11. One point made in the text is that a sphere has the smallest possible surface area for a given volume. Calculate the surface area of a sphere of volume = 1.00 cm^3. Calculate the surface area of a cube of the same volume.
12. Glassy substances are commonly formed by the supercooling of certain viscous organic liquids. For example, glycerol is extremely viscous and as it cools it forms a hard glass instead of a crystalline solid. Why should viscous liquids be more susceptible to supercooling than liquids of a lower viscosity?

15/ The Laws of Thermodynamics

The chemical bond exists in an H_2 molecule at room temperature because the bonded condition, compared with two separated hydrogen atoms, is one of lower energy. Our experience tells us that systems commonly transfer to a lower energy condition when given a choice between higher and lower energy. Figure 15–1(a) shows a 2-kg book in an unnatural state of suspension two meters above a floor. The potential energy of the book with respect to the floor is

$$\text{Potential energy} = mgh$$

$$= (2000\text{ g})\left(\frac{980\text{ cm}}{\text{sec}^2}\right)(200\text{ cm})$$

$$= 3.90 \times 10^8\text{ ergs} = 39.0\text{ joules}$$

Thus, 39.0 joules of work has to have been performed to lift the book two meters above the floor. Experience tells us that the state is unnatural because the book will fall spontaneously to the floor with the loss of 39.0 joules of potential energy. The book spontaneously moves to the condition of lower

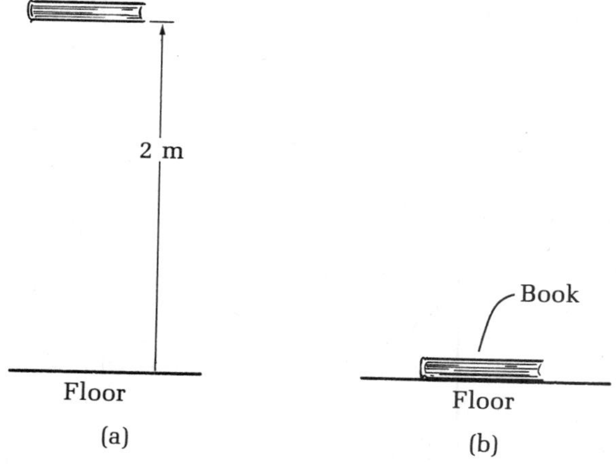

Figure 15–1
Conversion of the potential energy of a book to thermal energy.

energy just as hydrogen atoms do when they combine to form hydrogen molecules at room temperature.

From Figure 15–1 and other everyday experiences, there appears to be a law of nature that says that if anything is given a choice between existing at a high or lower energy, it will always choose the latter. Thus, hydrogen atoms should always be found as hydrogen molecules. However, this conclusion about atoms is erroneous. At elevated temperatures H_2 spontaneously dissociates into H atoms and the atomic state, still the one of higher energy, is the one found rather than the molecular state. Our conclusion about the inherent tendency of a given system to lose energy is not correct.

It is desirable to derive a criterion for predicting spontaneous change. Such a criterion might have been the one of energy minimization just discussed, but we have shown that to be erroneous in some instances. Heat plays a large role in most spontaneous changes including the hydrogen atom–hydrogen molecule change. After investigating heat as it pertains to changes, we shall be in a position to establish a general criterion for predicting spontaneous physical and chemical changes.

15–1. Thermochemistry

The First Law of Thermodynamics

Potential energy equal to 39.0 joules is apparently lost when the book in Figure 15–1(a) falls to the floor. However, extremely precise temperature measurements reveal that the temperature of the book and the floor in (b) are both slightly higher than they are in (a). Since the floor and the book have nonzero specific heats, energy must have been added to both to produce the temperature increases, and the amount of energy in each case is measurable. If this measurement is made in the case of the book and the floor, it is discovered that the energy associated with the temperature increase is exactly 39.0 joules. The potential energy of the book was not lost when the book fell to the floor; rather the energy was quantitatively converted to another form of energy that manifested itself in a temperature rise. This type of energy we shall call **thermal energy.** It includes all of the types of molecular motion we have so far examined in connection with heat capacities.

Temperature increases can be produced in other ways besides converting potential energy to thermal energy. The usual way to increase the temperature of an object is to bring it in contact with an object that is at a higher temperature. In common terminology, we say that "heat" passes from the hot substance to the cold substance. Actually the temperature of the hot substance is lowered and the temperature of the cold substance is raised. Thermal energy is lost by the hot substance and gained by the cold substance. This loss and gain is termed the transfer of "heat." Heat is put in quotes here because heat is not really a substance transferred. The term *heat* as used subsequently in the remainder of the text will denote transfer of thermal energy from a substance brought into contact with a substance of lower temperature. Thus, heat is a "bookkeeping" term. It is the amount of energy exchanged by the process of heat transfer.

In our elevated-book example illustrating conversion of potential energy to thermal energy (Figure 15–1), we calculated that 39.0 joules of potential energy were quantitatively converted to 39.0 joules = 39.0/4.185 cal = 9.32 cal of thermal energy. All experiments on energy conversion have shown, without exception, that energy is neither created nor destroyed in any change. It is merely converted from one form to another. This statement of the **conservation of energy** is known as the **first law of thermodynamics.** The name *thermodynamics* derives from words for "heat" and "motion" (or "change"), and thus the relations between heat and change form the science of thermodynamics.

The first law of thermodynamics is sufficient to disprove the proposal that all systems tend toward minimum energy in any spontaneous process. The book of our example loses energy, but the floor gains energy in the form of thermal energy. Because energy is conserved in any change, that energy lost by one system must be gained by another. If we are to derive a general criterion for spontaneity in physical and chemical changes, we must be able to predict which systems will gain and which systems will lose.

Enthalpy

Of particular interest to us are processes that take place in substances exposed to atmospheric pressure. This includes all of the chemistry of living systems, and all laboratory chemistry carried out in containers open to the atmosphere.

Figure 15–2(a) shows a mixture of $H_2(g)$ and $O_2(g)$ contained in a cylinder and enclosed with a piston, the whole immersed in a water bath. The gas-

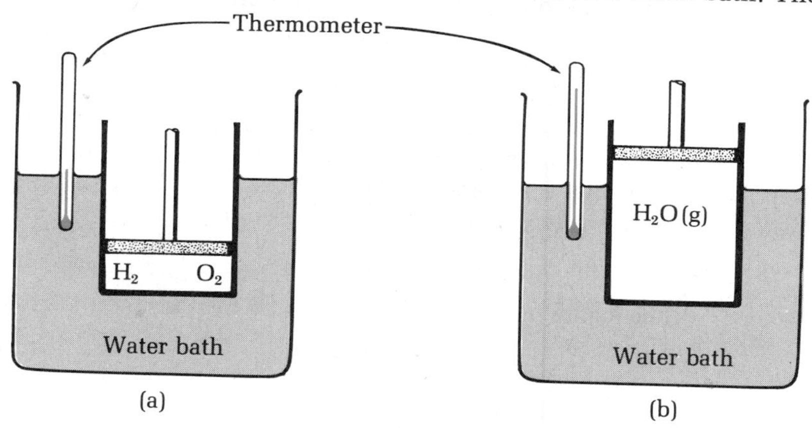

Figure 15–2
(a) Initial and (b) final states for the reaction of hydrogen and oxygen enclosed in a cylinder by a piston.

eous mixture can be ignited with a spark (not shown in the figure) and the reaction that takes place can be described as

$$2\,H_2(g) + O_2(g) \rightarrow 2\,H_2O(g) \qquad \textbf{15–1}$$

A large amount of energy is liberated in reaction 15–1. How does this energy appear? Experience tells us that a large amount of energy is converted to thermal energy in the water bath by heat transfer since the water bath experiences a temperature rise. Energy changes arising from heat transfer are given the symbol Q.

The temperature rise of the gas in the cylinder causes its pressure to rise, pushing the piston up and doing work W in the process. Thus work performance, like heat transfer, is a means by which energy is transferred. The change in the energy of the gas appears as the sum of energy lost in heat transfer and energy lost as work performed. Such is the case for all changes, ΔE, in the energy of any system.

The mathematical statement of the first law of thermodynamics is

$$\Delta E = Q - W \qquad \textbf{15–2}$$

Equation 15–2 says that the net change in energy of any system is the difference between the energy added to the system by heat transfer and the energy lost by the system in work performed. With equation 15–2 there is an implied definition that heat transfer to a system is positive and heat transfer from a system is negative. In addition, work done by a system is defined to be positive and work done on a system is negative. (You should be aware in reading other textbooks that equation 15–2 can also be written with a plus sign with a switch in the definitions of positive and negative work. In this textbook, we use equation 15–2 and the implied definitions.)

The book of Figure 15–1 lost 39.0 joules of energy, so that $\Delta E_{\text{book}} = -39.0$ joules. If this energy is all given to the floor in heat transfer, $Q_{\text{book}} = -39.0$ joules and $W_{\text{book}} = 0$. The same quantities for the floor are $\Delta E_{\text{floor}} = 39.0$ joules, $Q_{\text{floor}} = 39.0$ joules, and $W_{\text{floor}} = 0$. The combined ΔE of the book and the floor is $(-39.0 + 39.0)$ joules $= 0$. The first law of thermodynamics demands that the total ΔE for all systems taking part in a given change must be zero.

The work done by a system expanding against a constant pressure P is $P\Delta V$ where ΔV is the volume change of the system. Such is the case of a system expanding against the pressure of the atmosphere. Under this special condition of constant pressure, equation 15–2 becomes

$$\Delta E = Q_p - P\Delta V \qquad \textbf{15–3}$$

or the energy either lost or gained by the system by heat transfer is given by

$$Q_p = \Delta E + P\Delta V \qquad \textbf{15–4}$$

where the subscript p refers to constant pressure.

For processes carried out at constant pressure and in which the only work involved is given by $P\Delta V$, the energy involved in heat transfer is given by the change in a property of the system termed the **enthalpy** of the system. The enthalpy H is defined as follows:

$$H = E + PV \qquad \textbf{15–5}$$

With this definition ΔH at constant pressure is given by

$$\Delta H = \Delta E + P\Delta V = Q_p \qquad \textbf{15–6}$$

and the enthalpy change is equal to the energy either lost or gained due to the heat transfer involved in the process.

The Difference between ΔE and ΔH

How important is the work term $P\Delta V$ in typical processes? Relative to Figure 15–1, it was of no consequence at all since the volumes of the book and the floor do not change. Thus Q_p equals ΔE for both the book and the floor.

Let us consider reaction 15–1 carried out so that any work done is carried out by or against a pressure of 1 atm. At this relatively low pressure, $H_2(g)$, $O_2(g)$, and $H_2O(g)$ may be treated as ideal gases and PV for each is given by nRT, where n is the number of moles of each. Thus,

$$\begin{aligned} P\Delta V &= PV_{H_2O} - PV_{H_2} - PV_{O_2} \\ &= n_{H_2O}RT - n_{H_2}RT - n_{O_2}RT \\ &= RT\Delta n \end{aligned} \qquad \textbf{15–7}$$

If we begin with one mole of O_2 and two moles of H_2, Δn is -1 and $P\Delta V$ is calculated:

$$P\Delta V = (1.987)(298)(-1) = -592 \text{ cal at } 298\ ^\circ\text{K}$$

The measured ΔE for reaction 15–1 at 25 °C is -1.150×10^5 cal, and the term $P\Delta V$ is relatively insignificant in this case and in the case of most chemical reactions. Nevertheless, there are processes for which ΔE is relatively small, and for these $P\Delta V$ becomes very significant. As a consequence of its important relationship to heat transfer, the enthalpy is encountered in chemistry far more often than the energy, and enthalpy changes will be referred to frequently in the remainder of this textbook. In fact, this has been surreptitiously done even before this chapter. The values of the "bond energies" listed in Table 6–3 are in reality **bond enthalpies.** This liberty was taken because the differences between bond energies and bond enthalpies are relatively small, and at the time we were not ready for thermodynamics. Because of the small differences between bond energies and bond enthalpies, none of the conclusions of our intervening considerations are in any way affected by this small bit of deception. From this point on, we shall refer to the numbers in Table 6–3 by their true designation, bond enthalpies.

Standard Enthalpies of Formation

Since enthalpy changes in chemical transformations are very important, we should like to have them tabulated for easy referral. Unfortunately, such a tabulation would have to be massive, since the number of possible chemical reactions is astronomical. The task would be greatly simplified if we could tabulate an enthalpy value for one mole of each reactant and product. Then the enthalpy change for a reaction such as 15–1 would be given as follows:

$$\Delta(\text{enthalpy}) = 2(\text{enthalpy})_{H_2O} - (\text{enthalpy})_{O_2} - 2(\text{enthalpy})_{H_2} \qquad \textbf{15–8}$$

where $(\text{enthalpy})_{H_2O}$, for example, is the molar enthalpy of gaseous water.

Unfortunately, there is no such quantity as the absolute enthalpy of any substance since the absolute zero of enthalpy is not defined. Enthalpy is a function of energy E, and E is made up of the sum of the kinetic and potential energies of the particles of the system. The zero of potential energy is an arbitrary quantity, and consequently so are E and H.

Fortunately, we can construct a table of relative enthalpies that proves to be equally as useful as absolute enthalpies would be in the calculation of enthalpy changes. Consider the following reaction as an example:

$$CaO(s) + CO_2(g) \rightarrow CaCO_3(s) \qquad \textbf{15–9}$$

Enthalpies depend on both temperature and pressure. Thus, if we are going to tabulate them, we must be careful to specify both the temperature and the pressure. By agreement, the temperature and the pressure chosen for this tabulation are 298 °K and 1 atm. They are referred to as the **standard conditions,** and the enthalpies reported for these conditions are termed **standard enthalpies** (symbolized by $H°$).

The standard enthalpy change $\equiv \Delta H° \equiv$ (ΔH at 1 atm pressure and 298 °K) for reaction 15–9 is represented as follows:

$$\Delta H° = H°_{CaCO_3} - H°_{CaO} - H°_{CO_2} \qquad \textbf{15–10}$$

Each $H°$ of a compound is the sum of the $H°$ values of the component elements and the $\Delta H°$ of formation of the compound from the elements. Thus, the $\Delta H°$ in equation 15–10 may be reformulated:

$$\Delta H° = \Delta H°_{f(CaCO_3)} + H°_{Ca} + H°_{C} + (\tfrac{3}{2})H°_{O_2} - \Delta H°_{f(CaO)} - H°_{Ca}$$
$$- (\tfrac{1}{2})H°_{O_2} - \Delta H°_{f(CO_2)} - H°_{C} - H°_{O_2} \qquad \textbf{15–11}$$

where $\Delta H°_f$ symbolizes the **standard heat of formation.** The standard absolute enthalpies of the elements cancel in this reaction and any chemical reaction because the number and kind of each element is always conserved. Thus, equation 15–11 reduces to

$$\Delta H° = \Delta H°_{f(CaCO_3)} - \Delta H°_{f(CaO)} - \Delta H°_{f(CO_2)} \qquad \textbf{15–12}$$

and the standard enthalpies of formation may be used just as the standard enthalpies are in equation 15–10. In general, for any reaction,

$$\Delta H^\circ = \Sigma \Delta H^\circ_{f(\text{products})} - \Sigma \Delta H^\circ_{f(\text{reactants})} \qquad \textbf{15–13}$$

where Σ indicates the sum of the enthalpies of formation of the products or reactants.

Reactions for which ΔH° is negative spread energy by heat transfer to the surroundings and are termed **exothermic reactions,** while reactions for which ΔH° is positive are termed **endothermic reactions.**

Hess's Law

The measurement of ΔH°_f might seem like an easy process for a given compound. All that we should have to do is to mix the reactant elements in a cylinder, such as that shown in Figure 15–2, initiate the reaction, and measure the temperature change after the reaction proceeds to completion. The enthalpy change can be calculated from the temperature change. However, with a compound of any significant complexity, the component elements cannot be made to react spontaneously to produce the desired compound and only the desired compound. For example, consider acetylene, C_2H_2, for which the formation reaction is

$$H_2(g) + 2\,C(s) \rightarrow C_2H_2(g) \qquad \textbf{15–14}$$

There are thousands of known compounds that contain only carbon and hydrogen, and the reaction above can never be the exclusive one between carbon and hydrogen.

The solution of this problem is provided by the fact that any chemical reaction can be expressed as the combination of other chemical reactions. In the case of reaction 15–14, it is possible to relate the ΔH°_f to heats of reactions that do proceed spontaneously and cleanly to yield well-defined products. One type of reaction that proceeds quite well with most species is combustion, that is, reaction with gaseous oxygen, owing to the very high reactivity of gaseous oxygen. Similar combustions could be carried out with other reactive species such as $Cl_2(g)$ or $F_2(g)$. The following three combustion reactions proceed essentially to completion and their heats (the energies lost by heat transfer) are measurable under standard conditions.

1. $2\,H_2(g) + O_2(g) \rightarrow 2\,H_2O(l)$; $\quad \Delta H^\circ = 2\,\Delta H^\circ_{f(H_2O)}$
2. $C(s) + O_2(g) \rightarrow CO_2(g)$; $\quad \Delta H^\circ = \Delta H^\circ_{f(CO_2)}$
3. $2\,C_2H_2(g) + 5\,O_2(g) \rightarrow 4\,CO_2(g) + 2\,H_2O(l)$; $\quad \Delta H^\circ = 2\,\Delta H^\circ_{\text{comb.}(C_2H_2)}$

The symbol $\Delta H^\circ_{\text{comb.}}$ denotes the **standard enthalpy of combustion** of acetylene. Reaction 15–14 is just [reaction 1 + 4 (reaction 2) − reaction 3]/2.

In the nineteenth century, G. H. Hess concluded, from experimental work he performed, that the **heats of reactions may be combined in the same way as the reactions.** This conclusion is termed **Hess's law.** In the present case, Hess's law becomes

$$\Delta H^\circ_{f(C_2H_2)} = \frac{2\,\Delta H^\circ_{f(H_2O)} + 4\,\Delta H^\circ_{f(CO_2)} - 2\,\Delta H^\circ_{comb.(C_2H_2)}}{2} \quad \textbf{15–15}$$

Calorimetry

The construction of a table of standard enthalpies for most compounds is critically dependent on the precise measurement of heats of combustion, as for the case of acetylene. These combustion reactions are conveniently carried out in a **bomb calorimeter** as shown in Figure 15–3.

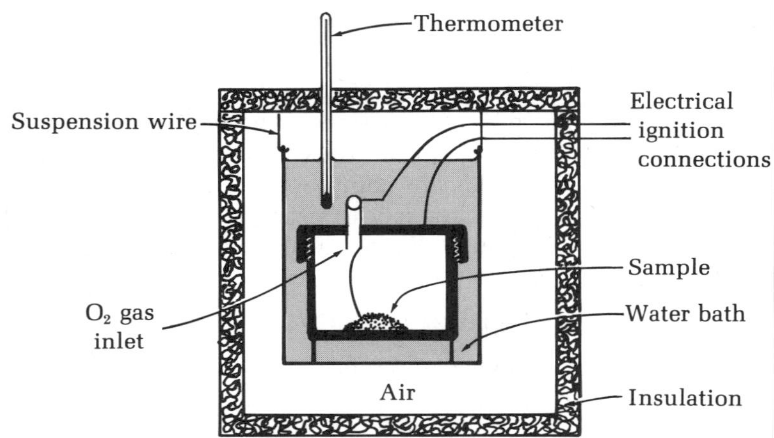

Figure 15–3
Bomb calorimeter for the measurement of heats of combustion.

The calorimeter is an excellent illustration of the application of the first law of thermodynamics. The sample to be combusted and a large excess of oxygen gas (to ensure complete reaction) are placed in a sealed steel bomb, which is then immersed in a bath of water. The bomb and bath are suspended by thin wires, as shown in the figure, so as to insulate them from their surroundings. The energy liberated in the combustion reaction is converted to thermal energy in the reaction products, the bomb, and the water bath. The reaction is initiated by an electrical current through the electrical leads, as shown in the figure, and after initiation the reaction proceeds spontaneously to completion. The thermal energy manifests itself in the temperature rise of the products, bomb, and water bath, and the temperature rise is recorded on the thermometer immersed in the water bath.

According to the first law of thermodynamics, the sum of ΔE_{bath}, ΔE_{bomb}, and $\Delta E_{chemicals}$ must be zero since energy is conserved. To be strictly correct, we should also add a fourth ΔE equal to the electrical energy added in the initiation process, but we shall ignore that small energy input here. No work is performed as the reaction proceeds, since the reaction is isolated from the atmosphere by the sealed bomb.

Let us suppose that the initial temperature of the reactants, bomb, and water is 298 °K and that the final, raised temperature is T_H. The overall change in the combustion is

$$\text{Bath at } 298\,°\text{K} + \text{Bomb at } 298\,°\text{K} + \text{Reactants at } 298\,°\text{K} \rightarrow \text{Bath at } T_H + \text{Bomb at } T_H + \text{Products at } T_H \qquad \textbf{15–16}$$

This overall process may be treated as the sum of several separate changes with well-defined ΔE values.

1. Reactants at 298 °K → products at 298 °K; $\Delta E = \Delta E^\circ_{\text{reaction}}$
2. Products at 298 °K → products at T_H; $\Delta E = (T_H - 298)\,\Sigma(nC_v)_{\text{products}}$
3. Bomb at 298 °K → bomb at T_H; $\Delta E = (\text{wt})_{\text{bomb}}(\text{sp ht})_{\text{bomb}}(T_H - 298)$
4. Bath at 298 °K → bath at T_H; $\Delta E = (T_H - 298)(nC_v)_{H_2O}$

Since the sum of all four ΔE's must be zero, we have

$$\Delta E = 0 = \Delta E^\circ_{\text{reaction}} + (T_H - 298)[\Sigma(nC_v)_{\text{products}} + (\text{wt})_{\text{bomb}}(\text{sp ht})_{\text{bomb}} + (nC_v)_{H_2O}] \qquad \textbf{15–17}$$

The stepwise nature of equation 15–17 is also shown in energy changes diagrammed in Figure 15–4.

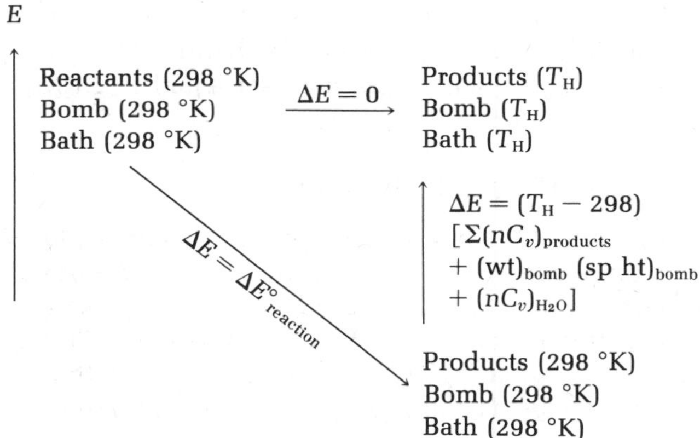

Figure 15–4
Energy changes accompanying a reaction carried out in a bomb calorimeter.

The temperature T_H is measurable and so are the numbers of moles n of all products formed and also their heat capacities C_v. The same is true of the weight and specific heat of the bomb and also the number of moles of bath water and the heat capacity of the bath water. The only unknown in equation 15–17 is $\Delta E^\circ_{\text{reaction}}$, and thus $\Delta E^\circ_{\text{reaction}}$ may be calculated from equation 15–17.

Let us apply equation 15–17 to an actual situation, namely a determination of the heat of combustion of liquid benzene. A 0.2530-g sample of benzene was completely combusted to CO_2(g) and H_2O(l) in a 1.000-kg steel bomb immersed in 1.000 liter of water. The temperature rise recorded for the bath, bomb, and contents was 2.29 °C. The specific heat of steel is 0.107 cal/g-°C, and thus ΔE_{steel} is (0.107)(1000)(2.29) = 245 cal; and ΔE_{H_2O} is $(\frac{1000}{18.0})(18.00)(2.29) = 2.29 \times 10^3$ cal, where 18.00 is C_v for liquid water. The products are 0.0195 mole of $CO_2(g)$ and 9.73×10^{-3} mole of $H_2O(l)$. The ΔE for the heating of these products is much less than one calorie and can be neglected in this calculation and in all bomb calorimeter calculations. From equation 15–17, we have

$$\Delta E^\circ_{reaction} = -245 - 2.29 \times 10^3 = -2.53 \times 10^3 \text{ cal}$$

This is the energy liberated by heat transfer by the combustion of (0.2530)/78 $= 3.24 \times 10^{-3}$ mole of benzene, so that the energy of combustion of one mole of liquid benzene is

$$\frac{(-2.53 \times 10^3)}{(3.24 \times 10^{-3})} = -7.81 \times 10^5 \text{ cal/mole}$$

The final step is to convert from $\Delta E^\circ_{reaction}$ to $\Delta H^\circ_{reaction}$. According to equation 15–7, the difference is $RT\Delta n$, where Δn is the change in the number of moles of gases in the course of the reaction. The balanced equation is

$$C_6H_6(l) + \tfrac{15}{2}\ O_2(g) \rightarrow 6\ CO_2(g) + 3\ H_2O(l) \qquad \textbf{15–18}$$

and Δn for the gases is $(-\frac{3}{2})$ so that $P\Delta V = RT\Delta n = -888$ cal at 298°. This is negligible compared with 7.81×10^5 cal, and thus the bomb calorimeter gives us a value of $\Delta E^\circ_{comb.}$, which we may set equal to $\Delta H^\circ_{comb.} = -7.81 \times 10^5$ cal/mole $= -781$ kcal/mole.

A Sample Calculation Using the First Law

In calorimetry we use the first law of thermodynamics to calculate an energy change from a measured temperature change. Eventually, we shall need to know temperature changes based on known or calculable energy changes. A 500-g block of iron at 525 °K is immersed in a 250-g bath of water at 278 °K. What is the final temperature? Energy is lost by the hot iron block and gained by the water as both go to some final temperature between 278 °K and 525 °K.

The first law of thermodynamics demands that the overall ΔE be zero; that is, the thermal energy that the iron loses must equal the thermal energy the water gains:

$$\Delta E = 0 = (nC_v)_{iron}(T_f - 525) + (nC_v)_{H_2O}(T_f - 278) \qquad \textbf{15–19}$$

The heat capacity of iron is 6.1 cal/mole-deg. With this value T_f, the final temperature, is calculated from equation 15–19:

$$\Delta E = 0 = \left(\frac{500}{55.85}\right)(6.1)(T_f - 525) + \left(\frac{250}{18.0}\right)(18.0)(T_f - 278)$$

$$T_f = 322\ °K = 49\ °C$$

First law calculations are always a matter of energy bookkeeping. The energy lost by one system must always equal the energy gained by all other systems, and we must see that the energy debits equal the energy credits.

Standard Heats of Formation and Bond Enthalpies

Table 15–1 lists the standard heats of formation ΔH_f° for a number of organic and inorganic compounds. (This table is also given in Appendix D.) In each case the state of the compound as a solid, liquid, or gas is listed along with the compound. The states of the elements that are the reactants in the stan-

Compound	ΔH_f° (kcal/mole)
Inorganic	
Al_2O_3(s)	−399.09
AgCl(s)	−30.36
B_2H_6(g)	7.5
$BaCl_2$(s)	−205.6
C(diamond)	0.45
CaO(s)	−151.79
$CaCO_3$(s)	−288.45
CO(g)	−26.42
CO_2(g)	−94.05
Fe_2O_3(s)	−196.5
HBr(g)	−8.66
HCl(g)	−22.06
HF(g)	−64.2
HI(g)	6.2
H_2O(g)	−57.80
H_2O(l)	−68.32
$HgCl_2$(s)	−55.0
KCl(s)	−104.18
$MgCl_2$(s)	−153.40
MgO(s)	−143.84
NH_3(g)	−11.04
NO(g)	21.60
NO_2(g)	8.09
N_2O_4(g)	2.31
NaCl(s)	−98.23
Na_2CO_3(s)	−270.3
O_3(g)	34.0
SO_2(g)	−70.76
SiO_2(quartz)	−205.4
$ZnCl_2$(s)	−99.4

Table 15–1
Standard Heats of Formation, ΔH_f°

Table 15–1 (continued)

Organic	
C_2H_2(g) (acetylene)	54.19
C_6H_6(l) (benzene)	11.72
CCl_4(l) (carbon tetrachloride)	−33.4
$CHCl_3$(l) (chloroform)	−31.5
C_2H_6(g) (ethane)	−20.24
C_2H_4(g) (ethylene)	12.50
CH_4(g) (methane)	−17.89
CH_3OH(l) (methanol)	−57.04

dard formation reactions are always the most commonly encountered ones at 298 °K and 1 atm. Thus, the standard states of some of the elements are H_2(g), Br_2(l), C(graphite), O_2(g), and some standard formation reactions at 25 °C and 1 atm are

$$\text{C(graphite)} + O_2(g) \rightarrow CO_2(g)$$

$$H_2(g) + Br_2(l) \rightarrow 2\,HBr(g)$$

$$2\,Al(s) + 3\,Cl_2(g) \rightarrow 2\,AlCl_3(s)$$

Let us illustrate the use of Table 15–1 by calculating the heat of combustion of benzene according to equation 15–18. The $\Delta H°$ value for the combustion is calculated:

$$\Delta H° = 3\Delta H^{\circ}_{f(H_2O,l)} + 6\Delta H^{\circ}_{f(CO_2,g)} - \Delta H^{\circ}_{f(C_6H_6,l)}$$

$$= 3(-68.32) + 6(-94.05) - (11.72) = -781 \text{ kcal}$$

The number agrees with the results of our bomb calorimeter experiment.

We can also use the data of Table 15–1 to calculate the bond enthalpies of Table 6–3, and this is how a table of bond enthalpies is constructed. For example, the OH bond enthalpy may be taken as half of the $\Delta H°$ for the following reaction that splits two OH bonds:

$$H_2O(g) \rightarrow 2\,H(g) + O(g) \qquad \textbf{15–20}$$

Reaction 15–20 is not the reverse of the standard formation reaction of H_2O(g), since H atoms and O atoms are not the standard states of the elements. In order to go from $\Delta H^{\circ}_{f(H_2O,g)}$ to $\Delta H°$ for reaction 15–20, we require values for the standard enthalpies of formation of hydrogen and oxygen atoms, that is, $\Delta H°$ values for the following two reactions:

$$\tfrac{1}{2}\,H_2(g) \rightarrow H(g) \qquad \textbf{15–21}$$

$$\tfrac{1}{2}\,O_2(g) \rightarrow O(g) \qquad \textbf{15–22}$$

These **atomization enthalpies** have been obtained for a number of atoms by a combination of calorimetry and spectroscopy, and several of the most important values are listed in Table 15–2. In the cases of H, O, and N, the enthalpies are simply half of the H_2 and O_2 bond enthalpies listed in Table 6–3. In the case of carbon, the enthalpy is for the formation of gaseous carbon atoms from graphite and this does not relate simply to any bond enthalpy listed in Table 6–3.

Atom	ΔH_f° (kcal/mole)
H	52.1
C	171.7
N	112.5
O	59.2
F	18.3
Cl	29.0
Br	26.7
I	25.5

Table 15–2
Standard Enthalpies of Formation of Gaseous Atoms

With the values from Tables 15–1 and 15–2, the ΔH° for reaction 15–20 is calculated as follows:

$$\Delta H^\circ = 2\Delta H^\circ_{f(H,g)} + \Delta H_{f(O,g)} - \Delta H^\circ_{f(H_2O,g)}$$

$$= 2(52.1) + 59.2 - (-57.8) = 221.2 \text{ kcal/mole}$$

and the average OH bond enthalpy in gaseous H_2O is $221.2/2 = 111$ kcal/mole. This is the value listed in Table 6–3.

15–2. A Criterion for Spontaneity

The Concept of Spread

Perhaps the most obvious everyday phenomenon is spontaneous change. Furthermore, spontaneous changes appear to have a natural sense of direction. We are not surprised to see an object roll down a hill, but the same object rolling up the hill is likely to make us investigate this unusual event. When a coin is dropped onto the floor, we expect it to bounce slightly and then sit there quietly. It would disturb us immensely if a stationary coin should suddenly jump six feet off the floor with no visible agent present. We all have an inherent sense, based on experience, of the natural way in which changes occur. When we have converted that sense into one general, simple concept that will tie together all changes, the unfamiliar as well as the familiar, we can predict qualitatively the sort of changes that must occur in unfamiliar situations such as chemical reactions.

Consider the simple process pictured in Figure 15–5(a). In the cylinder at the left, a gas is compressed to pressure P_1 by a piston held by a stop so that gas expansion cannot occur. The stop is then removed and the gas spontaneously expands, pushing out the piston until it hits the end of the cylinder. Why did the process occur, and why did it finally stop? An obvious explanation is that when the stop was removed, the forces on the piston were imbalanced and the gas expanded until the piston hit the cylinder end and the forces could again rebalance. Thus we can hypothesize that force imbalance results in spontaneous change and that **equilibrium,** the condition in which no change occurs, corresponds to a balance of forces. This consideration of forces is correct, but let us examine its usefulness in general.

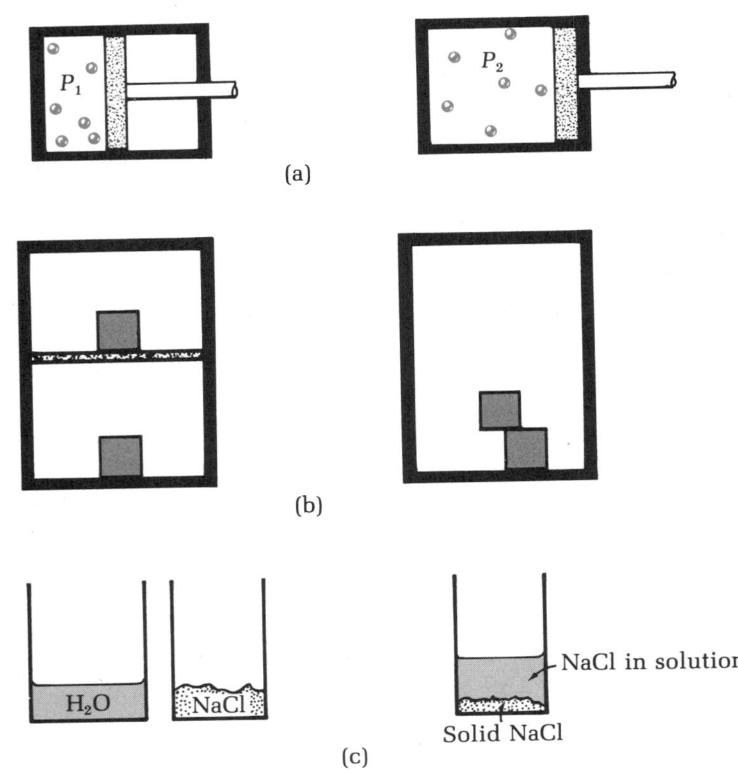

Figure 15–5
Three spontaneous processes: (a) the expansion of a gas against a lower pressure, (b) the transfer of energy from a hot substance to a cold substance, and (c) the dissolution of NaCl in water.

In Figure 15–5(b) we depict on the left a 100-g bar of iron at 100 °C separated by an insulating partition from a 100-g iron bar at 0 °C. The partition is then removed through a slot in the container wall. The hot iron bar falls on the cold bar, heat transfer takes place, and both bars reach a temperature of 50 °C. Why did spontaneous heat transfer take place? There is no obvious imbalance of forces between the hot and cold iron bars (though there is if we examine atomic collisions). The examination of forces on an atomic level is a difficult undertaking, in general, and we want a simple, useful explanation of spontaneity and equilibrium, if such indeed exists.

Figure 15-5(b) on the left also contains an imbalance, namely that of temperature, and equilibrium on the right corresponds to the temperature balance, or equality. Thus, from our observations it might be proposed that inequalities of intensive properties like pressure and temperature cause spontaneous changes and that equilibrium corresponds to equality of these same properties. This proposal certainly holds for Figures 15-5(a) and (b). But is it true in general?

Figure 15-5(c) shows, on the left, separate containers of water and sodium chloride. The solid NaCl is emptied into the beaker of water and a spontaneous dissolution occurs, resulting in the saturated solution shown on the right. Apparently spontaneity is caused by the inequality of NaCl concentration in solid NaCl versus pure water, but some inequality persists at equilibrium also. Although inequality persists, no spontaneous process results. Thus, the inequality of intensive parameters is not a general criterion for spontaneity, nor is the equality of intensive properties a general criterion for equilibrium.

Let us return to Figure 15-5(a) to see whether we can draw any other conclusions. The compressed gas on the left appears to be uniformly distributed in the volume to the left of the cylinder. It is, if you like, spread out to the maximum possible degree. A condition of uniform distribution, or maximum spread, is also observed in the cylinder on the right. However, because of the larger available volume, the spread of the gas on the right is greater than the spread for the gas on the left. Perhaps the general spontaneity and equilibrium criterion is that matter tends to spread out to the maximum possible degree.

In Figure 15-5(b), the matter-spread criterion fails because matter (the hot iron bar) is transported by gravity, but the concentration of matter stays constant. However, in this case, the uniform distribution of another quantity is achieved. The hot iron bar contains more energy (largely as the vibrational energy of iron atoms) than the cold bar. In the final state at 50 °C, this energy is uniformly distributed between the two bars and energy has been spread to the maximum possible degree within the constraints of the container. We can modify our criterion to read that matter and energy tend to spread out to the maximum possible degree; that is, both matter and energy tend toward as uniform a distribution as possible.

A simple conclusion from this idea is that any process that involves matter spread but no energy transfer is predicted to be spontaneous, as in Figure 15-5(a). Such processes involving no energy transfer (or very little) are not common, but those that do occur reinforce the conclusion. The mixing of gases is always spontaneous since it involves matter spread (each gas can occupy the original enclosing volume of the other as well as its own), and little energy transfer is involved because gaseous molecules interact only weakly. Any dissolution process should proceed well on matter-spread grounds if little or no energy transfer is involved. This idea finds verification in the rule that "like dissolves like": polar solvents dissolve polar materials, nonpolar solvents dissolve nonpolar materials, metals dissolve metals, and so forth. The key factor in like dissolving like is that the atoms or molecules in the mixture experience interactions with their surroundings similar to the interactions in the pure materials. Thus, the energy transfer on mixing is small, and dissolution is strongly driven by the matter-spread criterion. Alkali metal salts, we have found, are highly soluble in water because they dissolve with neither the absorption nor the liberation of a large amount of energy. Thus their solubility is accountable for by the matter-spread criterion.

Another simple conclusion to be tested is that any process that involves energy spread but no change in the spreadedness of matter is also predicted to be spontaneous. There are many examples of such spontaneity—that shown in Figure 15–1, for instance. Potential energy concentrated in the object is converted to thermal energy, which is free to spread where it can. If energy is not spread in this process, that is, if the collision with the floor is perfectly elastic, then the reverse process is also spontaneous and the book will bounce back to its original height. Equilibrium can be reached only when the original potential energy has been spread in the form of thermal energy.

Matter-spread and energy-spread criteria are useful concepts in simple cases involving only matter spread or energy spread, but most processes involve the two tendencies competively. In Figure 15–5(c) the dissolution of NaCl is favored on matter-spread grounds; why therefore isn't NaCl soluble in water in all proportions? If it were, even a trace of water added to a large sample of NaCl(s) would liquefy it, and this does not occur. No solid is infinitely soluble in any liquid, since as the solution becomes concentrated, the dissolution of further amounts of any solid becomes a highly endothermic process (even if the first solid added to pure water dissolves in a highly exothermic fashion). Thus the saturated solution represents a compromise between the matter spread of the dissolution and the energy spread of the precipitation of the solid from a saturated solution.

If we are to be able to predict when spontaneous changes will occur and where they will stop in cases where matter spread and energy spread oppose each other, we must be able to evaluate energy spread in terms of equivalent matter spread or both in terms of some third quantity so we can know which dominates in any given case and by how much. This should enable us to calculate the position of equilibrium for any proposed system or process.

Entropy and the Second Law of Thermodynamics

Let us begin our quantitative analysis of matter spread by reference to Figure 15–6(a). The figure shows two gases, A and B, separated by a partition.

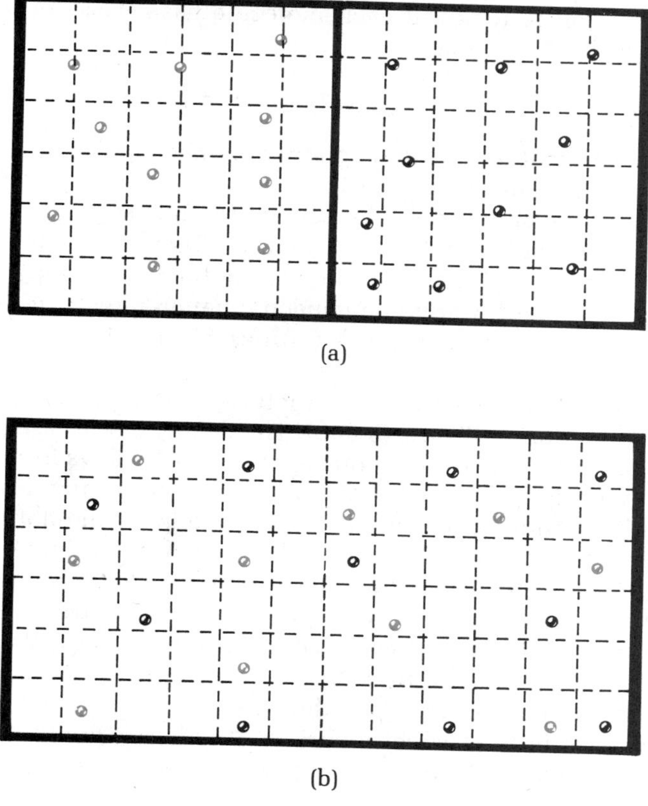

Figure 15-6 Molecular positions (a) before and (b) after spontaneous matter spreading in a gas.

Both gases are at the same temperature and pressure and they occupy the same volume. Therefore, the number of moles of A equals the number of moles of B. We have divided each of the two volumes into equal-sized cells by drawing in the dashed lines shown. Each cell on the left represents a position possibility for each of the A molecules shown in color, and each cell on the right is a position possibility for each B molecule.

What is the total number of positional options of the ten A molecules shown? There are 36 cells, and thus each molecule has 36 positional options. But the ten molecules are independent and their positional possibilities multiply. This multiplication of possibilities is exemplified by the throwing of two dice. Each die can come to rest with any one of six faces up. Thus, the total number of possibilities for the two dice together is (6)(6) = 36. In the case of the 36 cells and 10 molecules, we have an analogy to 10 dice and 36 faces, and the number of possibilities is $(36)^{10}$. This number of possibilities for molecules is commonly given the symbol Ω (Greek uppercase *omega*); thus Ω_A is $(36)^{10}$. Likewise, Ω_B is also $(36)^{10}$. Since A and B are independent, the combined number of possibilities is as follows:

$$\Omega_{AB} = \Omega_A\Omega_B = (36)^{10}(36)^{10} = (36)^{20} \qquad \textbf{15-23}$$

When the partition is removed each molecule has access to 72 cells, and Ω_{AB} changes to the following:

$$\Omega_{AB} = \Omega_A\Omega_B = (72)^{10}(72)^{10} = (72)^{20} \qquad \textbf{15–24}$$

The spontaneous matter spread was accompanied by an increase in positional possibilities, and the Ω_{AB} for the combined gases is $(2)^{20}$ times larger than Ω_{AB} for the separated gases. Furthermore, this factor of $(2)^{20}$ holds regardless of the size of the cells, as we should discover if we were to change the number and size of the cells in Figure 15–6. This type of position analysis can be made for any spontaneous matter-spreading process, and the total number of positional possibilities always increases in such a process.

Can we apply the same sort of Ω analysis to energy spreading? Let us consider the situation in Figure 15–5(b). The energy in the case of solid iron is the vibrational energy of iron atoms. Figure 15–7 shows iron-atom vibrational energy levels and an equal number of iron atoms (circles) in the hot bar (left) and the cold bar (right). The extra energy of the hot bar is manifested in the greater population of higher-energy vibrational levels. The levels in Figure 15–7 represent motional possibilities to the atoms in analogy with the positional possibilities of Figure 15–6. If the occupied levels in Figure 15–7 represent the only significantly populated ones, we can perform an Ω analysis, just as we did in connection with Figure 15–6. However, in this case Ω represents motional possibilities.

The Ω analysis in the case of Figure 15–7 is relatively complicated, but we can see what must happen by analogy with Figure 15–6. In Figure 15–6(b)

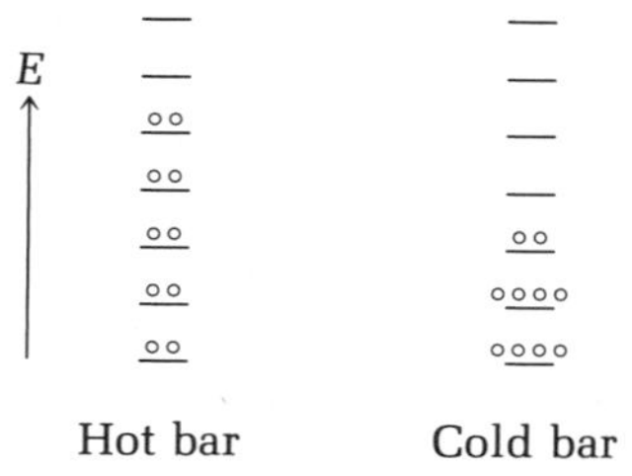

Figure 15–7 The occupations of vibrational energy levels in two iron bars at different temperatures.

each molecule is offered a choice between residing in any of the original 36 cells of Figure 15–6(a) or 36 new cells made available by the removal of the partition. It is these additional choices that led to the increase in Ω_A and in Ω_{AB}. On the left of Figure 15–5(b), extra energy is restricted to the hot bar. When the bars are brought in contact, the energy can remain as the vibrational energy of the original hot atoms or it can transfer to the newly accessible cold atoms. Just as the molecules in the gas distribute themselves uniformly between the cells, so the energy distributes itself uniformly between the original hot bar and the original cold bar. The Ω that is the total of motional possibilities in Figure 15–7 increases in the spontaneous energy transfer just as the positional Ω increases in Figure 15–6.

The atoms of any given substance have both motional and positional possibilities and the two are independent of each other; that is, the fact that an atom is in a given position gives no information whatever about its motion. Thus, all substances are characterized by a combined positional and motional Ω given by the equation

$$\Omega_{\text{substance}} = \Omega_{\text{position}}\Omega_{\text{motion}} \qquad \textbf{15–25}$$

If two or more substances interact as in Figures 15–6 and 15–7, the spontaneous change is such as to maximize the product of the Ω's of all of the substances involved within the constraints the system imposes. The tendency in nature toward maximum spread is equivalent to the tendency to maximize the combined Ω of all of the atoms in the universe. The tendency is to have as many atoms as possible share the positional and energetic wealth of the universe, and this tendency provides the only driving force for spontaneous change.

For any measurable amount of any substance, $\Omega_{\text{substance}}$ is huge because it gives the combined positional and motional possibilities of the order of Avogadro's number of atoms. Thus, it proves convenient to scale $\Omega_{\text{substance}}$ down in some manner so that we have manageable numbers to handle. For two interacting substances A and B, their combined Ω is represented as

$$\Omega_{\text{combined}} = \Omega_{\text{A}}\Omega_{\text{B}}$$

Suppose that we take the logarithm of each side of this expression and multiply by a positive constant:

$$(\text{Constant}) \log \Omega_{\text{combined}} = (\text{constant}) \log \Omega_{\text{A}} + (\text{constant}) \log \Omega_{\text{B}} \qquad \textbf{15–26}$$

Figure 15–8 shows a plot of log Ω as a function of Ω. Since log Ω consistently increases as Ω increases, log Ω maximizes in any physical or chemical change exactly where Ω maximizes. If the constant multiplying the logarithm has any positive value, (constant) log Ω will also maximize where Ω maximizes. We define the **entropy** S of any system as follows:

$$S \equiv (\text{constant}) \log \Omega \qquad \textbf{15–27}$$

The spontaneity criterion that we have deduced is that any process is spontaneous that increases the combined Ω of all substances in the universe. With equation 15–27, an equivalent criterion is that **any process is spontaneous that increases the total entropy of the universe,** where S_{universe} is the sum of the entropies of all substances in the universe. This statement of the spontaneity criterion in terms of entropy is the **second law of thermodynamics.** The principle inherent in equation 15–27 was first proposed at the beginning of this century by Ludwig Boltzmann, who was one of the

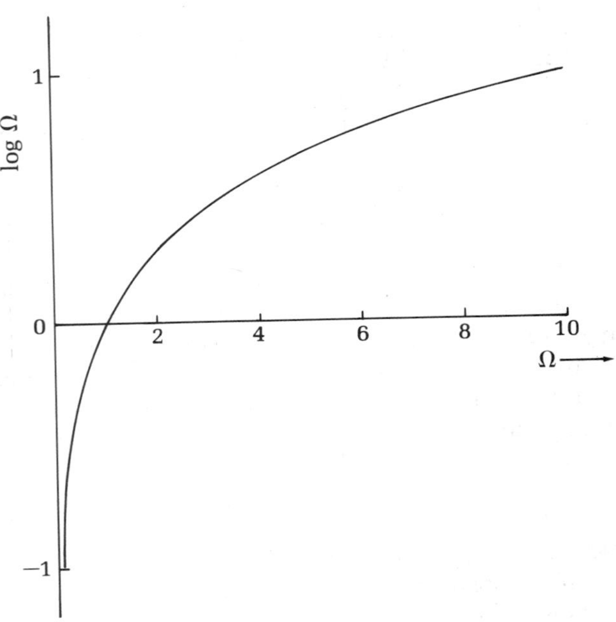

Figure 15–8
Plot of log Ω as a function of Ω.

first scientists to recognize a connection between spontaneity and the motions and positions of atoms.

In our subsequent investigation of entropies of various substances under a variety of conditions and the consideration of spontaneity in particular systems, it should always be kept in mind that entropy is only a convenient form of positional and motional spread. Any qualitative conclusions that develop about entropy in general or in specific systems should and will follow from the simple concepts of positional and motional spread.

Qualitative Conclusions about Entropy

The idea that entropy is a measure of atomic positions and motions within a system helps us to deduce how the entropy must depend qualitatively on such measurables as pressure, temperature, and concentrations of substances in solution. We can thus obtain a better grasp of this quantity and derive a way of measuring it. The following conclusions can be drawn immediately:

1. For a given amount of a pure substance at a given pressure and temperature, gaseous molecules possess a larger entropy than molecules in the liquid phase, which in turn possess a larger entropy than molecules in the solid phase. This is in direct relation to the greater variety of motional possibilities (free translation, free rotation, vibration) that gas molecules have over liquid molecules (restricted translation, restricted

rotation, vibration) and liquid over solid (vibration). In addition, the molar volume of a gas at a given P and T is much larger than the molar volume of the corresponding liquid or solid, and this adds a considerable position spread to the entropy of the gas-phase molecules.

If gas molecules contribute so much to the entropy of the universe, why isn't everything gaseous? That is, why doesn't everything spontaneously vaporize? Gas-phase molecules possess the highest entropy of the three phases, but they also possess the highest energy, since the intermolecular attractive forces responsible for the condensed liquid and solid phases must be overcome in producing the gaseous molecules. Thus, energy must be taken from another system (usually by heat transfer) in order to vaporize the first. Whether or not this will be spontaneous depends on what this energy extraction does to the entropy of the second system and not only to the first system, since spontaneity requires a total entropy increase in the universe. Thus, we must consider how the addition of energy to a substance or extraction therefrom affects its entropy.

2. The entropy of a pure substance increases if its thermal energy is increased at constant pressure or constant volume. The Ω of such a substance increases with the thermal energy increase due to an increase in motional possibilities. According to equation 15–27, the entropy of a system must increase if Ω increases for that system under any conditions.
3. A given thermal energy increase in a given amount of a substance at a given initial pressure and temperature produces a greater increase in the entropy of the substance if the initial temperature is low than if it is high. This follows directly from the consideration of the hot and cold iron bars of Figure 15–5(b). The total entropy change is positive (since the combined Ω increased) and is made up of the sum of a negative change for the hot bar (entropy loss) and a positive change for the cold bar (entropy gain). The gain must have exceeded the loss; but both entropy changes resulted from the transfer of the same amount of energy. The energy lost at higher temperature produced a smaller entropy loss than the gain in entropy produced by the same energy at a lower temperature.
4. For a pure substance at a given temperature, the entropy increases as the pressure decreases. As the pressure decreases, the volume increases, and because of positional spread, the entropy increases.
5. In a solution of any substance in any other substance at a given pressure and temperature, the entropy of a given amount of the dissolved substance increases as its concentration decreases. This is also a simple consequence of position spread.
6. The more complicated the structure of a molecule, the higher its entropy in a given phase at a given P and T. The more complex the molecule, the more the internal motions (vibrations and rotations) it can have and thus the higher the entropy.

15–3. Measurement of Entropy

The Boltzmann equation (equation 15–27) is useful in a qualitative way, but it is not useful quantitatively if we do not know how to construct the Ω function and how Ω depends on those variables like temperature and pressure that define changes in any system. The detailed construction of the Ω function necessitates a knowledge of the molecular structures, motions, and interactions in the system under investigation. Gathering such knowledge is, in general, a difficult task and is the province of the science of **statistical mechanics.**

Isothermal Expansion of an Ideal Gas

Fortunately, there are a few cases in which the entropy change can be calculated simply and directly from the Boltzmann equation. According to this equation any entropy change is given as follows:

$$\begin{aligned}\Delta S &= S_{\text{final}} - S_{\text{initial}} \\ &= (\text{constant})(\log \Omega_{\text{final}} - \log \Omega_{\text{initial}}) \\ &= (\text{constant}) \log \frac{\Omega_{\text{final}}}{\Omega_{\text{initial}}} \qquad \textbf{15–28}\end{aligned}$$

The term $\Omega_{\text{final}}/\Omega_{\text{initial}}$ is relatively simple to calculate if only position spread is involved in the process, that is, if there are no changes in molecular motions or intermolecular interactions. Such a process is the expansion of an ideal gas at constant temperature, as shown in Figure 15–9. The constancy of temperature fixes a constancy of molecular motion, and by definition, there are no intermolecular interactions in an ideal gas. Suppose that in the isothermal expansion the volume of the gas doubles; by what factor does Ω increase? The situation is very similar to that of Figure 15–6. In Figure

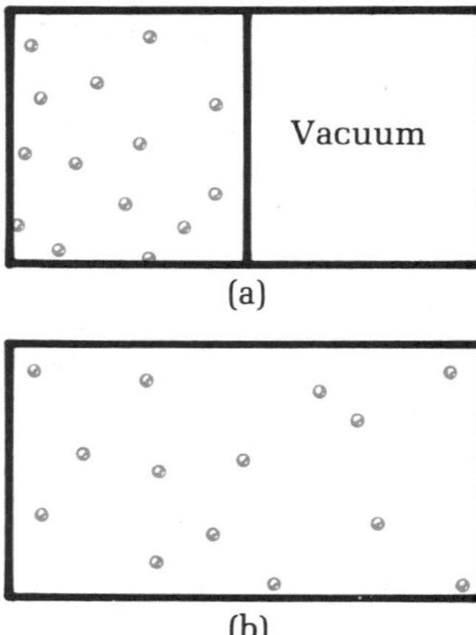

Figure 15–9
The isothermal expansion of an ideal gas into a vacuum: (a) before expansion and (b) after.

15–9(b) each molecule can be found in the original volume at any given time or it can be found in the new volume. If we have one mole of gas, there are $N =$ Avogadro's number such independent choices, and the number of position possibilities has increased by a factor of 2^N from the position possibilities in Figure 15–9(a). In Figure 15–6, each gas increases its Ω by a factor of $(2)^{10}$, so that the combined Ω increases by a factor of $(2)^{20}$. In the case of Figure 15–9,

$$\Omega_{\text{final}} = 2^N\, \Omega_{\text{initial}} \qquad \textbf{15–29}$$

Combined with equation 15–28, this yields the entropy increase

$$\Delta S = N(\text{constant}) \log 2 \qquad \textbf{15–30}$$

A general volume increase from V_i to V_f yields an entropy increase.

$$\Delta S = N\,(\text{constant}) \log \frac{V_f}{V_i} \qquad \textbf{15–31}$$

The process in Figure 15–9 involves pure matter spread; no transfer of energy is involved. Can we relate the ΔS of equation 15–31 to an equivalent energy spread? In other words, how much energy do we have to concentrate in the gas to balance the spontaneity due to the tendency of matter to spread? Figure 15–10 shows the same isothermal expansion carried out with the absorption of energy by the gas. The energy comes from a large water bath at the temperature T. The energy of an ideal gas depends only on its temperature, so ΔE_{gas} is zero in Figure 15–10. The absorbed energy is used to do work; this work is done by displacing a piston against an opposing force.

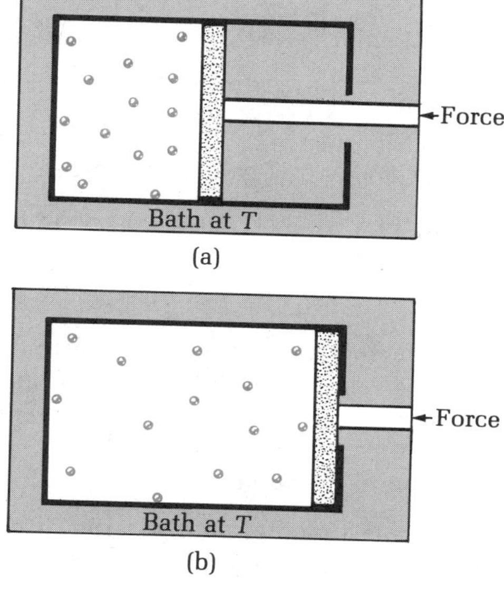

Figure 15–10
The isothermal expansion of an ideal gas under conditions in which energy is concentrated.

The gas pictured in Figure 15–10 expands just as the gas shown in Figure 15–9, but the expansion is slower in (b) because of the opposing force on the piston. How much energy must be absorbed for the speed of expansion to be zero, and the energy concentration exactly to balance the matter spread? The energy absorbed must equal the limit of the work done as the opposing force approaches the force on the piston that the gas exerts. By the calculus procedure of integration (a procedure beyond our scope) we find this limiting work to be the following:

$$W_{\text{limiting}} = RT \ln \frac{V_f}{V_i} = (2.303)\ RT \log \frac{V_f}{V_i} \qquad \textbf{15–32}$$

where ln denotes the **natural** logarithm. Appendix B at the end of the text contains a discussion of natural logarithms.

Since $\Delta E_{\text{gas}} = 0$, $Q_{\text{balancing}}$, the energy whose concentration balances the matter-spreading tendency, is equal to W_{limiting} in equation 15–32. A comparison of equations 15–31 and 15–32 reveals that

$$\Delta S = \frac{N\ (\text{constant})}{(2.303)RT}\ (Q_{\text{balancing}}) \qquad \textbf{15–33}$$

and we have a quantitative measure of an entropy change in terms of an energy taken by heat transfer from a bath at temperature T. Any ΔS can be measured in this balancing fashion, and equation 15–33 is the useful expression that we require for entropy changes. As a final touch, we can now choose a convenient value for the constant in equation 15–33:

$$(\text{Constant}) = \frac{(2.303)R}{N} = 2.303k \qquad \textbf{15–34}$$

where k is Boltzmann's constant, expressed in this case as follows:

$$k = \frac{1.987\ \text{cal/mole-deg}}{6.023 \times 10^{23}\ \text{molecules/mole}}$$

$$= 3.30 \times 10^{-24}\ \text{cal/molecule-deg}$$

With this substitution for the constant, equation 15–33 becomes

$$\Delta S = \frac{Q_{\text{balancing}}}{T} \qquad \textbf{15–35}$$

and the problem of measuring entropy changes becomes the problem of measuring $Q_{\text{balancing}}$ and the temperature of the bath from which $Q_{\text{balancing}}$ is taken.

Entropy Changes for Common Processes

A word used more commonly than *balancing* in the case of equation 15–35 is *reversible*. A balanced system is one that can be displaced in either of two directions by the slightest disturbance one way or the other. The balancing condition in Figure 15–10(a) occurs when the force that the gas exerts on the piston exactly equals the external force. Any slight decrease in the external force, and the gas will expand; and any slight increase will cause it to be compressed. Thus, the balanced state can be changed reversibly, and $Q_{balancing}$ in equation 15–35 is commonly termed $Q_{reversible}$ (Q_{rev}).

A common condition of reversibility occurs at the melting point of a solid (or freezing point of a liquid). Figure 15–11 shows a mixture of ice and liquid water in a beaker in a bath at exactly 0 °C. This is a condition of balance, or equilibrium. No change will be observed in the amounts of ice or water. If the slightest amount of energy is extracted from the bath causing its temperature to fall below 0 °C, some water will spontaneously freeze in the beaker. If a slight amount of energy is added to the bath, some ice will melt.

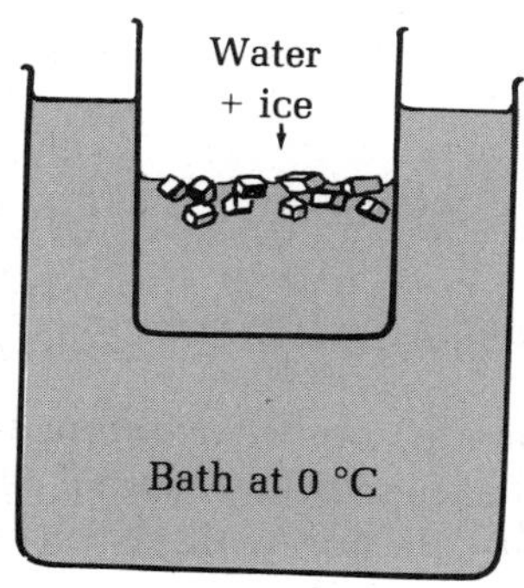

Figure 15–11
A condition of balance, liquid water and ice at 0 °C.

What is the balancing energy for the melting of ice at 0 °C? The driving force for the melting of any solid is the greater entropy of the liquid owing to the greater motion of the molecules of the liquid. On the other hand, the liquid state is also the one of higher energy and energy must be added to a solid in order to melt it. This energy (normally added by heat transfer) is termed the **enthalpy of fusion** (ΔH_{fusion}). At 0 °C the energy concentration just balances the tendency of the higher-entropy liquid to form, and the solid and liquid are in equilibrium. For any substance, the change of one mole of solid to one mole of liquid at its melting point produces an entropy increase in the substance:

$$\Delta S_{fusion} = \frac{\Delta H_{fusion}}{T_{fusion}} \qquad \textbf{15–36}$$

where T_{fusion} is the equilibrium melting point. If one mole of any substance freezes at the equilibrium freezing point, its entropy decreases:

$$\Delta S_{freezing} = \frac{-\Delta H_{fusion}}{T_{fusion}} \qquad \textbf{15–37}$$

A second balancing condition occurs at the boiling point of any substance. The vapor is the state of higher entropy, but energy equal to the enthalpy of vaporization (ΔH_{vap}) must be concentrated in the substance to make it vaporize. If one mole of a substance boils at its equilibrium boiling point, its entropy increases:

$$\Delta S_{vap} = \frac{\Delta H_{vap}}{T_{boiling}} \qquad \textbf{15–38}$$

where $T_{boiling}$ is the equilibrium boiling point. The conversion of one mole of vapor to liquid at $T_{boiling}$ decreases its entropy:

$$\Delta S_{condensation} = \frac{-\Delta H_{vap}}{T_{boiling}} \qquad \textbf{15–39}$$

A final important entropy change occurs when a substance is heated or cooled at constant pressure or constant volume. By the methods of calculus, we can obtain the two entropy changes for one mole of any substance as follows:

$$\Delta S_{(\text{temp change}, v)} = C_v \ln \frac{T_f}{T_i} \qquad \textbf{15–40}$$

$$\Delta S_{(\text{temp change}, p)} = C_p \ln \frac{T_f}{T_i} \qquad \textbf{15–41}$$

where equation 15–40 refers to constant volume and equation 15–41 refers to constant pressure. The temperatures T_f and T_i refer to the final and initial temperatures. In agreement with our qualitative conclusion in the preceding section, both equations show that an increase in temperature produces an entropy increase.

A Sample Entropy Calculation

We have used the first law of thermodynamics to show that when a 500-g block of iron at 525 °K is immersed in a 250-g bath of water at 278 °K, the final temperature attained is 322 °K. Since the energy transfer is spontaneous, the combined entropies of the iron and water must have increased in the process. Let us calculate ΔS to show that this is the case.

From equation 15–41, the entropy change is

$$\Delta S = \Delta S_{iron} + \Delta S_{H_2O} = (nC_p)_{Fe} \ln \frac{T_f}{T_i} + (nC_p)_{H_2O} \ln \frac{T_f}{T_i}$$

$$= \left(\frac{500}{55.85}\right)(6.1)(2.303) \log \frac{322}{525} + \left(\frac{250}{18}\right)(18)(2.303) \log \frac{322}{278}$$

$$= -26.7 + 36.7 = 10.0 \text{ cal/deg}$$

We note that ΔS is positive, as indeed it must be.

Absolute Entropies

Since entropies determine the direction of all spontaneous processes and the conditions for all equilibria, it would be extremely useful if we could have a table of standard entropies analogous to the table of enthalpies (Table 15–1). Unlike the enthalpy values that are given relative to an arbitrary base (the elements as zero), entropies can be obtained with respect to an absolute zero. From equation 15–27, an S of zero denotes an Ω of unity, or a single position possibility and motion possibility for each atom. This condition occurs at 0 °K where all atoms are in the lowest vibrational level and each has one motion possibility. In a crystal containing no vacancies (perfect crystal), each atom occupies a well-defined lattice site, and thus it has only one position possibility. These considerations are embodied in the **third law of thermodynamics,** which states that **the entropy of any perfect crystalline element or compound is zero at 0 °K.**

Since temperatures down to a small fraction of 1 °K have been obtained, it has been possible to measure heat capacities and enthalpies of phase transitions from essentially 0 °K up to any desired temperature. Suppose we wished to report the absolute entropy of nitrogen gas at 25 °C and 1 atm pressure (the standard conditions). Nitrogen melts at 63 °K and 1 atm pressure and boils at 77 °K and 1 atm. The heating of one mole of N_2 from 0 °K to 298 °K at 1 atm is constructed of the following steps:

1. $N_2(s, 0\ ^\circ K) \rightarrow N_2(s, 63\ ^\circ K)$
2. $N_2(s, 63\ ^\circ K) \rightarrow N_2(l, 63\ ^\circ K)$
3. $N_2(l, 63\ ^\circ K) \rightarrow N_2(l, 77\ ^\circ K)$
4. $N_2(l, 77\ ^\circ K) \rightarrow N_2(g, 77\ ^\circ K)$
5. $N_2(g, 77\ ^\circ K) \rightarrow N_2(g, 298\ ^\circ K)$

If the nitrogen is slowly heated from near 0 °K to 298 °K, all of the balancing heats can be measured at all temperatures and the absolute entropy of N_2 at 298 °K has been measured as 45.7 cal/mole-deg. The absolute entropies of a number of elements and compounds are listed in Table 15–3.

Note especially the increase in absolute entropy as the complexity of the molecules increases. This is illustrated in many entries in Table 15–3, but perhaps most clearly in the increment of 10 cal/mole-deg between CH_4 and C_2H_6 and again between C_2H_6 and C_3H_8. The increment of 10 is due to the additional motion of the new CH_2 unit added in each case.

Entropy Changes in Chemical Reactions

Suppose that we were to mix H_2 gas and N_2 gas at 25 °C and a partial pressure of each gas of 1 atm. Should we expect them to react to form ammonia, according to the following reaction?

Table 15–3 Absolute Entropies, $S°$, of Some Elements and Compounds (in cal/mole-deg)

Solids

Substance	$S°$	Substance	$S°$
Ag	10.20	Mg	7.77
AgCl	22.97	MgO	6.4
$AgNO_3$	33.68	$MgCl_2$	21.4
Be	2.28	Na	12.2
C(diamond)	0.58	NaCl	17.3
C(graphite)	1.36	$NaNO_3$	27.8
CaO	9.5	Pb	15.51
$CaCO_3$	22.2	$PbCl_2$	32.6
Cu	7.96	PbO	16.2
$CuSO_4$	27.1	Zn	9.95
Fe	6.49	$ZnCl_2$	25.9
Fe_2O_3	21.5	ZnS	13.8
HgO	17.2		
I_2	27.9		

Liquids

Inorganic		*Organic*	
BCl_3	50.0	C_6H_6 (benzene)	29.76
Br_2	36.4	CCl_4 (carbon tetrachloride)	51.25
CS_2	36.10		
H_2O	16.72	$CHCl_3$ (chloroform)	48.5
Hg	18.5	CH_3OH (methanol)	30.3

Gases

Inorganic		*Organic*	
Ar	36.98	C_2H_2 (acetylene)	48.00
Br_2	58.64	C_2H_6 (ethane)	54.85
Cl_2	53.29	C_2H_4 (ethylene)	52.45
CO	47.30	CH_4 (methane)	44.5
CO_2	51.06	CH_3OH (methanol)	56.8
H_2	31.21	C_3H_8 (propane)	64.5
HF	41.47		
HCl	44.62		
H_2O	45.11		
N_2	45.77		
NH_3	46.01		
O_2	49.00		
O_3	56.8		
SO_2	59.40		

$$3\ H_2(g) + N_2(g) \rightarrow 2\ NH_3(g) \qquad \textbf{15–42}$$

The coming together of four gaseous molecules to form two gaseous molecules is matter concentration, and it should yield a negative entropy change for the gas.

We can use the data of Table 15–3 to calculate $\Delta S°$ for reaction 15–42, as follows:

$$\begin{aligned}\Delta S° &= 2\ S°_{NH_3} - 3\ S°_{H_2} - S°_{N_2} \\ &= 2(46.01) - 3(31.21) - (45.77) \\ &= -47.38 \text{ cal/deg}\end{aligned}$$

The entropy change is large and negative as we predicted.

We might be tempted to predict that ammonia cannot be formed from hydrogen and nitrogen; this prediction is erroneous. The error comes because the entropy calculation above only accounts for part of the total entropy change. From Table 15–1, we calculate that reaction 15–42 liberates by heat transfer an amount of energy equal to

$$-\Delta H° = -(2\Delta H°_{f(NH_3)}) = 22.08 \text{ kcal}$$

This energy is gained by the surroundings at 298 °K, producing an entropy increase in the surroundings:

$$\Delta S_{surr} = \frac{Q_{to\ surr}}{T_{surr}} = \frac{22{,}080}{298} = 74.1 \text{ cal/deg}$$

The total entropy change of the system plus the surroundings is represented:

$$\begin{aligned}\Delta S_{total} &= \Delta S_{gas} + \Delta S_{surr} \\ &= -47.38 + 74.1 = 26.7 \text{ cal/deg}\end{aligned}$$

and the conversion is spontaneous.

SUGGESTIONS FOR ADDITIONAL READING

Bent, H. A., *The Second Law*, Oxford University Press, New York, 1965.

Mahan, B. H., *Elementary Chemical Thermodynamics*, W. A. Benjamin, New York, 1963.

Nash, L. K., *Elements of Chemical Thermodynamics*, 2nd ed., Addison-Wesley, Reading, Mass., 1970.

PROBLEMS

1. What is the rise in temperature of water as a consequence of falling in a waterfall from a height of 500 ft? Assume that all of the potential energy is converted to thermal energy.

2. One way to increase the temperature of water is to stir it. A force must be applied to the stirrer to make it move through the water. Suppose that this force is constant at 100 dynes (the stirrer is moving at a constant speed) and the stirrer moves a total distance of 100 meters. If 250 ml of water is stirred, what is its temperature rise as a result of the stirring?
3. A man walks up a flight of stairs. Has he performed work? Has his total energy increased? If it has, where did the energy come from? Is it possible that his total energy might have decreased in walking up the stairs? Explain.
4. A compressed iron spring is dissolved in acid. What happens to the potential energy of compression of the spring?
*5. A man stands holding a block of concrete. Is he doing work? If he is not doing work, why is he so tired? *Hint:* Consider an electromagnet suspending a car in a junkyard.
6. Each of the reactions below liberates energy (ΔE is negative) when they are carried out at 25 °C and 1 atm pressure. Predict in each case whether the energy liberated by heat transfer is greater than or less than $-\Delta E$ or whether $-\Delta E = Q_{released}$:
   a. $CO_2(g) + CaO(s) \rightarrow CaCO_3(s)$
   b. $H_2(g) + C_2H_4(g) \rightarrow C_2H_6(g)$
   c. $2\ NCl_3(l) \rightarrow N_2(g) + 3\ Cl_2(g)$
   d. $H_2(g) + Cl_2(g) \rightarrow 2\ HCl(g)$
7. The freezing of one mole of liquid water at 0 °C and 1 atm liberates 1440 cal of energy by heat transfer. What is ΔE_{H_2O} for the same process? The density of ice at 0 °C is 0.917 g/ml, while the density of liquid water is 1.000 g/ml.
8. After each of the following reactions is listed the ΔH at 25 °C and 1 atm pressure:
   a. $CaO(s) + H_2O(l) \rightarrow Ca(OH)_2(s)$; $\Delta H° = -15.3$ kcal
   b. $H_2(g) + \frac{1}{2}\ O_2(g) \rightarrow H_2O(l)$; $\Delta H° = -68.3$ kcal
   c. $Ca(s) + \frac{1}{2}\ O_2(g) \rightarrow CaO(s)$; $\Delta H° = -151.8$ kcal

   Employ Hess's law to calculate $\Delta H°$ for reaction d:

   d. $Ca(s) + O_2(g) + H_2(g) \rightarrow Ca(OH)_2(s)$
9. A 10-g ice cube at 0 °C is placed in an insulated container along with 250 ml of liquid water at 25 °C. What is the final temperature in the container? The melting of one mole of ice at 0 °C requires an energy input to the ice of 1440 cal/mole.
10. The extremely high stability of solid aluminum oxide can be utilized in the "thermite" reaction

$$2\ Al(s) + Fe_2O_3(s) \rightarrow Al_2O_3(s) + 2\ Fe(s)$$

From data given in Table 15–1, calculate the energy liberated by heat transfer by the reaction of two moles of Al(s) with one mole of Fe_2O_3(s) at 25 °C and 1 atm.

11. If the thermite reaction is carried out in a container open to the atmosphere, much of the liberated energy is lost to the surroundings. However, some of the energy goes into raising the temperature of the products. Suppose that 70 percent of the energy is lost to the surroundings. To what temperature does the remaining 30 percent of the energy convert the products? The C_p of Fe(s) is 6.1 cal/mole-deg, and that of Al_2O_3(s) is 20 cal/mole-deg. Iron melts at 1530 °C and the ΔH of fusion is 3560 cal/mole. The C_p of molten Fe may also be taken as 6.1 cal/mole-deg.

12. Suppose that we carry out the thermite reaction in the 1.000-kg bomb calorimeter referred to in the text. The bomb is surrounded by 1.000 l of water initially at 25 °C. If 0.1093 g of Al reacts in the bomb with a slight stoichiometric excess of iron oxide, what is the final temperature of the bomb and surrounding water?

13. The fuel value of a combustible substance is the energy evolved in heat transfer in the combustion of a unit weight of the substance. From data given in Table 15–1, calculate which has the greater fuel value, methane(g) or acetylene(g). Assume that both fuels are converted to H_2O(g) and CO_2(g).

14. How much energy is evolved by heat transfer when a 10.0-g piece of steel wool rusts according to the following reaction?

$$4\ Fe(s) + 3\ O_2(g) \rightarrow 2\ Fe_2O_3(s)$$

Does it surprise you that a hot piece of steel wool burns in air? Explain.

15. Use the data of Tables 15–1 and 15–2 to calculate the average C—H bond enthalpy in methane. Compare this value with the value in Table 6–3.

16. Use the data of Tables 15–1 and 15–2 to calculate the bond enthalpy in carbon monoxide. Are you surprised that it is larger than the N_2 bond enthalpy? Explain.

17. Indicate with each of the following processes whether matter is spread or concentrated and whether energy is spread or concentrated:
    a. The operation of an internal combustion engine.
    b. A person running.
    c. Rain formation in a cloud.
    d. $H_2(g) \rightarrow 2\ H(g)$.

18. When liquid H_2SO_4 dissolves in water, a large amount of energy is liberated by heat transfer. Likewise a sizable liberation of energy occurs when either gaseous HCl or solid NaOH dissolves in water. Liquid H_2SO_4 is soluble in water in all proportions, yet there is a limit to the

solubility of HCl(g) and NaOH(s). Why does this difference occur between H_2SO_4 and the other two species?

19. In the discussion of Figure 15–6, we calculated Ω_A, Ω_B, and Ω_{AB} before and after the removal of the partition. Suppose that there are five A molecules and twenty B molecules. Calculate Ω_A, Ω_B, and Ω_{AB} before and after the removal of the partition.

20. Benzene(l) and toluene(l) mix to form solutions with almost no heat transfer involved in the process. Can we use an analysis similar to that of Figure 15–6 to account for the spontaneity of the dissolution? Explain.

*21. We have stated that Ω_{motion} is greater in the hot bar in Figure 15–7 than in the cold bar. If we label the levels L_1, L_2, L_3, and so on, starting from the lowest up, and label the atoms A_1, A_2, A_3, and so on, a motional possibility for the combined atoms of one bar is the assignment of A_1 to L_1, A_2 to L_2, and so on. Other motional possibilities are obtained by permuting the atoms among the levels. Suppose that in the hot bar we have five atoms, one in each of five levels. Show that Ω_{motion} is (5)(4)(3)(2)(1) = 5! (called 5 factorial) = 120. In the cold bar, let us place the five atoms as two in L_1, two in L_2, and one in L_3. Show that Ω_{motion} is 5!/(2!)(2!) = 30, and indeed that $\Omega_{hot} > \Omega_{cold}$.

22. A friend of yours who has no scientific background has heard the word *entropy* and asks you what it means. What is your explanation of entropy?

23. Of the following pairs of substances, which is predicted to have the greater molar entropy? Explain your choices.
    a. $H_2O(s)$ at 0 °C and 1 atm or $H_2O(l)$ at 0 °C and 1 atm.
    b. $H_2O(l)$ at 25 °C and 24 torr or $H_2O(g)$ at 25 °C and 24 torr.
    c. $H_2O(g)$ at 25 °C and 24 torr or $H_2O(g)$ at 25 °C and 10 torr.
    d. Cu(s) at 25 °C and 1 atm or Cu(s) at 0 °C and 1 atm.
    e. Ethanol at 25 °C and 1 atm or methanol at 25 °C and 1 atm.

24. Show that equation 15–31 can also be expressed as

$$\Delta S = N(\text{constant}) \log \frac{P_i}{P_f} \qquad \textbf{15–43}$$

where P_i and P_f are the initial and final pressures.

25. One mole of an ideal gas expands at 25 °C and its volume increases tenfold. What is the minimum amount of work the gas can perform in this expansion? What is the limit to the maximum amount of work that can be performed (report a numerical answer)? Is this limit an attainable one? Explain.

26. Calculate ΔS_{total} for the spontaneous process that occurs in Figure 15–1. Assume that the floor and book are initially at 25 °C and the temperature increase is infinitesimal. Thus, the thermal energy is gained by the book and floor at a constant temperature of 25 °C.

27. Calculate the total entropy change in Figure 15–6 if A is 2 moles of N_2 initially at 10 atm pressure and B is 3 moles of He initially at 5 atm.

28. Calculate ΔS_{total} for the energy transfer in Figure 15–5(b). The C_p of iron is 6.1 cal/mole-deg.

29. Calculate ΔS_{total} for the spontaneous process of problem 9.

***30.** Suppose that the gases in problem 27 are initially at different temperatures: N_2 at 100 °C, and He at 0 °C. Calculate ΔS_{total} for the spontaneous process that occurs when the partition is removed. The C_v values of N_2 and He are 5.0 and 3.0 cal/mole-deg, respectively.

31. Certain crystalline compounds are found to possess nonzero entropies even at 0 °K. For instance, the entropy of CO(s) is $R \ln 2$ cal/mole-deg at 0 °K. The crystal structure of CO was found on examination to be one in which the CO molecules exist in a random mixture of CO and OC orientations. Use this fact and the Boltzmann equation to derive that indeed $R \ln 2$ should be the residual entropy.

32. Predict the sign of $\Delta S°$ for each of the following reactions from qualitative considerations. Test your intuition by calculating $\Delta S°$ from the data of Table 15–3.

a. $2\,H_2(g) + O_2(g) \rightarrow 2\,H_2O(g)$
b. $H_2(g) + C_2H_4(g) \rightarrow C_2H_6(g)$
c. $2\,CH_4(g) + O_2(g) \rightarrow 2\,CH_3OH(g)$
d. $2\,Mg(s) + O_2(g) \rightarrow 2\,MgO(s)$

***33.** Estimate the $\Delta S°$ for the following reaction from the change in the number of gas molecules:

$$H_2(g) + Cl_2(g) \rightarrow 2\,HCl(g)$$

Calculate $\Delta S°$ from the data of Table 15–3. Figure 15–6 shows you the source of the small error in your estimate. How?

34. Use the relation $S = k \ln \Omega$ to calculate Ω for Be(s) and Cu(s) under the standard conditions. Is this huge number due to a huge number of positional possibilities, motional possibilities, or both in these cases? If Ω is assumed to arise entirely from motional possibilities, what is the average number of motional possibilities for each atom in Cu(s) under the standard conditions? in Be(s)? Does this provide evidence in favor of the quantum hypothesis of discrete energy states? Explain. Copper obeys the law of Dulong and Petit and beryllium does not. Explain this result based on the calculated number of motional possibilities for the atoms of each metal.

35. The entropies of the solid elements typically increase with increasing atomic number, but there are some notable exceptions. For example, the $S°$ of Na is considerably larger than the $S°$ of Mg. Explain this reversal. *Hint:* The melting point of Mg is considerably higher than the melting point of Na.

36. Give two reasons why the $S°$ of $MgCl_2$ is greater than the $S°$ of MgO.

37. Calculate ΔS_{total} for the combustion of a 0.3592-g sample of liquid methanol. Assume that the products are $H_2O(g)$ and $CO_2(g)$ at 25 °C and that the energy liberated by heat transfer passes to the surroundings at 25 °C. Is this process favored by energy spread, matter spread, both, or neither?

16/ Free Energy and Equilibrium

The model of matter spread and energy spread, formulated in terms of entropy, can be used in quantitative analyses of chemical and physical changes. Any given spontaneous change eventually ceases and arrives at a state of balance, or equilibrium. Can we understand why a given change ceases where it ceases? Might we even be able to predict where it will cease? Both questions can be answered affirmatively.

All spontaneous changes increase the entropy of the universe. Thus, any change proceeds until it reaches a point at which the entropy change of the universe would be zero for the next infinitesimal progression of the change. At that point, the changing system has reached a state of balance or equilibrium. Suppose that we should want to find the temperature at which ice and liquid water are in equilibrium at a pressure of 1 atm. This temperature is not 25 °C because, as we know, ice will not exist in equilibrium at that temperature. This means that at 25 °C the $\Delta S_{universe}$ of the following process is positive:

$$H_2O(s) \rightarrow H_2O(l)$$

The $\Delta S_{universe}$ is the sum of an entropy increase for the solid water in converting to the liquid and an entropy decrease for the surroundings in supplying the energy required for fusion. The entropy increase is larger than the decrease and the ice spontaneously melts at 25 °C.

At 0 °C, the entropy of one mole of liquid water is 5.27 cal/deg larger than the entropy of one mole of solid water. Also at 0 °C, the enthalpy of fusion of ice is 1440 cal/mole. Thus, in the melting process at 0 °C, the entropy decrease of the surroundings is $1440/273 = 5.27$ cal/deg. The total entropy change of the water and surroundings is zero, and $H_2O(s)$ and $H_2O(l)$ are in equilibrium at 0 °C.

In chemistry we are most often concerned with conditions of equilibrium under two well-defined constraints, namely constant temperature and constant pressure. Consider the system in Figure 16–1. On the left are beakers containing a solution of NaCl in water in one and a solution of $AgNO_3$ in water in the other. Both solutions are at room temperature; both are exposed

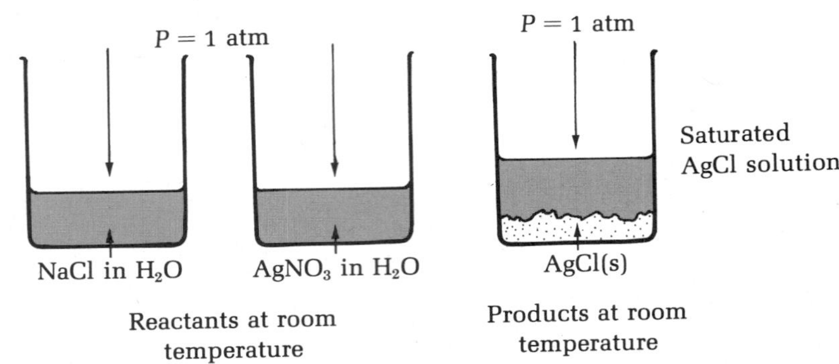

Figure 16–1
The formation of AgCl by mixing aqueous solutions of $AgNO_3$ and NaCl contained in uninsulated beakers exposed to a pressure of 1 atm.

to the atmosphere, and are at the same pressure of 1 atm. The two solutions are mixed, and since AgCl is only sparingly soluble in water, a voluminous white precipitate forms. The final equilibrium state is AgCl(s) in contact with a saturated solution of AgCl that also contains dissolved $NaNO_3$. The reaction will, in general, involve some heat transfer between system and surroundings; but if the system is allowed to stand, it will eventually reach the original room temperature. The final T and P are equal to the initial T and P as they will be in any such process in uninsulated systems exposed to the atmosphere. Many chemical processes and all biological processes may be considered to occur at constant T and P.

What we shall do in this chapter is to determine what special form the equation $\Delta S_{\text{universe}} = 0$ takes for constant T and P. This, we shall find, introduces a new quantity, the Gibbs free energy quantity. In succeeding chapters, the concept of this quantity will be applied to specific physical and chemical equilibria.

16–1. The Gibbs Free Energy

Spontaneous Changes at Constant Temperature and Pressure

The spontaneity criterion we have developed (Chapter 15) is $\Delta S_{\text{universe}} > 0$. For processes that take place at a constant temperature, like the melting of ice at 0 °C, we can calculate $\Delta S_{\text{universe}}$ as the sum of the entropy change in the chemical system of interest (the water-ice in that case) and a simple expression for the entropy change of the surroundings (the bath at 0 °C). The spontaneity criterion becomes

$$\Delta S_{\text{universe}} = \Delta S_{\text{system}} + \frac{Q_{\text{to surr}}}{T} > 0 \qquad \textbf{16–1}$$

where the second term is the simple expression for the entropy increase or decrease of the surroundings as a result of the acceptance ($Q > 0$) or the supplying ($Q < 0$) of energy by the surroundings by heat transfer.

We shall now add the additional restriction that the system change at constant pressure, that is, that the system do work on the surroundings by expanding against a constant pressure or have work done on it by a constant pressure. The only interactions with the surroundings in the process under investigation are heat transfer and expansion-compression work. Under these conditions (constant T, constant P, only $P\Delta V$ work), the energy delivered to the surroundings by heat transfer is $(-\Delta H_{system})$, and equation 16–1 becomes

$$\Delta S_{system} - \frac{\Delta H_{system}}{T} > 0 \qquad \textbf{16–2}$$

Equation 16–2 is customarily multiplied through by $-T$ and rearranged to yield

$$\Delta H_{system} - T\Delta S_{system} < 0 \qquad \textbf{16–3}$$

The spontaneity criterion in this special case of constant T and P is transformed to the change in a new and very important function of the system, G, the **Gibbs free energy,** defined as follows:

$$G \equiv H - TS \qquad \textbf{16–4}$$

At constant temperature ΔG_{system} is equal to $\Delta H_{system} - T\Delta S_{system}$. The spontaneity criterion is that **ΔG_{system} must be negative for any system undergoing any spontaneous change at constant T and P.** It must be emphasized that this is a special condition that holds **for and only for constant T and P.** There is no general tendency in nature towards minimum energy. The only general tendency is towards maximum spread.

Maximum Work

The term *Gibbs free energy* is a strange one and deserves some analysis. It honors J. Willard Gibbs, an American scientist who contributed an enormous amount to our understanding of the statistical basis of spontaneity and equilibrium through his theoretical studies during the latter part of the nineteenth century. What is the source of the word *free* in *free energy?* The criterion for a process to be spontaneous is that $S_{universe}$ increase for that process, but $\Delta S_{universe}$ need be only slightly greater than zero. A large positive $\Delta S_{universe}$ is unnecessary. As we saw depicted in Figure 15–10, any large, positive $\Delta S_{universe}$ can be decreased by obtaining work from the process. Thus, **any spontaneous process can be made to yield useful work.** However, there is an upper limit to this useful work set by $\Delta S_{universe}$ for the spontaneous process. For an isothermal process (as in Figure 15–9) the maximum useful work obtainable is $T\Delta S_{universe}$. In the case of Figure 15–9 $T\Delta S_{universe}$ is the work that could have been obtained, but was not.

In Chapter 15 we obtained $\Delta S_{universe}$ for the process depicted in Figure

15–9 as $\Delta S_{\text{universe}} = R \ln V_f/V_i$. The obtaining of useful work from the gas expansion is shown by Figure 15–10. The upper limit to the obtainable work is $RT \ln V_f/V_i$.

Let us consider the obtaining of work from a spontaneous chemical reaction, for example, reaction 15–42.

$$3\ H_2(g) + N_2(g) \rightarrow 2\ NH_3(g) \qquad \textbf{16–5}$$

This reaction concentrates matter ($\Delta S° = -47.38$ cal/deg) and spreads energy ($\Delta H° = -22.08$ kcal). If it proceeds under standard conditions with the performance of no work other than $P\Delta V$ the entropy of the universe increases by 26.7 cal/deg.

However, it is not necessary to have this much entropy produced. In order to have the reaction proceed it is necessary to spread enough energy to just counterbalance the matter concentration. This energy that must be spread is $(-T\Delta S°) = -(298)(-47.38) = 14.12$ kcal. The difference between the energy that is spread $(-\Delta H°)$ and the energy that must be spread $(-T\Delta S°)$ is $-(-22.08) - (14.12) = 7.96$ kcal, an amount of energy that we are *free* to convert to useful work performance. But $(-\Delta H°) - (-T\Delta S°) = (-\Delta G°)$, and thus $(-\Delta G°)$ is the *free energy* of the process.

The maximum useful work obtainable from reaction 15–42 under standard conditions is $(-\Delta G°)$, as we have just seen. In a preceding paragraph it was stated that the maximum useful work obtainable from any spontaneous process carried out at constant temperature is $T\Delta S_{\text{universe}}$, where $\Delta S_{\text{universe}}$ is that for the spontaneous process in which this useful work is not obtained. Let us examine $T\Delta S_{\text{universe}}$ for a spontaneous process occurring at constant T and P and involving only $P\Delta V$ work. From equations 16–1 and 16–2 we have

$$T\Delta S_{\text{universe}} = T\left(\Delta S_{\text{system}} - \frac{\Delta H_{\text{system}}}{T}\right)$$
$$= -\Delta G_{\text{system}}$$

Thus, **$(-\Delta G_{\text{system}})$ is the maximum useful work (work other than $P\Delta V$) that can be obtained from any spontaneous process operating at constant temperature and constant pressure.**

Engines and Refrigerators

Work can be derived from any spontaneous process. The most common examples of the use of stored energy to do work are the internal combustion engine and the heat engine. The combustion of gasoline is favored by both matter spread [the conversion of a liquid fuel to $CO_2(g)$ and $H_2O(g)$] and energy spread (the burning of gasoline is highly exothermic).

In a heat engine, such as a steam engine or steam turbine, work is obtained from the spontaneous heat transfer from a hot substance, the gases in the flame used to heat the steam, to a cold substance, the room containing the engine or turbine. In this case, the spreading of energy alone causes the process to be spontaneous since the steam is recycled and there is no net matter spread in a complete cycle. Because energy spread does provide the driving force, it is impossible to use all of the thermal energy of the flame to perform work. Some energy must be spread as thermal energy in the surroundings to supply the spontaneity of the heat engine. It was the consideration of the maximum possible work obtainable from heat engines that led Sadi Carnot, a French engineer, to arrive at the first formulation of entropy in the early nineteenth century. Carnot deduced the relation for the maximum possible **efficiency** of a heat engine:

$$\text{Maximum efficiency} = \frac{T_H - T_L}{T_H} \qquad \textbf{16–6}$$

where the efficiency is defined as the fraction of energy absorbed from the flame that can be used to do work, and T_H and T_L are the flame temperature and the temperature of the surroundings, respectively. For example, if the flame temperature is 1000 °K and the temperature of the surroundings is 298 °K, the maximum possible efficiency is as follows:

$$\text{Maximum efficiency} = \left(\frac{1000 - 298}{1000}\right) = 0.702$$

Thus, at best only 70.2 percent of the thermal energy of the gas molecules in the flame can be used to do work and the remaining 29.8 percent must be spread to the surroundings. In practice, the efficiency is much less than the maximum possible efficiency.

In the operation of a refrigerator, thermal energy is extracted from the substances inside the refrigerator and given to other substances (those outside the refrigerator). In the process, some substances become cold and others become warm. This is the opposite of the energy transfer in the heat engine. Work is not obtained; work must be performed to operate the refrigerator, and the energy to perform this work is purchased from the electric power company. The spontaneous operation of the refrigerator is obtained from the spreading of concentrated energy, electricity, to the surroundings as thermal energy. The minimum possible work required to pump a given amount of energy from the inside of a refrigerator was also deduced by Carnot:

$$\frac{\text{Work required}}{\text{Energy pumped}} = \frac{T_H - T_L}{T_L} \qquad \textbf{16–7}$$

where T_H is the room temperature and T_L is the temperature of the inside of the refrigerator. If T_H is 298 °K and T_L is 278 °K, it takes at least 0.072 cal = (0.072)(4.185) = 0.30 joule of electrical energy to expel one calorie of thermal energy from the inside of the refrigerator.

The Driving of One Process by Another

Nonspontaneous processes, that is, processes for which $\Delta S_{universe} \leq 0$, can be made to proceed; but they require that work be done directly or that work be done indirectly by coupling the nonspontaneous process to a spontaneous one, whose maximum work capability exceeds the minimum work requirement of the nonspontaneous process. An example of a direct performance of work is the refrigerator.

An example of an indirect performance of work is the reversal of the spontaneous process depicted in Figure 15–5(c). This reversal is represented in Figure 16–2 as a spontaneous distillation. In the initial condition shown, matter is spread as NaCl dispersed in water and energy is concentrated as the energy of natural gas (CH_4) molecules and oxygen molecules. In the final condition, matter has been reconcentrated as pure NaCl on the left and pure H_2O on the right, while energy has been spread to the ice bath on the right as thermal energy in the process of water condensation. The energy spreading exceeds in importance the concentrating of matter, and the process is spontaneous, in fact overwhelmingly so.

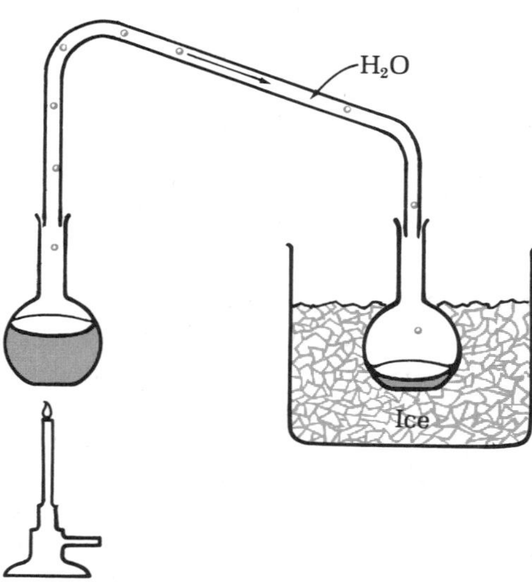

Figure 16–2 The spontaneous separation of a salt solution by distillation.

The separation of salt from seawater is a problem of enormous importance in a world desperately in need of fresh water for irrigation purposes. The process shown in Figure 16–2 is uneconomical since too much energy is spread, and the expense would be too great to supply the energy needed. In Chapter 17 we shall see a much more economical method to accomplish the separation of pure water from salt in seawater.

16–2. The Equilibrium Constant

Dependence of G on Pressure

Let us investigate the conditions under which the following reaction is in equilibrium at 25 °C:

$$CaCO_3(s) \rightleftharpoons CaO(s) + CO_2(g) \qquad \textbf{16–8}$$

The equilibrium situation is shown in Figure 16–3. A closed container holds a mixture of solid $CaCO_3$ and CaO in contact with gaseous CO_2, and the container is immersed in a bath at 25 °C.

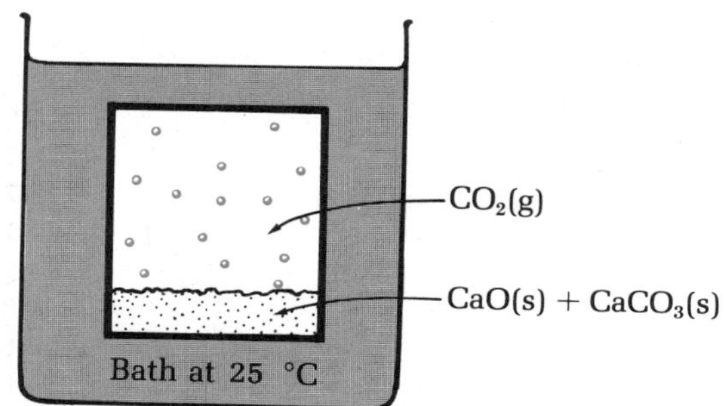

Figure 16–3 Equilibrium between the solids calcium oxide and calcium carbonate and gaseous carbon dioxide at 25 °C.

The double arrow in equation 16–8 indicate that both the formation and decomposition of $CaCO_3(s)$ are occurring at equilibrium. However, at equilibrium the rate of formation equals the rate of decomposition, and there is no net change in the amounts of any of the three substances.

Since CO_2 is a gas, its formation in reaction 16–8 is highly favored on matter-spread grounds. However, from the data of Table 15–1, we calculate $\Delta H°$ for the production of CaO(s) and $CO_2(g)$ from $CaCO_3(s)$ as follows:

$$\begin{aligned}\Delta H° &= \Delta H°_{f(CaO)} + \Delta H°_{f(CO_2)} - \Delta H°_{f(CaCO_3)}\\ &= -151.79 - 94.05 - (-288.45)\\ &= 42.61 \text{ kcal/mole}\end{aligned} \qquad \textbf{16–9}$$

Thus, the standard reaction concentrates energy, and this is unfavorable. This competition between $\Delta S°$ and $\Delta H°$ (actually a competition between $\Delta S°$ and $-\Delta H°/T$) is quite common in chemical and physical equilibria. The competing terms are embodied in the $\Delta G°$ of the reaction, where $\Delta G°$ is defined:

$$\Delta G° = \Delta H° - T\Delta S° \qquad \textbf{16–10}$$

As was the case with enthalpy, it proves highly convenient to have a table of standard Gibbs free energies of formation ($\Delta G_f°$) values for compounds, and ΔG° for any reaction can be calculated from these in the same way that ΔH° is calculated from the ΔH_f° values of Table 15–1. The ΔG_f° val-

Compound	ΔG_f° (kcal/mole)
Inorganic	
Al_2O_3(s)	−376.77
AgCl(s)	−26.22
B_2H_6(g)	19.8
$BaCl_2$(s)	−193.8
C(diamond)	0.69
CaO(s)	−144.4
$CaCO_3$(s)	−269.78
CO(g)	−32.81
CO_2(g)	−94.26
Fe_2O_3(s)	−177.1
HBr(g)	−12.72
HCl(g)	−22.77
HF(g)	−64.7
HI(g)	0.3
H_2O(g)	−54.64
H_2O(l)	−56.69
$HgCl_2$(s)	−44.4
KCl(s)	−97.59
$MgCl_2$(s)	−141.57
MgO(s)	−136.13
NH_3(g)	−3.98
NO(g)	20.72
NO_2(g)	12.39
N_2O_4(g)	23.49
NaCl(s)	−91.79
Na_2CO_3(s)	−250.4
O_3(g)	39.06
SO_2(g)	−71.79
SiO_2(quartz)	−192.4
$ZnCl_2$(s)	−88.3
Organic	
C_2H_2(g) (acetylene)	50.0
C_6H_6(l) (benzene)	41.30
CCl_4(l) (carbon tetrachloride)	−16.43
$CHCl_3$(l) (chloroform)	−17.1
C_2H_6(g) (ethane)	−7.86
C_2H_4(g) (ethylene)	16.28
CH_4(g) (methane)	−12.14
CH_3OH(l) (methanol)	−39.75

Table 16–1
Gibbs Free Energies of Formation, ΔG_f°

ues can be calculated from known values of ΔH_f° and ΔS_f° along with equation 16–10. Table 16–1 lists ΔG_f° values for the same compounds listed in Table 15–1. This table also appears in Appendix E.

In the case of reaction 16–8, the ΔG° is calculated:

$$\Delta G^\circ = \Delta G^\circ_{f(CaO)} + \Delta G^\circ_{f(CO_2)} - \Delta G^\circ_{f(CaCO_3)}$$
$$= -144.4 - 94.3 - (-269.78) \qquad \textbf{16–11}$$
$$= 31.1 \text{ kcal}$$

or the combined Gibbs free energy of one mole of CO_2 at 1 atm plus one mole of CaO at 1 atm is 31.1 kcal higher than the Gibbs free energy of one mole of $CaCO_3$ at 1 atm, all at 25 °C. The standard conditions are not equilibrium conditions in this case since for equilibrium, $\Delta G = -T\Delta S_{universe}$ must be zero ($\Delta S_{universe} = 0$ is the criterion for equilibrium).

If the three species are not in equilibrium at 1 atm, at what pressure are they in equilibrium at 25 °C? Surely it must be a lower pressure than 1 atm, because from equation 16–11 the production of $CaCO_3$(s) from CaO(s) and CO_2(g) at 1 atm has a ΔG of -31.1 kcal and the consumption of CO_2 is spontaneous.

How do the Gibbs free energies of the three species change as the pressure of CO_2(g) is lowered from 1 atm at constant T? First, consider $CaCO_3$(s). The change in the Gibbs free energy of a given amount of pure $CaCO_3$(s) due to a pressure change at constant temperature is the following:

$$\Delta G = \Delta H - T\Delta S = \Delta E + \Delta(PV) - T\Delta S \qquad \textbf{16–12}$$

Let us examine the three changes ΔE, $\Delta(PV)$, and $(-T\Delta S)$ separately. To an excellent approximation, solids and liquids are incompressible. Thus, no work is done in the pressure change and $\Delta E = 0$. No change in atomic positions and motions occurs and $\Delta S = 0$ as a consequence. The only nonzero term is $\Delta(PV)$. The molar volumes of most solids and liquids are quite small (that of $CaCO_3$(s) is 0.037 liter/mole). Since ΔP in the pressure drop from 1 atm can be no larger than 1 atm, $\Delta(PV) = V\Delta P$ is limited to 0.037 l-atm = 0.90 cal. This is negligible compared with the 31.1 kcal in equation 16–11. In general, in chemical reactions, the ΔG of nondissolved liquids and solids due to pressure changes at constant temperature may be neglected in the calculation of equilibrium positions.

If CO_2 at 1 atm or less and 298 °K can be treated as an ideal gas (an excellent approximation that we shall use throughout for gases in chemical and physical equilibria), we can easily obtain ΔG of CO_2(g) for the isothermal pressure change. For an ideal gas, E depends only on temperature, and so does PV. Thus, **ΔH is zero for an isothermal pressure change of any ideal gas.** The ΔS for an isothermal expansion of one mole of an ideal gas (Chapter

15) is given as follows:

$$\Delta S = R \ln \frac{V_f}{V_i} \qquad \textbf{16–13}$$

Through substitution of the ideal gas equation into equation 16–13, we obtain the ΔS in equation 16–13 as a function of initial and final pressures:

$$\Delta S = R \ln \frac{P_i}{P_f} \qquad \textbf{16–14}$$

The ΔG_{CO_2} due to pressure lowering from $P_i = 1$ atm to $P_f = P_e$, the equilibrium pressure, is given as

$$\Delta G_{CO_2} = -T\Delta S = -RT \ln \frac{P_i}{P_f} = -RT \ln \frac{1}{P_e} = RT \ln P_e \qquad \textbf{16–15}$$

which is a negative number. Matter is more spread in this low-pressure state, and the tendency toward spontaneous change in this state is diminished from that at 1 atm. Equilibrium occurs at that $P = P_e$ for which ΔG_{CO_2} just balances $\Delta G°$, that is,

$$\Delta G = \Delta G° + \Delta G_{CO_2} = 0$$

By substitution of equation 16–15 for ΔG_{CO_2}, we obtain the expression that allows us to calculate P_e:

$$\Delta G° = -RT \ln P_e \qquad \textbf{16–16}$$

and

$$\begin{aligned} P_e &= e^{-\Delta G°/RT} \\ &= e^{-\{31,100/(1.987)(298)\}} \\ &= 10^{-\{31,100/(1.987)(298)(2.3)\}} \\ &= 1.5 \times 10^{-23} \text{ atm} \end{aligned}$$

The attainment of the equilibrium situation of $\Delta G = 0$ through the pressure change is diagrammed in Figure 16–4. The level G_{CaCO_3} is shown as pressure-independent, while $G_{CaO} + G_{CO_2}$ increases linearly with log P_{CO_2} according to equation 16–15. Equilibrium occurs where the black line and colored line cross.

Equilibria Involving Gases

The simple relation given in equation 16–16 between $\Delta G°$ and the position of equilibrium at 298 °K suggests that such a simple expression might exist for all reactions involving gases. Let us consider the conditions of equilibrium at 298 °K for the following gas phase reaction:

$$3\ H_2(g) + N_2(g) \rightarrow 2\ NH_3(g) \qquad \textbf{16–17}$$

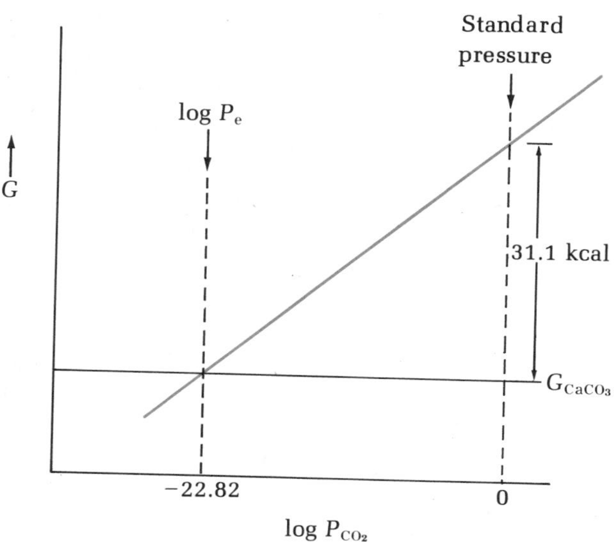

Figure 16-4 Comparison of the molar Gibbs free energy of solid $CaCO_3$ (solid black line) and the sum of the molar Gibbs free energies of solid CaO and gaseous CO_2 (colored line) at 298 °K as a function of the log CO_2 pressure. The standard pressure and equilibrium pressure are indicated by vertical dashed lines.

The left side is favored by the spreading of matter in this standard formation reaction of ammonia (end of Chapter 15). However, the formation of two moles of $NH_3(g)$ is exothermic by 22,080 cal/mole, and thus ammonia formation is favored by the spreading of energy. The ΔG_f° of −7960 cal for the formation of two moles of ammonia shows that energy spreading dominates in the standard reaction. Gaseous hydrogen, nitrogen, and ammonia cannot exist in equilibrium in contact with each other at partial pressures of 1 atm each. The spontaneous process that ensues is the formation of additional ammonia, since ΔG_f° for $NH_3(g)$ is negative.

Equation 16-15 may be generalized to the equation for the ΔG for the isothermal pressure change of any ideal gas, since the only assumption that went into the derivation of equation 16-15 was that of ideal behavior. We shall again treat H_2, N_2, and NH_3 as ideal gases. Thus, as we change the pressures from 1 atm, each single mole of gas experiences a ΔG of the form

$$\Delta G_{gas} = RT \ln P_{gas} \tag{16-18}$$

where P_{gas} is the partial pressure of the gaseous species in question.

Equilibrium is achieved when ΔG for reaction 16-17 is zero, that is, when the G of two moles of $NH_3(g)$ equals the combined G of 3 moles of $H_2(g)$ and one mole of $N_2(g)$, as shown in the following equation:

$$\begin{aligned} 2\,G_{NH_3} &= 2\,G^\circ_{NH_3} + 2\,RT \ln P_{e(NH_3)} = 3\,G_{H_2} + G_{N_2} \\ &= 3\,G^\circ_{H_2} + 3\,RT \ln P_{e(H_2)} + G^\circ_{N_2} + RT \ln P_{e(N_2)} \end{aligned} \tag{16-19}$$

Since $2\,G^\circ_{NH_3} - 3\,G^\circ_{H_2} - G^\circ_{N_2} = \Delta G^\circ = -7960$ cal, this equation may be rearranged to

$$\Delta G^\circ = -RT \ln \left(\frac{P^2_{e(NH_3)}}{P^3_{e(H_2)} P_{e(N_2)}} \right) = -7960 \text{ cal} \qquad \textbf{16–20}$$

Unlike reaction 16–8, which only involves one gaseous species, reaction 16–17 is not characterized by a unique equilibrium pressure for each of the three gases. Rather, equilibrium occurs for any pressure values that combine such that

$$\frac{P^2_{NH_3}}{P^3_{H_2} P_{N_2}} = e^{-\Delta G^\circ/RT} = 6.89 \times 10^5 \qquad \textbf{16–21}$$

The term $e^{-\Delta G^\circ/RT}$, which is a constant for any given reaction and temperature, is called the **equilibrium constant** of the reaction, and it is given the symbol K_e.

Construction of the Equilibrium Expression

According to equation 16–21, the equilibrium constant (symbolized by K_e) is equal to a combination of equilibrium pressures termed the **equilibrium expression.** How can we construct the equilibrium expression for any given reaction involving gases without having to go through the detailed consideration of G's as in equation 16–19? In the case of reaction 16–17, the ΔG° was calculated with the species on the right of the arrows as the products and those on the left as the reactants. We shall continue this convention for all equilibria that we write. The equilibrium expression in equation 16–21 contains the pressure of the product in the numerator and the pressures of the reactants in the denominator. Furthermore, each pressure is raised to the power of the coefficient of that species in the balanced reaction equation.

Two additional examples will further illustrate the construction of equilibrium expressions and the calculation of equilibrium constants. The first is reaction 16–8. There is only one gaseous species involved and it is a product. Thus,

$$K_e = e^{-\Delta G^\circ/RT} = 1.5 \times 10^{-23} = P_{CO_2}$$

and this is the result that we obtained earlier in this chapter.

The second example is

$$H_2(g) + Br_2(l) \rightleftharpoons 2\ HBr(g)$$

and the equilibrium expression is

$$K_e = e^{-\Delta G^\circ/RT} = \frac{P^2_{HBr}}{P_{H_2}} \qquad \textbf{16–22}$$

In this reaction, $\Delta G^\circ = 2\ \Delta G^\circ_{f(HBr)} = 2(-12{,}720)$ cal, so that $K_e = 4.6 \times 10^{18}$.

The alert reader might be puzzled at the fact that $e^{-\Delta G°/RT}$ is clearly a unitless quantity and yet equal to K_e, which appears to have the unit of pressure in both of the preceding examples. Actually K_e is also unitless, since the pressures enter the derivation through equation 16–15, which involves the ratio of the standard pressure of 1 atm to the equilibrium pressure. Thus, K_e in the first example really equals P_{CO_2} atm/1 atm, and the units cancel. However, for this cancellation of units to happen, and thus for the equilibrium expression to be properly related to $\Delta G°$, calculated from Table 16–1, **all gas pressures in the equilibrium expression must be in atmospheres.**

The Escaping Tendency

We have seen the importance of the molar Gibbs free energy of any substance. What is a molecular interpretation of molar G? We have also seen that entropy is a direct measure of positional and motional spreadedness in a system; but what does G of one mole of a substance measure? The definition of G is given as $G = H - TS$, so that G decreases if energy is lost from a system by heat transfer or if S for the system increases, or if both these events occur. Thus, a decrease in the molar G of a system is associated with the two spontaneous processes, the spreading of matter and the spreading of energy. Energy is spread to the surroundings if H_{system} decreases. Energy or matter or both are spread within the system if S_{system} increases. We can conclude that G of a given amount and kind of matter **measures the extent to which matter and energy are concentrated in that matter.** The larger the molar G of that matter, the greater the tendency of that matter to undergo spontaneous change.

Another commonly used interpretation of the molar G follows immediately—that the molar G is a measure of **escaping tendency.** Consider again reaction 16–8. Under standard conditions, the sum of the molar G's of CaO and CO_2 is greater than the molar G of $CaCO_3$. There is a spontaneous reaction in which CaO(s) and CO_2(g) escape from the state of high molar G to join as $CaCO_3$(s) in the state of lower molar G. As the reaction proceeds and the CO_2 pressure drops, the matter concentration on the right drops, and the escaping tendency on the right drops until it equals the escaping tendency on the left. Equilibrium is equality of escaping tendencies.

Throughout the remainder of this text references are made to the G of particular substances. This is always taken to be the G of one mole of the given substance, the escaping tendency of that substance.

Dependence of *G* on Solution Concentration

The solution phase is one of particular interest in chemistry since most chemical reactions are carried out by dissolving the reactants in some appropriate liquid solvent. In much of biological chemistry, this solvent is water, and the conditions of equilibrium in the blood and in the aqueous portions of cells are of critical importance. The quantitative measure of acid strength is an important factor in solution chemistry. Acetic acid, H_3CCOOH, is a typical carboxylic acid. Acetic acid is a Lowry-Brønsted acid that loses a proton to water according to the following reaction:

$$H_3CCOOH(aq) \rightleftharpoons H^+(aq) + H_3CCOO^-(aq) \qquad \textbf{16–23}$$

The deprotonation of acetic acid yields the acetate ion, H_3CCOO^-, and the acetate ion is capable of accepting protons from water, as the arrow from right to left in reaction 16–23 shows. At equilibrium, the rate of proton loss by acetic acid equals the rate of proton gain by acetate ion.

The condition of equilibrium universally is that the combined G's of the species on the two sides of the reaction equation must be equal; hence

$$G_{H_3CCOOH} = G_{H_3CCOO^-} + G_{H^+}$$

We proceed to find the conditions that produce this equality as we did with reactions 16–8 and 16–17. This means that we require ΔG_f° values for each of the three species in reaction 16–23. For molecules, atoms, or ions in solution, we must specify a standard concentration as well as a standard pressure and temperature, since concentration is an additional variable for solution species. This standard concentration is one-molar, that is, there is one mole of the given substance per liter of solution.

An additional problem in obtaining ΔG_f° values occurs with ions in solution, since we never obtain a solution containing only H^+ ions or only acetate ions. Ions always come in pairs. Thus it is impossible to obtain an absolute value for $\Delta G^\circ_{f(H^+,aq)}$ or for $\Delta G^\circ_{f(H_3CCOO^-,aq)}$. Fortunately, in reaction 16–23 and in all reactions involving ions in solution, we require only the sums or differences of ΔG_f° values for individual ions. This allows us to **define** the ΔG_f° value of any one ion to be zero; and all other ΔG_f° values for other ions are measured relative to that standard ion. This necessity of definition of a standard ion also holds for enthalpies and entropies of formation of ions in solution. Table 16–2 lists standard entropies, enthalpies of formation, and Gibbs free energies of formation of species in aqueous solution. (This table is also given in Appendix F.) (In Chapter 20 we shall see examples of the measurements of these numbers, and we shall see why the standard ion is chosen to be H^+.) We are now primarily concerned with showing how the data of Table 16–2 can be used to calculate the equilibrium concentrations of solution species. This table can be used to calculate ΔS°, ΔH°, and ΔG° values just as Tables 15–3, 15–1, and 16–1 have been used.

For reaction 16–23, ΔG° is calculated as follows:

$$\begin{aligned}\Delta G^\circ &= \Delta G^\circ_{f(H^+,aq)} + \Delta G^\circ_{f(H_3CCOO^-,aq)} - \Delta G^\circ_{f(H_3CCOOH,aq)}\\ &= 0 - 89.02 - (-95.51) = 6.49 \text{ kcal}\end{aligned}$$

Ion or Molecule	S° (cal/mole-deg)	ΔH_f° (kcal/mole)	ΔG_f° (kcal/mole)
H^+ (defined)	0	0	0
Ag^+	17.67	25.31	18.43
$Ag(NH_3)_2^+$	57.8	− 26.72	− 4.16
Al^{3+}	−74.9	−125.4	−115
BF_4^-	40	−365	−343
Br^-	19.29	− 28.90	− 24.57
Ca^{2+}	−13.2	−129.77	−132.18
CH_3COOH		−116.74	− 95.51
CH_3COO^-	20.8	−116.84	− 89.02
Cl^-	13.2	− 40.02	− 31.35
ClO_4^-	43.2	− 31.41	− 2.47
CN^-	28.2	36.1	39.6
CO_2	29.0	− 98.69	− 92.31
CO_3^{2-}	−12.7	−161.63	−126.22
Cu^{2+}	−23.6	15.39	15.53
F^-	− 2.3	− 78.66	− 66.08
Fe^{2+}	−27.1	− 21.0	− 20.30
Fe^{3+}	−70.1	− 11.4	− 2.53
H_2CO_3			−145.46
HCO_3^-	22.7	−165.18	−140.31
H_3PO_4	42.1	−308.2	−274.2
$H_2PO_4^-$	21.3	−311.3	−271.3
HPO_4^{2-}	− 8.6	−310.4	−261.5
H_2S	29.2	− 9.4	− 6.54
HS^-	14.6	− 4.22	3.01
I_2			3.93
I^-	26.14	− 13.37	− 12.35
K^+	24.5	− 60.04	− 67.46
Mg^{2+}	−28.2	−110.41	−108.99
Na^+	14.4	− 57.28	− 62.59
NH_3	26.3	− 19.32	− 6.36
NH_4^+	26.97	− 31.74	− 19.00
NO_3^-	35.0	− 49.37	− 26.43
OH^-	− 2.52	− 54.96	− 37.59
PO_4^{3-}	−52	−306.9	−245.1
S^{2-}	− 4	7.8	20.6
SO_4^{2-}	4.1	−216.90	−177.34
Zn^{2+}	−25.45	− 36.43	− 35.18

Table 16–2
Standard Entropies (S°), Heats of Formation (ΔH_f°), and Gibbs Free Energies of Formation (ΔG_f°) for Species in Aqueous Solution

Since ΔG° is not zero, equilibrium does not occur in a solution in which the concentrations of acetic acid, hydrogen ion, and acetate ion are all one-molar at 1 atm pressure and 25 °C. Instead, acetate ions must accept protons to form acetic acid, since ΔG for this process is −6.49 kcal. The equilibrium state will be characterized by much diminished H^+ and H_3CCOO^- concentrations.

What are the three equilibrium concentrations? We require an equation analogous to equation 16–15, but one that gives the change in G of a given amount of a dissolved substance with the change in the concentration of that substance in solution. As long as the molecules of dissolved substance do not significantly interact, the dilution of a solution is perfectly analogous to the isothermal expansion of an ideal gas. There is no enthalpy change and ΔS is positive because of the increased positional possibilities. The ΔG of one mole of a dissolved substance A is given by the equation

$$\Delta G_A = -T\Delta S_A = RT \ln \frac{[A]_f}{[A]_i} \qquad \textbf{16–24}$$

where [A] is the molarity of substance A in solution. If the expression $PV = nRT$ is substituted into equation 16–24 along with $[A] = n_A/V$, equation 16–24 is shown to be identical with ΔG for the isothermal expansion of an ideal gas. A solution that behaves according to equation 16–24 is termed an **ideal solution,** and all solutions approach ideal behavior as $[A] \rightarrow 0$. This is analogous to the approach of gases to ideal behavior as $P \rightarrow 0$.

The standard state for species in solution is not defined as an actual one-molar solution, but rather as a hypothetical ideal one-molar solution. This subtle difference allows us to use equation 16–24 in connection with Table 16–2 to calculate equilibrium concentrations of species in dilute solutions. The hypothetical standard state is always something of a mystery to students, but the mystery should not concern us at this point. The important thing now is that we can use Table 16–2 and equation 16–24 to accomplish for dilute solutions what we accomplished for gases by the combination of Table 16–1 and equation 16–15. (We shall be able to remove much of the mystery surrounding the standard state when we show in Chapter 20 how the values in Table 16–2 are obtained.)

We have avoided discussion of equilibria in gases at high pressures and solutions at high concentrations, and we shall continue to do so. In such systems, ΔG due to pressure or concentration changes arises from enthalpy changes as well as from entropy changes. The analysis of this additional complication is beyond the scope of this book.

The condition for equilibrium with reaction 16–23 is obtained by equating the total G's on the right and left sides of the reaction equation in analogy with equation 16–19:

$$\begin{aligned} G_{H_3CCOOH} &= G^\circ_{H_3CCOOH} + RT \ln [H_3CCOOH] \\ &= G_{H^+} + G_{H_3CCOO^-} \\ &= G^\circ_{H^+} + RT \ln [H^+] + G^\circ_{H_3CCOO^-} + RT \ln [H_3CCOO^-] \end{aligned} \qquad \textbf{16–25}$$

Since $G^\circ_{H^+} + G^\circ_{H_3CCOO^-} - G^\circ_{H_3CCOOH} = \Delta G^\circ = 6490$ cal, equation 16–25 can be rearranged to

$$\Delta G° = -RT \ln \frac{[H^+][H_3CCOO^-]}{[H_3CCOOH]} = 6490 \text{ cal} \qquad \textbf{16–26}$$

Equation 16–26 is completely analogous to equation 16–20 save that equilibrium molar concentrations appear instead of equilibrium gas pressures. (The subscript "e," indicating equilibrium, does not appear in equation 16–26, nor will it be used with pressures or concentrations in any subsequent equilibrium expression.) Again, the equilibrium is characterized by an equilibrium constant that is related to an equilibrium expression:

$$K_e = e^{-\Delta G°/RT} = 1.74 \times 10^{-5}$$

$$= \frac{[H^+][H_3CCOO^-]}{[H_3CCOOH]} \qquad \textbf{16–27}$$

Additional Examples

It is useful to consider a few additional examples of the construction of equilibrium expressions and the calculation of equilibrium constants. Many ionic solids display limited solubilities in water. One such solid is silver chloride and the dissolution occurs according to the equation

$$AgCl(s) \rightleftharpoons Ag^+(aq) + Cl^-(aq) \qquad \textbf{16–28}$$

We have seen how to construct equilibrium expressions from the pressures of gases. The procedure is the same for solution equilibria, save that solution concentrations appear instead of gas pressures. Hence, we have for reaction 16–28,

$$K_e = e^{-\Delta G°/RT} = [Ag^+][Cl^-] \qquad \textbf{16–29}$$

The value of $\Delta G°$ is calculable from the data of Tables 16–1 and 16–2:

$$\Delta G° = \Delta G°_{f(Ag^+,aq)} + \Delta G°_{f(Cl^-,aq)} - \Delta G°_{f(AgCl)}$$

$$= 18.43 - 31.35 - (-26.22)$$

$$= 13.30 \text{ kcal}$$

and K_e is 1.76×10^{-10}. With this small value, we see that AgCl is indeed only slightly soluble in water at 25 °C.

As our second example, we consider an oxidation-reduction reaction:

$$Mg(s) + 2\,H^+(aq) \rightleftharpoons Mg^{2+}(aq) + H_2(g) \qquad \textbf{16–30}$$

This reaction has been chosen because it involves both dissolved species and gaseous species. The equilibrium expression is

$$K_e = e^{-\Delta G°/RT} = P_{H_2}\frac{[Mg^{2+}]}{[H^+]^2} \qquad \textbf{16–31}$$

and it contains solution concentrations as well as the pressure of a gas. The ΔG° is calculable from the data given in Table 16–2:

$$\Delta G^\circ = \Delta G^\circ_{f(Mg^{2+},\,aq)} + \Delta G^\circ_{f(H_2)} - \Delta G^\circ_{f(Mg)} - 2\Delta G^\circ_{f(H^+,\,aq)}$$
$$= -108.99 + 0 - 0 - 2(0)$$
$$= -108.99 \text{ kcal}$$

and K_e is 8.68×10^{79}. This huge number shows that metallic magnesium easily dissolves in solutions of strong acids in water with the evolution of hydrogen gas.

16–3. Temperature Dependence of Equilibrium Constants

We have destroyed the concept of energy minimization as a general criterion for spontaneity by pointing out (beginning of Chapter 15) that in the case of the hydrogen atom–hydrogen molecule equilibrium, low temperatures favor hydrogen molecules and high temperatures favor hydrogen atoms. To discover the source of this effect of a temperature change, let us return to reaction 16–8, an equilibrium that involves a very low equilibrium pressure of CO_2 at 298 °K. What is the position of equilibrium for this system in a flame at 1000 °C? We could easily obtain K_e at 1000 °C if we could convert ΔG° from 298 °K and 1 atm to a corresponding value at 1000 °C = 1273 °K. We shall define ΔG°at temperatures other than 298 °K to mean ΔG for the conditions of 1 atm pressure in the case of gases and the hypothetical one-molar ideal solution in the case of solution species. Thus, K_e is calculable as $e^{-\Delta G^\circ/RT}$ at all temperatures, as long as the ΔG° used at a given temperature is the value at that temperature (rather than ΔG° at 298 °K). Since $\Delta G^\circ = \Delta H^\circ - T\Delta S^\circ$, the change in ΔG° with temperature arises from the obvious term T as well as from the temperature dependences of ΔH° and ΔS°.

Temperature Dependence of Enthalpy Changes

Let us investigate the temperature dependence of ΔH° first. The enthalpy of any substance increases with increasing temperature because energy must be added to produce the temperature increase. For a temperature change in one mole of any substance at constant pressure,

$$\Delta H = C_p \Delta T \qquad \textbf{16–32}$$

where we have approximated C_p as temperature-independent.

For the reaction of interest, we have

$$\Delta H^\circ_{1273} = H^\circ_{CO_2,298} + \Delta T(C_{p(CO_2)}) + H^\circ_{CaO,298} + \Delta T(C_{p(CaO)})$$
$$- H^\circ_{CaCO_3,298} - \Delta T(C_{p(CaCO_3)})$$

$$= \Delta H^\circ_{298} + \Delta T(\Delta C_p) \qquad \textbf{16–33}$$

The approximate heat capacities of $CaCO_3(s)$, $CaO(s)$, and $CO_2(g)$ in this temperature interval are 26, 15, and 11 cal/mole-deg respectively. Thus, ΔC_p is approximately zero for this reaction, a result that is a general one in chemical reactions. The result is that ΔH° for any reaction may be taken to a good approximation as temperature-independent in the calculation of the temperature dependence of ΔG°.

Temperature Dependence of Entropy Changes

The entropy change accompanying the constant-pressure temperature change of one mole of a pure substance with a temperature-independent C_p is

$$\Delta S = C_p \ln \frac{T_f}{T_i} \qquad \textbf{16–34}$$

This relation may be used to calculate an approximate value for ΔS°_{1273} in the reaction under investigation.

$$\begin{aligned}\Delta S^\circ_{1273} &= S^\circ_{CO_2,298} + C_{p(CO_2)} \ln \tfrac{1273}{298} \\ &\quad + S^\circ_{CaO,298} + C_{p(CaO)} \ln \tfrac{1273}{298} \\ &\quad - S^\circ_{CaCO_3,298} - C_{p(CaCO_3)} \ln \tfrac{1273}{298} \\ &= \Delta S^\circ_{298} + \Delta C_p \ln \tfrac{1273}{298}\end{aligned} \qquad \textbf{16–35}$$

Since $\Delta C_p \cong 0$, the standard entropy change in this reaction and all reactions may be taken to a good approximation as temperature-independent.

Spread and the Temperature Dependence of Gibbs Free Energy Changes

The temperature dependence of ΔG° arises almost entirely from the term T in the expression $\Delta G^\circ = \Delta H^\circ - T\Delta S^\circ$. We can easily interpret the role of this term T in our model of matter and energy spread, but first we shall finish the specific problem under investigation.

The ratio $K_{e(2)}/K_{e(1)}$ of equilibrium constants at temperatures T_2 and T_1 is defined as follows:

$$\frac{K_{e(2)}}{K_{e(1)}} = \frac{e^{-\Delta G^\circ_2/RT}}{e^{-\Delta G^\circ_1/RT}} = \frac{e^{-\Delta H^\circ_2/RT} e^{\Delta S^\circ_2/R}}{e^{-\Delta H^\circ_1/RT} e^{\Delta S^\circ_1/R}} = e^{-[(\Delta H^\circ/R)(1/T_2\ -\ 1/T_1)]} \qquad \textbf{16–36}$$

where ΔH° and ΔS° have been treated as temperature-independent. If the natural logarithm is taken for both sides of equation 16–36, we obtain

$$\ln K_{e(2)} - \ln K_{e(1)} = -\frac{\Delta H^\circ}{R}\left(\frac{1}{T_2} - \frac{1}{T_1}\right) \qquad \textbf{16–37}$$

Equation 16–37 is a general expression that applies to all reactions and all temperatures. In the case of reaction 16–8, it yields

$$\ln K_{e(1273)} = \ln (1.5 \times 10^{-23}) - \frac{42{,}610}{1.987}\left(\frac{1}{1273} - \frac{1}{298}\right)$$

$$= 2.56$$

and $K_{e(1273)} = 12.9$. The value 42,610 cal in this calculation is $\Delta H°$ for reaction 16–8, and it is calculable from the data given in Table 15–1. Since K_e in this case is the equilibrium pressure of CO_2, a sample of $CaCO_3(s)$ placed in an open container and heated to 1000 °C will be converted to $CaO(s)$ with the dispersal of the CO_2 to the atmosphere.

We see from equation 16–37 that $\Delta H°$ and T govern the temperature dependence of the equilibrium constant, and this is conceptually correct in terms of our matter spread–energy spread model. The K_e for the $CaCO_3(s)$ decomposition at 298 °K is small because the reaction is endothermic, and energy concentration predominates over matter spreading; that is, the entropy decrease of the surroundings is much larger than the entropy increase of the system in the standard reaction. If the same energy, equal to 42,610 cal, is taken from the surroundings at 1273 °K, a much smaller entropy decrease is produced in the surroundings than when the energy is taken at 298 °K, because the entropy change is inversely proportional to the temperature. **Thus, K_e for all endothermic reactions must increase as the temperature is increased.** Equation 16–37 also reveals that **K_e decreases with increasing T if $\Delta H° < 0$, that is, if the reaction is exothermic.** An exothermic reaction is favored by the spreading of energy. As the temperature is increased, the ΔS of the surroundings owing to the energy liberated by the reaction becomes smaller even though $\Delta H°$ remains constant.

What limit does K_e approach as T approaches infinity? Since

$$K_e = e^{-\Delta G°/RT} = e^{-\Delta H°/RT}e^{\Delta S°/R}$$

the term $\Delta H°$ becomes less and less important as the temperature is increased, regardless of whether the reaction is endothermic or exothermic. At T = infinity (∞), K_e becomes

$$K_{e(\infty)} = e^{\Delta S°/R} \quad \textbf{16–38}$$

Equation 16–38 provides the explanation of why hydrogen molecule dissociation is favored by high temperature. The separated atoms have a higher entropy than the molecules, and this state of higher entropy must eventually be favored over the lower-energy molecular state as the temperature is increased. This also provides the explanation for the progression from solid to liquid to gas as the predominant phases for pure substances as the temperature is increased from 0 °K. Since $S_{gas} > S_{liquid} > S_{solid}$ for any pure substance at a given pressure, the predominance of each of these phases in the various temperature ranges is understandable. The solid phase is the one of lowest energy and entropy and must be the most stable phase at the lowest temperatures. The gas phase must predominate at the highest temperatures because it is the phase of highest entropy; and the liquid exists at intermediate temperatures.

Equation 16–37 reveals that a plot of ln K_e versus $1/T$ is linear for all reactions. Figure 16–5 shows plots of ln K_e versus $1/T$ for an exothermic reaction and for an endothermic reaction, both of which possess the same $\Delta S°$.

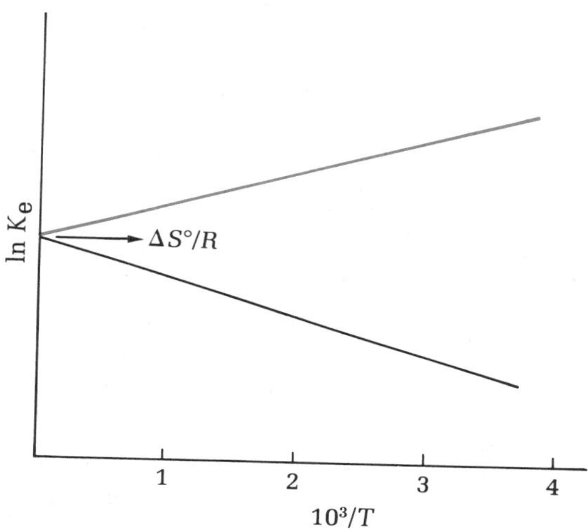

Figure 16–5
Plots of the natural logarithm of the equilibrium constant *versus* reciprocal absolute temperature for an endothermic reaction (black line) and an exothermic reaction (colored line). Both reactions have the same $\Delta S°$ value.

The plots show the independence of K_e on $\Delta H°$ at $T = \infty (1/T = 0)$, and they also reveal a convenient way to measure $\Delta H°$. Equation 16–37 reveals the slope of the straight lines of Figure 16–5 to be $-\Delta H°/R$. Thus, $\Delta H°$ values are easily obtained from the temperature dependences of equilibrium constants.

SUGGESTIONS FOR ADDITIONAL READING

Bent, H. A., *The Second Law*, Oxford University Press, New York, 1965.

Nash, L., *Elements of Chemical Thermodynamics*, Addison-Wesley, Reading, Mass., 1970.

PROBLEMS

1. Does equation 16–1 apply to the process depicted in Figure 15–5(b)? If not, why not?

***2.** What is the analog of equation 16–2 for a process that occurs at constant temperature and constant volume?

3. Show that the maximum work obtainable from the process shown in

Figure 15–1 is $T\Delta S_{universe}$, where $\Delta S_{universe}$ is that for the process in which no work is obtained.

4. If we assume that all of the original potential energy in Figure 15–1 is given to the floor as thermal energy, what is ΔG_{book}? Is $-\Delta G_{book}$ equal to the maximum amount of work that can be obtained from the process?
5. Work must be performed to electrolyze liquid water to gaseous hydrogen and gaseous oxygen. What is the minimum amount of work that must be performed to electrolyze 100 g of liquid water at 25 °C to produce hydrogen gas and oxygen gas, both at 1 atm partial pressure?
6. Consider the burning of liquid methanol to produce gaseous CO_2 and gaseous H_2O at 25 °C. Show from the data of Tables 15–1 and 15–3 that the maximum energy available from this process to do work exceeds the available energy for heat transfer, that is, $-\Delta G° > -\Delta H°$. Why is this the case? Isn't this a violation of the first law of thermodynamics? If not, what is the source of the energy that makes up the difference between $\Delta G°$ and $\Delta H°$?
7. When solid silver nitrate dissolves in water, the resultant solution is quite cool, indicating that ΔH for the dissolution is positive and energy is concentrated in the process. Yet $AgNO_3$ is highly soluble in water. How can that be?
8. Why do electric power plants use superheated steam (steam heated to over 300 °C) in power generation instead of steam at 100 °C?
9. The oceans contain an enormous amount of energy in the form of the motional energy of water molecules. Why doesn't New York City set up a power plant in which ocean water is taken in, thermal energy converted to electrical energy, and cold water expelled? The oceans appear to be a free source of energy.
10. Unlike the oceans, the sun really is a useful, free source of energy, which can be trapped and converted to electrical energy. Why is sunlight useful in this regard and ocean water not?
11. One of the immediate conclusions that can be drawn from equation 16–7 is that absolute zero (0 °K) is unattainable. Derive this conclusion from the equation.
12. You can spontaneously produce matter concentration, for instance, by raking leaves and putting them in a basket. Is this an exception to our general spontaneity criteria? Explain.
13. The carbon isotope ^{14}C is radioactive. Does this suggest a way to prove that both the left-to-right and right-to-left reactions in equation 16–8 occur even at equilibrium? How would you carry out the experiment to prove it?
14. Calculate the ΔG_{N_2} for the expansion of two moles of N_2 gas at 100 °C from 1.00 atm pressure to 0.100 atm. Since the ΔG is calculated to be negative, is this expansion a spontaneous process?

15. Why doesn't Br_2 appear in the equilibrium expression for reaction 16–22?

16. Construct the equilibrium expression and calculate the equilibrium constant at 25 °C for each of the following reactions:
 a. $2\ NO_2(g) \rightleftharpoons N_2O_4(g)$
 b. $3\ O_2(g) \rightleftharpoons 2\ O_3(g)$
 c. $CH_4(g) + \frac{1}{2}\ O_2(g) \rightleftharpoons CH_3OH(l)$

17. The standard formation reaction of benzene is

$$3\ H_2(g) + 6\ C(\text{graphite}) \rightarrow C_6H_6(l)$$

From the structures and bondings of graphite and benzene presented in earlier chapters of this text, explain why ΔH_f° is positive for benzene. Explain why ΔG_f° is more positive than ΔH_f°.

18. The interactions between the dissolved species are much greater in a one-molar NaCl aqueous solution than in gaseous HCl at 1 atm pressure and 25 °C. Give two reasons for the greater interaction.

19. From the data given in Table 16–2, calculate the relative strengths of $H_2S(aq)$ in going to $H^+(aq)$ and $HS^-(aq)$, and acetic acid in going to $H^+(aq)$ and $H_3CCOO^-(aq)$, both at 25 °C. That is, which is the stronger acid and which the weaker?

20. From the data given in Table 16–2, find which is more soluble in water at 25 °C, AgCl or AgI. The ΔG_f° for AgI(s) is −15.83 kcal/mole.

21. Vinegar is a solution of acetic acid in water. Use the data of Table 16–2 to show whether or not solid zinc is expected to dissolve appreciably in vinegar.

22. For each of the following equilibria, predict which species—those on the left or those on the right—are favored by an increase in temperature:
 a. $N_2O_4(g) \rightleftharpoons 2\ NO_2(g)$
 b. $NO(g) + \frac{1}{2}\ O_2(g) \rightleftharpoons NO_2(g)$
 c. $N_2(g) + O_2(g) \rightleftharpoons 2\ NO(g)$

23. Does the solubility of AgCl in water increase or decrease with increasing temperature? Explain.

24. The English scientist Joseph Priestley first prepared molecular oxygen in the laboratory by heating red, solid mercuric oxide to achieve the reaction

$$2\ HgO(s) \rightarrow 2\ Hg(l) + O_2(g)$$

The ΔG_f° and ΔH_f° values for HgO at 25 °C are −13.99 and −21.68 kcal/mole, respectively. At what temperature is the equilibrium pressure of O_2 equal to 1 atm?

17/ Equilibrium between Phases

The matter spread–energy spread model of physical and chemical changes together with the quantitative framework of Gibbs free energy changes and equilibrium constants will help us in making predictions and proposing explanations for phenomena associated with physical equilibria between solids, liquids, and gases.

17–1. Phase Equilibria in Pure Substances

Liquid-Vapor Equilibrium

The two intensive physical properties pressure and temperature are all that must be specified to determine completely all intensive thermodynamic properties of any pure element or compound. These thermodynamic properties include the molar energy, enthalpy, entropy, and Gibbs free energy. If G values are uniquely determined at each P and T for the solid, liquid, and gaseous phases of any pure substance, the phase or phases most stable at that P and T are also determined, since that phase or those phases have the lowest Gibbs free energy per mole. Normally there will be a single phase. However, at certain pressures and temperatures, two or more phases possess the same molar G. Where this occurs, these two or more phases coexist in equilibrium.

A diagram can be constructed for any pure substance with P as the ordinate and T as the abscissa. Every point on such a diagram corresponds to either a pure phase or one or more phases in equilibrium, and consequently it is termed a **phase diagram.** Such a phase diagram for pure water is shown in Figure 17–1, where pressure is plotted as log P on the ordinate and temperature as $10^3/T$ on the abscissa. Log P was chosen for the ordinate because the range of P to be included in the discussion covers many powers of ten, and the ordinate scale must be compressed. The reason for the choice of $10^3/T$ for the abscissa will be apparent shortly.

Certain points on the diagram correspond to one easily identified phase. High values of temperature and low pressures yield the vapor phase uniquely,

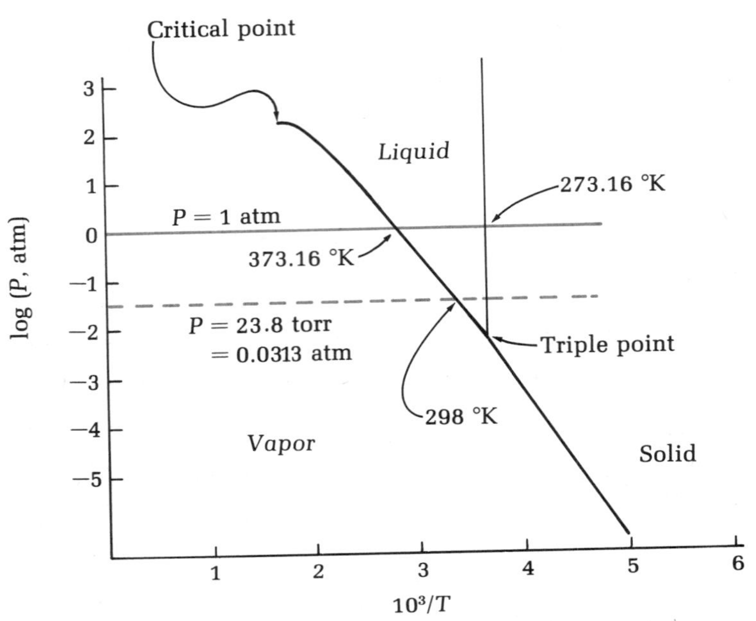

Figure 17–1 Phase diagram for water.

and the region is labeled "vapor" in Figure 17–1. The unique region labeled "solid" occurs at low temperatures and all but the lowest pressures. The liquid range is much more limited and hard to define. This is not surprising in terms of our previous findings (Chapter 16). The solid phase is the one of lowest H and must dominate the phase diagram at low temperatures. The vapor phase is the one of highest S and must predominate at high temperatures. The liquid phase represents a compromise between low H and high S, and it is not expected to exist over a wide range of P and T.

Let us consider first the measurement and then the calculation of one of the border points between the vapor and the liquid regions, namely the pressure at which $H_2O(l)$ and $H_2O(g)$ are in equilibrium at 298 °K. Such an equilibrium pressure exists over a range of temperatures, and it is termed the **equilibrium vapor pressure.** The measurement of the vapor pressure is straightforward and is depicted in Figure 17–2. The liquid water and its vapor are both maintained at 25 °C by immersion in a water bath at 25 °C. The pressure produces a displacement h of the manometer fluid, which is chosen to have a low density to maximize h, and a low vapor pressure, so it does not contribute its own pressure to the total pressure in the bulb. Mineral oil is a good candidate as a manometer fluid. The vapor pressure P is given by the equation

$$P = \frac{h(d)}{13.6} \tag{17–1}$$

where d is the density of the manometer fluid, 13.6 g/ml is the density of

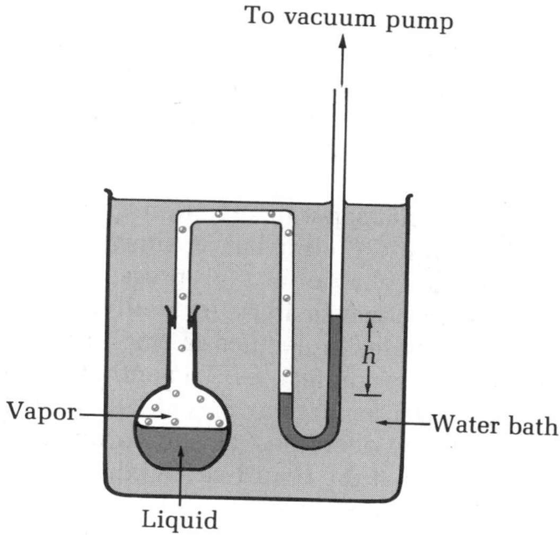

Figure 17–2
Apparatus for measurement of the equilibrium vapor pressure of a liquid.

liquid mercury, h is measured in mm, and P is in torr.

The equilibrium in question is

$$H_2O(l) \rightleftharpoons H_2O(g) \qquad \textbf{17–2}$$

for which ΔG° is obtainable from Table 16–1 as follows:

$$\Delta G^\circ = \Delta G^\circ_{f(H_2O,\,g)} - \Delta G_{f(H_2O,\,l)}$$
$$= -54.64 - (-56.69) = 2.05 \text{ kcal/mole}$$

The equilibrium constant for equilibrium 17–2 is calculated:

$$K_e = P_{H_2O} = e^{-\Delta G^\circ/RT} = 0.0313 \text{ atm} = 23.8 \text{ torr}$$

This equilibrium point has been located by an arrow at 298 °K on Figure 17–1. Water vapor at 1 atm is not in equilibrium with $H_2O(l)$ at 25 °C. However, the entropy increase of the water vapor owing to the pressure decrease (matter spreading) in going to 23.8 torr from 760 torr (the standard state) results in equilibrium.

What is the equation of the line in Figure 17–1 that represents liquid-vapor equilibrium? It is simply one example of the temperature dependence of an equilibrium constant, and equation 16–37 applies. In this case, the equation is

$$\ln P_T = (2.303) \log P_T$$
$$= (2.303) \log P_{298} - \frac{\Delta H^\circ_{\text{vap}}}{R}\left(\frac{1}{T} - \frac{1}{298}\right) \qquad \textbf{17–3}$$

where P_T is the vapor pressure at temperature T and $\Delta H^\circ_{\text{vap}}$ is the standard heat of vaporization (which is always positive). A plot of $\log P_T$ versus $1/T$

is predicted to be linear with a slope of $(-\Delta H^\circ_{vap}/2.303\ R)$, and ΔH°_{vap} for water can be calculated from Table 15–1 to be 10.52 kcal/mole. The temperature at which the vapor pressure equals 1 atm is the normal boiling point, and this temperature is indicated in Figure 17–1 as 373.16 °K. The linearity of the plot of log P_T versus $1/T$ is the reason $1/T$ was chosen as the abscissa scale in Figure 17–1.

The line in Figure 17–1 that separates the liquid and the vapor regions is straight below approximately 1 atm; but it shows a significant negative deviation from straight-line behavior at high pressures. At pressures of 100 atm or more, water vapor behaves in a decidedly nonideal fashion. The average distance between water molecules is small enough that significant attraction exists between water molecules. The enthalpy difference between liquid and vapor is no longer as large as it is at 298 °K, and the slope of the liquid-vapor equilibrium line decreases, as shown in the figure.

Figure 17–1 also shows that the liquid-vapor equilibrium line terminates at $T = 647$ °K and $P = 218$ atm. Such a termination must occur with all pure substances, and the point at which it occurs is termed the **critical point.** We have already mentioned that as P increases, the vapor molecules are located closer together on the average; that is, the vapor density increases. At the same time, the density of the equilibrium liquid decreases with increasing temperature. Eventually the two densities must equalize, and this happens at the critical point. It is absurd to refer to a vapor-liquid equilibrium past this point since there is only one phase present.

A common display of the behavior of a substance near the critical point is drawn in Figure 17–3. For CO_2, the critical point is located at 31 °C and 73 atm. In the display, a piece of dry ice (solid CO_2) is placed in a small,

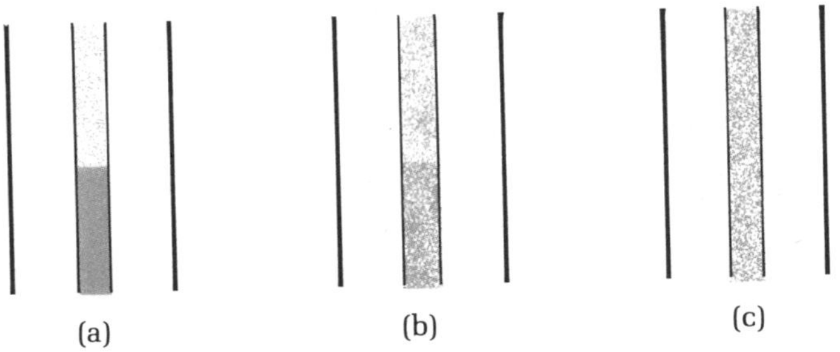

Figure 17–3
Magnifications of a tube containing CO_2 near its critical point. The liquid portion is darkened in order to distinguish it from the vapor. Part (a) is at a temperature ~ 0.1 °C below the critical point, (b) is ~0.01 °C below the critical point, and (c) is 0.1 °C above the critical point.

thick-walled glass tube and the tube is sealed by melting the glass with a torch. A properly sealed tube can easily withstand a pressure of 100 atm. The sealed CO_2 is allowed to warm to room temperature, where it exists in well-defined liquid and vapor phases with a clearly observable meniscus between the two. A small heater is used to warm the tube slowly from room temperature. Near 31 °C, the meniscus begins to shimmer like "heat waves" on a hot summer day, and over a range of approximately 0.1 °C the meniscus disappears. The shimmering is caused by the nearly equal densities of liquid and vapor. Any convection will disturb the liquid-vapor interface and cause shimmering.

Solid-Vapor Equilibrium

If the bath in Figure 17–2 is cooled from 25 °C, the measured vapor pressure will drop along the vapor-liquid equilibrium line of Figure 17–1. Eventually it must reach a temperature at which the low-enthalpy solid phase possesses a G equal to that of the liquid and that of the vapor. At this temperature, three phases (solid, liquid, and vapor) are simultaneously in equilibrium. This unique point for any pure substance is termed the **triple point,** and for water it is 0.0098 °C and 4.579 torr.

Let us now replace the water bath in Figure 17–2 with a lower-freezing liquid such as methanol, and continue the cooling. At the triple point, the molar G_{solid} = the molar G_{liquid} = the molar G_{vapor}. As we lower the temperature from this point in Figure 17–2, the liquid can no longer exist. The solid has become the condensed phase of lowest molar G below the triple point at a pressure of 4.579 torr. The molar G of the vapor can be lowered by decreasing its pressure from 4.579 torr. Such a pressure drop has virtually no effect on the molar G_{liquid}, however, and the molar G_{liquid} remains larger than the molar G_{solid} and the liquid phase disappears from the phase diagram.

The solid-vapor equilibrium line is another example of the application of equation 16–37, which becomes

$$\begin{aligned} \ln P_T &= (2.303) \log P_T \\ &= (2.303) \log P_{\text{tp}} - \frac{\Delta H^\circ_{\text{sub}}}{R}\left(\frac{1}{T} - \frac{1}{T_{\text{tp}}}\right) \end{aligned} \qquad \textbf{17–4}$$

where the subscript "tp" refers to the triple point. A plot of $\log P_T$ versus $1/T$ is linear, as shown in Figure 17–1, and the slope is $-\Delta H^\circ_{\text{sub}}/2.303\,R$. The ΔH° in equation 17–4 is that for the conversion of solid to vapor, a process called **sublimation.** We can calculate $\Delta H^\circ_{\text{sub}}$ by using Hess's law. We can view the process as the sum of two steps (even though it does not actually take place in two steps):

$$H_2O(s) \rightarrow H_2O(l); \qquad \Delta H = \Delta H^\circ_{\text{fusion}}$$

$$H_2O(l) \rightarrow H_2O(g); \qquad \Delta H = \Delta H^\circ_{\text{vap}}$$

We then have

$$\begin{aligned} \Delta H^\circ_{\text{sub}} &= \Delta H^\circ_{\text{fusion}} + \Delta H^\circ_{\text{vap}} \\ &= 1440 + 10{,}520 \\ &= 11{,}960 \text{ cal/mole} \end{aligned} \qquad \textbf{17–5}$$

Thus $\Delta H^\circ_{\text{sub}}$ is only slightly larger than the enthalpy of vaporization of the liquid. As shown in Figure 17–1, the slope of the solid-vapor equilibrium line is only slightly more negative than the slope of the liquid-vapor line.

Solid-Liquid Equilibrium

Liquid water cannot exist in the apparatus of Figure 17–2 at temperatures below the triple point. In the experiment, P is always the equilibrium vapor pressure, since there is no external pressure source. Suppose that we set up the experiment shown in Figure 17–4. In Figure 17–4(a), one mole of water has been introduced into a previously evacuated cylinder and frozen in a bath at −5 °C. The ice is contained in a volume larger than the molar volume

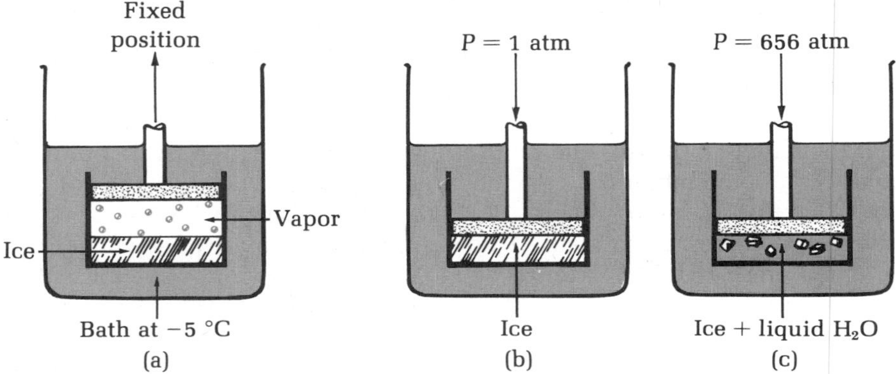

Figure 17–4 Apparatus for demonstrating the pressure dependence of the melting point of ice.

of ice by a piston that is anchored by some external device. The equilibrium state is the point on the solid-vapor line of Figure 17–1 at 268 °K. Now the piston anchor is removed, and the piston moves down, compressing the vapor. Any compression of the vapor raises its molar G and causes the vapor to escape to the solid phase by condensation, leading to the new equilibrium state of Figure 17–4(b), that is, only solid water at a P equal to 1 atm plus the pressure due to the weight of the piston.

What would happen if even more pressure were applied to the piston? Let us consider what happens to the molar G of the ice:

$$G_{ice} = E_{ice} + PV_{ice} - TS_{ice} \qquad \textbf{17–6}$$

Again we shall treat the ice as incompressible, as we did with the two solids in reaction 16–8, and thus E_{ice} and S_{ice} are pressure independent; but PV_{ice} increases with increasing pressure. Thus, the molar G of any condensed phase rises with increasing pressure, although the rate of increase is much smaller than for gases, as we have seen (Chapter 16).

The escaping tendency (molar G) of ice increases with increasing pressure, but is there any phase to escape to? It cannot be the vapor because the molar G of the vapor rises with pressure even more rapidly than the molar G of the solid; but what of the molar G of the liquid?

$$G_{liquid} = E_{liquid} + PV_{liquid} - TS_{liquid} \qquad \textbf{17–7}$$

The liquid is also essentially incompressible, and E_{liquid} and S_{liquid} can be taken as pressure-independent. Because of an increase in PV_{liquid}, G_{liquid}

increases with pressure. However, $V_{liquid} < V_{solid}$, that is, ice floats in liquid water; thus, the molar G_{liquid} increases with P at a slower rate than the molar G_{solid} does. If the pressure is large enough, the molar G_{solid} must exceed the molar G_{liquid}, and the ice must melt, as shown in Figure 17–4(c).

What is the value of P at which liquid water and ice are in equilibrium at -5 °C? First we need to know what ΔG is in going from one mole of solid water to one mole of liquid water at -5 °C and 1 atm pressure. The ΔH_{H_2O} for the melting of one mole of ice at -5 °C and 1 atm pressure is 1395 cal, and ΔS_{H_2O} for the same process is 5.11 cal/deg. Thus, ΔG at a pressure of 1 atm is an unfavorable 25.5 cal/mole, and the melting does not occur. The pressure at which the solid and liquid are in equilibrium is calculated from

$$\begin{aligned} P(V_{solid} - V_{liquid}) &= 25.5 \text{ cal/mole} \\ &= 1.053 \text{ liter-atm/mole} \end{aligned} \qquad \textbf{17–8}$$

The densities of ice and $H_2O(l)$ are 0.917 and 1.000 g/ml respectively, so that $V_{solid} = 0.0196$ and $V_{liquid} = 0.0180$ liter/mole. These values, together with equation 17–8, yield an equilibrium pressure of 656 atm. In general, equilibrium in and between condensed phases is relatively insensitive to pressure, and very large pressures must be applied to produce significant shifts in such equilibria. This is represented in Figure 17–1 by the steepness of the solid-liquid equilibrium line, even on a logarithmic scale. It is also of some interest to note that the positive slope of the solid-liquid line is not representative of most substances. The melting points of most substances increase with increasing pressure because $V_{solid} < V_{liquid}$; that is, most substances contract as they freeze.

All pure elements and compounds have a phase diagram like that of Figure 17–1, with well-defined solid, liquid, and vapor regions. However, some substances will not pass through a liquid phase as they are heated in an open container, that is, at a constant pressure of 1 atm. Carbon dioxide is such a substance. In Figure 17–1 for water, the solid, colored horizontal line at $P = 1$ atm crosses the vapor-liquid line at $T = 373.16$ °K and the liquid-solid line at $T = 273.16$ °K, and there is a fairly broad temperature range over which the liquid is observable at this pressure. For CO_2, the 1-atm line crosses only the solid-vapor line at -78 °C, and the liquid is never observable at this pressure. If, however, CO_2 is heated in a closed container so that the pressure increases with increasing temperature, the solid is seen to melt at -56.6 °C and a pressure of 5.1 atm.

17–2. Phase Equilibria Involving Solutions

Entropy of Formation of Solutions

Most people are aware that the equilibrium lines in Figure 17–1 can be shifted by adding another substance. For example, solid calcium chloride added to water lowers the freezing point of the water; this substance is used, therefore, to melt ice on the streets in winter. Sugar added to water raises its boiling point, and simmering syrup is at a much higher temperature than boiling water.

For pure water, the molar G of liquid water equals the molar G of solid water at 0 °C at 1 atm pressure. If adding calcium chloride lowers the freezing point, the ΔG of the conversion of water from the salt solution to solid water must be positive at 0 °C. Thus, ice spontaneously melts to form the solution at 0 °C. If ΔG for the freezing of ice from the salt solution at 0 °C is positive, the salt must have increased the molar G of *ice* or decreased the molar G of water in the salt solution relative to pure liquid water (or both changes occur). When a solution of calcium chloride freezes completely, the products are crystals of pure water and crystals of the salt. The molar G of the ice remains unaffected by the presence of calcium chloride. Thus, the depression of the freezing point must be due to a decrease in the molar G of liquid water as a consequence of the dissolution of the salt. A decrease in G_{liquid} of 25.5 cal/mole is sufficient to lower the freezing point of water to −5 °C, as we have seen.

We can show in an analogous fashion that in the boiling point elevation of water by sugar, the increase must reflect an increase in the molar G of water vapor, a decrease in the molar G of liquid water, or both as a result of the addition of sugar. The vapor phase is approximately pure water since sugar has a negligible vapor pressure over boiling syrup. Thus, the boiling-point elevation, like the freezing-point depression, is due to the lowering of the molar G of the liquid water through the dissolution of some substance in it.

A lowered molar G can arise from a lowered molar enthalpy or a raised molar entropy, or both. First we consider systems in which G changes as a consequence of entropy changes alone.

Are there substances that form solutions by processes in which the enthalpy change of each substance is very nearly zero? Yes there are, and an example is the mixing of toluene (methylbenzene) and benzene. Toluene and benzene are very similar substances, and the energy of the toluene-benzene interaction is nearly equal to the energies of the toluene-toluene and benzene-benzene interactions. Thus, the molar enthalpy of each substance has the same value in the solution as it has in the pure substance.

There are entropy changes accompanying the formation of benzene-toluene solutions. Let us suppose, for the sake of calculation, that a solution contains equal numbers of benzene and toluene molecules. The presence of the toluene opens up new position possibilities for the benzene since toluene molecules and benzene molecules can exchange positions in the solution. Thus, the volume that toluene molecules occupy at any given time is volume that is available to benzene at other times. Both the toluene and the benzene in the solution have been spread compared with pure toluene and pure benzene, in analogy with the gases in Figure 15–6.

The entropy increase of one mole of the benzene in going from pure benzene to the solution is given by

$$\Delta S_{\text{benzene}} = R \ln 2$$

Any solution contains a liquid termed the **solvent,** which is usually the substance present in the solution in the greatest molar amount, and one or more other dissolved species termed **solutes.** We can designate either substance in the toluene-benzene solution as the solvent. The general expression for $\Delta S_{\text{solvent}}$ for a dissolution that occurs with no enthalpy changes is

$$\Delta S_{\text{solvent}} = R \ln \left(\frac{x_i}{x_f}\right)_{\text{solvent}} \quad \text{cal/deg-mole} \qquad \textbf{17–9}$$

where x is the **mole fraction** of the solvent, defined as follows:

$$x_{\text{solvent}} = \frac{n_{\text{solvent}}}{n_{\text{solvent}} + n_{\text{solute}}} \qquad \textbf{17–10}$$

with n indicating number of moles. In the case of the toluene-benzene solution containing one toluene molecule per benzene molecule,

$$x_{\text{solvent}} = x_{\text{solute}} = 0.5$$

and equation 17–9 yields a $\Delta S_{\text{solvent}}$ of $R \ln 2$ for the solution process.

Raoult's Law

Equation 17–9 represents a starting point for our analysis of various equilibria involving solutions. Vapor-liquid equilibria can be studied by considering solutions containing methanol and ethanol. Methanol and ethanol are similar substances, and equation 17–9 is valid for solutions of one substance in the other. The equilibrium vapor pressure of pure ethanol is 44.5 torr at 20 °C, and the equilibrium vapor pressure of pure methanol is 88.7 torr at the same temperature. What is the vapor pressure over a solution of ethanol and methanol of a given ratio of numbers of moles of the two alcohols? We are concerned with two equilibria:

$$\text{Ethanol(l)} \rightleftharpoons \text{ethanol(g)} \qquad \textbf{17–11}$$

$$\text{Methanol(l)} \rightleftharpoons \text{methanol(g)} \qquad \textbf{17–12}$$

whose equilibrium constants and equilibrium expressions in the two pure substances are

$$P^\circ_{\text{ethanol}} = e^{-\Delta G^\circ/RT} = K_{e(\text{ethanol})} \qquad \textbf{17–13}$$

$$P^\circ_{\text{methanol}} = e^{-\Delta G^\circ/RT} = K_{e(\text{methanol})} \qquad \textbf{17–14}$$

where P° indicates the equilibrium vapor pressure over the pure liquid in each case, and ΔG° is the Gibbs free energy difference at 20 °C between pure liquid and pure vapor at 1 atm pressure.

The molar entropy of ethanol in solution is greater than the molar entropy of pure ethanol by an amount given by equation 17–9. Thus, the molar G of ethanol in the solution is lower than the molar G in pure liquid ethanol, and the difference is given by the equation

$$\begin{aligned} \Delta G_{(\text{ethanol, l})} &= G_{(\text{ethanol, solution})} - G_{(\text{ethanol, pure})} \\ &= -T\Delta S_{(\text{ethanol, l})} \\ &= -RT \ln \left(\frac{x_i}{x_f}\right)_{\text{ethanol}} \\ &= RT \ln x_{(\text{ethanol, l})} \end{aligned} \qquad \textbf{17–15}$$

where $x_{(\text{ethanol, l})}$ is the mole fraction of ethanol in the solution. The molar G of methanol is also lowered in the change from pure methanol to methanol in solution, and the change is given as follows:

$$\Delta G_{(\text{methanol, l})} = RT \ln x_{(\text{methanol, l})} \qquad \textbf{17–16}$$

Both ΔG values are negative. Thus, the equilibrium vapor pressure of each liquid must decrease in each case from the equilibrium vapor pressure of the pure liquid because the escaping tendency of each liquid has decreased from the standard value. For equilibrium to be maintained with a liquid whose molar G has decreased, the molar G of the vapor must decrease by an equal amount through a pressure decrease. This yields

$$\Delta G_{(\text{ethanol,g})} = \Delta G_{(\text{ethanol,l})}$$

$$RT \ln \frac{P_{\text{ethanol}}}{P^\circ_{\text{ethanol}}} = RT \ln x_{(\text{ethanol,l})} \qquad \textbf{17–17}$$

where we have employed equation 16–15 for $\Delta G_{(\text{ethanol, g})}$ and equation 17–15 for $\Delta G_{(\text{ethanol, l})}$. By cancelling RT and taking the antilogarithm of equation 17–17, we obtain

$$P_{\text{ethanol}} = P^\circ_{\text{ethanol}} x_{(\text{ethanol, l})} \qquad \textbf{17–18}$$

Likewise for methanol we derive

$$P_{\text{methanol}} = P^\circ_{\text{methanol}} x_{(\text{methanol, l})} \qquad \textbf{17–19}$$

Equations analogous to 17–18 and 17–19 are obeyed to a good approximation by a number of solutions, and this behavior was discovered experimentally in the last century by the French scientist François Raoult. For solutions that obey **Raoult's law**, the vapor pressure of each substance in solution is given by the equation

$$P_{\text{substance}} = P^\circ_{\text{substance}} x_{(\text{substance, l})} \qquad \textbf{17–20}$$

where $P_{\text{substance}}$ is the equilibrium vapor pressure above the solution and $P^\circ_{\text{substance}}$ is the vapor pressure of the pure substance.

Distillation

Equations 17–19 and 17–20 are plotted in Figure 17–5. The black, dashed line is the vapor pressure of ethanol as a function of solution composition, while the colored, dashed line is the vapor pressure of methanol. The solid black line is the total vapor pressure over the solution, and it is the sum of the two dashed lines. This vapor pressure–composition plot suggests a way to separate methanol from ethanol in an essentially quantitative fashion.

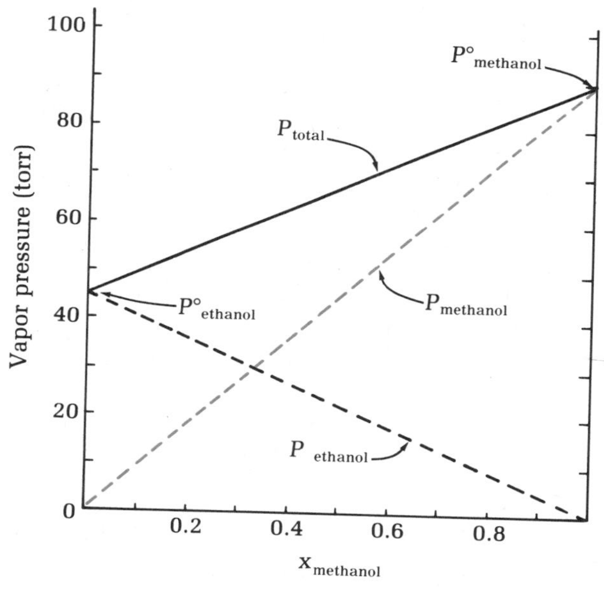

Figure 17–5
Vapor pressure as a function of composition for methanol-ethanol solutions.

Consider a solution in which $x_{(\text{methanol, l})}$ is 0.5. At 20 °C for this solution

$$P_{\text{ethanol}} = P^\circ_{\text{ethanol}} x_{(\text{ethanol, l})}$$
$$= (44.5)(0.5) = 22.2 \text{ torr}$$

$$P_{\text{methanol}} = P^\circ_{\text{methanol}} x_{(\text{methanol, l})}$$
$$= (88.7)(0.5) = 44.4 \text{ torr}$$

so that $P_{\text{total}} = 66.6$ torr. This low-pressure vapor is an excellent approximation to an ideal gas, for which Dalton's law of partial pressures holds. Thus, the mole fractions in the equilibrium vapor are

$$x_{(\text{ethanol, g})} = \frac{P_{\text{ethanol}}}{P_{\text{total}}}$$

$$= \frac{22.2}{66.6} = 0.333$$

$$x_{(\text{methanol, g})} = \frac{P_{\text{methanol}}}{P_{\text{total}}}$$

$$= \frac{44.4}{66.6} = 0.667$$

In any vapor-liquid equilibrium that obeys Raoult's law, the result above will be obtained at any composition and at any temperature; that is, **the equilibrium vapor is always enriched in the component of greater $P°$ (the more volatile component) as compared with the mole fraction of that component in the equilibrium liquid.**

Suppose that we start with a liter of methanol-ethanol solution at 20 °C and condense 1 ml of the equilibrium vapor in a separate container. The condensate is highly enriched ($x_{(\text{methanol, l})} \cong 0.667$) in methanol compared with the remaining 999 ml ($x_{(\text{methanol, l})} \cong 0.5$). If we remove the condensate to a container where it exists in equilibrium with its own vapor, the vapor pressures in this container at 20 °C are as follows:

$$P_{\text{ethanol}} = (44.5)(0.33) = 14.8 \text{ torr}$$

$$P_{\text{methanol}} = (88.7)(0.67) = 59.2 \text{ torr}$$

and the mole fractions in the equilibrium vapor are given as

$$x_{(\text{ethanol, g})} = \frac{14.8}{14.8 + 59.2} = 0.20$$

$$x_{(\text{methanol, g})} = \frac{59.2}{14.8 + 59.2} = 0.80$$

Each successive distillation leads to a higher methanol fraction in the vapor, which suggests that perhaps twenty such steps would yield a small amount of very nearly pure methanol.

The procedure of such successive distillations and removals would be exceedingly laborious and time-consuming in the separation of an entire liter of solution. Fortunately, the process can be automated in the "fractionating" apparatus shown in Figure 17–6. The liquid is heated to boiling in the vessel on the left, and vapor enriched in methanol rises in the column. In the cool column the vapor condenses onto glass beads or some other inert material that can hold the condensed vapor. Further hot vapor rising in the column heats the condensed liquid and revaporizes it, with the resulting vapor even more enriched in methanol. Thus, the vapor is subjected to successive distillations as it rises in the slowly heating column, and an efficient column of the type might easily provide the equivalent of twenty successive distillations. By the time the vapor reaches the thermometer it is essentially pure methanol, and this fact can be verified since the thermometer will read

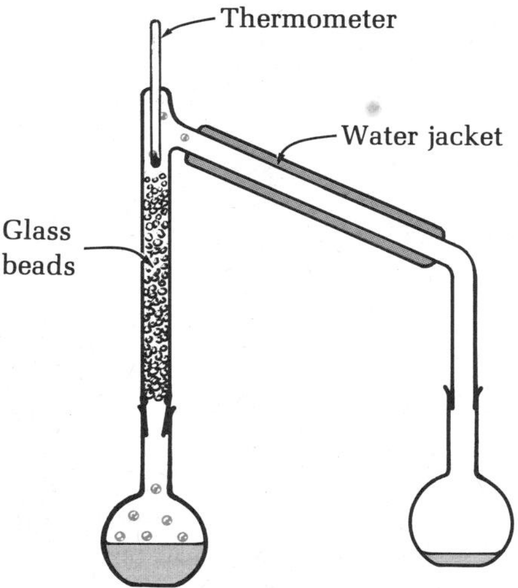

Figure 17-6
Fractionation apparatus for the separation of volatile liquids by distillation.

64.6 °C, the boiling point of pure methanol. The thermometer will continue to record this temperature as the methanol is condensed and collects in the vessel on the right. When virtually all of the methanol has been distilled, the thermometer will quickly rise to 78.4 °C, the boiling point of pure ethanol. The separation of species that obey Raoult's law in solution is easily accomplished by distillation.

Determining Molecular Weights by Applying Raoult's Law

A common application of Raoult's law is in determining molecular weights, particularly of nonvolatile substances, which are difficult to study by mass spectrometry, or of substances whose molecular weights may well change from phase to phase because of the aggregation of simple molecular units. An example of the latter is benzoic acid (C_6H_5COOH), and we shall carry through a calculation of its molecular weight through use of Raoult's law.

We must choose a volatile solvent in which benzoic acid dissolves with very little enthalpy change, and a good candidate is found to be benzene (C_6H_6), which has a vapor pressure of 75 torr at 20 °C. A sample of 1.000 g of benzoic acid is dissolved in 20.000 g of benzene and the measured vapor pressure over the solution is found to be 1.18 torr lower than for pure benzene. Since benzoic acid has a negligible vapor pressure at 20 °C, this ΔP is given as

$$\begin{aligned}\Delta P &= P^\circ_{\text{benzene}} - P_{\text{benzene}} \\ &= P^\circ_{\text{benzene}} - P^\circ_{\text{benzene}} x_{(\text{benzene, l})} \\ &= P^\circ_{\text{benzene}}(1 - x_{(\text{benzene, l})}) \\ &= P^\circ_{\text{benzene}} x_{\text{benzoic acid}} \qquad \textbf{17-21}\end{aligned}$$

Thus,

$$x_{\text{benzoic acid}} = \frac{1.18}{75} = 1.58 \times 10^{-2}$$

By definition,

$$x_{\text{benzoic acid}} = \frac{n_{\text{benzoic acid}}}{n_{\text{benzoic acid}} + n_{\text{benzene}}}$$

$$= \frac{\text{wt}_{\text{benzoic acid}}/MW_{\text{benzoic acid}}}{\text{wt}_{\text{benzoic acid}}/MW_{\text{benzoic acid}} + \text{wt}_{\text{benzene}}/MW_{\text{benzene}}}$$

The weights of benzoic acid and benzene are 1.000 g and 20.000 g, and the molecular weight of benzene is 78.00. With these values, the molecular weight of benzoic acid in benzene is calculated as 244 g/mole. Since the unit C_6H_5COOH has a weight of 122 g/mole, we must conclude that benzoic acid forms a dimer in benzene by aggregation of two formula units of molecular weight 122 each. The structure of this dimer is shown in Figure 17–7. The carboxylic acid group is highly polar and the two benzoic acid molecules associate by hydrogen-bonding.

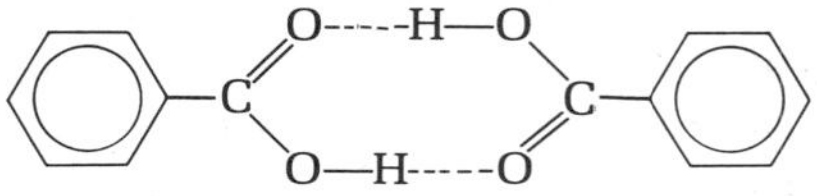

Figure 17–7 The structure of the dimer of benzoic acid.

Boiling-Point Elevation

Another way to use vapor-pressure lowering in the molecule-weight determination is through the accompanying elevation of the solution boiling point through dissolution of the nonvolatile, solid benzoic acid.

The boiling point of pure benzene may be calculated from the fact that $\Delta G^\circ_{\text{vap}}$ must be zero at the boiling point, or

$$\Delta G^\circ_{\text{vap}} = \Delta H^\circ_{\text{vap}} - T_b \Delta S^\circ_{\text{vap}} = 0 \qquad \textbf{17–22}$$

where T_b is the boiling point. From equation 17–22, T_b is calculable if $\Delta H^\circ_{\text{vap}}$ and $\Delta S^\circ_{\text{vap}}$ are known.

The ΔS_{vap} of the solvent from the solution is smaller than ΔS_{vap} from pure benzene because the entropy of benzene has been increased in the solution relative to the entropy of pure liquid benzene, making the entropy of benzene in solution nearer the entropy of the vapor. If the temperature is held constant at the normal boiling point of benzene, the benzene vapor pressure must be less than 1 atm over the solution for the molar G of benzene vapor to equal the molar G of benzene in solution.

Instead of lowering the pressure of the vapor, there is a second way in which one can reestablish the vapor-liquid equilibrium. Equation 17–23 gives the expression from which the equilibrium vapor pressure of pure benzene is calculable at any temperature:

$$\ln P_{\text{benzene}} = \frac{-\Delta G^\circ_{\text{vap}}}{RT} = \frac{-\Delta H^\circ_{\text{vap}}}{RT} + \frac{\Delta S^\circ_{\text{vap}}}{R} \qquad \textbf{17–23}$$

In going from pure benzene to benzene in solution, a second entropy term must be added to account for the entropy increase of the benzene in solution:

$$\ln P_{\text{benzene}} = \frac{-\Delta H^\circ_{\text{vap}}}{RT} + \frac{\Delta S^\circ_{\text{vap}}}{R} + \ln x_{(\text{benzene, l})} \qquad \textbf{17–24}$$

This extra term, which is negative, can be balanced either by lowering P_{benzene} according to Raoult's law or by making $-\Delta H^\circ_{\text{vap}}/T$ less negative. The latter is done by raising the temperature. The appropriate balancing occurs when

$$\ln x_{(\text{benzene, l})} + \Delta\left(\frac{-\Delta H^\circ_{\text{vap}}}{RT}\right) = \ln x_{(\text{benzene, l})} - \frac{\Delta H^\circ_{\text{vap}}}{R}\Delta\left(\frac{1}{T}\right) = 0 \qquad \textbf{17–25}$$

and $\Delta(1/T)$ must be negative (the boiling point must increase). Another way to view the boiling point increase is that raising the entropy of liquid benzene by making a solution lowers the favorable matter spreading in going from liquid to vapor. In order to return to equilibrium with the vapor state at the same pressure (1 atm), the unfavorable energy concentrating in the positive ΔH_{vap} must be performed at a higher temperature so as to yield a less unfavorable ΔS of the surroundings.

If $\Delta H^\circ_{\text{vap}}$ is known for benzene, equation 17–25 can be used as it stands to calculate x_{benzene} and the molecular weight of benzoic acid from the measured boiling points of the solution and of pure benzene. However, several excellent approximations can generally be used to arrive at a very common form of equation 17–25. Let us investigate the validity of the approximations and arrive at the common form by employing our specific benzoic acid–benzene solution.

The $\Delta H^\circ_{\text{vap}}$ value for benzene is 7,600 cal/mole at its boiling point and

$$\begin{aligned}\ln x_{(\text{benzene, l})} &= \ln (1 - 1.58 \times 10^{-2}) \\ &= -1.59 \times 10^{-2} \cong -x_{\text{benzoic acid}}\end{aligned}$$

which illustrates the first approximation commonly employed. With these values and equation 17–25,

$$\begin{aligned}\Delta\left(\frac{1}{T}\right) &= \left(\frac{1}{T_{\text{raised}}} - \frac{1}{T_{\text{b}}}\right) = \frac{-\Delta T}{T_{\text{b}}T_{\text{raised}}} \\ &\cong (-x_{\text{benzoic acid}})R/\Delta H^\circ_{\text{vap}} \\ &= \frac{(-1.58 \times 10^{-2})(1.987)}{(7600)} = -4.13 \times 10^{-6}\ \text{deg}^{-1}\end{aligned}$$

where $\Delta T = (T_{raised} - T_b)$ is the boiling point elevation from T_b, the boiling point of the pure solvent. The boiling point of pure benzene is 80.099 °C, and from $\Delta(1/T)$ the elevation ΔT of the boiling point in going to the solution is approximately 0.5 °C. This typical ΔT value corresponds to approximately a 0.1 percent increase in T_b, the boiling point of the pure solvent. Thus, to an excellent approximation $\Delta(1/T)$ may be equated to $-\Delta T/T_b^2$.

With our two approximations, equation 17–25 reduces to

$$\frac{\Delta T\, \Delta H^\circ_{vap}}{RT_b^2} \cong x_{benzoic\ acid}$$

The mole fraction can also be approximated as

$$x_{benzoic\ acid} = \frac{n_{benzoic\ acid}}{n_{benzoic\ acid} + n_{benzene}} = \frac{4.09 \times 10^{-3}}{4.09 \times 10^{-3} + 0.256}$$

$$\cong \frac{n_{benzoic\ acid}}{n_{benzene}}$$

and this is our third and last approximation.

We may now relate the mole ratio, $n_{benzoic\ acid}/n_{benzene}$, to a commonly employed measurement of concentration by multiplying both numbers of moles by 50 to scale our 20.000 g of benzene to 1 kg of benzene. Thus, $n_{benzoic\ acid}$ becomes the number of moles of benzoic acid in one kilogram of solvent, which is by definition $m_{benzoic\ acid}$, the **molality** of benzoic acid in solution. In addition, $n_{benzene}$ becomes $1000/MW_{benzene}$, and our final equation for the boiling point elevation in a general case is

$$\Delta T \cong \frac{(MW_{solvent})RT_b^2 m_{solute}}{1000\Delta H^\circ_{vap}} = k_b m_{solute} \qquad \textbf{17–26}$$

where k_b is termed the **boiling-point elevation constant;** it is a function of constants characteristic of the solvent. For benzene, k_b is calculated as

$$k_b = \frac{(78.0)(1.987)(353.3)^2}{(1000)(7600)} = 2.54 \text{ deg-kg/mole}$$

Freezing-Point Depression

Let us consider the effect of a factor like dissolved solid NaCl on the phase diagram for water given in Figure 17–1. The addition of another substance (NaCl) means that P and T no longer suffice to determine the phases in the system. A concentration variable must be added, such as x_{NaCl}, which can be plotted on an axis extending perpendicular to the page. The variable x_{NaCl} refers to the mole fraction of NaCl in the entire system and not in any particular phase. Figure 17–1 is one of an infinite number of projections of the P–T behavior of given H_2O–NaCl systems of fixed x_{NaCl}. The positions of the equilibrium lines vary continuously from those of pure water at

$x_{NaCl} = 0$ (Figure 17–1) to those of pure NaCl at $x_{NaCl} = 1$. Figure 17–8 shows the diagram for pure water along with the projection for a 2.0-molal sodium chloride solution. The black lines are for pure water and the colored lines are for the sodium chloride solution.

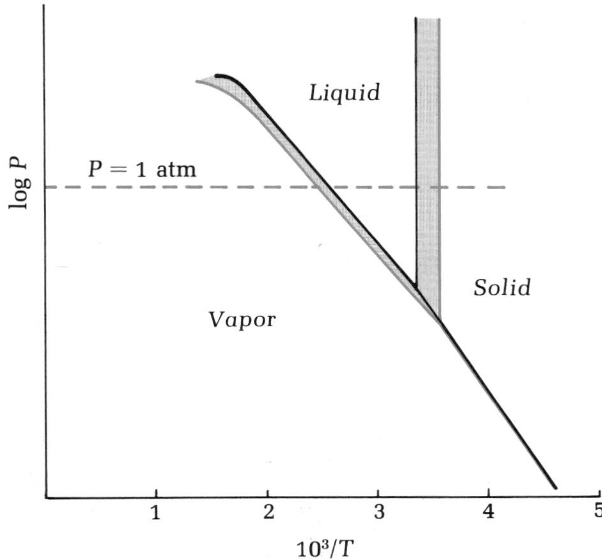

Figure 17–8
Phase diagram for pure water (black lines) and a projection of the phase diagram for a 2.0 molal sodium chloride solution (colored lines).

Sodium chloride possesses a negligible vapor pressure over the liquid-vapor equilibrium temperature range shown in Figure 17–8. Thus, P is the equilibrium vapor pressure of H_2O along this line, and the colored, vapor-liquid line is under the black line because of vapor pressure lowering. The solid phase is a mixture of separate and identifiable crystals of ice and solid sodium chloride. Therefore, the solid-vapor equilibrium remains unchanged by the presence of NaCl, and the black and colored lines for this equilibrium are coincident.

As shown in Figure 17–8 (and discussed previously in this chapter), the lowering of the equilibrium vapor pressure of the liquid as compared with the solid means a lower tendency of the liquid to escape to the solid phase. Thus, the liquid-solid line is displaced to lower temperatures, and both the normal freezing point of water and the triple point are lowered by the same temperature, ΔT. The effect of the lowered molar G of liquid water as a result of the dissolution of NaCl is to expand the liquid region of the phase diagram in all possible directions, as shown by the colored area in Figure 17–8.

We can relate the freezing-point depression to the molality of the NaCl just as we did the boiling-point elevation in the preceding section. At 273.16 °K and 1 atm pressure, the molar $G_{(H_2O,s)}$ = the molar $G_{(H_2O,l)}$ in pure water. In the solution, the molar $G_{(H_2O,l)}$ is different from the molar G of pure liquid water by $RT \ln x_{(H_2O,l)}$ as in the liquid-vapor equilibrium. This change favors the liquid state and increases the already positive ΔS_{H_2O} of the melting of ice, and the melting of ice mixed with salt becomes spontaneous at 0 °C. To counteract the effect of the increased ΔS_{H_2O} for melting, we can supply the necessary energy of fusion, ΔH°_{fusion}, at a lower temperature, thereby increasing the negative ΔS_{surr}. This is almost exactly the reasoning used in obtaining equation 17–25, and the corresponding equation here is

$$-\ln x_{(H_2O,l)} + \Delta\left(\frac{-\Delta H^\circ_{fusion}}{RT}\right) = 0$$

$$-\ln x_{(H_2O,l)} - \frac{\Delta H^\circ_{fusion}}{R}\,\Delta\left(\frac{1}{T}\right) = 0 \qquad \textbf{17–27}$$

Equations 17–25 and 17–27 are identical in form and differ only in the substitution of ΔH°_{fusion} for ΔH°_{vap} and in the sign of $\Delta(1/T)$, which must be positive in equation 17–27. We can follow through the same set of approximations with equation 17–27 as with equation 17–25 to obtain the commonly employed approximate equation

$$\Delta T \cong \frac{(MW_{solvent})RT_f^2 m_{solute}}{1000\Delta H^\circ_{vap}} = k_f m_{solute} \qquad \textbf{17–28}$$

where $\Delta T = T_f - T_{lowered}$, the freezing point of the pure solvent minus the lowered freezing point. In the case of water,

$$k_f = \frac{(18.00)(1.987)(273.16)^2}{(1000)(1440)} = 1.86 \text{ deg-kg/mole}$$

Salts dissolved in water can produce quite sizable lowerings of the freezing point. A solution, for example, containing 20 g of NaCl in 100 g of water shows a freezing point lowering of 12.73 °C. From equation 17–28, this corresponds to a sodium chloride molality of 6.84 and a molecular weight for NaCl of 29, or half the formula weight of NaCl. The reason for this strange result is that a solution that is 3.42 molal in NaCl is 3.42 molal in Na^+ and 3.42 molal in Cl^- for a total molality of dissolved material of 6.84. The number 29 is the average atomic weight of sodium and chlorine, whose ions are the solute in this case. The basis for vapor pressure and freezing-point lowering and boiling-point elevation, as we have discussed them, is the increased entropy of the solvent due to its competition with solute species for specific positional sites. In this competition, the nature of the solute particles does not enter; only their number is important. As a consequence, the three properties we have examined (and any other property that depends only on the number of solute particles and not their nature) are commonly termed **colligative properties.**

There are several other important topics to be considered under the general heading of solid-liquid equilibria involving solutions. However, at this point, we shall complete our introduction of colligative properties by examining osmotic pressure.

Osmotic Pressure

Consider the situation depicted in Figure 17-9. A cylinder containing a dilute sugar solution is enclosed on top by a piston and at the bottom by a semipermeable membrane that has pores small enough to allow the passage of water but not large enough to pass the sugar molecules. Such membranes are commonly available and their synthesis has become an important area of current research. The membranes that enclose living cells are also semipermeable, and this selective permeability is the basis of such processes as nerve impulse conduction.

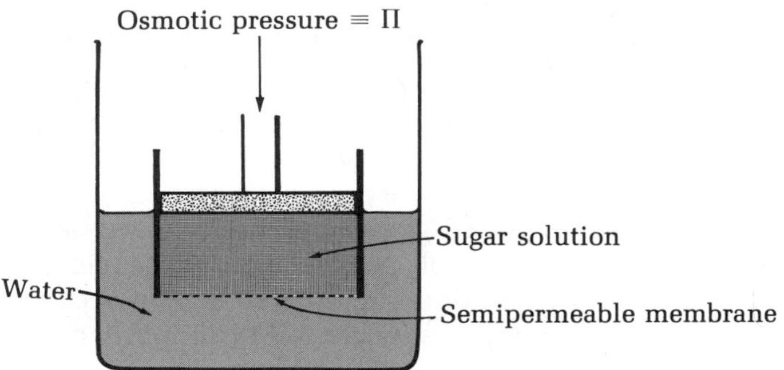

Figure 17-9 Apparatus for the demonstration of osmotic pressure.

The water in the sugar solution has a lower molar G than the molar G of pure water by a ΔG equal to $RT \ln x_{(H_2O,l)}$ (as discussed previously). The escaping tendency of water is higher from the pure water side, and water will tend to flow through the membrane into the solution, a process termed **osmosis.** To prevent the water flow, some means must be found to increase the molar G_{H_2O} in the solution by an amount equal to $-RT \ln x_{(H_2O,l)}$. This is done by applying the pressure Π (Greek uppercase *pi*) to the piston, increasing the molar G_{H_2O} in the incompressible solution by $\Delta G_{H_2O} = \Delta H_{H_2O} = \Pi V_{H_2O}$ (compare discussion of Figure 17-4). If the solution contains n_{H_2O} moles of water, the equation that gives the balancing of changes in G is

$$\Pi V_{solution} = -n_{H_2O}RT \ln x_{(H_2O,l)} \qquad \textbf{17-29}$$

where $V_{solution}$ is the volume of the solution.

If the solution is dilute, we can apply two of the approximations used in the derivation of equation 17-26:

$$-n_{H_2O}RT \ln x_{(H_2O,l)} \cong n_{H_2O}RTx_{solute}$$

$$\cong n_{H_2O}RT\left(\frac{n_{solute}}{n_{H_2O}}\right)$$

These approximations convert equation 17–29 to the approximate equation

$$\Pi \cong \left(\frac{n_{solute}}{V_{solution}}\right)RT = [\text{solute}]RT \qquad \textbf{17–30}$$

where Π is termed the **osmotic pressure** and [solute] is the solute concentration in moles of solute per liter of solution (molarity).

Determination of the Molecular Weights of Large Molecules

Since Π is a function only of temperature and the concentration of solute (but not its nature), it is also a colligative property; and it may be used in molecular-weight determinations. A little thought suggests that the osmotic pressure should provide a much more precise measure of molecular weight than any of the other three colligative properties discussed previously, since the molar G in a condensed phase is relatively insensitive to pressure. Thus, even a quite dilute solution should be characterized by a sizable osmotic pressure. This fact finds its greatest application in determining the molecular weights of very large molecules.

Suppose that we have a solution containing 0.100 g of hemoglobin in 5.00 ml of solution. Normally we should think of this as a concentrated solution. If the 0.100 g were NaCl instead, the ion concentration would be 0.67 molar. This would yield a freezing-point depression for water of approximately 1.2 °C, and the molecular weight of NaCl is easily measurable from this sizable depression. The molecular weight of hemoglobin is 6.45×10^4 and its molar concentration in our solution is 3.1×10^{-4}. The resultant freezing-point depression of water is 5.8×10^{-4} degree, and this is too small to be measured. Freezing-point depression cannot be used to determine the molecular weights of large molecules.

Let us now calculate the osmotic pressure for our hemoglobin solution.

$$\Pi = [\text{hemoglobin}]RT = (3.1 \times 10^{-4})(0.082)(298)$$
$$= 7.6 \times 10^{-3}\ \text{atm} = 5.8\ \text{torr}$$

This seems like a small pressure, but it can be measured quite precisely through use of an apparatus that involves a small modification of Figure 17–9; this apparatus is shown in Figure 17–10. In Figure 17–10, the osmotic pressure is produced hydrostatically by allowing a very small amount of water to enter the solution through the membrane, causing the solution to rise a distance h in the capillary. In the case of the 3.1×10^{-4} molar hemoglobin solution, h is (13.6)(5.8) = 79 mm, an easily and precisely measurable distance.

Within this same vein, it is interesting to calculate what h would be if

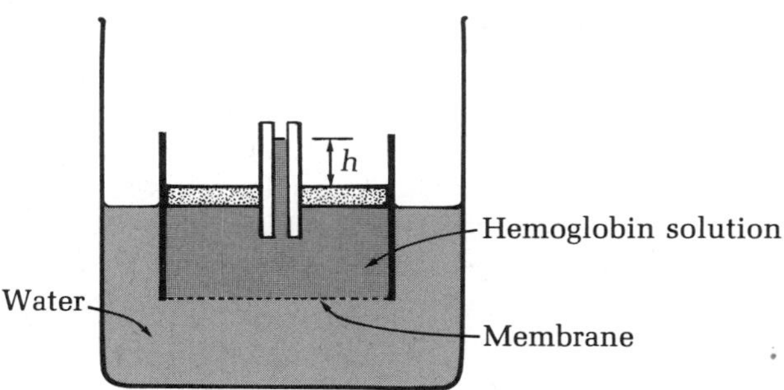

Figure 17–10
Apparatus for the measurement of small osmotic pressures.

the solution were 0.100 molar. The osmotic pressure at 298 °K is 2.44 atm = 2.52×10^4 mm of water = 82.7 ft of water. Thus, osmotic pressure can be (and is) one of the principal reasons water can be raised from the ground to great heights in trees.

Dialysis and Reverse Osmosis

Figure 17–9 suggests an important application of the osmotic effect, and this is its reversibility. If the pressure applied to the piston exceeds Π, then the molar G_{H_2O} in the solution will exceed the molar G in pure water, and pure water can literally be squeezed from a solution. The work involved in this process, termed **reverse osmosis,** is very nearly the minimum possible, namely the positive ΔG of separating a solution into pure water and a more concentrated solution. This is exactly the type of process required for the most economical obtaining of fresh water from seawater. The practical application of this concept, however, depends on the development of membranes that are strong enough to withstand pressures of many atmospheres, and that will permit the passage of water across them but not the passage of dissolved ions or organic species.

Figure 17–10 suggests another common application of semipermeable membranes. Suppose that the hemoglobin solution also were 0.1 molar in NaCl, and that the membrane were chosen to be permeable to water, Na^+, and Cl^-, but not the large hemoglobin molecules. The Na^+ and Cl^- will tend to spread to the maximum possible degree, and this means equal salt concentrations on each side of the membrane. When this condition is reached, the external solution can be removed and replaced with pure water, whereupon more salt is removed from the hemoglobin solution. The process described is **dialysis,** and it can be used in this case to make a virtually salt-free hemoglobin solution. Dialysis is the process by which ions and small organic molecules are removed from the bloodstream in the kidneys.

Deviations from Raoult's Law

Up to this point in this chapter, equilibria involving solutions have been restricted to those that obey Raoult's law, that is, where the molar enthalpy of each dissolved substance is equal to the molar enthalpy of the pure substance, and for which the entropy changes of the species are given by equation 17–9. While Raoult's law is a good approximation for many systems there are also many common solutions that are formed with considerable evolution or absorption of energy by heat transfer or for which equation 17–9 does not apply to the entropy changes. For instance, water added to sulfuric acid evolves much energy, giving an energy-spread driving force in addition to a matter-spread driving force to the solution process. We have seen that the matter spreading of solution formation lowers the escaping tendency of the solvent with respect to that of pure solvent. The energy spreading when water dissolves in sulfuric acid must lead to an additional lowering of the molar G of water, and the vapor pressure of H_2O over a sulfuric acid solution should be considerably lower than the vapor pressure predicted by Raoult's law; such is indeed the case.

If the deviation from Raoult's law behavior is sufficiently large, a serious complication arises in the attempted separation of the two solution species by simple distillation. We shall illustrate this with the solution of dioxane ($C_4H_8O_2$) in water. The structure of dioxane is

```
H₂C—CH₂
 /     \
O       O
 \     /
H₂C—CH₂
```

The hydrocarbon portion of the dioxane molecule yields a dioxane-water interaction that opposes solution formation (though not strongly enough to prevent solution formation at all proportions).

Figure 17–11 shows the equilibrium vapor pressures of water and dioxane over the solution as a function of composition at 35 °C. The vapor pressures of both water and dioxane (black lines) are above those (colored lines) predicted by Raoult's law, as they should be since the unfavorable dioxane-water interaction adds to the escaping tendency of both species. The vapor-pressure elevation is so large that, unlike the Raoult's law case in which the solution vapor pressure increases linearly with $x_{(\text{dioxane, l})}$, the total of water and dioxane vapor pressures goes through a maximum at approximately $x_{(\text{dioxane, l})} = 0.6$. Let us consider a solution in which $x_{(\text{dioxane, l})}$ is 0.2. What is the composition of the vapor in equilibrium with this solution at 35 °C? From Figure 17–11, the vapor pressures of water and dioxane are both 37 torr at this concentration. Thus, using Dalton's law of partial pressures, we calculate that x_{dioxane} in the vapor is 0.5, and again the vapor is enriched in the more volatile component, just as it was in the case of the methanol-ethanol solutions treated previously.

In the case of methanol-ethanol solutions, a series of successive distillations finally yields essentially complete separation of the two liquids. Can

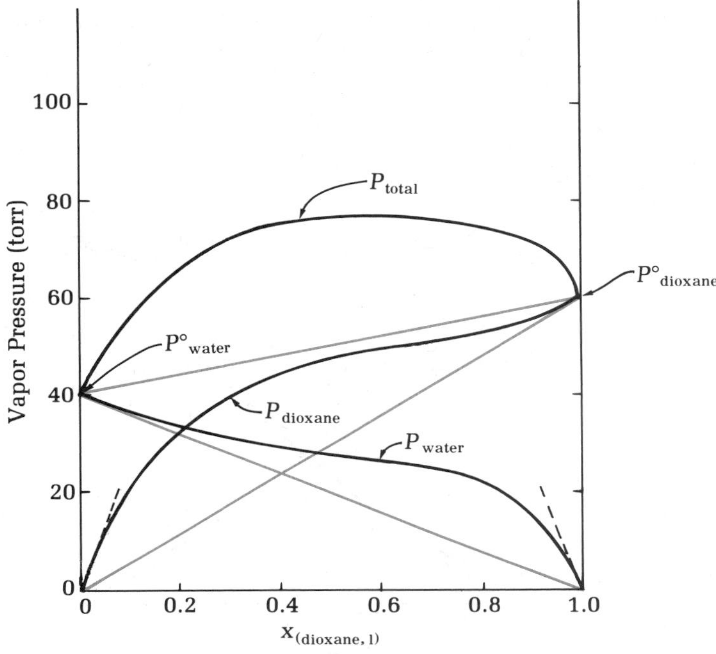

Figure 17-11 Vapor pressure as a function of composition for dioxane-water solutions at 35 °C. The black lines refer to the solution and its components. The colored lines are predicted by Raoult's law, and the dashed lines are predicted by Henry's law.

the same thing be done with dioxane-water? If we start with $x_{(\text{dioxane, l})} = 0.2$, the first distillate has $x_{(\text{dioxane, l})} = 0.5$, as we have just calculated. Distillation of this distillate yields a second distillate even more enriched in dioxane. Let us consider what happens when the liquid distillate in the fractionating column reaches $x_{\text{dioxane}} = 0.6$. Here the two vapor pressures are 49 torr and 27 torr for dioxane and water, respectively. The equilibrium vapor has x_{dioxane} approximately equal to 0.6, the same as the liquid. Further calculations show for this system **and for any other system for which the total vapor pressure as a function of composition goes through a maximum, the compositions of equilibrium liquid and vapor are identical at the composition corresponding to the maximum vapor pressure.** Fractional distillation will not proceed beyond this point and pure dioxane cannot be prepared by this method. The greatest concentration of dioxane obtainable by distillation from aqueous solution at 35 °C is $x_{(\text{dioxane, l})} \cong 0.6$. This solution, which has the same concentration as its equilibrium vapor, is termed an **azeotrope.** Perhaps the best-known azeotrope is ethanol and water, which contains 95 percent ethanol by weight and is the common product of fermentation of sugar followed by simple distillation.

Systems like water–sulfuric acid that form in an exothermic fashion show vapor pressure–composition curves in which the vapor pressure goes through a minimum. The curve for a commonly quoted example, the vapor pressures over solutions of acetone (CH_3COCH_3) and chloroform ($CHCl_3$), is shown in Figure 17–12. Unlike most C—H bonds, the C—H bond in chloroform is relatively polar because of the withdrawal of electron density by

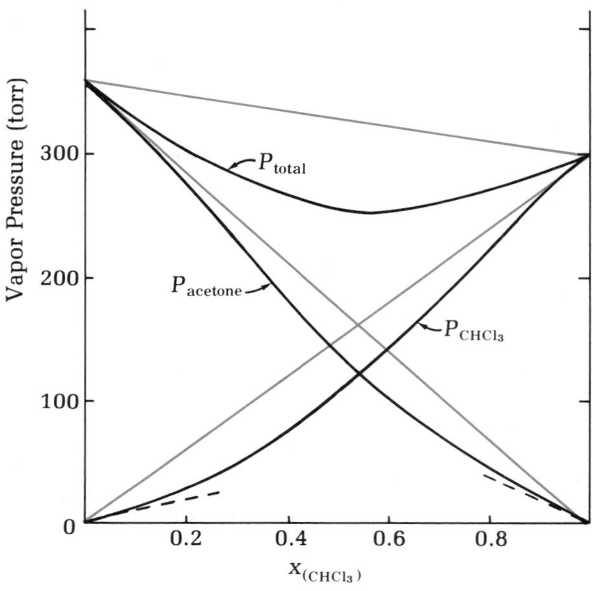

Figure 17–12 Vapor pressure as a function of composition for chloroform-acetone solutions at 35 °C. The black lines refer to the solution and its components. The colored lines are predicted by Raoult's law, and the dashed lines by Henry's law.

the three chlorines. As a consequence, chloroform hydrogen-bonds with acetone, as shown in Figure 17–13. The solution forms with a negative ΔH and this is reflected in the vapor pressures for each component, which are lower than those predicted from Raoult's law.

Figure 17–13 The hydrogen-bonded interaction of chloroform and acetone.

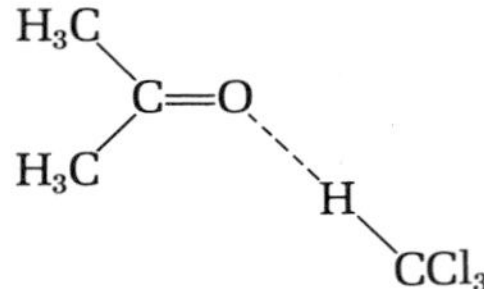

Let us consider an acetone-chloroform solution in which x_{CHCl_3} is 0.5. The vapor pressures of acetone and chloroform are nearly identical at this composition. Thus the vapor also has $x_{CHCl_3} = 0.5$ and a vapor pressure minimum also results in an azeotrope. Acetone and chloroform, like dioxane and water, cannot be separated by simple distillation.

Vapor-Liquid Chromatography

The separation and identification of species in solution forms an important part of analytical chemistry. Fractional distillation would appear to be a very easy way to accomplish the separation save for the problem of azeotropes. In any complex mixture, the molecular interactions are so numerous that many azeotropes can and will prevent a clean separation of components. To prevent the formation of azeotropes, we must prevent the interactions that form them. Thus, we must achieve a fractional distillation in such a way that the molecules of all volatile species are always kept at a large average distance from each other. One way to keep molecules of two or more substances apart is to make a dilute solution of those substances. If the solvent chosen has a very low vapor pressure, we shall not have to worry about solute-solvent azeotropes. Furthermore, we should prefer that the solution not form in an exothermic fashion, because we wish to distill the solute from the solution, and we do not want its vapor pressure lowered unnecessarily. Excellent solvents for these purposes are nonpolar, involatile mineral oil, or equally nonpolar, involatile paraffin wax.

Suppose we start with a very dilute solution of chloroform and acetone in mineral oil, in which the chloroform and acetone concentrations are equal. What will be the composition of the equilibrium vapor above this solution at 35 °C? Let us suppose for the sake of calculation that Raoult's law is obeyed by both acetone and chloroform in the mineral oil solution. At 35 °C, P°_{acetone} is 345 torr, while $P^\circ_{\text{CHCl}_3}$ is 290 torr; the equilibrium vapor has $x_{\text{acetone}}/x_{\text{CHCl}_3} = \frac{345}{290}$, or $x_{\text{acetone}} = 0.544$. As with simple distillation, the equilibrium vapor is enriched in the more volatile of the two species. Suppose the equilibrium vapor is now dissolved in a second sample of mineral oil. The equilibrium vapor over this second solution will be even more enriched in acetone than the first vapor was, again in complete analogy with simple fractional distillation. By a succession of such distillations and dissolutions, acetone and chloroform could be completely separated.

The process we have outlined is called **vapor-liquid chromatography**, and chemists routinely use it for separating and identifying volatile species. The chromatography apparatus is diagrammed in Figure 17–14. The separation is accomplished by passing the heated, vaporized substances to be

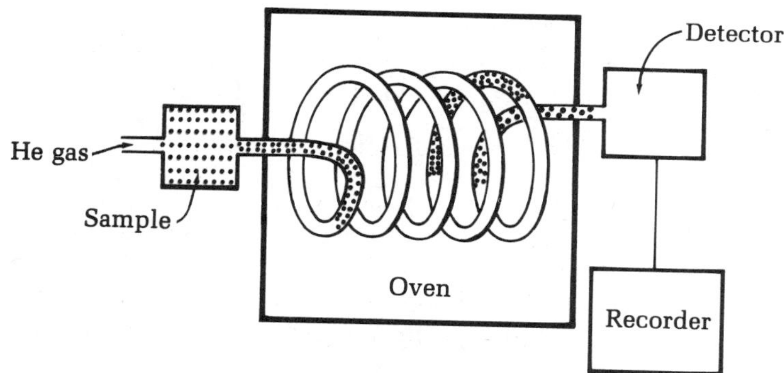

Figure 17–14 Schematic representation of a vapor-liquid chromatographic separation of two species.

separated through a small diameter copper tube of length 20–30 ft. The tube is packed with some porous inert solid that is coated with the involatile solvent. Helium is typically used as a carrier gas to sweep the vapor through the tube, which is preheated to ensure a sufficiently high vapor pressure of the volatile species being separated. If the tube temperature is carefully controlled, each distinct chemical species in the mixture separated will possess a highly predictable and reproducible retention time for passing through the tube, and this measured retention time is very useful in identifying unknown species. Detecting species at the end of a column is usually accomplished by measuring the thermal conductivity of the gas emerging from the tube. Helium is an excellent thermal conductor. As any substance other than helium emerges from the tube in significant amounts, the thermal conductivity decreases, and this decrease is easily detected and recorded.

The figure shows two species indicated in color and in black entering the tube as a vapor mixture. The repeated dissolutions and distillations in the tube result in the separation of the two species into bands shown in color and in black in the right portion of the tube. Figure 17–15 shows the thermal conductivity as a function of time as the two bands pass through the detector. These two inverted peaks are displayed on the recorder shown in Figure 17–14.

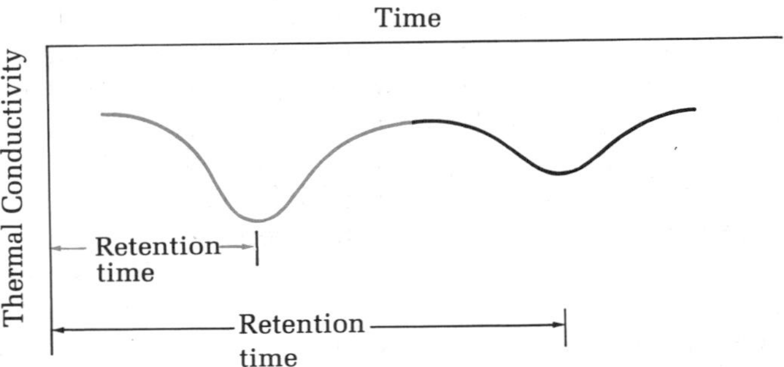

Figure 17–15 Recorder trace from the separated species in Figure 17–14.

The chromatography principle of employing dilute solutions to effect separations has been applied to other phase equilibria in addition to vapor-liquid. Thus, chromatography is not restricted to volatile species; it is, in fact, the principal procedure for separating large, involatile biological molecules.

Henry's Law

In our calculation on the chromatographic separation of acetone and chloroform, we assumed that Raoult's law is obeyed. This will not be true in general, and we must see how deviations from Raoult's law influence the separation. We are concerned with the following two equilibria:

$$\text{Acetone (mineral oil)} \rightleftharpoons \text{acetone (g)}$$

$$\text{Chloroform (mineral oil)} \rightleftharpoons \text{chloroform (g)}$$

Each of these two equilibria is characterized by an equilibrium constant:

$$K_{\text{acetone}} = \frac{P_{\text{acetone}}}{[\text{acetone}]}$$

$$K_{\text{chloroform}} = \frac{P_{CHCl_3}}{[CHCl_3]}$$

and these two equilibrium constant expressions are two examples of **Henry's law**, which is that the concentration of a dissolved gas is proportional to the pressure of that gas. This relationship is the reason champagne does not bubble in the sealed bottle but does bubble when the cork is removed. When the gaseous pressure of CO_2 decreases on removal of the cork, the solubility of CO_2 also decreases, and bubbles of gaseous CO_2 form.

Henry's law may also be stated as: The equilibrium vapor pressure of a dissolved substance is directly proportional to its concentration in solution. The equilibrium constant expression involving solution concentrations is only valid for dilute solutions (Chapter 16). For more concentrated solutions, equation 16–24 is invalid, and the equilibrium constant is not a valid concept. Thus, Henry's law is valid only for dilute solutions. However, we are dealing with dilute solutions in gas-liquid chromatography, and we may employ Henry's law. In a dilute solution, the volume of solution is, to an excellent approximation, the volume of solvent since the number of moles of solute is negligible in comparison with the number of moles of solvent. By the following set of excellent approximations, we can convert Henry's law to an expression that closely resembles Raoult's law:

$$
\begin{aligned}
P_{\text{solute}} &= K_{\text{solute}}\left(\frac{n_{\text{solute}}}{V_{\text{solution}}}\right) \\
&\cong K_{\text{solute}}\left(\frac{n_{\text{solute}}}{V_{\text{solvent}}}\right) \\
&= K_{\text{solute}}\rho_{\text{solvent}}\left(\frac{n_{\text{solute}}}{\text{wt}_{\text{solvent}}}\right) \\
&= \frac{K_{\text{solute}}\rho_{\text{solvent}}}{MW_{\text{solvent}}}\left(\frac{n_{\text{solute}}}{n_{\text{solvent}}}\right) \\
&\cong \left(\frac{K_{\text{solute}}\rho_{\text{solvent}}}{MW_{\text{solvent}}}\right) x_{\text{solute}}
\end{aligned}
$$

where ρ_{solvent} is the density of the solvent.

The term $K_{\text{solute}}\ \rho_{\text{solvent}}/MW_{\text{solvent}}$ is an effective $P°$ for the solute, and it will equal the actual vapor pressure of the pure solute if Raoult's law is obeyed. If the vapor pressure-composition curve displays a maximum, P_{solute} will be higher than the Raoult's law value at all concentrations. A plot of P_{solute} versus x_{solute}, at small values of x_{solute}, will be linear, with a steeper slope than the Raoult's law line. This is shown by the dashed lines both for water at high x_{dioxane} and for dioxane at low x_{dioxane} in Figure 17–11. Both species possess high effective vapor pressures because of the relatively unfavorable interaction between them. The opposite situation is shown for chloroform and acetone in Figure 17–12. In chloroform-acetone solutions, the interaction is relatively favorable, and the effective vapor pressures are lower than the Raoult's law values.

Liquid-Liquid Extraction

An interesting and useful example of the competition between the tendency toward the uniform distribution of matter and the tendency toward the uniform distribution of energy is the distribution of a solute between two solvents. Suppose we have an aqueous solution containing 0.1 *M* NaCl and 0.1 *M* acetic acid (CH_3COOH), and we wish to obtain pure acetic acid. There are several ways in which we could proceed, but one very convenient way is to add a roughly equivalent volume of diethyl ether ($C_2H_5OC_2H_5$) to the aqueous solution and shake the mixture in a closed container.

What is the new equilibrium situation? If matter spreading were the only criterion of spontaneity, the water, NaCl, and acetic acid would spread into the ether, and the ether would spread into the aqueous solution until all species were uniformly distributed in one solution phase. This does not happen, since NaCl is ionic, H_2O is highly polar, ether is only slightly polar, and acetic acid has a polarity intermediate between that of water and that of ether. Water and NaCl have higher molar enthalpies when dissolved in ether than when they are dissolved in water. Energy is concentrated in H_2O and NaCl as these species dissolve in ether and in ether as ether dissolves in water. At equilibrium we have two distinct phases as shown in Figure 17–16, one solution in which water is the solvent, and a second solution in which ether is the solvent. The densities of diethyl ether and water are 0.7 and 1.0 g/ml, respectively, and the ether solution floats on the aqueous solution, as shown in the figure.

The equilibrium situation shown in Figure 17–16 actually consists of four separate and simultaneous equilibria:

Figure 17–16
A mixture of water and diethyl ether in equilibrium.

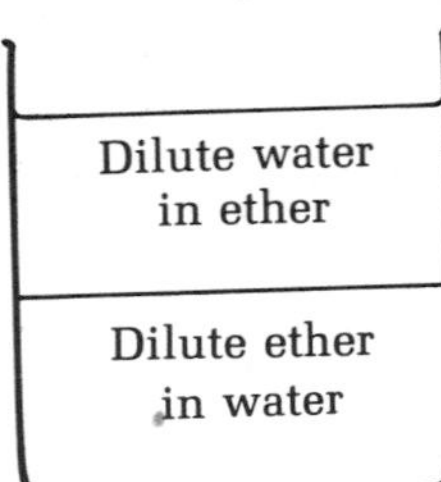

1. Ether (ether) ⇄ ether (aq).
2. H_2O (ether) ⇄ H_2O (aq).
3. NaCl (ether) ⇄ NaCl (aq).
4. Acetic acid (ether) ⇆ acetic acid (aq).

The lengths of the arrows connecting the two sides of these equilibria indicate which side predominates. For example, NaCl is found predominantly in the aqueous phase.

At equilibrium, the concentration of each of the four species in each of the two phases remains constant with time. In each case the escaping tendency of a given species in ether equals the escaping tendency of that species in water. Since the water-ether interaction is energetically unfavorable, the molar *H* of water and the molar *H* of ether increase as each substance dissolves in the other. This increase must be balanced by an entropy increase due to the spreading of matter; that is, the more unfavorable the energy of the ether-water interaction, the smaller the solubility of water in ether and ether in water. We may use these ideas to investigate equilibria (3) and (4), the partitioning of NaCl and acetic acid between water and ether.

These are equilibria between dilute solutions, and we may write equilibrium constant expressions for them:

$$K_3 = \frac{[\text{NaCl}]_{\text{aq}}}{[\text{NaCl}]_{\text{ether}}}$$

$$K_4 = \frac{[\text{acetic acid}]_{\text{aq}}}{[\text{acetic acid}]_{\text{ether}}}$$

where the equilibrium constants in this case of solution-solution equilibrium are termed **distribution constants.**

If the energy of the NaCl-water interaction were equal to that of the NaCl-ether interaction, K_3 would be unity. Maximum matter spreading would be the only equilibrium criterion, and this would occur at equality of concentrations in the two phases. However, NaCl interacts much more favorably with water than with ether. Thus, K_3 is, in fact, a very large number and NaCl is virtually insoluble in ether. The intermediate polarity of acetic acid gives rise to a relatively high solubility in both water and ether, and K_4 is of the order of unity. Suppose, for the sake of simplified calculation, that we set K_4 at 1.0 and K_3 at 1.0×10^{10}. Let us start with 100 ml of an aqueous solution containing 0.10 M NaCl and 0.10 M acetic acid. We add 100 ml of ether and shake until equilibrium is established. At equilibrium we have $[\text{NaCl}]_{\text{ether}} = 1.0 \times 10^{-11}$, $[\text{NaCl}]_{\text{aq}} = 0.10$, $[\text{acetic acid}]_{\text{ether}} = [\text{acetic acid}]_{\text{aq}} = 5 \times 10^{-2}$ M. Now we decant the ether solution, add another 100 ml of pure ether to the aqueous solution, and shake. Then we decant the second ether solution and add it to the first. After ten such equilibrations and decantings, the aqueous solution will have $[\text{NaCl}]_{\text{aq}} = 0.10$ and $[\text{acetic acid}]_{\text{aq}} = 0.10/2^{10} = 1.0 \times 10^{-4}$ M. We shall also have one liter of ether in which $[\text{acetic acid}]_{\text{ether}} = 1.0 \times 10^{-2}$ M. Thus, the process of **extraction** provides a quick and nearly quantitative separation of dissolved acetic acid from dissolved sodium chloride. Diethyl ether boils at 34.6 °C and acetic acid boils at 118.1 °C. Diethyl ether and acetic acid are easily separated by simple distillation. Since most chemical preparative reactions are carried out in solution, extraction is very commonly used in isolating and purifying reaction products.

Enthalpy Changes and Colligative Properties

We have considered previously the role of a dissolved substance in the depression of the freezing point of the solvent. In the derivation of equation 17–28, we assumed that the molar G_{solvent} in solution was changed only by the positional spreading caused by dissolving the solute in it. Thus, we might think that equation 17–28 is invalid for a dilute solution of water in dioxane or dioxane in water (or any other system that displays a marked deviation from Raoult's law). Such is not the case, however, and equation 17–28 is an excellent approximation for the freezing-point depression of the solvent in **any** dilute solution. A clue to why this is so is provided by

Figures 17–11 and 17–12. Consider the equilibrium vapor pressure of dioxane as a function of $x_{(\text{dioxane},l)}$ in Figure 17–11. At low $x_{(\text{dioxane},l)}$ Henry's law is obeyed. The dioxane pressure is a linear function of $x_{(\text{dioxane},l)}$, but the slope differs markedly from the Raoult's law slope. At $x_{(\text{dioxane},l)}$ near unity, P_{dioxane} again approaches a linear dependence on $x_{(\text{dioxane},l)}$; but the slope now approaches the Raoult's law slope. Examination of Figures 17–11 and 17–12 for the behavior of the vapor pressures of water, dioxane, chloroform, and acetone shows that in each case, as the mole fraction of a species in solution approaches unity, the vapor pressure of the species approaches that predicted by Raoult's law.

Why should the unfavorable dioxane-water interaction be very important for dioxane in a dilute solution of dioxane in water, such that Raoult's law is not obeyed by dioxane; yet in the same solution the same interaction is of little consequence for water, and Raoult's law is nearly obeyed by water? Figure 17–17 shows schematic representations of pure water (all W's), pure dioxane (all D's), and a dilute solution of dioxane in water (W's and

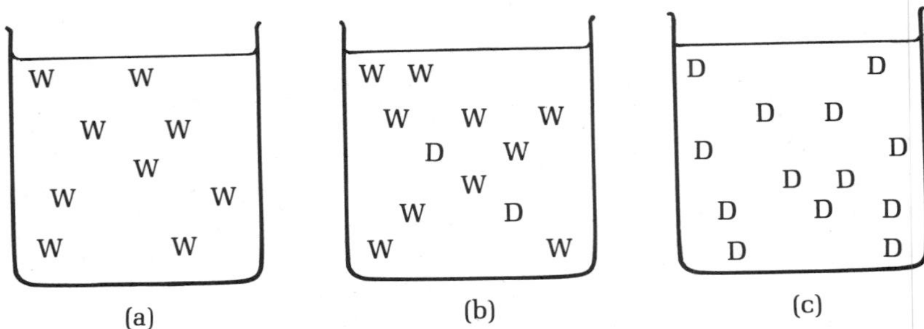

Figure 17–17 Representation of the locations of molecules of water (W) and dioxane (D) in (a) pure water, (c) pure dioxane, and (b) a dilute solution of dioxane in water.

D's). Raoult's law results from the additional positional possibilities in Figure 17–17(b) over those in either 17–17(a) or 17–17(c). Each W is free to exchange positions with each D. Since there are few D's, the relative effect on W of the additional positional possibilities is small; but the effect on D is very large. Thus, according to Raoult's law, the vapor pressure of W is only slightly less over (b) than over (a), whereas the vapor pressure of D is much less over (b) than over (c).

Now let us consider the effect of the D–W interaction on the two vapor pressures over (b). Every D has W's as nearest neighbors in this solution and the elevation of the molar G_{dioxane} due to this interaction is the maximum possible at this temperature and pressure. Thus, the Henry's law slopes of Figure 17–11 are the steepest tangents to each curve in the figure; that is, they represent the largest deviation in the figure from Raoult's law behavior. Most of the W's in Figure 17–17(b) are completely surrounded by other W's and the elevation of the molar G_{water} due to the D–W interaction is nearly zero. Thus, P_{water} approaches Raoult's law behavior as x_{water} approaches unity.

We saw that any of the so-called colligative properties (boiling point elevation, osmotic pressure, and so on) are useful in molecular-weight determinations. In the case of each of these properties, we have derived a quantitative relation, which can be used in the appropriate molecular-weight determination. All of these equations were derived from Raoult's law, and they are valid only to the extent that Raoult's law is obeyed. We have just showed that the colligative property equations, for example equation 17–28, are valid for **any dilute solution,** regardless of the interaction between solute and solvent. It is this fact that makes colligative properties so widely useful in molecular-weight determinations in chemistry.

The Freezing of Solutions

A complete examination of freezing involves, besides the freezing of pure liquids (Chapter 14), the freezing of solutions. Consider a mixture of the two elements antimony (Sb) and lead (Pb). The phase diagram for such a two-component system is three-dimensional, like the H_2O–NaCl diagram previously considered. In Figure 17–8 we investigated one of a family of projections of this three-dimensional diagram at various mole fractions of sodium chloride. These projections are particularly illustrative of how phases of one component are influenced by the presence of another component.

Suppose that we wish to display the composition dependence of the position of some equilibrium such as that between liquid and solid at 1 atm pressure. This is part of the projection of the three-dimensional diagram at a constant pressure of 1 atm. In Figure 17–8, it is a plane perpendicular to the page and including the dashed line shown at $P = 1$ atm. A strict projection of Figure 17–8 would plot $10^3/T$ on the ordinate and x_{NaCl} on the abscissa. However, in this case the ordinate is commonly transformed to T, so that equilibrium temperature is directly displayed as a function of composition. Such a plot will display the composition dependence of both the boiling point and freezing point. We have already gone through an extensive discussion of vapor-liquid equilibrium as a function of solution composition, and we shall restrict our attention here to the solid-liquid portion of the plot of T versus x_{Pb} at $P = 1$ atm for the Pb–Sb system as shown in Figure 17–18. The presence of Pb lowers the melting point of Sb, and the presence of Sb lowers the melting point of Pb—the general phenomenon discussed previously. The melting point of pure Sb is 904 °K and all Sb–Pb solutions exist in equilibrium with solid Sb at temperatures lower than 904 °K. As shown in the figure, any mixture of two solids must, because of the mutual lowering of melting points, have a well-defined minimum possible melting temperature. This is termed the **eutectic temperature,** and the composition that yields it is the **eutectic composition.** In the case of Pb and Sb, the **eutectic point** occurs at $x_{\mathrm{Pb}} = 0.71$ and $T = 519$ °K. Common solder is a

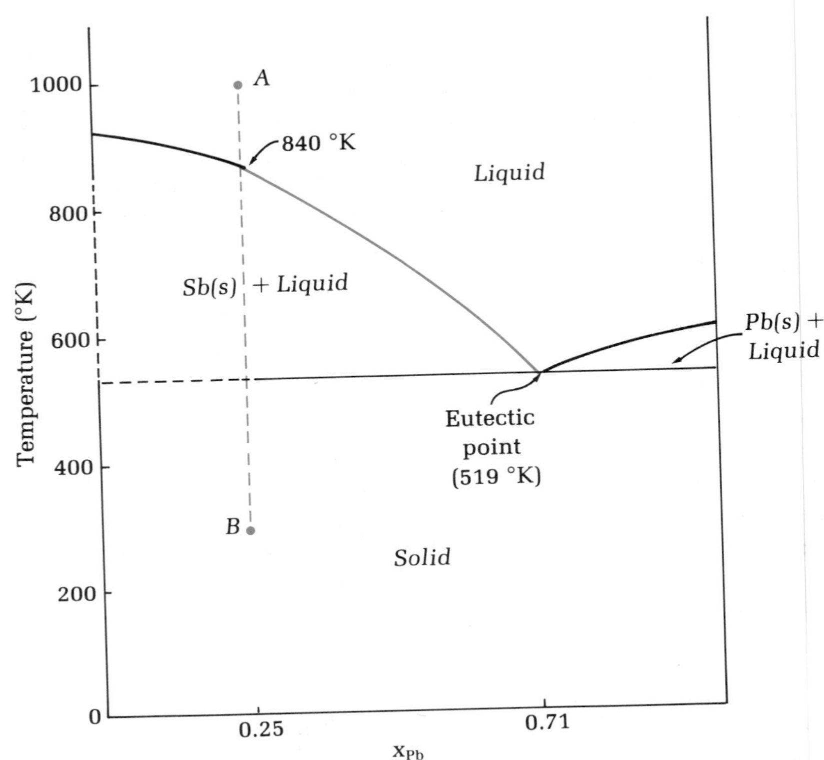

Figure 17–18
Solid-liquid portion of the antimony-lead phase diagram at $P =$ 1 atm.

eutectic mixture of tin and lead with a melting point sufficiently low that it is easy to liquefy and maintain in a shapeable liquid form.

There are several well-defined regions in Figure 17–18. To understand better how these regions arise, let us consider a process that connects regions. Let us take a mixture of solid Sb and solid Pb in which x_{Pb} is 0.25 and heat it to 1000 °K (point *A* in Figure 17–18). Now let the sample cool by letting it stand in a room at 300 °K. First the liquid cools along the colored vertical line below *A* in the diagram. At 840 °K the system reaches the Sb(l)–Sb(s) equilibrium line, and a small amount of Sb(s) forms. The removal of this Sb from solution results in an increased x_{Pb} for the solution. This more concentrated solution has a lower freezing point than the original solution. The antimony freezes out of solution over a temperature range beginning at 840 °K. Since there are two phases present, Sb(s) and solution, we can represent the cooling, as the freezing out of Sb(s) occurs, by two lines (one for each phase) in Figure 17–18. The solid colored line shows the cooling of the solution, which becomes more concentrated in lead as the temperature falls. The black dashed line shows the cooling of solid antimony. The system as a whole follows the vertical colored dashed line during this process, and every point on the colored dashed line corresponds to solid antimony in equilibrium with a solution whose composition is given by the solid colored line at that temperature.

At the eutectic temperature, the Sb equilibrium line meets the Pb equilibrium line. The new solid formed at the eutectic temperature is no longer pure Sb but contains Pb at $x_{Pb} = 0.71$, the same composition as the equilibrium liquid. Like a pure substance, the eutectic mixture will freeze, with the temperature remaining constant until all liquid has been converted to solid at the eutectic temperature. The composition of the total solid formed in the entire cooling process slowly increases from $x_{Pb} = 0$ to $x_{Pb} = 0.25$, and this is shown by the horizontal black dashed line in the figure. Along this horizontal line, both Sb(s) and Pb(s) exist in equilibrium with liquid of the eutectic composition. When all liquid has solidified, the system reaches the junction of the black dashed and colored dashed lines at the eutectic temperature, and the final process is the cooling of the solid along the vertical colored dashed line to 300 °K (point B).

The cooling process described can be used to locate two important temperatures in the construction of Figure 17–18, namely the eutectic temperature and the Sb(s)-Sb(l) equilibrium temperature at $x_{Pb} = 0.25$. Figure 17–19 shows the plot obtained for temperature as a function of time as the cooling takes place from point A to point B. At first the temperature drops steeply from $T_A = 1000$ °K as the liquid cools. As soon as Sb(s) begins to form, the rate of temperature decrease slows markedly as energy equal to $\Delta H_{\text{fusion}} = -\Delta H_{\text{freezing}}$ is spread by heat transfer. We have also indicated a supercooling of the liquid before the first Sb(s) forms. Supercooling occurs in general with solutions as well as pure liquids. As shown in Figure 17–19, the equilibrium temperature at which the first Sb(s) would have appeared had there been no supercooling is obtainable by the dashed extrapolation. At the eutectic temperature, the temperature decrease stops completely as eutectic liquid freezes to a solid of the same composition, and the eutectic temperature is easily identified. Finally, when the entire sample has solidified, the temperature asymptotically approaches 300 °K.

SUGGESTIONS FOR ADDITIONAL READING

Bent, H. A., *The Second Law*, Oxford University Press, New York, 1965.
Porterfield, W. W., *Concepts of Chemistry*, W. W. Norton, New York, 1972.

PROBLEMS

1. The ΔG_f° value of gaseous benzene at 25 °C is 1.24 kcal/mole more positive than ΔG_f° for liquid benzene. What is the equilibrium vapor pressure of benzene at 25 °C?
2. The normal boiling point of any liquid is that temperature at which the

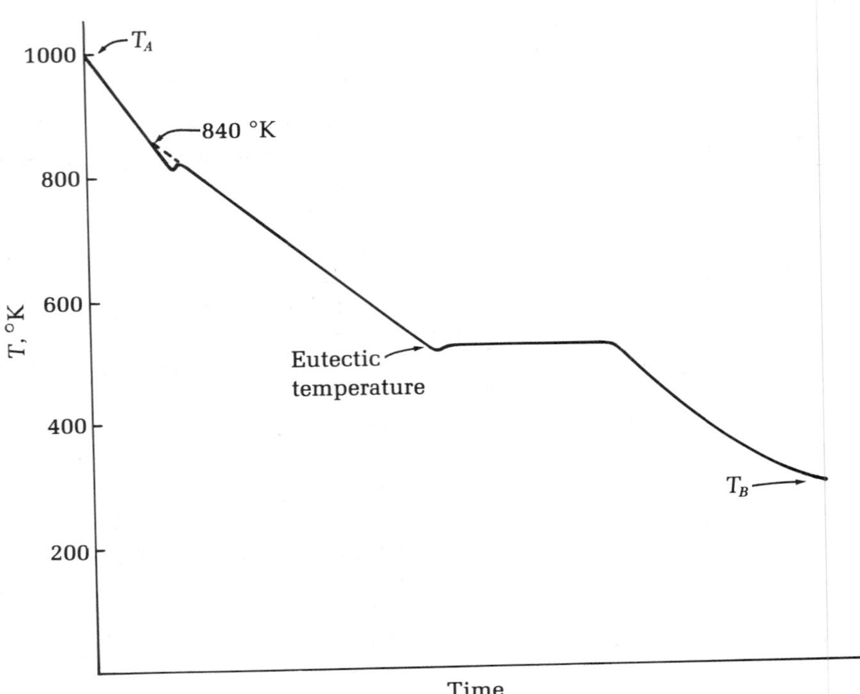

Figure 17–19
Temperature as a function of time for the cooling of an antimony-lead mixture of x_{Pb} = 0.25.

molar G_{liquid} = the molar G_{vapor} at a pressure of 1 atm. This temperature can be obtained to a good approximation from the $\Delta S°$ and $\Delta H°$ of vaporization calculable from Tables 15–1 and 15–3 by assuming that these values are temperature-independent. Use the data in these tables to calculate an approximate value of the normal boiling point of water.

*3. Consider the arrangement shown in Figure 17–20. Liquid water is in a closed container under a pressure of argon gas. If we assume that argon is insoluble in water and that the vapor can be treated as an ideal gas,

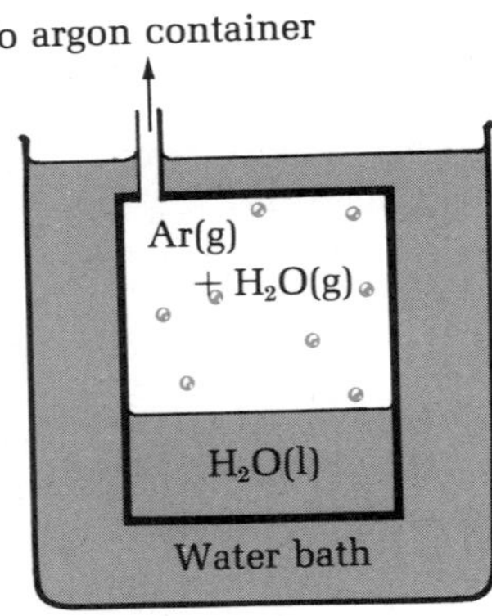

Figure 17–20
Apparatus for measuring the pressure dependence of the vapor pressure of water.

what can we predict for the relative vapor pressure of water in this container—is it higher, lower, or the same compared with the vapor pressure of water at the same temperature in the absence of argon? In other words, what is the pressure dependence of vapor pressure? Explain. *Hint:* Consider the effect of the argon on the molar $G_{(H_2O,l)}$ and the molar $G_{(H_2O,g)}$.

4. Criticize the following definition of the critical temperature. The critical temperature is that temperature above which a gas cannot be converted to a liquid no matter how much pressure is applied to it.
5. The slope of the vapor-liquid equilibrium line in Figure 17–1 is zero at the critical point. Propose an explanation. Would you predict that this is true for all pure substances? Why?
6. Consider a mixture of ice and solid calcium chloride at −30 °C. Is the vapor pressure of water over this mixture, in comparison with the vapor pressure over pure ice, higher, lower, or the same at this temperature? Explain.

*7. Figure 17–1 shows that ice, liquid water, and water vapor cannot coexist at 1 atm pressure. Yet an open beaker of water at 0 °C will definitely involve all three phases in coexistence. Explain this paradox.

8. Normally the sliding of one solid across another is greatly impeded by friction. As a consequence, moving parts of machinery must be lubricated. Yet, in ice-skating, a steel surface slides easily over ice. How does the solid-liquid equilibrium for water lead to this phenomenon? Benzene melts at a temperature near that of water. Would you expect to be able to skate on a surface of solid benzene? Explain.
9. Show that equation 17–9 holds for each gas A and B in the isothermal mixing of the two gases in Figure 15–6.
10. Solutions of benzene in toluene obey Raoult's law to an excellent approximation. At 20 °C the vapor pressures of pure benzene and pure toluene are 75 torr and 22 torr, respectively. What is the composition of the vapor in equilibrium with a benzene-toluene solution in which $x_{benzene} = 0.40$?
11. How many successive distillations and condensations are required to obtain an $x_{benzene}$ of 0.90 or higher starting with a benzene-toluene solution of $x_{benzene} = 0.40$ and maintaining the temperature at 20 °C?
12. Naphthalene is a solid at room temperature with a vapor pressure much lower than that of benzene. The vapor pressure of pure benzene at 25 °C is 93.60 torr. A solution of 1.000 g of naphthalene in 10.000 g of benzene has a vapor pressure of 88.22 torr. Calculate the molecular weight of naphthalene.
13. Calculate the boiling point of the solution in problem 12. This is a relatively concentrated solution. Test the validity of equation 17–26 in this

case by calculating the boiling point from both equations 17–25 and 17–26.

14. Solutions of HCl in water are commercially available in large quantities. These solutions contain 36 percent HCl by weight and the solution density is 1.18 g/ml. Calculate the molality, molarity, and mole fraction of HCl in these solutions.

15. For the concentrated HCl solutions of problem 14, at what temperature does ice begin to form when they are cooled? Remember that HCl is a strong acid. Should you use equation 17–27 or equation 17–28 in this calculation?

16. So called "permanent" antifreeze for car radiators consists principally of ethylene glycol ($HOCH_2CH_2OH$). What is the percentage of ethylene glycol by weight required in a car radiator to give antifreeze protection to 0 °F? You may use equation 17–28 as a sufficient approximation.

17. If salt is spread on a lawn, the grass quickly dies. The reason for this lies in the phenomenon of osmosis. How?

18. If red blood cells are placed in pure water, they burst. Propose an explanation. What would you put into the water to prevent this?

19. Fruit juices very quickly undergo fermentation when left exposed to air. Microorganisms convert the sugar in the juice to ethanol and CO_2. However, honey may be stored for long periods of time without being subject to fermentation. Propose an explanation.

20. A 0.2520-g sample of an organic polymer (plastic) is dissolved in sufficient CCl_4 to yield 10.0 ml of solution. In an osmotic pressure measurement such as that shown in Figure 17–10, the height h for this solution was measured to be 32 mm at 25 °C. The density of CCl_4 is 1.59 g/ml. What is the molecular weight of the polymer?

21. You have a solution of a protein in water containing a number of inorganic ions. You wish to study the effect of dissolved KCl on this protein. This means that you must remove all other ions so as to have only K^+ and Cl^-. How would you do it?

22. Is the boiling point of a dioxane-water solution with $x_{dioxane} = 0.60$ higher or lower than the boiling point of either pure H_2O or pure dioxane? Explain.

23. Liquid sulfuric acid is sometimes used to remove water vapor from the air in closed systems. Why is sulfuric acid used rather than some other liquid?

*24. If pure helium is passed through the apparatus shown in Figure 17–14, a horizontal line is obtained that serves as a base or reference line from which the peaks can be measured. The area between any peak and the baseline is proportional to the mole fraction of the species yielding the peak. Thus, the ratio of peak areas is the ratio of moles of species in the mixture that was separated. Show that this is the case only if the values for thermal conductivity of the separated species, as compared with helium, are both negligible.

25. This chapter contains a lengthy discussion of vapor-liquid chromatography, a separation technique based on the solution-vapor equilibrium. Most separations of biochemical species use the equilibrium between dissolved species and species adsorbed to the surfaces of solids. How does such an equilibrium provide a basis for separations?

26. Show (using Henry's law) how it is possible for a liquid that has a higher boiling point than a second liquid to have a shorter retention time in a vapor-liquid chromatography column.

27. What is the approximate vapor pressure of diethyl ether over a saturated solution of diethyl ether in water at 35 °C? Pure diethyl ether boils at 34.6 °C.

***28.** The solubilities of most substances in most solvents increase with increasing temperature. All solids show a limited solubility in any given solvent for reasons that we considered in Chapter 15. Suppose that we have a mixture of organic solids, A and B, which are chemically similar species. After some experimenting, we locate a solvent in which B is twice as soluble as A and in which the solubilities of both A and B increase with increasing temperature. Show that these facts provide a basis for a simple procedure with which to obtain a sample of pure A.

29. Barbiturates are the salts of derivatives of barbituric acid, whose structure is

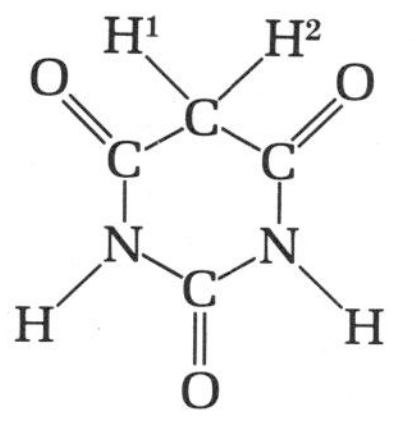

The derivatives are produced by substituting hydrocarbon radicals for the hydrogen atoms labeled 1 and 2.

The drugs act on or within cell membranes of the central nervous system. Up to a limit, the potency of the drug increases as the sizes of these hydrocarbon substituents increase. From our previous discussion of the structure of phospholipid membranes (Chapter 10), can you propose a reason for this relation between drug structure and drug activity?

30. Figure 17–21 shows the solid-liquid portion of the phase diagram for phenol (C_6H_5OH) and aniline ($C_6H_5NH_2$) at $P = 1$ atm. This figure differs markedly from Figure 17–18 in that there are two eutectic points and an intermediate maximum melting temperature at $x_{aniline} = 0.5$. What is the source of these new features? *Hint:* Phenol is an acid and aniline is a base. Label each of the lettered regions between the lines drawn in Figure 17–21 by indicating the species and phases that exist in those regions. In analogy with the path traced on Figure 17–18 for the cooling of an Sb–Pb melt from point A to point B, trace paths on Figure 17–21 for the cooling of two different melts from 310 °K to 250 °K. The mole fractions of aniline in the two melts are 0.5 and 0.9 (the eutectic composition to the right).

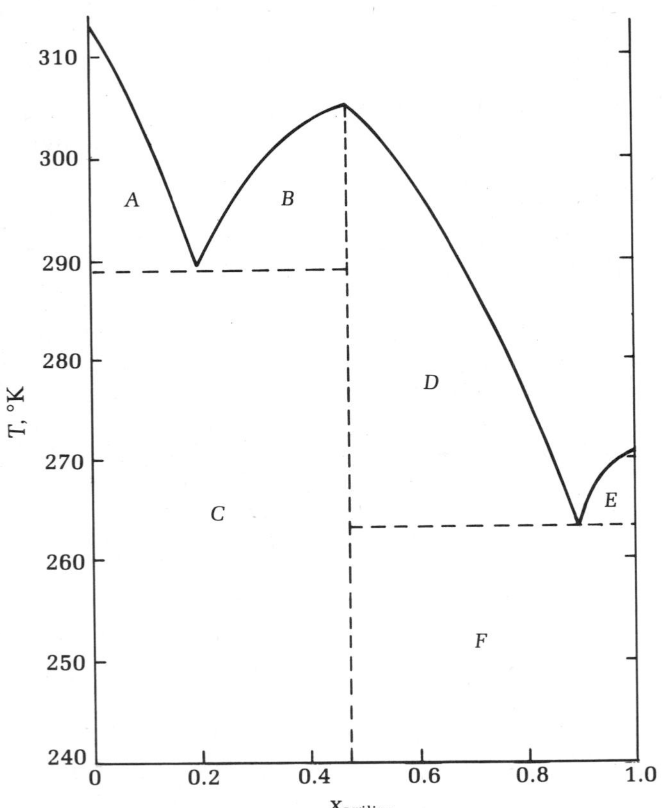

Figure 17–21
Liquid-solid portion of the phenol-aniline phase diagram at P = 1 atm.

18/ Equilibrium in Gaseous Reactions

We now have a model comprising the factors that determine the conditions of equilibrium in any system, namely maximum spread of matter and maximum spread of energy. Equilibrium in gases and dilute solutions can be described through an equilibrium constant, which is uniquely related to the standard Gibbs free energy change for the chemical or physical transformation under study. The equilibria we have so far examined have involved substances whose identities remain unchanged. Thus, the equilibrium $H_2O(l) \rightleftharpoons H_2O(g)$ involves the substance water on both sides of the equilibrium.

There are other processes in which chemical species exist in equilibrium with different chemical species. They are represented by reactions we have considered briefly, such as

$$CaCO_3(s) \rightleftharpoons CaO(s) + CO_2(g)$$

$$3\ H_2(g) + N_2(g) \rightleftharpoons 2\ NH_3(g)$$

Some important factors are related to chemical equilibrium. The pressure of a given gas or the concentration of a given dissolved species is often of great importance. For example, if we are synthesizing a compound from other species, we want to maximize the yield of this product in the most economical way possible. If the species is a noxious pollutant, we want to minimize its production or maximize its consumption in a reaction. In other instances, it may be necessary to hold the pressure or concentration of a species nearly constant, even though many reactions are proceeding which involve this species. This is an important factor in living systems.

To control pressures or concentrations in a rational way, we must be able to relate them to equilibrium constants and see how they are manipulated within a given equilibrium constant. The development of this manipulative ability is the principal aim of this chapter and Chapter 19. This chapter centers on reactions in the gas phase, and Chapter 19 centers on the solution phase.

18.1. Equilibria and Nitrogen Oxide Air Pollution

The reaction

$$N_2(g) + 2\ O_2(g) \rightleftharpoons 2\ NO_2(g) \tag{18–1}$$

is potentially troublesome. Nitrogen dioxide (NO_2) is a poisonous, malodorous brown gas (at room temperature) that is an important constituent of urban air pollution. What is its equilibrium partial pressure in the atmosphere at 298 °K? The equilibrium partial pressures of N_2 and O_2 are 0.8 and 0.2 atm, respectively. From Table 16–1, $\Delta G°$ for reaction 18–1 is $2\ \Delta G_{f(NO_2)} = 2(12.39) = 24.78$ kcal/mole. Thus,

$$K_e = \frac{P^2_{NO_2}}{P_{N_2}P^2_{O_2}} = e^{-\Delta G°/RT} = 6.7 \times 10^{-19} \tag{18–2}$$

This small number shows that the equilibrium pressure of NO_2 must be small compared with the equilibrium pressures of N_2 and O_2. From equation 18–2, we obtain this pressure as

$$\begin{aligned} P_{NO_2} &= (K_e P_{N_2} P^2_{O_2})^{1/2} \\ &= [(6.7 \times 10^{-19})(0.8)(0.2)^2]^{1/2} \\ &= 1 \times 10^{-10} \text{ atm} \end{aligned}$$

From this calculation we should conclude that NO_2 does not pose a pollution problem. However, both the production and the decomposition of NO_2 according to reaction 18–1 proceed slowly under normal circumstances in the atmosphere. Up to this point, we have had very little to say about the rates of chemical reactions. We have shown that certain reactions such as the combination of gaseous hydrogen and oxygen to produce water are spontaneous, but we have not associated spontaneity with rate. The reason that we have not is that there is no such association. The hydrogen-oxygen reaction is imperceptibly slow at room temperature in spite of the fact that $\Delta G°$ for it is huge and negative. The question of reaction rate and the factors that influence it is complex (see Chapter 21). In the case of reaction 18–1, any NO_2 introduced to the atmosphere must decompose until P_{NO_2} falls to 1×10^{-10} atm; but the slowness of the reaction at 298 °K means that P_{NO_2} can be maintained at a dangerous level.

Nitrogen dioxide is introduced into the atmosphere in large quantities from internal combustion engines. From Table 15–1, we see that the formation of $NO_2(g)$ is endothermic with $\Delta H_f° = 8.09$ kcal/mole, and K_e for reaction 18–1 has a higher value at elevated temperatures than the value at 25 °C. Equilibrium between N_2, O_2, and NO_2 at high temperatures is achieved in internal combustion engines. The temperature within a cylinder of an operating internal combustion engine is very nonuniform. Small regions, particularly near the spark, can have extremely high temperatures. For purposes of calculation, we shall use a temperature of 3000 °K for these extremely hot regions.

The temperature dependence of K_e is given by equation 16-37, which becomes in this case

$$(2.303) \log K_{e(3000)} = (2.303) \log K_{e(298)} - \frac{\Delta H^\circ}{R}\left(\frac{1}{3000} - \frac{1}{298}\right) \qquad \textbf{18-3}$$

The ΔH° for reaction 18-1 is $2\Delta H_f^\circ$ for $NO_2 = 16{,}180$ cal, and K_e at 3000 °K is calculated to be 3.3×10^{-8}. This is still a small number, and NO_2 does not exist in significant amounts in hot automobile exhaust gases. Actually, we should have realized this before performing the calculation. The production of $NO_2(g)$ is disfavored at 298 °K on both matter-spread and energy-spread grounds ($\Delta H_f^\circ = 8.09$ kcal and $\Delta S_f^\circ = -14.4$ cal/mole-deg). The unfavorable entropy is obvious from inspection of reaction 18-1 since three gaseous molecules go to two. Elevated temperatures favor the state of greater entropy (Chapter 16), and this is $N_2(g) + 2\ O_2(g)$.

Unfortunately, NO_2 is not the only nitrogen oxide that can be formed from $N_2(g)$ and $O_2(g)$. Nitric oxide (NO) is formed according to the reaction

$$N_2(g) + O_2(g) \rightleftharpoons 2\ NO(g) \qquad \textbf{18-4}$$

By inspection, this reaction, which involves two gaseous molecules on both sides, should have ΔS° near zero and K_e near unity at 3000 °C. Let us check this prediction. From Table 16-1, $\Delta G^\circ_{f(NO)} = 20.72$, and thus K_e for reaction 18-4 is 4.0×10^{-31} at 298 °K. The source of the small K_e is a large $\Delta H^\circ_{f(NO)}$ of 21.60 kcal/mole. The value of $\Delta S^\circ_{f(NO)}$ is 2.95 cal/mole-deg, a small number, as we predicted. From equation 18-3, K_e for reaction 18-4 is calculated to be 0.013 at 3000 °K. While K_e increases by a huge factor between 298 °K and 3000 °K, it is still much smaller than the infinite temperature value of $e^{\Delta S^\circ/R} = 19.5$. Nevertheless, K_e at 3000 °C is large enough that a significant pressure of NO is formed.

Let us calculate what fraction of $N_2(g)$ is converted to NO in a typical region at $T = 3000$ °K. Air is drawn into the cylinder along with fuel, and caused to be compressed and to react, and the temperature is raised to 3000°. Most of the oxygen gas is consumed in the combustion of gasoline, a reaction with a large negative ΔG° (and at high temperatures a rapid reaction also). A typical equilibrium oxygen pressure within a cylinder at 3000 °K might be 1.00 atm. Nitrogen is a much less reactive gas than oxygen and will not react with gasoline to any significant extent, even at 3000 °K. The N_2 pressure in the cylinder is quite high, and 150 atm might be taken as a typical value for the condition of 3000 °K.

The equilibrium expression for reaction 18-4 is

$$K_e = 0.013 = \frac{P_{NO}^2}{P_{N_2}P_{O_2}} \qquad \textbf{18-5}$$

from which

$$P_{NO} = [(150)(1.00)(0.013)]^{1/2} = 1.4 \text{ atm}$$

Thus, approximately 1 percent of the $N_2(g)$ is converted to NO(g) in a typical high-temperature region, and the NO is quickly cooled to atmospheric temperature by being expelled in the exhaust. Reaction 18–4, like reaction 18–1, proceeds slowly in both directions at 298 °K, and NO gas will only very slowly decompose to $N_2(g)$ and $O_2(g)$ once it is formed.

The presence of NO(g) and $O_2(g)$ at room temperature allows the following reaction to proceed:

$$2\ NO(g) + O_2(g) \rightarrow 2\ NO_2(g) \qquad \textbf{18–6}$$

From Table 16–1, the $\Delta G°$ of reaction 18–6 is 2(12.39) − 2(20.72) = −16.66 kcal/mole, and K_e for reaction 18–6 is 1.6×10^{12}. The oxidation of nitric oxide is rapid at 298 °K, and thus the NO in the automobile exhaust is quickly converted to NO_2 in the atmosphere.

What can be done technically to minimize the production of NO? We might think that high-temperature regions should be eliminated since reaction 18–4 is extremely temperature-sensitive, and even a relatively small temperature decrease would result in a dramatic decrease in NO production. However, high-temperature regions result from the extremely rapid combustion of fuel and they are nearly impossible to eliminate.

A second way to minimize NO production is to lower the O_2 or N_2 pressure in the cylinder. For example, if the typical P_{O_2} in equation 18–5 were changed from 1.00 atm to 0.100 atm, the equilibrium pressure of NO_2 would be

$$P_{NO} = [(150)(0.100)(0.013)]^{1/2} = 0.44 \text{ atm}$$

and two-thirds less NO is produced than in the case in which $P_{O_2} = 1.00$ atm. This calculation illustrates an important (and perhaps obvious) conclusion about chemical equilibria. **Lowering the equilibrium concentration or pressure of any of the reactants results in smaller equilibrium concentrations or pressures of the products. Likewise, an increase in reactant concentrations or pressures in equilibrium also raises the equilibrium concentrations or pressures of the products.**

In the case of reaction 18–4, it appears that the best way to eliminate NO production would be to eliminate N_2 entirely, since it is not needed for fuel combustion. However, this would eliminate air as the source of O_2, and the use of pure oxygen would greatly add to the cost of operating an automobile.

Lowering P_{O_2} in a gasoline engine in order to lower P_{NO} also has drawbacks. While an increased ratio of fuel to air leads to more complete use of oxygen and a lower P_{O_2}, it also leads to a less complete combustion of fuel. Nitrogen oxide pollution is traded for pollution from carbon monoxide and uncombusted fuel. An engine has recently been developed that alleviates this problem. It is based on the fact that the high-temperature regions are located largely near the spark. The cylinder has two chambers, a small one enclosing the spark, and a larger one bounded by the piston. Two different fuel-air mixtures are introduced to the cylinder through two different intake valves. The small chamber in which the spark occurs contains a mixture rich in fuel. Thus, in the high-temperature regions, P_{O_2} is low and NO production is suppressed. The flame from the small chamber ignites an air-rich mixture in the larger chamber. The larger chamber contains a high P_{O_2}, but the high-temperature regions are avoided. In addition, the largest part of the complete cylinder contains an air-rich mixture, so that pollution from incomplete combustion of fuel is also minimized.

18–2. Obtaining Equilibrium Partial Pressures

Dependence of Equilibrium Partial Pressures on Total Pressures

The discussion of nitrogen oxide pollution illustrates the great utility of Table 16–1 in solving practical problems and demonstrates some of the calculations involving gas-phase equilibria. We now know two ways to change an equilibrium pressure of a product, that is, by changing the temperature or by changing the pressures of one or more of the reactants. What remains is to investigate the dependence of product partial pressure on the total pressure of the system. This dependence can be shown by examining another equilibrium in the nitrogen-oxygen system:

$$2\ NO_2(g) \rightleftharpoons N_2O_4(g) \qquad \textbf{18–7}$$

(We encountered the species dinitrogen tetroxide (N_2O_4) in Chapter 8.)

From Table 16–1 the $\Delta G°$ for reaction 18–7 is $23.49 - 2(12.39) = -1.29$ kcal/mole, and $K_e = 8.83$ kcal/mole. The $\Delta G°$ is near zero because the unfavorable matter concentration of the dimerization of NO_2 is balanced by the favorable energy spreading from the exothermic formation of a new chemical bond in N_2O_4.

What is the composition of the gas in a container of NO_2 and N_2O_4 at 298 °K in which the measured total pressure is 1.00 atm? This is a very common type of calculation in gas-phase equilibrium systems since it is the total pressure that is very easily controlled. The equilibrium constant expression for reaction 18–7 is

$$K_e = \frac{P_{N_2O_4}}{P^2_{NO_2}} \qquad \textbf{18–8}$$

To calculate the two individual partial pressures in equation 18–8, we must relate the pressures to each other. In this case, that relation is through the total pressure, that is $P_{N_2O_4} + P_{NO_2} = 1.00$ atm. We have

$$K_e = 8.83 = \frac{1.00 - P_{NO_2}}{P^2_{NO_2}}$$

which yields $P_{NO_2} = 0.285$ atm and $P_{N_2O_4} = 0.715$ atm.

What fraction of the nitrogen in the system exists as N_2O_4? We can calculate this easily from the equilibrium partial pressures, since by Dalton's law of partial pressures, the partial pressure of a gas is proportional to the number of moles of the gas present. Thus, the fraction of N in the form of N_2O_4 is represented as follows:

$$\text{Fraction} = \frac{2\ P_{N_2O_4}}{P_{NO_2} + 2\ P_{N_2O_4}} = 0.834$$

Suppose that we lower the total pressure of NO_2 and N_2O_4 in the container from 1.00 atm to 0.100 atm. Now we follow essentially the same procedure as that for a total pressure = 1.00 atm to calculate a new P_{NO_2} of 0.064 and a new $P_{N_2O_4}$ of 0.036 atm. The fraction of nitrogen in the form of N_2O_4 is now 0.529. As the total pressure is reduced, the state of greater matter spread, that is, $NO_2(g)$, is the preferred one, and the fraction of N in the form of N_2O_4 decreases. **In any equilibrium that involves only gaseous species, diminished total pressure favors the side of the equilibrium having the greater number of gaseous species.**

The Principle of Le Châtelier

We have seen three ways in which the position of an equilibrium can be shifted, that is, by change of temperature, change of pressure, or adding or removing one or more of the reactants or products. We have also seen by thermodynamic analysis or calculation or both the direction of the shifts in equilibrium each of these three changes produces. These shifts can also be deduced from a very useful generalization termed the **Le Châtelier Principle.** The principle states that **if a stress is placed on any equilibrium system, the equilibrium position will shift in the direction that relieves the stress.**

Let us apply the Le Châtelier Principle to some of the equilibrium shifts already considered to show that it is indeed consistent with the conclusions that we have previously drawn. We first analyzed the effect of temperature, and showed that for an endothermic reaction the equilibrium shifts in the direction of the products as the temperature is increased. The stress in this case is the energy input necessary to raise the temperature. An endothermic reaction will absorb this energy and remove the stress as it proceeds. By similar reasoning we conclude that an exothermic reaction is promoted by lowering the temperature. The stress in this case is the removal of energy, and the exothermic reaction removes the stress by supplying energy by heat transfer.

We have illustrated the effect of another stress (Chapter 17) by showing that the equilibrium melting point of water is lowered by increasing the pressure. In that case, the pressure is the stress and it drives the equilibrium in the direction of lower volume. Since the molar volume of $H_2O(l)$ is less than the molar volume of $H_2O(s)$, the equilibrium shifts in the direction of the liquid. Increasing the pressure at a given temperature will always favor the state of lower molar volume, according to the Le Châtelier Principle. This provides a quick interpretation of the pressure dependence of the NO_2–N_2O_4 system previously discussed. At a given pressure and temperature, the molar volume of pure $NO_2(g)$ is twice the volume of one-half mole of pure N_2O_4. Thus, $NO_2(g)$ is favored by diminished total pressure, and N_2O_4 by an increased total pressure.

The third stress is the addition of further reactant or product to an equilibrium system. The stress due to the addition of a reactant is relieved by the formation of additional product. The stress due to the addition of a product is relieved by the formation of reactants.

Sample Calculations Involving Gas-Phase Equilibria

We have used K_e to calculate individual pressures for the species involved in reaction 18–7. Calculations of that type are common in the treating of gas-phase equilibria; two further illustrations are presented.

At elevated temperatures, H_2 and I_2 exist in equilibrium with HI according to the following reaction:

$$H_2(g) + I_2(g) \rightleftharpoons 2\ HI(g) \qquad \textbf{18–9}$$

The equilibrium constant is 55.2 at 699 °K. Suppose that we introduce into a sealed container an amount of pure HI sufficient to yield a pressure of 0.500 atm of HI at 699 °K if none of it decomposes. Of course some of it will decompose, and the equilibrium gas will be a mixture of H_2, I_2, and HI. What are the equilibrium partial pressures of H_2, I_2, and HI? The equilibrium expression is

$$K_e = \frac{P_{HI}^2}{P_{H_2}P_{I_2}} \qquad \textbf{18–10}$$

and we must relate the three pressures to each other in such a way as to change equation 18–10 to one containing one variable instead of three. In this case, this can be accomplished through the balanced reaction equation. Let us define x as the equilibrium partial pressure of H_2. According to the balanced reaction equation, x must also equal the equilibrium partial pressure of I_2, since one I_2 molecule is produced with one H_2 molecule in the HI decomposition. In order to make one H_2 molecule, two HI molecules must decompose so that the equilibrium HI partial pressure must be $(0.500 - 2x)$.

With these substitutions, equation 18–10 becomes

$$K_e = 55.2 = \frac{(0.500 - 2x)^2}{x^2}$$

or by taking the square root of both sides of the equation, we have

$$7.43 = \frac{0.500 - 2x}{x}$$

This expression yields $x = 0.053$ atm. The equilibrium partial pressures are

$$P_{H_2} = P_{I_2} = 0.053 \text{ atm} \qquad \text{and} \qquad P_{HI} = 0.394 \text{ atm}$$

Let us now shift the equilibrium for reaction 18–9 by adding a quantity of iodine sufficient to yield an I_2 partial pressure of 0.200 atm in the container at 699 °K if there were no other source of I_2 and if none of the I_2 decomposed. What are the new equilibrium partial pressures of HI, H_2, and I_2? Again, we must relate the three partial pressures through one x variable. If none of the newly introduced iodine were to decompose, the partial pressures would be

$$P_{I_2} = 0.053 + 0.200 = 0.253 \text{ atm}$$

$$P_{H_2} = 0.053 \text{ atm}$$

$$P_{HI} = 0.394 \text{ atm}$$

If these numbers are substituted into equation 18–10, we obtain

$$\frac{P_{HI}^2}{P_{H_2}P_{I_2}} = \frac{(0.394)^2}{(0.253)(0.053)} = 11.5$$

This number is smaller than the equilibrium value, and additional HI must be formed in order to reestablish equilibrium.

Let us define x as the partial pressure of I_2 that is lost in the reestablishment of equilibrium. This must also be the partial pressure of H_2 lost; and the HI partial pressure must increase by the amount 2x. The new equilibrium expression is

$$55.2 = \frac{(0.394 + 2x)^2}{(0.253 - x)(0.053 - x)}$$

The expanded expression is the following quadratic equation:

$$51.2x^2 - 18.47x + 0.585 = 0$$

which can be solved for x by means of the expression

$$x = \frac{-b \pm \sqrt{b^2 - 4ac}}{2a} \qquad \textbf{18–11}$$

In equation 18–11, a is the coefficient of x^2 in the quadratic equation, b is the coefficient of x, and c is the constant term. In the present case, a is 51.2, b is −18.47, and c is 0.585, so that x is 0.035 atm. The equilibrium partial pressures are

$$P_{HI} = 0.464 \text{ atm} \qquad P_{H_2} = 0.018 \text{ atm} \qquad P_{I_2} = 0.218 \text{ atm}$$

Thus, the addition of the reactant I_2 shifted the equilibrium in the direction of production of additional HI.

SUGGESTIONS FOR ADDITIONAL READING

Sienko, M. J., *Chemistry Problems*, Benjamin, Menlo Park, Calif., 1972.

PROBLEMS

1. Two sources of atmospheric pollution from internal combustion engines are CO(g) and NO(g). It is possible to cause these species to react with each other according to the following reaction equation:

$$2\,NO(g) + 2\,CO(g) \rightleftharpoons 2\,CO_2(g) + N_2(g)$$

From data given in Table 16–1, calculate the equilibrium constant for this reaction at 25 °C. Why doesn't the reaction occur in automobile exhaust?

2. The Haber process for the production of ammonia employs the following equilibrium:

$$3\,H_2 + N_2(g) \rightleftharpoons 2\,NH_3(g)$$

If we wish the ratio P_{NH_3}/P_{N_2} to be maximized, should this reaction be carried out at a high total pressure or a low total pressure? Should it be carried out at a high temperature or low temperature? Explain.

3. A reaction that has been used for some time to produce gaseous fuel is the "water gas" reaction,

$$C(s) + H_2O(g) \rightleftharpoons CO(g) + H_2(g)$$

Should this reaction be run at room temperature or elevated temperature? Should it be run at high or low total pressure? Explain.

*4. At what temperature will the hydrogen exist 90 percent in the form of H_2 in the water gas reaction at a total pressure of 1.00 atm?

5. A 5.00-g sample of solid N_2O_4 is introduced into a 10.0-liter container and allowed to warm to 25 °C. Calculate the equilibrium partial pressures of NO_2 and N_2O_4.

6. Suppose that the container in problem 5 also contains helium gas at

1.00-atm partial pressure. What effect does the helium have on P_{NO_2} and $P_{N_2O_4}$? Assume ideal behavior for all gases.

7. Suppose that we introduce 10.0 g of I_2 and 0.113 g of H_2 into a 10.0-liter container, seal the container, and heat to 699 °K. What are the equilibrium partial pressures of H_2, I_2, and HI in this container? Assume that no solid I_2 exists under these conditions.

8. Suppose that we had started with 4.52 g of HI in addition to the other quantities in problem 7. What are the new equilibrium partial pressures at 699 °K?

*9. For convenience, equilibrium expressions for gas-phase reactions are sometimes expressed in moles/liter of the constituent gases. What is the new standard state for the gases in this case? Under what condition will $\Delta G°$ for the new standard state equal $\Delta G°$ for the more conventional standard state of 1 atm pressure?

10. The ΔG_f° of Cl atoms is 25.2 kcal/mole and the ΔH_f° is 29.0 kcal/mole at 25 °C. Calculate the equilibrium partial pressure of Cl atoms in chlorine gas at 1000 °C and 1 atm total pressure.

19/ Equilibrium in Aqueous Solution

We have encountered four important examples of equilibria involving species in aqueous solution. The first involves the addition of acids or bases to water and centers around the competition for protons. The second is donor-acceptor competition between water molecules and other nucleophiles for association with metal ions in solution. The third involves the solubility of sparingly soluble salts, such as AgCl. And the fourth is the competition for electrons in redox reactions. The first three of these types can be considered together in a quantitative investigation.

19–1. Acid-Base Equilibria

The Self-Dissociation of Water

In our study of the concepts of acids and bases (Chapter 7), it was pointed out that water is both a Lowry-Brønsted acid and a Lowry-Brønsted base. Water serves as an acid in the donation of a proton to ammonia (NH_3) to produce ammonium ions (NH_4^+) and it serves as a base in accepting a proton from HCl to form the species H_3O^+. Water also displays its dual character by self-ionizing, according to the following equilibrium:

$$H_2O(l) \rightleftharpoons H^+(aq) + OH^-(aq) \qquad \textbf{19–1}$$

The equilibrium expression is

$$K_e = [H^+][OH^-] \qquad \textbf{19–2}$$

and K_e at 25 °C can be evaluated from the data in Tables 16–1 and 16–2:

$$\Delta G^\circ = \Delta G^\circ_{f(H^+,\,aq)} + \Delta G^\circ_{f(OH^-,\,aq)} - \Delta G^\circ_{f(H_2O,\,l)}$$

$$= (0) + (-37.59) - (-56.69)$$

$$= 19.1 \text{ kcal}$$

$$K_e = e^{-\Delta G^\circ/RT} = e^{-[19{,}100/(1.987)(298)]}$$

$$= 1.0 \times 10^{-14}$$

Pure water at 25 °C possesses an aquated hydrogen-ion concentration equal to the hydroxide-ion concentration equal to 1.0×10^{-7} mole/liter. The K_e in equation 19–2 is commonly encountered in combination with other equilibrium constants, and it is given its own symbol, K_w.

These equilibrium concentrations of H^+ and OH^- are surprisingly large when we consider that the OH bond in the water molecule has bond enthalpy of 111 kcal/mole, and that this bond is broken to make H^+ and OH^- in reaction 19–1. From Tables 16–1 and 16–2, we can calculate $\Delta H°$ for reaction 19–1:

$$\begin{aligned} \Delta H° &= \Delta H°_{f(H^+,\,aq)} + \Delta H°_{f(OH^-,\,aq)} - \Delta H°_{f(H_2O,\,l)} \\ &= (0) + (-54.96) - (-68.32) \\ &= 13.36 \text{ kcal/mole} \end{aligned}$$

Heterolytic bond-breaking (both bonding electrons leave with the same nucleus) in solution requires an energy input of only 13.36 kcal/mole, while the homolytic cleavage (one bonding electron leaves with each nucleus) requires an energy input of 111 kcal/mole. The source of this huge difference is indicated by the value of $\Delta S°$ for reaction 19–1. The $\Delta S°$ is easily obtainable from $\Delta H°$ and $\Delta G°$ and it is calculated:

$$\begin{aligned} \Delta S° &= \frac{\Delta H° - \Delta G°}{T} \\ &= \frac{13{,}360 - 19{,}100}{298} = -19.26 \text{ cal/deg-mole} \end{aligned}$$

This $\Delta S°$ value is completely contrary to our intuition concerning $\Delta S°$ and the spread of matter. The splitting of water appears to spread matter. Furthermore, a production of two species from one as in reaction 19–1 will invariably yield a positive $\Delta S°$ for reactions in the gas phase or in nonpolar or weakly polar solvents. The explanation of this apparent contradiction lies in the nature of H^+ and OH^- in water. The proton associates with a water molecule to form H_3O^+, an ion that hydrogen-bonds with water molecules in a considerably stronger fashion than water itself does. Figure 19–1 shows the structure of H_3O^+ strongly hydrogen-bonded to three water

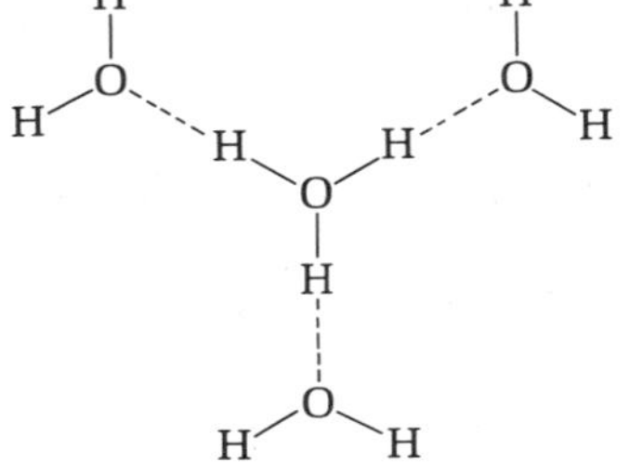

Figure 19–1
The structure of H_3O^+ and part of its aquation. The dashed lines indicate hydrogen bonds.

molecules. These three water molecules are, in turn, hydrogen-bonded to other water molecules, and the proton concentrates a sizable number of water molecules around it in aqueous solution. A similar picture can be drawn for OH^- and a portion of its hydration is shown in Figure 19–2. Thus, the production of H^+ and OH^- in water involves a binding together of water molecules by the ions. This binding results in the negative entropy change.

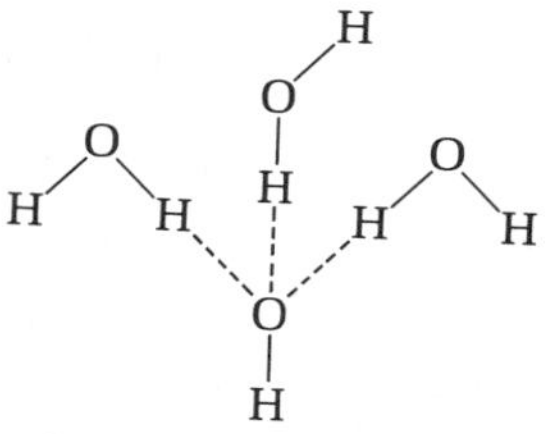

Figure 19–2 A part of the aquation of the hydroxide ion in aqueous solution. The dashed lines indicate hydrogen bonds.

Weak Acids and pH

We have defined an acid as any substance that increases $[H^+]$ in water when added to water. Up to this point, we have only discussed this increase in $[H^+]$ in qualitative terms. For example, it was pointed out that HCl and HNO_3 are strong Lowry-Brønsted acids, acetic acid is a weak Lowry-Brønsted acid, and Na^+ is an extremely weak Lewis acid. To treat one of these acids in quantitative terms, let us weigh out 6.00 g of acetic acid and add it to enough water to make 1.00 l of solution. What is the hydrogen-ion concentration in this solution? We have treated the ionization of acetic acid (Chapter 16), showing that the equilibrium expression and equilibrium constant at 25 °C are

$$K_e = 1.74 \times 10^{-5} = \frac{[H^+][H_3CCOO^-]}{[H_3CCOOH]} \qquad \textbf{19–3}$$

Equation 19–3 is the form of the equilibrium expression for all Lowry-Brønsted acids in water; that is, the expression has the undissociated acid in the denominator and $[H^+]$ and the deprotonated form of the acid in the numerator. The K_e for the dissociation of an acid is commonly given the symbol K_a, and the general expression for K_a is

$$K_a = \frac{[H^+][\text{deprotonated acid}]}{[\text{acid}]} \qquad \textbf{19–4}$$

Table 19–1 lists a number of K_a values for weak acids. Strong acids such as HNO_3, HCl, and $HClO_4$, are quantitatively dissociated in dilute solution in water and K_a is very much greater than unity. Because of the large range of K_a values listed in the table, it proves convenient to define a scaled version of K_a, the pK_a, defined as

$$pK_a = -\log K_a \qquad \textbf{19–5}$$

Table 19–1 Dissociation Constants of Some Acids in Water at 298 °K

Acid	K_a	pK_a
HSO_4^- (hydrogen sulfate)	1.2×10^{-2}	1.92
H_3PO_4 (orthophosphoric acid)	7.5×10^{-3}	2.12
HF (hydrogen fluoride)	3.5×10^{-4}	3.45
H_2CO_3 (carbonic acid)	1.7×10^{-4}	3.77
H_3CCOOH (acetic acid)	1.8×10^{-5}	4.76
H_2S (hydrogen sulfide)	1.0×10^{-7}	7.00
$H_2PO_4^-$ (dihydrogen orthophosphate)	6.2×10^{-8}	7.21
NH_4^+ (ammonium)	5.6×10^{-10}	9.25
HCO_3^- (hydrogen carbonate)	5.6×10^{-11}	10.25
HPO_4^{2-} (hydrogen orthophosphate)	1×10^{-12}	12
HS^- (hydrogen sulfide ion)	1.1×10^{-13}	12.96

The table lists the pK_a of each acid in the column adjacent to the K_a.

Let us return to the acetic acid solution that contains 6.0 g of acetic acid per liter of solution. If no acetic acid ionizes,

$$\frac{[H^+][H_3CCOO^-]}{[H_3CCOOH]} = \frac{(1.0 \times 10^{-7})(0)}{0.100} = 0$$

and this is a nonequilibrium situation. Some acetic acid must ionize, and we can use equation 19–3 to calculate how much ionizes. Let x equal the acetate-ion concentration produced in the ionization, equal to the equilibrium concentration of acetate ion. From the balanced reaction equation, we can relate the other equilibrium concentrations to x:

$$[H^+] = 1.0 \times 10^{-7} + x; \qquad [H_3CCOOH] = 0.100 - x$$

and the equilibrium equation is

$$1.74 \times 10^{-5} = \frac{(1.0 \times 10^{-7} + x)(x)}{0.100 - x}$$

The resultant quadratic equation yields $x = 1.31 \times 10^{-3}$. Only a little more than 1 percent of the acetic acid is ionized in equilibrium, and we are certainly justified in calling acetic acid a weak acid.

The weak acid has produced over a ten-thousandfold increase in $[H^+]$ over that in pure water, however, since the equilibrium value of $[H^+]$ is 1.31×10^{-3} and the value in pure water is 1.0×10^{-7}. As with the conversion of K_a to pK_a it proves convenient to define and use a scaled version of $[H^+]$, the **pH,** defined:

$$pH = -\log [H^+] \qquad \textbf{19–6}$$

In the case of our acetic acid solution,

$$pH = -\log(1.31 \times 10^{-3})$$
$$= -\log(1.31) - \log(10^{-3})$$
$$= -0.12 + 3.00 = 2.88$$

The most obvious physiological effect of a pH as low as 2.88 is a sour taste, and solutions of even quite weak acids taste sour. Lemon juice, tomatoes, and unripe fruits derive their sour taste from the organic acids they contain.

Ionization of a Weak Acid as a Function of Concentration

Suppose that our one-liter acetic acid solution had been prepared with 0.60 g of acetic acid instead of 6.0 g. Again let us define x as the equilibrium concentration of acetate ion, so that the equilibrium expression is

$$1.74 \times 10^{-5} = \frac{(1.0 \times 10^{-7} + x)(x)}{0.0100 - x}$$

The value of x in this more dilute solution is 4.08×10^{-4} and the solution pH is 3.38.

In this case, 4.08 percent of the original acetic acid has ionized in equilibrium as compared with only 1.31 percent of the acetic acid in the more concentrated solution treated in the preceding section. This result is the solution analog of the shift in the $2\ NO_2 \rightleftharpoons N_2O_4$ equilibrium with a decrease in total pressure. A lowered total concentration is the analog of a lowered total pressure, and the lowered concentration favors the dissociated form of the acid, just as a lowered pressure favors $2\ NO_2$ as compared with N_2O_4.

Weak Bases

Let us now consider a solution of a base in water. Such a base is acetate ion, obtained by the deprotonation of acetic acid. Acetate ion reacts with water according to the following equation:

$$H_3CCOO^-(aq) + H_2O(l) \rightleftharpoons H_3CCOOH(aq) + OH^-(aq) \qquad \textbf{19–7}$$

and the equilibrium expression is

$$K_e = \frac{[OH^-][H_3CCOOH]}{[H_3CCOO^-]} \qquad \textbf{19–8}$$

All bases react with water according to the same fundamental equation,

$$\text{Base(aq)} + H_2O(l) \rightleftharpoons \text{protonated base(aq)} + OH^-(aq) \qquad \textbf{19–9}$$

for which the equilibrium expression is

$$K_e \equiv K_b = \frac{[OH^-][\text{protonated base}]}{[\text{base}]} \qquad \textbf{19–10}$$

Every base possesses its own characteristic K_b value, but it is not necessary to list a table of K_b values, since they are easily obtained from other values. An acid and the deprotonated form of the acid constitute what is termed a **conjugate acid-base pair.** For example, acetic acid is the conjugate acid of acetate ion, and acetate ion is the conjugate base of acetic acid. We can relate K_b for acetate ion to K_a for acetic acid by multiplying the right side of equation 19–8 by unity $= [H^+]/[H^+]$:

$$K_b = \frac{[H^+][OH^-][H_3CCOOH]}{[H_3CCOO^-][H^+]} = \frac{K_w}{K_a} \qquad \textbf{19–11}$$

For this conjugate pair and for any conjugate acid-base pair, we have

$$K_w = K_a K_b \qquad \textbf{19–12}$$

For acetate ion,

$$K_b = \frac{K_w}{K_a} = \frac{1.0 \times 10^{-14}}{1.74 \times 10^{-5}}$$
$$= 5.7 \times 10^{-10}$$

and acetate ion is quite a weak base.

Suppose that we add 0.100 mole of solid sodium acetate to enough water to prepare 1.00 l of solution. What is the pH of the resultant solution? As we have seen, the sodium ion is an extremely weak Lewis acid and exerts no influence on the solution pH. The predominant effect comes from the basic acetate ion. Let x be the concentration of acetic acid formed from acetate ion by reaction 19–7. This is also the equilibrium concentration of acetic acid. The equilibrium values of $[H_3CCOO^-]$ and $[OH^-]$ are given as

$$[H_3CCOO^-] = 0.100 - x$$
$$[OH^-] = 1.0 \times 10^{-7} + x$$

and the equilibrium expression is

$$5.7 \times 10^{-10} = \frac{(1.0 \times 10^{-7} + x)(x)}{0.100 - x}$$

The value of x is 7.5×10^{-6}. The equilibrium hydroxide ion concentration is 7.6×10^{-6} and the $[H^+]$ value given as

$$[H^+] = \frac{1.0 \times 10^{-14}}{7.6 \times 10^{-6}} = 1.3 \times 10^{-9}$$

The solution pH is 8.9. **When a base is added to pure water, the resultant pH is always greater than 7.0. When an acid is added to pure water, the resultant pH is always less than 7.0.**

Titration

Suppose we take a 50-ml sample of the 0.100 molar acetic acid solution and add dropwise a 0.100 molar solution of NaOH from a graduated buret, as shown in Figure 19–3. As the strong base is added, we monitor the solution pH with a pH meter. (The pH meter is a relatively simple device discussed

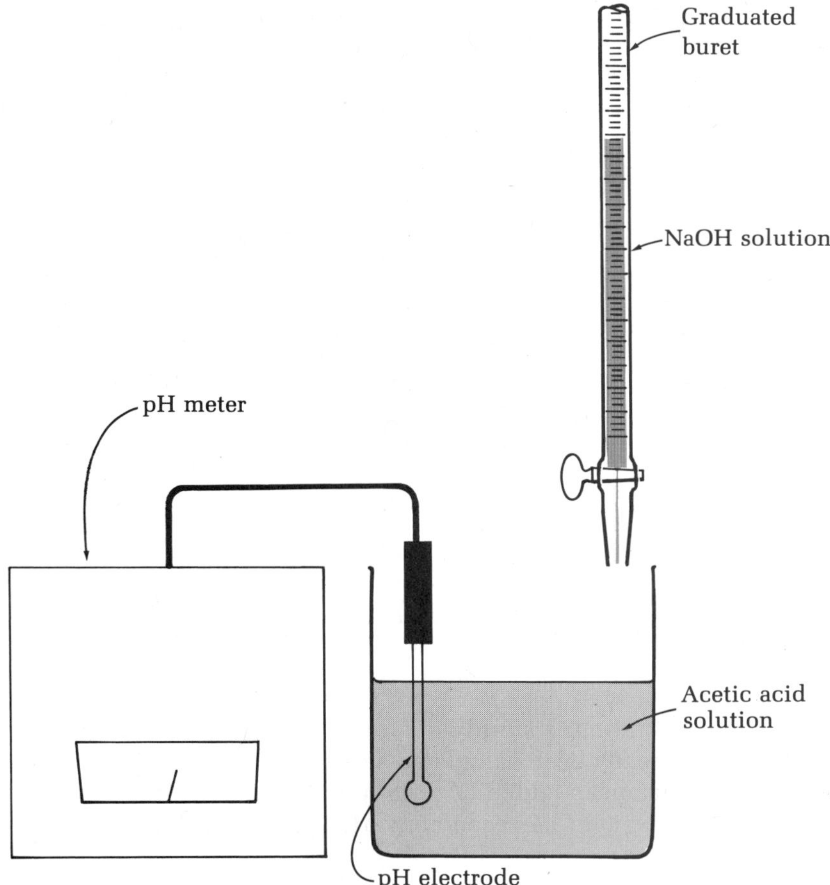

Figure 19–3 Arrangement of apparatus for an acid-base titration.

in some detail in Chapter 20.) The reaction that takes place is

$$H_3CCOOH(aq) + OH^-(aq) \rightleftharpoons H_3CCOO^-(aq) + H_2O(l) \qquad \textbf{19–13}$$

for which the equilibrium expression is

$$K_e = \frac{[H_3CCOO^-]}{[H_3CCOOH][OH^-]}$$

$$= \frac{K_a}{K_w} = 1.8 \times 10^9 \qquad \textbf{19–14}$$

This large equilibrium constant ensures that the reaction proceeds essentially quantitatively to the right; that is, the number of moles of hydroxide ion added equals the number of moles of acetate ion formed. The process of carrying out a monitored quantitative reaction by adding one solution to another is termed **titration.**

Figure 19–4 shows the plot of solution pH as a function of the volume of NaOH solution added. This plot is termed a **titration curve.** The pH at the beginning of the titration is simply that of a 0.100 M solution of acetic acid, and this pH has already been calculated as 2.88. What is the pH when 25 ml of OH^- solution has been added? At that point in the titration, exactly half of the original acetic acid has been converted to acetate ion, so that

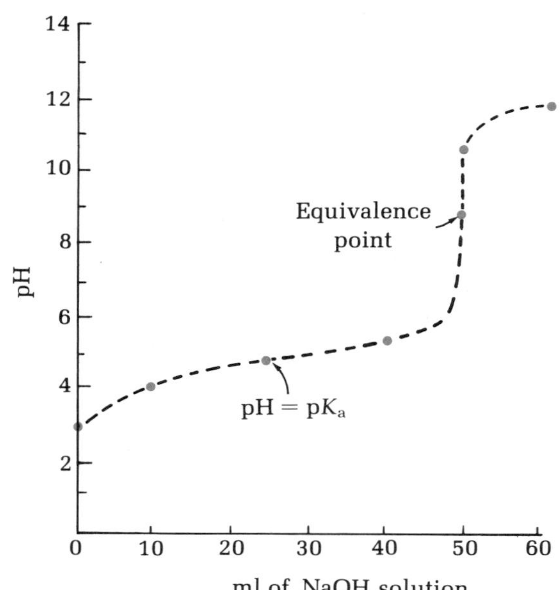

Figure 19–4 Titration curve for the titration of 50.00 ml of 0.100 M acetic acid with 0.100 M NaOH.

$$[H_3CCOOH] = [H_3CCOO^-]$$

We can obtain a very useful relation by taking the negative logarithm of both sides of equation 19–3 and rearranging:

$$pH - pK_a = \log \frac{[H_3CCOO^-]}{[H_3CCOOH]} \qquad \textbf{19–15}$$

Equation 19–15 can be generalized to any acid, where it is known as the **Henderson-Hasselbalch equation;** it takes the form

$$pH - pK_a = \log \frac{[\text{conjugate base}]}{[\text{acid}]} \qquad \textbf{19–16}$$

This equation is easily interpreted in a commonsense fashion. When the pH is larger than the pK_a, the basic form of the conjugate pair predominates over the acid form and the difference between the pH and pK_a is the number of powers of 10 of this dominance. When the pH is less than the pK_a, the acid form predominates by the number of powers of 10 indicated by $pK_a - pH$. When the pH equals the pK_a, both acid and base forms are present in equal concentrations. Likewise, when the acid and conjugate base concentrations are equal, as they are at the point in Figure 19–4 where 25 ml of base have been added, the pK_a is read as the pH value at that point. That pH value is 4.76, and 4.76 is the pK_a of acetic acid.

The Henderson-Hasselbalch equation allows us to estimate very quickly how a given species is partitioned between its acid and base forms at a given pH. For example, HF has a pK_a of 3.45 and blood has a pH of 7.4. This pH is higher than the pK_a by $7.4 - 3.45 = 4.0$. Fluoride ion must predominate over HF in blood by a factor of 10^4; that is,

$$\frac{[F^-]}{[HF]} = 10^4$$

Equation 19–15 allows us to fill in quickly any points in Figure 19–4 for which $[H_3CCOOH]$ and $[H_3CCOO^-]$ are comparable in magnitude. For example, at 10 ml of added base solution $[H_3CCOO^-]/[H_3CCOOH] = 0.25$, and the pH is calculated to be 4.16. At 40 ml of added base $[H_3CCOO^-]/[H_3CCOOH] = 4.0$ and $pH = 5.36$, as shown in Figure 19–4.

The **equivalence point** in the titration plotted in Figure 19–4 is that point at which a number of moles of OH^- equal to the original number of moles of acid has been added. In the case of Figure 19–4, this occurs at 50 ml of added base, and the solution at that point is a 0.050 molar solution of sodium acetate. The pH of this solution of a weak base is calculated by the method previously given herein to be 8.7 and the equivalence point is indicated in Figure 19–4. If we didn't know the number of moles of acetic acid originally present, the titration will serve as a quantitative analysis for acetic acid if the equivalence point can be located precisely. The equivalence point is usually easy to locate in a precise fashion.

Buffers and Indicators

Suppose that we added 1.0 ml of base in excess of the equivalent amount in Figure 19–4. There is no more acetic acid left for reaction and the pH is simply that of 1.0 ml of 0.100 molar NaOH diluted to 101 ml. The $[OH^-]$ is calculated to be

$$[OH^-] = \frac{0.100}{101} = 9.9 \times 10^{-4}$$

The $[H^+]$ is $1.0 \times 10^{-14}/9.9 \times 10^{-4} = 1.01 \times 10^{-11}$, and the pH is 10.99. This is a very sharp rise in the pH from the pH at the equivalence point as shown in the figure, and it contrasts markedly with the relatively small slope of the titration curve away from the equivalence region.

Throughout most of the titration, acetic acid and its conjugate base are present in solution in comparable concentrations. In the case of Figure 19–4, the strong base, OH^-, which would normally increase the pH steeply when added to water, is converted to the weak base, acetate ion, and the pH rise is small. A comparable resistance to large pH change occurs with the addition of strong acid to the solution because of the following reaction:

$$H^+(aq) + H_3CCOO^-(aq) \rightleftharpoons H_3CCOOH(aq) \qquad \textbf{19–17}$$

for which the equilibrium expression and constant are

$$K_e = \frac{[H_3CCOOH]}{[H^+][H_3CCOO^-]} = \frac{1}{K_a} = 5.5 \times 10^4 \qquad \textbf{19–18}$$

Thus, the acetate ion converts the strong acid, H^+, to the weak acid, acetic acid, in an efficient fashion. The resistance to large pH changes shown by the acetic acid–acetate ion system is termed **pH buffering,** and it occurs any time a weak acid and its conjugate base are present in comparable concentrations in solution.

The pH change between 50 and 51 ml of base in Figure 19–4 shows the effect of the addition of strong base to an unbuffered system. The buffering effect is lost at the equivalence point with the depletion of acetic acid. The loss of the buffering effect at the equivalence point leads to maximizing of the slope of the titration curve at that point, and thus the equivalence point is obtainable with high precision.

The sharp changes in pH near the equivalence point in titration curves allow the use of so-called **pH indicators** in titrations carried out for analytical purposes. These indicators are dyes that can exist in either an acid or a conjugate base form, with the two forms differing markedly in color. A common indicator is **phenolphthalein,** a dye of pK_a approximately 9, which is colorless in the acid form but red in the base form. Canonical resonance structures of the acid and base forms are shown in Figure 19–5. The conjugation extends over the entire anion in the base form, while it is

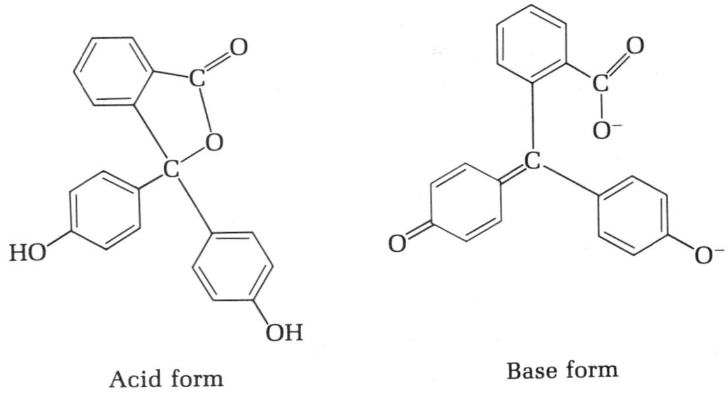

HO, OH, O — Acid form

O, O⁻, O⁻ — Base form

Figure 19–5
Canonical resonance structures of the acid and base forms of phenolphthalein.

restricted to individual benzene rings in the acid form. The greater conjugation of the base form results in its color (reasons presented in Chapter 12).

Suppose that the 0.100 *M* acetic acid solution titrated in Figure 19-4 also contained 1.0×10^{-5} *M* phenolphthalein. At the initial pH of 2.88, the colorless acid form of the dye predominates over the base form by a factor of approximately 10^6, so that the solution is colorless. Suppose that the titration is one drop (~0.05 ml) of added base below the equivalent amount of 50.00 ml of 0.100 molar NaOH. At this point 99.9 percent of the original acid has been titrated and $[H_3CCOO^-]/[H_3CCOOH] = 1000$. From the Henderson-Hasselbalch equation, we obtain the pH as 7.76, and the dye is still predominantly in the acid form by a factor of approximately 20. With the addition of one more drop of base, the pH rises to 8.7, at which the acid and base forms of the dye exist in comparable concentrations; and the solution displays a marked red color, indicating the equivalence point.

Since the pH value at the equivalence point varies considerably depending on the acid or base titrated, different indicators covering a wide range of applicable pH values are commonly available. Table 19-2 lists some of these, together with the equivalence point pH values that fall within the range of applicability of each. Also listed are the colors of the acid and base forms in each case. Perhaps the most common application of pH indicators is litmus paper, containing the dye azolitmus. This dye has a pH range of 4.5 to 8.5 and is red at acid pH and blue at basic pH. It has long been used as a quick test for acidity or basicity of aqueous solutions. A more recent version is "universal" pH indicator paper, containing a mixture of several indicators covering different ranges and having different colors. Each pH yields a different partitioning of the various dyes in acid and base forms and a different overall color as a function of pH; the pH can be determined within a range of approximately one unit.

Indicator	pH Range	Color	
		Acid	*Base*
Cresol red	0.2–1.8	Red	Yellow
Thymol blue	1.2–2.8	Red	Yellow
Methyl yellow	2.8–4.0	Red	Yellow
Methyl orange	3.1–4.4	Red	Yellow
Methyl red	4.2–6.3	Red	Yellow
Bromthymol blue	6.0–7.6	Yellow	Blue
Phenol red	6.8–8.4	Yellow	Red
Phenolphthalein	8.3–10.0	Colorless	Red
Alizarin yellow	10.1–12.1	Yellow	Lilac

Table 19-2
Common pH Indicators

Polyprotic Acids

As the last topic in our consideration of acids and bases, we treat the important topic of **polyprotic acids,** that is, those molecules that can supply more than one proton to the solvent. These include many molecules of biological interest, one of which, CO_2, is the principal cause of the pH buffering effect in blood. Consequently, we center our attention on CO_2 to demonstrate equilibria involving polyprotic acids.

Actually CO_2 is a Lewis acid but forms the diprotic Lowry-Brønsted acid H_2CO_3, carbonic acid, by reaction with water:

$$CO_2(aq) + H_2O(l) \rightleftharpoons H_2CO_3(aq) \qquad \textbf{19–19}$$

From Tables 16–1 and 16–2, the $\Delta G°$ for reaction 19–19 is calculated as follows:

$$\Delta G° = \Delta G°_{f(H_2CO_3,\,aq)} - \Delta G°_{f(CO_2,\,aq)} - \Delta G°_{f(H_2O,\,l)}$$

$$= -145.46 - (-92.31) - (-56.69)$$

$$= 3.54 \text{ kcal}$$

and K_e for reaction 19–19 is 2.5×10^{-3}. Carbonic acid exists in equilibrium with its conjugate base, hydrogen carbonate ion, through the reaction

$$H_2CO_3(aq) \rightleftharpoons H^+(aq) + HCO_3^-(aq) \qquad \textbf{19–20}$$

for which K_a is listed in Table 19–1 as 1.7×10^{-4}. Reactions 19–19 and 19–20 may be added to yield the effective reaction for CO_2 serving as an acid:

$$CO_2(aq) + H_2O(l) \rightleftharpoons HCO_3^-(aq) + H^+(aq) \qquad \textbf{19–21}$$

The K_a for CO_2 is given by

$$K_{a(CO_2)} = \frac{[HCO_3^-][H^+]}{[CO_2]} \qquad \textbf{19–22}$$

Equation 19–22 may be expressed in terms of the equilibrium expressions for reactions 19–19 and 19–20:

$$K_{a(CO_2)} = \frac{[H_2CO_3]}{[CO_2]} \times \frac{[H^+][HCO_3^-]}{[H_2CO_3]} \qquad \textbf{19–23}$$

Thus, K_a for CO_2 is the product of K_a for H_2CO_3, and the K_e for reaction 19–19 $= (2.5 \times 10^{-3})(1.7 \times 10^{-4}) = 4.3 \times 10^{-7}$, and CO_2 is an acid of $pK_a = 6.37$. (When reactions add, their equilibrium constants multiply.) At the pH of blood (7.4), the CO_2, which is a product of respiration, exists 91 percent as hydrogen carbonate and 9 percent as carbon dioxide. This makes blood a buffered solution and this buffering is critically important to us. Any significant change in blood pH is fatal. (The chemistry of blood is quite complex, and blood contains other buffers in addition to carbon dioxide and hydrogen carbonate.)

Equilibria involving polyprotic acids are complicated by more than one ionization step. In the case of CO_2, the second step is

$$HCO_3^-(aq) \rightleftharpoons CO_3^{2-}(aq) + H^+(aq) \qquad \textbf{19–24}$$

for which K_a is listed in Table 19–1 as 5.6×10^{-11}. We term this constant K_{a2} to signify the second ionization step of H_2CO_3, and the K_a of reaction 19–20 is termed K_{a1}. There is a large ratio between these successive ionization constants as well as between the successive constants for ionization of H_2S and H_3PO_4 listed in Table 19–1. With each ionization, the succeeding ionization step becomes more difficult. We might think that this effect was owing to the energy required to separate positive and negative charges. Thus, for example, it should be more difficult to separate a proton from the negatively charged $H_2PO_4^-$ than from the uncharged H_3PO_4. Examination of Table 16–2 shows that this reasoning is fallacious. There is relatively little difference in the ΔH_f° of the successive species involved in ionization; for example, the ΔH_f° values for H_3PO_4, $H_2PO_4^-$, HPO_4^{2-} and PO_4^{3-} are -308.2, -311.3, -310.4, and -306.9 kcal/mole, respectively. There are, however, huge differences in the absolute entropies of the successive species; for example, for orthophosphoric acid, the S° values decrease successively from 42.1 cal/mole-deg for H_3PO_4 to -42 cal/mole-deg for PO_4^{3-}. This entropy decrease is the principal cause of the large ratios between successive ionization constants, and it arises from two sources. The first is the production of a new H^+ in each ionization and the accompanying binding of solvent to it. The second and more important is the increase in negative charge on the anion. This results in a large increase in the solvent association with the anion and a large decrease in solvent entropy.

The large ratios between successive ionization constants greatly simplify the interpretation of titration curves for polyprotic acids. We illustrate this with the titration of 50 ml of a 0.100 molar solution of sodium carbonate, Na_2CO_3, with 0.100 molar HCl, for which the titration curve is given in Figure 19–6. Carbonate ion is a base ($K_b = 1.8 \times 10^{-4}$) and the pH at the start of the titration is 11.62. The first additions of strong acid convert an equivalent amount of CO_3^{2-} to HCO_3^-. The concentration of CO_2 is insignificantly small, and it remains so during the first half of the titration. At 25 ml of added acid, $[CO_3^{2-}] = [HCO_3^-]$, and the pK_a of HCO_3^- can be read from the titration curve as the pH value at this point (10.25).

There are two equivalence points in the titration of CO_3^{2-} because CO_3^{2-} can accept two protons. At the first equivalence point (50 ml of added acid), we have a 0.050 molar solution of sodium hydrogen carbonate. What is the pH of this solution? The HCO_3^- can serve as either an acid or as a base. It behaves as an acid according to reaction 19–24, with a K_a of 5.6×10^{-11},

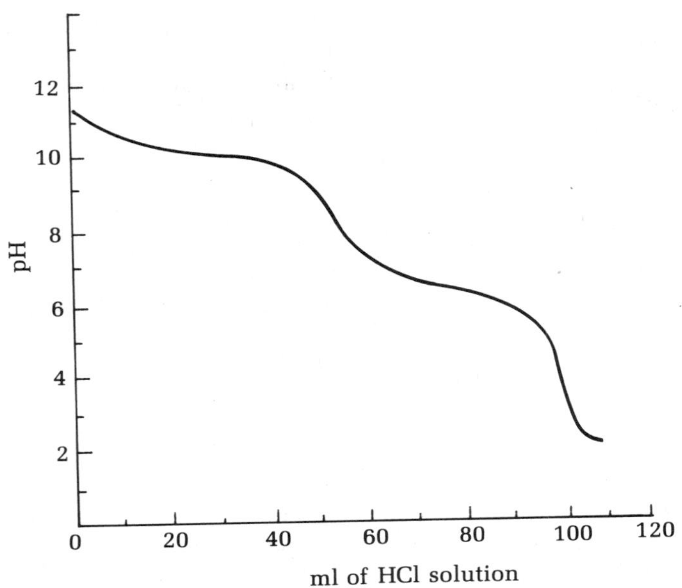

Figure 19–6
Titration curve for the titration of 50.00 ml of 0.100 *M* Na_2CO_3 with 0.100 *M* HCl.

and as a base according to reaction 19–25:

$$HCO_3^-(aq) + H_2O(l) \rightleftharpoons CO_2(aq) + OH^-(aq) + H_2O(l) \qquad \textbf{19–25}$$

The K_b for reaction 19–25 is given by the equation

$$K_b = \frac{K_w}{K_{a(CO_2)}} = \frac{1.0 \times 10^{-14}}{4.3 \times 10^{-7}} = 2.3 \times 10^{-8}$$

Thus, hydrogen carbonate is a much stronger base than acid, and we can center on reaction 19–25 exclusively in calculating the solution pH. The procedure for calculating the pH at the equivalence point is identical to that previously used with the solution of sodium acetate, and the result is a pH of 9.50.

As additional HCl is added past the first equivalence point, HCO_3^- is converted to CO_2. At 75 ml of added acid, $[HCO_3^-]$ equals $[CO_2]$, and the pH equals the pK_a of $CO_2 = 6.37$. At the second equivalence point (100 ml of added acid), we have a 0.033 molar solution of CO_2 and a pH of 3.92. Because the second equivalence region is longer and steeper than the first equivalence region, it is used for analysis of CO_3^{2-}, and methyl orange would be a suitable indicator.

19–2. Metal Ion Complexes

Stability Constants

Ions such as Ni^{2+}, Fe^{3+}, Zn^{2+}, and Cu^{2+} are good electrophiles that normally exist in water as hydrated species with water molecules bonded to the metal ions (Chapter 11). If another nucleophile is added to the solution, a competition is set up between the nucleophile and water for bonding to the electrophilic metal ion.

One such competition is that involving silver ion and ammonia:

$$Ag^+(aq) + 2\,NH_3(aq) \rightleftharpoons Ag(NH_3)_2{}^+(aq) \qquad \textbf{19–26}$$

We should predict that if the diammine silver(I) ion is favored in this equilibrium over aquated Ag^+ and NH_3, it surely must be because of energy spreading; that is, $\Delta H°$ must be significantly negative. The process of taking ammonia molecules from a solution in which they have positions and motions independent of each other and binding them together with a silver ion is surely the concentrating of matter, and $\Delta S°$ is predicted to be negative.

The $\Delta G°$ is obtainable from Table 16–2:

$$\Delta G° = \Delta G°_{f(Ag(NH_3)_2{}^+,\,aq)} - \Delta G°_{f(Ag^+,\,aq)} - 2\Delta G°_{f(NH_3,\,aq)}$$
$$= -4.16 - 18.43 - 2\,(-6.36)$$
$$= -9.87 \text{ kcal}$$

The equilibrium constant for the formation of a metal ion complex is termed the **stability constant** for the complex (symbolized as K_{st}). In the case of $Ag(NH_3)_2{}^+$, the K_{st} is 1.7×10^7, and the complex is a relatively strong one. The $\Delta H°$ for reaction 19–26 is also calculable from the data of Table 16–2 as −13.39 kcal, and $\Delta S°$ is −11.8 cal/deg. Thus, our prediction concerning energy spread and matter concentration is correct and it will be correct for most complexation reactions.

Table 19–3 lists the stability constants of several common metal ion complexes. Complexes like those listed in Table 19–3 are very important in the qualitative and quantitative analysis of metal ions in solution. These analyses almost always involve a coupling of complex-ion equilibria with dissolution equilibria of sparingly soluble salts. (We shall treat this subject in a subsequent section of this chapter.)

Complex	K_{st}	Complex	K_{st}
$Ag(NH_3)_2{}^+$	1.7×10^7	$Fe(CN)_6{}^{4-}$	1×10^{37}
$Ag(CN)_2{}^-$	1×10^{21}	$Fe(CN)_6{}^{3-}$	1×10^{44}
$Co(NH_3)_6{}^{2+}$	8×10^4	$Ni(NH_3)_6{}^{2+}$	1×10^9
$Co(NH_3)_6{}^{3+}$	5×10^{33}	$Ni(CN)_4{}^{2-}$	1×10^{14}
$Co(CN)_6{}^{3-}$	1×10^{64}	$Zn(CN)_4{}^{2-}$	1×10^{19}
$Cu(NH_3)_4{}^{2+}$	4×10^{12}	$Zn(NH_3)_4{}^{2+}$	3×10^9

Table 19–3
Stability Constant Values of Metal Ion Complexes

Chelation

The stability constant of $Ni(NH_3)_6^{2+}$ is given in Table 19–3 as 1×10^9. The $\Delta H°$ for the formation of this complex in solution from $Ni^{2+}(aq)$ and six aqueous ammonia molecules is -19 kcal, and $\Delta S°$ is -22 cal/deg. The $\Delta H°$ and $\Delta S°$ values have the signs characteristic of the formation of many complexes, a situation we considered in connection with $Ag(NH_3)_2^+$.

The ammonia molecule contains only one electron donor site, that is, the nitrogen region. Suppose that we incorporate two such sites in a single molecule such as ethylenediamine, $H_2NCH_2CH_2NH_2$. Such a molecule constitutes a **bidentate ligand** since it can complex with a single metal ion at two sites, as depicted in Figure 11–8. Figure 19–7 is a comparison of the structures of $Ni(NH_3)_6^{2+}$ and $Ni(en)_3^{2+}$. The ethylenediamine ligand is symbolized by N—N.

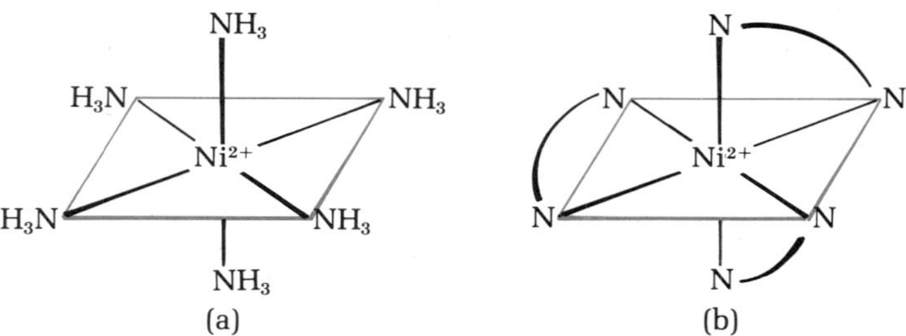

Figure 19–7 The structures of (a) hexammine nickel(II) ion and (b) tris-ethylenediamine nickel (II) ion. The ethylenediamine ligand is symbolized by N—N.

Ethylenediamine is a slightly stronger nucleophile than ammonia, so we should predict that $Ni(en)_3^{2+}$ has a slightly larger stability constant than $Ni(NH_3)_6^{2+}$. The measured K_{st} of $Ni(en)_3^{2+}$ is 1×10^{18}, which represents an enormous increase over K_{st} for $Ni(NH_3)_6^{2+}$. The large surprise comes in the $\Delta S°$ value for the formation of $Ni(en)_3^{2+}$, which is -3 cal/deg as compared to -22 cal/deg for the formation of $Ni(NH_3)_6^{2+}$.

The source of the less unfavorable entropy for $Ni(en)_3^{2+}$ formation is not difficult to deduce. With the monodentate ligand NH_3, six molecules must be taken from a one-molar solution (the standard state), where their positional and motional possibilities are independent of each other, and concentrated together into $Ni(NH_3)_6^{2+}$, where the positions are tied together and much more limited. With ethylenediamine, this process of matter concentration must be performed with only three molecules and it is much less unfavorable (though still unfavorable). The formation of complexes involving more than one binding site in the same molecule is termed **chelation,** and $Ni(en)_3^{2+}$ is an example of a **chelate complex.** From this entropy analysis, we should expect all chelate complexes to show a considerably greater stability than the corresponding monodentate complexes.

If a bidentate ligand forms stronger complexes than a corresponding monodentate ligand, we might expect a hexadentate ligand to form very strong complexes with most metal ions. Such a ligand is the **ethylenediaminetetraacetate ion** (symbolized as EDTA), whose structure is shown in Figure 19–8. The ion possesses six good electron donor sites, that is, the two nitrogens and four acetate groups (the ion may be viewed as a fusion of four

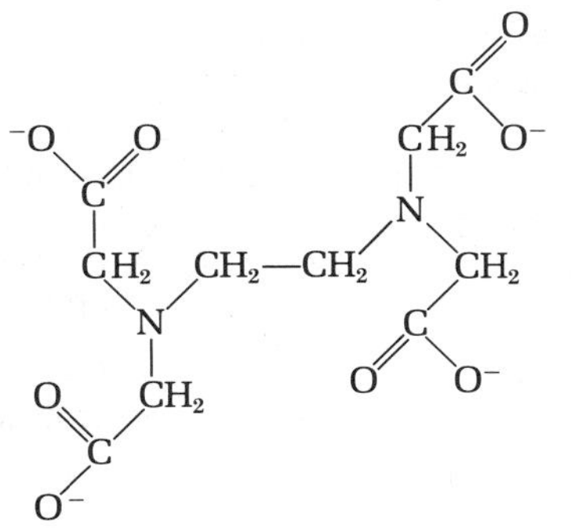

Figure 19–8 The structure of ethylenediaminetetraacetate ion.

acetate ions onto a central ethylenediamine molecule); and the ion is large enough and flexible enough to wrap itself around most metal ions so that each of these six sites is serving as an electron donor. As an example of the magnitude of the chelate effect with EDTA, the stability constant for formation of $Mg(EDTA)^{2-}$ is 5×10^8. Yet neither ammonia nor acetate ion displays any significant affinity for Mg^{2+}. It is even doubtful whether a complex of the formula $Mg(NH_3)_2(acetate)_4{}^{2-}$, the monodentate analog of $Mg(EDTA)^{2-}$, could even be formed in detectable amounts in solution.

The ability of EDTA to form complexes with Mg^{2+} as well as Ca^{2+} has been commercially exploited for some time. "Hard water" is hard because of the presence of metal ions such as Mg^{2+} and Ca^{2+}, which form precipitates with the anions of soaps. EDTA "softens" water by lowering the concentrations of aquated Ca^{2+} and Mg^{2+} by converting them to the EDTA complexes.

Chelation is responsible for very high stability constants for a number of metal ion complexes of great biological importance. Two examples are hemoglobin, a complex of Fe^{2+}, and chlorophyll, a complex of Mg^{2+}.

19–3. The Solubilities of Salts

A Thermodynamic Analysis

In Chapter 15 it was stated that a salt such as NaCl is very soluble in water because of the matter spreading in the process of dissolution. However, as we have seen previously in this chapter, the introduction of ions into water causes a binding together of water molecules by the ions; so perhaps our statement in Chapter 15 was incorrect. Let us check by calculating $\Delta G°$,

$\Delta H°$, and $\Delta S°$ for the following change:

$$NaCl(s) \rightleftharpoons Na^+(aq) + Cl^-(aq) \qquad \textbf{19–27}$$

These standard thermodynamic changes are calculable from the data given in Tables 15–1, 16–1, and 16–2. The values of $\Delta G°$, $\Delta H°$, and $\Delta S°$ are -2.25 kcal, 0.93 kcal, and 10.33 cal/deg, respectively. These numbers reveal that our statement was correct and NaCl is very soluble because of the spreading of Na^+ and Cl^-. The $\Delta H°$ is nearly zero because the energies of the aquated ions are approximately equal to the energies of the ions in the solid.

Let us now look at $\Delta H°$ and $\Delta S°$ for the dissolution of a relatively insoluble salt. In Chapter 16, we calculated the $\Delta G°$ for the dissolution of AgCl as 13.30 kcal. The $\Delta H°$ and $\Delta S°$ are 15.65 and 7.88 kcal, respectively. Again, as with NaCl, the entropy is favorable for the dissolution; but in contrast with NaCl, a large amount of energy must be concentrated, and AgCl is relatively insoluble.

Solubility Products

The formula of any salt is expressible in the form C_nA_m where C is a cation, A is an anion, and n and m are integers. The dissolution equilibrium of this salt is

$$C_nA_m(s) \rightleftharpoons n\ C^{m+}(aq) + m\ A^{n-}(aq) \qquad \textbf{19–28}$$

and the equilibrium expression, termed the **solubility product constant** (K_{sp}), is

$$K_{sp} = [C^{m+}]^n[A^{n-}]^m \qquad \textbf{19–29}$$

A number of such solubility products are given in Table 19–4.

What is the relation between the solubility product and the solubility of a salt, where the solubility is defined as the number of moles of the salt that are contained in one liter of a saturated solution? Let us take CaF_2 as an example, and let x be the solubility. The relation between K_{sp} and x is given as

$$K_{sp} = 2.8 \times 10^{-11} = [Ca^{2+}][F^-]^2 = (x)(2x)^2$$

where K_{sp} was obtained from Table 19–4. The solubility of calcium fluoride in water at 25 °C is 1.9×10^{-4} mole/liter.

The Separation of Metal Ions by Selective Precipitation

Suppose that we have a solution that is 1.00×10^{-3} M in $Fe(NO_3)_2$, and 1.00×10^{-3} M in $Cu(NO_3)_2$, and we want a quantitative separation of the two cations. Both cations form relatively insoluble solids when combined with S^{2-}, as shown by the K_{sp} values in Table 19–4. However, the ratio between the two K_{sp} values is quite high, and this provides the means of sep-

Solid	K_{sp}	Solid	K_{sp}
AgBr	5.0×10^{-13}	FeS	1×10^{-19}
AgCl	1.8×10^{-10}	HgS	1×10^{-52}
AgI	8.5×10^{-17}	$Mg(OH)_2$	1×10^{-11}
Ag_2CrO_4	1.9×10^{-12}	$MgCO_3$	2×10^{-5}
Ag_2S	1×10^{-51}	MgF_2	7×10^{-9}
$Al(OH)_3$	2×10^{-32}	MnS	2×10^{-13}
BaF_2	1.7×10^{-6}	NiS	3×10^{-19}
$BaCO_3$	1.6×10^{-9}	PbF_2	3×10^{-8}
$BaSO_4$	1.1×10^{-10}	$PbCO_3$	1×10^{-13}
CaF_2	2.8×10^{-11}	$PbSO_4$	2×10^{-8}
$CaCO_3$	1×10^{-8}	PbS	1×10^{-28}
$CaSO_4$	3×10^{-5}	SnS	1×10^{-28}
$Cu(OH)_2$	1.6×10^{-19}	$Zn(OH)_2$	5×10^{-17}
CuS	6×10^{-36}	$ZnCO_3$	2×10^{-10}
$Fe(OH)_2$	1.6×10^{-15}	ZnS	3×10^{-24}
$Fe(OH)_3$	3×10^{-38}		

Table 19-4
Solubility Products for Ionic Species in Water at 25 °C

aration, as we shall see. The separation of species is the first important step in identifying the constituents of any mixture. Sulfide ion plays an important role in the **qualitative analysis** of solutions of metal ions.

Let us take the solution containing the two cations and add sufficient HCl to it to make it 0.100 molar in H^+. Now we saturate it with H_2S by bubbling gaseous H_2S through it. This saturated solution is 0.10 molar in H_2S. What is the S^{2-} concentration in this solution? Sulfide ion is produced from H_2S by the following two successive ionizations:

$$H_2S(aq) \rightleftharpoons H^+(aq) + HS^-(aq) \qquad \textbf{19–30}$$

$$HS^-(aq) \rightleftharpoons H^+(aq) + S^{2-}(aq) \qquad \textbf{19–31}$$

with equilibrium constants of $K_{a1} = 1.0 \times 10^{-7}$ and $K_{a2} = 1.1 \times 10^{-13}$. The overall reaction of S^{2-} production from H_2S is the sum of reactions 19–30 and 19–31; and the equilibrium expression and K_e value for the overall reaction are

$$K_e = \frac{[H^+]^2[S^{2-}]}{[H_2S]} = K_{a1}K_{a2} = 1.1 \times 10^{-20} \qquad \textbf{19–32}$$

Since H_2S is a weak acid, only a small fraction of it ionizes, and the equilibrium values of $[H^+]$ and $[H_2S]$ are both 0.10 *M*. This yields $[S^{2-}]$ as 1.1×10^{-19} *M*.

The product of the Fe^{2+} and S^{2-} concentrations is

$$[Fe^{2+}][S^{2-}] = (1.0 \times 10^{-3})(1.1 \times 10^{-19}) = 1.1 \times 10^{-22}$$

Table 19–4 lists the K_{sp} of FeS as 1×10^{-19}. No FeS will form unless the product of concentrations in the equilibrium expression exceeds K_{sp}. In this case, 1.0×10^{-22} is much less than K_{sp}, and the Fe^{2+} ions remain in solution. The product of $[Cu^{2+}]$ and $[S^{2-}]$ is also 1.1×10^{-22}, but this far exceeds the K_{sp} of 6×10^{-36}. The Cu^{2+} ions will be quantitatively removed from solution and the desired separation is obtained.

Precipitation of Salts in Quantitative Analysis

The precipitation of sparingly soluble salts also commonly serves as a basis for quantitative analysis. Suppose that we have a solution containing NaCl and Na_2SO_4, each at 0.0100 M concentration, and we wish to determine the sulfate ion concentration. The salt $BaCl_2$ is quite soluble in water, although Table 19–4 lists K_{sp} for $BaSO_4$ as 1.1×10^{-10}. If our NaCl–Na_2SO_4 solution is made 0.10 M in Ba^{2+}, the equilibrium concentration of SO_4^{2-} will be given as

$$[SO_4^{2-}] = \frac{K_{sp}}{[Ba^{2+}]} = \frac{1.1 \times 10^{-10}}{0.10}$$

$$= 1.1 \times 10^{-9}$$

and the SO_4^{2-} is almost quantitatively precipitated as $BaSO_4$. The precipitate can then be filtered, dried, and weighed to determine the number of moles of SO_4^{2-} in the original solution.

A particularly convenient form of quantitative analysis can be carried out for chloride ion, based on the selective precipitation of sparingly soluble salts. Suppose that we want to determine the Cl^- concentration in a sodium chloride solution, and that the unknown concentration will turn out to be 0.200 molar. For reasons that will become apparent, we add enough of the solid yellow salt potassium chromate, K_2CrO_4, to make the solution 5×10^{-3} M in CrO_4^{2-}.

We now add dropwise from a graduated buret a 0.100 molar solution of $AgNO_3$. The K_{sp} of AgCl is 1.8×10^{-10}, and therefore AgCl forms when $[Ag^+]$ exceeds $1.8 \times 10^{-10}/(0.200) = 9 \times 10^{-10}$ M. The first drop of $AgNO_3$ solution would produce a $[Ag^+]$ higher than 9×10^{-10} M if no precipitation occurred, and therefore solid AgCl must form. Table 19–4 shows that Ag_2CrO_4 is also sparingly soluble, with a K_{sp} of 1.9×10^{-12}. Precipitation of Ag_2CrO_4 will occur when $[Ag^+]$ reaches $[(1.9 \times 10^{-12})/5 \times 10^{-3}]^{1/2} = 2 \times 10^{-5}$ M. Thus, it might appear that the first drop of $AgNO_3$ solution would also precipitate Ag_2CrO_4. Such is not the case, however, because the Cl^- removes Ag^+ as AgCl. The $[Ag^+]$ cannot rise to 2×10^{-5} M until the $[Cl^-]$ reaches $1.8 \times 10^{-10}/2 \times 10^{-5} = 9 \times 10^{-6}$ M. The chloride ion is almost quantitatively precipitated before any Ag_2CrO_4 is formed. However, one drop of $AgNO_3$ over the equivalence point (where the number of moles of Ag^+ added equals the number of moles of Cl^- in the original solution) will result in the formation of the bright red solid Ag_2CrO_4. The formation of the red solid serves as an equivalence point indicator in this precipitation titration, just as the pH indicators serve in acid-base titrations.

Competition between Complexation and Precipitation

The large stability constant of $Ag(NH_3)_2^+$, and the high solubility of $Ag(NH_3)_2Cl$, together provide a commonly used qualitative test for Ag^+ in solution. We have already shown that the addition of Cl^- to Ag^+ in aqueous solution results in the precipitation of the white solid AgCl. Suppose that we have 0.100 g (7.0×10^{-4} mole) of solid AgCl, and we add it to 10.0 ml of 2.00 *M* ammonia in water. Both Ag^+ and Cl^- must dissolve to some extent in this new solution because of the nonzero solubility product of AgCl. Once in solution, most of the silver ion will exist in the highly soluble form $Ag(NH_3)_2^+$ because of the high stability constant of this complex. Therefore, it is possible that all of the AgCl could dissolve in the ammonia solution. Let us test that possibility by calculation.

If the 0.100 g of AgCl dissolves, the resultant solution has $[Cl^-] = [Ag^+] + [Ag(NH_3)_2^+] = 0.070\ M$. The equilibrium expression for reaction 19–26 is

$$K_{st} = \frac{[Ag(NH_3)_2^+]}{[Ag^+][NH_3]^2} = 1.7 \times 10^7 \qquad \textbf{19–33}$$

The number of moles of NH_3 in solution greatly exceeds the number of moles of either Ag^+ or $Ag(NH_3)_2^+$ so that, to a good approximation, $[NH_3]$ in equation 19–26 may be taken as 2.00 molar. With this value, equation 19–33 shows that $[Ag(NH_3)_2^+] \gg [Ag^+]$, so that $[Ag(NH_3)_2^+]$ may be taken as 0.070 molar. Substitution of these values for $[Ag(NH_3)_2^+]$ and $[NH_3]$ into equation 19–33 yields $[Ag^+] = 1.0 \times 10^{-9}$. The product of $[Ag^+]$ and $[Cl^-]$ is 7×10^{-11}, a number less than the AgCl solubility product of 1.8×10^{-10}, and indeed the entire precipitate will dissolve. The common qualitative test for silver ion in aqueous solution is the formation of a white precipitate on addition of chloride ion and the redissolving of the precipitate on addition of concentrated aqueous ammonia.

SUGGESTIONS FOR ADDITIONAL READING

Sienko, M. J., *Chemistry Problems*, Benjamin, Menlo Park, Calif., 1972.
Butler, J. N., *Ionic Equilibrium*, Addison-Wesley, Reading, Mass., 1964.
Butler, J. N., *Solubility and pH Calculations*, Addison-Wesley, Reading, Mass., 1964.

PROBLEMS

1. Hydrochloric acid is a strong acid, which quantitatively dissociates in water to produce H^+ and Cl^-. What is the hydroxide ion concentration in a 0.100 *M* solution of HCl in water?

2. An American chemist, George Olah, has recently shown that an acid many orders of magnitude stronger than H_2SO_4 or HNO_3 can be prepared by combining a strong Lowry-Brønsted acid, HSO_3F (fluorosulfuric acid), with a good Lewis acid, SO_2. Write an equilibrium expression to show how the presence of the Lewis acid produces an enormous enhancement in the strength of the Lowry-Brønsted acid. Would water be chosen as the solvent to test the acid strength of this "superacid"? Name a proton acceptor that might be more suitable.
3. Complete and balance each of the following reaction equations. In each case in which liquid water is one of the species involved, the liquid water is present in great excess.
   a. $Na_2O(s) + H_2O(l) \rightarrow$
   b. $NaOH(s) + H_2O(l) \rightarrow$
   c. $SO_3(g) + H_2O(l) \rightarrow$
   d. $Na_2O(s) + SO_3(g) \rightarrow$
   e. $N_2O_5(s) + H_2O(l) \rightarrow$
   f. $HNO_3(l) + H_2O(l) \rightarrow$
4. In the text, we showed that the acid dissociation of acetic acid is disfavored by a large negative $\Delta S°$, a common occurrence for weak acids. From the data in Table 16–2, show that the dissociation of NH_4^+ is an exception to this rule. Why should we have been able to predict this exception?
5. By the combination of equations 19–2, 19–4, and 19–5, calculate the pK_a of water.
6. Convert the following concentrations to pH values:
   a. $[H^+] = 7.4 \times 10^{-6}$
   b. $[H^+] = 5.6 \times 10^{-12}$
   c. $[OH^-] = 2.3 \times 10^{-2}$
7. Calculate the pH of a 0.0200 *M* solution of ammonium chloride (NH_4Cl). (The Cl^- has a negligible effect on the pH.)
8. Refer to the qualitative discussion of acids and bases in previous chapters and indicate which of the acids in each of the following pairs possesses the larger pK_a.
   a. HCl or HI.
   b. Methanol or phenol.
   c. HOI or HOCl.
   d. $HOClO_3$ or HOCl.
   e. PH_3 or NH_3.
   f. NH_3 or H_2O.
   g. $Mg^{2+}(aq)$ or $Zn^{2+}(aq)$.
   h. $Fe^{2+}(aq)$ or $Fe^{3+}(aq)$.
9. Which solution will possess the higher pH, a 0.100 *M* solution of NH_3 in water or a 0.100 *M* solution of sodium acetate in water? Explain.

10. Calculate the pH of a 0.0100 *M* solution of NaF.

11. A 0.3081-g sample of an unknown acid was dissolved in water and titrated with 0.1090 *M* NaOH solution. The single equivalence point was reached at 22.61 ml of acid. What is the molecular weight of the unknown acid?

12. In the titration in problem 11, the pH was found to be 5.12 after 15.16 ml of NaOH solution had been added. What is the pK_a of the unknown acid?

13. Of the four possibilities, NH_4^+, NH_3, H_3CCOO^-, and H_3CCOOH, which two predominate in a 0.100 *M* ammonium acetate solution at the following pH values: (a) 7.0; (b) 2.0; (c) 12.0.

14. What weights of what substances would you add to water to make one liter of a buffer solution at pH = 4.00?

15. Most drugs enter the bloodstream from the gastrointestinal tract in nonionic form since they must pass across nonpolar membranes to do so. Aspirin is a weak acid with the structure:

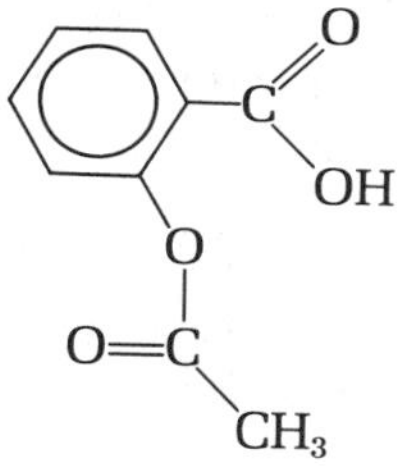

The stomach has a pH of 2–3 and aspirin quickly passes through the stomach wall into the bloodstream. Weak bases are usually much slower in entering the bloodstream. Explain this difference.

16. Construct a titration curve for the titration of 50 ml of a 0.100 *M* solution of NH_4Cl with a 0.100 *M* NaOH solution. This titration is never carried out in practice, nor is any other very weak acid analyzed by titration in aqueous solution. Does your titration curve show why? Explain.

17. In the titration of a 0.100 *M* ammonia with 0.100 *M* HCl, what is the pH at the equivalence point? What indicator would you use in this titration?

18. Finish and balance the following reactions to indicate the predominant reaction that occurs in each:
    a. $Na_2S(s) + H_2O \rightarrow$
    b. $NaHSO_4(s) + H_2O \rightarrow$
    c. $Na_2HPO_4(s) + H_2O \rightarrow$
    d. $Na_2CO_3(s) + H_2O \rightarrow$
    e. $NaHCO_3(s) + H_2O \rightarrow$

*19. Seawater is a buffered system whose pH is held nearly constant by the dissolution of a weak acid, CO_2, from the air, and a weak base, CO_3^{2-}, from rocks at the ocean bottom. If the concentrations of dissolved CO_2 and CO_3^{2-} in seawater are equal, what is the pH of seawater?

20. Of the three possible species, H_2S, HS^-, and S^{2-}, which predominates in a buffered solution at pH = 10.0? In the titration of S^{2-} with HCl solution, where does pH 10.0 lie on the titration curve (first buffer region, first equivalence region, second buffer region, and so on)?

21. Equal volumes of a 0.100 *M* $AgNO_3$ solution and a 0.300 *M* NH_3 solution are mixed. Calculate the concentration of uncomplexed Ag^+ in the resultant solution.

22. Equal volumes of a 0.100 *M* $AgNO_3$ solution, a 0.400 *M* NH_3 solution, and a 1.000 *M* NaCN solution are mixed. What is the concentration of uncomplexed Ag^+ in the resultant solution? (See Problem 23.)

23. Silver ion will react with cyanide ion in aqueous solution to add a maximum of two cyanide ions according to the reaction

$$Ag^+(aq) + 2\ CN^-(aq) \rightleftharpoons Ag(CN)_2^-$$

for which the stability constant is 1×10^{21}. This complex is so stable that it has long been used in silver-ore refining. The reaction that gives dissolution of silver from the ore is

$$4\ Ag(s) + 8\ CN^-(aq) + O_2(g) + 2\ H_2O(l) \rightarrow 4\ Ag(CN)_2^-(aq) + 4\ OH^-(aq)$$

Use the stability constant given, together with data in Tables 16–1 and 16–2, to calculate $\Delta G°$ for this dissolution reaction to show that it is indeed a favorable reaction.

Consider this oxidation reaction in the absence of cyanide:

$$4\ Ag(s) + O_2(g) + 2\ H_2O(l) \rightarrow 4\ Ag^+(aq) + 4\ OH^-(aq)$$

Use Tables 16–1 and 16–2 to show that this is an unfavorable reaction, so that the cyanide complexation is necessary for dissolution of a significant amount of the metal.

24. Gold will not dissolve in either concentrated nitric acid or concentrated hydrochloric acid, but it will dissolve in a mixture of the two acids, termed **aqua regia.** The dissolution reaction is

$$Au(s) + 4\ Cl^-(aq) + 3\ NO_3^-(aq) + 6\ H^+(aq) \rightarrow AuCl_4^-(aq) + 3\ NO_2(g) + 3\ H_2O(l)$$

In view of this reaction why will neither acid dissolve gold by itself?

25. The standard dissolution reactions of many salts containing highly charged ions show negative $\Delta S°$ values and also positive $\Delta H°$ values. Can you explain a negative $\Delta S°$ for a dissolution process?

26. From the K_{sp} listed in Table 19–4 for each of the following species, calculate how many grams of each dissolve in 100 ml of water at 25 °C: (a) AgCl; (b) Ag_2CrO_4; (c) $Mg(OH)_2$.

27. How many grams of $CaCO_3$ will dissolve in one liter of a 0.0100 *M* solution of Na_2CO_3?

*28. From the solubility product of HgS, calculate the number of grams of HgS present in one liter of a saturated solution. How many Hg^{2+} ions are in this liter of solution? With a number this small, how could K_{sp} ever be determined?

29. What is the maximum pH at which we can prepare a 0.100 *M* solution of $Mg(NO_3)_2$ in water without formation of solid $Mg(OH)_2$?

30. A suspension of $Mg(OH)_2$ in water is the common stomach antacid "milk of magnesia." Magnesium hydroxide undergoes the same quantitative neutralization reaction with strong acid as sodium hydroxide does. The reaction is

$$Mg(OH)_2(s) + 2\ H^+(aq) \rightarrow Mg^{2+}(aq) + 2\ H_2O(l)$$

However, NaOH would never be used as an antacid. Why can one of these bases be used and not the other, when both react similarly with strong acid?

31. A drop of silver nitrate solution is added to a solution containing 0.100 *M* NaCl, 0.0100 *M* NaBr, and 0.00100 *M* NaI. What solid, if any, forms?

32. At what pH will FeS begin to precipitate from a 0.0100 *M* solution of Fe^{2+} that is saturated with H_2S?

33. The stability constant of the complex $Cu(NH_3)_4^{2+}$ is 4×10^{12}. Ammonia is a base and $Cu(OH)_2$ has a solubility product of 1.6×10^{-19}, as listed in Table 19–4. If 10 ml of 0.100 *M* $CuCl_2$ solution is mixed with 10 ml of 1.00 *M* ammonia solution, will a precipitate of $Cu(OH)_2$ form?

34. Limestone is largely $MgCO_3$ and $CaCO_3$, whose solubility products are listed in Table 19–4. Both $Mg(HCO_3)_2$ and $Ca(HCO_3)_2$ are highly soluble in water. Can 10.0 grams of $CaCO_3$ dissolve in one liter of an aqueous solution whose pH is maintained at 2.0? Assume that the CO_2 produced remains in solution.

20/ Electrochemistry

All spontaneous processes can be made to yield useful work, and this obviously includes spontaneous chemical reactions. The conversion of matter and energy concentration to the performance of work may be carried out in a very inefficient fashion, as in an internal combustion engine, or in a very efficient fashion. One of the most efficient procedures is the operation of an **electrical cell,** in which the reaction yields electricity as it proceeds. **Electrochemistry** is that branch of chemistry concerned with the relation between electricity and chemical reactions. Its study includes how electrical cells are constructed, how given cell voltages are generated, and how cell voltages are used to obtain very valuable thermodynamic data. A number of other practical applications of electrochemistry exist, such as the measurement of pH, the obtaining of pure elements by electrolysis, and the transfer of metals in the process of electroplating.

20–1. Electrical Cells

Maximum Work from Chemical Reactions

For any process occurring at constant P and T, $-\Delta G$ is the maximum useful work that is obtainable. Most processes that are commonly used to perform useful work are operated in a very inefficient fashion. For instance, most of the potential energy of water in Lake Erie is converted to thermal energy in Lake Ontario in passing over Niagara Falls. Only a small fraction of the potential energy is converted to electrical energy. The combustion of natural gas (CH_4) is another spontaneous process with a $\Delta G°$ of -195.5 kcal/mole at 25 °C. Work is derived from this process, as in a gas-powered engine, from the energy liberated. Essentially no work is done during the combustion, which occurs much more rapidly than piston motion occurs. The product of combustion is a gas at high pressure and high temperature. The second spontaneous process, which provides the useful work, is the expansion of the high-pressure products against the lower external pressure. A necessary by-product of the operation is a large heat transfer to the surroundings.

Is it necessary to have a large amount of energy pass to the surroundings by heat transfer in the combustion of methane? For the combustion to occur, $\Delta S_{universe}$ must increase. The $\Delta S°$ of the reaction

$$CH_4(g) + 2\ O_2(g) \rightleftharpoons CO_2(g) + 2\ H_2O(l) \qquad \textbf{20–1}$$

is calculable from Table 15–3 as −58.00 cal/deg, and the $\Delta H°$ is −212.78 kcal. If all of this energy, $-\Delta H°$, is delivered to the surroundings at 298 °K by heat transfer, the entropy increase of the universe is $-58.00 + (212{,}780)/298 = 656.02$ cal/deg-mole. However, it is not necessary to spread 212.78 kcal of energy to have reaction 20–1 proceed spontaneously. Any energy in excess of $(58.00)(298) = 17.2$ kcal will suffice to make $\Delta S_{universe} > 0$, and the remaining energy, 195.5 kcal, can be used to do useful work. One measure of the efficiency of obtaining work is (work)/($-\Delta H$), that is, the ratio of work obtained to the energy available for heat transfer. The maximum possible efficiency under standard conditions is $(-\Delta G°)/(-\Delta H°)$. For reaction 20–1 this maximum efficiency is $195.5/212.78 = 0.918$, or 91.8 percent.

To obtain any efficiency near this limit, we must control reaction 20–1 so that it liberates little more energy by heat transfer than is essential to produce spontaneity. It should also be pointed out that efficiencies defined in this way can be greater than 100 percent. The maximum efficiency in obtaining work from reaction 20–1 is less than 100% because the reaction concentrates matter (three gaseous molecules are converted to one gaseous molecule). If a process spreads matter (an example is the combustion of gasoline), no energy need be spread to supply spontaneity. In fact, thermal energy from the surroundings can be concentrated in work performance to a limit of $T\Delta S_{system}$. In such a case $(-\Delta G_{system})$ is greater than $(-\Delta H_{system})$ and the efficiency, as we have defined it, exceeds 100%.

The natural question that arises concerns our ability to control such an obviously spontaneous process as the combustion of methane. Won't it always proceed at its own pace, quite independent of what is convenient for us? To answer this question, we center first on a more elementary reaction than reaction 20–1.

Half-Cells

Consider the following reaction:

$$Zn(s) + Cu^{2+}(aq) \rightleftharpoons Zn^{2+}(aq) + Cu(s) \qquad \textbf{20–2}$$

which is a simple redox reaction (Chapter 7). The $\Delta G°$ for this reaction is calculable from Table 16–2 as $-35.18 - 15.53 = -50.71$ kcal, while $\Delta H°$ and $\Delta S°$ are −51.82 kcal and −3.72 cal/deg respectively. The reaction is highly spontaneous, and 50.71 kcal of useful work can be obtained from it for each mole of zinc consumed. However, the process can be carried out in such a way that no work is obtained. The reaction can be carried out in an open beaker in which solid zinc is added to a solution of copper(II), such as $CuCl_2$. The reaction proceeds rapidly in an exothermic fashion, and no useful work is obtained.

Let us now see how work can be obtained from reaction 20–2. The reaction involves the transfer of two electrons from Zn(s) to Cu^{2+}. The reaction could be controlled if we were to keep the reactants physically separated and force the electron transfer to occur through a wire. In this transfer, chemical energy would be directly converted to electrical energy. Each of the two separate compartments is termed a **half-cell** and the combination of the two half-cells constitutes the electrical cell. Figure 20–1 shows a cell based on reaction 20–2. All half-cells contain an electrical conductor in contact with a solution and the conductors (Cu(s) and Zn(s) in this case) are termed **electrodes.** The electrode that supplies electrons to the wire is termed the **anode** (Zn(s) here) and the electron-accepting electrode is the **cathode** (Cu(s) here).

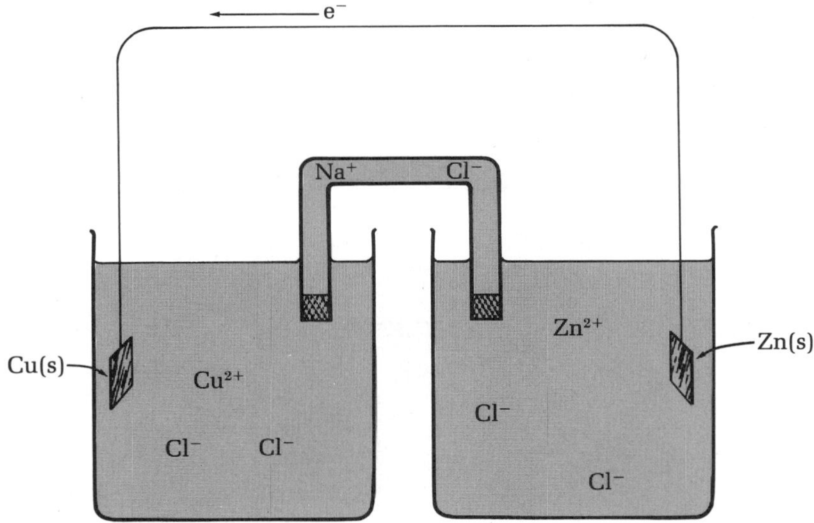

Figure 20–1
Copper–zinc electrical cell.

The second connection in addition to the wire between the two half-cells in Figure 20–1 is a **salt bridge.** Without the salt bridge the copper half-cell would build up an excess of negative ions in solution since the Cu^{2+} ions are depleted in the reaction. Likewise, Zn^{2+} ions would build up in solution in the other half-cell. This charge buildup opposes further electron transfer from the Zn half-cell to the Cu half-cell, and the reaction would stop very shortly after it started. The salt bridge is an aqueous solution containing two ions that do not participate in the cell reaction (Na^+ and Cl^- in Figure 20–1). With the salt bridge present, sodium ions can move through the porous plug into the copper half-cell to replace the Cu^{2+} ions that react, and chloride ions move into the zinc half-cell to balance the charges of the Zn^{2+} ions produced, and excess charges do not build up in the half-cells.

Unlike the reaction between zinc and copper (II) ions in the beaker, the cell reaction can be controlled by controlling the electric current through the wire. The only heat transfer that occurs accompanies the rise in temperature of the wire and of the two solutions caused by the motion of ions through them. As the electrical current is made to approach zero, the useful work derived (for example, to operate an electric motor) approaches $-\Delta G$ for the cell reaction.

Fuel Cells

Reaction 20–1 is also a redox reaction, in which methane is oxidized and O_2 is reduced. We can, in principle, construct an electrical cell based on this reaction also, and such a cell is shown in Figure 20–2. The electron transfers in both half-cells take place at an electrode constructed of a very nonreactive metal (platinum is commonly used as this metal). The gases, CH_4 and O_2, are bubbled through the half-cells at the left and right, respectively. The gases come into simultaneous contact with the electrode and the solution so that electron transfer can take place with the electrode and atom or ion transfer can take place with the solution.

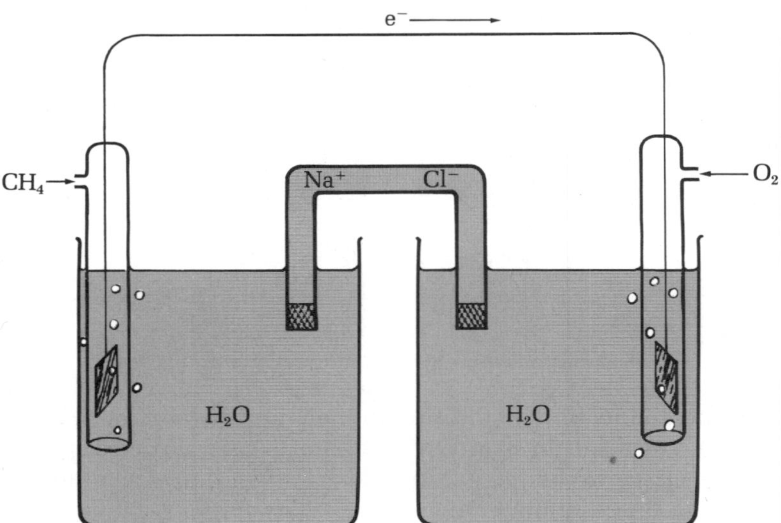

Figure 20–2
A proposed methane-oxygen fuel cell.

Electrical cells that use combustion as the chemical reaction for supplying electrons are termed **fuel cells.** If fuel cells are potentially very efficient, why do inefficient internal combustion engines still exist? The answer lies in the fact that in spite of much research effort, no cell such as that of Figure 20–2 has ever been made to function in an economically satisfactory fashion for large-scale use. The problem is that while reaction 20-1 is highly spontaneous, it proceeds at a very slow rate in the cell. This is yet another example of the lack of correlation between the $\Delta G°$ of a reaction and the rate at which the reaction proceeds. A reaction that has been successfully incorporated into a fuel cell is that between gaseous hydrogen and oxygen, and this cell has been used for electrical-power generation on space flights (a relatively small-scale application).

The Half-Cell Method for Balancing Redox Equations

You will recall the procedure based on oxidation numbers for the recognition of redox reactions and the determination of the numbers of electrons lost or gained by the reactant species (Chapter 7). The advantage of the oxidation-number procedure lies in the speed with which it can be applied. This speed is lost if complicated polyatomic reactants and products are involved because the assignment of oxidation numbers becomes complicated and often ambiguous.

An alternative procedure can be devised based on **half-cell reactions,** and this procedure applies equally unambiguously to complicated reactions and simple ones. The steps in this half-cell method are:

1. Imagine that the oxidizing agent and the reducing agent are in two separate containers in contact with water (even if the reaction is not carried out in aqueous solution). These two reaction halves are the half-cells, and we start by dealing with them separately.
2. Pick out a suspected oxidizing or reducing agent and the product formed from it. If the reactant must obtain H in the reaction, supply it as $H^+(aq)$. If it must obtain O, supply it from $H_2O(l)$. If it must lose either H or O, let them be given to the water as $H^+(aq)$ and $H_2O(l)$ respectively. If any other atom must be added, obtain it from one of the other reactants; and if another atom must be lost, let it appear in one of the products.
3. Once the proper atoms appear on both sides of the equation, balance the equation in numbers of each atom.
4. At this point the half-cell reaction will be mass-balanced (the nuclear masses on both sides of the equation will be equal), but the combined electrical charges on both sides will not be equal. Balance the charge by adding the proper number of electrons to the proper side.
5. At this point we have balanced one half-cell. Now repeat the operation with the remaining reactant(s).
6. To balance the complete equation, multiply one or both half-cell reactions by an integer so that the number of electrons lost in one half-reaction equals the number gained in the other half-reaction. Now add the two half-reactions to get the balanced redox equation.

Let us illustrate the procedure with reaction 20-1. One reactant is CH_4, which is converted to CO_2. We require oxygens from H_2O, and hydrogens are given to water as $H^+(aq)$ to yield the mass-balanced equation

$$CH_4(g) + 2\ H_2O(l) \rightarrow CO_2(g) + 8\ H^+(aq)$$

The equation is electrically balanced by adding eight electrons to the right:

$$CH_4(g) + 2\ H_2O(l) \rightarrow CO_2(g) + 8\ H^+(aq) + 8\ e^- \qquad \textbf{20–3}$$

Equation 20–3 reveals the same fact as the oxidation number method; namely, that the conversion of CH_4 to CO_2 involves the loss of eight electrons. In a similar manner, the second half-reaction is obtained as

$$O_2(g) + 4\ H^+(aq) + 4\ e^- \rightarrow 2\ H_2O(l) \qquad \textbf{20–4}$$

The balanced reaction 20–1 is obtained by multiplying reaction 20–4 by 2 and adding it to reaction 20–3. During the addition of the two half-cell reactions, the species H^+ cancels out, showing that the reaction does not have to be carried out in aqueous solution.

Let us illustrate the method further with a reaction that is carried out in aqueous solution, namely the following unbalanced reaction:

$$Fe^{2+}(aq) + MnO_4^-(aq) \rightarrow Fe^{3+}(aq) + Mn^{2+}(aq)$$

The first half-cell reaction is

$$Fe^{2+}(aq) \rightarrow Fe^{3+}(aq) + e^- \qquad \textbf{20–5}$$

and the second half-cell reaction is

$$5\ e^- + MnO_4^-(aq) + 8\ H^+(aq) \rightarrow Mn^{2+}(aq) + 4\ H_2O(l) \qquad \textbf{20–6}$$

The complete balanced reaction is obtained by multiplying reaction 20–5 by five and adding it to reaction 20–6.

$$5\ Fe^{2+}(aq) + MnO_4^-(aq) + 8\ H^+(aq) \rightarrow Mn^{2+}(aq) + 5\ Fe^{3+}(aq) + 4\ H_2O(l) \qquad \textbf{20–7}$$

Reaction equation 20–7 reveals that the reaction consumes acid; as a consequence, it and any other reaction in which acid is consumed are normally carried out in the solution of a strong acid. Likewise, reactions in which OH^- is consumed are normally carried out in a solution of a strong base.

20–2. Cell Voltages

You are certainly aware that all electrical cells are characterized by a voltage, which is a function of the chemical composition of the cell. For instance, the typical lead storage battery generates either six or twelve volts depending on the number of cells it contains. There is a quantitative relation between voltage and cell composition. Examining this relation will give us a very useful measure of the standard oxidizing strength or reducing strength or both of various chemical reagents.

Standard Reduction Potentials

We now know that $-\Delta G$ for the cell reaction equals the maximum work that can be performed by electrons moving between the two half-cells. This electrical work is defined as the product of the electrical charge moved times the electrical driving force (voltage) moving it. Let us consider reaction 20–2 as an example. The $\Delta G°$ is -50.71 kcal, and two moles of electrons are transferred in the reaction to which this $\Delta G°$ refers. The electrical charge of one mole of electrons is called a **Faraday** of charge (symbolized by $\mathscr{F}$) and it is 96,500 coulombs. (The coulomb in this case is not the statcoulomb (esu) defined in Chapter 3.) The electronic charge in terms of the coulomb to be used in this chapter is $(96{,}500)/6.023 \times 10^{23} = 1.602 \times 10^{-19}$ coulomb/electron.

The product of voltage (symbolized by $\mathscr{E}$) and charge in coulombs is energy in joules. Since we have been using ΔG values in cal, we must employ the conversion factor

$$1 \text{ cal} = 4.185 \text{ joules}$$

The result of the combination of all of these factors is

$$\Delta G = \frac{-n\mathscr{F}\mathscr{E}}{4.185} \qquad \textbf{20–8}$$

where n is the number of moles of electrons transferred in the balanced reaction (two in the case of reaction 20–2).

Cell voltages are measured with a **potentiometer,** since $-\Delta G$ equals the electrical work only at the limit where no current flows. The potentiometer supplies an adjustable, measurable voltage in opposition to the cell voltage. When the two voltages are equal, no current flows, and the cell voltage is read as the potentiometer voltage. In the Cu–Zn cell the negative side of the external potential would have to face the zinc half-cell and the positive side faces the copper half-cell, as shown in Figure 20–3.

Since the cell voltage is directly related to $-\Delta G$ of the cell reaction, the cell voltage varies with all of the properties on which ΔG depends, that is, temperature and pressure and also the nature of the reactants and products and their concentrations. Just as we report standard ΔG values for reactions, we can also report $\mathscr{E}°$, the standard voltage = the voltage of the cell containing all reactants and products in their standard states. Such a tabulation of standard cell voltages would be enormous because of the enormous number of redox reactions. It would be extremely convenient if we could tabulate the voltages of standard half-cells so that the voltage of any standard cell could be easily obtained by combination of the half-cell voltages. That is done by defining and measuring the **standard reduction potentials** of all reductions of all oxidizing agents. The standard reduction potential is defined as $-4.185\Delta G°/n\mathscr{F}$ for the reaction

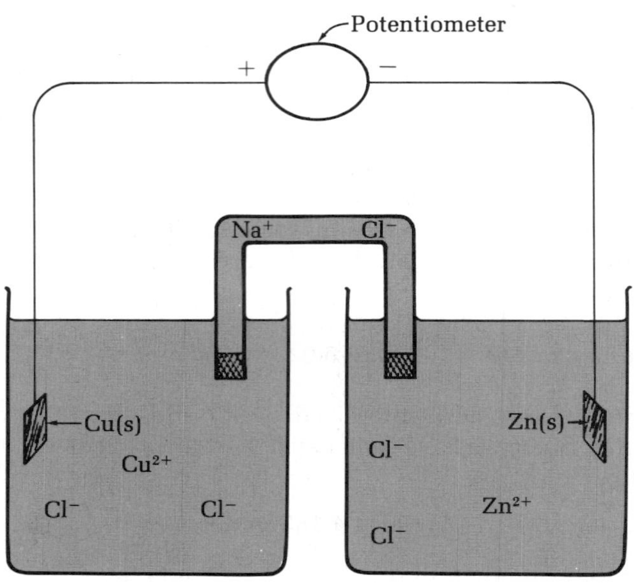

Figure 20–3 Representation of the measurement of the voltage of a copper-zinc cell with a potentiometer.

$$\text{Oxidizing agent} + H_2(g) \rightarrow \text{reducing agent} + 2\ H^+(aq) \qquad \textbf{20–9}$$

where we have used equation 20–8 in the definition. The standard reduction potential is the standard ability to oxidize gaseous hydrogen.

Let us consider a specific case, namely the reaction

$$Zn^{2+}(aq) + H_2(g) \rightarrow Zn(s) + 2\ H^+(aq) \qquad \textbf{20–10}$$

It is a relatively easy matter to obtain $\Delta G°$ for this reaction from Table 16–2 since $\Delta G° = -\Delta G°_{f(Zn^{2+},aq)} = 35.18$ kcal/mole and $\mathscr{E}° = -(4.185)(35{,}180)/(2)(96{,}500) = -0.76$ volt. The negative voltage indicates that in the standard cell the reaction will not proceed spontaneously in the direction indicated by equation 20–10 with the Zn half-cell as the cathode and the hydrogen half-cell as the anode. Rather, the Zn half-cell is the anode and Zn(s) will be oxidized to Zn^{2+}.

Now consider a second standard reduction,

$$Cu^{2+}(aq) + H_2(g) \rightarrow Cu(s) + 2\ H^+(aq) \qquad \textbf{20–11}$$

The $\Delta G° = -\Delta G°_{f(Cu^{2+},aq)} = -15.53$ kcal/mole, and $\mathscr{E}° = +0.34$ volt. Thus, Cu^{2+} is indeed reduced in the standard cell. We now have two standard reduction potentials, that of Zn^{2+} reduced to Zn(s) as −0.76 volt and that of Cu^{2+} to Cu(s) as +0.34 volt. Reaction 20–2 is reaction 20–11 − reaction 20–10 so that

$$\Delta G°_{(20\text{-}2)} = \Delta G°_{(20\text{-}11)} - \Delta G°_{(20\text{-}10)} \qquad \textbf{20–12}$$

and by substitution of equation 20–8

$$\frac{-n_{(20\text{-}2)}\mathscr{F}\mathscr{E}^{\circ}_{(20\text{-}2)}}{4.185} = \frac{-n_{(20\text{-}11)}\mathscr{F}\mathscr{E}^{\circ}_{(20\text{-}11)}}{4.185} \frac{-n_{(20\text{-}10)}\mathscr{F}\mathscr{E}^{\circ}_{(20\text{-}10)}}{4.185} \qquad \textbf{20–13}$$

Each of the three n values is 2, and equation 20–13 reduces to

$$\mathscr{E}^{\circ}_{(20\text{-}2)} = \mathscr{E}^{\circ}_{(20\text{-}11)} - \mathscr{E}^{\circ}_{(20\text{-}10)} = +0.34 - (-0.76)$$
$$= +1.10 \text{ volts}$$

This illustrates a general rule for cells. **The standard cell voltage is equal to the standard reduction potential of the oxidized species that appears on the left side of the reaction equation minus the standard reduction potential of the oxidized species on the right side.** If the calculated cell voltage is positive, the cell reaction proceeds as written. If the cell voltage is negative, the reaction proceeds in the opposite direction. In the present case, the positive voltage of 1.10 volts indicates that Cu^{2+} oxidizes Zn(s), as written in reaction 20–2.

Table 20–1 lists standard reduction potentials for a number of species and reactions.

Table 20–1 Standard Reduction Potentials

Half-Cell Reaction	$\mathscr{E}^{\circ}$ (volts)
$F_2 + 2\,e^- \rightarrow 2\,F^-$	2.87
$O_3 + 2\,H^+ + 2\,e^- \rightarrow O_2 + H_2O$	2.07
$PbO_2 + SO_4^{2-} + 4\,H^+ + 2\,e^- \rightarrow 2\,H_2O + PbSO_4$	1.69
$MnO_4^- + 8\,H^+ + 5\,e^- \rightarrow Mn^{2+} + 4\,H_2O$	1.49
$ClO_4^- + 8\,H^+ + 8\,e^- \rightarrow Cl^- + 4\,H_2O$	1.36
$Cl_2 + 2\,e^- \rightarrow 2\,Cl^-$	1.36
$Cr_2O_7^{2-} + 14\,H^+ + 6\,e^- \rightarrow 2\,Cr^{3+} + 7\,H_2O$	1.33
$O_2 + 4\,H^+ + 4\,e^- \rightarrow 2\,H_2O$	1.23
$Br_2 + 2\,e^- \rightarrow 2\,Br^-$	1.08
$2\,Hg^{2+} + 2\,e^- \rightarrow Hg_2^{2+}$	0.91
$ClO^- + H_2O + 2\,e^- \rightarrow Cl^- + 2\,OH^-$	0.90
$Ag^+ + e^- \rightarrow Ag$	0.80
$Hg_2^{2+} + 2\,e^- \rightarrow 2\,Hg$	0.79
$Fe^{3+} + e^- \rightarrow Fe^{2+}$	0.77
$I_2 + 2\,e^- \rightarrow 2\,I^-$	0.54
$Cu^+ + e^- \rightarrow Cu$	0.52
$O_2 + 2\,H_2O + 4\,e^- \rightarrow 4\,OH^-$	0.40
$ClO_4^- + H_2O + 2\,e^- \rightarrow ClO_3^- + 2\,OH^-$	0.36
$Cu^{2+} + 2\,e^- \rightarrow Cu$	0.34
$Cu^{2+} + e^- \rightarrow Cu^+$	0.15
$2\,H^+ + 2\,e^- \rightarrow H_2$ (defined)	0
$Pb^{2+} + 2\,e^- \rightarrow Pb$	−0.13
$Ni^{2+} + 2\,e^- \rightarrow Ni$	−0.25

Half-Cell Reaction	$\mathscr{E}°$ (volts)
$Co^{2+} + 2\,e^- \rightarrow Co$	−0.28
$PbSO_4 + 2\,e^- \rightarrow Pb + SO_4^{2-}$	−0.36
$Cr^{3+} + e^- \rightarrow Cr^{2+}$	−0.41
$Fe^{2+} + 2\,e^- \rightarrow Fe$	−0.44
$Zn^{2+} + 2\,e \rightarrow Zn$	−0.76
$Al^{3+} + 3\,e^- \rightarrow Al$	−1.66
$Mg^{2+} + 2\,e^- \rightarrow Mg$	−2.36
$Na^+ + e^- \rightarrow Na$	−2.71
$Ca^{2+} + 2\,e^+ \rightarrow Ca$	−2.87
$K^+ + e^- \rightarrow K$	−2.92
$Li^+ + e^- \rightarrow Li$	−3.04

Table 20–1 (continued)

The Use of Standard Reduction Potentials

Let us use Table 20–1 to calculate the voltages of some additional standard cells. The first is based on the oxidation of iron(II) ions by permanganate ion in acidic solution according to equation 20–7:

$$5\,Fe^{2+}(aq) + 8\,H^+(aq) + MnO_4^-(aq) \rightarrow 5\,Fe^{3+}(aq) + Mn^{2+}(aq) + 4\,H_2O(l)$$

This reaction shows MnO_4^- being reduced and Fe^{2+} being oxidized. Thus, the cell voltage is the $\mathscr{E}°$ for the $MnO_4^- \rightarrow Mn^{2+}$ half-cell minus the $\mathscr{E}°$ for the $Fe^{3+} \rightarrow Fe^{2+}$ half-cell. From Table 20–1, this is $1.49 - 0.77 = +0.72$ volt and the positive sign indicates that permanganate oxidizes Fe^{2+} ions under the standard conditions.

The second example involves the following reaction:

$$2\,Ag(s) + Cu^{2+}(aq) \rightarrow 2\,Ag^+(aq) + Cu(s) \qquad \textbf{20–14}$$

In the reaction as written, Cu^{2+} is reduced, so that $\mathscr{E}°$ for reaction 20–14 is

$$\mathscr{E}°_{(20-14)} = \mathscr{E}°_{(Cu^{2+} \rightarrow Cu)} - \mathscr{E}°_{(Ag^+ \rightarrow Ag)}$$

$$= 0.34 - 0.80 = -0.46 \text{ volt}$$

Thus, reaction 20–14 does not proceed in the standard cell as written. Rather, Ag^+ is a stronger oxidizing agent than Cu^{2+}, and Ag^+ oxidizes metallic copper.

It is also clear that Table 20–1 is very useful in chemical syntheses and analyses, since the good oxidizing agents are neatly grouped at the top of the table and the good reducing agents at the bottom. Suppose that we wish to analyze a solution for Fe^{2+} by quantitatively oxidizing it to Fe^{3+}. Table 20–1 reveals a list of oxidizing agents that cannot accomplish this reaction because they are poorer oxidizing agents than Fe^{3+}, that is, all those below Fe^{3+} in the table. Those species immediately above Fe^{3+} in the table will oxidize Fe^{2+}, but not quantitatively. Only the strongest oxidizing agents in the table will suffice, and this is why MnO_4^- was chosen in reaction 20–7.

Suppose that we have a solution of silver nitrate and we wish to reclaim the silver as Ag(s). Table 20–1 lists many species that could accomplish this reduction, that is, the reduced forms of all species significantly below Ag^+ in the table. For example, metallic zinc would serve this purpose quite well.

The Measurement of ΔG_f° for Ions in Solution

The $\mathscr{E}^\circ$ values in Table 20–1 can be calculated from ΔG_f° values given in Table 16–2, and we performed this calculation in the cases of the $Cu^{2+} \rightarrow Cu$ and $Zn^{2+} \rightarrow Zn$ half-cells. This calculation can be reversed, and in practice, most of the ΔG_f° values for ions in Table 16–2 were calculated from measured cell voltages. We shall illustrate how this is done by using Zn^{2+} as an example.

Let us consider reaction 20–10:

$$Zn^{2+}(aq) + H_2(g) \rightarrow Zn(s) + 2\,H^+(aq)$$

If we could construct the standard cell based on this reaction so that $\mathscr{E}^\circ$ is directly measurable, $\Delta G^\circ_{f(Zn^{2+},\,aq)}$ is obtainable from equation 20–8 as

$$\Delta G^\circ_{(20-10)} = -\Delta G^\circ_{f(Zn^{2+},aq)} = \frac{-2\mathscr{F}\mathscr{E}^\circ}{4.185}$$

The value of $\mathscr{E}^\circ$ is obtainable from Table 20–1 as -0.76 volt and $\Delta G^\circ_{f(Zn^{2+},\,aq)}$ is calculated as -35.18 kcal/mole.

However, the standard cell is a hypothetical one because the standard states of Zn^{2+} and H^+ are hypothetical ones, and no measurements can be made with a hypothetical cell. However, we can calculate ΔG_f° for Zn^{2+} from the voltage of a cell in which each of the ions is present at a low concentration. Consider the cell depicted in Figure 20–4. The measured voltage is 0.585 volt with the zinc half-cell serving as the anode in this cell that contains both $[Zn^{2+}]$ and $[H^+]$ at low values (the hydrogen ion concentration in the hydrogen half-cell can be held nearly constant by buffering the solution). The spontaneous reaction in the cell is the reverse of reaction 20–10, so that reaction 20–10 is characterized by a voltage of -0.585 volt in the cell.

The ΔG for reaction 20–10 under the cell conditions is given by the equation

$$\Delta G = \frac{-n\mathscr{F}\mathscr{E}}{4.185} = \frac{-(2)(96{,}500)(-0.585)}{4.185}$$

$$= 27.00 \text{ kcal/mole}$$

We can change this value to that at the standard state, however, because we know (Chapter 16) that the molar Gibbs free energy of a dissolved species A

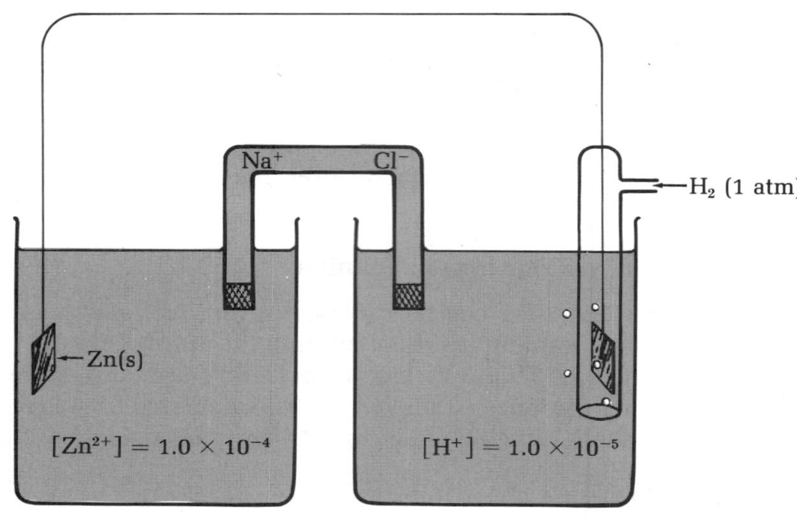

Figure 20–4
Cell for measuring a voltage related to the standard reduction potential of Zn^{2+} in aqueous solution.

changes with concentration in an ideal solution by

$$\Delta G = RT \ln \frac{[A]_f}{[A]_i} \qquad \textbf{20–15}$$

In the case of Figure 20–4, the change in ΔG of the cell in going from the concentrations given to the standard state concentrations is given as

$$\begin{aligned}\Delta(\Delta G) &= \Delta G_{\text{product}} - \Delta G_{\text{reactant}} \\ &= 2\,RT \ln \frac{[H^+]_f}{[H^+]_i} - RT \ln \frac{[Zn^{2+}]_f}{[Zn^{2+}]_i} \\ &= RT \ln \frac{[1/(1.0 \times 10^{-5})]^2}{1/(1.0 \times 10^{-4})} \\ &= 8.18 \text{ kcal}\end{aligned}$$

Thus, the $\Delta G°$ of reaction 20–10 is 27.00 + 8.18 = 35.18 kcal. This $\Delta G°$ yields $\mathscr{E}°$, the standard reduction potential of Zn^{2+}, as −0.76 volt and $\Delta G°_{f(Zn^{2+},\,aq)}$ as −35.18 kcal/mole. An entire table of $\Delta G_f°$ values for ions in solution can be constructed from cell voltages as we have illustrated with Zn^{2+}.

The Nernst Equation

In the case of the zinc-hydrogen cell, the relation of $\Delta G°$ of reaction 20–10 to the ΔG in Figure 20–4 was given as

$$\Delta G° = \Delta G + RT \ln \frac{[Zn^{2+}]}{[H^+]^2} \qquad \textbf{20–16}$$

where $[Zn^{2+}]$ is 1.0×10^{-4} and $[H^+]$ is 1.0×10^{-5} *M* in the cell. The ΔG's are related to cell voltages through equation 20–8, so that equation 20–16 may also be expressed as

$$\frac{-n\mathscr{F}\mathscr{E}^\circ}{4.185} = \frac{-n\mathscr{F}\mathscr{E}}{4.185} + RT \ln \frac{[Zn^{2+}]}{[H^+]^2} \qquad \textbf{20–17}$$

By rearrangement of equation 20–17 and conversion to the common logarithm, we obtain an expression for the voltage of the cell as a function of reactant and product concentrations:

$$\mathscr{E} = \mathscr{E}^\circ - \frac{(4.185)(2.303)RT}{n\mathscr{F}} \log \frac{[H^+]^2}{[Zn^{2+}]} \qquad \textbf{20–18}$$

At a T of 298 °K, equation 20–18 becomes

$$\mathscr{E} = \mathscr{E}^\circ - \frac{0.0592}{n} \log \frac{[H^+]^2}{[Zn^{2+}]} \qquad \textbf{20–19}$$

We recognize the term $[H^+]^2/[Zn^{2+}]$ in equation 20–19 as part of the equilibrium expression for reaction 20–10. The entire equilibrium expression is $[H^+]^2/[Zn^{2+}]P_{H_2}$, and P_{H_2} does not appear in equation 20–19 because its value is unity in Figure 20–4. For a general cell reaction

$$a\,A + b\,B \rightarrow c\,C + d\,D$$

the cell voltage is

$$\mathscr{E} = \mathscr{E}^\circ - \frac{0.0592}{n} \log \frac{f_C^c f_D^d}{f_A^a f_B^b} \qquad \textbf{20–20}$$

where f is partial pressure in atm for a gas, concentration in moles/liter for a dilute solute, and unity for a pure solid or liquid. Equation 20–20 is known as the **Nernst equation** after the German scientist who first formulated it.

Use of the Nernst Equation

Let us see how the Nernst equation is constructed in an example and how it is used in that case. Suppose that we construct a cell that uses the reaction

$$Ag^+(aq) + Fe^{2+}(aq) \rightarrow Ag(s) + Fe^{3+}(aq) \qquad \textbf{20–21}$$

In one half-cell, $[Ag^+]$ is 1.00×10^{-2} M, and in the other half-cell $[Fe^{2+}] = 1.00 \times 10^{-2}$ and $[Fe^{3+}] = 5.00 \times 10^{-3}$ M. What is the voltage of the cell? The equilibrium expression is $[Fe^{3+}]/[Ag^+][Fe^{2+}]$, and n is 1. The standard voltage is $0.80 - 0.77 = 0.03$ volt. The Nernst equation in this case is

$$\begin{aligned}\mathscr{E} &= 0.03 - (0.0592) \log \frac{[Fe^{3+}]}{[Ag^+][Fe^{2+}]} \\ &= 0.03 - (0.0592) \log \frac{(5.00 \times 10^{-3})}{(1.00 \times 10^{-2})^2} \\ &= -0.07 \text{ volt}\end{aligned} \qquad \textbf{20–22}$$

The negative voltage shows that the spontaneous cell reaction is the reverse of reaction 20–21, and the silver half-cell functions as the anode. As the cell reaction proceeds, the Ag^+ and Fe^{2+} concentrations increase, and the Fe^{3+} concentration decreases. Equilibrium occurs when the cell voltage is zero. What are $[Ag^+]$, $[Fe^{2+}]$, and $[Fe^{3+}]$ at equilibrium? Let the concentration of Fe^{3+} converted to Fe^{2+} at equilibrium be x, so that at equilibrium we have

$$\mathscr{E} = 0 = 0.03 - (0.0592)\log\frac{(5.00 \times 10^{-3} - x)}{(1.00 \times 10^{-2} + x)^2}$$

This equation yields an x of 4.3×10^{-3}, so that equilibrium is reached at $[Fe^{3+}] = 7 \times 10^{-4}$ molar, $[Fe^{2+}] = [Ag^+] = 1.43 \times 10^{-2}$ molar.

Concentration Cells

The most common application of the Nernst equation is in the determination of ion concentrations in solution, particularly when these ions are present in extremely low concentration and would be difficult to measure by other means. We shall illustrate this application by designing a pH meter, which could then be used in acid-base titrations, as discussed in Chapter 19. We shall also use this discussion to illustrate a point previously made, namely that useful work can be obtained from any spontaneous process.

Figure 20–5(a) shows two beakers, each containing 100 ml of HCl solution, one at 0.100 *M* and the other at 1.00×10^{-3} *M*. If the two solutions are mixed, a spontaneous matter spreading occurs, and the result is 200 ml of solution at $[H^+] = 5.05 \times 10^{-2}$ molar. This process can be made to generate useful work by constructing the **concentration cell** shown in Figure 20–5(b). Both half-cells have hydrogen gas at the standard pressure of 1 atm, but neither is a standard half-cell because of the nonstandard $[H^+]$ values.

Can a spontaneous process occur in Figure 20–5(b)? Surely there is a tendency for the $[H^+]$ in the two solutions to equalize, and this can occur by electron transfer, as indicated in the figure. In the half-cell containing the lower acid concentration, $H_2(g)$ is oxidized to two H^+, and in the other half-cell two H^+ are reduced to H_2. As these oxidations and reductions proceed, the $[H^+]$ values in the two solutions approach equality.

What is the voltage of the cell in Figure 20–5? The chemical reaction in the cell is

$$H_2 \text{ (low-acid half-cell)} + 2\,H^+ \text{ (high-acid half-cell)} \rightarrow$$
$$2\,H^+\text{(low-acid half-cell)} + H_2 \text{ (high-acid half-cell)} \quad \textbf{20–23}$$

In this case the Nernst equation is

$$\mathscr{E} = \mathscr{E}^\circ - \frac{0.0592}{2}\log\frac{P_{H_2\text{(high-acid)}}[H^+]^2_{\text{(low-acid)}}}{P_{H_2\text{(low-acid)}}[H^+]^2_{\text{(high-acid)}}} \quad \textbf{20–24}$$

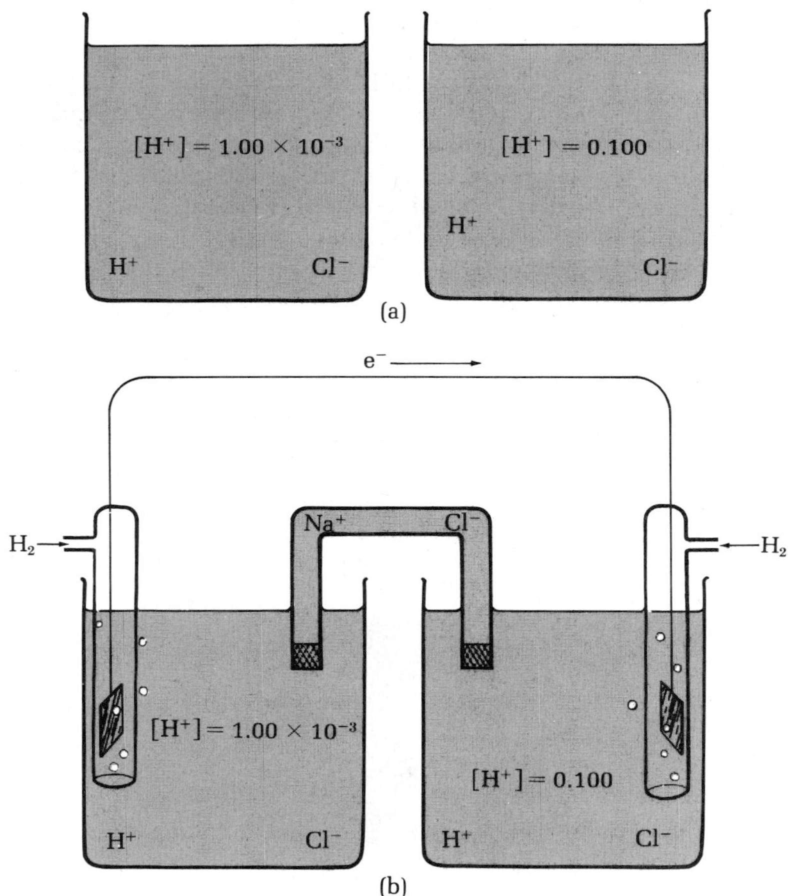

Figure 20–5
(a) Two solutions that are the bases of a spontaneous process. (b) The deriving of work from the spontaneous process.

where $\mathscr{E}^\circ = 0$ (identical standard half-cells) and $P_{H_2(\text{low-acid})} = P_{H_2(\text{high-acid})} =$ 1.0 atm to yield

$$\mathscr{E} = -\frac{0.0592}{2}\log\frac{(1.00 \times 10^{-3})^2}{(0.100)^2} = 0.1184 \text{ volt}$$

This voltage is easily and precisely measurable. Suppose that we know the larger H^+ value because we made up the solution to be 0.100 molar in HCl. The other acid solution may be one of unknown pH. In this case equation 20–24 becomes

$$\mathscr{E} = 0.0592 \text{ pH}_{(\text{unknown})} - 0.0592 \qquad \textbf{20–25}$$

Since the voltage is proportional to the pH of the solution under investigation, a pH meter is simply a high-impedance voltmeter with a scale calibrated in pH instead of volts.

It is clear from the Nernst equation that the hydrogen gas–hydrogen ion half-cell does not have to be used in the construction of a pH meter; rather, any half-cell in which either H^+ or OH^- appears in the balanced half-cell reaction can be used. Nor does the entire cell have to be a concentration cell, as in Figure 20–5; that is, the two standard half-cells need not be identical. A very suitable pH meter could be based on reaction 20–7 because $[H^+]$ is contained in the equilibrium expression for the reaction.

The Glass Electrode pH Meter

In practice, pH is measured much more simply than it would be in the arrangement shown in Figure 20–5(b). The practical pH meter takes advantage of the properties of thin glass membranes. Suppose that we arranged the two solutions of Figure 20–5(a) in the manner shown in Figure 20–6, where they are separated by a thin glass membrane. The glass membrane in this

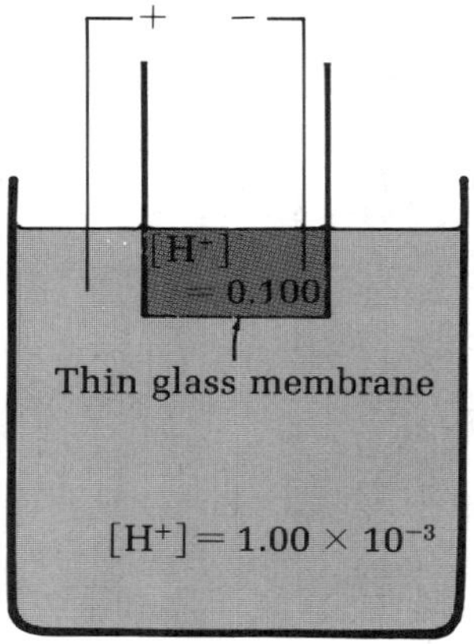

Figure 20–6 The generation of a voltage between two acid solutions through use of a glass membrane.

case can effectively allow hydrogen ions to pass, but not other ions. (We shall see a mechanism for this proton transmission shortly.) The passage of hydrogen ions occurs from the more concentrated solution to the less concentrated solution in the spontaneous matter-spreading process. However, since the anions cannot pass across the membrane, an excess of negative electrical charge builds in the 0.100 molar solution and an excess of positive charges builds in the dilute solution. After the passage of only a very small number of hydrogen ions, the ion passage will stop, and an equilibrium is reached with a voltage between the two solutions of 0.1184 volt. In practice, no electrode is inserted in the solution of unknown pH. Rather, the electrical potential generated in the half-cell of known $[H^+]$ is coupled to another half-cell of known voltage and the voltage of this complete cell is measured. Since the potential generated in the solution of known acid concentration on one side of the glass membrane is a function of the unknown acid concentration, so is the measured voltage, which can be calibrated to read directly the previously unknown pH. The glass membrane procedure has the advantage that essentially no contamination of the unknown solution occurs during the pH measurement. As shown in Figure 19–3, a small glass tube containing the glass membrane and reference half-cell is lowered into the solution whose pH is to be measured. Only glass surfaces come in contact with the unknown acid solution.

The glass membrane has been shown to transmit hydrogen ions in a novel way that is useful in analyzing for other ions. The hydrogen nucleus, tritium, made up of a proton and two neutrons, is radioactive and this radioactivity makes its location easy to detect. Furthermore, even a single radioactive nuclear decay is detectable, and consequently extremely small quantities of radioactive tracers can be detected. When tritiated HCl is used in the preparation of a 0.100 *M* solution, as shown in Figure 20–6, no radioactivity is detected in the dilute solution. Therefore, no hydrogen ions actually pass through the glass membrane. Apparently the glass membrane effectively transmits protons by displacement, as shown in Figure 20–7. Glass surfaces contain Si–OH groups that are weak acids. These groups lose protons on the

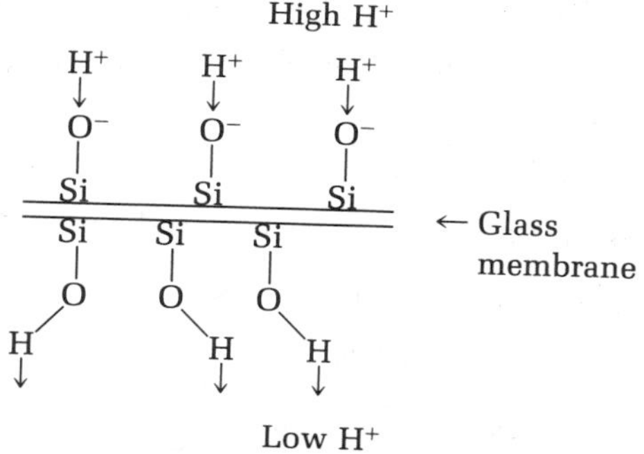

Figure 20–7
The protonation (high acid side) and deprotonation (low acid side) of a thin glass membrane in a glass electrode pH meter.

low-acid side of the membrane to form SiO^- groups. This negative charge is detectable by the species at the opposite glass surface and these species accept protons from the 0.100 molar $[H^+]$ solution to a greater degree than they would if the first surface had not been deprotonated. The changes at the two surfaces may be expressed as

$$H^+\text{(high-acid side)} + \text{glass surface(high-acid side)} \rightarrow \text{protonated surface(high-acid side)}$$

$$\text{Glass surface(low-acid side)} \rightarrow H^+\text{(low-acid side)} + \text{deprotonated surface(low-acid side)}$$

The net result of these two steps is to lower $[H^+]$ in the high-acid solution and to increase $[H^+]$ in the low-acid solution. This is effectively transmission of protons across glass surfaces. In fact, glass surfaces can bind many cations other than protons and effectively transmit them by the same mechanism as with protons. Thus, a pH meter can be adapted with the proper membrane to measure any one of a number of other cations (Na^+, K^+, and Ca^{2+} are examples of cations whose concentrations have been measured with the use of such membranes).

Quantitative Analysis

We can carry out the analysis of acetic acid in aqueous solution by titration with sodium hydroxide solution because the equilibrium constant for the following reaction is quite large:

$$H_3CCOOH(aq) + OH^-(aq) \rightarrow H_3CCOO^-(aq) + H^+(aq)$$

and the reaction between acetic acid and hydroxide ions is nearly quantitative.

Many redox reactions also have large equilibrium constants and they are the basis for quantitative analyses. An example is the oxidation of Fe^{2+} by MnO_4^- according to equation 20–7:

$$5\ Fe^{2+}(aq) + MnO_4^-(aq) + 8\ H^+(aq) \rightarrow Mn^{2+}(aq) + 5\ Fe^{3+}(aq) + 4\ H_2O(l)$$

The $\mathscr{E}°$ for this reaction was calculated previously in this chapter to be +0.72 volt, and this yields a $\Delta G°$ as follows:

$$\Delta G° = \frac{-n\mathscr{F}\mathscr{E}°}{4.185} = \frac{-(5)(96{,}500)(0.72)}{(4.185)} = -83 \text{ kcal}$$

The equilibrium constant at 298 °K is calculated as:

$$K_e = e^{-\Delta G°/RT} = 10^{-[(-83{,}000)/(1.987)(298)(2.303)]}$$
$$= 1 \times 10^{61}$$

This huge value shows that MnO_4^- oxidizes Fe^{2+} quantitatively to Fe^{3+}, and Fe^{2+} can be titrated with a solution of MnO_4^- (in the form of $KMnO_4$). Potassium permanganate is highly colored and the equivalence point is easy to locate since it is the point at which the solution titrated shows the first permanent color.

Commercial Cells

From the Nernst equation, we see that the voltages of cells normally change with time as the concentrations of the products of the cell reaction increase and the concentrations of reactants decrease. A varying voltage is a severe drawback in a cell that is to be put to practical use. The solution to this problem is to use reactants and products that are either pure liquids or solids and consequently do not appear in the equilibrium expression, or to buffer the concentrations of dissolved species so that they change only slightly as the cell reaction proceeds.

The lead storage battery is perhaps the most common example of the use of solids and buffered solutes in electrical cells. The chemical reaction that serves as the basis of the lead storage battery is

$$Pb(s) + PbO_2(s) + 4\ H^+(aq) + 2\ SO_4^{2-}(aq) \rightarrow 2\ PbSO_4(s) + 2\ H_2O(l) \qquad \textbf{20–26}$$

As this reaction stands, the voltage should decrease steadily in operation as $[H^+]$ and $[SO_4^{2-}]$ decrease. Both concentrations must be buffered if the voltage is to remain constant in normal operation. Both bufferings occur by adding a very high concentration of H_2SO_4. If both $[H^+]$ and $[SO_4^{2-}]$ are initially very high, a large quantity of lead and lead dioxide can be consumed without a large fractional change in $[H^+]$ or $[SO_4^{2-}]$. The high $[H^+]$ and $[SO_4^{2-}]$ also increase the voltage of the cell since both are reactants, and an increase in reactant concentrations makes the ΔG of the reaction more negative. An approximate voltage of the lead storage cell is obtainable from the following two half-cell potentials listed in Table 20–1.

$$PbO_2 + SO_4^{2-} + 4\ H^+ + 2\ e^- \rightarrow 2\ H_2O + PbSO_4 \qquad \mathscr{E}° = 1.69$$

$$PbSO_4 + 2\ e^- \rightarrow Pb + SO_4^{2-} \qquad \mathscr{E}° = -0.36$$

Combination of these two values yields $\mathscr{E}° = 1.69 - (-0.36) = 2.05$ as the voltage of the standard cell. The actual cell has a higher voltage (2.15 volts) because of the high values of $[H^+]$ and $[SO_4^{2-}]$. A "battery" is a collection of such cells linked in series so that the voltages add. A "twelve-volt battery" for an automobile contains six such cells.

A second common cell that employs solids is the dry cell used in flashlights. A cross section of the cell is shown in Figure 20–8. The cell really is not dry at all but contains a paste of water, MnO_2, NH_4Cl, $ZnCl_2$, and

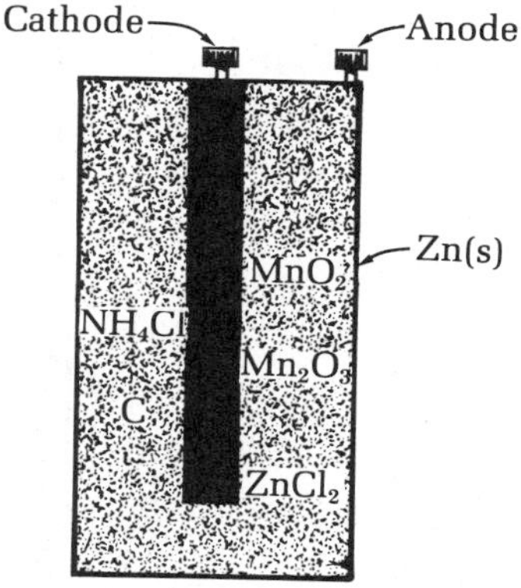

Figure 20–8
Cross section of a dry cell.

carbon between a carbon rod in the center of the cell and a zinc can that serves to enclose the cell. The cell reaction is

$$2\ MnO_2(s) + Zn(s) + 2\ NH_4^+(aq) \rightarrow$$
$$Zn^{2+}(aq) + Mn_2O_3(s) + 2\ NH_3(aq) + H_2O(l) \qquad \textbf{20–27}$$

Thus, the zinc can is the anode, as shown in Figure 20–8, and the carbon rod serves as the cathode. The manganese product is a complex mixture of Mn(III) species, which is represented in reaction 20–27 by a single species, Mn_2O_3.

20–3. The Reversing of Cell Reactions

Electrolysis

We have shown that spontaneous processes can be used efficiently in cells to obtain useful work. The reaction used in a hydrogen-oxygen fuel cell is

$$2\ H_2(g) + O_2(g) \rightarrow 2\ H_2O(l) \qquad \textbf{20–28}$$

What is the voltage of the standard hydrogen-oxygen fuel cell? The two half-cell reactions and the standard potentials are

$$2\ H^+ + 2\ e^- \rightarrow H_2 \qquad \mathscr{E}^\circ = 0$$

$$O_2 + 4\ H^+ + 4\ e^- \rightarrow 2\ H_2O \qquad \mathscr{E}^\circ = 1.23$$

If the fuel cell is set up as shown in the simplified fashion of Figure 20–9, the voltage is 1.23 volts. In this case, the salt bridge is produced by adding NaCl to the common solution for both half-cells.

Suppose that we inserted a battery on the wire connecting the two half-cells shown in Figure 20–9, with the negative junction connected to the hydrogen half-cell and the positive junction to the oxygen half-cell. If the voltage of this battery were to exceed 1.23 volts, the spontaneous reaction in the hydrogen-oxygen cell would be the reverse of reaction 20–28. This reversal of any natural cell reaction by the input of electrical energy is termed **electrolysis.**

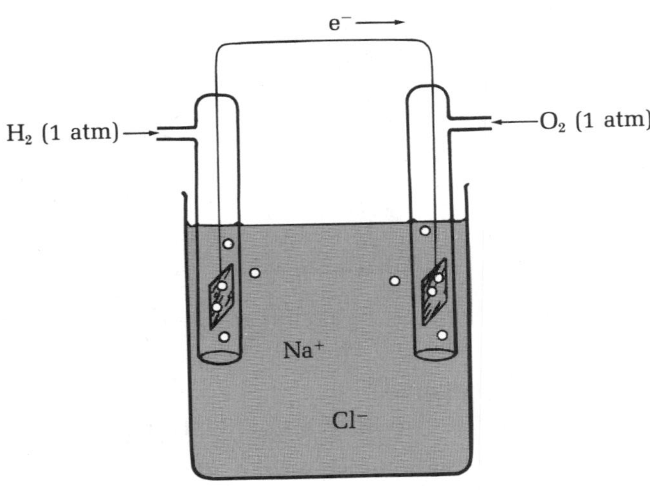

Figure 20–9
Simplified representation of a hydrogen-oxygen fuel cell.

Electrolysis has long been used as the principal means (and in some cases the only means) for the preparation of very strong oxidizing agents and reducing agents. Some substances commonly prepared by electrolysis are F_2, Cl_2, Li, Na, K, Ca, Mg, and Al. Figure 20–10 shows the electrolytic process developed in the last century by the American inventor Charles M.

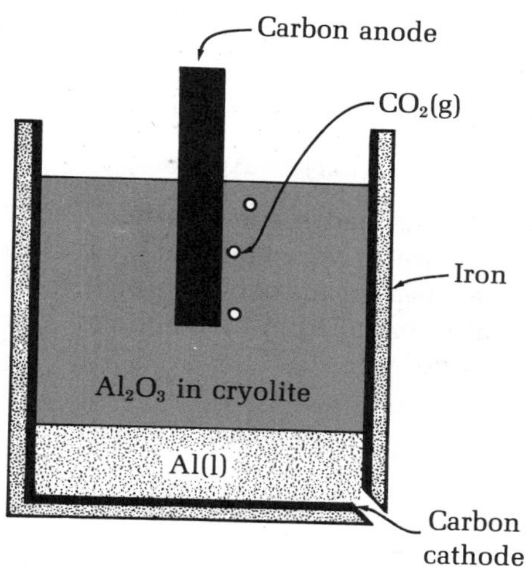

Figure 20–10 Electrolysis apparatus for the production of metallic aluminum.

Hall, for the production of metallic aluminum from Al_2O_3. Aluminum oxide dissolves in cryolite, Na_3AlF_6, at elevated temperature, and it is this solution that is electrolyzed. The carbon electrodes also participate in the reaction and the reaction equation is

$$2\ Al_2O_3(\text{in cryolite}) + 3\ C(s) \rightarrow 4\ Al(l) + 3\ CO_2(g) \qquad \textbf{20–29}$$

Electroplating

Consider the arrangement in Figure 20–11. We should like to transfer silver from the bar at the left to the tray on the right. This can be accomplished by

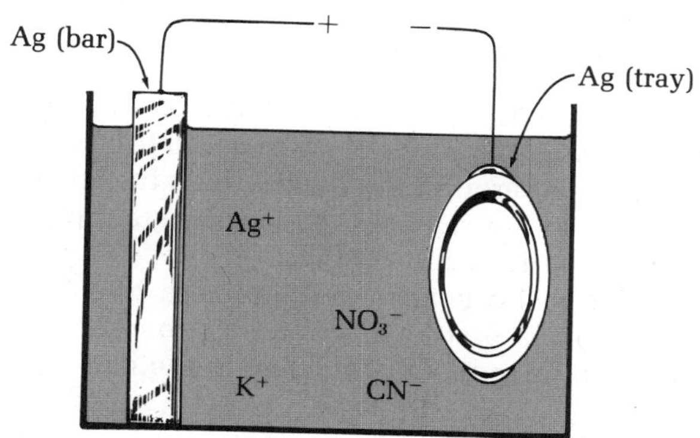

Figure 20–11 Apparatus for silver plating.

applying the voltages as shown in the figure. Electrons are extracted from the bar at the left and the anode reaction is

$$Ag\ (bar) \rightarrow Ag^+(aq) + e^-\ (to\ wire)$$

These electrons are accepted by Ag^+ ions at the cathode tray in the following reaction:

$$Ag^+(aq) + e^-\ (from\ wire) \rightarrow Ag\ (tray)$$

The sum of these two half-cell reactions is the desired reaction

$$Ag\ (bar) \rightarrow Ag\ (tray)$$

The transfer process is termed **electroplating**, and it is commonly performed with relatively nonreactive metals such as silver, gold, copper, and chromium. It is also used as a means of obtaining these metals in high purity by transferring the desired metal from the anode to the cathode and leaving the impurities at the anode.

20–4. Corrosion

The Rusting of Iron

Oxygen is undoubtedly the most important oxidizing agent to us because it is the oxidation of foodstuffs by oxygen that supplies the energy with which we function. Unfortunately, oxygen also participates in other, less desirable redox reactions. One such reaction is

$$4\ Fe(s) + 3\ O_2(g) + 6\ H_2O(l) \rightarrow 4\ Fe(OH)_3(s) \qquad \textbf{20–30}$$

The product of the reaction is rust. The rusting of iron is an example of **corrosion**, a widespread and expensive consequence of the reactivity of molecular oxygen. Rusting occurs by a rather unusual process. It actually occurs in two steps, and these are the following:

$$2\ Fe(s) + O_2(g) + 2\ H_2O(l) \rightarrow 2\ Fe(OH)_2(s) \qquad \textbf{20–31}$$

$$4\ Fe(OH)_2 + O_2(g) + 2\ H_2O(l) \rightarrow 4\ Fe(OH)_3(s) \qquad \textbf{20–32}$$

The first of these two steps occurs electrochemically, which can be shown by placing a drop of water on a steel surface and letting it slowly evaporate. A ring of rust will be left around the place where the edge of the drop was and a pit will be left in the steel at the place where the center of the drop was. This is represented in Figure 20–12.

Let us see how the rust ring is formed in Figure 20–12. Oxygen dissolves in water and it dissolves first at the water-air interface. As O_2 dissolves, oxygen molecules can be reduced according to the half-cell reaction

$$O_2(g) + 2\ H_2O(l) + 4\ e^- \rightarrow 4\ OH^-(aq)$$

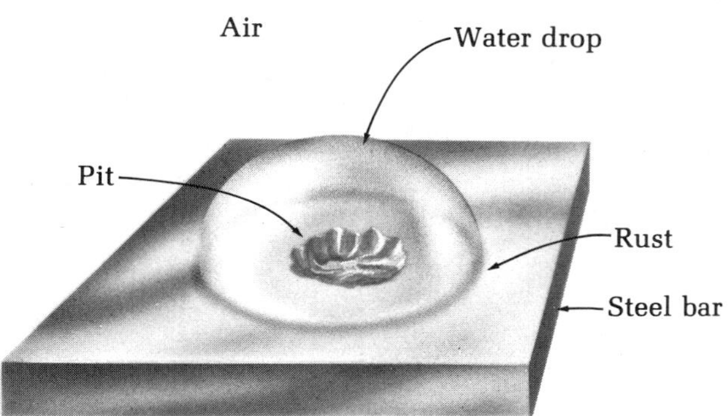

Figure 20–12
The rusting of steel caused by oxygen and a water drop.

If these electrons cannot be supplied by some reducing agent, this half-cell reaction cannot occur. However, the iron in the steel bar is a reducing agent and the iron near the center of the drop is oxidized to Fe^{2+} with the electrons transferring through the bar to O_2 molecules near the edge of the drop. The Fe^{2+} ions and the hydroxide ions (which are the product of oxygen reduction) then diffuse together through the water drop to combine to form insoluble $Fe(OH)_2$. The $Fe(OH)_2$ then reacts with gaseous oxygen to produce the red brown $Fe(OH)_3$. The cell that is equivalent to the arrangement in Figure 20–12 is shown in Figure 20–13.

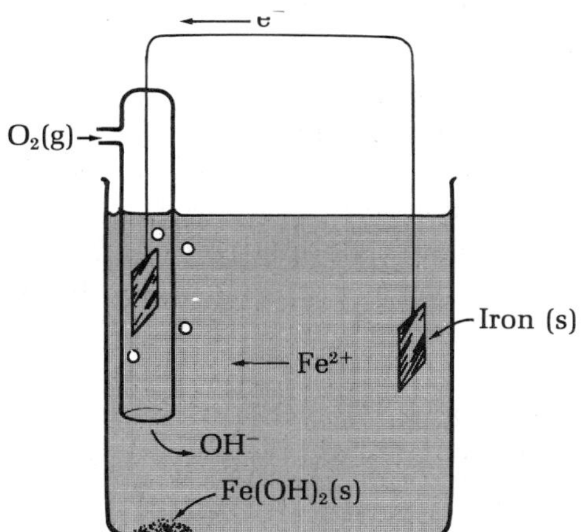

Figure 20–13
A cell based on the electrochemical reaction involved in the rusting of iron.

Protection against Corrosion

The most obvious way to protect a metal such as iron from corrosion is to prevent its contact with water and oxygen. This can be done by painting or plating with another metal (Cr, Cu, Zn, and Sn are commonly used).

The electrochemical nature of corrosion is particularly well illustrated by a type of protection from corrosion known as **cathodic protection.** This method of protection is particularly useful for iron objects that must be located underground, where they are in constant contact with moist earth. The iron object is connected by a wire to a block of a metal such as Mg, which is more easily oxidized than the iron. The oxidation of iron is completely replaced by the oxidation of Mg, and the magnesium can be periodically replaced to maintain the protection.

SUGGESTIONS FOR ADDITIONAL READING

Masterton, W. L., and Slowinski, E. J., *Chemical Principles*, W. B. Saunders, Philadelphia, 1973.

Bauman, R. P., *An Introduction to Equilibrium Thermodynamics*, Prentice-Hall, Englewood Cliffs, N.J., 1966.

PROBLEMS

1. From the data given in tables in Chapters 15 and 16 calculate the maximum possible efficiency for obtaining work from each of the following reactions at 25 °C. Assume standard conditions.
   a. $2\ H_2(g) + O_2(g) \rightarrow 2\ H_2O(g)$
   b. $2\ Na(s) + Cl_2(g) \rightarrow 2\ NaCl(s)$
   c. $2\ C_6H_6(l) + 15\ O_2(g) \rightarrow 12\ CO_2(g) + 6\ H_2O(g)$
2. Write a balanced equation for the anode reaction in the methane-oxygen fuel cell. Propose an explanation to account for the slowness of this reaction at room temperature.

*3. A functioning methane-oxygen fuel cell has been developed that uses molten KOH as the solvent in place of water. In view of the answer to problem 2, explain the advantage of using this extremely strong base as the electrolyte.

4. Balance each of the following reactions by using the half-cell method:
   a. $Fe^{2+}(aq) + Cr_2O_7^{2-}(aq) \rightarrow Fe^{3+}(aq) + Cr^{3+}(aq)$
   b. $CH_4(g) + Cl_2(g) \rightarrow CCl_4(l) + HCl(g)$
   c. $F_2(g) + H_2O(l) \rightarrow HF(g) + O_2(g)$
   d. $ClO^-(aq) + I_3^-(aq) \rightarrow Cl^-(aq) + I_2(s)$
5. What is *n* in equation 20–8 in each of the following balanced reaction equations?
   a. $2\ Fe^{3+}(aq) + 2\ I^-(aq) \rightarrow 2\ Fe^{2+}(aq) + I_2(s)$
   b. $3\ H_2(g) + N_2(g) \rightarrow 2\ NH_3(g)$
   c. $Cu(s) + 4\ H^+(aq) + 2\ NO_3^-(aq) \rightarrow Cu^{2+}(aq) + 2\ NO_2(g) + 2\ H_2O(l)$
   d. $2\ C_6H_6(l) + 15\ O_2(g) \rightarrow 12\ CO_2(g) + 6\ H_2O(g)$
6. From the ΔG_f° values listed in Tables 16–1 and 16–2, verify the standard

reduction potentials in Table 20–1 for the following reactions:
a. $Mg^{2+} + 2\ e^- \rightarrow Mg$
b. $Fe^{3+} + e^- \rightarrow Fe^{2+}$
c. $O_2 + 4\ H^+ + 4\ e^- \rightarrow 2\ H_2O$

7. Calculate the voltage of the standard cells based on the following reactions:
a. $Cu(s) + 2\ H^+(aq) \rightarrow Cu^{2+}(aq) + H_2(g)$
b. $Hg^{2+}(aq) + Hg(l) \rightarrow Hg_2{}^{2+}(aq)$
c. $Cr_2O_7{}^{2-}(aq) + 6\ Fe^{2+}(aq) + 14\ H^+(aq) \rightarrow$
$2\ Cr^{3+}(aq) + 6\ Fe^{3+}(aq) + 7\ H_2O(l)$
Which of the reactions proceed in the standard cell as written, and which proceed in the opposite direction?

8. Complete and balance each of the following reaction equations:
a. $Cr^{2+}(aq) + H^+(aq) \rightarrow$
b. $Cl_2(g) + Br^-(aq) \rightarrow$
c. $F_2(g) + H_2O(l) \rightarrow$

9. Suppose that the half-cell at the right in Figure 20–4 were buffered at pH 6.0. What would the cell voltage be in that case?

10. Why is it convenient to choose H^+ as the standard ion in Table 16–2 rather than some other ion?

11. Consider a cell based on reaction 20–14. The Ag^+ and Cu^{2+} concentrations are both 0.0100 *M*. What is the cell voltage? Which half-cell is the anode?

12. Draw cells in the manner shown in Figures 20–1 and 20–2 that use each of the following reactions:
a. $Fe^{3+}(aq) + Ag(s) + Cl^-(aq) \rightarrow Fe^{2+}(aq) + AgCl(s)$
b. $H_2(g) + Br_2(l) \rightarrow 2\ H^+(aq) + 2\ Br^-(aq)$
c. $Ni(s) + Cu^{2+}(aq) \rightarrow Cu(s) + Ni^{2+}(aq)$

13. Design a cell that uses the following reaction:

$$2\ Ag^+(aq) + Ni(s) \rightarrow Ni^{2+}(aq) + 2\ Ag(s)$$

and that has a potential of exactly 1.00 volt. This requires that you specify values for $[Ni^{2+}]$ and $[Ag^+]$.

14. Consider the cell shown in Figure 20–14. What is its voltage? Which half-cell is the anode?

***15.** What is the ΔG of the process that occurs when the two solutions of Figure 20–5(a) are mixed? The volume of each solution is one liter.

16. Suppose that the solution in the half-cell at the right in Figure 15–5(b) is known to be 0.100 *M* in H^+ but the pH of the other solution is unknown. The cell voltage is found to be 0.3120 volt with the unknown solution half-cell serving as the anode. What is the pH of the unknown solution?

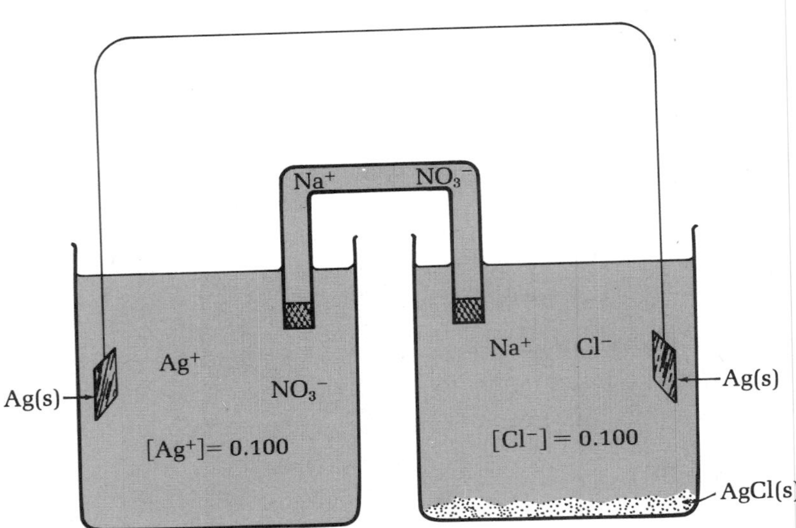

Figure 20–14
A silver concentration cell.

17. Molecular oxygen is the oxidizing agent in the oxidation of foodstuffs in living systems. Table 20–1 lists the standard reduction potential of O_2 in going to H_2O. What is this reduction potential at $P_{O_2} = 0.20$ atm and pH = 7.4?

18. If the glass membrane electrode of a pH meter is lowered into a beaker containing 0.0100 *M* HCl in water, the meter will read pH = 2.00. Suppose that the HCl were dissolved in liquid ammonia instead of water to make a 0.0100 *M* solution. What would you expect the meter to record if the glass membrane electrode were lowered into this new solution—pH = 2, or a number higher or lower than this value? Explain.

19. The electrical charge stored in a lead storage battery is commonly checked with a hydrometer, which measures the density of the solution in the battery. How does this density relate in a qualitative fashion to the degree of charge of the battery?

20. If a very reactive light metal such as lithium were used as the anode material in a storage battery, a much greater amount of energy could be stored in a lighter battery than is the case with the lead storage battery. What is a serious drawback with Li that has precluded its use in batteries?

21. Why do the concentrations of $Zn^{2+}(aq)$ and $NH_3(aq)$ not build up as the spontaneous reaction in a dry cell proceeds? *Hint:* Consider the electrophilic character of Zn^{2+}.

22. Aluminum has been referred to as "congealed electricity." An ampere is an electrical current of one coulomb per second, and a current of 1000 amperes is quite large. In the production of metallic aluminum by electrolysis, how long must a 1000-amp current flow through the

electrolysis cell to produce one ton of aluminum? Remember that the electrons in one mole have a combined charge of 96,500 coulombs.

23. Why is aluminum not produced by electrolyzing an aqueous solution of an aluminum salt?

24. In silver and gold plating, it is desirable to have the plating process occur slowly so that the coating is smooth and uniform. This is accomplished by dissolving KCN in the electrolyte solution, as shown in Figure 20–11. How does the presence of cyanide ion in solution slow the plating process?

***25.** The rusting of iron in Figure 20–12 occurs much more rapidly if the drop is one of salt water rather than pure water. Propose an explanation.

21/ Chemical Kinetics

In the discussion of fuel cells it was pointed out that there is no correlation between $\Delta G°$ for a reaction and the rate of that reaction. If the rate does not depend on $\Delta G°$, on what factors does the rate depend? To answer this question, we must be able to construct a model for the process by which any given set of reactants yields products. The model chemists use most commonly is a series of discrete steps involving the formation and subsequent reaction of molecules that are **intermediates** between reactants and products. The listing of these intermediates and the steps for their formation and reaction is the formulation of a **reaction mechanism.**

The construction of reaction mechanisms is quite likely the most difficult task that is attempted in this text. Particular reaction mechanisms are probably the subject of more heated debate than any other models in chemistry. Nevertheless, if we are to acquire any insight into the reasons why some reactions are rapid and others are slow, we must do so within a framework of particular reaction mechanisms, even though these mechanisms are subject to debate.

Our aim is twofold. We first want to display the types of measurements and reasoning that are involved in formulating reaction mechanisms. Secondly, we want to use these mechanisms to gain insight into reaction rates and the factors that determine them. The "correctness" of any of the mechanisms presented is not an issue. The mechanisms presented are plausible ones, and open to debate as all mechanisms are open to debate.

The study of reaction rates and of the factors that influence reaction rates is termed **chemical kinetics.**

21-1. Activation—Why Some Reactions Are Slow

Three Factors That Determine Reaction Rates

The following two reactions are good examples of the lack of correlation between $\Delta G°$ and reaction rate:

$$H_2(g) + Cl_2(g) \rightarrow 2\,HCl(g) \tag{21-1}$$

$$H^+(aq) + OH^-(aq) \rightarrow H_2O(l) \tag{21-2}$$

The $\Delta G°$ of reaction 21–1 is $2\,\Delta G°_{f(HCl,g)} = -45.54$ kcal, while $\Delta G°$ for reaction 21–2 is $\Delta G°_{f(H_2O,l)} - \Delta G°_{f(OH^-,aq)} = -19.10$ kcal. While the $\Delta G°$ of reaction 21–1 is more negative than $\Delta G°$ for reaction 21–2, reaction 21–1 proceeds at an imperceptibly slow rate at 298 °K in the absence of light. Reaction 21–2 proceeds as rapidly as a solution of an acid can be mixed with a solution of a base.

To understand how this huge difference in reaction rates arises, we must picture the way in which the two reactions proceed on a molecular level. We shall do such an analysis of reaction 21–2 to show that there are three key factors that determine whether a reaction is rapid or slow. Reaction 21–2 is one of many extremely rapid reactions that the German scientist Manfred Eigen has studied by using methods devised in the 1950s. The mechanism to be discussed has been proposed for reaction 21–2 from Eigen's investigation of the reaction.

The first factor that enters when two or more species such as $H^+(aq)$ and $OH^-(aq)$ react is **encounter**, since the species cannot react if they do not encounter each other. Both H^+ and OH^- are associated with water molecules in aqueous solution. Thus, the ions that encounter each other in reaction 21–2 each contain a certain number of bound water molecules. One such encounter in the case of reaction 21–2 occurs when H^+ and OH^- are separated by two water molecules, as shown in Figure 21–1.

Figure 21–1 also shows the second important aspect of a reaction mechanism. The probability of reaction depends not only on encounter but also on the relative orientation of the two species. One can suggest that in the case of Figure 21–1, an optimum orientation might involve the formation of a hydrogen bond at the horizontal dashed line separating $H_9O_4^+$ from $H_7O_4^-$. **Orientation** is our second important factor.

Reaction 21–1 appears to involve the simple formation of an O—H bond from H^+ and OH^-. However, Figure 21–1 shows that, as is almost invariably the case in reaction mechanisms, one or more bonds must be broken so that others may be formed. In Figure 21–1, three proton transfers occur (as indicated by the arrows), three O—H bonds (indicated by the asterisks) are broken, and three new O—H bonds (indicated by the daggers) are formed. **Bond breaking** is the third (and usually the most important) of the factors that determine reaction rates.

The processes of encounter and orientation are matter concentrating and the process of bond breaking, which must precede or accompany bond making, involves energy concentrating. Both concentratings decrease $S_{universe}$ (and increase G_{system} if they occur at constant temperature and pres-

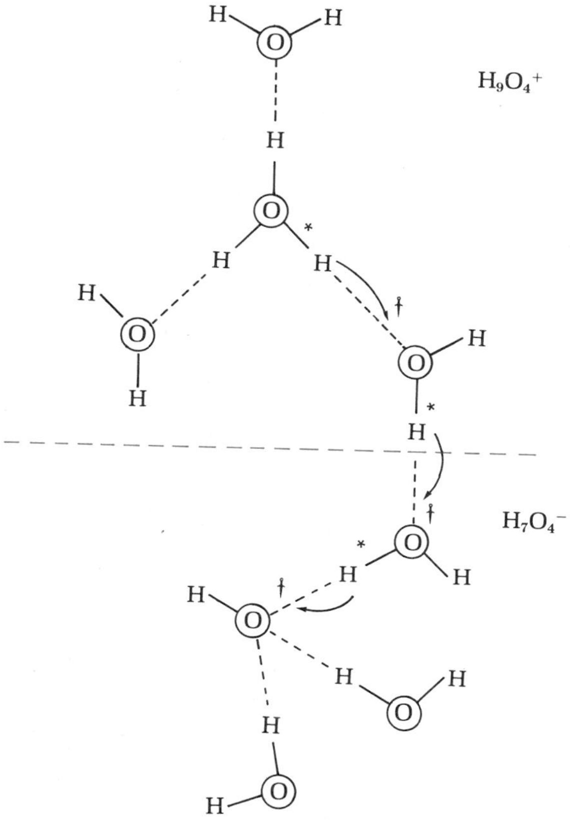

Figure 21-1
A proposed mechanism for the reaction of H^+ and OH^- in aqueous solution. The asterisks indicate O—H bonds broken and the daggers indicate O—H bonds formed.

sure). Thus, all reactants face a barrier in the Gibbs free energy that must be surmounted for reaction to occur. The situation is diagrammed in Figure 21-2 in a plot of the Gibbs free energy as a function of the distance along the reaction path for a typical reaction. The reaction path is some appropriate set of nuclear position coordinates that describe encounter, orientation, bond breaking, and bond making. Every reaction path contains a point indicated by the double dagger in Figure 21-2 that corresponds to the maximum G, the maximum matter and energy concentration. The point is termed the **activated state** (or transition state), and the difference between $G^{\ddagger}$ and $G_{\text{reactants}}$ is termed the **Gibbs free energy of activation** (commonly symbolized by $\Delta G^{\ddagger}$). It is the value of $\Delta G^{\ddagger}$ rather than the value of ΔG° for the reaction that determines the rate of reaction. Subsequently, we shall develop a quantitative relation between reaction rate and $\Delta G^{\ddagger}$.

A Key to Rapid Reactions—Simultaneous Bond Making and Bond Breaking

In the case of reaction 21-2, the activation barrier ($\Delta G^{\ddagger}$) must be quite low

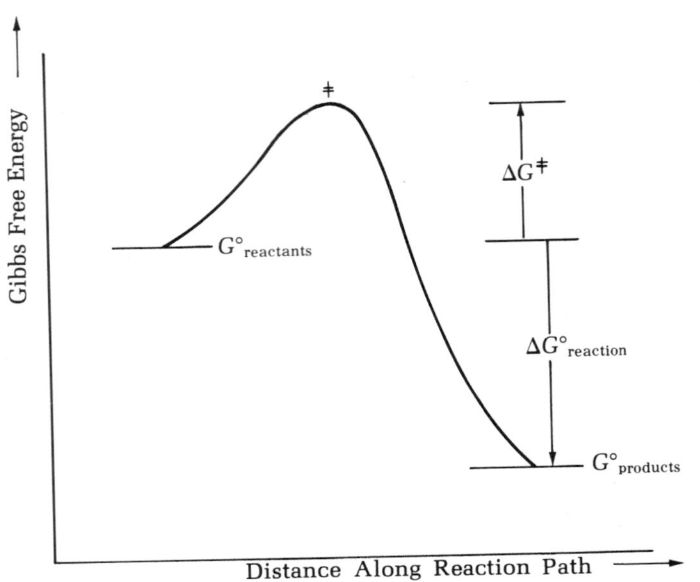

Figure 21–2
The variation of Gibbs free energy between reactants and products along a typical reaction path.

because the reaction is so rapid. As we shall see, the encounter between two species present at their standard concentrations (or pressures in the case of gases) does not present a significant reaction barrier at room temperature since molecular motion is quite rapid. The problem of orientation does not arise with reaction 21–2 since water molecules normally hydrogen-bond to each other in water, and consequently $H_9O_4^+$ and $H_7O_4^-$ easily associate with each other in the manner shown in Figure 21–1. Finally, there is the matter of bond breaking and making. The bonds shown by the asterisks in Figure 21–1 are stretched in the bond-breaking process and the energy increases, as shown in Figure 12–9. However, as one O—H bond is stretched, the corresponding O—H bond, indicated by a dagger in Figure 21–1, is formed. The energy along the reaction path for proton transfer between two oxygen atoms of Figure 21–1 is displayed in Figure 21–3. The oxygen nuclei labeled O* and $O^†$ are anchored at a fixed distance from each other. The proton can move anywhere on the line between them, and the distance along the line is plotted in the figure as r_{O^*-H}, the distance from the proton to O*. This process of proton transfer between O* and $O^†$ does not involve either matter spread or matter concentration, and ΔG for it arises strictly from energy changes. The colored curves in Figure 21–3 are potential energy curves (of the type shown in Figure 12–9) for O*—H and H—$O^†$ bonds in the absence of hydrogen-bonding to the other oxygen atom. The black curve is the sum of the two colored curves (remember that potential energies of bonded species are negative). The black curve shows two minima. The minimum at the left corresponds to O*—H · · · $O^†$ and that at the right cor-

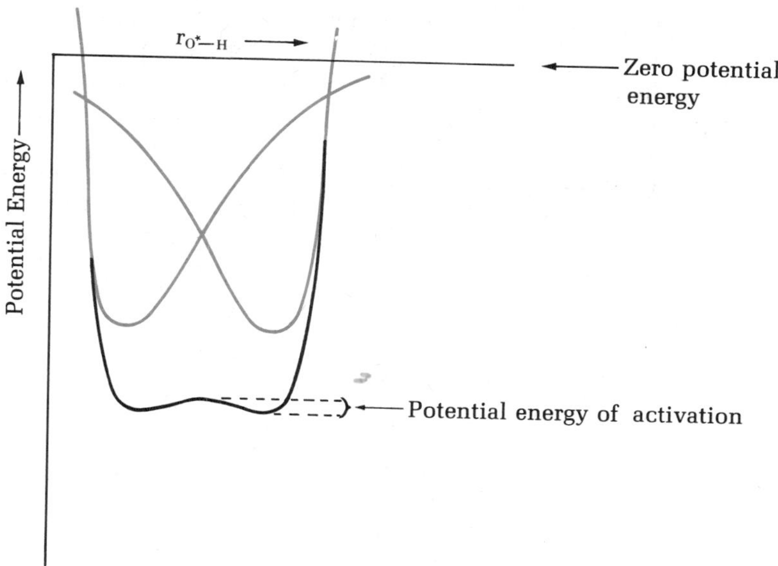

Figure 21–3
Potential energy as a function of O—H distance for a proton located between two stationary oxygen atoms.

responds to $O^* \cdots H—O^+$. The highest potential energy between these two minima occurs when the proton is exactly halfway between the two oxygen atoms, and the energy difference between this maximum and the two equal minima is the potential energy of activation for proton transfer. This activation barrier can be very small and if it is small, the proton can rapidly transfer from one minimum to the other.

A Key to Slow Reactions—Isolated Bond Breaking

Now that we have seen the three factors—encounter, orientation, and bond breaking—involved in activation, let us try to discern the reason for the slowness of reaction 21–1 at room temperature. As with the encounter of H^+ and OH^-, the encounter of H_2 and Cl_2 molecules when both are present at the standard pressure of 1 atm offers no significant barrier to the reaction. Both molecules are simple in structure, with a relatively small number of significantly different orientational possibilities, and the orientational factor offers no significant obstacle to the reaction. The bond-breaking barrier depends on the reaction mechanism, and the reaction will surely proceed by the mechanism that offers the smallest barrier due to bond breaking. One possible mechanism is

$$H_2(g) + Cl_2(g) \rightarrow 2\,HCl(g) \qquad \textbf{21–3}$$

You might ask what is the difference between equations 21–1 and 21–3, since they appear to be the same. The difference is that we have identified equation 21–1 as an overall reaction equation and equation 21–3 as an elementary reaction step. A reaction equation gives the reactants and products in an overall reaction, but it gives no information about the steps through which the reaction proceeds. A reaction step such as equation 21–3 shows which species encounter each other in that step and which species result from that encounter.

One picture of the one-step mechanism 21–3 is given in Figure 21–4. If the H_2 and Cl_2 molecules encounter each other in the proper orientation,

H H → H H

Cl Cl → Cl Cl

(a)

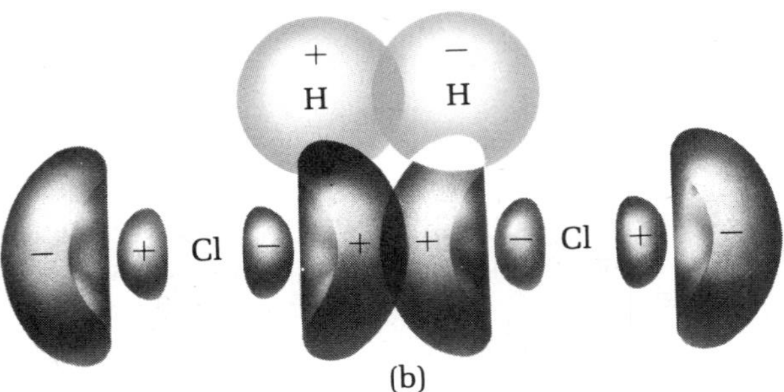

(b)

Figure 21–4
A proposed electron transfer for the conversion of H_2 and Cl_2 to 2 HCl (a). One of the orbital overlaps involved in the electron transfer (b).

as shown, it would appear that four electrons can easily change positions as shown in Figure 21–4(a); and two HCl bonds form as the H_2 and Cl_2 bonds break. This situation of simultaneous bond making and breaking would appear to be similar to that in Figure 21–3, and we should predict that reaction 21–1 is very rapid. Since this conclusion is incorrect, we must investigate the electron transfer in Figure 21–4(a) more closely. The electron transfer from Cl_2 results in the breaking of the H—H bond at the same time that H—Cl bonds are being formed. This requires that there be bonding overlap between the $3p\sigma$ orbital of Cl_2 and the $1s\sigma^*$ orbital of H_2. Such an overlap would allow the transfer of two electrons from the region between the Cl atoms to the regions between H and Cl. The acceptance of these electrons by the $1s\sigma^*$ orbital would cancel the bonding of the two electrons in the $1s\sigma$ orbital of H_2, and the H_2 bond would break. A similar picture must be drawn of electron transfer from the $1s\sigma$ orbital of H_2 to the $3p\sigma^*$ orbital of Cl_2 to result in the rupture of the Cl—Cl bond. Unfortunately, the overlap shown in Figure 21–4(b) yields no net H—Cl bonding, since there is reinforcing overlap at the left and cancelling overlap at the right. The electron transfer in Figure 21–4(a) simply cannot occur to yield a low-energy barrier. Reactions are almost invariably slow at room temperature unless the bonds that must be broken are very weak or unless the breaking of strong bonds is accompanied by the simultaneous formation of bonds equally strong, or stronger.

The overlap problem can be overcome in the case of reaction 21-1 by splitting either H_2 or Cl_2 into constituent atoms. Since the Cl_2 bond enthalpy is approximately half the H_2 bond enthalpy, it is the breaking of the Cl_2 bond that initiates the reaction. However, the $\Delta G°$ for the breaking of the Cl_2 bond is 50.4 kcal/mole and this large activation free energy causes the reaction to be very slow at room temperature. As we saw in equations 12-9, the reaction rate is accelerated by exposing the Cl_2 molecules to radiation of a frequency sufficient to split the Cl—Cl bond. Figure 21-5 shows a Cl atom colliding with an H_2 molecule in such a way as to create an HCl molecule and an H atom. The overlap shown in Figure 21-5(b) is that between the $1s\sigma^*$ orbital of H_2 and one of the $3p$ orbitals of a Cl atom. If an electron transfers from the H_2 $1s\sigma$ orbital to the $1s\sigma^*$ orbital as shown in Figure 21-5(a), the H—Cl bond is formed as the H—H bond is broken. As a consequence, the reaction between Cl and H_2 is very rapid. In a similar fashion, we can show that the reaction of an H atom with a Cl_2 molecule to regenerate the Cl atom and produce an HCl molecule is rapid also.

$$\text{Cl} \downarrow \quad \text{H} \uparrow\downarrow \text{H} \rightarrow \text{Cl} \uparrow\downarrow \text{H} \qquad \downarrow \text{H}$$

(a)

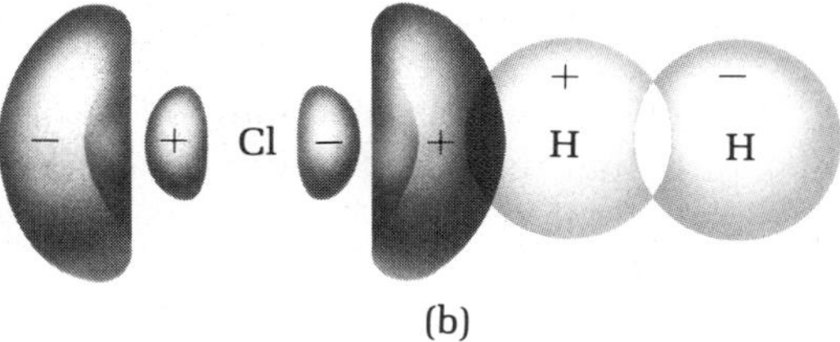

(b)

Figure 21-5
(a) Formation of HCl by the transfer of an electron from H_2 to the region between H and Cl. (b) Orbital overlap involved in (a).

The detailed pictures we have so far considered for the steps through which two reactions might proceed are both plausible. Other plausible pictures can also be drawn. Whatever form they take, they will necessarily involve an activation barrier, with the three factors—encounter, orientation, and bond breaking—that contribute to such a barrier.

21-2. Mechanisms and Rate Laws

Encounter Control in the Gas Phase—Explosions

It was stated that encounter would not prevent reaction 21-1 from being rapid if encounter represented the only barrier to reaction. Let us see how rapidly reaction 21–1 would occur if each H_2 molecule reacts whenever it encounters a Cl_2 molecule. We must return to the consideration of collisions between gas molecules (Chapter 2). Equation 2–25 gives the number of collisions per second involving a given gas molecule and its gaseous neighbors. For collisions between a given H_2 and a collection of Cl_2 molecules, the expression is

$$\text{Collisions of given H}_2\text{ molecule/sec} = \frac{\pi D^2 c_{rms} n_{Cl_2} N}{V} \qquad \textbf{21–4}$$

where D is the average of the H_2 and Cl_2 molecular diameters, n_{Cl_2}/V is the number of moles of Cl_2 per unit volume, c_{rms} is the root mean square speed of the H_2 molecules, and N is Avogadro's number.

The ideal gas law gives

$$\frac{n_{Cl_2}}{V} = \frac{P_{Cl_2}}{RT}$$

so that equation 21–4 can be rewritten as

$$\text{Collisions of given H}_2\text{ molecule/sec} = \left(\frac{\pi D^2 c_{rms} N}{RT}\right) P_{Cl_2} \qquad \textbf{21–5}$$

We can convert this expression to the number of collisions experienced by all H_2 molecules in one cm^3 in one second by multiplying the right side of equation 21–5 by $n_{H_2}N/V = P_{H_2}N/RT$.

$$\text{Collisions of all H}_2\text{ in one cm}^3\text{/sec} = \left(\frac{\pi D^2 c_{rms} N^2}{R^2T^2}\right) P_{Cl_2} P_{H_2} \qquad \textbf{21–6}$$

If each of these collisions results in reaction, the collision rate represents a loss of Δn_{H_2} moles of H_2 per unit time in the total volume, V.

$$\text{Collisions of all H}_2\text{ in one cm}^3\text{/sec} = -\frac{\Delta n_{H_2} N/V}{\text{sec}} \qquad \textbf{21–7}$$

In equation 21–7, the minus sign is included because Δn_{H_2} is negative. From the ideal gas equation,

$$\frac{\Delta n_{H_2}}{V} = \frac{\Delta P_{H_2}}{RT}$$

so that equations 21–6 and 21–7 may be combined with the ideal gas law to give

$$\frac{-\Delta P_{H_2}}{\text{sec}} = \left(\frac{\pi D^2 c_{rms} N}{RT}\right) P_{Cl_2} P_{H_2} \qquad \textbf{21–8}$$

If the one-second time interval is reduced to an infinitesimally small time interval, we arrive at an expression for the reaction rate at any given time during the course of the reaction:

$$\text{Rate} = \left(\frac{\pi D^2 c_{rms} N}{RT}\right) P_{Cl_2} P_{H_2} = k_{rate} P_{Cl_2} P_{H_2} \qquad \textbf{21–9}$$

(In the terminology and notation of calculus, the rate is $-dP_{H_2}/dt$, the negative of the derivative of the hydrogen pressure with respect to time.)

Equation 21–9 is termed the **rate law** of the reaction for the postulated mechanism, and k_{rate} is termed the **rate constant** of the reaction. If we estimate D in equation 21–9 as 4×10^{-8} cm and c_{rms} as 2×10^5 cm/sec, and use R as 82.06 cm³-atm/mole-°K, the value of k_{rate} at 298 °K is calculated as follows:

$$k_{rate} = \frac{(3.14)(4 \times 10^{-8})^2(2 \times 10^5)(6.023 \times 10^{23})}{(82.06)(298)}$$

$$= 2 \times 10^{10} \text{ atm}^{-1}\text{sec}^{-1}$$

Let us see what this rate constant yields for a time dependence of P_{H_2} during this reaction. For simplicity, let us assume that the initial pressures, that is, the pressures of H_2 and Cl_2 before any reaction takes place, are the same. The dependence of P_{H_2} on time is then obtained by the calculus procedure of integration, as

$$\frac{1}{P_{H_2}} - \frac{1}{P_{O(H_2)}} = k_{rate}t \qquad \textbf{21–10}$$

In equation 21–10, $P_{O(H_2)}$ is the H_2 pressure at the start of the reaction ($t = 0$) and P_{H_2} is the pressure at some later time t. Suppose that $P_{O(H_2)}$ is 1 atm. The pressure P_{H_2} will fall to 1×10^{-3} atm in the time

$$t = \frac{1/P_{H_2} - 1/P_{O(H_2)}}{k_{rate}} = \frac{1000 - 1}{2 \times 10^{10}}$$

$$= 5 \times 10^{-8} \text{ sec}$$

The $\Delta H°$ of reaction 21–1 is −44.12 kcal. In 5×10^{-8} sec, very little of this energy liberated can be lost to the surroundings. Almost all of it goes into raising the temperature of the HCl product (and to increasing the pressure of that gaseous product, if the reaction is carried out in a closed container). It is the combination of this temperature and pressure increase that results in explosions for rapid gas phase reactions. As we can see from our calculation, if encounter between reaction pairs were the only barrier to gas phase reactions, all such reactions with any sizably negative $\Delta H°$ would occur explosively.

The General Form of Rate Laws

The terms P_{H_2} and P_{Cl_2} enter the rate law (equation 21–9) for the reaction for a postulated mechanism because it is the collision of an H_2 molecule and a

Cl_2 molecule that defines encounter for this postulated mechanism. Part of our task in deducing the mechanism of any reaction is specifying the atoms, ions, or molecules that encounter each other in the various steps of the reactions. The nature of these encounters is necessarily reflected in the rate law, and there is always a unique relation between a postulated mechanism and the resultant rate law for any reaction. The experimental determination of the rate law is an invaluable first step in proposing a mechanism of any reaction.

Consider the general reaction

$$\alpha\, A + \beta\, B \rightarrow \gamma\, C + \delta\, D \qquad \textbf{21–11}$$

which may occur either in the gaseous phase or in solution. The rate law is always an expression for both the rate of disappearance of reactants and the rate of appearance of products. For reactions in the gas phase and for reactions in the solution phase, rates are usually reported as the loss or gain of molar concentration with time.

We have encountered one rate law in equation 21–9, that is, that for the gas phase reaction of H_2 and Cl_2 according to the hypothetical mechanism given in reaction 21–3. A second reaction (whose mechanism actually involves the simple encounter between two species) is reaction 21–2 and its rate law is shown experimentally to be

$$\text{Rate of disappearance of } H^+ = \text{rate of disappearance of } OH^- = k_{rate}[H^+][OH^-] \qquad \textbf{21–12}$$

The rate constant in equation 21–12 is 2×10^{11} liters/mole-sec, and this large value shows that the reaction between H^+ and OH^- is extremely rapid.

In the cases of both reactions 21–1 and 21–2, the proposed mechanisms lead to rate laws in which the reactants appear raised to the powers given by the corresponding coefficients in the balanced reaction equations. This occurrence is purely coincidental, and no such relation between the rate law and the reaction equation exists in general. However, there is a relation between the rate law and the coefficients of the reactant species in one or more elementary reaction steps, as we shall illustrate subsequently.

Rate laws can be quite complex, but our stated purposes are served by limiting ourselves to rate laws of the following form, which is written for equation 21–11:

$$-\left(\frac{1}{\alpha}\right)\frac{d[A]}{dt} = -\left(\frac{1}{\beta}\right)\frac{d[B]}{dt} = \left(\frac{1}{\gamma}\right)\frac{d[C]}{dt} = \left(\frac{1}{\delta}\right)\frac{d[D]}{dt} = k_{rate}[A]^a[B]^b[C]^c[D]^d[X]^x \qquad \textbf{21–13}$$

where, for purposes of simplified notation, we have used the calculus symbol for a rate. Thus, for example, $-d[A]/dt$ is the rate of disappearance of A, and $d[C]/dt$ is the rate of appearance of C. The exponents a, b, and so on may be positive or negative integers or fractions, or a given exponent will be zero if that species is not involved in the encounters leading to activation. Often a solvent is involved in the encounters leading to activation, but the functional form of the solvent concentration in the rate law is seldom determined. The species X in equation 21–13 denotes any substance involved in activation but not appearing in the reaction equation. Such substances are termed **catalysts,** and we have already encountered some examples. If the exponent x is negative, the substance X is termed an **inhibitor.** A number of drugs and poisons owe their activities in living systems to their ability to inhibit specific biochemical reactions. The appearance of reaction products in a rate law is relatively rare; that is, the exponents c and d in equation 21–13 are typically zero. When products do appear in the rate law, the phenomenon is termed **autocatalysis.** Rate laws more complex than equation 21–13 do arise in some instances. However, those with the form of equation 21–13, on which we are concentrating, are found most commonly.

Obtaining a Rate Law from Experimental Data

Let us now consider a specific reaction to demonstrate a general procedure for obtaining a rate law from experimental data. The reaction is the oxidation of iodide ion by hypochlorite ion in basic solution to yield chloride ion and hypoiodite ion:

$$I^-(aq) + ClO^-(aq) \rightarrow IO^-(aq) + Cl^-(aq) \qquad \textbf{21–14}$$

The reaction is carried out by mixing basic solutions of I^- and ClO^- and then measuring the concentration of one of the reactants or one of the products as a function of time. In a series of qualitative experiments, we discover that the rate of appearance of IO^- at the beginning of the reaction appears to depend in some manner on the initial concentrations of I^- and ClO^-, and on the solution pH. The rate is also found to be totally unaffected by the addition of IO^- and Cl^- to the initial solutions of the experiment. Thus, the rate law appears to be of the form

$$\frac{d[IO^-]}{dt} = k_{rate}[I^-]^a[ClO^-]^b[OH^-]^c \qquad \textbf{21–15}$$

It would be relatively difficult to determine the values of a, b, and c if all three concentrations in equation 21–15 were allowed to undergo large fractional changes simultaneously. This problem is handled by the technique of **flooding** the system with all of the species in the rate law save one, and in our case this one is I^-. Thus, we prepare solutions so that the initial reaction mixture contains $[I^-] = 1.0 \times 10^{-3}\ M$, $[ClO^-] = 0.100\ M$, and $[OH^-] = 1.00\ M$. When the iodide ion has completely reacted, the hypochlorite ion concentration is reduced only by 1 percent. Thus $[ClO^-]^b$ and $[OH^-]^c$ may be treated as constants to yield

$$\frac{d[IO^-]}{dt} = k'_{rate}[I^-]^a \qquad \textbf{21–16}$$

where $k'_{rate} = k_{rate}[ClO^-]^b[OH^-]^c$ is termed a **pseudo rate constant.**

Table 21–1 shows some values of $[I^-]$ as a function of time for three experiments that differ in the initial concentrations of the species appearing in equation 21–15. The time intervals given in this table are very short and thus a special apparatus must be constructed for the rapid mixing of the reactant solutions. Furthermore, some method must be devised to determine the progress of the reaction over very short times. This latter task can typically be performed by the rapid recording of a spectroscopic absorption characteristic of one of a reactant or product, since the intensity of such absorption varies predictably with the concentration of the absorbing species.

Table 21–1
Iodide Ion Concentrations as a Function of Time in the Production of IO^- from I^- by Oxidation with ClO^-

$[ClO^-]_0 = 0.100\ M$, $[OH^-]_0 = 1.00\ M$		$[ClO^-]_0 = 0.050\ M$, $[OH^-]_0 = 1.00\ M$		$[ClO^-]_0 = 0.050\ M$, $[OH^-]_0 = 2.00\ M$	
t, units of 10^{-2} sec	$[I^-]$, units of $10^{-4}\ M$	t, units of 10^{-2} sec	$[I^-]$, units of $10^{-4}\ M$	t, units of 10^{-2} sec	$[I^-]$, units of $10^{-4}\ M$
0	10.0	0	10.0	0	10.0
1.00	9.4	1.00	9.7	5.0	9.3
2.50	8.6	5.0	8.6	10.0	8.6
5.0	7.4	10.0	7.4	20.0	7.4
10.0	5.5	20.0	5.5	50	4.7
20.0	3.0	50	2.2	100	2.2
50	0.5	100	0.5	200	0.5

In the case of reaction 21–14, the species IO^- absorbs visible light strongly in the region near $\bar{\nu} = 25{,}000\ cm^{-1}$. The need for rapid mixing of solutions commonly arises in studying the rates of reactions (particularly biochemical reactions), and commercial equipment is available that can accomplish the mixing in a few milliseconds. Of course, there are a large number of solution reactions that proceed much more slowly than reaction 21–14 and rapid mixing is not necessary for them. Furthermore, the progress of these slow reactions can be followed by extracting samples of the reacting mixture and analyzing for one of the reactants or products by conventional chemical means.

Our next task is to determine a, b, and c in equation 21–15 from the data of Table 21–1. We have adjusted the initial concentrations so that equation 21–16 applies, and thus we first determine values for the exponent a and for k'_{rate} in equation 21–16. The most common value of a nonzero exponent in a rate law is unity, and that possibility is our first study. Since $d[IO^-]/dt = -d[I^-]/dt$, we can rewrite equation 21–16 as the following equation for the case in which $a = 1$:

$$\frac{-d[I^-]}{dt} = k'_{\text{rate}}[I^-] \qquad \textbf{21–17}$$

If we know $[I^-]$ at $t = 0$ and have equation 21–17 for the rate of change of $[I^-]$ with time, we can obtain an expression for $[I^-]$ as a function of time. This expression for $[I^-]$ is obtained by the calculus procedure of integration, and the result is

$$\ln [I^-] - \ln [I^-]_0 = -k'_{\text{rate}}\, t \qquad \textbf{21–18}$$

where $[I^-]_0$ is the iodide ion concentration at $t = 0$.

From equation 21–18, we see that if $a = 1$, a plot of $\ln [I^-]$ versus t is linear with $-k'_{\text{rate}}$ as the slope. Figure 21–6 shows this plot with the data from the left column of Table 21–1. As shown, the plot is indeed linear and a is unity. The value of a is termed the **order** of the rate law with respect to I^-, and the rate law for reaction 21–14 is said to be **first-order** in I^-.

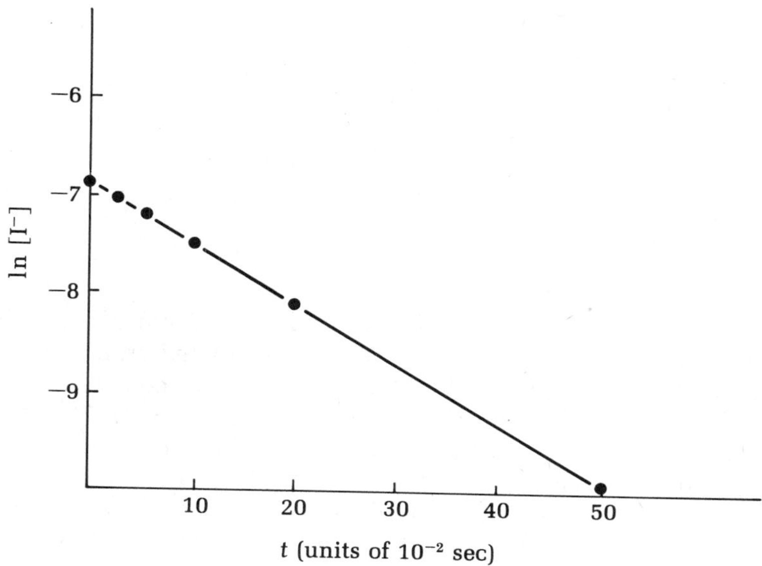

Figure 21–6
The natural logarithm of the iodide ion concentration as a function of time for data from the left column of Table 21–1.

Let us also test the same data of Table 21–1 for **second-order** behavior ($a = 2$). According to the derivation leading to equation 21–10, the second-order rate law in this case would be

$$\frac{-d[I^-]}{dt} = k'_{\text{rate}}[I^-]^2 \qquad \textbf{21–19}$$

and the dependence of $[I^-]$ on time is

$$\frac{1}{[I^-]} - \frac{1}{[I^-]_0} = k'_{rate}\, t \qquad \textbf{21–20}$$

Figure 21–7 shows a plot of $(1/[I^-])$ versus t that should be linear if $a = 2$. Clearly it is not linear, and if we had plotted Figure 21–7 before Figure 21–6, we should have rejected the possibility that a is two, based on that nonlinearity.

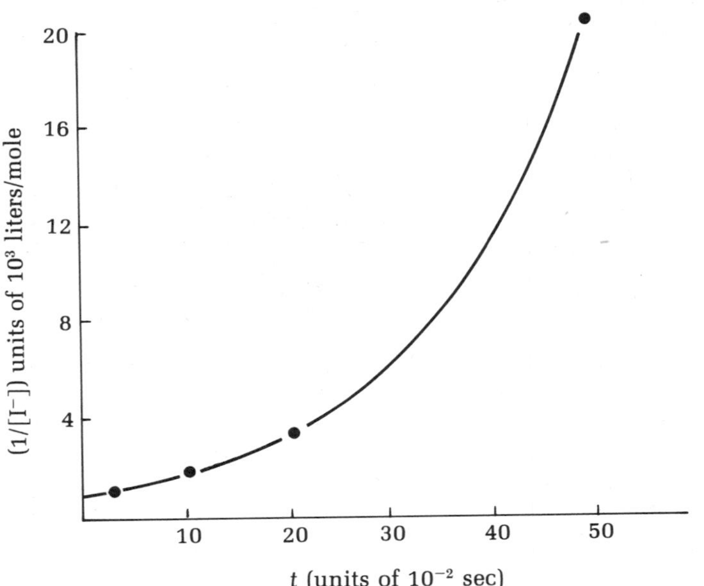

Figure 21–7 The reciprocal of the iodide ion concentration as a function of time for data from the left column of Table 21–1.

Now that we have obtained a from the data of Table 21–1, let us also obtain b, c, and k_{rate}. The value of k'_{rate} is calculated from the slope of the straight line in Figure 21–6 as 6.0 sec^{-1}. We can construct analogous linear plots from the data in the other two columns of Table 21–1, and the two k'_{rate} values obtained from those plots are 3.0 sec^{-1} and 1.50 sec^{-1} for the middle and right columns, respectively. Since k'_{rate} is given by $k_{rate}[ClO^-]_0^b[OH^-]_0^c$, and only $[ClO^-]_0$ changes, of the three initial concentrations in the left and middle columns, b must be unity. When $[ClO^-]_0$ was halved, k'_{rate} was halved also. The only difference between the middle and right columns is the doubling of $[OH^-]_0$ at the right. This doubling halved k'_{rate}, so c must be -1, and the complete rate law is

$$\frac{d[IO^-]}{dt} = k_{rate}\frac{[I^-][ClO^-]}{[OH^-]} \qquad \textbf{21–21}$$

with k_{rate} being 60 sec^{-1}.

The Half-Time of a Reaction

We obtained the value of unity for the order of the rate law with respect to I^- from the linearity of Figure 21–6. Actually the data of Table 21–1 yield a value of unity for a almost by inspection through a method derivable from the antilogarithmic form of equation 21–18:

$$\frac{[I^-]}{[I^-]_0} = e^{-k'_{\text{rate}}t} \qquad \textbf{21–22}$$

Equation 21–22 is an expression for the fraction of the concentration of the original I^- that remains unreacted at time t. For example, in the reaction for which the data are given in the left column of Table 21–1, 0.86 of the original $[I^-]$ remains at $t = 2.5 \times 10^{-2}$ sec. What fraction of this 0.86 remains after the next 2.5×10^{-2} sec? From the table, we obtain this fraction as $(7.4 \times 10^{-4})/(8.6 \times 10^{-4}) = 0.86$ again. **Equal time intervals yield equal fractional decreases in a first-order rate law.** This behavior is completely consistent with equation 21–22, since $[I^-]/[I^-]_0$ is a function only of k'_{rate} and the elapsed time t in that equation.

This unique property of first-order rate laws is commonly stated in an implicit fashion by quoting a **half-time** for the reaction. The half-time (symbolized by $t_{1/2}$) is the time required for the limiting reactant at some reference time to decrease in concentration or pressure to half of the value at the reference time. From equation 21–22, we can derive that $t_{1/2}$ is related to k'_{rate} by the equation

$$t_{1/2} = \frac{\ln 2}{k'_{\text{rate}}} = \frac{0.693}{k'_{\text{rate}}} \qquad \textbf{21–23}$$

Table 21–2 reinforces the concept of the constant $t_{1/2}$ by showing the fraction of initial reactant remaining at times measured in units of $t_{1/2}$.

Time	Fraction Remaining
0	1
$t_{1/2}$	$\frac{1}{2}$
$2t_{1/2}$	$\frac{1}{4}$
$3t_{1/2}$	$\frac{1}{8}$
$4t_{1/2}$	$\frac{1}{16}$

Table 21–2
Fraction of Limiting Reactant Remaining in a First-Order Reaction at Various Multiples of the Half-Time

By way of contrast, let us obtain the expression for $t_{1/2}$ for a second-order rate law from equation 21–20. The result is

$$t_{1/2} = \frac{1}{k'_{\text{rate}}[I^-]_0} \qquad \textbf{21–24}$$

so that for a second-order reaction, the half-time increases as the reaction proceeds.

From Rate Law to Proposed Mechanisms

Let us now try to deduce a mechanism for reaction 21–14 consistent with the rate law of equation 21–21. Before we do this, we must consider some general features of reaction mechanisms. The simple encounter mechanisms that we considered for reactions 21–1 and 21–2 involve only one step. Most reaction mechanisms involve a sequence of several steps, in which one or more products of one step become the reactants in the next step. In such a sequence, it is often found to be the case that one step is much slower than all the others. This step is termed the **rate-determining step** (rds), since the overall reaction can proceed no more rapidly than this step proceeds.

Steps that precede the rate-determining step have sufficient ample time to reverse themselves before the rate-determining step occurs, in which case these preceding steps involve equilibria. All steps that follow the rate-determining step proceed to yield products at a rate controlled by the supply rate from the rate-determining step. A general mechanism has the features outlined as follows:

$$\begin{aligned} &\text{Reactants} \rightleftharpoons \text{intermediates} && \text{(equilibria)} \\ &\text{Intermediates} \rightarrow \text{intermediates}' && \text{(rds)} \\ &\text{Intermediates}' \rightarrow \text{products} && \text{(rapid)} \end{aligned} \qquad \textbf{21–25}$$

The reactant atoms are rearranged in some set of equilibria, the steps of which occur more rapidly than the rate-determining step (rds). These rearranged species are termed reaction intermediates. In the rate-determining step, one or more intermediates (or intermediates plus reactants) produce new intermediates or products or both. If new intermediates are formed in the rate-determining step, they rapidly convert to products in subsequent reactions.

Since the rate-determining step controls the reaction rate, the rate law must reflect the atoms and electrical charges that encounter each other in the rate-determining step. This statement has great application in the proposing of mechanisms, and it is best shown by example.

The atoms and electrical charges that encounter each other in the rate-determining step of a reaction mechanism are the sum of those in the numerator of the right side of the rate law minus those in the denominator of the right side. In the case of equation 21–21, it is impossible to subtract the denominator atoms from the numerator atoms since there are no hydrogen atoms in the numerator. This problem is solved by recalling that the reaction is carried out in aqueous solution in which $[OH^-]$ is inversely related to $[H^+]$ by

$$[H^+][OH^-] = K_w = 1.0 \times 10^{-14}$$

Thus, equation 21–21 may be rewritten as

$$\frac{d[IO^-]}{dt} = \frac{k_{rate}}{k_w}[H^+][I^-][ClO^-] \qquad \textbf{21–26}$$

and we see that H^+ catalyzes the reaction. From equation 21–26, we obtain the atoms and charges that encounter each other in the rate-determining step as

$$H^+ + I^- + ClO^- = HIClO^-$$

The rate law does not reveal how the atoms and charges of $HIClO^-$ are arranged as they encounter each other in the rate-determining step. Some possibilities are

$$HI + ClO^-$$

$$HClO + I^-$$

$$H^+ + ClO^- + I^-$$

and there are other possibilities. Our task is to choose the most plausible combination of species for the rate-determining step. For this, we call on our knowledge of the chemistries of the proposed species.

It is very unlikely that the reaction occurs by the simultaneous encounter of H^+, ClO^-, and I^-, as shown in Figure 21–8. While encounters between two species occur quite commonly, the probability that three or more separate species undergo a simultaneous encounter is very low, unless one or more of the species are solvent molecules. While such encounters do occur in some reactions the occurrence is relatively rare, and we shall exclude the possibility here (and in the problems at the end of the chapter). The number of separate species that encounter each other in the rate-determining step is termed the **molecularity** of the step. Figure 21–8 depicts a **termolecular** step, and Figures 21–1 and 21–4 show **bimolecular** steps.

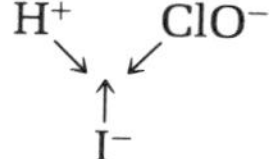

Figure 21–8
A highly improbable encounter in one possible mechanism for reaction 21–14.

Two possible bimolecular encounters are encounters between HI and ClO^- and between HClO and I^-. The acid HI is a very strong acid and it is unlikely that any HI molecules exist in the highly basic solutions of Table 21–1. However, HClO is a weak acid of $K_a = 2 \times 10^{-8}$, and a small but significant concentration of HClO is present even when $[OH^-]$ is quite large. Thus, a quite reasonable mechanism is the following:

$$\begin{aligned} H^+ + ClO^- &\rightleftharpoons HClO && \text{(equilibrium)} \\ HClO + I^- &\rightarrow HIO + Cl^- && \text{(rate-determining step)} \qquad \textbf{21–27} \\ HIO &\rightarrow IO^- + H^+ && \text{(rapid)} \end{aligned}$$

Let us check this mechanism to see that it yields the rate law, equation 21–26. Since HClO and I^- encounter each other in the rate-determining step, the rate of production of Cl^- is given by

$$\frac{d[Cl^-]}{dt} = \frac{d[IO^-]}{dt} = k_{rds}[I^-][HClO]$$

where k_{rds} is the rate constant for the rate-determining step. Notice that the rate expression for the rate-determining step contains the concentrations of the species that encounter each other in that step. Furthermore, each of these species appears in the rate expression raised to the power equal to its coefficient in the balanced equation for the step. The concentration of HClO is

$$[HClO] = \frac{[H^+][ClO^-]}{K_a}$$

so that the rate law derived from mechanism 21–27 is

$$\frac{d[IO^-]}{dt} = \frac{k_{rds}}{K_a}[I^-][H^+][ClO^-] \qquad \textbf{21–28}$$

Equations 21–26 and 21–28 are identical if $k_{rate}/K_w = k_{rds}/K_a$, and mechanism 21–27 is consistent with the experimentally determined rate law.

Further Examples of the Deduction of Mechanisms from Rate Laws

Constructing reaction mechanisms from rate laws requires considerable practice, for which some additional examples will be helpful.

Chlorine and carbon monoxide react in the gas phase to yield the poisonous gas phosgene, according to the reaction

$$Cl_2(g) + CO(g) \rightarrow \begin{matrix} Cl \\ \quad \diagdown \\ \quad\; C{=}O \\ \quad \diagup \\ Cl \end{matrix} \qquad \textbf{21–29}$$

and the rate law (expressed in terms of partial pressures) has been found to be

$$\frac{dP_{OCCl_2}}{dt} = k_{rate}P_{CO}P_{Cl_2}{}^{3/2} \qquad \textbf{21–30}$$

The exponent 3/2 indicates that $\frac{3}{2}$ Cl_2 molecules, or three chlorine atoms, are involved in the rate-determining step; and the combined formula of the species involved in the rate-determining step is $COCl_3$.

We shall again limit the possible mechanisms to ones involving the encounter of two species or fewer in the rate-determining step. One such example would be the encounter of Cl_3 with CO, which is the combination $COCl_3$; but there is no evidence for the existence of Cl_3 in gaseous chlorine. A second possibility is the encounter of Cl with CO to form OCCl. A complete mechanism incorporating this step is

$$\begin{aligned} Cl_2 &\overset{K_1}{\rightleftharpoons} 2\,Cl && \text{(equilibrium)} \\ Cl + CO &\overset{K_2}{\rightleftharpoons} OCCl && \text{(equilibrium)} \\ OCCl + Cl_2 &\rightarrow OCCl_2 + Cl && \text{(rds)} \end{aligned} \qquad \textbf{21–31}$$

where K_1 and K_2 are the equilibrium constants for the two equilibria preceding the rate-determining step.

The rate of the bimolecular rate-determining step is given by

$$\frac{dP_{OCCl_2}}{dt} = k_{rds}P_{OCCl}P_{Cl_2} \qquad \textbf{21–32}$$

The pressure of the intermediate P_{OCCl} is related to P_{CO} and P_{Cl_2} through the equilibrium constants K_1 and K_2:

$$P_{OCCl} = K_2P_{Cl}P_{CO}$$

$$P_{Cl} = K_1^{1/2}P_{Cl_2}^{1/2}$$

so that $P_{OCCl} = K_1^{1/2}K_2P_{CO}P_{Cl_2}^{1/2}$, and the rate law for the reaction is

$$\frac{dP_{OCCl_2}}{dt} = k_{rds}K_1^{1/2}K_2P_{CO}P_{Cl_2}^{3/2} \qquad \textbf{21–33}$$

Equation 21–33 is the experimental rate law, with $k_{rate} = k_{rds}K_1^{1/2}K_2$. We should keep in mind that the agreement between equations 21–30 and 21–33 does not prove that mechanism 21–31 is correct, since we may be able to write other mechanisms that will also yield the experimental rate law. A further test is the plausibility of the steps and independent evidence for the intermediate species proposed in the mechanism.

Now let us consider another reaction. Nitrogen dioxide is produced rapidly in the following reaction:

$$2\,NO(g) + O_2(g) \rightarrow 2\,NO_2(g) \qquad \textbf{21–34}$$

The rate law for this reaction is found experimentally to be

$$\frac{dP_{NO_2}}{dt} = k_{rate}P_{NO}^2P_{O_2} \qquad \textbf{21–35}$$

According to this rate law, the combination of atoms that encounter each other in the rate-determining step have the formula N_2O_4. We can form the combination N_2O_4 in several different ways; the following are two possible mechanisms, both of which yield the experimental rate law:

Mechanism 1

$$2\,NO \overset{K_1}{\rightleftharpoons} N_2O_2 \qquad \text{(equilibrium)}$$

$$N_2O_2 + O_2 \rightarrow N_2O_4 \qquad \text{(rds)}$$
$$N_2O_4 \rightarrow 2\ NO_2 \qquad \text{(rapid)}$$

Mechanism 2

$$NO + O_2 \overset{K_2}{\rightleftharpoons} NO_3 \qquad \text{(equilibrium)}$$
$$NO_3 + NO \rightarrow 2\ NO_2 \qquad \text{(rds)} \qquad \textbf{21–37}$$

The rate law derivable from mechanism 21–36 is

$$\frac{dP_{NO_2}}{dt} = 2k_{rds}K_1P_{NO}^2P_{O_2} \qquad \textbf{21–38}$$

and the rate law from mechanism 21–37 is

$$\frac{dP_{NO_2}}{dt} = 2k_{rds}K_2P_{NO}^2P_{O_2} \qquad \textbf{21–39}$$

In this case it is even difficult to make a distinction between the two mechanisms based on the plausibility of the two intermediates, N_2O_2 and NO_3. Nitric oxide contains an unpaired antibonding electron, and two NO molecules can combine to form the structure shown in Figure 21–9(a). All four nuclei are in the same plane and the two original antibonding electrons form the N—N bond. The O_2 molecule can accept the four electrons shown by colored dots in Figure 21–9(a), two electrons into the O_2 2 $p\sigma^*$ orbital and two electrons into the $2p\pi^*$ orbital. This destroys the O═O bonding as it forms the nitrogen-oxygen bonding shown in Figure 21–9(b). The N_2O_4, shown formed in Figure 21–9(b), can then decompose to 2 NO_2.

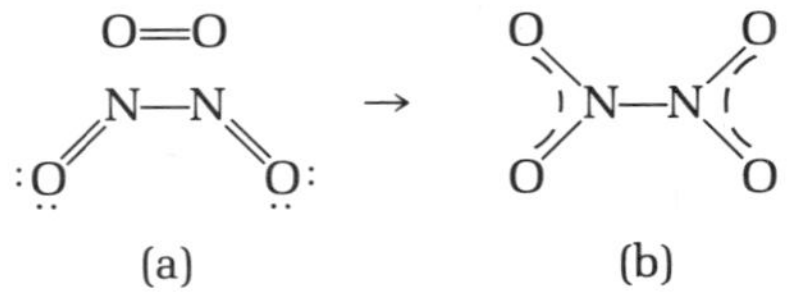

Figure 21–9
The rate determining step in one proposed mechanism for the formation of 2 NO_2 from 2 NO and O_2.

(a) (b)

The oxygen molecule contains two antibonding, unpaired electrons, and bond formation between NO and O_2 to form NO_3 is possible, as proposed in mechanism 21–37. The NO_3 molecule can then be attacked by NO in a process of simultaneous bond making and bond breaking to yield two NO_2 molecules directly. In order to discover which of the two mechanisms, 21–36 or 21–37, is actually dominant, additional evidence is required, such as the spectroscopic detection of either N_2O_2 or NO_3 during the reaction.

21–3. Activation

Unimolecular Reactions

At a low pressure of O_2, the decomposition of ozone to give O_2 follows the rate law:

$$\frac{-dP_{O_3}}{dt} = k_{rate}P_{O_3} \qquad \textbf{21–40}$$

It is proposed that this reaction does not involve an encounter as the rate-determining step. Instead, the rate-determining step involves isolated O_3 molecules. We shall investigate this **unimolecular** reaction in some detail because it is a convenient vehicle for visualizing the process of activation in general.

The proposed mechanism of decomposition of ozone is the following:

$$\begin{aligned} O_3 &\rightarrow O_2 + O \qquad &\text{(rds)} \\ 2\,O &\rightarrow O_2 \qquad &\text{(rapid)} \end{aligned} \qquad \textbf{21–41}$$

The $\Delta H°$ for the rate-determining step is calculable from the O═O bond enthalpy listed in Table 6–3 and $\Delta H°_{f(O_3)}$ listed in Table 15–1.

$$\Delta H° = \frac{(\text{bond enthalpy})_{O_2}}{2} - \Delta H°_{f(O_3)}$$

$$= (\tfrac{118}{2}) - 34 = 25 \text{ kcal/mole}$$

$$= 1.7 \times 10^{-12} \text{ erg/molecule}$$

For an O_3 molecule to split into an O_2 molecule and an O atom, it must acquire an energy of 1.7×10^{-12} erg over and above its average energy. This is an enthalpy barrier to the reaction, and it is the principal contribution to $\Delta G^{\ddagger}$ of the reaction.

How does a molecule gain additional energy in a gas mixture at a fixed temperature? Clearly it does so by collisions with its neighbors and with the walls, as we pictured in Figure 2–18. Because of these collisions, all collections of molecules at a given temperature contain a distribution of molecules over a wide range of energies. One such distribution involves translational kinetic energies related to Figures 2–16 and 2–17. Others are the relative populations of rotational and vibrational energy states (Chapter 12). Thus, the O_3 molecules in the rate-determining step of mechanism 21–41 constitute that fraction of the O_3 molecules that have energies at least 1.7×10^{-12} erg above the average energy. With this in mind, we can write the overall decomposition mechanism as

$$\begin{aligned} O_{3(normal)} &\rightleftharpoons O_{3(energetic)} \qquad &\text{(equilibrium)} \\ O_{3(energetic)} &\rightarrow O_2 + O \qquad &\text{(rds)} \\ 2\,O &\rightarrow O_2 \qquad &\text{(rapid)} \end{aligned} \qquad \textbf{21–42}$$

The rate law derivable from mechanism 21–42 is

$$\frac{-dP_{O_3}}{dt} = k_{rds}K_eP_{O_3} \qquad \textbf{21–43}$$

where K_e is the equilibrium constant for the formation of the sufficiently energetic O_3 molecules, and k_{rds} is the rate constant for the unimolecular decomposition of the energetic molecules.

Before we proceed further, we must ask how the second step of mechanism 21–42 can really be rate-determining. Should not O_3 molecules decompose as soon as they acquire enough energy to decompose? Figure 21–10(a) shows an O_3 molecule with a very large translational energy but only a small vibrational energy. This molecule cannot react, because in order to break an O═O bond, the large energy must be concentrated in stretching vibrations or in rotation. The conversion of translational energy to vibrational energy can occur through molecule-molecule collisions in the gas or by collisions at a wall, as depicted in Figure 21–10(b). In the figure, one of the O═O bonds is highly compressed as a result of the collision. If the potential energy in this compressed state is greater than the bond energy, the bond will break as the molecule leaves the wall, as shown in Figure 21–10(c). Of course, the collision might well have occurred with a different orientation with no bond breaking and with translational energy loss. Thus, the rate-determining step can be relatively slow because the rate of this step reflects the probability that the excitation energy is located in a vibrational mode that will break a bond before that energy is lost in ineffective collisions. This consideration is particularly important in the unimolecular decompositions of relatively complex molecules.

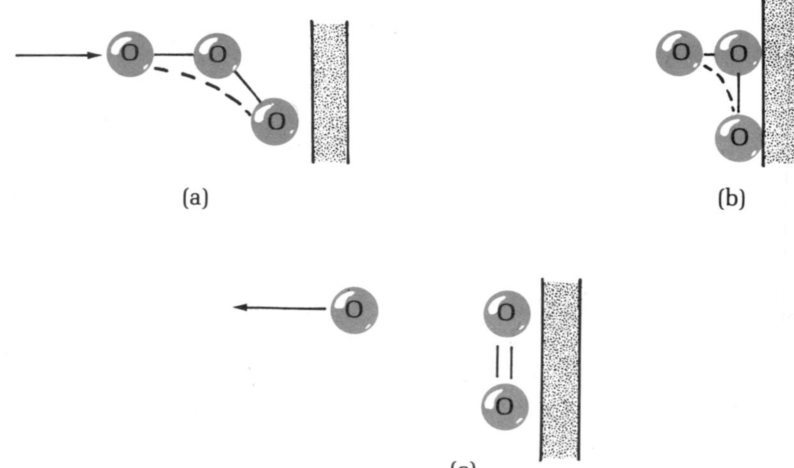

Figure 21–10
A mechanism for the dissociation of a translationally energetic O_3 molecule by collision with a wall. (a) Before collision, (b) during collision, and (c) after collision.

A General Model of Reaction Rates

The mechanism of decomposition of ozone has some important features in common with the mechanisms of almost all reactions. Up to this point we have used only the concentration or the pressure terms in rate laws in deducing reaction mechanisms. However, the magnitude of k_{rate} also derives from the mechanism, and we should be able to extract valuable information about the mechanism from the magnitude of k_{rate}.

A comparison of equations 21–40 and 21–43 shows that k_{rate} for the ozone decomposition can be expressed as follows:

$$k_{rate} = k_{rds} K_e$$

if the postulated mechanism is correct. The term K_e is the equilibrium constant for the production of sufficiently energetic molecules, and as such it is related to a $\Delta G°$ for the production of energetic molecules:

$$K_e = e^{-\Delta G_1^\circ/RT} = e^{\Delta S_1^\circ/R} e^{-\Delta H_1^\circ/RT} \qquad \textbf{21–44}$$

In equation 21–44, the term ΔG_1° is the component of the activation free energy due to Step 1 of mechanism 21–42.

We now require an expression for k_{rds}, and this requires the construction of a model for conversion of energetically excited molecules to products. The American scientist Henry Eyring has constructed such a model, in which he pictures an equilibrium between energetically excited species and species at the top of the Gibbs free energy barrier. The model is depicted in Figure 21–11.

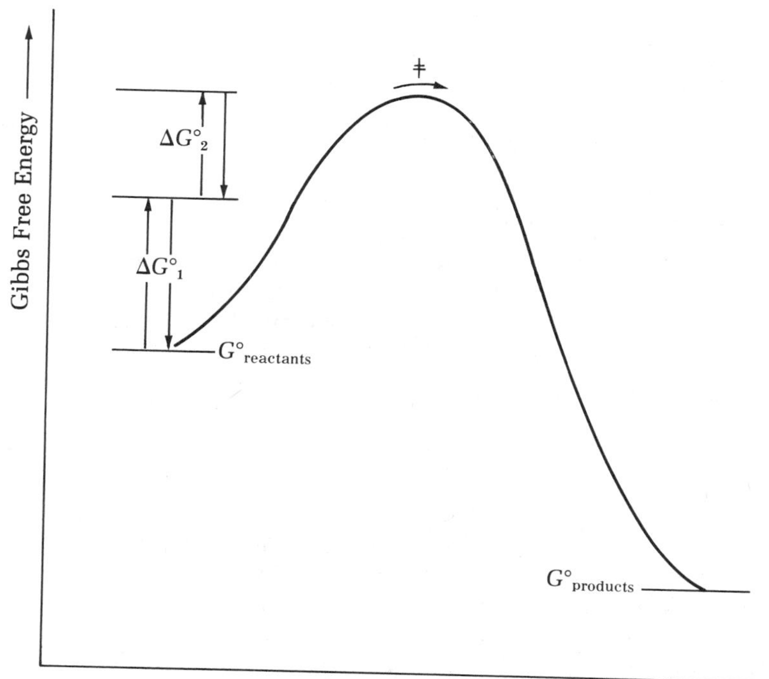

Figure 21–11
Gibbs free energy changes involved in the Eyring equilibrium model of reaction rates. The change ΔG_1° is that between normal reactants and energetically excited reactants, and ΔG_2° is that between energetically excited reactants and species at the top of the activation barrier.

Although we have introduced the Eyring model in the context of one reaction, it may be applied to many other reactions also. In the model, reactant species must acquire excess energy to be able to react. This energy must be located in the molecular motions required to bring about reaction, and thus there is a second component to the barrier to reaction. The merits of the Eyring model have been the subject of much debate, but it remains a model that is very commonly applied to the analysis of rate constants.

We shall not develop the Eyring expression for k_{rds}, but merely state it, as follows:

$$k_{rds} = \frac{kT}{h} e^{-\Delta G_2^\circ/RT} = \frac{kT}{h} e^{\Delta S_2^\circ/R} e^{-\Delta H_2^\circ/RT} \qquad \textbf{21–45}$$

where k is Boltzmann's constant and h is Planck's constant.

If we combine equations 21–44 and 21–45, we obtain the following proposed expression for the rate constant for the ozone decomposition. The expression should apply to the rate constant of any other reaction to the extent that the Eyring equilibrium picture is a valid one:

$$\begin{aligned} k_{rate} &= \frac{kT}{h} e^{-(\Delta G_1^\circ + \Delta G_2^\circ)/RT} \\ &= \frac{kT}{h} e^{(\Delta S_1^\circ + \Delta S_2^\circ)/R} e^{-(\Delta H_1^\circ + \Delta H_2^\circ)/RT} \\ &= \frac{kT}{h} e^{-\Delta G^\ddagger} \\ &= \frac{kT}{h} e^{\Delta S^\ddagger/R} e^{-\Delta H^\ddagger/RT} \end{aligned} \qquad \textbf{21–46}$$

Temperature Dependence of Reaction Rates

We have stated that certain reactions are imperceptibly slow at room temperature and extremely rapid at elevated temperatures. One such reaction is that between hydrogen and oxygen. Let us consider the temperature dependence of k_{rate} predicted by equation 21–46. The equation can be reformulated logarithmically to yield

$$\ln \frac{k_{rate}}{T} = 2.303 \log \frac{k_{rate}}{T} = \ln \frac{k}{h} + \frac{\Delta S^\ddagger}{R} - \frac{\Delta H^\ddagger}{RT} \qquad \textbf{21–47}$$

Thus, a plot of log (k_{rate}/T) versus $1/T$ is predicted to be linear with a slope of $-\Delta H^\ddagger/2.303R$. Figure 21–12 shows a plot of log (k_{rate}/T) versus reciprocal temperature for the gas-phase decomposition of HI to give H_2 and I_2. Within the experimental uncertainty of the data, the relation between log (k_{rate}/T) and $1/T$ shown in Figure 21–12 is indeed linear.

Most reaction rate constants have a temperature dependence such that log (k_{rate}/T) is very nearly a linear function of $1/T$. Thus, this prediction of

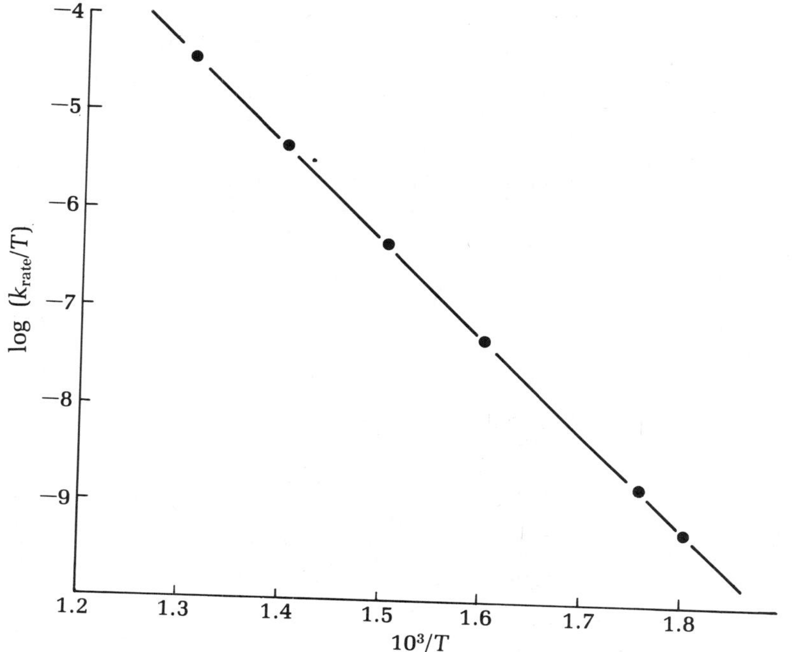

Figure 21–12 Variation with reciprocal temperature of the log rate constant divided by temperature for the decomposition of HI. The values of k_{rate}/T are in units of liters per mole per second.

the equilibrium model is correct. However, other models also yield this prediction of the temperature dependence of rate constants. If more precise data are plotted for the HI decomposition, a better approximation to linearity is obtained by plotting log $(k_{rate}/T^{1/2})$ versus $1/T$, and this is not in agreement with equation 21–47. Activation is an extremely complex process in general and no model now exists that allows us to picture it in some simple way of general applicability. The Eyring model perhaps comes the closest to this ideal at present and this is why it is widely used.

Values of $\Delta H^{\ddagger}$ and $\Delta S^{\ddagger}$ obtained from applying equation 21–47 to rate data are commonly reported. Since equation 21–47 is only an approximate expression, these values of $\Delta H^{\ddagger}$ and $\Delta S^{\ddagger}$ are only approximations to the true activation parameters that combine to give the barrier depicted in Figure 21–2. Nevertheless, the approximate $\Delta H^{\ddagger}$ and $\Delta S^{\ddagger}$ values can be quite useful.

Let us obtain values for $\Delta H^{\ddagger}$ and $\Delta S^{\ddagger}$ from Figure 21–12. The slope of the straight line is equal to $-\Delta H^{\ddagger}/2.303R$, so that

$$\Delta H^{\ddagger} = -\left(\frac{\Delta \log (k_{rate}/T)}{\Delta(1/T)}\right) R(2.303)$$

With the data in Figure 21–12, this yields a calculated $\Delta H^{\ddagger}$ for the HI decomposition of 44.5 kcal/mole. The $\Delta S^{\ddagger}$ value can then be obtained by substituting this $\Delta H^{\ddagger}$ value into equation 21–47. The result is a calculated $\Delta S^{\ddagger}$ of −10 cal/mole-deg.

The gaseous decomposition of HI is just one of many reactions for which the rate constant shows a dramatic increase with increasing temperature. The rate constant for the HI decomposition increases by a factor of approximately 1000 between 600 °K and 750 °K. Even relatively large activation barriers can be overcome by carrying out reactions at elevated temperatures. It is very common to carry out reactions at elevated temperatures both in the laboratory and in industrial processes, so that the reactions proceed at convenient and economical rates.

Application of $\Delta H^{\ddagger}$ and $\Delta S^{\ddagger}$ to the Postulation of Reaction Mechanisms

The calculated $\Delta H^{\ddagger}$ value for the decomposition of ozone is 25 kcal/mole. This value is equal to the value previously calculated as ΔH° for the rate-determining step in mechanism 21–41. There is no necessity for such an exact agreement, but the agreement definitely adds weight to the case for mechanism 21–41.

As an additional example, let us consider the formation of HI from H_2 and I_2:

$$H_2(g) + I_2(g) \rightarrow 2\,HI(g) \qquad \textbf{21–48}$$

The calculated $\Delta H^{\ddagger}$ value for this reaction is 40 kcal/mole. This is approximately equal to the I_2 bond energy of 36 kcal/mole, and it is reasonable to propose that the most important step in the mechanism of formation of HI is the splitting of I_2 to form iodine atoms.

The $\Delta S^{\ddagger}$ value for a given reaction is also of value in proposing a mechanism. For example, the unimolecular splitting of a simple molecule should be expected to display a positive $\Delta S^{\ddagger}$ since matter is spread in the splitting process. An example is the decomposition of azomethane to yield ethane and nitrogen according to the reaction

$$H_3C{-}N{=}N{-}CH_3 \rightarrow H_3C{-}CH_3 + N{\equiv}N \qquad \textbf{21–49}$$

The $\Delta H^{\ddagger}$ value is 53 kcal/mole and $\Delta S^{\ddagger}$ is 17 cal/mole-deg. The positive $\Delta S^{\ddagger}$ indicates that the splitting of the azomethane molecule plays a large role in the activation. Two possible mechanisms, both of which involve the splitting of the azomethane molecule in the rate-determining step, are the following:

$$\begin{aligned} H_3C{-}N{=}N{-}CH_3 &\rightarrow N_2 + 2\,CH_3 && \text{(rds)} \\ 2\,CH_3 &\rightarrow C_2H_6 && \text{(rapid)} \end{aligned} \qquad \textbf{21–50}$$

$$\begin{aligned} H_3C{-}N{=}N{-}CH_3 &\rightarrow H_3C{-}N{=}N + CH_3 && \text{(rds)} \\ CH_3 + H_3C{-}N{=}N &\rightarrow C_2H_6 + N_2 && \text{(rapid)} \end{aligned} \qquad \textbf{21–51}$$

21–4. Catalysis

Since the rates of reactions are limited by barriers arising from encounter, orientation, and bond breaking, we have only two choices if we wish to increase the rate of a reaction that proceeds by a given mechanism. First, we can increase the concentrations or pressures of the species that appear in the numerator of the rate law. However, there is a limit to these concentrations and pressures and we cannot produce really dramatic changes in the rate unless we increase the rate constant. As we saw in Figure 21–12, this can be done in dramatic fashion by increasing the temperature, and this is our second choice. However, the production and maintenance of elevated temperatures is expensive, and we should like to avoid having to use them.

A third possibility for increasing a reaction rate is to alter the mechanism, and this is why catalysts are employed. The role of a catalyst is to change the reaction mechanism to one with a lower activation barrier and consequently a greater rate constant. One of the great advantages in knowing a reaction mechanism is that the nature of the activation barrier becomes apparent and a catalyst can be designed in a rational fashion.

We have already encountered several examples of catalysis in the text. One example is reaction 21–14, which is catalyzed by H^+. A significant contribution to the reaction barrier between I^- and ClO^- is the potential energy of repulsion between these two negatively charged species. The protonation of ClO^- presumably removes this barrier and markedly increases the reaction rate.

Formation of Polyethylene

In Figure 11–15 we saw the steps proposed for the catalysis of the polymerization of ethylene. Let us examine why the polymerization reaction requires a catalyst by investigating the uncatalyzed reaction in the gas phase. The first step in the uncatalyzed polymerization is

$$2\ H_2C{=}CH_2 \rightarrow {-}H_2C{-}CH_2{-}CH_2{-}CH_2{-} \qquad \textbf{21–52}$$

and two C=C bonds are replaced by three C—C bonds. The $\Delta H°$ for reaction 21–52 can be estimated from the bond enthalpies of Table 6–3 as follows:

$$\Delta H° = -3\ (\text{bond enthalpy})_{C-C} + 2\ (\text{bond enthalpy})_{C=C}$$

$$= [-3(83) + 2(146)]\ \text{kcal/mole}$$

$$= 43\ \text{kcal/mole}$$

This large activation enthalpy causes ethylene molecules to be quite unreactive toward each other at room temperature. The activation barrier can be overcome by a temperature increase. However, the product of reaction 21–52 must react with another ethylene molecule for the chain to continue to lengthen. A highly favored reaction that competes with chain lengthening is the cyclization of the product of reaction 21–52 to form cyclobutane. In fact, the formation of the cyclic product is the principal reaction of ethylene in the gas phase, and polyethylene cannot be made in that fashion.

Figure 21–13 shows the structures involved in the catalyzed polymerization that we considered in Figure 11–15. Figure 21–13(a) shows reactants C_2H_4 and C_2H_5 in the first chain-lengthening step. One role of the titanium atom that is manifested in Figure 21–13(a) is overcoming the encounter and orientation barriers to reaction by binding the two reactants to itself. Another important role is that the Ti—C* and C—C bonds in Figure 21–13(b) can form simultaneously with the breaking of the C═C pi bond and the Ti—C bond in Figure 21–13(a), and this leads to a relatively low activation energy. Finally, no species like the product of reaction 21–52 ever forms, and thus the competing cyclization reaction cannot occur.

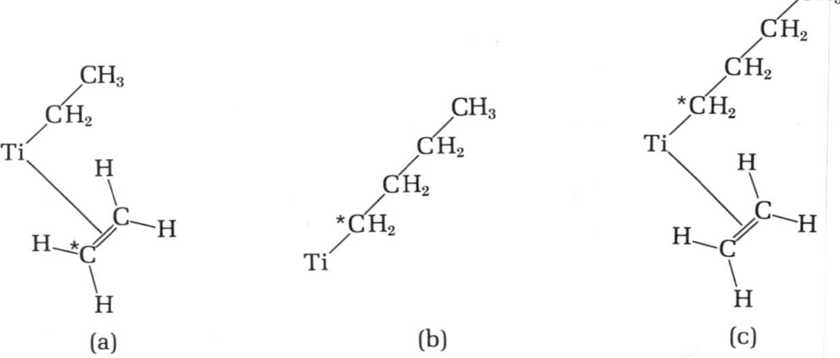

Figure 21–13 Successive structures formed in the catalyzed polymerization of ethylene.

Acid Catalysis of Ester Hydrolysis

Another example of catalysis occurs in the hydrolysis of an ester to yield a carboxylic acid and an alcohol:

$$RC(=O)OR' + H_2O \rightarrow RC(=O)OH + R'OH \qquad \textbf{21–53}$$

This reaction is catalyzed by H^+, a fact that we could have predicted based on our previous considerations of the source of reactivity of compounds containing the carbonyl group. The C═O group is polarized as $C^{\delta+}{=}O^{\delta-}$, and the $C^{\delta+}$ portion shows a Lewis acid character that can be markedly strengthened by protonation of the $O^{\delta-}$ to form the ion in Figure 21–14(a). This strong Lewis acid can then accept a pair of electrons from a water molecule to form the structure shown in Figure 21–14(b). This structure can decompose to re-form the original reactants or it can lose an $R'O^-$ ion and two protons to yield the carboxylic acid and the alcohol as products.

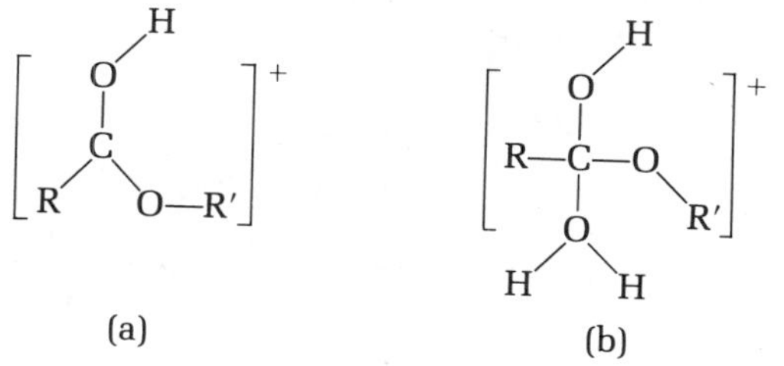

$$\left[\begin{array}{c} \text{H} \\ \diagup \\ \text{O} \\ | \\ \text{C} \\ \diagup \quad \diagdown \\ \text{R} \qquad \text{O—R}' \end{array}\right]^+ \qquad \left[\begin{array}{c} \text{H} \\ \diagup \\ \text{O} \\ | \\ \text{R—C—O} \\ | \qquad \diagdown \\ \text{O} \qquad \text{R}' \\ \diagup \quad \diagdown \\ \text{H} \qquad \text{H} \end{array}\right]^+$$

(a) (b)

Figure 21–14 Two intermediates formed in the H^+ catalyzed hydrolysis of an ester.

Catalysis by Solvent

Many reactions in the laboratory and in industrial processes are carried out in solution. The solvent can also serve to catalyze the reaction and an important choice in any reaction is the solvent in which it is to be carried out. A dramatic example occurs in the hydrolysis of tert-butyl chloride to form tert-butyl alcohol, H^+, and Cl^-.

$$H_3C-\underset{\underset{CH_3}{|}}{\overset{\overset{CH_3}{|}}{C}}-Cl + H_2O \rightarrow H_3C-\underset{\underset{CH_3}{|}}{\overset{\overset{CH_3}{|}}{C}}-OH + H^+ + Cl^- \qquad \textbf{21–54}$$

The reaction is typically carried out in a mixed solvent of ethanol and water, since tert-butyl chloride is only sparingly soluble in water. The rate constant for reaction 21–54 shows a dramatic dependence on the percentage of water in the solvent. When this percentage is increased from 10 percent to 50 percent, the reaction rate constant increases by a factor of approximately 200. To understand this huge enhancement, we must know something about the reaction mechanism. A very important piece of information about the mechanism comes from the hydrolysis of a compound similar to tert-butyl chloride, as shown in Figure 21–15. The alkyl chloride shown in Figure 21–15(a) has four different substituents bonded to the central carbon, and thus the structure shown is one of two enantiomers. The hydrolysis products of this enantiomer are the two enantiomeric alcohols, (b), and (c) in nearly a 50–50 mixture. This indicates that (a) is converted in the course of the reaction to an intermediate that does not possess an asymmetric carbon atom. The most likely candidate for such an intermediate is the cation shown in Figure 21–16(a). All of the nuclei in the ion, with the exception of the protons of the methyl group, lie in one plane, and the ion is optically inactive. It can be attacked from either side of the plane to yield the two enantiomeric alcohols shown.

The formation of the cation in Figure 21–16(a) requires the breaking of a C—Cl bond, and this presents a sizable activation barrier. This barrier is very significantly lowered by a solvent that can solvate the ions as they are being formed. This requires a polar solvent. Both ethanol and water are

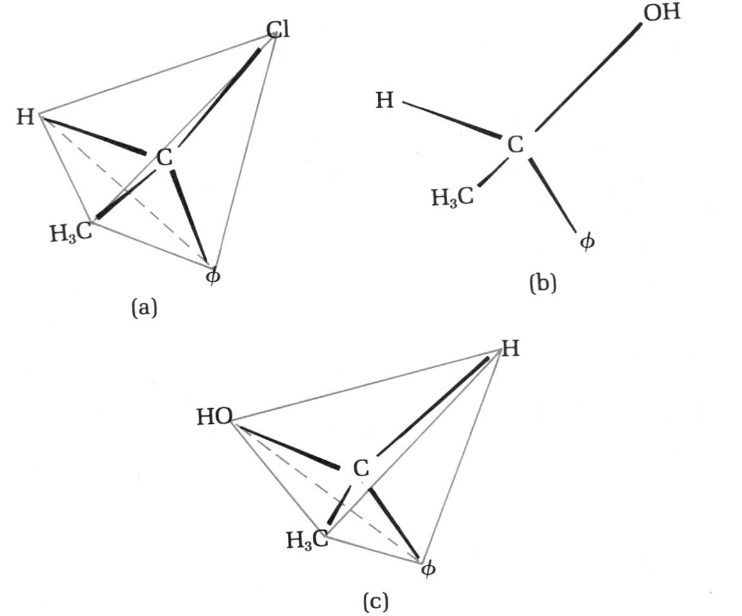

Figure 21–15
Reactant (a) and products, (b) and (c), in the hydrolysis of one enantiomer of an alkyl chloride. The symbol ϕ signifies a benzene ring.

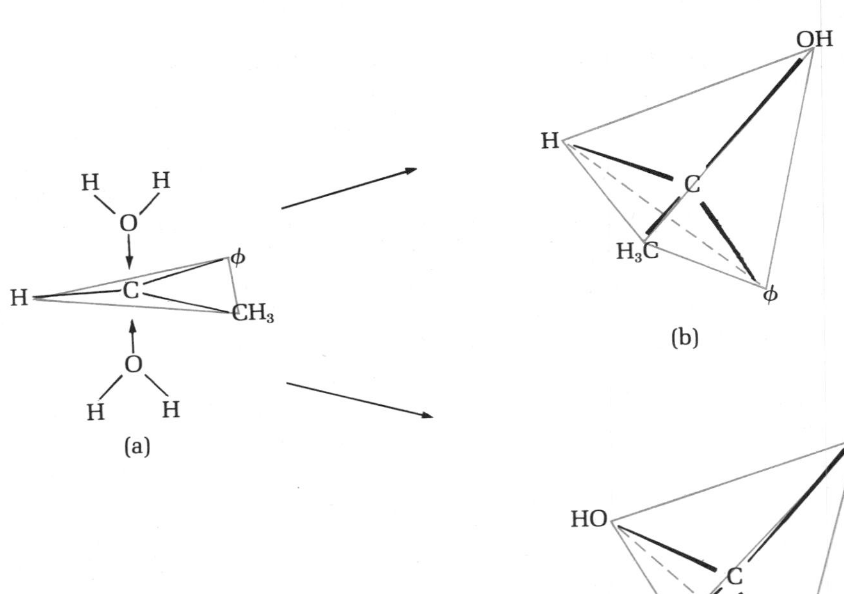

Figure 21–16
The formation of either of two enantiomeric alcohols, (b) and (c), by the attack of a water molecule on a planar cation (a).

polar; but most salts are considerably more soluble in water than in ethanol, indicating that the C—Cl bond splitting becomes much easier as the solvent is made richer in water.

Heterogeneous Catalysis

Most of the examples of catalysis considered to this point have represented **homogeneous** catalysis; that is, the catalyst is present in the same phase as the reactants and products. Industrial processes very commonly involve **heterogeneous catalysis,** in which a solid catalyst influences the reaction of gaseous or solution species. One common reaction subject to such catalysis is the hydrogenation of C—C double bonds. The hydrogenation of ethylene is an example:

$$H_2(g) + H_2C{=}CH_2(g) \rightarrow H_3C{-}CH_3(g) \qquad \textbf{21–55}$$

Figure 21–17 shows the bimolecular encounter between H_2 and C_2H_4 molecules that one might believe would lead to the rapid formation of ethane. Again we might propose, as in Figure 21–4, that four electrons transfer, as shown by the black arrows in Figure 21–17(a), to break the C—C pi bond and the H—H bond as two C—H bonds form. However, if we were to analyze this transfer as in Figure 21–4(b), we should see that it cannot occur. As with the reaction between H_2 and Cl_2, one bond must break completely before any other bond can form, and this results in a large activation barrier.

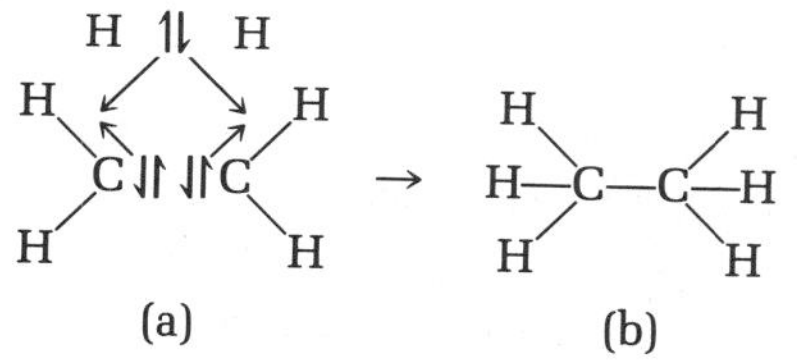

(a) (b)

Figure 21–17
A proposed mechanism for hydrogenation of ethylene.

The most commonly used catalyst in hydrogenation reactions is finely divided metallic nickel. The fine division yields a large surface area, and the Ni can react with H_2 at this surface to form Ni—H bonds as shown in Figure 21–18. The Ni—H bonds shown extend up from the planar surface of the metal. Transition metals can bind to compounds containing C═C bonds

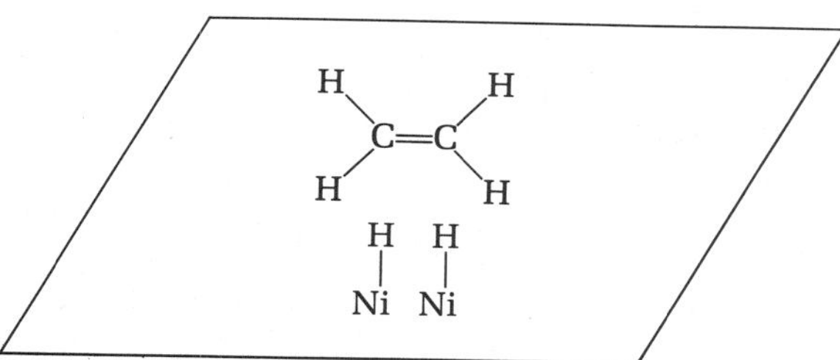

Figure 21–18
A portion of the surface of metallic nickel that is in contact with a gaseous mixture of hydrogen and ethylene.

(Chapter 11). Thus, ethylene molecules will also bind to the nickel surface, as shown in Figure 21–18. The plane of the ethylene molecule is parallel to the nickel surface. The nickel plays roles that are very similar to those played by the titanium in the polymerization of ethylene. The binding of hydrogen atoms and ethylene molecules overcomes the barriers of encounter and orientation. Furthermore, the orbital overlaps are such that C—H bonding can form at the same time as Ni—H bonding breaks. Thus, $\Delta H^{\ddagger}$ is small. Transition metals and their compounds are the most common catalysts in many different industrial processes largely because of their ability to bind simultaneously to a wide variety of substances. The binding of molecules to solid surfaces is a subject under heavy investigation by industrial researchers since the detailed mechanisms of action of many important industrial catalysts are only very poorly understood. This area of research is very important, and one in which there will be increasing activity for many years to come.

SUGGESTIONS FOR ADDITIONAL READING

King, E. L., *How Chemical Reactions Occur*, Benjamin, New York, 1963.
Frost, A. A., and Pearson, R. G., *Kinetics and Mechanism*, Wiley, New York, 1961.
Dence, J. B., Gray, H. B., and Hammond, G. S., *Chemical Dynamics*, Benjamin, New York, 1968.

PROBLEMS

1. A given reaction possesses a ΔG° of -55.3 kcal and a $\Delta G^{\ddagger}$ of 25.1 kcal. What is the $\Delta G^{\ddagger}$ for the reverse of this reaction?

*2. It is impossible for two hydrogen atoms to encounter each other in the gas phase to form a stable H_2 molecule. Some third atom or molecule must be present in the encounter. Why?

3. Figure 21–4(b) shows one of the two orbital overlaps required in the mechanism for HCl formation depicted in Figure 21–4(a). Draw the second overlap picture to show that it is also zero.

4. Which of the two following reactions would you predict to proceed more rapidly at room temperature in the absence of a catalyst? Explain.
   a. $H_2(g) + Cl_2(g) \rightarrow 2\ HCl(g)$
   b. $3\ H_2(g) + N_2(g) \rightarrow 2\ NH_3(g)$

5. Draw the orbital overlap that provides one explanation why the following reaction is rapid.

$$H + Cl_2 \rightarrow HCl + Cl$$

*6. A mechanism that we could propose for reaction 21–1 involves the H_2 and Cl_2 molecules meeting end-on to bring about the following reaction:

$$\text{H—H} + \text{Cl—Cl} \rightarrow \text{H} + \text{H—Cl} + \text{Cl}$$

Show by orbital overlap that this reaction can occur. Calculate the enthalpy barrier to the reaction. Would you predict a significant role for this mechanism in the reaction of H_2 and Cl_2? Explain.

7. The rates of a large number of chemical reactions increase by many powers of ten in raising the temperature from 298 °K to 1000 °K. Do any of the terms in equation 21–9 show such a dramatic temperature dependence? Is the dramatic temperature dependence due to an increase in molecular encounters at elevated temperatures? Explain.

8. The compound di-tert-butyl peroxide decomposes in the gas phase at elevated temperature according to the following equation:

$$(CH_3)_3C\text{—O—O—}C(CH_3)_3 \rightarrow 2\ O{=}C(CH_3)_2 + C_2H_6$$

The reactant and both products of the reaction are gases and may be treated as ideal gases in the analysis of the data. Table 21–3 lists some values of the total pressure of reactants and products as a function of time (time zero corresponds to pure reactant). Test these data to see if the rate law is first- or second-order in di-tert-butyl peroxide. Determine the value of the reaction-rate constant.

Time (min)	Pressure (torr)
0	173.5
5	205.3
9	228.6
15	259.2
20	282.3

Table 21–3
Total Pressure as a Function of Time in the Decomposition of Di-*tert*-butyl Peroxide

Hint: Calculate the partial pressure of di-*tert*-butyl peroxide from the total pressure and then plot it as in Figures 21–6 and 21–7.

9. Nitric oxide can be reduced by hydrogen in the gas phase at elevated temperature according to the reaction

$$2\ NO(g) + H_2(g) \rightarrow N_2O(g) + H_2O(g)$$

Table 21–4 lists some half-times for the reaction as a function of the initial pressures of the reactants. From these data deduce the order of

Table 21–4
Half-Times as a Function of Initial Pressures in the Gas Phase Reaction of NO and H_2

Experiment	P°_{NO} (torr)	$P^\circ_{H_2}$ (torr)	$t_{1/2}$ (sec)
1.	600	10	19.2
2.	600	20	19.2
3.	20	600	835
4.	40	600	417

the reaction with respect to both NO and H_2 and evaluate the rate constant.

10. An example of a rate law that is more complex than equation 21–13 occurs with the reaction

$$H_2O_2(aq) + 2\ H^+(aq) + 3\ I^-(aq) \rightarrow I_3^-(aq) + 2\ H_2O(l)$$

for which the rate law is

$$\frac{d[I_3^-]}{dt} = k_1[H_2O_2][I^-] + k_2[H_2O_2][I^-][H^+]$$

This rate law indicates that two mechanisms proceed with comparable rates. Propose two such mechanisms, each to be consistent with one of the two terms in the rate law above.

11. For the reaction in problem 9, propose a mechanism that is consistent with the rate law deduced in that problem.

12. Propose a mechanism for the decomposition of di-tert-butyl peroxide in the gas phase. Organic reactions usually yield small amounts of products in addition to the principal product, and the identification of these species is very useful in identifying intermediate species in the mechanism of formation of the principal product. Two of these minor products in the case of the decomposition of di-tert-butyl peroxide are methane and methylethyl ketone. Write reaction steps for the formation of these products from intermediates in the reaction mechanism.

13. When ozone is present in a small amount compared with O_2, as in air, the rate law for the decomposition of ozone to yield oxygen is

$$\frac{-dP_{O_3}}{dt} = k_{rate}\frac{P^2_{O_3}}{P^2_{O_2}}$$

Propose a mechanism that is consistent with this rate law.

14. The oxidation of iodide ion by persulfate ion proceeds according to the reaction

$$S_2O_8^{2-}(aq) + 2\ I^-(aq) \rightarrow 2\ SO_4^{2-}(aq) + I_2(s)$$

The following mechanism has been proposed:

$$S_2O_8^{2-} + I^- \overset{K_e}{\leftrightarrows} SO_4I^- + SO_4^{2-} \qquad \text{(equilibrium)}$$

$$SO_4I^- + I^- \rightarrow SO_4^{2-} + I_2 \qquad \text{(rds)}$$

What rate law does this mechanism yield? The actual rate law shows a first-order dependence on both $[I^-]$ and $[S_2O_8^{2-}]$. Propose a plausible mechanism.

15. The species OCCl has been proposed as an intermediate in the formation of phosgene from CO and Cl_2. Propose a structure and describe the bonding in OCCl. The bond enthalpy of carbon monoxide is 257 kcal/mole. Refer to the bond enthalpies in Table 6–3 to calculate a $\Delta H°$ for the formation of OCCl from CO and Cl. Is OCCl a plausible reaction intermediate? Explain.

***16.** Show that the orbital overlap is favorable in Figure 21–9(a) for the transfer of electrons from the nonbonding nitrogen orbitals to the antibonding σ and π orbitals of O_2 in order to break the O=O bonds as the two nitrogen-oxygen bonds are formed.

17. Methyl ethyl ether can be synthesized in the following reaction:

$$H_3CI + H_3CCH_2O^- \rightarrow H_3COC_2H_5 + I^-$$

Rate constants for this reaction have been measured as a function of temperature, and two values are 2.4×10^{-4} at 285 °K and 1.0×10^{-3} at 297 °K. Calculate the values of $\Delta H^{\ddagger}$ and $\Delta S^{\ddagger}$ for this reaction. It has been proposed that the mechanism involves a bimolecular rate-determining step. Are the values of $\Delta H^{\ddagger}$ and $\Delta S^{\ddagger}$ consistent with a bimolecular rate-determining step? What solvent would you choose in which to carry out this reaction? Why is methyl iodide used instead of methyl chloride?

18. Electron exchange can be studied by isotopic substitution. For example, a solution of one isotope of Ce^{3+} can be mixed with the solution of another isotope in the form of Ce^{4+}. As electron transfer occurs between Ce^{3+} and Ce^{4+}, isotopic scrambling occurs, and after a sufficient time has elapsed, both the Ce^{3+} and the Ce^{4+} have the same isotopic composition. Thus, a study of the isotopic composition as a function of time yields the rate law for the electron exchange

$$^*Ce^{3+} + Ce^{4+} \rightarrow {}^*Ce^{4+} + Ce^{3+}$$

In this reaction the asterisk indicates one of the two isotopes. The $\Delta S^{\ddagger}$ for this electron exchange has been found to be −40 cal/deg. It has been proposed that the reaction occurs by a simple bimolecular step, so how do you account for this huge entropy barrier?

19. The activation enthalpy for reaction 21–48 is 40 kcal/mole. Refer to Figure 21–2, and then employ the value of $\Delta H^\circ_{f(HI)}$ to calculate $\Delta H^{\neq}$ for the reverse of reaction 21–48. The rate law for reaction 21–48 shows a first-order dependence on both P_{H_2} and P_{I_2}. At equilibrium, the rates of reactions 21–48 and its reverse must be equal and $P^2_{HI}/P_{H_2}P_{I_2} = K_e$, the equilibrium constant for the reaction. Use this information to derive the rate law for the reverse of reaction 21–48.

20. Two possible mechanisms for the reaction of NO and ozone to yield NO_2 and O_2 are

Mechanism 1

$$NO + O_3 \rightarrow NO_2 + O_2 \text{ (rds)}$$

Mechanism 2

$$O_3 \rightarrow O_2 + O \text{ (rds)}$$

$$O + NO \rightarrow NO_2 \text{ (rapid)}$$

The rate of the reaction shows only a small temperature dependence, and $\Delta H^{\neq} = 2$ kcal/mole. Which of these two mechanisms (if either) is consistent with a small activation enthalpy? Explain.

*21. Is it possible for a catalyst to affect a reaction rate without affecting the rate of the reverse of that reaction? *Hint:* Is it possible for a catalyst to change the equilibrium constant for a reaction?

22. The following reaction is catalyzed by H^+:

$$H_3CCH_2OH + Br^- \rightarrow H_3CCH_2Br + OH^-$$

Propose a mechanism that shows how the catalyst operates.

23. The reaction of $CO + O_2$ to produce CO_2 is very slow at room temperature in spite of the large negative ΔG° for it. Why is this reaction slow, whereas the reaction between NO and O_2 is extremely rapid?

24. The reaction between CO and O_2 is catalyzed by Cu_2O. Propose a mechanism for this catalysis. *Hint:* The ΔG°_f and ΔH°_f for solid Cu_2O are -34.98 and -39.84, respectively. Is calcium oxide also likely to catalyze this reaction? Explain.

22/ The Chemistry of Life Processes

Examining life processes is a fitting conclusion to the important concepts we have been investigating. If we are to understand how living organisms function we must apply many of those concepts. The molecules of living systems have complex structures and perform complex functions. Our aim is to understand the relation between the structures and functions.

22–1. The Driving Force for Life Processes

The only two spontaneous processes are the spreading of matter and the spreading of energy. Ultimately everything that occurs on this planet must be a consequence of the net spreading of matter or energy or both. Let us imagine ourselves sitting on the moon taking a detached look at the earth. Is there a net spreading of matter occurring there? There would be if the atmosphere were being lost to outer space and the oceans were evaporating to replace it. Such a process could yield the driving force for life processes, but it does not occur. Matter is trapped to the earth by gravity. The matter is rearranged in spontaneous processes, but net spreading does not appear to be occurring as we view it from the moon.

Is energy spreading occurring? The earth is obviously receiving sunlight because we see the earth by reflected sunlight from our position on the moon. We discern colors on the earth, and we conclude that much of the sunlight is absorbed and ultimately converted to thermal energy. Conversion of sunlight to thermal energy is the spreading of energy. The energy concentrated in a single photon of visible light can be divided into the thermal energy of a large number of molecules, and thus the process of light to thermal energy is definitely energy spreading.

The conversion of sunlight to thermal energy provides the ultimate driving force for all processes on this planet. Consider as an example a refrigerator in Buffalo, New York. Its driving force is the conversion of electricity to thermal energy. The output of this process is obvious if we touch the compressor, and the input is obvious if we neglect to pay the power company. What spontaneous process produces the electricity? In Buffalo, electricity is produced by a hydroelectric plant that takes advantage of the process

$$H_2O(\text{Lake Erie}) \rightarrow H_2O(\text{Lake Ontario}) + \text{thermal energy}$$

Water passing over Niagara Falls has its potential energy converted into thermal energy in Lake Ontario. The spontaneity of this energy spreading drives the hydroelectric turbines.

How does water come to be in Lake Erie? Lakes are collected rainwater. What is the driving force for rain? Hot, moist air encounters cool, dry air and, in the process of energy spreading, thermal energy is taken from the hot air and given to the cold. One way for energy to be liberated from the hot, moist air is for rain to form so that energy equal to the enthalpy of vaporization can be given to the cold air. How does some air become hot and moist? Air is heated by the absorption of sunlight, and the moisture comes principally from the evaporation of ocean water. Thus, the steps involved in the operation of the refrigerator are

1. Sunlight + water at low potential energy (ocean) → water vapor + (thermal energy)$_1$.
2. Water vapor → water at high potential energy (Lake Erie) + (thermal energy)$_2$.
3. Water at high potential energy (Lake Erie) → water at low potential energy (ocean) + (thermal energy)$_3$ + electricity. **22–1**
4. Electricity + energy inside refrigerator → energy outside refrigerator + (thermal energy)$_4$.

The sum of all these steps is

Sunlight + energy inside refrigerator → energy outside refrigerator + (thermal energy)$_1$ + (thermal energy)$_2$ + (thermal energy)$_3$ + (thermal energy)$_4$ **22–2**

Every step of the process involves energy spreading, and process 22–1 derives its spontaneity from conversion of sunlight to the four thermal energies.

We can outline a process analogous to 22–1 for the operation of living systems. The elements required for this process are

1. A trap for sunlight.
2. A large storehouse of Gibbs free energy, analogous to the water in Lake Erie.
3. A conveniently usable form of concentrated energy, analogous to electricity.

The primary storehouse of high *G* for living systems is **sugar** and its polymeric form **starch.** The sunlight trap is **chlorophyll,** and the analog of steps (1) and (2) of process 22–1 is **photosynthesis.** In photosynthesis, carbon dioxide and water are converted to sugar. Thus CO_2 and H_2O are the analogs of the water at low potential energy in process 22–1. The analog of step (1) of 22–1 is the absorption of light to produce an energetically excited chlorophyll. In the analog of step (2), this excited molecule brings about photosynthesis. The analog of step (3) of 22–1 is **respiration,** in which sugar is reconverted to CO_2 and H_2O with the concomitant production of a more conveniently usable type of energy-rich molecule. Finally, the energy-rich molecules react to provide the driving force for the myriad of processes that characterize the functioning of living systems. In all of the steps in living systems that match the steps in 22–1, thermal energy is produced. However, living systems operate in a much more efficient fashion than the scheme of 22–1 and the amount of thermal energy produced per amount of sunlight trapped is much smaller than in 22–1. Figure 22–1 diagrams the

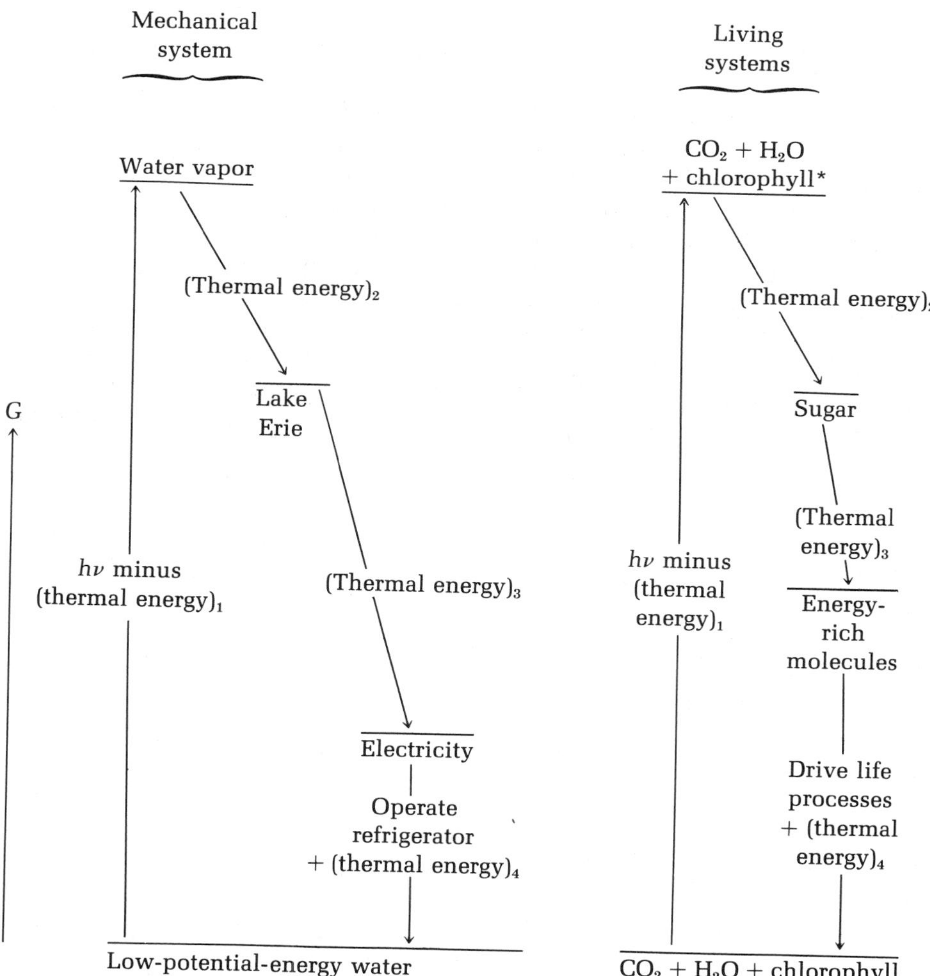

Figure 22–1
Analogy between the driving of a mechanical process and the driving of living processes by the conversion of sunlight to thermal energy.

analogy between living systems and process 22–1. In the mechanical process, water is the material cycled between high-energy and low-energy states, and in the living systems carbon, hydrogen, and oxygen are cycled between high- and low-energy forms.

22–2. Photosynthesis

Carbohydrates

The products of photosynthesis are composed of molecules with the general formula $C_nH_{2n}O_n$. Thus, each molecule contains n carbon atoms and the atoms of n water molecules. The substances are termed **carbohydrates** because they may be decomposed to carbon and water. The most common example of a carbohydrate is **glucose,** whose molecular structure is shown in Figure 22–2. Glucose is a simple sugar (the common term is *monosaccharide*). Simple sugar molecules contain either five or six carbons, and the functional groups are hydroxyl groups and a single carbonyl group, forming either a ketone or an aldehyde, as shown in Figure 22–2(a).

The asterisks in Figure 22–2(a) denote asymmetric carbon atoms (recall Chapter 10). Since each asymmetric carbon yields two optical isomers, there are fifteen other molecules that are related by optical isomerism to the molecule depicted in Figure 22–2(a). Only two of these isomers are termed *glucose*, and the structure in Figure 22–2(a) is D-glucose. A small capital D or L in the name of an optical isomer does not indicate the sense of the rotation of the plane of polarized light, as *d* and *l* do. Rather D and L refer to specific molecular structures, and a D isomer may well be levorotatory, for example.

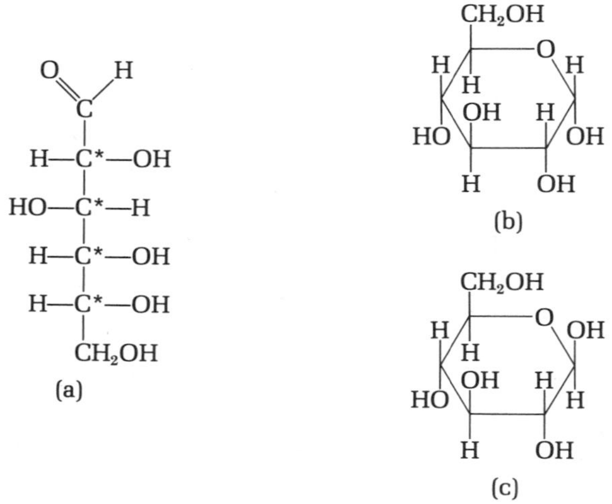

(a)

(b)

(c)

Figure 22–2
The structure of D-glucose: (a) open-chain structure; the ring forms, (b) α-D-glucose and (c) β-D-glucose.

Aldehydes and ketones can react with water by having H^+ add to the carbonyl oxygen and OH^- add to the carbonyl carbon. Aldehydes and ketones can add alcohols in analogous fashion. Monosaccharides in aqueous solution exist as an equilibrium mixture of the open-chain form shown in Figure 22–2(a) and ring forms in which the hydroxyl group on the asymmetric carbon farthest removed from the aldehyde or ketone group adds to the carbonyl bond. For D-glucose, this addition yields one or both of two compounds α-D-glucose and β-D-glucose, and their structures are shown in Figures 22–2(b) and (c). The two ring forms differ in the relative positions of the hydroxyl groups.

The difference between the α and β isomers of glucose is very considerable in biological systems. Monosaccharides can polymerize by forming ether linkages, as shown in Figure 22–3. Starch is a high-molecular-weight polymer involving the linkage of α-glucose molecules, as shown in Figure 22–3. Animals contain substances that can split the linkage (termed the

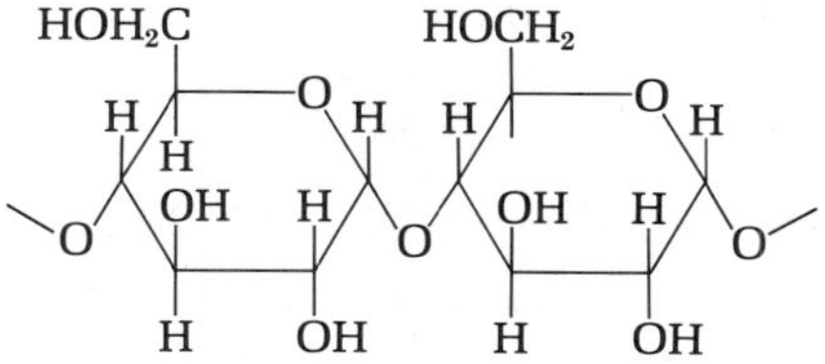

Figure 22–3
A portion of the structure of starch.

glycoside bond) between α-glucose molecules and they are able to use starch as a foodstuff. **Cellulose** is a high-molecular-weight polymer involving the corresponding linkage of β-glucose molecules. Most animal digestive systems (termites are an exception) cannot split the linkage in cellulose, and consequently cellulose is not a foodstuff for most animals.

The glycoside bond also occurs in substances of much smaller molecular weight than starch and cellulose. For example, **sucrose** (common table sugar) is a disaccharide formed from the linkage of α-D-glucose and β-D-fructose, as shown in Figure 22–4.

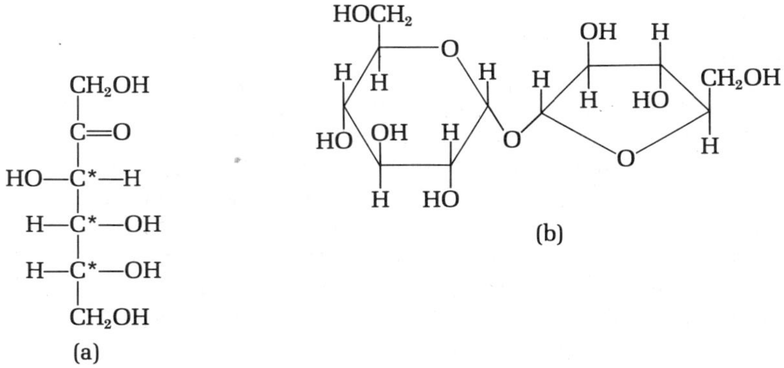

Figure 22–4
The structures of (a) D-fructose and (b) sucrose.

The Chemical Result of Photosynthesis

The carbohydrates are the primary energy storehouse molecules in living systems. They are primary in the sense that they are most closely related to photosynthesis. As we shall see, there are other important energy-concentrating molecules as well, but they are not direct products of photosynthesis. Let us investigate the photosynthesis of glucose.

The oxidation of glucose according to the reaction

$$C_6H_{12}O_6(s) + 6\ O_2(g) \rightarrow 6\ CO_2(g) + 6\ H_2O(l) \qquad \textbf{22–3}$$

possesses a $\Delta G°$ of -686 kcal/mole. The huge negative $\Delta G°$ for the combustion of glucose is typical of a six-carbon sugar or of a six-carbon portion of starch. Thus, we are justified in placing the carbohydrates at their high level in Figure 22–1. The photosynthesis of glucose is the reverse of reaction 22–3, and it requires a Gibbs free energy **input** of 686 kcal/mole of glucose produced. This energy is derived from sunlight.

We can use the half-cell method of balancing redox equations (recall Chapter 20) to show that each glucose molecule loses 24 electrons in reaction 22–3. The balanced half-reactions are

$$C_6H_{12}O_6(s) + 6\ H_2O(l) \rightarrow 6\ CO_2(g) + 24\ H^+(aq) + 24\ e^- \qquad \textbf{22–4}$$

$$24\ e^- + 6\ O_2(g) + 24\ H^+(aq) \rightarrow 12\ H_2O \qquad \textbf{22–5}$$

Thus, the photosynthesis of glucose, which involves the reversal of these half-reactions, is the reduction of CO_2 by H_2O, with each CO_2 gaining 4 electrons and each water molecule losing two electrons. Very few species are sufficiently strong oxidizing agents to oxidize H_2O to O_2 (elemental fluorine is one). Thus, the trapping of sunlight must lead to the formation of an extremely strong oxidizing agent. Likewise CO_2 is a very poor oxidizing agent, and the trapping of light must also produce a very strong reducing agent in order to accomplish the reduction of CO_2. Finally, the strong oxidizing agent and strong reducing agent must be kept apart or they will react with each other in preference to H_2O and CO_2, respectively.

The Role of Chlorophyll

How are these strong oxidizing and reducing agents produced simply by the absorption of light? A strong reducing agent is a substance like elemental sodium, which contains one or more electrons in an orbital or orbitals of relatively high energy. Such an electron can be produced in any substance by the absorption of electromagnetic radiation, in which the electron passes from a low-energy orbital to an excited orbital. With most molecules, the absorption of light does not bring about reduction because the energy absorbed from the radiation is quickly reradiated or converted to thermal energy. To bring about the desired reduction, the excited molecule must be able to hold its excess energy long enough to encounter the molecule to be reduced and to pass the electron to it.

How can an electron be trapped in an excited state for a relatively long time? We have seen that electronic excitation occurs with no change in the electron-spin orientation (Chapter 12). If the electron-spin orientation of the electron can change by some mechanism in the excited state, the return of the reoriented electron is greatly impeded, and the electron is held in the excited state for a relatively long time. Figure 22–5 diagrams this scheme for the trapping of an electron in an excited state. If no electron acceptor is available for the excited electron, the energy will eventually be lost by reradiation. The slow reemission of light following its absorption is the reason that some substances glow in the dark; this phenomenon of the slow reemission of light is termed **phosphorescence.** Thus, the molecule that traps light in photosynthesis must be phosphorescent when there are no molecules to be reduced.

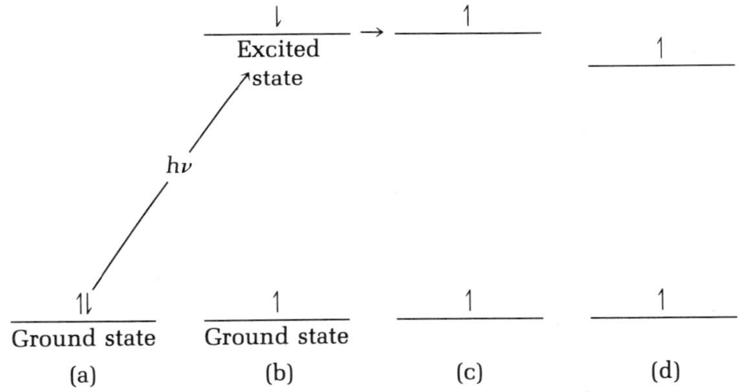

Figure 22–5
Steps involved in producing the strong oxidizing agent and strong reducing agent required in photosynthesis. (a) Ground state of unexcited chlorophyll, (b) chlorophyll states immediately after excitation, (c) chlorophyll states after spin reorientation, (d) excited state electron and ground state electron in different chlorophyll molecules.

Figure 22–5(c) depicts both a strong reducing agent and a strong oxidizing agent. The high-energy electron is the source of the reducing power, and the vacancy in the ground level is the source of the oxidizing character. An extremely important requirement is that the excited electron should not be located near the low-energy vacancy in the same molecule. If this happens to be the case, when a molecule is reduced by accepting the high-energy electron, the molecule immediately loses the electron to the low-energy vacancy. The excited electron should be transferred to the excited level of another molecule that is fixed in some way so that it cannot transfer the electron back to the vacancy. Figure 22–5(d) illustrates the excited electron in one molecule and the vacancy in the ground level of the molecule originally excited.

The molecule that performs all of the functions outlined is chlorophyll, whose structure is shown in Figure 22–6. Most of the molecules we shall study are rather complex. The molecular structures are presented not to be

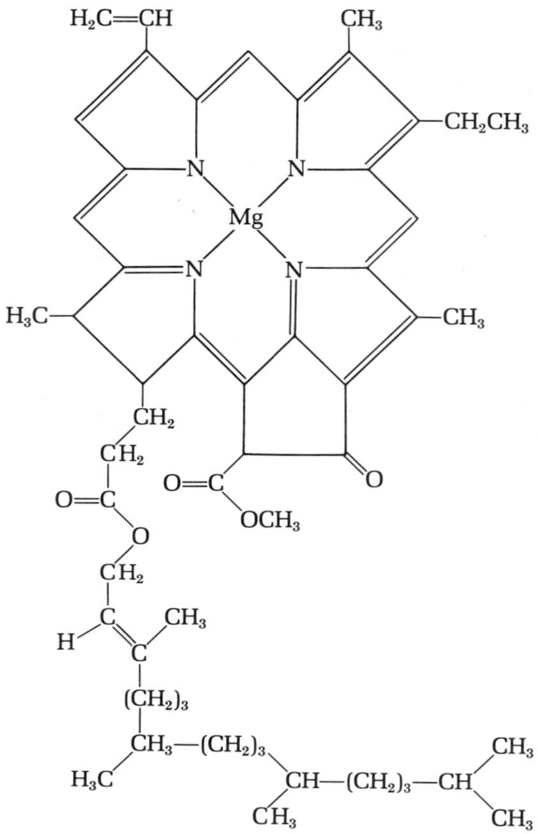

Figure 22–6
The structure of chlorophyll a.

memorized but to present something definite to refer to in examining structure-function relations. The ring portion of the chlorophyll molecule is a large conjugated system and absorbs light strongly in the visible range by essentially the same mechanism as that for vitamin A aldehyde (Chapter 12). The molecule is a donor-acceptor complex of a Mg^{2+} ion and a ligand with a charge of -2, so that the chlorophyll molecule possesses no net charge. The lack of a charge together with the long hydrocarbon "tail" of the molecule enables it to reside in cell membranes, and this is an important aspect of its functioning. In the membrane, chlorophyll molecules can be located near each other, but their motion relative to each other is highly restricted. This allows electrons to pass to certain chlorophyll molecules, where they yield the desired reducing character (Figure 22–5(d)), but it prevents those electrons from returning to the low-energy vacancies in the chlorophyll molecules originally excited.

Another important feature of photosynthesis can be deduced if we try to contemplate a detailed mechanism for the reverse of the reaction 22–3.

The reduction of CO_2 can lead to many possible products, including among others any of the hydrocarbons. If photosynthesis is to produce carbohydrates as the product exclusively, it must involve a great deal more than simple electron transfer to CO_2. Preferably the strong reducing agent should be unable to reduce CO_2 directly, either because the two species cannot encounter each other or because there is another large activation barrier to reaction. Thus, electron transfer should occur through a system of courier molecules, each of which can bring about only a small number of changes of the CO_2 and the intermediates between CO_2 and glucose. This courier mechanism occurs by the reduction of the courier by our strong reducing agent. The courier then either reduces another courier or reduces CO_2 or one of the intermediate species. (We shall examine some common courier molecules in a subsequent section.)

The following scheme outlines the sequence of steps in photosynthesis:

1. Chlorophyll + sunlight → strong reducing agent.
2. Strong reducing agent + oxidized couriers → reduced couriers + strong oxidizing agent. **22–6**
3. Reduced couriers + CO_2 → reduced CO_2 + oxidized couriers.
4. Strong oxidizing agent + H_2O → chlorophyll + oxygen.

The sum of the steps in this scheme is sunlight combining with water and carbon dioxide to produce reduced CO_2 (carbohydrates) and oxygen. All of the other species are recycled and consequently serve as catalysts.

22–3. Metabolic Oxidation of Carbohydrates

The Electron Transfer Chain

The chemical reactions in living systems make up what is termed **metabolism.** Metabolism is divided into those reactions in which complex molecules are produced from simpler ones (termed **anabolism**) and those in which complex molecules are degraded to simpler molecules (termed **catabolism**). Anabolism produces energy-storehouse molecules, molecules that provide the structure of the living system, and molecules capable of carrying out complex functions (the analog of the refrigerator in Figure 22–1). Catabolic reactions supply the energy to drive the anabolic reactions and also to operate the complex functions of the complex molecules that result from anabolic reactions. In analogy with Figure 22–1, they supply the energy to build the refrigerator at the factory as well as to operate it in the home.

The fundamental structural unit of living systems is the cell. Figure 22–7 shows a simplified drawing of an animal cell. Oxidative degradation of molecules such as glucose takes place in **mitochondria,** and two mitochondria are shown in Figure 22–7.

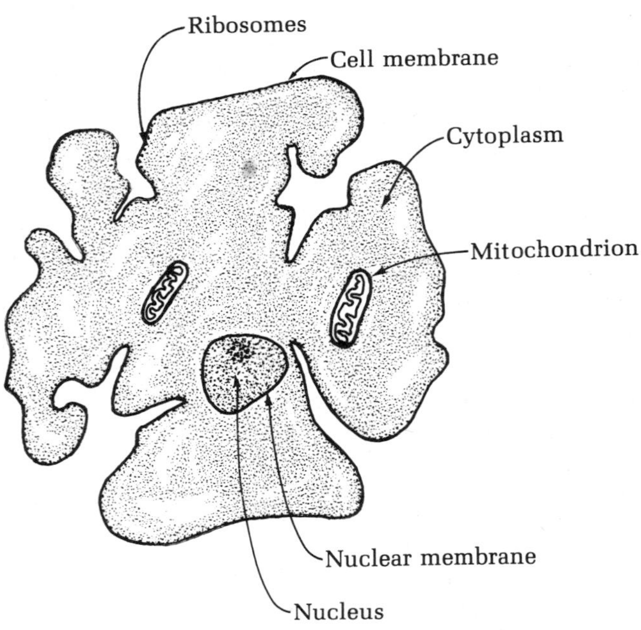

Figure 22–7
Simplified drawing of an animal cell.

Catabolic oxidation of glucose is the prime catabolic reaction for the production of energy-rich molecules in Figure 22–1. The oxidation of glucose to carbon dioxide and water is extremely slow at room temperature in the absence of catalysts. In living systems the catalysts are protein molecules, termed **enzymes,** that operate in conjunction with courier molecules of the type mentioned previously. (The structures and enzymatic activities of proteins are the subject of a subsequent section of this chapter.) In this section we shall focus on courier electron transfer molecules and the intermediates in the oxidation of glucose.

Figure 22–8 shows the chain of electron transfers between foodstuff molecules (such as glucose molecules as they undergo oxidation) and molecular oxygen. The courier molecules are the seven species from NAD to cytochrome oxidase inclusive. The reduction potential of oxygen gas at pH 7.4 is 0.80 volt. The reduction potentials of the four cytochromes are 0.05, 0.22, 0.25, and 0.28 for cytochromes b, c_1, and c, and cytochrome oxidase, respectively. Thus, molecular oxygen accepts electrons from cytochrome oxidase, which then accepts electrons from cytochrome c, which oxidizes cytochrome c_1, and so forth. The weakest oxidizing agent of the seven is NAD, which has a reduction potential of −0.33 volt. However, NAD is a sufficiently strong oxidizing agent to oxidize foodstuffs.

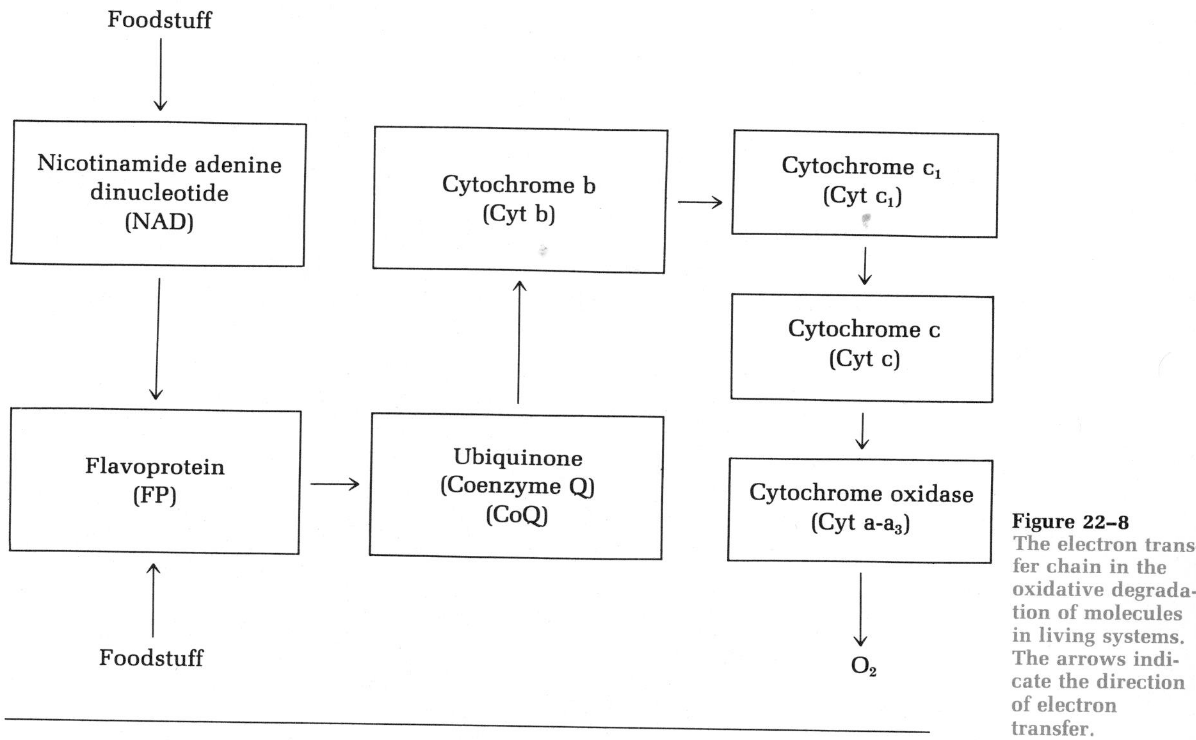

Figure 22–8 The electron transfer chain in the oxidative degradation of molecules in living systems. The arrows indicate the direction of electron transfer.

OPTIONAL MATERIAL

The Cytochromes

Cytochrome oxidase, one of the courier electron transfer agents in Figure 22–8, is a large molecule of molecular weight 240,000. Most of the molecule consists of protein, but the redox character is due to iron and copper ions complexed within the molecule. These ions can exist in the Fe(III), Fe(II), Cu(II), and Cu(I) states, and one molecule of cytochrome oxidase can supply the four electrons required to reduce an O_2 molecule to two water molecules. The nature of the binding of the copper ions in cytochrome oxidase is not known, but the iron binding involves a structure that is similar to that found in many biologically important molecules; it is shown in Figure 22–9. The large ring system containing four nitrogen atoms was also encountered in the structure of chlorophyll (Figure 22–6). The molecule depicted in Figure 22–9 is called **porphin,** and both chlorophyll and cytochrome oxidase (and also several other important molecules) are derived from it. The porphin molecule can lose both N—H protons to produce a tetradentate ligand that can associate with an Mg^{2+} as in Figure 22–6 or with an Fe^{2+} as in Figure 22–10. Iron complexes derived from porphin are termed **hemes,** and the structure shown in Figure 22–10 is termed heme A. The two carboxylate groups at the bottom of Figure 22–10 provide the points of

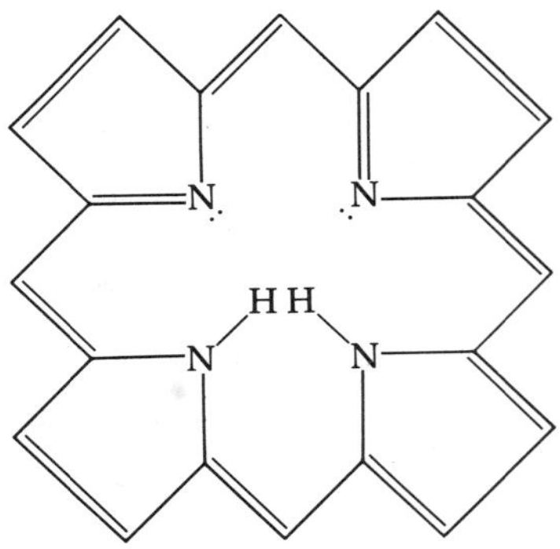

Figure 22–9
The structure of porphin.

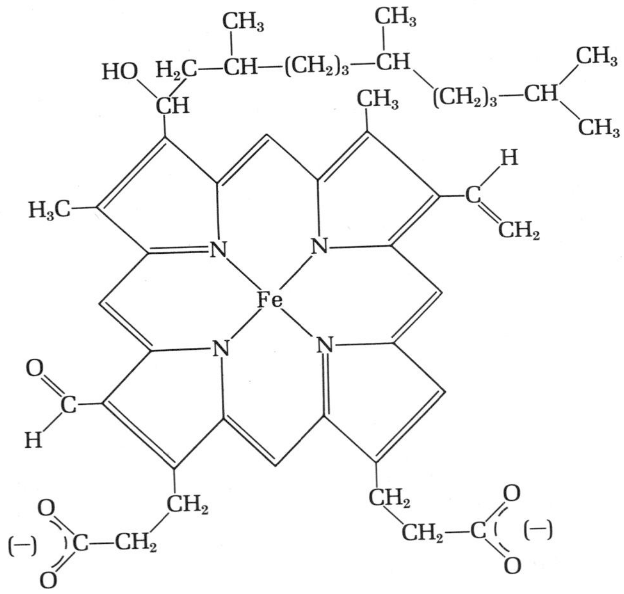

Figure 22–10
The structure of heme A.

attachment of the heme to the protein portion of cytochrome oxidase. The name *cytochrome* derives from the fact that these species are colored; this color comes from the extensive conjugation in the heme.

All four cytochromes in Figure 22–8 contain heme groups, but the detailed structures of the hemes are different for the different cytochromes, and the nature of the protein is also different. These secondary changes account for the differing reduction potentials among the cytochromes. Figure 22–11 shows the heme portion of cytochrome c. The groups on the periphery of the heme differ from those of heme A. The two sulfur atoms at the top of the figure provide the link to the protein portion of the cytochrome.

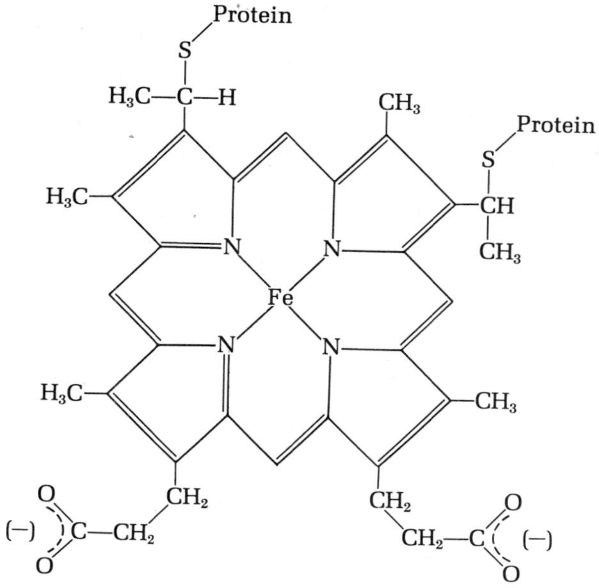

Figure 22–11
The heme portion of cytochrome c.

Ubiquinone

The structure of ubiquinone (or coenzyme Q; CoQ, it is commonly termed) is shown in Figure 22–12(a). The name *quinone* is associated with the structure

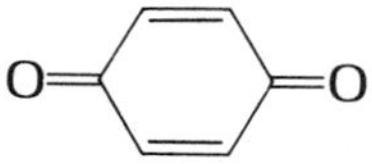

and quinones are mild oxidizing agents. The reduced form is obtained by the addition of two electrons and two protons to the quinone, and it is termed a hydroquinone. The hydroquinone form of CoQ is shown in Figure 22–12(b).

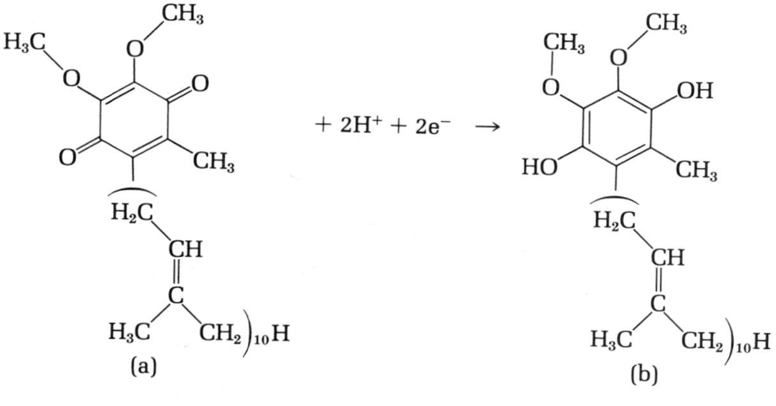

$$+ 2H^+ + 2e^- \rightarrow$$

(a) (b)

Figure 22–12
(a) The oxidized and (b) reduced forms of ubiquinone.

Flavoproteins

The flavoproteins (FP) are derivatives of riboflavin (vitamin B_2). The vitamins, in general, are relatively small molecules (the molecular weights are of the order of several hundred amu) that serve in conjunction with enzymes to catalyze specific biochemical reactions. Figure 22–13(a) shows the structure of riboflavin in its oxidized form. The prefix *ribo-* comes from the polyhydric alcohol chain in the figure, which is a derivative of the sugar D-ribose. This sugar is a constituent of several important biochemical molecules. The —OH on the terminal carbon in the chain provides the link to the remainder of the flavoprotein. The oxidized form of riboflavin can accept two electrons and two protons to form the reduced species shown in Figure 22–13(b). The asterisks indicate the two protons accepted.

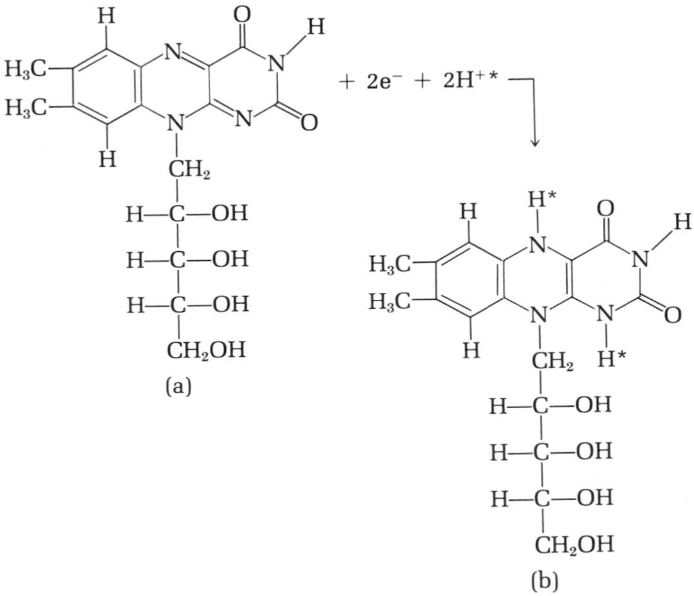

$+ 2e^- + 2H^{+*}$

(a)

(b)

Figure 22–13
The oxidized (a) and reduced (b) forms of riboflavin (Vitamin B_2).

Nicotinamide Adenine Dinucleotide

Nicotinamide adenine dinucleotide is also the derivative of a B vitamin, in this case the vitamin niacinamide (more commonly termed nicotinamide). Figure 22–14(a) shows the oxidized form of NAD. The dashed box locates the nicotinamide portion of the molecule. The two sugar units are both derived from β-D-ribose and are linked by a pyrophosphate bridge. The final portion of the molecule is the nitrogen-containing fused ring system termed **adenine.** Adenine is an important constituent of several biochemical molecules.

The reduction of NAD occurs in the nicotinamide portion through the acceptance of two electrons and one proton to yield the structure of the

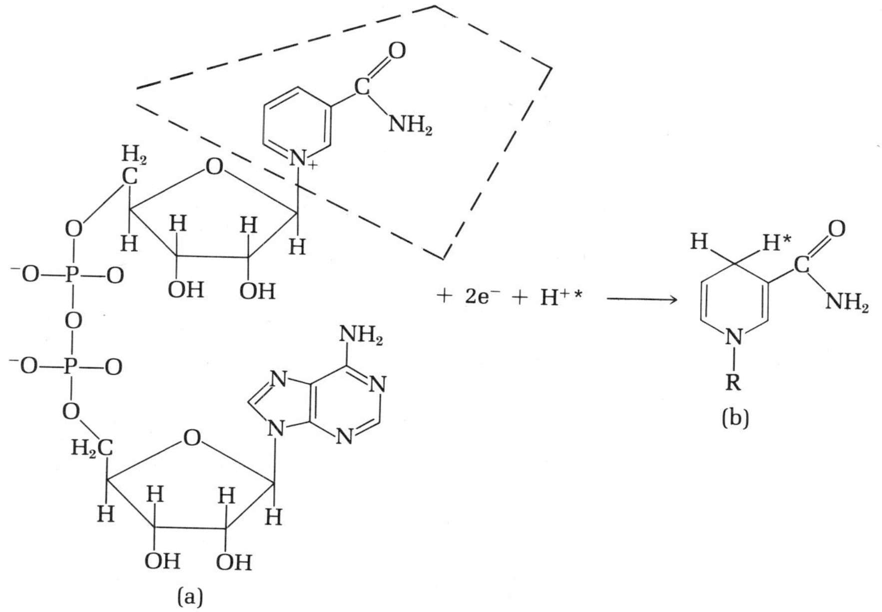

Figure 22–14 (a) The oxidized and (b) the reduced forms of nicotinamide adenine dinucleotide. The symbol *R* in (b) denotes the portion of (a) outside the dashed enclosure.

reduced molecule shown in Figure 22–14(b). The oxidized form of NAD is commonly symbolized as NAD^+, because of the positive charge on the nicotinamide ring, and the reduced form is symbolized by NADH.

Adenosine Triphosphate

According to the scheme to the right in Figure 22–1, an important feature in the oxidation of carbohydrates is the production of energy-rich molecules. The energy-rich species is **adenosine triphosphate** (ATP); its structure is shown in Figure 22–15(b). During the oxidative degradation of molecules such as glucose, ATP is synthesized from **adenosine diphosphate** (ADP), whose structure is shown in Figure 22–15(a), by the lengthening of the polyphosphate chain. (The molecule ADP is identical in structure with the portion of NAD shown in the lower part of Figure 22–14(a)). The $\Delta G°$ for this production of ATP from ADP is 7.3 kcal/mole, and it arises almost exclusively from the matter concentration in making the larger ion and solvating it. Thus, the production of a mole of ATP represents a storage of 7.3 kcal of Gibbs free energy, and this energy can be reclaimed in any subsequent reaction that reverses this synthesis.

Since the synthesis of ATP possesses a positive $\Delta G°$, it must be driven by a process having a ΔG that is more negative than −7.3 kcal so that the overall process possesses a negative ΔG. Examples of such processes are

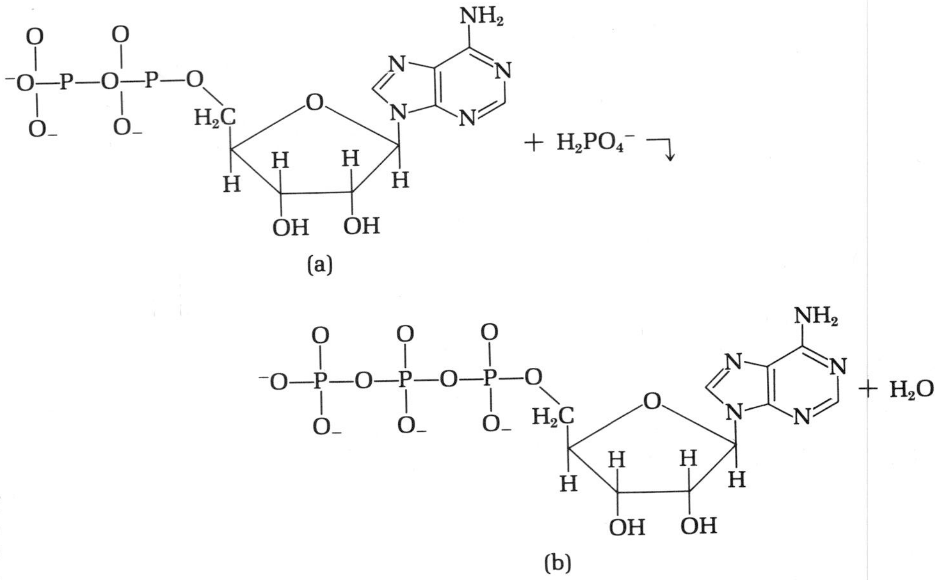

Figure 22–15
(a) The conversion of adenosine diphosphate to (b) adenosine triphosphate.

some of the electron transfers, as shown in Figure 22–8. Adenosine triphosphate molecules are synthesized during the processes in which electrons are transferred from NADH to oxidized FP, from reduced cytochrome b to oxidized cytochrome c_1, and from reduced cytochrome c to oxidized cytochrome oxidase.

Energy Storage in the Oxidation of Glucose

There are two distinct parts to the reaction path by which a molecule of glucose is converted to six CO_2 molecules and six water molecules. In the first part of the path, glucose is converted to **pyruvate,** a process termed **glycolysis.** The structure of the pyruvate anion is

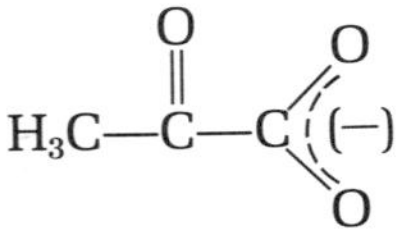

$$H_3C—\overset{\overset{\displaystyle O}{\|}}{C}—C\overset{O}{\underset{O}{\lessgtr}}(-)$$

The production of pyruvate represents only a small degree of oxidation of glucose, since both glucose and pyruvate contain one oxygen atom per carbon atom, and CO_2 contains two oxygen atoms per carbon atom. In glycolysis with one molecule of glucose, eight molecules of ATP are synthesized for an energy storage of (8)(7.3) = 58.4 kcal/mole of Gibbs free energy. This is a rather small quantity of energy compared with the $-\Delta G°$ of reaction 22–3 (686 kcal/mole).

Pyruvate is a common intermediate in the biological oxidation of several types of foodstuff molecules. The conversion of pyruvate to CO_2 is the principal source of ATP in the oxidation of foodstuffs. This conversion occurs in a cyclic mechanism termed the **tricarboxylic acid cycle,** or **Krebs cycle,** named for the scientist who discovered it. In the conversion of two pyruvate anions (derived from one glucose molecule) to carbon dioxide and water, thirty additional ATP molecules are synthesized. Thus, the complete oxidation of glucose results in the storage of $(38)(7.3) = 277$ kcal/mole of Gibbs free energy. Furthermore, this energy is in a form in which it can bring about very specific chemical reactions required for the functioning of living systems. In animals one of the most important such functions is muscle contraction, and the energy required for muscle contraction is derived from the conversion of ATP to ADP. The detailed molecular mechanism of muscle contraction is not known, although much is known about the process. (For an extended treatment of the topic, see Chapter 26 of the textbook by Lehninger, in the list at the end of this chapter.)

OPTIONAL MATERIAL

Glycolysis

The steps and intermediates of glycolysis are shown in Figure 22–16. The first step in Figure 22–16 involves the transfer of a phosphate group from ATP to glucose. Phosphate transfer from ATP to substrate molecules is one of the most common ways in which the free energy stored in ATP brings about chemical reactions of other molecules. The input and output of molecules that are not substrate molecules are shown by the curved arrows. Thus, Step 1 involves ATP as a reactant and ADP as a product. The numbering of the carbons of glucose is shown in color in the second structure in the sequence, so that the product of the first step is termed α-D-glucose-6-phosphate.

In the second step, the glucose phosphate converts to the isomeric fructose phosphate; and in Step 3, the fructose phosphate accepts another phosphate. In the fourth step, the molecule splits in half to produce two molecules of glyceraldehyde-3-phosphate. (The numbering of the glyceraldehyde carbons is given in color.)

Step 5 is the first oxidation step in the sequence and it is the oxidation of two aldehyde molecules to two phosphocarboxylic acid molecules with the transfer of four electrons to two NAD^+ molecules. (The symbol P_i in the figure stands for $H_2PO_4^-$.) Then 1,3-diphosphoglyceric acid successively undergoes phosphate transfer to ADP, isomerization, dehydration, and a second phosphate transfer to ADP to yield two pyruvate anions as the product.

Two ATP molecules are directly netted in Figure 22–16 (two each in steps 6 and 9 minus one each in steps 1 and 3). Six of the eight ATP mole-

α-D-Glucose

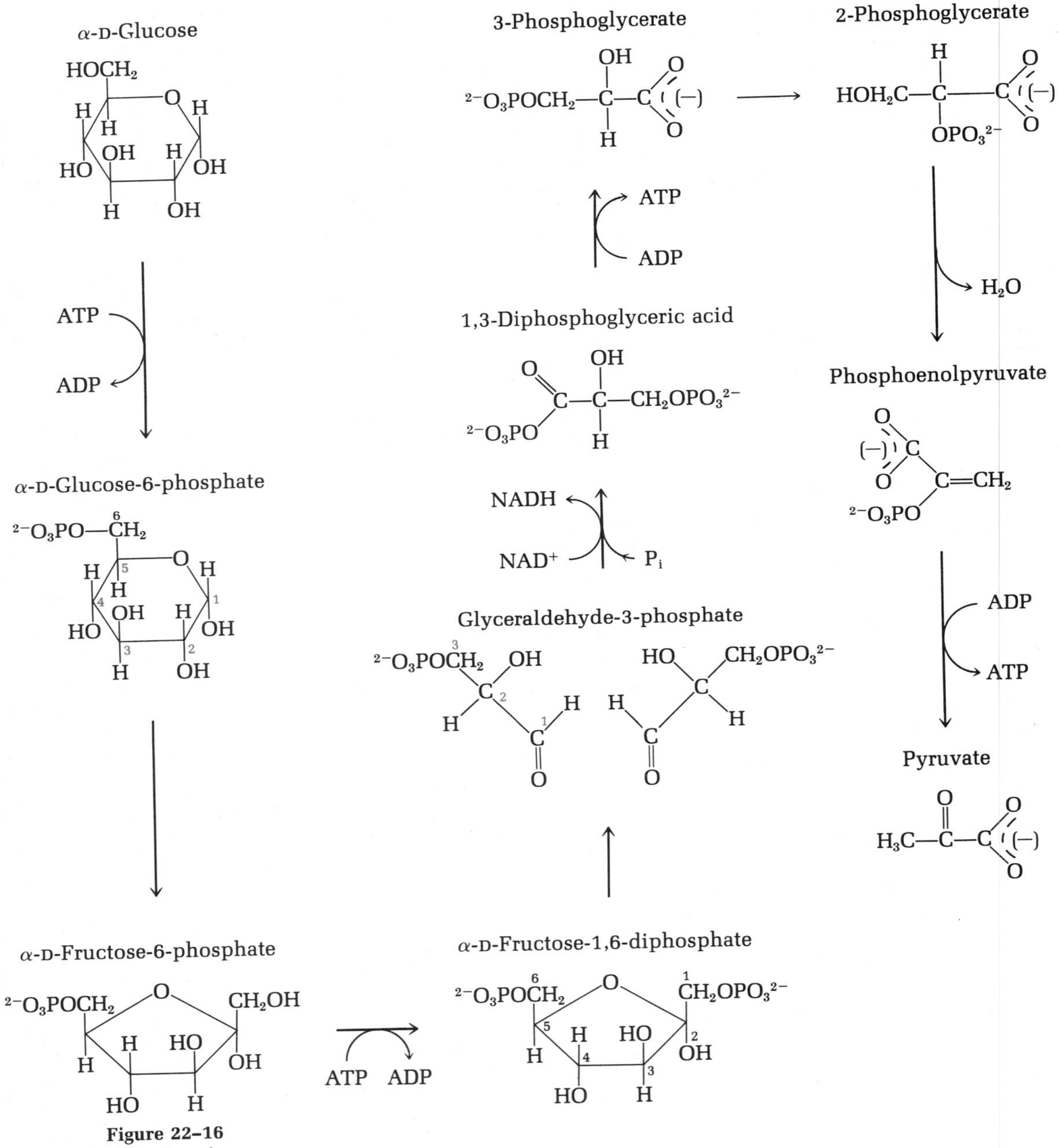

ATP → ADP

α-D-Glucose-6-phosphate

α-D-Fructose-6-phosphate

ATP ADP

α-D-Fructose-1,6-diphosphate

Glyceraldehyde-3-phosphate

NAD^+ P_i → NADH

1,3-Diphosphoglyceric acid

ADP → ATP

3-Phosphoglycerate

2-Phosphoglycerate

H_2O

Phosphoenolpyruvate

ADP → ATP

Pyruvate

Figure 22–16
Series of reactions in the conversion of glucose to pyruvate.

cules produced in glycolysis are produced in the electron transport chain of Figure 22–8 when four electrons pass from the two NADH molecules produced in Step 5 to an oxygen molecule.

The Krebs Cycle

The conversion of a pyruvate anion to water and carbon dioxide is diagrammed in Figure 22-17. The first CO_2 molecule is actually lost before entry

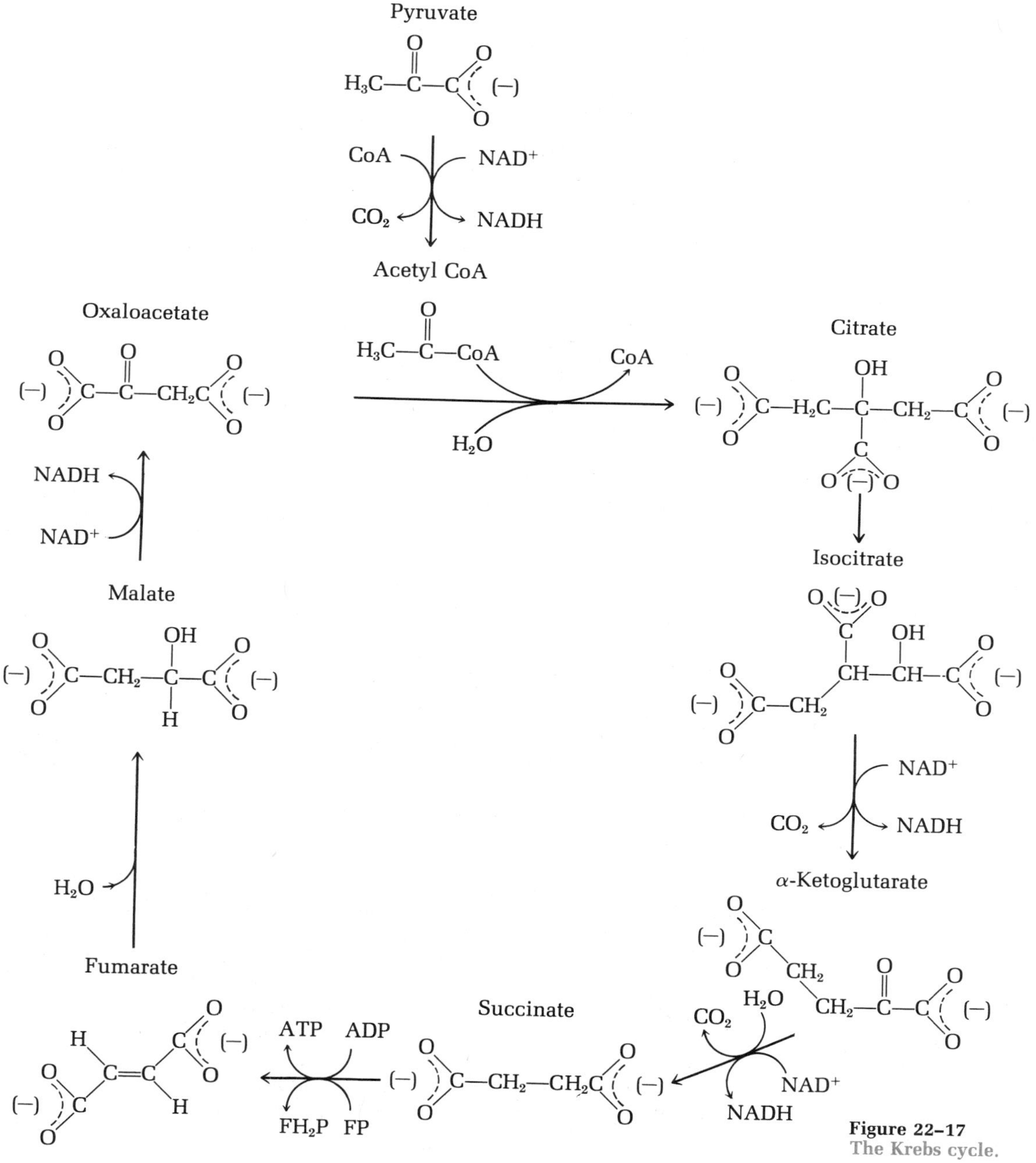

Pyruvate

CoA, NAD+ → CO_2, NADH

Acetyl CoA

Oxaloacetate

Citrate

Isocitrate

NAD+ → CO_2, NADH

α-Ketoglutarate

H_2O, NAD+ → CO_2, NADH

Succinate

ATP, ADP, FH_2P, FP

Fumarate

H_2O

Malate

NAD+ → NADH

Figure 22-17
The Krebs cycle.

into the cyclic part of the mechanism, as shown at the top of Figure 22–17. The molecule CoA shown entering this step is **coenzyme A,** which is derived from the B vitamin, **pantothenic acid.** Figure 22–18(a) shows the structure of coenzyme A and identifies the portion derived from pantothenic acid. The derivative of pyruvate that enters the Krebs cycle is acetyl CoA, and Figure 22–18(b) shows the point of attachment to the coenzyme of the acetyl group

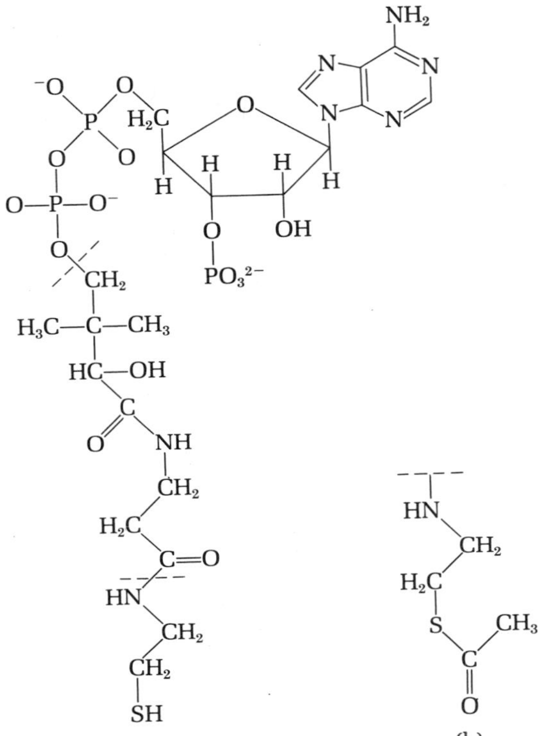

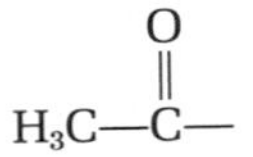

Figure 22–18
(a) The structure of coenzyme A and (b) acetyl CoA. The part of the CoA molecule derived from pantothenic acid is contained between the dashed lines in (a). The part of (b) above the dashed line is identical to that in (a).

$$H_3C-\overset{\overset{\displaystyle O}{\|}}{C}-$$

The entry into the cycle occurs with the transfer of the acetyl group from acetyl CoA to oxaloacetate to produce citrate and to regenerate coenzyme A. At this point, the six original carbons of glucose have been converted to two CO_2 molecules and two carbons in each of two citrate ions. A total of eight electrons have been lost to four NAD^+ molecules and fourteen ATP molecules have been synthesized. As Figure 22–17 shows, two additional CO_2 molecules are produced per citrate ion in the cycle and three NAD^+ molecules and one FP molecule are reduced per citrate ion. Each of these reductions involves the transfer of two electrons, so that eight electrons are lost per citrate ion in the cycle. Since each glucose molecule yields two citrate ions, the cycle results in a sixteen-electron loss per glucose molecule. This number of electrons plus the eight lost up to the entry into the cycle accounts for the twenty-four-electron loss in reaction 22–3.

There are twelve ATP molecules synthesized per citrate ion in the cycle, or twenty-four ATP molecules per original glucose molecule. The twelve ATP molecules arise from nine produced in the electron transfer chain from three NADH molecules, two produced in the electron transfer chain from the reduced flavoprotein and one produced directly in the conversion of succinate to fumarate. This yields a total of thirty-eight ATP molecules per glucose molecule in the conversion of glucose to CO_2 and water.

22–4. Metabolism of Fats

Biosynthesis of Fatty Acids

Animals advanced in the evolutionary scale possess only a limited capacity to store carbohydrates, and thus carbohydrate molecules are not the convenient energy storehouse in those systems that they are in plants. Advanced animals convert excess carbohydrates to fats, which can be stored in large amounts in fatty tissue. Fats exceed carbohydrates in being an even more concentrated storehouse of energy. For example, consider the fatty acid, palmitic acid, whose structure is given in Table 10–6. This molecule is far less oxidized than glucose since it contains only two oxygen atoms per sixteen carbon atoms while glucose contains six oxygen atoms per six carbon atoms. Approximately twice as much energy per carbon atom is stored in a fat as in a carbohydrate.

You may have noticed that each of the fatty acids listed in Table 10–6 contains an even number of carbon atoms. This occurs because fatty acid molecules are synthesized two carbons at a time in living systems. The energy for the synthesis is obtained from the conversion of ATP to ADP. The two-carbon starting material for the synthesis is acetyl CoA, and a large number of electrons must be supplied to form the highly reduced fatty acid molecules. These electrons are supplied by NADPH, a molecule identical in structure to NADH, save for one additional phosphate group. The complete reaction for the synthesis of palmitic acid is

$$\text{8 acetyl CoA} + \text{14 NADPH} + 14\ H^+ + \text{7 ATP} + H_2O \rightarrow \text{palmitic acid} + \text{8 CoA} + 14\ NADP^+ + \text{7 ADP} + 7\ P_i \qquad \textbf{22–7}$$

OPTIONAL MATERIAL

Mechanism of Biosynthesis of Fatty Acids

Figure 22–19 diagrams the scheme by which palmitic acid is synthesized

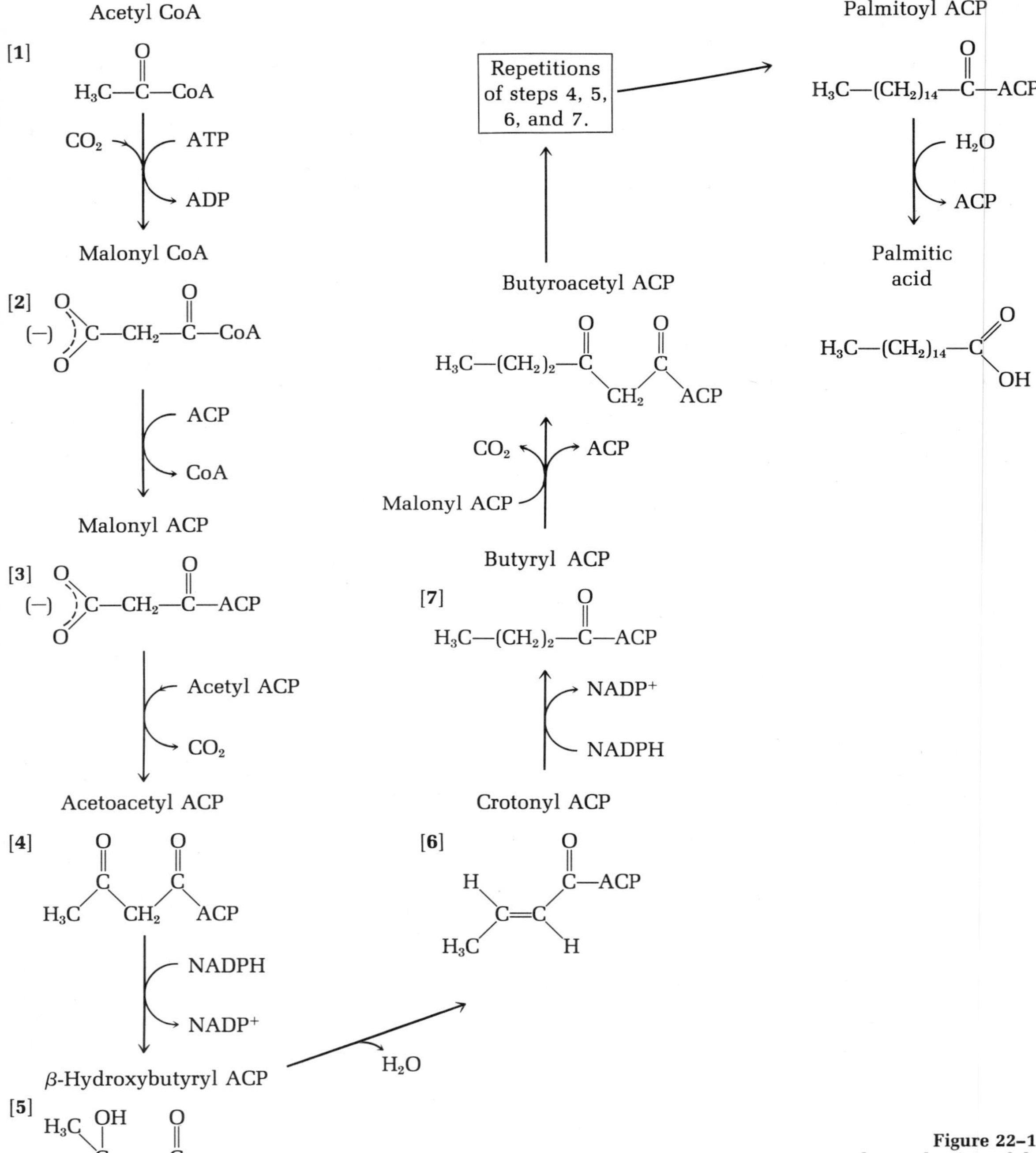

Figure 22–19
The mechanism of the biosynthesis of palmitic acid.

from acetyl CoA, with energy derived from the conversion of ATP to ADP. The designation ACP refers to **acyl carrier protein.** ACP consists of a protein portion and a nonprotein part that contains the site of attachment of acyl groups (acetyl, malonyl, and butyryl are examples of acyl groups). The nonprotein part of ACP is identical with the lower portion (including the first phosphate) of coenzyme A shown in Figure 22–18(a), and the site of acyl attachment is the sulfur atom, as in acetyl CoA. Acetyl ACP and malonyl ACP are produced by acyl transfer from the corresponding CoA derivatives, as shown in Step 2 of Figure 22–19 for the malonyl transfer.

The figure shows each step and intermediate species formed up to the six-carbon chain of butyroacetyl ACP. At that point steps 4, 5, 6, and 7 are repeated to yield the analogous eight-carbon chain. In this fashion, the chain is lengthened by two carbons at a time until the sixteen carbon chain of the palmitoyl group is reached. In the final step, palmitoyl ACP is hydrolyzed to produce palmitic acid and regenerate ACP.

Steps 4, 5, 6, and 7 must be carried out seven times to convert eight acetyl CoA molecules to one palmitic acid molecule. Steps 4 and 6 both involve the oxidation of one NADPH molecule to $NADP^+$. This accounts for the fourteen NADPH molecules in equation 22–7.

Oxidation of Fatty Acids

Just as fatty acids are synthesized from acetyl CoA molecules, the degradation of fatty acids occurs two carbons at a time to yield acetyl CoA molecules. However, there are differences in the details between the synthesis and the degradation of fatty acids. For example, the oxidizing agents in the degradation are FP and NAD^+ instead of $NADP^+$, the oxidizing agent produced in Figure 22–19. The overall reaction for the conversion of the sixteen-carbon chain of palmitic acid to the acetyl groups of eight acetyl CoA molecules is

$$\text{Palmitic acid} + 8\ \text{CoA} + 7\ \text{FP} + 7\ \text{NAD}^+ + 6\ \text{H}_2\text{O} + \text{ATP} \rightarrow$$
$$8\ \text{acetyl CoA} + 7\ \text{FH}_2\text{P} + 7\ \text{NADH} + 7\ \text{H}^+ + \text{ADP} + \text{P}_i \qquad \textbf{22–8}$$

Since each FH_2P yields two ATP molecules and each NADH yields three ATP molecules in the electron-transfer chain, reaction 22–8 results in the synthesis of thirty-four molecules of ATP and the storage of 248 kcal of Gibbs free energy. This is a large amount of energy, but it is small compared to the $\Delta G°$ for the complete oxidation of palmitic acid according to reaction 22–9:

$$\text{Palmitic acid} + 23\ \text{O}_2 \rightarrow 16\ \text{CO}_2 + 16\ \text{H}_2\text{O} \qquad \textbf{22–9}$$

The $\Delta G°$ for this complete oxidation is $-2{,}340$ kcal/mole. As in the oxidation of carbohydrates, most of the energy from reaction 22–9 is obtained from the oxidation of the acetyl group of acetyl CoA in the Krebs cycle. Since the Krebs cycle produces twelve ATP molecules per acetyl group entering it, the total number of ATP molecules synthesized in the course of reaction 22–9 is 8(12) + 34 = 130 molecules for an energy storage of (130)(7.3) = 949 kcal/mole of palmitic acid.

Let us use this value along with the value of 277 kcal/mole for the complete oxidation of glucose to compare the energy storage per gram of fats versus carbohydrates. The molecular weights of palmitic acid and glucose are 256 and 180 amu, respectively. Thus, the oxidation of palmitic acid results in the storage of 3.7 kcal/gram, while glucose yields only 1.5 kcal/gram. Fats cause more of a problem to weight-conscious people than carbohydrates do, certain diet faddists notwithstanding.

OPTIONAL MATERIAL

Mechanism of Oxidation of Fatty Acids

The mechanism for conversion of palmitic acid to acetyl CoA is given in Figure 22–20. The process is very similar to the reverse of that in Figure 22–19, but there are some key differences. Acyl carrier protein (ACP) is not involved in Figure 22–20, nor is $NADP^+$ or malonyl CoA. In every other way, however, the steps in Figure 22–20 are the reverses of corresponding steps in Figure 22–19.

22–5. The Structures and Functions of Proteins

There are many chemical reactions in a process such as the complete oxidation of glucose. The direct reaction of oxygen with glucose is slow at room temperature. Thus, each step in the electron transfer must be catalyzed. The principal catalysts in biochemical reactions are the large molecules called enzymes, which typically have molecular weights between 10^4 and 10^6 amu. Enzymes belong to the class of compounds termed proteins. The proteins are analogous to the refrigerator of Figure 22–1 in that they form the structures that carry out the functions of living systems. For example, muscle tissue consists largely of protein and the structures of muscle protein molecules provide the basis for muscle contraction.

Our approach to the study of proteins will be as follows. First, we shall consider their chemical compositions and important structural features. Next, we shall consider the structure of a specific enzyme and a proposed mechanism of action of that enzyme. Finally, we shall treat the mechanism by which proteins are synthesized within living cells.

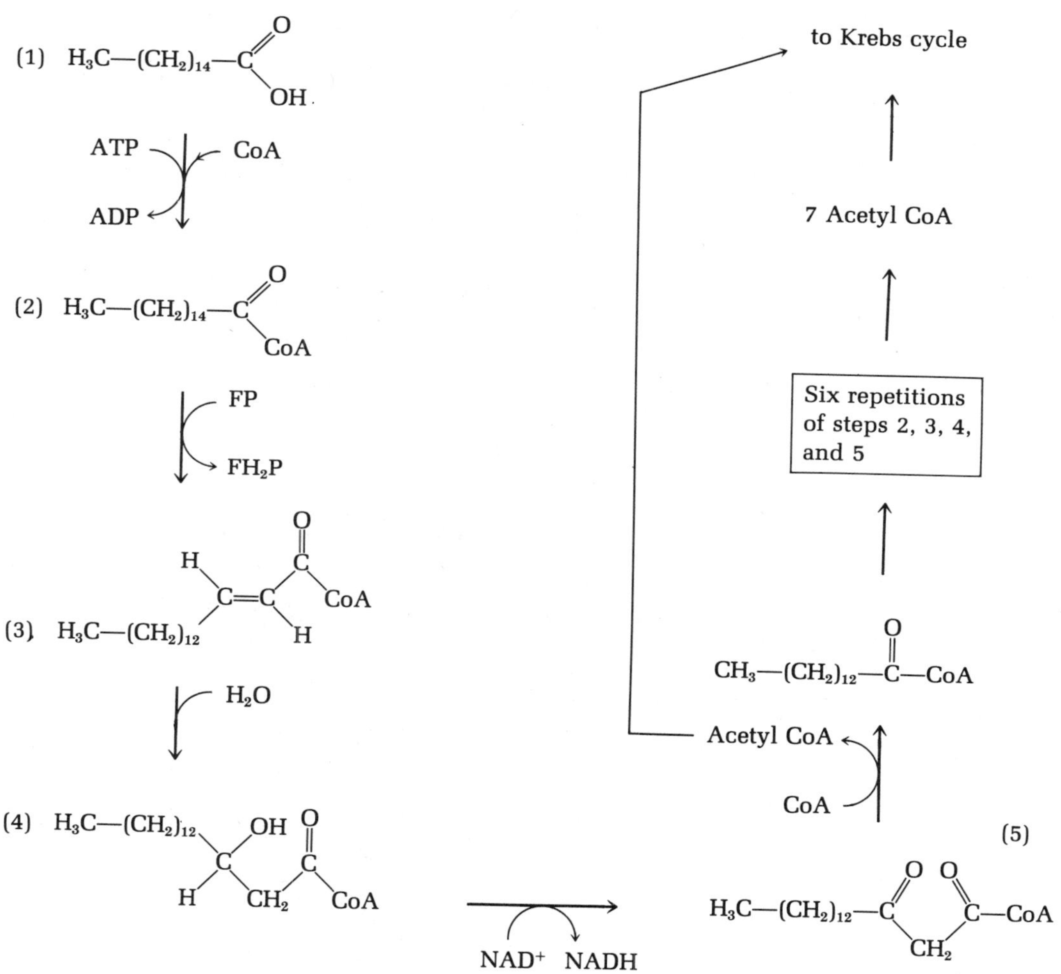

Figure 22–20
The sequence of steps in the oxidation of palmitic acid to eight acetyl CoA molecules.

Amino Acids

Proteins are polymeric molecules that have as monomeric units the twenty **amino acids** whose structures are shown in Figure 22–21. All amino acids contain at least one carboxylic acid group and at least one nucleophilic nitrogen atom (located on the carbon next to the carboxylic acid group in all cases except proline). In addition, each of the amino acids contains a characteristic group (termed the R group), and the R groups show a wide variation in chemical behavior from one amino acid to another. These R groups yield the catalytic activities of enzymes. The structure shown in the figure for each amino acid is the protonated or deprotonated form that predominates in aqueous solution at pH 7.4 (the pH of blood). Also given with each amino acid is the abbreviated name that commonly represents the acid.

Amino Acid	Abbreviated Name	Structure
Glycine	Gly	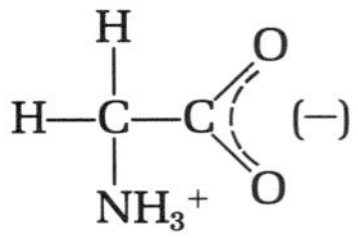 $H{-}CH(NH_3^+){-}COO^{(-)}$
Alanine	Ala	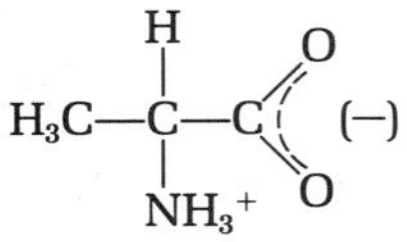 $H_3C{-}CH(NH_3^+){-}COO^{(-)}$
Valine	Val	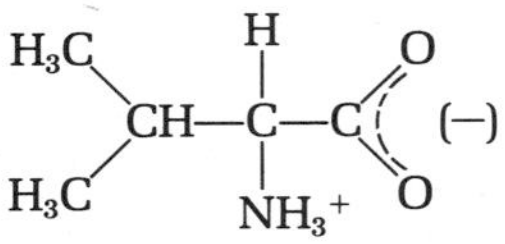 $(H_3C)_2CH{-}CH(NH_3^+){-}COO^{(-)}$
Leucine	Leu	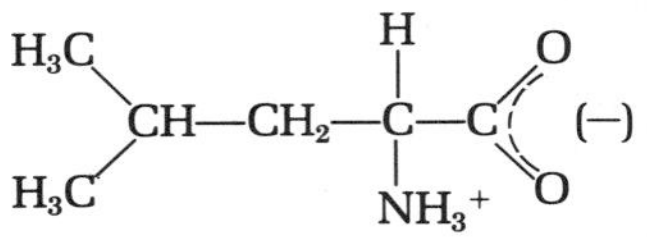 $(H_3C)_2CH{-}CH_2{-}CH(NH_3^+){-}COO^{(-)}$
Isoleucine	Ile	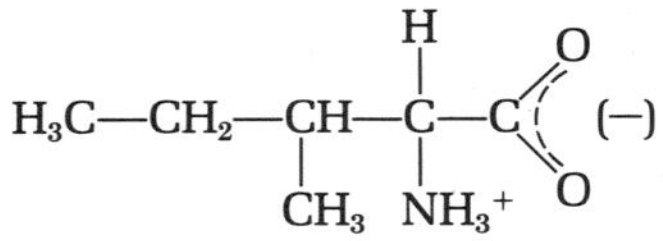 $H_3C{-}CH_2{-}CH(CH_3){-}CH(NH_3^+){-}COO^{(-)}$
Proline	Pro	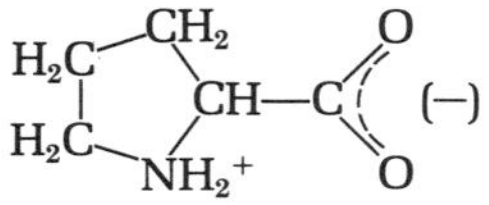 $H_2C{-}CH_2{-}CH(COO^{(-)}){-}NH_2^+{-}CH_2$ (ring)
Phenylalanine	Phe	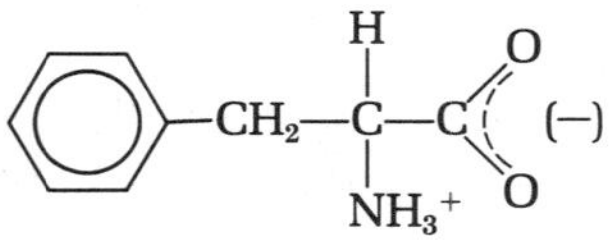 $C_6H_5{-}CH_2{-}CH(NH_3^+){-}COO^{(-)}$
Tryptophan	Trp	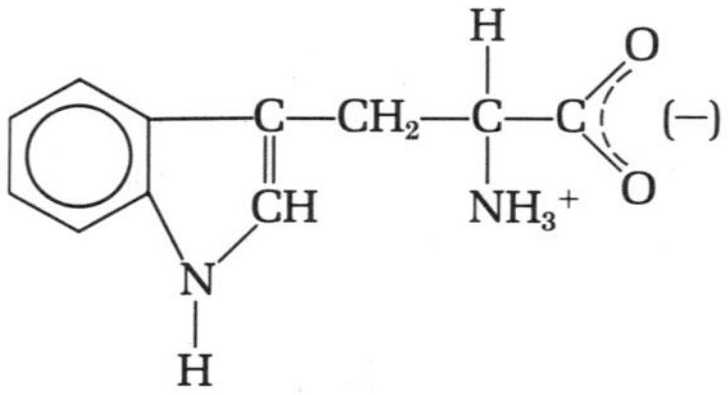 indolyl ($C{=}CH{-}N{-}H$ ring fused to benzene)${-}CH_2{-}CH(NH_3^+){-}COO^{(-)}$

Figure 22–21
The twenty common amino acids. The structure shown for each amino acid is either the protonated or deprotonated form that predominates in aqueous solution at pH 7.4.

Amino Acid	Abbreviated Name	Structure
Tyrosine	Tyr	$HO-C_6H_4-CH_2-CH(NH_3^+)-COO^{(-)}$
Serine	Ser	$HO-CH_2-CH(NH_3^+)-COO^{(-)}$
Threonine	Thr	$H_3C-CH(OH)-CH(NH_3^+)-COO^{(-)}$
Asparagine	Asn	$H_2N-C(=O)-CH_2-CH(NH_3^+)-COO^{(-)}$
Glutamine	Gln	$H_2N-C(=O)-CH_2-CH_2-CH(NH_3^+)-COO^{(-)}$
Cysteine	Cys	$HS-CH_2-CH(NH_3^+)-COO^{(-)}$
Methionine	Met	$H_3C-S-CH_2-CH_2-CH(NH_3^+)-COO^{(-)}$
Lysine	Lys	$H_3N^+-CH_2-CH_2-CH_2-CH_2-CH(NH_3^+)-COO^{(-)}$
Arginine	Arg	$(H_2N)_2C^{(+)}-NH-(CH_2)_3-CH(NH_3^+)-COO^{(-)}$

Figure 22-21 (continued)

Amino Acid	Abbreviated Name	Structure
Histidine	His	$HC{=}C(-NH-CH{=}N-)-CH_2-CH(NH_3^+)-COO^{(-)}$
Aspartic acid	Asp	$^{(-)}OOC-CH_2-CH(NH_3^+)-COO^{(-)}$
Glutamic acid	Glu	$^{(-)}OOC-(CH_2)_2-CH(NH_3^+)-COO^{(-)}$

Figure 22–21 (continued)

Primary Structures of Proteins

In forming proteins, the amino acids join in an amide linkage termed the **peptide bond.** For example, a glycine molecule may unite with an alanine molecule, as shown in reaction 22–10.

$$H_2C(NH_3^+)-COO^{(-)} + H_3C-CH(NH_3^+)-COO^{(-)} \rightarrow H_2C(NH_3^+)-C(=O)-N(H)-CH(CH_3)-COO^{(-)} + H_2O \qquad \textbf{22–10}$$

The peptide bond is the bond between the carbonyl carbon of one amino acid and the amine nitrogen of a second amino acid.

The NH_3^+ group of the **dipeptide** that is the product of reaction 22–10 may form a second peptide bond with the carboxylate group of a third amino acid to form a **tripeptide,** or the free carboxylate group of the dipeptide may react with the NH_3^+ group of a third amino acid, to form a different tripeptide. Proteins typically contain hundreds of amino acids in the form of one or more **polypeptide** chains. Since there are twenty choices for each link in the chain, the possible number of sequences in a chain containing a hundred amino acids is $(100)^{20} = 10^{40}$, an astronomical number. An example of a specific sequence in a rather small molecule is that of beef insulin, shown in Figure 22–22. The molecule contains two polypeptide chains, termed A and B. The free NH_3^+ groups at the two chain

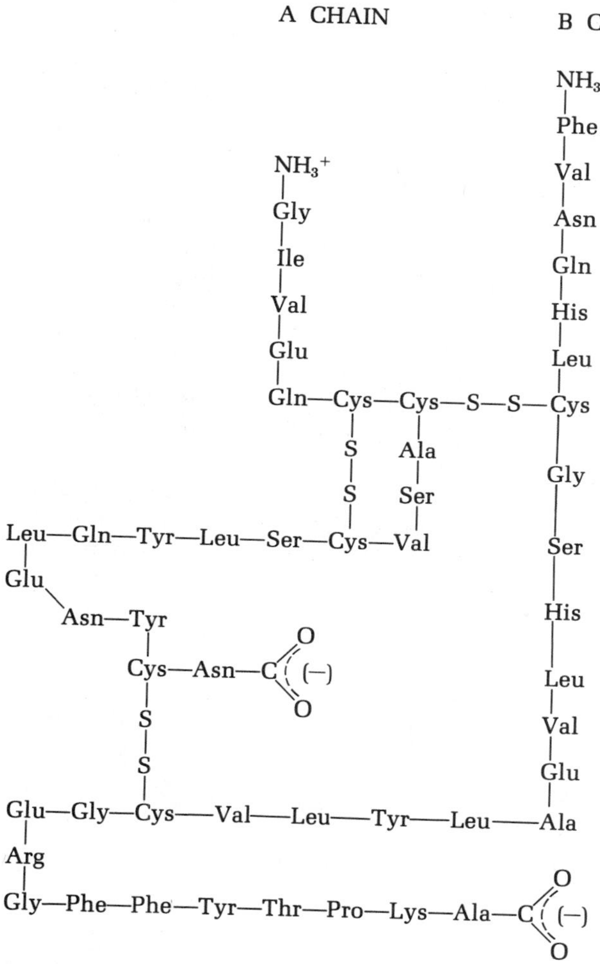

A CHAIN B CHAIN

NH_3^+ (B chain) — Phe — Val — Asn — Gln — His — Leu — Cys

NH_3^+ (A chain) — Gly — Ile — Val — Glu

Gln—Cys—Cys—S—S—Cys

S Ala Gly

S Ser

Leu—Gln—Tyr—Leu—Ser—Cys—Val Ser

Glu

Asn—Tyr His

Cys—Asn—$C(O)O^{(-)}$ Leu

S Val

S Glu

Glu—Gly—Cys—Val—Leu—Tyr—Leu—Ala

Arg

Gly—Phe—Phe—Tyr—Thr—Pro—Lys—Ala—$C(O)O^{(-)}$

Figure 22–22
The primary structure of beef insulin.

terminals belong to a linked glycine molecule (termed a glycine **residue**) in the A chain and to a phenylalanine residue in the B chain. The two free carboxylate terminals occur in an asparagine residue in the A chain and an alanine residue in the B chain.

A common role of the amino acid cysteine is the formation of disulfide (—S—S—) bridges between chains, a process that occurs by the oxidation of the (—SH) groups of the two cysteine residues to be joined. Figure 22–22 shows three such disulfide bridges, one within the A chain and two between the A and B chains. The structure shown in Figure 22–22 is termed the **primary structure** of the polypeptide. The primary structure is specified by the sequence of amino acids in the chains plus the locations of the disulfide bridges.

Secondary and Tertiary Structures of Proteins

The structure of insulin shown in Figure 22–22 is incomplete since it does not show how the chains are arranged in space. The R groups in Figure 22–21 are such that associations between different R groups in the same chain or in different chains can occur. For example, the first several amino acids in the figure contain nonpolar R groups, and these can associate for the same reason that hydrocarbon "tails" associate in soap micelles.

A very important type of association between amino acid residues in protein chains is hydrogen bonding. A very important role of hydrogen bonding in the structures of proteins was deduced by the American scientists Linus Pauling and R. B. Corey in 1951. From examining physical models of polypeptide chains, they concluded that a regularly spaced array of hydrogen bonds could be formed between amino acid residues of the same chain if the chain were coiled into the structure termed an **α-helix.** This helical structure is shown in Figure 22–23. It has the form of a spring that is wound in a right-handed fashion; that is, if the spring is observed end-on, clockwise motion along the spring also involves motion away from the observer. The hydrogen bonds occur between the carbonyl oxygen of one

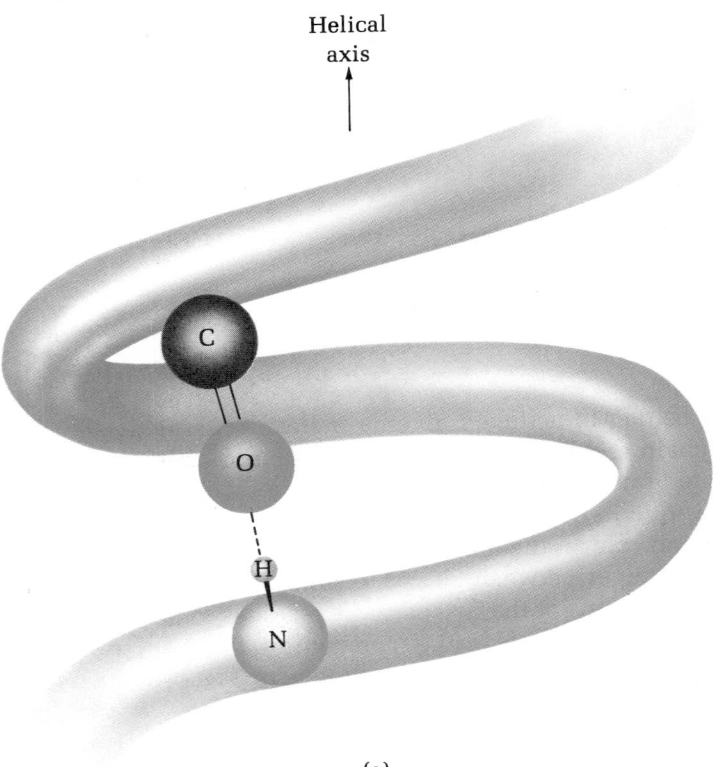

Figure 22–23
A portion of an α-helical peptide chain: simplified representation showing one hydrogen bond (a); a more detailed representation (b).

amino acid residue and the N—H hydrogen of another residue further along the chain. In the α-helical form, every N—H hydrogen and every carbonyl oxygen in the chain is involved in a hydrogen bond, and this results in an extraordinary stability of the helical structure. In Figure 22–23(b), atoms other than the carbonyl oxygens, N—H hydrogens, and those that form the polypeptide backbone have been omitted so that the regular spacing of the hydrogen bonds is more readily apparent. The hydrogen bonds are indicated by dashed lines. The distance along the helical axis (vertical distance in the figure) in one helical turn is 5.4 Å, and 3.6 amino acid residues are contained along the chain in this one turn. Each N—H proton is hydrogen-bonded to the carbonyl oxygen of the third peptide linkage along the chain from it.

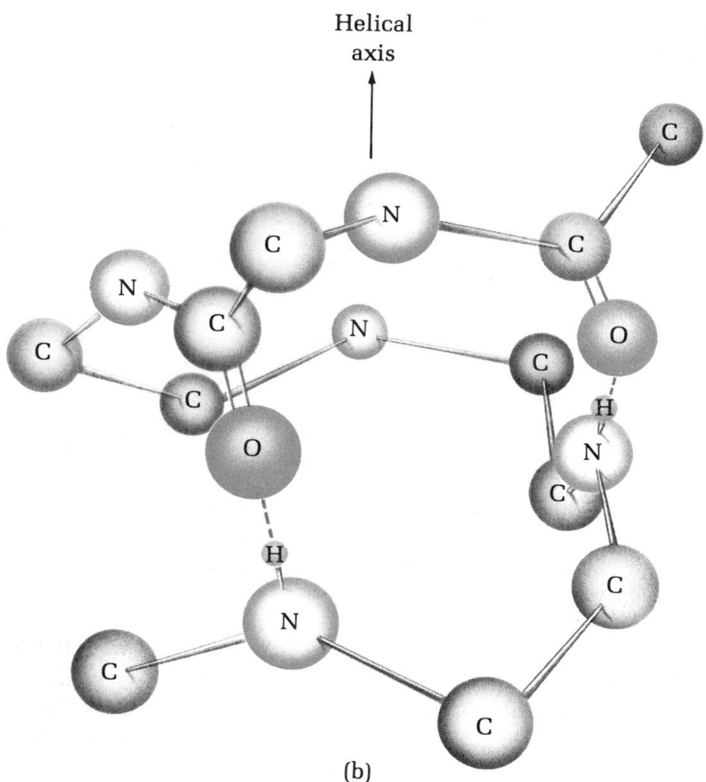

Figure 22–23 (continued)

The α-helix forms the **secondary structure** of many proteins or of segments of those proteins. There are, however, very important structural features over and above even the secondary structure, and they make up the **tertiary structure.** For example, the fibrous nature of hair derives from the coiling of α-helical chains to form supercoils, whose structure resembles that of rope. Figure 22–24 shows such a supercoiled structure formed from three α-helical chains. Fibrous proteins like those found in hair, tendon, muscle, and bone possess an orderly array of polypeptide chains, even in those cases in which the secondary structure is not the α-helix. The principal protein found in advanced animals is the fibrous protein **collagen,** which composes over one-third by weight of all body protein. The collagen molecular unit consists of three polypeptide chains, none of which is α-helical. However, the three chains are wound together to form a three-stranded "rope."

Figure 22–24 The supercoiling of three α-helices (colored, white, and black) in the fibrous protein of hair. The α-helix is shown in the white strand.

Enzymes belong to the class of molecules known as **globular proteins,** in which the chains wind in a much more irregular fashion than in the fibrous proteins. In spite of their irregularity the secondary and tertiary structures of these proteins are extremely important in relation to their functions. One of the most exciting advances in science over the past quarter century has been the development of X-ray crystallographic techniques for determining the complete molecular structures (including secondary and tertiary structure) of globular proteins. It is absolutely essential, to have such structural information if we are to understand how any given protein molecule functions.

Hemoglobin and Myoglobin

Some large and important globular proteins consist of discrete subunits each of which contains polypeptide chains. Each subunit possesses its own primary, secondary, and tertiary structures, and the arrangement of the subunits in relation to each other is termed the **quaternary structure** of the molecule. One such large globular molecule is **hemoglobin,** the molecule that transports oxygen in the bloodstream of advanced animal forms. This molecule contains four peptide chains, two of which (termed the α-chains) contain 141 amino acid residues and the other two of which (termed the β-chains) contain 146 residues. In addition, each chain is weakly bound to a heme ring system so that the hemoglobin molecule contains four Fe^{2+} ions.

The four subunits of hemoglobin are closely related to the simpler molecule **myoglobin,** which stores molecular oxygen in muscle cells. Figure 22–25 represents the structure of myoglobin as deduced by J. C. Kendrew and his associates in England from X-ray crystallographic data. Myoglobin contains one heme group, the iron and four nitrogens of which are shown in black in the upper center of the figure. The protein chain contains 153 amino acid residues and its tertiary structure is shown by the sausage-shaped out-

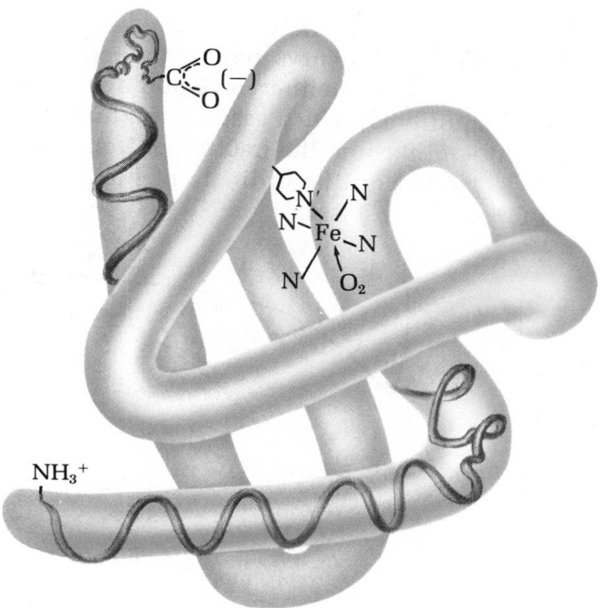

Figure 22–25
The tertiary structure of myoglobin. Also shown are two α-helical segments and the location of the heme portion of the molecule.

line in the figure. The segments of chain between the bends in the sausage are α-helical, and two of these segments are shown in the figure. The iron ion (Fe^{2+}) can bond to six electron-donor atoms. Four of these donor atoms are the four nitrogens of the heme. The fifth donor is one of the two ring nitrogens of the amino acid histidine. The sixth position is the one to which molecular oxygen binds, as indicated by the arrow in the figure. In other important heme proteins such as cytochrome c, the six donors are the four heme nitrogens and two atoms from amino acids. Consequently the cytochromes are not transport and storage agents but rather electron transfer agents.

We might ask quite logically why the hemoglobin molecule contains four heme groups when the single heme of myoglobin serves perfectly well to bind molecular oxygen. The key difference between the molecules occurs in the functions demanded of them. Hemoglobin is required to associate with as many oxygen molecules as possible under conditions in which O_2 is abundant, that is, at the lung surface. It is then required to give up those O_2 molecules under conditions in which oxygen is in demand, that is, in cells where it is required for oxidation of foodstuffs. Since myoglobin provides the O_2 storage and transport within cells, under conditions of a low availability of O_2 it must bind O_2 more strongly than hemoglobin does. Myoglobin thus accepts all of the oxygen bound to hemoglobin so that the hemoglobin can return to the lung surface for a new supply.

Figure 22–26 shows the competition for O_2 between myoglobin and hemoglobin in a graphic fashion. Each curve refers to an aqueous solution containing a given concentration of either myoglobin or hemoglobin. This solution is allowed to equilibrate with a gaseous mixture containing a given O_2 partial pressure. At equilibrium, the fraction of iron ions bound to O_2 is measured and this is plotted on the ordinate of Figure 22–26 versus the partial pressure of O_2 on the abscissa.

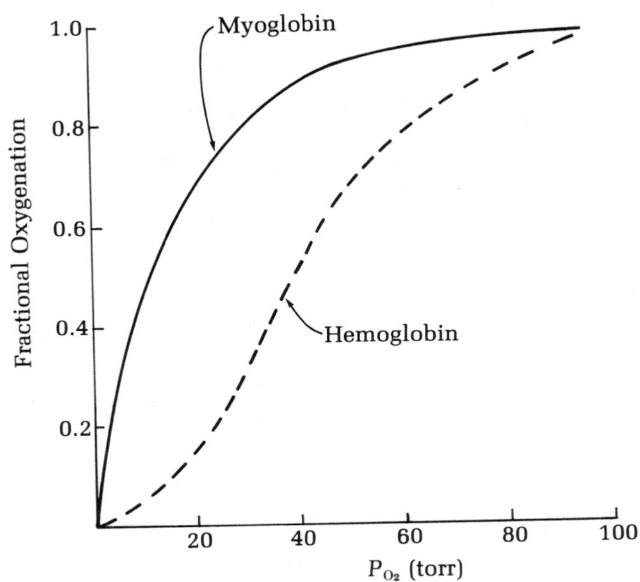

Figure 22–26 The fraction of aqueous myoglobin (solid line) and hemoglobin (dashed line) oxygenated as a function of the equilibrium partial pressure of O_2.

The figure shows that hemoglobin is a very poor binder of O_2 at low P_{O_2} as we require, but its oxygen-binding capability increases markedly as P_{O_2} increases until it becomes a strong oxygen binder at P_{O_2} of the order of that present in the lung. The myoglobin curve in Figure 22–26 is the normal one for the equilibrium in reaction 22–11:

$$\text{Myoglobin(aq)} + O_2(g) \rightleftharpoons \text{myoglobin—}O_2\text{(aq)} \qquad \textbf{22–11}$$

However, such a simple equilibrium does not account for the S-shaped curve for hemoglobin in Figure 22–26. The S-shaped curve arises because the binding of an O_2 to one heme group of hemoglobin markedly enhances the ability of the remaining three heme groups to bind oxygen. This enhancement would be relatively easy to explain if the four hemes of the hemoglobin molecule were in close proximity, but X-ray crystallographic studies have shown that they are well separated. Thus, the explanation must lie in the tertiary and quaternary structure of the molecule. It is obvious from the comparison between hemoglobin and myoglobin at low P_{O_2} that the protein conformation must play an important role in oxygen binding since this conformation is the only significant difference between myoglobin and the subunits of hemoglobin. The binding of an O_2 molecule apparently causes a change in the tertiary or quaternary structure or both, which results in an increased tendency toward oxygen binding in mono-oxygenated hemoglobin. The effect is a dramatic demonstration of the importance of protein conformation to protein function, and it underscores our point about the necessity of having the detailed molecular structure in mind in order to account for protein function.

Structure and Function of an Enzyme

Because of the complexity of enzyme molecules, detailed structural information from X-ray crystallography is available for only a few enzymes (though this number grows each year). One of these few enzymes is **α-chymotrypsin,** an enzyme that is secreted into the small intestine and that catalyzes the hydrolysis of polypeptides into individual amino acids. This molecule was first isolated in the 1930s, and its name is an exception to the naming of most enzymes. The typical suffix for an enzyme is *-ase*, and the prefix indicates the function of the enzyme. Thus, for example, **carboxypeptidase** hydrolyzes polypeptides into their constituent amino acids by attacking the peptide bond between the amino acid at the free carboxylate terminus and the next residue in the chain. Both α-chymotrypsin and carboxypeptidase are **peptidases,** indicating that they catalyze the hydrolysis of peptide bonds. However, α-chymotrypsin is not restricted to the carboxylate terminus of the polypeptide chain but can split peptide links at several sites along the chain.

The α-chymotrypsin molecule consists of 241 amino acid residues in three chains containing five disulfide bridges. It is difficult to represent the tertiary structure of the molecule in a two-dimensional projection because it coils in a complex fashion to form a nearly spherical outline. In addition, only a small portion of α-helical region is observed, located near the free carboxylate terminus of one of the three chains. This complicated chain winding is characteristic of enzymes, and the precise conformation is absolutely essential to their activity. (We shall see this shortly in the case of α-chymotrypsin.)

Because of the presence of the R groups, the different amino acids can play a variety of roles in proteins. The first thing that the R groups can do is to bind the substrate molecule that the enzyme is to attack. The twenty amino acids of Figure 22–21 represent a wide variety of abilities to associate with various portions of substrate molecules. For example, there are three electrical charges represented by R groups. Most of the amino acids have uncharged R groups at pH 7.4; but lysine and arginine have positively charged R groups, and the R groups of aspartic acid and glutamic acid are negatively charged. The positively charged and negatively charged R groups can associate with oppositely charged groups on substrate molecules.

The neutral R groups may be divided into polar and nonpolar groups. The saturated hydrocarbon groups of alanine, valine, leucine, and isoleucine tend to form "hydrophobic pockets" in the tertiary structure of the enzyme by associating with each other and excluding water. Hydrocarbon portions of substrate molecules are bound in these pockets in much the same way that an oil droplet is bound in a detergent micelle. Those polar R groups containing hydroxyl groups (tyrosine, serine, and threonine) can form hydrogen bonds to substrate molecules.

A particularly important amino acid in enzymes is histidine. Figure 22–21 shows that the nitrogen-containing ring (termed the imidazole ring) of histidine is not protonated at pH 7.4. The pK_a of this ring in histidine is 6.0, and both the protonated and unprotonated rings of histidine are present in significant amounts under physiological conditions. Histidine can be both a proton donor and a proton acceptor in enzymes, and this is a main reason that histidine is a key amino acid. The imidazole ring can also associate with metal ions, as shown with myoglobin in Figure 22–25. Metal ions are often required to produce the catalytic activity of enzymes. Several R groups in Figure 22–21 are capable of strong association with metal ions. In addition to histidine, they are aspartic acid, glutamic acid, and the two sulfur-containing amino acids (which associate with metal ions through the sulfur atoms). Metal ions provide catalytic activation of enzymes most commonly by binding to one or more of the amino acid residues to convert the enzyme structure to one that is active, or by forming a bridge between the enzyme and substrate that permits strong binding of the enzyme and substrate.

The amino acid residues that directly attack the substrate molecule define the **active site** of the enzyme. In the case of α-chymotrypsin, these residues are a histidine residue and a serine residue that are widely separated in distance along the chains but neighbors in the folded structure of the active enzyme. In the numbering sequence for the residues of this enzyme, the key active-site residues are His 57 and Ser 195, and these numbers give a good indication of the separation of the two residues along the chains.

The barriers that must be overcome in any reaction are encounter, orientation, and bond breaking, with the largest of these usually bond breaking. The bond-breaking barrier can be substantially lowered if it is accompanied by simultaneous bond making. However, it is often the case that encounter or orientational restrictions or both prevent the simultaneous bond making and breaking. For example, simultaneous bond making and breaking might be possible in a given termolecular rate-determining step; but termolecular encounters are very improbable. The ideal situation for reaction occurs if the reactants can be brought together with a number of catalysts in just the proper orientation to permit the simultaneous bond making and breaking. Such multispecies encounters are normally far more improbable than even a termolecular encounter; but in an enzyme the catalysts are amino acid R groups, and their encounter with each other is forced by the enzyme structure. Thus, enzymes offer a potential for catalysis that no other system offers.

The requirement for specific catalysts in specific locations with respect to a given substrate molecule yields an important characteristic of enzymes, namely their **specificity.** Each enzyme can catalyze only a very few reactions, and some are capable of catalyzing only one reaction of a single substrate species. This high specificity is a key requirement in the orderly functioning of a complex biological system.

Figure 22–27 shows two proposed mechanisms for the action of α-chymo-

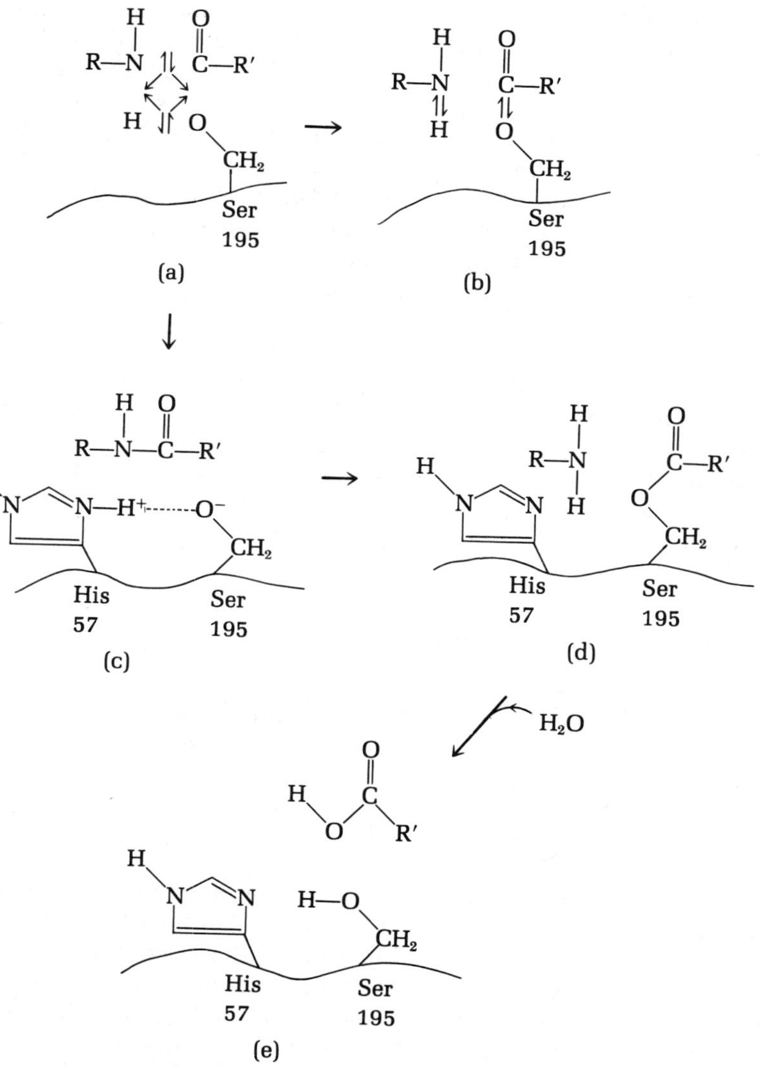

Figure 22–27
Two proposed mechanisms for the splitting of a peptide bond by α-chymotrypsin. The direct conversion of (a) to (b) as shown is an incorrect mechanism. A more reasonable mechanism involves the sequence of structures (a) → (c) → (d) → (e).

trypsin on a peptide link. Figure 22–27(a) shows a substrate molecule

$$\mathrm{R{-}\overset{\displaystyle H}{\overset{|}{N}}{-}\overset{\displaystyle O}{\overset{\|}{C}}{-}R'}$$

at the active site of the enzyme. The peptide bond could be split if electrons (as shown by the half-arrows) could transfer as shown by the black arrows in Figure 22–27(a). But this transfer is analogous to the transfer shown in Figure 21–4, and it does not occur. Thus, the structure shown in Figure 22–27(b) does not form as shown. However, the peptide bond of the substrate can be split if the serine O—H bond is split first, and this can occur by proton transfer to His 57, as shown in Figure 22–27(c). This proton transfer is analogous to Cl—Cl bond breaking in the H_2–Cl_2 reaction, and conversion as shown from Figure 22–27(c) to 22–27(d) occurs easily and rapidly because it involves simultaneous bond making and bond breaking. Histidine is a proton acceptor in the formation of the structure in Figure 22–27(c) and a proton donor in the formation of the species in Figure 22–27(d). This dual character of histidine is very useful in enzyme catalysis.

The final step of the second mechanism shown in Figure 22–27 is the hydrolysis of the ester link involving Ser 195 (again assisted by His 57) to regenerate the enzyme and free the carboxylate portion of the polypeptide. The catalytic action of α-chymotrypsin is so effective that it results in a reaction rate constant that is approximately 10^9 times as large as that for the reaction in the absence of the enzyme. This enhancement factor is typical of enzymes in general, and it is essential for the functioning of complex biological systems. There are so many steps that occur during a reaction such as 22–3 that the organism could not function unless each of these steps occurs very rapidly.

22–6. Biosynthesis of Proteins

What is quite probably the greatest triumph of science to date in unraveling the mysteries of life processes is the determination of the mechanism by which genetic information is stored and then passed from one generation to the next in reproduction. Protein synthesis is the key to this process.

Storage of Genetic Information

The number of possible sequences of twenty amino acids in a polypeptide chain of any significant length is astronomical. Yet the union of a single egg cell and sperm cell contains all of the information necessary to dictate the structures of all of the proteins that will eventually control the form and function of an adult human being. Because of the huge number of possibilities for protein structures, the information concerning how they are to be synthesized must be contained in molecules that also must have an astronomical number of structural possibilities. This information storage molecule has been identified within the past few decades as **deoxyribonucleic acid** (DNA), which is the principal constituent of cell nuclei.

DNA is a polymer with a molecular weight that may reach over 10^9 in some cases. Unlike proteins, which are polymers containing a maximum of twenty different kinds of monomer units (the twenty amino acids), DNA contains only four different kinds of monomer units. The monomers of DNA are termed **nucleotides** and their structures are shown in Figure 22–28. The structures of these nucleotides are closely related to the structures of ATP and ADP, given in Figure 22–15. The generic name *nucleotide* is given to any molecule possessing the three structural entities shown in Figures 22–15 and 22–28. The first is an organic ring termed the **base** (because it contains N atoms that are good proton acceptors), the second is a sugar, and the third is one or more phosphates. The sugar in DNA is 2′-deoxy-β-D-ribose, which differs from the β-D-ribose of ATP in the absence of the oxygen bonded to the 2′ carbon (the numbering of the sugar carbons is shown in Figure 22–28(a)).

Two of the organic bases, termed **adenine** and **guanine**, are derived from the organic base **purine**, and the nucleotides derived from these bases are **purine nucleotides.** Thus, dAMP, dGMP, ATP, and ADP are purine nucleo-

Figure 22–28 The four nucleotides of DNA.

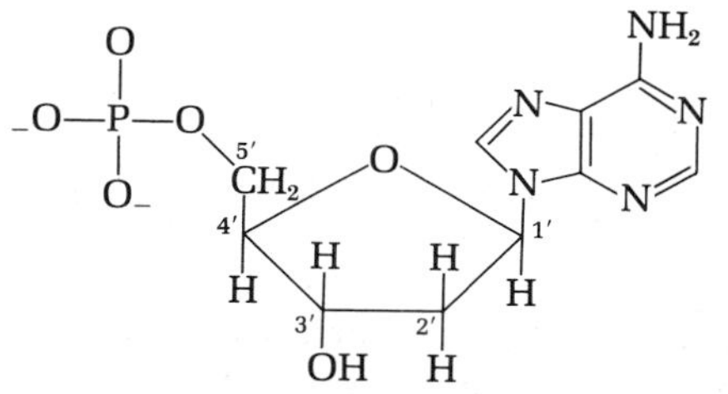

2′-Deoxyadenosine-5′-monophosphate (dAMP)
(a)

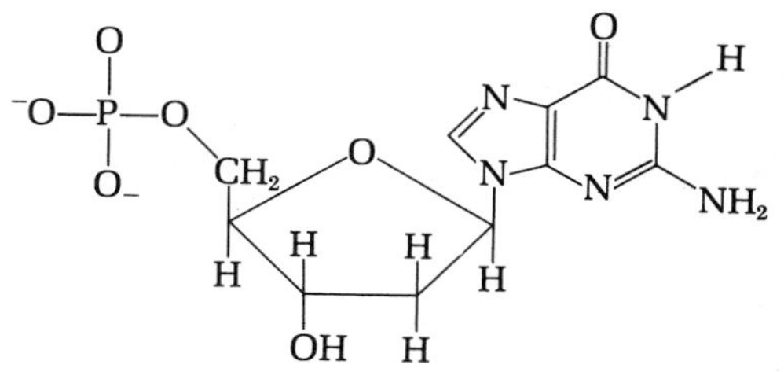

2′-Deoxyguanosine-5′-monophosphate (dGMP)
(b)

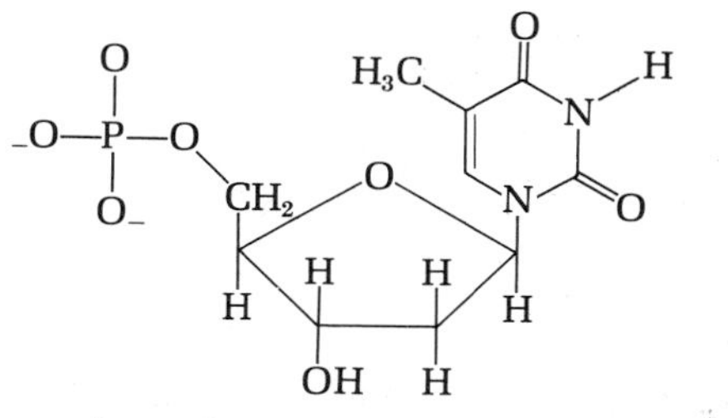

2′-Deoxythymidine-5′-monophosphate (dTMP)
(c)

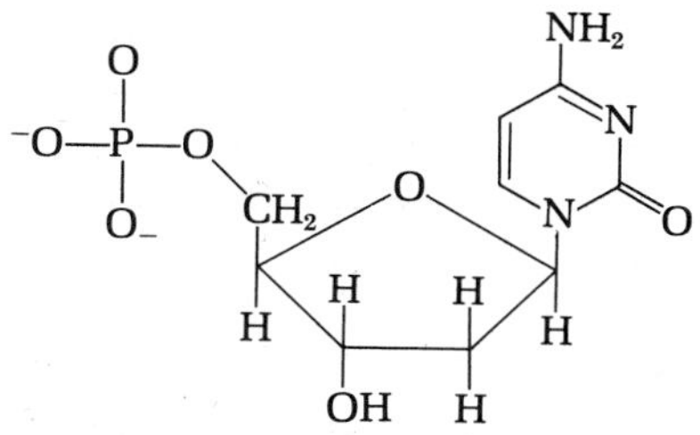

2′-Cytidine-5′-monophosphate (dCMP)
(d)

tides. The bases **thymine** and **cytosine** are derived from the organic base **pyrimidine,** and dTMP and dCMP are **pyrimidine nucleotides.** Figure 22–29 shows the structures of purine and pyrimidine, along with the structures of the four DNA bases.

Chemical analysis of DNA molecules reveals that the bases adenine and thymine occur in a mole ratio of unity, and the same is true of the mole ratio of guanine to cytosine. However, the ratio of adenine to guanine varies from one type of DNA molecule to another, depending on the source. This information along with the X-ray crystallographic data of the English scientists Maurice Wilkins and Rosalind Franklin led to the proposal of the structure

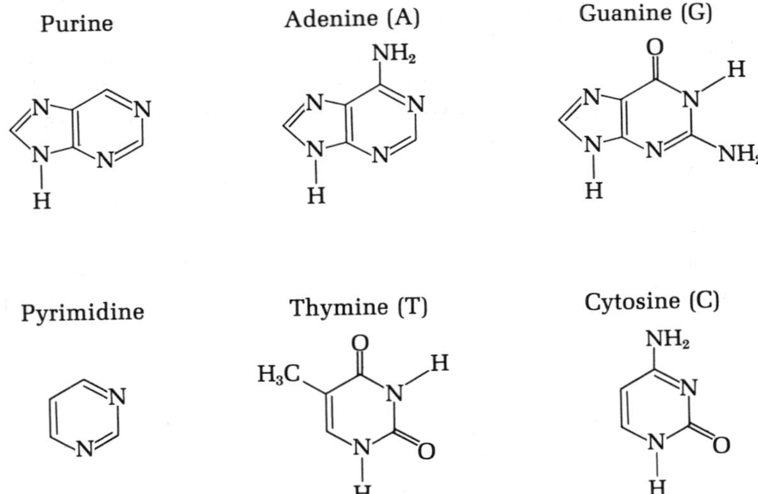

Figure 22–29
The structures of purine and pyrimidine and the DNA bases derived from them.

of DNA in 1953 by the American scientist James Watson and the English scientist Francis Crick. In the Watson-Crick model, which is the model molecular biologists universally employ, the nucleotides are polymerized into two chains, each of which possesses the primary structure shown in Figure 22–30. The phosphates bridge the 3′ carbon of one sugar molecule to the 5′ position of the adjacent molecule. The ordering of the bases along the chain shows no predictable pattern just as the sequence of amino acids in a given protein is not predictable. The bases in Figure 22–30 are symbolized by the letters G, A, and T for guanine, adenine, and thymine. We shall use these symbols plus C for cytosine in subsequent representations of the DNA structure.

The two polynucleotide chains of a given DNA molecule associate to form the helical structure depicted in Figure 22–31. The sugar and phosphate groups in the backbone are symbolized by S and P, respectively. The bonding between the two strands is provided by hydrogen bonds between the bases adenine and thymine, and between guanine and cytosine. Figure

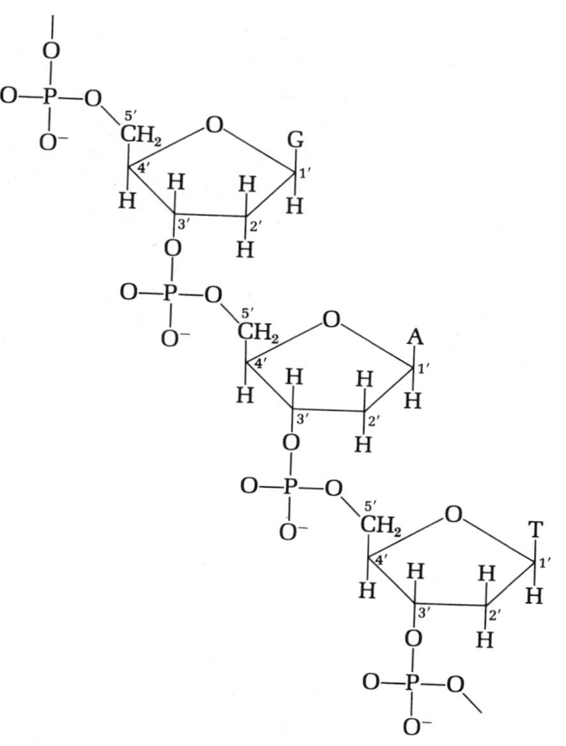

Figure 22-30
A portion of one polynucleotide chain of DNA. The symbols, G, A, and T stand for guanine, adenine, and thymine, respectively.

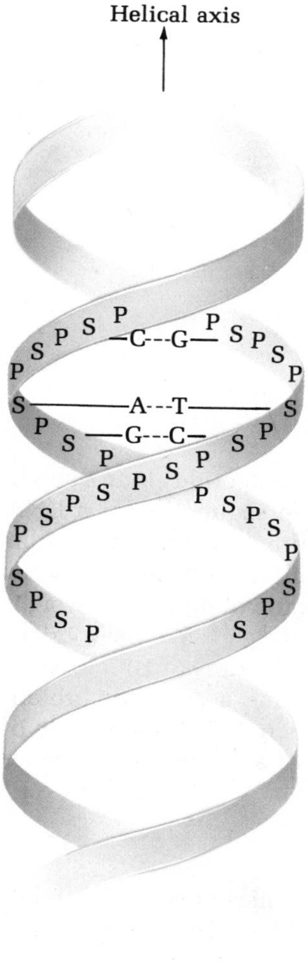

Figure 22-31
The two-stranded helical structure of DNA.

22–32 shows the hydrogen bonding between the **complementary base pairs.** It is because of this hydrogen bonding that the mole ratio for A and T and for G and C in DNA are each unity. No other hydrogen-bonded pairs, such as G–T, A–C, G–A, and so on, are found in DNA. The DNA structure is essentially like a spiral staircase. The purine and pyrimidine bases are planar, and in Figure 22–31 these planes lie perpendicular to the helical axis so that they form the "steps" of the "spiral staircase."

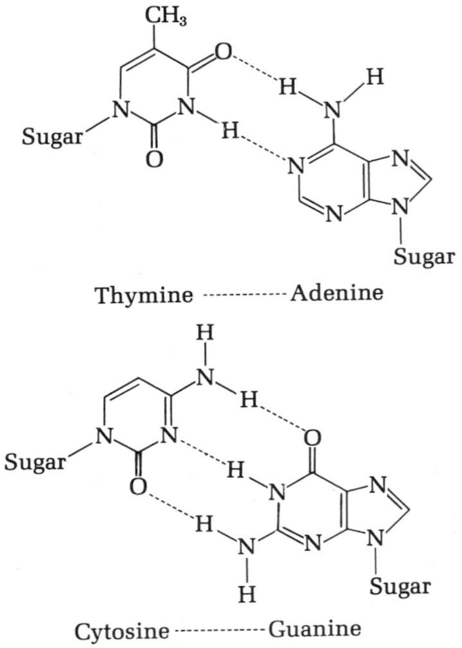

Thymine ············ Adenine

Cytosine ············ Guanine

Figure 22–32
The hydrogen bonding between the complementary base pairs of DNA.

The choice of any one of four bases for any given nucleotide along a single DNA strand gives an enormous number of possible sequences for a DNA molecule of molecular weight 10^9. Since the approximate average molecular weight of a nucleotide is 350 amu, there are on the order of one million bases that can be selected independently in a DNA molecule of molecular weight 10^9. The number of possible sequences is $(4)^{10^6} = 10^{(6 \times 10^5)}$: We have calculated that a protein containing 100 amino acids can have any of 10^{40} possible sequences. If the DNA sequence somehow represents a code for protein synthesis, there is enough information stored in a single DNA molecule of molecular weight 10^9 to determine the sequences of an enormous number of proteins.

Transfer of Genetic Information

The mechanism by which genetic information is transferred is easily understood in terms of the double-helical structure of DNA. The mechanism is depicted in Figure 22–33. The (a) part of the figure represents the intact helix, which unwinds to form the separated strands shown in (b). Past Figure 22–33(b), the mechanism resembles that for the hydrogenation of ethylene, shown in Figure 21–18. In the hydrogenation process, the nickel serves as a template for bringing the reactants together in the proper orientation for reaction. As depicted in Figure 22–33(c), each DNA strand binds free nucleotides by hydrogen bonding, and the sequence of these new complementary

```
P             P      P                 P
 \           /        \               /
  S—A····T—S           S—A       T—S
 /           \        /               \
P             P      P                 P
 \           /        \               /
  S—G····C—S           S—G       C—S
 /           \        /               \
P             P      P                 P
 \           /        \               /
  S—C····G—S           S—C       G—S
 /           \        /               \
P             P      P                 P
      (a)                   (b)

P                                          P
 \                                        /
  S—A····T—S—P3        P3—S—A····T—S
 /                                        \
P                                          P
 \                                        /
  S—G····C—S—P3        P3—S—G····C—S
 /                                        \
P                                          P
 \                                        /
  S—C····G—S—P3        P3—S—C····G—S
 /                                        \
P                                          P
                   (c)

P                                          P
 \                                        /
  S—A····T—S               S—A····T—S
 /           \            /               \
P             P          P                 P
 \           /            \               /
  S—G····C—S               S—G····C—S
 /           \            /               \
P             P          P                 P
 \           /            \               /
  S—C····G—S               S—C····G—S
 /           \            /               \
P             P          P                 P
                   (d)
```

Figure 22–33
The mechanism of replication of DNA.

partners is the same as in the original DNA molecule. The nucleotides are then joined through the action of the enzyme **DNA polymerase,** and we arrive at (d), the replication of (a). The symbol P_3 in Figure 22–33(c) indicates that the free nucleotides that form the new chains are triphosphates, as in ATP. Energy is required in the chain-making process, and this is derived from the splitting of the triphosphates into the monophosphates of the DNA chain and free pyrophosphate.

The Genetic Code

What is the nature of the DNA code? Surely the sequence of base pairs is a code for the amino acid sequences of specific proteins; but does each base on a given chain represent the code for a specific amino acid? A moment's thought convinces us that this cannot be the case since there are only four different bases and twenty different amino acids. Nor can adjacent bases in a strand represent individual amino acids, since there are only $(4)^2 = 16$ such possibilities. The minimum number of adjacent bases that can code for an amino acid is three, since there are $(4)^3 = 64$ possibilities for such base triplets. Thus, the AGC sequence on the left of Figure 22–33(a) could code for one of the twenty amino acids.

The translation of the DNA code is not direct, since DNA resides in the cell nucleus and protein synthesis occurs in **ribosomes,** extremely small granular particles that are well removed from the nucleus (Figure 22–7). Thus, a messenger is required to carry the code from DNA to the ribosomes, and this messenger is termed **messenger ribonucleic acid** (*m*-RNA). A ribonucleic acid molecule is similar in many respects to DNA, but there are some characteristic differences. First, RNA molecules are smaller than DNA, and second, they commonly occur as single polynucleotide strands instead of the double-stranded helix. In addition, the sugar unit is β-D-ribose instead of 2′-deoxy-β-D-ribose. Finally, the bases encountered in RNA are adenine, guanine, and cytosine as in DNA, but the fourth base is a modification of thymine, termed **uracil** (symbolized by U). Figure 22–34 shows that uracil is just thymine with a hydrogen atom in place of the methyl group. Uracil is capable of pairing with adenine just as thymine does.

Figure 22–34
The structure of uracil.

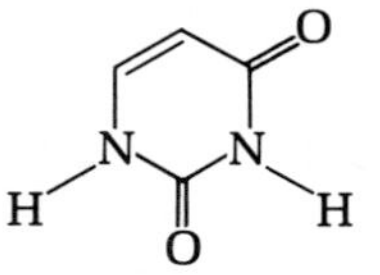

Messenger RNA is synthesized in nearly the same way as DNA, depicted in Figure 22–33. A portion of a DNA molecule unwinds, and the nucleotides ATP, GTP, CTP, and UTP form hydrogen bonds to the complementary bases along one of the DNA strands. These nucleotides are then polymerized to form *m*-RNA, and the energy is derived as in DNA formation by the splitting off of pyrophosphate.

The genetic code contained in the DNA molecule is then contained in the sequences of bases in *m*-RNA. Experiments have shown that a sequence of three base pairs in the *m*-RNA chain is the code for a specific amino acid, and such base triplets are termed **codons.** Figure 22–35 shows the 64 possible codons and the amino acid represented by each. Each codon contains the sequence of groups

$$\text{—P—}\underset{\displaystyle B_1}{\underset{|}{\text{S}}}\text{—P—}\underset{\displaystyle B_2}{\underset{|}{\text{S}}}\text{—P—}\underset{\displaystyle B_3}{\underset{|}{\text{S}}}$$

where B_1, B_2, and B_3 are the three bases listed in the order in which they appear from top to bottom in Figure 22–30. Each of the 20 amino acids appears at least once in Figure 22–35. Three codons (UAA, UAG, and UGA)

	U		C		A		G	
U	UUU	Phe	UCU	Ser	UAU	Tyr	UGU	Cys
	UUC	Phe	UCC	Ser	UAC	Tyr	UGC	Cys
	UUA	Leu	UCA	Ser	UAA		UGA	
	UUG	Leu	UCG	Ser	UAG		UGG	Trp
C	CUU	Leu	CCU	Pro	CAU	His	CGU	Arg
	CUC	Leu	CCC	Pro	CAC	His	CGC	Arg
	CUA	Leu	CCA	Pro	CAA	Gln	CGA	Arg
	CUG	Leu	CCG	Pro	CAG	Gln	CGG	Arg
A	AUU	Ile	ACU	Thr	AAU	Asn	AGU	Ser
	AUC	Ile	ACC	Thr	AAC	Asn	AGC	Ser
	AUA	Ile	ACA	Thr	AAA	Lys	AGA	Arg
	AUG	Met	ACG	Thr	AAG	Lys	AGG	Arg
G	GUU	Val	GCU	Ala	GAU	Asp	GGU	Gly
	GUC	Val	GCC	Ala	GAC	Asp	GGC	Gly
	GUA	Val	GCA	Ala	GAA	Glu	GGA	Gly
	GUG	Val	GCG	Ala	GAG	Glu	GGG	Gly

Figure 22–35 The genetic code. Each triplet of RNA bases is a code for the amino acid to the right of the triplet in the box.

do not represent amino acids. These codons signal the termination of the polypeptide chain in protein synthesis.

Mechanism of Protein Synthesis

Since protein synthesis occurs at the ribosomes, there must be a mechanism for transporting amino acids to ribosomes and also a means for their binding to *m*-RNA, which in turn must be bound to the ribosome. Each of these functions is carried out by one of two additional types of RNA. Ribosomes contain large RNA molecules as permanent constituents, and these molecules provide the binding of messenger RNA. The transfer of amino acids is provided by relatively small RNA molecules, of the order of 100 nucleotides; they are called **transfer RNA** (*t*-RNA). Each amino acid binds to its own *t*-RNA with its own nucleotide sequence. The amino acid becomes linked to the 3′-phosphate of one end of the *t*-RNA chain, as shown in Figure 22–36, and the product is termed an **amino acyl-*t*-RNA.** The energy for the linking comes from the splitting of ATP into adenosine monophosphate (AMP) and free pyrophosphate. The terminal nucleotide that binds the amino acid always contains the base adenine, as shown in the figure.

We still have the problem of explaining how a given amino acid recognizes its signal from messenger RNA so that it takes its proper position in the polypeptide chain to be formed. This key step arises from the structure adopted by all of the amino acyl-*t*-RNA molecules, and this characteristic

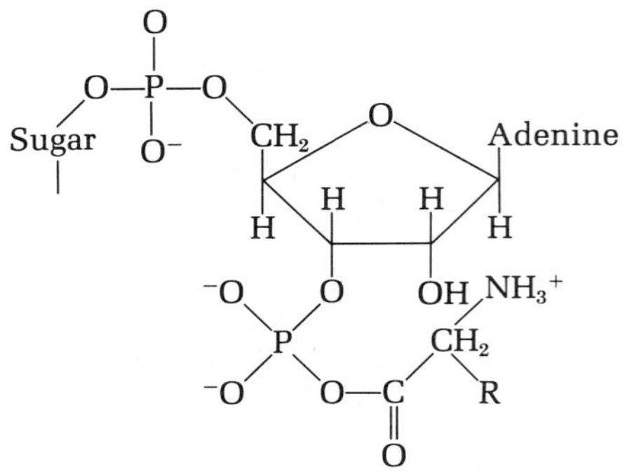

Figure 22–36 The binding of an amino acid residue in an amino acyl *t*-RNA molecule.

structure is shown in Figure 22–37. Many of the bases of the *t*-RNA are complementary, and the *t*-RNA forms three helical regions by base pairing, as shown by the dashed lines in the figure. There are three loops, however, in which base pairing does not occur, and the bases in these loops are free to associate with bases in other molecules. Three adjacent bases in one of these arms are termed the **anticodon** because these bases bind to the *m*-RNA at the codon corresponding to the amino acid carried by the *t*-RNA.

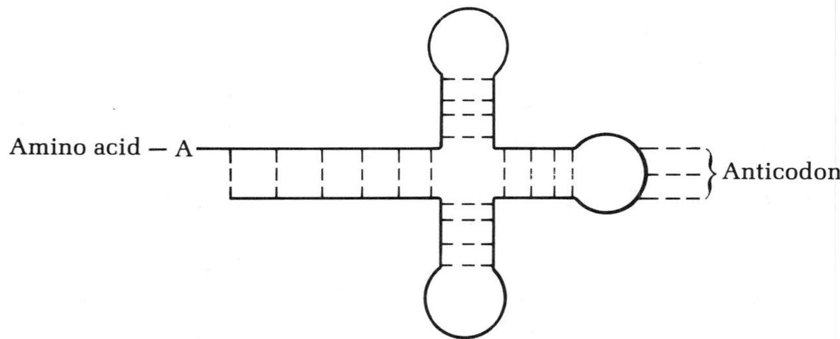

Figure 22–37 Representation of the structure of an amino acyl *t*-RNA molecule. The dashes inside the cross indicate the hydrogen bonding in two-stranded helical regions. The three dashed lines outside the cross locate the three bases constituting the anticodon.

Figure 22–38 shows a segment of *m*-RNA and the binding of an amino acyl-*t*-RNA of the amino acid leucine. One of the leucine codons from Figure 22–35 is CUG, and the complementary anticodon is GAC. Thus, the codon and anticodon bind, and leucine is held in the proper position. Experiments have shown that the protein grows along the *m*-RNA chain from the 5′-phosphate end to the 3′-phosphate end, although it is not known how and where the polypeptide chain is initiated in the cells of the higher animals.

The next codon on the *m*-RNA chain in Figure 22–38 in the direction of polypeptide chain growth is UGC, which is a cysteine codon with a complementary anticodon of ACG. The figure shows the binding of cysteine-*t*-RNA at this codon. (One of the problems at the end of the chapter asks you

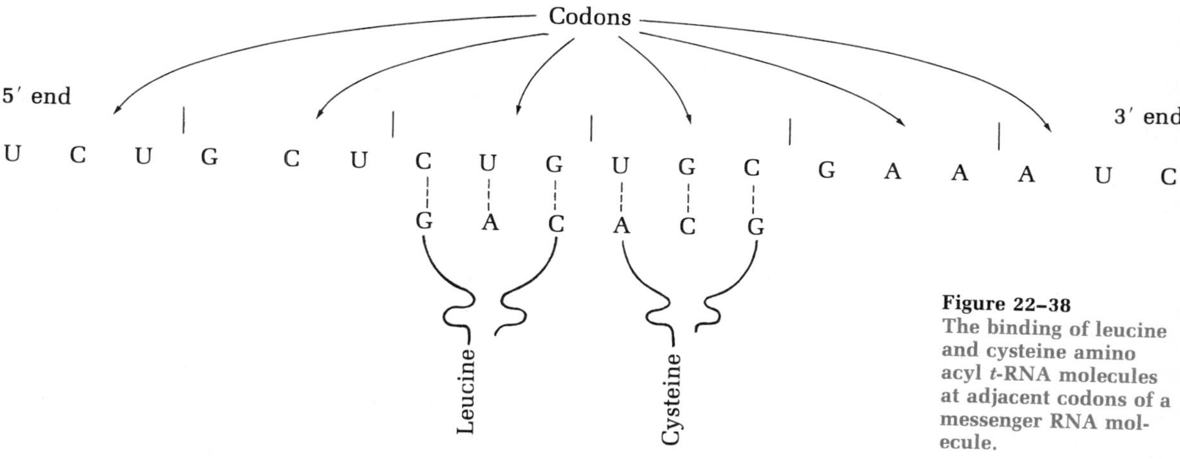

Figure 22-38
The binding of leucine and cysteine amino acyl *t*-RNA molecules at adjacent codons of a messenger RNA molecule.

to add the remaining two amino acyl-*t*-RNA molecules to the right of the two in the figure.)

Figure 22-39 shows a more detailed view of the two adjacent amino acyl-*t*-RNA molecules of Figure 22-38. Protein chains have been shown to grow from the free $—NH_3^+$ end, so the $—NH_3^+$ of cysteine is shown directed toward the carbonyl group of leucine. The amino acyl-*t*-RNA bond of the leucine compound is then replaced by the leucine-cysteine peptide bond to yield the structure shown in Figure 22-40. The empty *t*-RNA molecule then leaves and a new amino acyl-*t*-RNA enters to the right in the figure, and the polypeptide chain grows until a termination codon is reached.

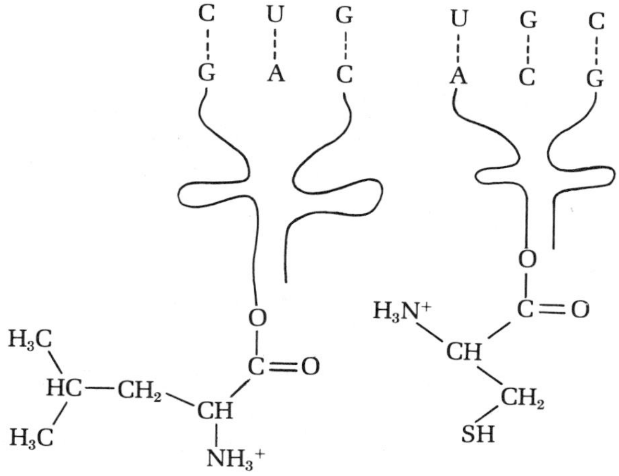

Figure 22-39
The relative orientations of two amino acid residues prior to peptide bond formation, as part of the biosynthesis of a protein.

This extremely complex process of genetic information transfer and protein synthesis brings this text to its conclusion. It should be apparent to the

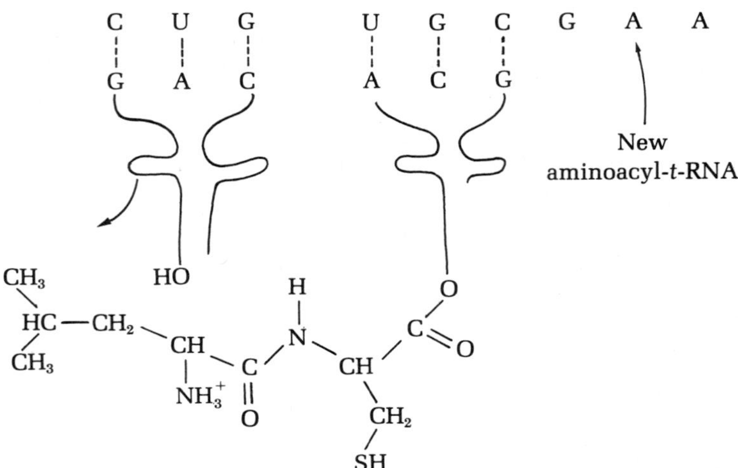

Figure 22–40
Structure resulting from peptide bond formation between the two amino acid residues of Figure 22–39.

reader that many other processes occur in living systems that are even more complex; they are, however, potentially understandable and can be understood in terms of specific molecular structures and reactions. Perhaps the foremost among these is the human nervous system. This research area is attracting a growing number of the world's most renowned scientists, and it promises to be one of the most dynamic areas in science for decades to come.

SUGGESTIONS FOR ADDITIONAL READING

Lehninger, A. L., *Biochemistry,* Worth, New York, 1970.

Watson, J. D., *Molecular Biology of the Gene,* Benjamin, New York, 1970.

Dickerson, R. E., and Geis, I., *The Structure and Action of Proteins,* Harper and Row, New York, 1970.

PROBLEMS

1. Trace the spontaneity involved in the operation of a windmill back to the process sunlight → thermal energy.
2. How many electrons must be transferred to six CO_2 molecules to reduce them to one molecule of cyclohexane?
3. Some organisms accomplish CO_2 reduction with H_2S instead of water. What oxidation state would you predict for the sulfur as it appears in the product? Is there any advantage to the organism to derive the electrons for reduction from H_2S rather than from H_2O? Explain.
4. If chlorophyll is phosphorescent, why doesn't grass glow in the dark?
5. The chlorophyll in plants is located in structures termed chloroplasts. It has been demonstrated that aqueous suspensions of chloroplasts will generate O_2 from H_2O when illuminated in the presence of $Fe(CN)_6^{3-}$

in solution, even in the absence of CO_2. Explain this oxygen evolution in an answer based on the mechanism for electron transport in photosynthesis.

6. One of the most interesting areas of investigation in biochemistry is chemical evolution, the evolution of biochemical molecules. Do chlorophyll and heme A show any evidence in their structures that they possess a common ancestor molecule? Explain.

7. Is there a molecular structural resemblance between ubiquinone and rubber? Explain.

8. It has been proposed that the reduction of species such as NAD^+ occurs by the addition of a hydride ion to the vitamin derivative. Refer to the structures given in Figure 22–14 and show by arrows the electron shifts involved in the conversion of 22–14(a) to 22–14(b) accompanying the addition of H^- to 22–14(a).

9. The first step in Figure 22–16 is the conversion of glucose to glucose 6-phosphate. If the phosphate comes from $H_2PO_4^-$ (P_i), the reaction

$$\text{Glucose} + P_i \rightarrow \text{glucose 6-phosphate} + H_2O \qquad \textbf{22–12}$$

possesses an equilibrium constant of 5.8×10^{-3}. However, the phosphate in Figure 22–16 does not come from P_i but rather from ATP. Use the value of $\Delta G°$ for the splitting of ATP together with the equilibrium constant for reaction 22–12 to calculate the equilibrium constant for the reaction

$$\text{Glucose} + \text{ATP} \rightarrow \text{glucose 6-phosphate} + \text{ADP} \qquad \textbf{22–13}$$

10. Show that the conversion of a molecule of glyceraldehyde 3-phosphate to 1,3-diphosphoglyceric acid in Figure 22–16 involves the loss of two electrons.

***11.** Reaction 22–3 produces a net incorporation of 6 oxygen atoms per glucose molecule in the conversion of $C_6H_{12}O_6$ to six CO_2. Specify the steps in Figures 22–16 and 22–17 in which these oxygen atoms are incorporated.

***12.** In Step 1 in Figure 22–19, CO_2 is added to acetyl CoA and in Step 3 this CO_2 is expelled again. What value is there in its initial incorporation when it is only going to be expelled shortly?

***13.** How many electrons does a palmitic acid molecule lose during its oxidation to CO_2 and H_2O? Account for all of these electrons in Figure 22–20 and in the steps of the Krebs cycle.

14. Draw structures for each of the following tripeptides. Show all of the atoms and all of the bonds.

a. H_3N^+—His—Cys—Gly—C(=O)O(−)

b. H_3N^+—Trp—Ser—Met—C(=O)O(−)

15. All of the enantiomeric amino acids in proteins are of the L form. If a D amino acid is synthetically incorporated into the structure, the α-helix of Figure 22–23 cannot form in that region. What does this indicate about the orientations of the R groups in Figure 22–23? What sort of secondary structure do you predict for a polypeptide containing only D amino acids?
16. A certain α-helical polypeptide chain contains 100 amino acid residues. If the energy of an average hydrogen bond is 5 kcal/mole, what is the total hydrogen bonding energy contributing to the stability of this α-helical structure?
17. The active site of α-chymotrypsin is a region known to contain few water molecules. It has been proposed that this exclusion of water is very important in contributing to the catalytic strength of enzymes in general. Explain how lack of water would promote a reaction step such as the conversion of the structure in Figure 22–27(a) to the structure in Figure 22–27(c).
18. What are two possible anticodons for the amino acid proline?
19. Complete Figure 22–38 by adding the two aminoacyl-*t*-RNA molecules that are dictated by the two codons to the right of the figure.

Appendices

Appendix A / Commonly Used Physical Constants

Avogadro's number	$N = 6.023 \times 10^{23}$/mole
Electron mass	$m_e = 9.108 \times 10^{-28}$ g
Electron charge	$e = 4.803 \times 10^{-10}$ esu
	$e = 1.602 \times 10^{-19}$ coul
Ideal gas constant	$R = 1.987$ cal/mole-°K
	$R = 0.08206$ l-atm/mole-°K
	$R = 8.313 \times 10^{7}$ erg/mole-°K
	$R = 8.313$ joule/mole-°K
Boltzmann's constant	$k = 1.380 \times 10^{-16}$ erg/molecule-°K
	$k = 3.299 \times 10^{-24}$ cal/molecule-°K
Faraday constant	$\mathscr{F} = 96{,}500$ coul/mole of electrons
Planck's constant	$h = 6.625 \times 10^{-27}$ erg/Hz
Speed of light	$c = 2.998 \times 10^{10}$ cm/sec

Energy Conversion Factors

1 cal = 4.185 joule
1 joule = 1.000×10^{7} erg
1 kcal/mole = 6.95×10^{-14} erg/molecule
1 ev/molecule = 23.06 kcal/mole
1 ev/molecule = 1.602×10^{-12} erg/molecule
1 l-atm = 24.2 cal

Appendix B / Exponents and Logarithms

B–1. Exponents

Exponential Notation

In Chapter 1 it is pointed out that scientists use a notation for numbers that separately denotes the significant figures and the power of ten in the number.

Some examples are 3.25×10^7 and 6.21×10^{-4}. A negative exponent indicates 1 divided by 10 to the power indicated by the number in the exponent. Thus, for example

$$10^{-4} = \frac{1}{10^4} = 0.0001$$

To convert a decimal such as 0.0023 to its equivalent exponential form, move the decimal point to the right of the first significant figure and count how many numbers (counting zeros as numbers) have been passed in doing so. This is the number in the exponent. If the decimal is less than 1, the exponent is negative; and if it is greater than 1, the exponent is positive. Thus, 0.0023 is represented as follows:

$$0.0023 = 2.3 \times 10^{-3}$$

because three numbers are passed (two zeros and the 2) in moving the decimal point to between the 2 and 3. A sample number greater than 1 is

$$1743 = 1.743 \times 10^3$$

Again, three numbers are passed (3, 4, and 7) in moving the decimal point from the right of the 3 to between the 1 and 7.

Multiplication and Division of Exponents

The multiplication or division of numbers expressed in exponential form is performed in two steps. Let us consider the example

$$(6.523 \times 10^{10})(1.69 \times 10^3)$$

The first operation is to multiply the two pre-exponential numbers to obtain 11.0, where three significant figures are reported, as in 1.69. The second operation is to **add** the two exponents to obtain 10^{13}. Thus, the product of the two numbers is

$$(6.523 \times 10^{10})(1.69 \times 10^3) = 11.0 \times 10^{13} = 1.10 \times 10^{14}$$

Now consider the division of two numbers in exponential form.

$$\frac{1.23 \times 10^4}{9.12 \times 10^{-2}}$$

Again, there are two operations, and the first is the division of the two pre-exponential numbers to yield 0.135. The second is the **subtraction** of the denominator exponent from the exponent in the numerator to yield $10^4/10^{-2} = 10^{[4-(-2)]} = 10^6$. The quotient of the two numbers is

$$\frac{1.23 \times 10^4}{9.12 \times 10^{-2}} = 0.135 \times 10^6 = 1.35 \times 10^5$$

Raising Exponential Numbers to Powers

Suppose that we wish to square the number 6.19×10^5. As with multiplication and division, raising a number in this form to a power is accomplished in two operations. The first is to raise the pre-exponential number to the desired power. In this case, $(6.19)^2 = 38.3$. The second operation is the multiplication of the exponent by the power to obtain in this case $(10^5)^2 = 10^{10}$. The overall operation is

$$(6.19 \times 10^5)^2 = 38.3 \times 10^{10} = 3.83 \times 10^{11}$$

If the power is less than 1, we are extracting a root of the number, but the operation is the same as raising the number to a power greater than 1. For example, the square root of a number is obtained by raising the number to the power $\frac{1}{2}$. For example,

$$\begin{aligned}(6.19 \times 10^5)^{1/2} &= (61.9 \times 10^4)^{1/2} = (61.9)^{1/2}(10^4)^{1/2} \\ &= 7.87 \times 10^2\end{aligned}$$

The number was converted to the form 61.9×10^4 so that the exponent times $\frac{1}{2}$ yields an integer.

B-2. Logarithms

Obtaining Common Logarithms

On some occasions it is necessary to display in a plot the magnitude of some quantity that changes over many powers of ten. It is impossible to display this change on a page if the quantity is plotted simply as itself. For example, if one unit of the quantity is represented by 1 mm, 1000 units would be a length of one meter, and that does not fit on a page.

One way to compress the scale so that many powers of ten can be shown on the same plot is to plot the **common logarithm** (symbolized as log) of the quantity. The common logarithm of a number is defined as the power to which 10 must be raised to give the number. For example, 1000 is 10^3, so the common logarithm of 1000 is 3. Likewise 0.01 is 10^{-2}, and the common logarithm of 0.01 is -2. By plotting the logarithm of a number instead of the number itself, five units of length represent five powers of ten, and huge ranges are easily plotted.

Numbers that are not simply integral powers of 10 have common logarithms related to the exponential form of the number. For example, consider the number 623. In exponential form, the number is 6.23×10^2. Now we wish to know x in the equation

$$10^x = 10^{\log(6.23 \times 10^2)} = 6.23 \times 10^2$$

This equation can be written

$$6.23 \times 10^2 = 10^{\log 6.23} 10^{\log 10^2} = 10^{\log 6.23 + \log 10^2}$$

Thus,

$$x = \log 10^2 + \log 6.23 = 2 + \log 6.23$$

The common logarithm of any number is the sum of an integer that gives the power of ten of that number and the logarithm of a number between 1 and 10. The logarithms of numbers between 1 and 10 are given in "log tables" in all mathematics handbooks, most physics and chemistry handbooks, and many textbooks. For example, the log table lists log 6.23 as 0.794, and thus log 623 is 2.794.

Let us consider some additional examples:

$$\begin{aligned} \log 12.1 &= \log (1.21 \times 10) \\ &= \log 1.21 + \log 10 \\ &= 0.083 + 1.000 \\ &= 1.083 \end{aligned}$$

$$\begin{aligned} \log 0.0544 &= \log (5.44 \times 10^{-2}) \\ &= \log (5.44) + \log (10^{-2}) \\ &= 0.736 - 2.000 \\ &= -1.264 \end{aligned}$$

Obtaining an Antilogarithm

Several times in the text you are required to obtain a number from the logarithm of that number. This is the process of obtaining an antilogarithm. This is easily done with the use of a log table. Let us consider the logarithm −4.471; what number n is the antilogarithm of −4.471? From the definition of a logarithm, we have

$$\begin{aligned} n &= 10^{-4.471} = 10^{(-5+0.529)} \\ &= 10^{0.529} \times 10^{-5} \end{aligned}$$

We refer to a log table to see which number between 1 and 10 has the logarithm 0.529, and this number is 3.38. Thus, n is 3.38×10^{-5}.

Operations with Logarithms

Since common logarithms are exponents of 10, they can be combined as exponents are. Thus we have the following relations:

$$\log xy = \log x + \log y$$

$$\log \frac{x}{y} = \log x - \log y$$

$$\log \frac{1}{y} = -\log y$$

Since one step in raising a number to a power is multiplying the power by the exponent of the number, we have the following additional relation involving logarithms:

$$\log x^n = n \log x$$

Natural Logarithms

In Chapters 15 and 21 we encounter relations between quantities through the logarithm of one of the quantities. The logarithm that naturally appears in these relations is defined somewhat differently from the common logarithm. The number 10 is called the **base** of the common logarithm system. The base of the **natural logarithm** system is the number e, which is written to six significant figures.

$$e = 2.71828$$

The natural logarithm of a number x (symbolized as ln x) is easily obtained from the common logarithm (log x). The conversion factor is ln 10 = 2.303. We have

$$x = 10^{\log x} = (e^{\ln 10})^{\log x} = e^{2.303 \log x} = e^{\ln x}$$

Thus,

$$\ln x = 2.303 \log x$$

The corresponding antilogarithmic relation is

$$e^y = 10^{(y/2.303)}$$

Appendix C/ Nomenclature of Inorganic Compounds

In Chapter 10 the system of nomenclature devised by the International Union of Pure and Applied Chemistry (IUPAC) is presented as applied to organic compounds. In Chapter 11 the nomenclature of metal-ion complexes is presented in some detail. This Appendix includes the nomenclature of simple inorganic species encountered in the text, particularly in Chapters 8 and 9.

Cations

The most common positive ions (cations) in inorganic chemistry are derived by electron extraction from metal atoms. In cases in which only one charge type is encountered with a given metal, the cation is simply given the name of the metal:

K^+	potassium	Mg^{2+}	magnesium
Ca^{2+}	calcium	Al^{3+}	aluminum

Some elements, particularly the transition metals, form more than one type of cation. In these cases, a Roman numeral indicates the charge:

Cu^+	copper(I)	Fe^{2+}	iron(II)
Cu^{2+}	copper(II)	Fe^{3+}	iron(III)

In some texts these ions are named according to an older system based on the Latin names of the elements. These Latin names are used as prefixes coupled with the suffixes *-ous* and *-ic*. These suffixes represent the lower and higher charges, respectively:

Cu^+	cuprous	Fe^{2+}	ferrous
Cu^{2+}	cupric	Fe^{3+}	ferric

The Roman numeral representation is used exclusively in this text.

Anions

Several common negative ions (anions) are derived by adding one or more electrons to an atom of a nonmetallic element. The names of these anions are obtained by combining the suffix *-ide* with prefixes derived from the names of the elements. Some examples are

H^-	hydride	F^-	fluoride
N^{3-}	nitride	Cl^-	chloride
O^{2-}	oxide	Br^-	bromide
S^{2-}	sulfide	I^-	iodide

There are some common inorganic anions that contain more than one atom. Some of these are

OH^-	hydroxide	ClO_4^-	perchlorate
CO_3^-	carbonate	ClO_3^-	chlorate
NO_3^-	nitrate	ClO_2^-	chlorite
NO_2^-	nitrite	ClO^-	hypochlorite
O_2^{2-}	peroxide	IO_4^-	periodate
PO_4^{3-}	orthophosphate	IO^-	hypoiodite
$P_2O_7^{4-}$	pyrophosphate	MnO_4^-	permanganate
SO_4^{2-}	sulfate	CrO_4^{2-}	chromate
$S_2O_7^{2-}$	pyrosulfate	$Cr_2O_7^{2-}$	dichromate
SO_3^{2-}	sulfite		

Some of the anions in this group can be protonated to yield hydrogen-containing anions. Some examples are

HCO_3^-	hydrogen carbonate
HPO_4^{2-}	hydrogen orthophosphate
$H_2PO_4^-$	dihydrogen orthophosphate
HSO_4^-	hydrogen sulfate
HS^-	hydrogen sulfide
HSO_3^-	hydrogen sulfite

Ionic Compounds

Ionic compounds are named by combining the name of the cation with the name of the anion. Some examples are

NaCl	sodium chloride
K_2SO_4	potassium sulfate
$Fe(OH)_3$	iron(III) hydroxide
$NaHCO_3$	sodium hydrogen carbonate
NaH_2PO_4	sodium dihydrogen orthophosphate

Binary Compounds

A compound such as NaCl is a binary compound consisting of ions. Many compounds exist that contain two different elements and are not ionic. Their naming is like the naming of ionic compounds, however. Some examples are

HCl	hydrogen chloride
CO_2	carbon dioxide
SO_3	sulfur trioxide
N_2O	dinitrogen oxide
NO	nitrogen oxide
N_2O_4	dinitrogen tetroxide

There are also in wide usage several older names for common nonionic inorganic compounds. These are also used in the text, and some of them are

H_2O	water
NH_3	ammonia
N_2O	nitrous oxide
NO	nitric oxide
PH_3	phosphine
B_2H_6	diborane
SiH_4	silane

Oxyacids

The oxyacids are a common and important group of inorganic compounds. Each oxyacid contains hydrogen, oxygen, and one additional element. Their names derive from the names of the additional elements. When only one oxyacid is formed by a given element, the suffix employed in the name is *-ic*. Some examples are

H_2CO_3	carbonic acid
H_3BO_3	boric acid

Some elements form more than one oxyacid. When two compounds are formed, the one containing the greater number of oxygens takes the suffix *-ic* and the other takes the suffix *-ous*. Thus, we have

HNO_3	nitric acid
HNO_2	nitrous acid

Chlorine forms four oxyacids; the prefix *per-* indicates the highest number of oxygens and *hypo-* indicates the lowest number. The four acids are

$HClO_4$	perchloric acid
$HClO_3$	chloric acid
$HClO_2$	chlorous acid
$HClO$	hypochlorous acid

Appendix D/ Standard Heats of Formation

Compound	ΔH_f° (kcal/mole)
Inorganic	
$Al_2O_3(s)$	−399.09
$AgCl(s)$	−30.36
$B_2H_6(g)$	7.5
$BaCl_2(s)$	−205.6
C(diamond)	0.45
$CaO(s)$	−151.79
$CaCO_3(s)$	−288.45
$CO(g)$	−26.42
$CO_2(g)$	−94.05
$Fe_2O_3(s)$	−196.5
$HBr(g)$	−8.66
$HCl(g)$	−22.06

Compound	ΔH_f°(kcal/mole)
$HF(g)$	−64.2
$HI(g)$	6.2
$H_2O(g)$	−57.80
$H_2O(l)$	−68.32
$HgCl_2(s)$	−55.0
$KCl(s)$	−104.18
$MgCl_2(s)$	−153.40
$MgO(s)$	−143.84
$NH_3(g)$	−11.04
$NO(g)$	21.60
$NO_2(g)$	8.09
$N_2O_4(g)$	2.31
$NaCl(s)$	−98.23
$Na_2CO_3(s)$	−270.3
$O_3(g)$	34.0
$SO_2(g)$	−70.76
SiO_2(quartz)	−205.4
$ZnCl_2(s)$	−99.4
Organic	
$C_2H_2(g)$ (acetylene)	54.19
$C_6H_6(l)$ (benzene)	11.72
$CCl_4(l)$ (carbon tetrachloride)	−33.4
$CHCl_3(l)$ (chloroform)	−31.5
$C_2H_6(g)$ (ethane)	−20.24
$C_2H_4(g)$ (ethylene)	12.50
$CH_4(g)$ (methane)	−17.89
$CH_3OH(l)$ (methanol)	−57.04

Appendix D (continued)

Appendix E / Gibbs Free Energies of Formation

Compound	ΔG_f° (kcal/mole)
Inorganic	
Al_2O_3(s)	−376.77
AgCl(s)	−26.22
B_2H_6(g)	19.8
$BaCl_2$(s)	−193.8
C(diamond)	0.69
CaO(s)	−144.4
$CaCO_3$(s)	−269.78
CO(g)	−32.81
CO_2(g)	−94.26
Fe_2O_3(s)	−177.1
HBr(g)	−12.72
HCl(g)	−22.77
HF(g)	−64.7
HI(g)	0.3
H_2O(g)	−54.64
H_2O(l)	−56.69
$HgCl_2$(s)	−44.4
KCl(s)	−97.59
$MgCl_2$(s)	−141.57
MgO(s)	−136.13
NH_3(g)	−3.98
NO(g)	20.72
NO_2(g)	12.39
N_2O_4(g)	23.49
NaCl(s)	−91.79
Na_2CO_3(s)	−250.4
O_3(g)	39.06
SO_2(g)	−71.79
SiO_2(quartz)	−192.4
$ZnCl_2$(s)	−88.3
Organic	
C_2H_2(g) (acetylene)	50.0
C_6H_6(l) (benzene)	41.30
CCl_4(l) (carbon tetrachloride)	−16.43
$CHCl_3$(l) (chloroform)	−17.1
C_2H_6(g) (ethane)	−7.86
C_2H_4(g) (ethylene)	16.28
CH_4(g) (methane)	−12.14
CH_3OH(l) (methanol)	−39.75

Appendix F/ Standard Entropies ($S°$), Heats of Formation ($\Delta H°_f$), and Gibbs Free Energies of Formation ($\Delta G°_f$), for Species in Aqueous Solution

Ion or Molecule	$S°$ (cal/mole-deg)	$\Delta H_f°$ (kcal/mole)	$\Delta G_f°$ (kcal/mole)
H^+ (defined)	0	0	0
Ag^+	17.67	25.31	18.43
$Ag(NH_3)_2^+$	57.8	−26.72	−4.16
Al^{3+}	−74.9	−125.4	−115
BF_4^-	40	−365	−343
Br^-	19.29	−28.90	−24.57
Ca^{2+}	−13.2	−129.77	−132.18
CH_3COOH		−116.74	−95.51
CH_3COO^-	20.8	−116.84	−89.02
Cl^-	13.2	−40.02	−31.35
ClO_4^-	43.2	−31.41	−2.47
CN^-	28.2	36.1	39.6
CO_2	29.0	−98.69	−92.31
CO_3^{2-}	−12.7	−161.63	−126.22
Cu^{2+}	−23.6	15.39	15.53
F^-	−2.3	−78.66	−66.08
Fe^{2+}	−27.1	−21.0	−20.30
Fe^{3+}	−70.1	−11.4	−2.53
H_2CO_3			−145.46
HCO_3^-	22.7	−165.18	−140.31
H_3PO_4	42.1	−308.2	−274.2
$H_2PO_4^-$	21.3	−311.3	−271.3
HPO_4^{2-}	−8.6	−310.4	−261.5
H_2S	29.2	−9.4	−6.54
HS^-	14.6	−4.22	3.01
I_2			3.93
I^-	26.14	−13.37	−12.35
K^+	24.5	−60.04	−67.46
Mg^{2+}	−28.2	−110.41	−108.99
Na^+	14.4	−57.28	−62.59
NH_3	26.3	−19.32	−6.36
NH_4^+	26.97	−31.74	−19.00
NO_3^-	35.0	−49.37	−26.43
OH^-	−2.52	−54.96	−37.59
PO_4^{3-}	−52	−306.9	−245.1
S^{2-}	−4	7.8	20.6
SO_4^{2-}	4.1	−216.90	−177.34
Zn^{2+}	−25.45	−36.43	−35.18

Answers to Selected End-of-Chapter Problems

Chapter 1

1. 5.00 g
2. three
3. a. 0.154 b. 27.96 c. 3.7
4. a. 21.5 l c. 1.61 km e. 11 m/sec
5. 28.3 l; 62.4 lb/ft^3
6. 39.34 percent
13. 1.0×10^{25} molecules
15. $K_3FeC_6N_6$
16. 14.1 percent
17. 52.2 percent
18. CH_2O
19. 84 percent sulfur left unreacted
20. 0.4489
21. 54 g
22. 15.2 ml; 3.736×10^{-3} mole of NH_3
23. 87.9 percent

Chapter 2

1. 772.7 torr, 1.017 atm
2. 5.30×10^{-3} torr
4. a. Extensive b. Intensive c. Intensive d. Extensive
5. $-40\,°C = -40\,°F$
6. 0.55 cal/g-°F
8. 6.3 l
9. 55 psi
10. 0.282 l
11. 33.4 percent
12. 58 amu, C_3H_6O
13. 29 amu
14. There are approximately eight oxygen molecules to every water molecule.
18. 3.72×10^3 joules/mole
21. The probability is 8×10^{-28}.
25. 3.8×10^{-8} cm

Chapter 3

1. 0.234 g of Na
2. 3.4×10^{-38} dyne
5. 77.5 percent ^{35}Cl
6. 9.45×10^{17} cm/yr; 5.87×10^{12} miles/yr
8. 2.56 ft
9. $n_1 = 1,\ n_2 = 3$
10. $n_1 = 2,\ n_2 = 3$
15. $r_1 = 5.3 \times 10^{-9}$ cm; $r_2 = 2.1 \times 10^{-8}$ cm
17. 2.63×10^5 cm/sec

Chapter 4

2. Electron density in each segment is 0.100 electron.
4. $n = 4,\ l = 3$
   Three planar nodes; no spherical nodes
   Seven functions
5. -54.4 ev
6. 49.2 ev
7. 2.50, 0.62
8. $A = 1.1 \times 10^5\ cm^{-1}$
11. Rb and Cs
12. As, Sb, and Bi
15. C > F; H > He; Na > Li
18. Li, N, S, Fe, Cr^{2+}, Cu^{2+}

Chapter 5

2. -4.7×10^{-11} erg

10. $1s\sigma^2 1s\sigma^{*2} 2s\sigma^2 2s\sigma^{*2} 2p\sigma^2 2p\pi^4 2p\pi^{*3}$
Bond order = $\frac{3}{2}$

13. 0.18 electron

16. Na_2O: Na^+ $(1s^2 2s^2 2p^6)$,
O^{2-} $(1s^2 2s^2 2p^6)$
Na_2S: Na^+, S^{2-} $(1s^2 2s^2 2p^6 3s^2 3p^6)$
$MgCl_2$: Mg^{2+} $(1s^2 2s^2 2p^6)$
Cl^- $(1s^2 2s^2 2p^6 3s^2 3p^6)$
MgO: Mg^{2+}, O^{2-}
Al_2O_3: Al^{3+} $(1s^2 2s^2 2p^6)$, O^{2-}

18. The two ion pair models yield the following bond energies.

H^+—Cl^-, 31 kcal/mole $\qquad$ H^-—Cl^+, −22 kcal/mole

Chapter 6

3. N, O, and F

4. a.

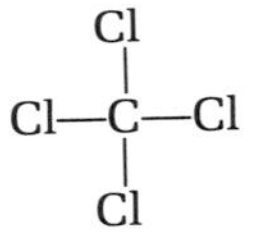

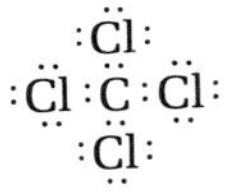

```
   Cl                :Cl:
   |               :Cl:C:Cl:
Cl—C—Cl              :Cl:
   |
   Cl
```

b.

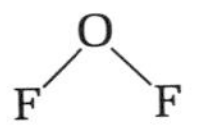

```
  O                  :O:
 / \               :F: :F:
F   F
```

c. H—Cl $\qquad$ H:Cl:

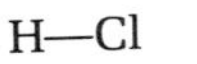

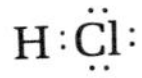

d.

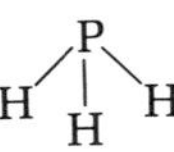

```
H:P:H
  H
```

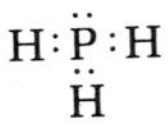

5. a. Tetrahedral b. Tetrahedral c. Linear
d. Triangular e. Bent f. Linear

7. a. Four b. Three c. Two

9. a. sp b. sp^2 c. sp^2
d. sp e. sp

10. sp^3

11. sp^2

12. −373 kcal/mole

15. 0.33 electron

18. a. −71 kcal b. −21 kcal

20. 70 kcal/mole

Chapter 7

1. a. Sc_2O_3 b. Ca_3P_2 c. $SiCl_4$ d. CS_2

5. A structure similar to that of NH_3

7. Four-electron loss per ethyl alcohol molecule

9. a. Two-electron loss b. Six-electron gain c. Eight-electron gain

11. $N{\equiv}\overset{+}{N}{-}\overset{-}{O}$

Chapter 8

1. −22 kcal

3. Δ*E* per mole of ammonia is −78 kcal.

8. Tetrahedral, sp^3

11. 180°

12. :Ö–N:N–Ö: (Lewis structure, O–N–N–O)

13. :Ö:, :Ö: (Lewis structure, $O_2N{:}NO$)

17.

```
   :Ö:
H:Ö:P:H
    H
```

20.

```
    :Ö:
:F:S:Ö:H
    :Ö:
```

Chapter 9

7. $ClO^-(aq) + Cl^-(aq) + 2\ H^+(aq) \rightarrow Cl_2(g) + H_2O(l)$

9. The O—Xe—O angle is near the tetrahedral angle in XeO_3. XeO_4^{4-} is square-planar.

12.

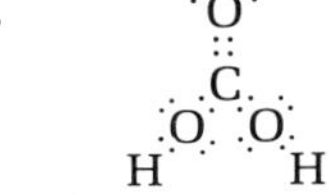

Carbonic acid

Hydrogen carbonate

14. The structure is analogous to that of B_2H_6.

15.

HO—B, O, B—OH (cage structure of four B and five O)

$H_2B_4O_7$

[HO—B, O, B—OH, with OH on upper and lower B]$^{2-}$

$H_4B_4O_9^{2-}$

Chapter 10

1. Hexane; 2-methylpentane; 3-methylpentane; 2,2-dimethylbutane; 2,3-dimethylbutane

5. The chair form predominates.

6. 1-pentene; *cis*-2-pentene; *trans*-2-pentene; 2-methyl-2-butene; 2-methyl-1-butene; cyclopentane; methylcyclobutane; 1,2-dimethylcyclopropane; 1,1-dimethylcyclopropane

7. $\Delta E = -23$ kcal for reaction 10–4.
$\Delta E = -88$ kcal for reaction 10–18.

9. There are four structures.

10. 2-methylbutadiene

12. 1-butyne; 2-butyne; 1,3-butadiene; 1,2-butadiene; cyclobutene; methylcyclopropene; bicyclobutane; methylenecyclopropane,

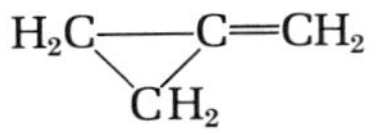

H_2C—C=CH_2 (ring: H_2C, C, CH_2)

15. The positions shown by the asterisks are positions in which the second nitro group will not be found.

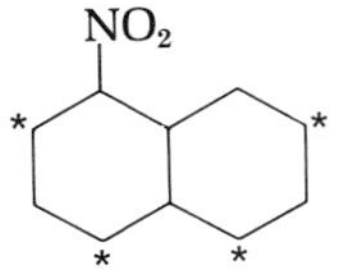

NO_2 * * * *

16. One structure is

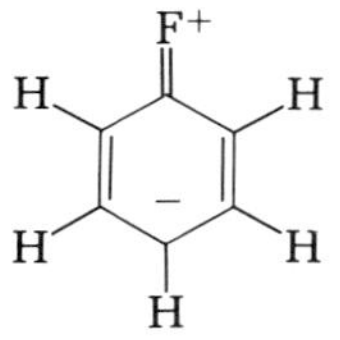

F^+ H H H H H −

20. Butanal (butyraldehyde)
Butanone (methyl ethyl ketone)
2-Methylpropanal (isobutyraldehyde)

22. OH
H_3C—C—CH_3
CN

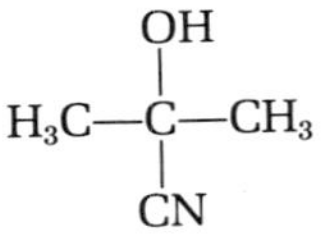

23. Eight electrons per *cis*-2-butene

26. One resonance structure is

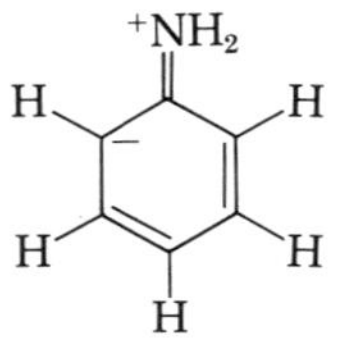

$^+NH_2$ H H H H H −

29. The asterisks indicate the asymmetric carbon atoms.

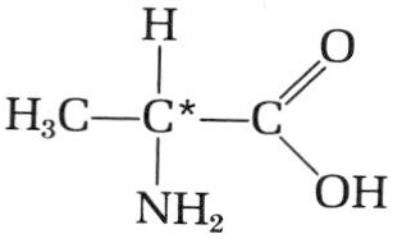

$H_3C-C^*H(NH_2)-C(=O)OH$

(a)

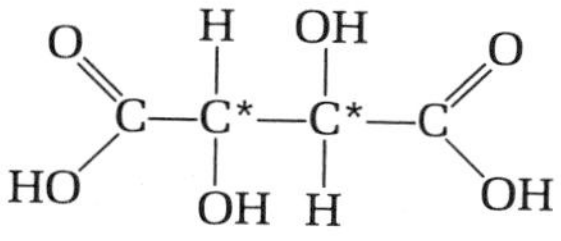

$HO(O=)C-C^*H(OH)-C^*H(OH)-C(=O)OH$

(b)

$HO-{}^*CH$ (ring: $CH_2-CH_2-CH_2-CH=CH$)

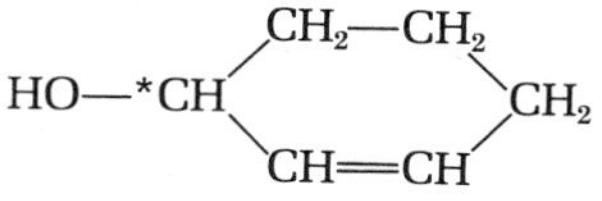

(c)

Chapter 11

1.

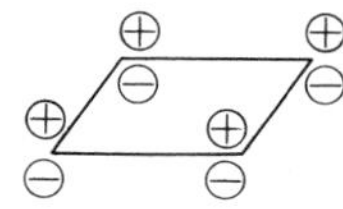

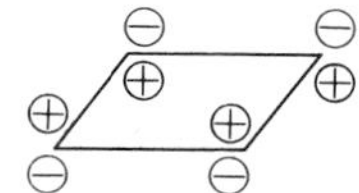

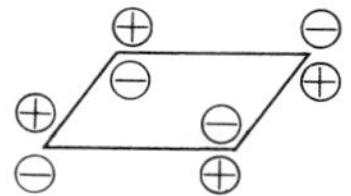

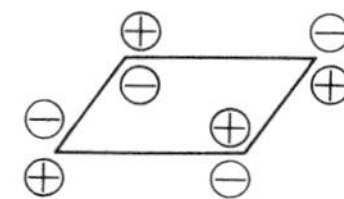

17. 180°

18. Two bridged tetrahedra

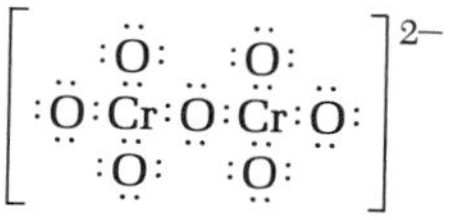

$$\left[:\ddot{O}:\ddot{Cr}:\ddot{O}:\ddot{Cr}:\ddot{O}:\right]^{2-}$$ (each Cr also bonded to two $:\ddot{O}:$, above and below)

19. Tetrahedral

$$\left[:\ddot{O}:\ddot{Mn}:\ddot{O}:\right]^{-}$$ (Mn also bonded to $:\ddot{O}:$ above and below)

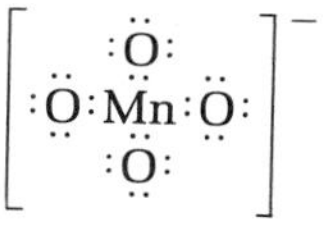

24. $Cu(NH_3)_4{}^{2+}$ is square-planar.

26. a. Chloropentammine cobalt(III) tetrafluoroborate
b. Sodium tetrachlorocuprate(II)
c. Potassium hexacyanocobaltate(III)
d. Dichloro ethylenediamine platinum(II)

27.

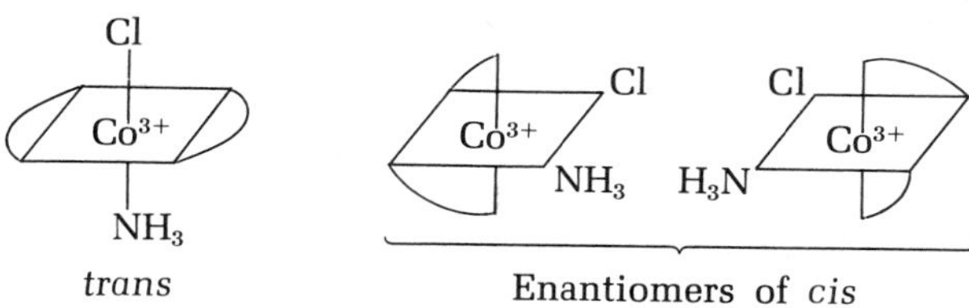

28. Co^{3+} and Ni^{2+}

30. $Ni(CO)_4$, tetrahedral
$Cr(CO)_6$, octahedral

33. Tetrahedral

Chapter 12

4. a. 3 translations, 3 rotations, 9 vibrations
   b. 3 translations, 3 rotations, 12 vibrations
   c. 3 translations, 2 rotations, 4 vibrations
5. Motion (a), symmetric stretch is not excited. Motion (b), out-of-plane bend is excited.

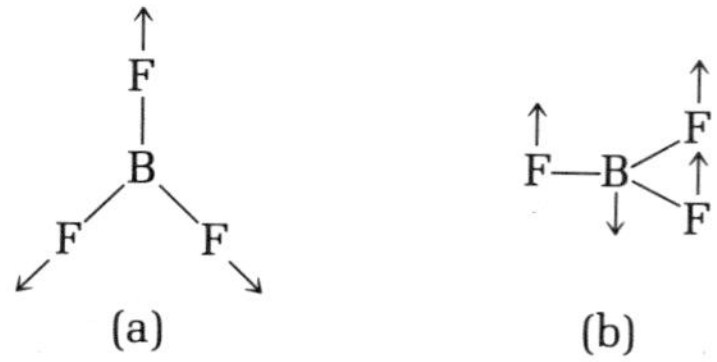

11. The smaller vibration energy and greater bond energy are characteristic of DCl.
19. $1s\sigma^2 1s\sigma^{*2} 2s\sigma^2 2s\sigma^{*2} 2p\sigma^2 2p\pi^3 2p\pi^{*1}$
22. Nitrogen single, nonbonding electron → antibonding pi orbital.
23. 1.0×10^8 Hz
26. $H_3C—CH_2—CH_2—NO_2$
27. 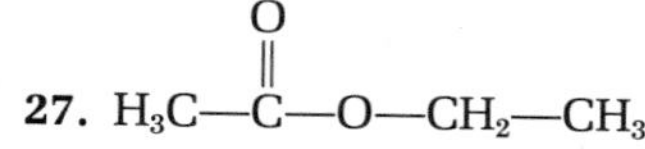

Chapter 13

1. The angle θ is smaller for $n = 1$.
5. It is the same as body-centered tetragonal.
7. 4.79 Å
8. Gold
9. Twelve; twelve
10. The two layers between spheres 3 and 10 contain spheres 2, 4, 5, 7, 8, and 12 in one layer and 1, 6, 9, 11, 13, and 14, is the other layer.
13. $r_2/r_1 = 0.155$.
14. Cell length = 4.13 Å. Cs—Cl distance = 3.57 Å.

Chapter 14

8. 4.81×10^{-5} cm
11. Area of sphere = 4.8 cm^2. Area of cube = 6.0 cm^2.

Chapter 15

1. 0.356 °C
2. 9.6×10^{-5} °C
6. a. $Q_{released} > (-\Delta E)$. b. $Q_{released} > (-\Delta E)$.
   c. $Q_{released} < (-\Delta E)$. d. $Q_{released} = (-\Delta E)$.
7. $\Delta E_{H_2O} = -1479$ cal.
8. −235.4 kcal
9. 294.2 °K
10. $\Delta H^\circ = -202.6$ kcal.
11. 1691 °C
12. 25.370 °C
13. CH_4, 11.96 kcal/g C_2H_2, 11.52 kcal/g
14. 18 kcal
15. 100 kcal

16. 257 kcal/mole
25. $W_{limiting} = 5.71 \times 10^3$ joules
26. 0.0312 cal/deg
27. 6.89 cal/deg
28. 0.26 cal/deg
29. 0.29 cal/deg
30. 7.12 cal/deg
37. 6.16 cal/deg

Chapter 16

5. 1.32×10^6 joules
14. -3.42×10^3 cal
16. a. $K_e = P_{N_2O_4}/P^2_{NO_2} = 8.83$ b. $K_e = P^2_{O_3}/P^3_{O_2} = 5.0 \times 10^{-58}$
c. $K_e = 1/P_{CH_4}P^{1/2}_{O_2} = 1.8 \times 10^{20}$
19. The K_e for the dissociation of H_2S is 9.9×10^{-8}. Acetic acid is stronger.
20. The K_e for the dissolution of AgI is 8.5×10^{-17}. AgCl is more soluble.
21. The K_e for the reaction is 1.9×10^{16}. Zinc dissolves appreciably.
23. Increases
24. 840 °K.

Chapter 17

1. 93 torr
2. 370.6 °K
10. $x_{benzene} = 0.70$
11. Three
12. 128
13. From equation 17–25, we obtain $T = 355.19$ °K; and from equation 17–26, we obtain $T = 355.24$ °K.
14. Molarity = 12; molality = 15; mole fraction = 0.22.
15. 233.8 °K
16. 37 percent ethylene glycol
20. 1.3×10^5
27. 1 atm

Chapter 18

1. 3×10^{120}
4. 1075 °K
5. $P_{NO_2} = 0.097$ atm; $P_{N_2O_4} = 0.085$ atm.
7. $P_{HI} = 0.403$ atm; $P_{H_2} = 0.121$ atm; $P_{I_2} = 0.025$ atm.
8. $P_{HI} = 0.570$ atm; $P_{H_2} = 0.139$ atm; $P_{I_2} = 0.043$ atm.
10. 6.5×10^{-3} atm.

Chapter 19

1. 1.0×10^{-13} mole/l
3. a. $Na_2O(s) + H_2O(l) \rightarrow 2\ Na^+(aq) + 2\ OH^-(aq)$
b. $NaOH(s) + H_2O(l) \rightarrow Na^+(aq) + OH^-(aq) + H_2O(l)$
c. $SO_3(g) + H_2O(l) \rightarrow H^+(aq) + HSO_4^-(aq)$
d. $Na_2O(s) + SO_3(g) \rightarrow Na_2SO_4(s)$
e. $N_2O_5(s) + H_2O(l) \rightarrow 2\ H^+(aq) + 2\ NO_3^-(aq)$
f. $HNO_3(l) + H_2O(l) \rightarrow H^+(aq) + NO_3^-(aq) + H_2O(l)$
5. 15.7
6. a. 5.13 b. 11.25 c. 12.37
7. 5.47
10. 7.77
11. 125.0
12. 4.81
13. a. H_3CCOO^-, NH_4^+ b. H_3CCOOH, NH_4^+ c. H_3CCOO^-, NH_3
17. 5.26, methyl red

18. a. $Na_2S(s) + H_2O(l) \rightarrow 2\ Na^+(aq) + HS^-(aq) + OH^-(aq)$
b. $NaHSO_4(s) + H_2O(l) \rightarrow Na^+(aq) + H^+(aq) + SO_4^{2-}(aq) + H_2O(l)$
c. $Na_2HPO_4(s) + H_2O(l) \rightarrow 2\ Na^+(aq) + H_2PO_4^-(aq) + OH^-(aq)$
d. $Na_2CO_3(s) + H_2O(l) \rightarrow 2\ Na^+(aq) + HCO_3^-(aq) + OH^-(aq)$
e. $NaHCO_3(s) + H_2O(l) \rightarrow Na^+(aq) + CO_2(aq) + OH^-(aq) + H_2O(l)$

19. 8.31
21. $1.2 \times 10^{-6}\ M$
22. $5 \times 10^{-22}\ M$
26. a. 1.9×10^{-4} g b. 2.6×10^{-3} g c. 1×10^{-3} g
27. 1×10^{-4} g
29. 9.0
31. AgI
32. 2.0

Chapter 20

1. a. 0.945 b. 0.934 c. 1.0285
4. a. $6\ Fe^{2+}(aq) + Cr_2O_7^{2-}(aq) + 14\ H^+(aq) \rightarrow 6\ Fe^{3+}(aq) + 2\ Cr^{3+}(aq) + 7\ H_2O(l)$
b. $CH_4(g) + 4\ Cl_2(g) \rightarrow CCl_4(l) + 4\ HCl(g)$
c. $2\ F_2(g) + 2\ H_2O(l) \rightarrow O_2(g) + 4\ HF(g)$
d. $2\ I_3^-(aq) + ClO^-(aq) + 2\ H^+(aq) \rightarrow 3\ I_2(s) + Cl^-(aq) + H_2O(l)$
5. a. 2 b. 6 c. 2 d. 60
7. a. −0.34 volt b. 0.12 volt c. 0.56 volt
8. a. $2\ Cr^{2+}(aq) + 2\ H^+(aq) \rightarrow 2\ Cr^{3+}(aq) + H_2(g)$
b. $Cl_2(g) + 2\ Br^-(aq) \rightarrow 2\ Cl^-(aq) + Br_2(l)$
c. $2\ F_2(g) + 2\ H_2O(l) \rightarrow O_2(g) + 4\ H^+(aq) + 4\ F^-(aq)$
9. −0.52 volt
11. −0.40 volt
13. One pair of concentrations is $[Ni^{2+}] = 0.100\ M$; $[Ag^+] = 0.045\ M$.
14. 0.46 volt
15. −3.87 cal
16. 6.27
17. 0.78 volt
22. 113 days

Chapter 21

1. 80.4 kcal
6. $\Delta H^{\ddagger} = 59$ kcal
8. First-order; $k_{rate} = 0.0188\ min^{-1}$.
9. The reaction is first-order in H_2 and second-order in NO; $k_{rate} = 1.0 \times 10^{-7}\ sec^{-1}\ torr^{-2}$.
11. A possible mechanism is

$$2\ NO \rightleftharpoons N_2O_2 \quad \text{(equilibrium)}$$
$$N_2O_2 + H_2 \rightarrow H_2O + N_2O \quad \text{(rds)}$$

12. A possible mechanism is

$$(CH_3)_3C\text{—}O\text{—}O\text{—}C(CH_3)_3 \rightarrow 2\ (H_3C)_2C\text{=}O + 2\ CH_3 \quad \text{(rds)}$$

$$2\ CH_3 \rightarrow C_2H_6 \quad \text{(rapid)}$$

The following two steps would account for the two minor products:

$$CH_3 + (H_3C)_2C{=}O \rightarrow CH_4 + (H_3C)(H_2C)C{=}O$$

$$CH_3 + (H_3C)(H_2C)C{=}O \rightarrow (H_3C)(H_3C{-}CH_2)C{=}O$$

13. A plausible mechanism is

$O_3 \rightleftharpoons O_2 + O$ (equilibrium)

$2\ O \rightarrow O_2$ (rds)

14. A plausible mechanism is

$S_2O_8^{2-} + I^- \rightarrow SO_4I^- + SO_4^{2-}$ (rds)

$SO_4I^- + I^- \rightarrow SO_4^{2-} + I_2$ (rapid)

17. $\Delta H^{\ddagger} = 20.1$ kcal; $\Delta S^{\ddagger} = -4.6$ cal/mole-deg

19. $\Delta H^{\ddagger}_{reverse} = 28$ kcal

$$\frac{-dP_{HI}}{dt} = k_{reverse}P^2_{HI}$$

24. A plausible mechanism is

$CO + Cu_2O \rightarrow CO_2 + 2\ Cu$ (rds)

$4\ Cu + O_2 \rightarrow 2\ Cu_2O$ (rapid)

Chapter 22

2. 36 electrons

13. 92 electrons

18. GGA, GGG, GGU, and GGC

9. 1.3×10^3.

16. 500 kcal/mole

Index

Index

1 2 3 4 5 6 7 8 9 10